1983

Sourcebook of Experiments for the Teaching of Microbiology

Special Publications of the Society for General Microbiology

Publications officer: Colin Ratledge

Sourcebook of Experiments for the Teaching of Microbiology

edited by

S. B. Primrose

Department of Biological Sciences
University of Warwick
Coventry

A. C. Wardlaw

Department of Microbiology
University of Glasgow
Glasgow

1982

ACADEMIC PRESS

A Subsidiary of Harcourt Brace Jovanovich, Publishers
London · New York
Paris · San Diego · San Francisco · São Paulo
Sydney · Tokyo · Toronto

ACADEMIC PRESS INC. (LONDON) LTD.
24–28 Oval Road, London NW1 7DX

United States Edition published by
ACADEMIC PRESS INC.
111 Fifth Avenue, New York, New York 10003

British Library Cataloguing in Publication Data

Sourcebook of experiments for the teaching of
microbiology.—(SGM Special Publications; 8)
1. Microbiology—Study and teaching (Higher)
2. Microbiology—Experiments
I. Primrose, S. B. II. Wardlaw, A. C. III. Series
576′.0724   QR61

ISBN 0–12–565680–7

LCCCN 81–69579

Typeset by
Macmillan India Ltd., Bangalore.

Printed in Great Britain by Page Bros. (Norwich) Ltd.

Preface

The stimulus for this book was our belief that some parts of many undergraduate laboratory courses in microbiology—including our own—are unnecessarily dull! At the same time, most of us who teach the subject have a few favourite experiments—experiments which are a delight to put on and which the students enjoy doing. So we thought that a collection of such experiments would be useful both for ourselves and other teachers.

We therefore wrote to a large number of department heads and other colleagues (mainly in the UK) explaining what we had in mind, i.e. that we were **not** looking for the standard bread-and-butter type experiments or exercises which tend, perhaps unavoidably, to make up the bulk of undergraduate teaching manuals. The sourcebook is not intended to replace such compilations. Instead, what we have tried to provide here is a reservoir of experiments and exercises which

(a) highlight interesting and important areas of microbiology;
(b) are known from experience to stimulate the students;
(c) are satisfying from the teachers' point of view because of their relative simplicity, reliability, robustness towards student error, perhaps a certain dramatic quality and because they are open-ended in the sense of easily leading into mini-projects by ringing the changes on the experimental variables.

The 111 experiments eventually selected are a testimonial to the enthusiastic response we received to our letter of request. Inevitably we have disappointed, and perhaps offended, some of our colleagues who submitted material in good faith only to have most or all of it rejected. For this and to them we apologize, but to keep the book to a reasonable size we had to be quite ruthless in what was chosen. In fact what you see here comes from less than ten per cent of the material submitted.

We have tried to cover the whole field of microbiology as broadly as possible and in a reasonably balanced way. Thus we have tried to get good representation of both the pro- and eukaryotes and the viruses; of the basic processes of microbial growth and death; of the biochemical, molecular and genetical aspects of microbes, and of the various interactive features of microbiology, such as microbial relations with animals and plants and industrial microbiology. Inevitably there are some areas where the representation is rather thin, or even non-existent. Usually this is because no suitable material was made available to us.

In conclusion, may we thank all the colleagues who have helped in the production of this book—not only the authors whose work is represented, but everyone else who generously sent us samples from laboratory manuals and provided ideas on content and format. We hope that you, the teacher, will find the sourcebook useful and that your students will have their practical work supplemented by some of the experiments we have collected.

January 1982   S.B.P.*
                                                                                                                 A.C.W.

---

* Dr. Primrose is currently Director of Microbiology and Process Research, at G. D. Searle & Co. Ltd., Lane End Road, High Wycombe, Bucks, HP12 4 HL.

How to Use this Book

THE MAIN SECTIONS

This book is intended mainly for teachers rather than undergraduates, so there are no instructions on how to do a Gram stain or spread an agar plate. We have assumed that each department where microbiology is taught has its own house manual of basic techniques and introductory exercises and therefore there is no point in repeating them. But perhaps you, the teacher, recognize that there is a part of your course where the experiments are not as interesting as they might be. Alternatively, you might want to introduce an area of microbiology not represented on your course but need some suggestions for experiments. This is where the sourcebook should help.

The 111 experiments and exercises have been segregated under nine main headings. The titles of those sections are somewhat unconventional and were selected to show that the commonly used subdivisions of microbiology are artificial and may stifle imagination. The contents of the sections are as follows.

1 **Simple and Diverse**   A collection of ten experiments which are simple to do and cover many aspects of microbiology.
2 **The Micro-organism Feeds**   Eleven experiments on nutrient uptake, transport of metabolites, enzyme induction and predation.
3 **The Micro-organism Grows**   Thirteen experiments on the growth of bacteria, fungi and protozoa including quantitative aspects and cell differentiation.
4 **The Micro-organism Excretes**   Eight experiments on the release of metabolites during microbial growth.
5 **The Micro-organism Dies**   Sixteen experiments on the effects of antimicrobial agents of various kinds tested in a variety of different ways.
6 **Micro-organisms in the World Around Us**   Fourteen experiments on the activities and interactions of viruses, bacteria, algae and fungi in soil and natural aquatic habitats and including mathematical modelling and numerical taxonomy.
7 **Micro-organism Meets Animal**   Twelve experiments on the interactions of micro-organisms with the body or body fluids of man and marine animals and with the hen's egg. This section also includes experiments on bacterial toxins.
8 **Micro-organism Meets Plant**   Sixteen experiments on the interactions of viruses, bacteria and fungi with whole plants and with fruits and seeds, and including material on the degradation of the plant polymers cellulose and lignin as well as on the nitrogen cycle.

9 Microbial Genes   Eleven experiments on mutagens, colicins, plasmids and gene exchange.

Within each of the above sections the experiments are arranged approximately in increasing order of difficulty or complexity—at least in the judgement of the editors!

THE SUBJECT INDEX

An alternative point of access to the contents of the sourcebook is provided by the subject index. For example, if you wish to see all the experiments dealing with **fungi**, reference to this entry in the index reveals that experiments 7, 12, 13 etc. are all on this topic. Likewise all experiments involving the use of **radioactive tracers** can be located under that heading.

THE FORMAT OF THE INDIVIDUAL EXPERIMENTS

Many of our contributors queried the particular format that we demanded for presenting the experiments. However, this format was chosen only after careful thought and with the primary intention of assisting the user to find and use suitable material with a minimum of effort.

The Information Panel   This gives a summary of the essential details of the experiment and contains three pieces of information: the level of undergraduate for whom the experiment is intended, the conventional subject areas to which the experiment is relevant, and any special features of the experiment which may favour its selection. With regard to suitability for students at different stages we have used the following categories:

All undergraduate years: can be used at any stage of an undergraduate course and no special skills are required.

Beginning undergraduates: suitable for students in their first year of exposure to microbiology.

Advanced undergraduates: suitable for students who have already had an introductory course in microbiology.

Introduction.   This section gives the background to the experiment and concludes with a concise statement of the objectives of the experiment. No attempt is made to give a comprehensive introduction to the literature of the subject but usually a few key references are provided.

Experimental.   This section gives detailed instructions, in the form of flow diagrams, for performing the experiment. At the outset is given the time required for an average class to complete the experiment. This is to enable teachers to decide if the experiment can be fitted in to the time available in a particular class. Each part of the flow diagram is subdivided into Day 1, Day 2, etc. In many cases it is not essential that the second part of an experiment be done on the day immediately following Day 1. However, where it is essential to have a daily follow-up this is indicated.

Materials. Here are listed any special materials or equipment required for the experiment. This should enable the teacher to decide at a glance whether the experiment can be mounted with the existing departmental resources or whether particular strains, reagents or equipment, etc. would have to be ordered or prepared in advance. Where necessary the names and addresses of suppliers of the listed items are given here, together with diagrams of specific apparatus. This section also contains the recipes for culture media and reagents.

Specific Requirements. This is a summarized check-list for each day of the experiment. It contains all of the special items listed under **Materials** together with the ordinary items like test tubes and Petri dishes which all microbiology departments possess. We have tried to impress our contributors with the need to make the list fully comprehensive, i.e. to assume a "bare bench" situation and to include items such as test tube racks which may be mundane but are essential for the experiment. However, we have assumed that students have available at all times a microscope, slides and cover glasses, staining set, Bunsen burner and a bacteriological loop. The class technician should find this section very useful.

Further Information. This section contains sample results and their interpretation and suggestions as to how the experiments might be modified or extended.

AVAILABILITY AND STORAGE OF CULTURES

Where no specific strain of an organism has been indicated, then any strain can be used. Where a department does not already possess the organism in question it should be available from one of the culture collections listed. Alternatively, the students might be set the task in an earlier experiment of isolating that organism from its usual habitat.

When a specific strain of an organism is listed, then it is recommended that only this strain be used and no substitutions be made until the experiment has been thoroughly class-tested. All the specific strains required for experiments described in this book have been deposited in one or more of the major culture collections and should be obtained from them. The addresses of these culture collections are given on page 763. Arrangements have been made with the National Collection of Industrial Bacteria to supply those strains which are listed in this book as being held only in the culture collection of individual authors or departments. Therefore readers requiring such strains are requested to write to the NCIB (quoting the SGM Source Book) rather than to the individual authors themselves.

Preservation of Cultures

The best way to store most micro-organisms is by lyophilization. Where such facilities are not available a suitable alternative is storage on glass beads at $-20°$C. In fact, using this method one of us has maintained over 300 cultures for 10 years with no losses or contamination problems. The method is particularly simple and is described below.

Half-fill a 25 ml screw-cap bottle (e.g. Universal bottle) with glass beads ($\sim$ 4–6 mm diam.) and add 2 ml glycerol. Note that the glycerol is easier to dispense if it is warmed to $50–60°$C. Sterilize the bottles of beads by autoclaving. Take either 8 ml of an overnight broth culture or aseptically wash the organisms off a Petri dish with 8–10 ml saline, buffer or both and add to the glycerol plus beads. Shake vigorously to mix the glycerol with the culture and then with a Pasteur pipette suck off any excess liquid above the level of the beads. Store the bottles at $-20\,°C$ in a deep freeze or in the freezer compartment of an ordinary refrigerator. To resuscitate a culture, remove the bottle from the freezer and warm gently in the hand while shaking vigorously. Very soon a few beads will fly loose from the surface and these should be deposited aseptically on the surface of an appropriate nutrient medium. The container is returned immediately to the freezer so that the remainder of the culture is not subjected to an unnecessary cycle of thawing and freezing.

A FEW WORDS ABOUT SAFETY

Apart from the common laboratory hazards there are three classes of hazard more specifically associated with experiments in microbiology: physical (e.g. radiation), chemical and microbiological. What constitutes a physical or chemical hazard is generally agreed upon by the scientific community but there is still considerable disagreement among microbiologists about which organisms are unsuitable for class use. For example, some teachers consider that *Serratia marcescens* and *Salmonella typhimurium* are too dangerous for undergraduates to handle, whereas others would have no reservations about using these organisms. In our initial screening of possible contributions for this manual we did not exclude any experiment because it involved micro-organisms which some people might consider hazardous. We have left judgement on microbiological safety to the individual reader. Where a possible chemical or physical hazard is involved the word "**caution**" has been inserted at the appropriate place in the text. In most countries there are legally enforceable regulations concerning the safe handling of radioactive material and we trust that these will be observed by all concerned.

Contents

Part Three
The Micro-organism Grows

Part Four
The Micro-organism Excretes

Part Five
The Micro-organism Dies

Part Six
Micro-organisms in the World around Us

Part Nine
Microbial Genes

Part One
Simple and Diverse

1

A Synthetic Epidemic

S. B. PRIMROSE

*Department of Biological Sciences, University of Warwick,
Coventry, CV4 7AL, England*

Level: Beginning undergraduates
Subject area: General microbiology
Special features: Extremely enjoyable introduction
to basic microbiological
technique

INTRODUCTION

During the first few laboratory sessions in microbiology, students are normally introduced to basic techniques such as streaking plates for isolated colonies, subculturing, etc. Important as these are, they do not constitute very exciting practical experiments. The aim of this exercise is to introduce a "fun" element into the introductory laboratory sessions in microbiology.

Basically, the experiment consists of presenting each student with a sweet (piece of candy), one of which has been contaminated with the red-pigmented *Serratia marcescens* and the remainder with the yellow-pigmented *Sarcina lutea* (*Micrococcus luteus*). The students rub the candy between their hands and then indulge in a handshaking exercise in order to start an "epidemic". After the handshaking is over, the students swab their hands and test the swabs for the presence of *Serratia*. When the results of the class are pooled it is possible to do some detective work and identify the student who initiated the epidemic. You could say he is caught red-handed!

The results obtained in this experiment provide an insight into the ease with which microorganisms are spread by contact, as well as the methods used by public health officials in tracking down the source of an epidemic. More important, it provides colourful material which is used for practising basic microbiological procedures. Thus such techniques become part of a fun experiment instead of being rather dull exercises in their own right.

EXPERIMENTAL

Day 1 *(45 min)*

Each student is presented with a sticky sweet contained in a numbered Petri dish (*see* note 1). One of the sweets has been moistened with a culture of *S. marcescens* (*see* note 2), the rest with a culture of *S. lutea* (*M. luteus*).

↓

Each student dons a pair of disposable plastic gloves (*see* note 3). On a signal from the instructor all the students rub their sweet between the palms of their gloved hands. The sweets are then returned to the Petri dishes for disposal later.

↓

On a signal from the instructor the student with Petri dish number 1 shakes hands with another member of the class, preferably not a neighbouring student (*see* note 4). When student number 1 has returned to his place, the student with Petri dish number 2 then shakes the hand of another member of the class. The handshaking continues in this manner until each person has shaken hands with a person of their choice. The instructor records on the blackboard the sequence of handshakes.

↓

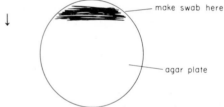

Students with Petri dishes numbers 5, 10, 15, 20, etc., swab their right palm with a sterile cotton swab (*see* note 5) and then rub the swab on an agar plate as shown here.

When the swabbing is completed, a second round of handshaking is again initiated by the student with Petri dish number 1. Each student should endeavour to shake hands with someone they did not shake hands with in round 1. The instructor again records the sequence of handshakes.

↓

When the 2nd round of handshaking is complete, all the students swab their right palm with a sterile swab (*see* note 5) and then rub the swab on an agar plate, as shown above. Students with Petri dishes numbers 5, 10, 15, 20, etc. should use fresh swabs and nutrient agar plates.

↓

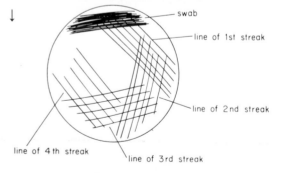

Discard the gloves, sweets and swabs. Each student then takes his/her inoculated plate(s) and with a flamed loop makes streaks of the initial inoculum. Although there are many methods of streaking plates, the one shown here is the best method. The plates are then incubated for 2–3 days at room temperature in the presence of light (*see* note 6).

Day 2 *(15–60 min*, see *note 7)*
The plates are examined for the red colonies of *S. marcescens* and the students pool their results and try to determine who started the epidemic.

Notes and Points to Watch

1. The experiment can be used with classes containing 15–30 students. If more than 30 students are present the class should be divided into suitably sized groups. In our classes we normally have 60–65 students and split them into three groups of approximately 20.
2. Approximately 0·5 ml of an overnight broth culture should be poured over the sweets in the Petri dishes. The best results are obtained when sweets numbered 5, 6 or 7 (for classes of 15 students) or sweets 7, 8, 9 or 10 (for larger classes) are inoculated with *S. marcescens*. This ensures that the epidemic is initiated early in the round of handshaking.
3. *S. marcescens* and *Sarcina lutea* may occasionally cause infections, particularly if introduced into wounds. For this reason all students should wear gloves to avoid direct skin contact.
4. Students seem naturally lazy and have a natural temptation to shake hands with a neighbour. This does not assist the spread of the "infection".
5. Most people are right-handed and therefore have a tendency to swab their left palm. However, people shake with their right hand so it is the right palm which must be swabbed. Failure to pay attention to this fact can result in failure of the experiment.
6. The red pigment of *S. marcescens* develops best at temperatures below 30 °C in the presence of light. If the plates are incubated at 37 °C in the absence of light no pigmentation will be observed. Pigmentation is also affected by the composition of the medium. Nutrient agar is satisfactory but peptone–glycerol medium containing 5 g peptone and 10 ml glycerol per litre of distilled water gives even better results.
7. The time spent on day 2 depends on whether the students simply record the results or actually attempt to analyse the data during the laboratory session.

MATERIALS

1. *Serratia marcescens*. Any strain will do but if Experiment 38 is to be done later it would be useful to use strain STM 38.
2. *Sarcina lutea* (called *Micrococcus luteus* in the 8th edn of Bergey's Manual). Any strain will do.

Any other pigmented bacterium would be a satisfactory alternative.

3. Sterile, moist, cotton swabs are best prepared by autoclaving, in test tubes, swabs which have been moistened with distilled water or nutrient broth.

SPECIFIC REQUIREMENTS

Day 1
10 ml overnight broth culture of *Serratia marcescens*

20 ml overnight broth culture of *Sarcina lutea* sterile, moist cotton swabs (6 swabs for every 5 students)

cont.

cont.

nutrient agar plates (6 plates for every 5 students)	disposable gloves (1 pair per student)
sticky sweets in numbered Petri dishes (1 per student)	***Day 2*** none

FURTHER INFORMATION

Yellow-pigmented colonies of *Sarcina lutea* should be seen on all the plates and red-pigmented colonies of *Serratia* on many of them. With two such readily distinguishable colony types it is very easy to demonstrate to students the need for selecting well-isolated colonies when purifying micro-organisms. In addition, the two organisms are particularly suitable for a Gram-staining exercise since one is a Gram-negative rod and the other a Gram-positive coccus.

Table 1

Student initiating handshake	Student whose hand was shaken		Presence of *S. marcescens* after	
	Round 1	Round 2	Round 1	Round 2
1	9	10		−
2	8	3		−
3	13	4		+
4	12	6		+
5	7	3	+	+
6	14	11		+
7	11	8		+
8	10	9		+
9	2	10		+
10	15	11	−	+
11	13	14		+
12	7	13		+
13	14	15		+
14	4	8		+
15	14	11	+	+

Typical results from a class experiment are shown in Table 1. The results in Table 1 are interpreted as follows:

No. 10 was free of *Serratia* at the beginning of the 2nd round.

No. 10 was free of *Serratia* when he shook the hand of no. 9.

No. 10 got it from no. 9.

No. 9 is not the culprit for he shook the hand of no. 1 who is free.

No. 9 must have caught *Serratia* from no. 8.

No. 8 is not the culprit for no. 8 shook the hand of no. 2 who is free.

No. 8 must have got it from no. 7. Is the culprit no. 7 or no. 11?

No. 5 shook hands with no. 7 before no. 7 shook hands with no. 11 and no. 5 is infected at the end of the 1st round. Therefore, no. 11 is not the culprit.

Did no. 5 or no. 7 initiate the epidemic? Only the instructor can tell!

Occasionally a student who should be positive appears to be free of *Serratia*. This is usually due to the student swabbing the wrong palm. However, this does not matter for it mimics the situation encountered in epidemics where some individuals never develop infection despite exposure.

ACKNOWLEDGEMENTS

The exact origin of this experiment is not clear, but I originally discovered it in an old laboratory manual used in the Department of Bacteriology, University of California, Davis.

2

Assessing the Efficiency of Toilet Paper as a Barrier against Microbes

M. C. ALLAN

Department of Applied Microbiology, University of Strathclyde, Glasgow, G1 1XW, Scotland

> *Level*: All undergraduate years
> *Subject areas*: Hygiene
> *Special features*: Relevance to everyday hygiene

INTRODUCTION

The following short experiment was introduced to reinforce the maxim "now wash your hands" after using the toilet, by showing the porosity of toilet paper and perhaps the inadequacy of normal hand-washing.

The experiment is carried out by microbiology students and also by students of food science and nutrition, and environmental health. However, it is also appropriate for application in schools and for food handlers, etc.

EXPERIMENTAL

Day 1 *(30 min)*

Pour the test organism (*see* note 1) in a liquid medium on to a sterile nutrient agar plate, so that the surface is covered. Leave it for a few minutes. Pour off the excess liquid into a disinfectant solution. The surface of the agar is then taken as imitating a specimen of moist faeces.

↓

Wrap finger(s) in one or more layers of toilet paper and rub over the surface of the nutrient agar plate to simulate the normal "cleaning" in using toilet paper.

↓

Discard the toilet paper by dropping it into a bowl of disinfectant.

↓

Imprint the contaminated finger(s) onto, or rub over, half of a MacConkey agar plate (where *Escherichia coli* is the test organism) or over half of a malt extract or malt wort agar plate (where yeast is the test organism).

↓

Wash the finger(s) as carefully as would normally be done after using the toilet and dry on a towel.

↓

Again imprint the finger(s) onto, or rub over, the second half of the MacConkey or malt extract or malt wort agar plate. This will show a "before and after" result on the same plate (*see* note 2).

↓

Incubate the plate for an appropriate time at a suitable temperature, e.g. 24 h at 37 °C for *E. coli*, or 2–3 days at 30° C if the yeast is used.

- -

Day 2 *(10 min)*

Inspect the plates and compare the pattern of growth with those obtained by other students using different types of, and number of sheets of, toilet paper.

- -

Notes and Points to Watch

1. The best comparison with everyday life would be to use *E. coli* as test organism. This also permits the use of MacConkey agar as the medium for culturing the contamination from the finger(s). The selectivity of this medium cuts out the growth of many normal skin organisms which makes the degree of contamination through the toilet paper easier to appreciate, since the students' attention can be directed to the rose-pink, lactose-fermenting colonies. Similarly the reduction in the number of such colonies after washing is obvious even though the washing may increase the number of skin bacteria available for transfer to the medium.
2. If *E. coli* is being used, it may be advisable to dip the hands into a suitable antiseptic solution and wash them thoroughly at the end of the experiment.

MATERIALS

1. As test organism, a 24 h nutrient broth culture of *Escherichia coli* is normally used. However, if this is felt inappropriate (e.g. in schools) a 48–72 h malt extract, or malt wort broth, culture of *Saccharomyces cerevisiae* (baker's yeast) will serve equally well. One culture should be provided per student or pair of students.

2. Nutrient agar plates; one per student or pair of students.
3. MacConkey agar or malt agar plates; one per student.
4. Bowl of disinfectant.
5. Toilet paper, various types.

SPECIFIC REQUIREMENTS

Day 1   *Day 2*
all the items listed above   none
37 °C incubator

FURTHER INFORMATION

This is a very simple experiment, yet provides a very dramatic illustration of the need for careful personal hygiene (Figure 1). It is capable of considerable elaboration and might be subjected to statistical analysis. Although *E. coli* is a more typical organism, the yeast cell, being much larger than the bacterium, is perhaps a more powerful illustration of the point which the experiment seeks to convey. This could be illustrated by microscope slides of a mixture of the two organisms.

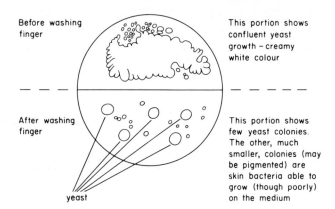

Before washing finger

This portion shows confluent yeast growth – creamy white colour

After washing finger

This portion shows few yeast colonies. The other, much smaller, colonies (may be pigmented) are skin bacteria able to grow (though poorly) on the medium

yeast

Figure 1   Typical result of toilet paper experiment with *Saccharomyces cerevisiae* as test organism on malt agar medium.

The experiment can be varied in several ways, e.g.

(a) comparison of the "shinier" vs. the "softer" type of toilet paper,
(b) a comparison of medicated vs. non-medicated toilet papers,
(c) increasing the number of sheets of toilet paper used to prevent heavy contamination of the finger(s). The results here can be surprising, and
(d) by allowing the nutrient agar plate to dry off so a comparison can be made of "wet" vs. "dry" faeces.

3

An Illuminating Demonstration of Bacterial Luminescence

C. M. TURLEY

Department of Marine Biology, Marine Science Laboratories, Menai Bridge, Anglesey, LL59 5EH, Wales

> *Level*: All undergraduate years
> *Subject areas*: Marine microbiology and ecology
> *Special features*: Provides "light" relief!

INTRODUCTION

Light is produced by some insects, fungi and algae as well as by bacteria. Luminous bacteria are present in all marine environments and develop abundantly on the surface of dead fish. Perhaps

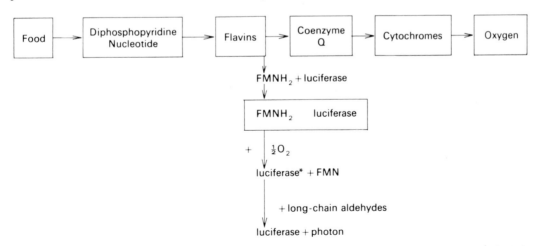

Figure 1 A schematic diagram of the reactions involved in bacterial luminescence as a side branch of the electron-transport system.

the most interesting of the luminous bacteria are those that live symbiotically in special organs of some fish and squids. The bacteria are supplied with nutrients and in return luminesce continuously, thereby supplying the hosts with devices for recognition, defence and the attracting of prey. Some fish (e.g. the Indonesian fish *Photoblepharon*) can flash its lights on and off by operating a shutter mechanism or "eyelid" across the pouches which contain the luminous bacteria. Some highly specialized organs have been developed to contain the luminous bacteria and are described by Dales (1966). Some squids have even developed reflectors and lenses around the areas which contain the luminous bacteria. McElroy and Seliger (1962) dramatically illustrate luminescing organisms with photographs for which the light source was only that produced by the luminescing organisms themselves.

The light-emitting reaction is a side branch of the oxidative, or electron-transport, processes of the bacteria cell (Figure 1). Reduced flavin mononucleotide ($FMNH_2$) reacts with an enzyme luciferase. The resulting complex is oxidized by molecular oxygen with formation of oxidized FMN and excited luciferase. The return of the luciferase to its ground state results in the emission of a photon.

The aim of this exercise is to illuminate the processes of bacterial luminescence and to act as a few minutes of "light" relief during one of the more mentally strenuous experiments incorporated in this book.

EXPERIMENTAL

Day 1 *(20 min plus 20 min with 7 h in between)*

Liquid culture (*see* note 1) of a luminescent bacterium (*see* note 2).

↓

Early in the morning, paint a message (*see* note 3) on a large, nutrient marine-agar plate with a sterile paint brush and incubate at 20°C.

↓

Early in the morning, inoculate two 20 ml amounts of a marine liquid medium (*see* note 1) with 2 ml of the liquid culture. Aerate one culture with a bubbler and incubate both at 20°C.

Late in the afternoon, inoculate a second marine agar plate as a duplicate of the one inoculated in the morning, and incubate at 20°C.

Day 2 *(20 min)*

In a dark room examine (*see* note 4) the large plates for bioluminescence and inspect the aerated (*see* note 5) and non-aerated liquid cultures (*see* note 6).

If luminescence is not detectable in the morning (*see* note 7), re-examine the cultures in the afternoon.

Day 3 (20 min)

If necessary (*see* note 7) re-examine the cultures.

‒ ‒

Notes and Points to Watch

1. Bacteria isolated from marine fish or squid (*see* note 2) should be grown in glycerol marine broth, while *Photobacterium phosphoreum* (NCIMB 844) should be grown in sea-water yeast peptone broth.

2. Luminescent bacteria can be isolated from marine fish or squid provided the animals have not been washed in fresh water or stored in ice. If such animals are available, they should be incubated overnight at 15–20 °C. Next day they will usually be covered with colonies of luminous bacteria which can be transferred to glycerol marine agar. Alternatively a suitable, reliable strain from the National Collection of Industrial and Marine Bacteria is *P. phosphoreum* (NCIMB 844). This strain should be grown on sea-water yeast peptone agar.

3. A large metal tray containing the appropriate agar medium gives a dramatic demonstration of bioluminescence. If the demonstration is required for a few minutes of "light entertainment" for the students a witticism such as "And God said let there be light—so I obliged literally, signed Photobacterium" can be painted on the agar dish. However, if a more serious approach is required something like the following can be written

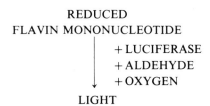

REDUCED
FLAVIN MONONUCLEOTIDE
+ LUCIFERASE
+ ALDEHYDE
+ OXYGEN
LIGHT

4. The demonstration requires a dark room to view the luminescence. The observers should spend five minutes in the dark room to allow for the accommodation of the eyes before examining the cultures.

5. The role of oxygen in the light-emitting processes can be easily demonstrated by aerating one suspension of luminous bacteria and leaving another suspension unaerated. The bacteria in the unaerated suspension exhaust the dissolved oxygen and cease to luminesce except at the surface where they are exposed to air, whereas the aerated suspension luminesces brightly.

6. If a barely luminous bacteria, e.g. a dark mutant, has been isolated, the low luminosity may be because the isolate cannot synthesize a long-chain aldehyde (dodecanal) which is essential for high luminosity. If this aldehyde is added to a suspension of the mutant it glows brightly, thus demonstrating the requirement for aldehydes in bacterial light emission.

7. Strains of luminous bacteria may grow at different rates, and light emission usually occurs within 18–48 h of subculture. Therefore, if this experiment is prepared purely as a demonstration for the students, inoculate the "plates" at staggered intervals (as insurance) at approximately 48, 36, 26 and 18 h before the demonstration is required.

MATERIALS

1. A luminous bacterium such as *Photobacterium phosphoreum* (NCIMB 844), or organisms isolated from marine fish or squid.
2. (a) *Sea-water yeast peptone broth* (for *P. phosphoreum*)

yeast extract (Oxoid or Difco),	3 g
peptone,	5 g
"aged" filtered sea-water,[a]	750 ml
tap or distilled water,	250 ml

This and the other media are sterilized by autoclaving at 121 °C for 15 min.

   [a] sea-water stored in the dark for a minimum of 3 weeks to age it.
   Artificial sea-water may be used in place of the "aged" sea-water.

   (b) *Sea-water yeast peptone agar* (for *P. phosphoreum*)
   This is the above medium solidified by incorporating 15 g agar per litre.

   (c) *Glycerol marine broth* (for isolates from fish or squid)

nutrient broth (Oxoid or Difco) powder,	25 g
glycerol,	3 ml
sodium chloride,	25 g
distilled water,	to 1 litre

   Adjust the pH to 7·8 and sterilize by autoclaving.

   (d) *Glycerol marine agar* (for isolates from fish or squid) This is the above medium solidified by incorporating 15 g agar per litre.

3. Large plates or trays of the appropriate solid medium (*see* note 1). Four plates should be prepared for demonstration purposes (*see* note 7), and more if groups of students are to inoculate their own plates.
4. Paintbrush, wrapped in aluminium foil and sterilized by autoclaving.
5. Darkened room.
6. Source of compressed air, or an aquarium pump.

SPECIFIC REQUIREMENTS

cultures of luminous bacteria
plates, or trays, of solid medium and tubes, bottles or flasks of liquid medium
compressed air
darkened room
sterile paintbrush
2 ml sterile pipettes

FURTHER INFORMATION

The light produced by luminous bacteria is confined to a fairly narrow band of wavelengths in the blue–green region of the visible spectrum with a maximum near 490 nm. Emission is continuous and is always associated with the consumption of molecular oxygen. It therefore only occurs under aerobic conditions. Light emission by a suspension of luminous bacteria provides one of the most sensitive methods for detecting traces of dissolved oxygen.

Luminous bacteria have been favoured organisms for studying the action of drugs and other inhibitors of cell respiration because the effects are easily observed with a photocell. Dark mutant strains of luminous bacteria have been used to examine the ability of various chemicals to restore luminescence, e.g. the addition of dodecanal can increase light emission. The rate at which the aldehyde penetrates the cell membrane can be determined by measuring with a photocell the rate at which the light intensity increases.

The US Navy has funded much research on bioluminescence in the sea (caused by dinoflagellates). Such bioluminescence may be of great embarrassment during military strategic operations (Johnson, 1975). Another interesting note about luminous bacteria is that Japanese families used them as a source of household illumination during "blackouts" in the Second World War (Haneda, 1955).

Further details of bacterial luminescent systems are given by Harvey (1952) McElroy (1961) and Hastings (1978).

REFERENCES

R. P. Dales (1966). Symbiosis in marine organisms. *In* "Symbiosis", (S. M. Henry, ed.), Vol. I, pp. 299–326. Academic Press, London and New York.

Y. Haneda (1955). Luminous organisms of Japan and the Far East. *In* "The Luminescence of Biological Systems", (F. H. Johnson, ed.), pp. 337–8. American Association for the Advancement of Science, Washington D.C.

E. N. Harvey (1952). "Bioluminescence", pp. 1–95. Academic Press, London and New York.

J. W. Hastings (1978). Bacterial and dinoflagellate luminescent systems. *In* "Bioluminescence in Action", (P. J. Herring, ed.), pp. 129–70. Academic Press, London and New York.

F. H. Johnson (1975). Final Report: Research on bioluminescence, p. 5. Task number NR 108–860. Office of Naval Research, Biological Department Princeton University, Princeton, New Jersey.

W. D. McElroy (1961). Bacterial luminescence. *In* "The Bacteria", (I. C. Gunsalus and R. Y. Stanier, eds), Vol. II, pp. 479–508. Academic Press, London and New York.

W. D. McElroy and H. H. Seliger (1962). Biological luminescence. *Scientific American*, **207** (6), 76–91.

4

Use of Triple Loops for Colony Counts of Bacteria

A. G. SCHAFFER

*Department of Science, Bristol Polytechnic,
Frenchay, Bristol, BS16 1QY, England*

Level: All undergraduate years
Subject areas: General microbiology
Special feature: Economy and improvisation. Data
obtained suitable for statistical
analysis

INTRODUCTION

Triple wire loops which deliver about 20 μl of liquid provide a cheap and effective alternative to the surface drop (Miles and Misra) technique for colony ("viable") counting of bacteria. The loops can also be used for serial dilutions, with a substantial saving of sterile pipettes and diluent.

Recent investigations (Jarvis *et al.*, 1977; Kramer, 1977; Hedges *et al.*, 1978) have shown that there is no significant difference between any of the methods commonly used for colony counts. The newer methods (Droplette, Spiral Plate Maker and the use of a precision micro-pipette for dilutions) all saved time, labour and/or materials when compared with a standard pour-plate method or a conventional dilution procedure. However, these new methods are too expensive for large classes. Thus the Miles and Misra technique continues to be used, but even this can be expensive if a large class is to carry out an elaborate exercise.

The aim of the experiment is to show that colony counts can be carried out both cheaply and effectively with triple loops.

EXPERIMENTAL

The procedure for calibrating and using a triple loop for a colony count on a bacterial culture is described. This procedure may be incorporated into a more elaborate experiment.

Day 1 *(1–3 h, depending on experiment)*

Contruct a triple loop by winding nickel–chromium alloy wire (26 s.w.g.) three times round a loop handle (4·8 mm diam.). Twist the free ends together so that the loops of wire come into close contact (*see* note 1).

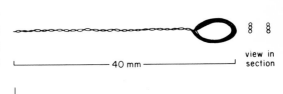

↓

Calibrate loop (*see* note 2).

↓   ↓   ↓

Spectrophotometric method
Transfer one loopful of a solution of a water soluble dye to 3 ml diluent. Repeat to give at least 5 replicates. Measure absorbance and compare with standard curve for dye (*see* note 8).

Weighing method
Add 20 loopfuls of water to preweighed filter paper. Weigh paper.

Spot method
Deliver 20 μl of dye solution on to filter paper with a micro-pipette. Compare spots from loop with standard spot.

↓

To make a serial dilution of a broth culture, first prepare sterile tubes (10 × 40 mm) (*see* note 3) containing 0·18 ml of diluent (e.g. use 9 × 20 μl drops). Transfer one triple loopful of culture to the first tube to give 10^{-1} dilution. Flame loop and repeat procedure to give 10^{-2} dilution. Repeat as necessary.

↓

Use the loop to spread replicate samples of the diluted culture onto one-sixth or one-quarter of the dry surface of a suitable nutrient agar medium (*see* note 4).

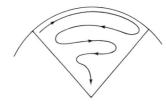

↓

Incubate the plates at the lowest temperature and/or shortest time for the organism under study (*see* note 5).

- -

Day 2 *(15–60 min)*

Count colonies and calculate the number of c.f.u. ml^{-1} in the original culture (*see* note 6).

- -

Notes and Points to Watch

1. The triple loops hold about 20 μl of aqueous solutions. Usually the drop of liquid is stable but the loops do become distorted with use (*see* Further Information).

2. The method of calibration depends on the purpose of the exercise. The spectrophotometric method is only of value if the variation in the capacity of an individual loop is of interest as in an experiment to compare loops and droppers (*see* Further Information). We have used bromocresol purple ($0.4\,gl^{-1}$) as the dye, $0.1\,M$ sodium hydroxide as the diluent and wavelength of 590 nm. The weighing method is rapid and the spot method is convenient if a number of loops need to be checked in preparation for a class.
3. The tubes used for the dilutions must be wide enough to allow easy access to the diluent. The preparation time needed for the sterile tubes can be greatly reduced if they are sterilized in an inverted position and not plugged or capped.
4. A useful experiment is to compare the counts obtained (and the materials used) using conventional dilution and plating techniques with those using the triple loop (*see* Further Information).
5. The accuracy of all colony counting techniques is improved if the colonies are not allowed to spread too far. The bacteria commonly used for teaching (e.g. *Escherichia coli, Beneckea natriegens, Bacillus megaterium, Staphylococcus aureus*) give small but countable colonies after 12–18 h incubation at 20 °C or less.
6. The calculation of the number of c.f.u. ml^{-1} should take account of the volume of the loop employed. But it is a useful exercise to estimate the error which would result if the loop was assumed to deliver exactly 20 μl and whether this error is within the limits expected of a colony counting method.
7. The major disadvantage of using the triple loops for teaching introductory classes in microbiology is that many (full-time) students seem to accept the technique as the "standard" method and not as just one of the methods which can be used.
8. The data gathered by a class could provide material for a discussion of the meaning of precision and accuracy (which are often confused). Statistical analysis could be done to calculate standard deviation and standard error of the mean, to check normality of the data, to do simple analysis of variance and tests of significance, e.g. t-test, to see if different loops differ significantly.

MATERIALS

1. Nickel-chromium alloy wire, 26 s.w.g., 0.457 mm diam. Griffin and George Ltd.

SPECIFIC REQUIREMENTS

Day 1
wire (150 mm lengths) and loop holders
dye solution (e.g. bromocresol purple, $0.4\,gl^{-1}$)
diluent (e.g. $0.1\,M$ NaOH)
spectrophotometer and cuvettes
50 ml volumetric flask (for dilution of the dye)
1 ml and 5 ml graduated pipettes
filter paper
balance (automatic, weighing to 10 mg)
micro-pipette (20 μl), optional

broth culture (5 or 10 ml) of suitable organism
sterile saline (10 ml)
sterile tubes (10 mm × 50 mm) and racks
sterile glass "50 dropper" pipettes (fine-bore disposable Pasteur pipettes from Bilbate Ltd., 16 Middlemarch, Daventry, England)
plates of suitable agar medium

Day 2
equipment for colony counting

FURTHER INFORMATION

Results of a class experiment showed that the volume of liquid delivered by 20 different triple loops was no more variable than that delivered by 20 commercial 50 dropper pipettes. The loops delivered (mean ± s.e.m.) 21·3 ± 4·8 μl and the droppers 23·9 ± 5·0 μl. The individual droppers were less variable than the individual loops. Repeated heating of a loop has no significant effect upon the volume it holds: one loop was shown to hold 20·7 ± 3·4 μl (mean ± s.d.) before heating, and 22·7 ± 2·2 μl after heating to red heat 20 times, in an experiment where ten replicate determinations were made by the spectrophotometric method.

The loops are of particular value in experiments designed to study the death of bacteria where the conventional sampling procedure requires many sterile pipettes. In such experiments, a flamed triple loop can be used for all the samples and each sample transferred to a diluent or directly to an agar plate.

The saving of pipettes used for serial dilutions and droppers for colony counting is exemplified by a student's results of an experiment designed to investigate the interaction between nalidixic acid and rifampicin when added to a growing culture of *E. coli* (Figure 1). These results were obtained using 1 triple loop, 5 droppers, 50 dilution tubes and 10 nutrient agar plates. A more conventional procedure would have used about 20 droppers, 50 1-ml pipettes and 50 bottles of diluent.

The reaction of students to the use of triple loops is of interest. In general, part-time students, schooled in conventional techniques and used to a liberal supply of sterile materials, reject the

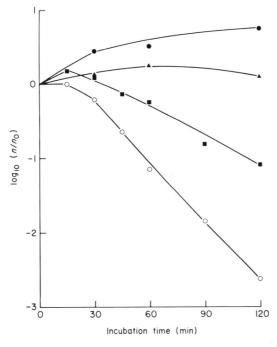

Figure 1   The effect of nalidixic acid (5 μg ml^{-1}) and rifampicin (20 μg ml^{-1}) on the survival of *Escherichia coli* in nutrient broth at 37°C: ●, control; ▲, rifampicin; ■, nalidixic acid; ○, nalidixic acid and rifampicin.

methods as being crude and inaccurate. In contrast, the best full-time students, who are aware of the limitations of all colony counting techniques and are used to a limited supply of sterile materials, really appreciate how the loops permit much more elaborate and interesting experiments to be performed.

REFERENCES

A. J. Hedges, R. Shannon and R. P. Hobbs (1978). Comparison of the precision obtained in counting viable bacteria by the Spiral Plate Maker, the Droplette and the Miles and Misra methods. *Journal of Applied Bacteriology*, **45**, 57–65.

B. Jarvis, V. H. Lach and J. M. Wood (1977). Evaluation of the Spiral Plate Maker for the enumeration of micro-organisms in foods. *Journal of Applied Bacteriology*, **43**, 149–57.

J. Kramer (1977). A rapid microdilution technique for counting viable bacteria in food. *Laboratory Practice*, **26**, 675–6.

5

The Efficiency of the Autoclave

L. MULLENGER

*Department of Biological Sciences, Lanchester Polytechnic,
Coventry, CV1 5FB, England*

> *Level*: Beginning undergraduates
> *Subject area*: Introductory microbiology
> *Special features*: Very simple and works well

INTRODUCTION

The working principle of an autoclave is that pure saturated steam at a pressure above atmospheric pressure is used to heat the load in the autoclave. Initially the load is cool, therefore the steam condenses on it liberating the latent heat of condensation and thus raising the temperature of the load. For the proper working of the autoclave it is necessary to ensure that the steam is fully saturated; that is, any air in the autoclave must be removed. Pure saturated steam has a temperature of 121 °C at a pressure of 103·4 kPa (15 lbf in^{-2}) or 115 °C at a pressure of 69 kPa (10 lbf in^{-2}). However, the temperature will be lower if any air remains mixed with the steam. This follows from Dalton's law of partial pressures which states that the total pressure of a mixture of steam and air will equal the sum of their individual pressures. Thus the more air there is present, the lower will be the partial pressure of the steam and thus the lower the overall temperature of the mixture.

In modern autoclaves air is removed either by evacuating the chamber before admission of the steam, or the air is displaced by gravity; steam being lighter than air. In the latter case the autoclave is fitted with a temperature-sensitive vent in the lowest part of the chamber. In a simple portable, pressure-cooker type of autoclave it is not possible to displace all of the air through the top vent although sufficient is removed to make them useable in practice.

This experiment is designed to illustrate the effect of leaving varying mixtures of steam and air in a portable autoclave on its ability to sterilize cultures of *Bacillus stearothermophilus*. This organism is often used as a biological indicator of the satisfactory operation of an autoclave.

Those wishing to photocopy this experiment must follow the instructions given on page v.

EXPERIMENTAL

Day 1 *(3 h)*

With the lower bung in position, place the aspirator in a sink of water. Fill the aspirator to the top with water. Seal with the top bung and remove the lower bung. The aspirator should remain full of water and will act as a gas jar to collect displaced air from the portable autoclave.

↓

Set up the portable autoclave on a gas ring (*see* note 2). Attach a suitable length of rubber gas tubing to the outlet vent on the autoclave lid and insert the free end into the lower outlet of the aspirator (*see* Figure 1).

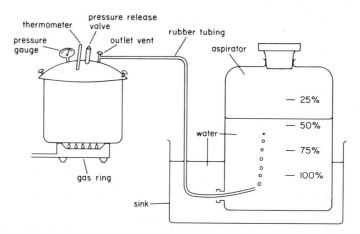

Figure 1

↓

Make sure the autoclave outlet vent is fully open. Light the gas and continue to heat the autoclave until all the air has been expelled into the aspirator, i.e. when the water level in the aspirator ceases to fall. Remove the tubing from the aspirator and turn off the gas supply.

↓

Lower the water level in the sink until it is about 2·5 cm below that in the aspirator. Dry the outside of the aspirator and mark the internal water level using a spirit-based marker pen.

↓

With a metre-ruler mark the positions where the water level would be if the autoclave had 25, 50 and 75 % of the air expelled.

↓

Now do 5 separate runs. With each, place a Universal bottle containing $5\,cm^3$ of a suspension of *B. stearothermophilus* in the autoclave. For the different runs expel either no air, 25, 50, 75 or 100 % of the air. When the correct amount has been expelled close the outlet vent and continue to heat until the pressure in the autoclave rises to 69 kPa (*see* note 3). By manipulating the gas level maintain this pressure for 10 min. If the autoclave lid is fitted with a thermometer note the temperature attained in each case.

↓

After 10 min turn off the gas supply, allow the autoclave to cool and remove the specimen (*see* note 4).

↓

Make three decimal dilutions in saline of all autoclaved samples and 10 decimal dilutions of the unautoclaved stock sample of *B. stearothermophilus*. Using molten nutrient agar held at 60 °C make triplicate pour plates of each autoclaved dilution (include three plates of undiluted material) and plate the 10^{-6}–10^{-10} dilutions of the control. Incubate plates overnight at 37 °C (*see* note 1).

Day 2 *(30 min)*

Count the number of colonies obtained and compute the % survival.

Notes and Points to Watch

1. *Bacillus stearothermophilus* has an optimum growth temperature of between 50 and 60 °C, and can grow at temperatures up to 70 °C. However, for the purposes of this experiment it has proved quite feasible to grow the organism at 37 °C.
2. Students should be closely supervised when using the autoclave which can be rather unstable on a gas ring. An electric autoclave does not permit sufficient temperature control to maintain a steady 69 kPa.
3. The automatic pressure valve on the lid of the autoclave would allow the air in the autoclave to escape. This must be set to open at 103·4 kPa. If the autoclave is then run at 69 kPa, as suggested, this valve will not operate and will not interfere with the result.
4. Autoclaves, steam and samples are all very hot and students should wear protective gloves.

MATERIALS

1. *Bacillus stearothermophilus* (NC1B 8157). Grow for at least one week in nutrient broth at 37 °C to induce spore production. Cell concentration should be as high as possible but at least 10^7 cells cm^{-3}.
2. A portable gas autoclave. A suitable model (J/S118) can be obtained from R. W. Jennings and Co Ltd., Scientech House, Main Street, East Bridgford, Nottingham, NG13 8PQ, England.
3. A 20 litre glass aspirator.

SPECIFIC REQUIREMENTS

These items are needed for each student.

6 Universal bottles containing 5 cm³ *B. stearothermophilus*
25 dilution bottles containing 9 cm³ saline
75 empty Petri dishes

4×250 cm³ molten nutrient agar held at 60 °C
30 sterile 1 cm³ pipettes with pipette bulb
1 Whirlimixer (for mixing dilution bottles)
1 clock
disposal jar for used pipettes

FURTHER INFORMATION

Table 1 presents a set of typical class results with *B. stearothermophilus*. If a non-spore forming organism is tested at the same time (e.g. *E. coli.*) no survivors will be found in any of the runs. A large proportion of the initial sample may be expected to still be present as vegetative cells. Although not tested experimentally the impact of this experiment might be increased by attempting to obtain a pure spore preparation by some suitable means such as pasteurization.

Table 1

% air removed	Temperature (°C)	% survival
Control sample	Not autoclaved	100^a
0	89	3.3×10^{-2}
25	97	1×10^{-2}
50	105	3.6×10^{-4}
75	111	4.8×10^{-6}
100	115	nil^b

[a] viable count of 2.5×10^7 colonies
[b] no survivors from 2.5×10^7 bacteria

6

Spectrophotometric Examination of Prodigiosin

D. B. DRUCKER

Department of Bacteriology and Virology, University of Manchester, Stopford Building, Oxford Road, Manchester, M13 9PT, England

> *Level*: Beginning undergraduates
> *Subject area*: General microbiology; microbial biochemistry
> *Special features*: Colourful experiment

INTRODUCTION

All bacteria excrete solutes from their cells. Occasionally a solute is coloured and therefore instantly noticeable. *Serratia marcescens* excretes the red soluble product prodigiosin, pigment production of which depends both on the genotype of the strain and on experimental conditions, e.g. concentration of certain cations, temperature, illumination. Prodigiosin (Figure 1) is a heterocyclic compound which acts both as an Eh and a pH indicator and is soluble in acetone.

Production of prodigiosin is an important characteristic in diagnostic microbiology for the identification of *S. marcescens*. Its colour indicates that it absorbs certain wavelengths of visible light more than other visible wavelengths, i.e. it has an **absorption spectrum**. The absorption spectrum is measured by determining the absorbance of a series of wavelengths of light by a

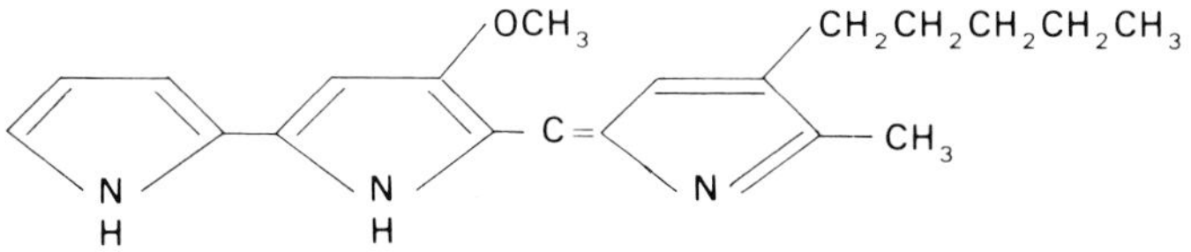

Figure 1   Structure of prodigiosin.

solution of pigment. Also, using dilutions of pigment solution, absorbance can be determined over a series of concentrations. A calibration graph can then be plotted (according to Beer's law), where concentration is proportional to absorbance, A, over a limited range of values.

The spectrophotometer is a widely used instrument in microbiology. Like a thermometer it is useless unless carefully calibrated. The two "ends of the scale" are $0\% \ T$ (no light transmitted) and $100\% \ T$ (all light transmitted). Concentration is proportional to absorbance. Absorbance, $A = -\log T$, where $T =$ transmittance, or the proportion of light transmitted through a sample.

The student must understand the relationship of A and T, and the significance of $0\% \ T$ and $100\% \ T$. Once these ideas are mastered the student can use any spectrophotometer and, more important, use it intelligently instead of "driving by numbers".

The aims of this experiment are: (1) to extract the pigment; (2) to illustrate the use of the spectrophotometer for determining an absorption spectrum; and (3) to illustrate Beer's law.

EXPERIMENTAL

Place a loopful of a few red colonies in a Universal bottle. Gently add 5 ml of acetone. The prodigiosin will dissolve in the acetone leaving an intact colourless smear of organisms.

$\downarrow$

Evaporate the acetone solution under a stream of air in a fume cupboard.

$\downarrow$

Dissolve the residue in 5 ml 0·1 M sodium phosphate buffer pH 5·0 (see note 3).

$\downarrow$

Obtain the absorption spectrum by measuring A (absorbance) at 20 nm increments from 440 to 700 nm (see notes 1 and 2) (Figure 2). A blank of phosphate buffer should be used.

$\downarrow$

Beer's law may be verified by measuring absorbances of a known series of dilutions of prodigiosin.

- -

Notes and Points to Watch

1. If available, a double beam instrument could also be used.
2. *Practice.* The instrument should be demonstrated to the student. Bad practices should be discouraged, e.g. pipetting liquid into cuvettes, rather than decanting; or spilling liquids inside the instrument.
3. *Plastic cuvettes.* Plastic cuvettes may be used if the acetone extract is dried and the residue dissolved in water. This is important, for acetone renders plastic cuvettes opaque. The acetone extract may also be examined directly if glass cuvettes are used.

SPECIFIC REQUIREMENTS

These items are needed for each group of students.

S. *marcescens* on an agar plate
spectrophotometer

acetone
Universal bottles
0·1 M sodium phosphate buffer, pH 5·0
cuvettes

ADDITIONAL INFORMATION

Figure 2 depicts an absorption spectrum for prodigiosin in acid buffer. This experiment may be combined with Experiment 38 to examine pigments of mutants. *See also* Experiment 1 on spread of infection.

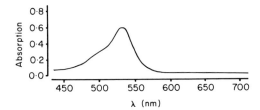

Figure 2   Absorption spectrum of prodigiosin at acid pH.

A Model System for Gradient Centrifugation

L. MULLENGER

*Department of Biological Sciences, Lanchester Polytechnic,
Coventry, CV1 5FB, England*

> *Level*: Beginning undergraduates
> *Subject area*: Miscellaneous techniques
> *Special features*: Simple, always works, good fun

INTRODUCTION

The fractionation and purification of subcellular components, e.g. ribosomes, mitochondria, nucleic acids or proteins may be achieved on density gradients of such substances as sucrose sorbitol or caesium chloride. The caesium ion has such a large mass that a caesium chloride solution will spontaneously form into a gradient of increasing density if a centrifugal force in the region of $500\,000\,g$ is applied over a period of approximately 24 h. Where gradients of sucrose or sorbitol are required these have to be artificially manufactured with gradient-making machines (for details *see* Williams and Wilson, 1975).

Although density gradient centrifugation is an extremely important biochemical technique it requires highly sophisticated equipment not readily available to undergraduate classes, e.g. an ultracentrifuge. Also the resulting gradients are usually small in volume and the banding of subcellular components is not readily visualized.

The experiments described below scale up the whole exercise by allowing the student to manufacture a large gradient ($100\,cm^3$) and separate a mixture of bacteria and yeast cells which are readily visualized when banded. Furthermore, the ultracentrifuge can be replaced by a simple bench centrifuge and fractionation of the gradient can be performed easily.

EXPERIMENTAL

Experiment 1 (a class demonstration; 5 min)

A model for isopycnic centrifugation

Isopycnic (equal density) centrifugation depends upon particles sedimenting in a gradient (which may or may not be preformed) until the buoyant density of the particle and the density of the

---

gradient are equal. Thus particles are separated by their buoyant density and not by their size and shape.

In the model system proposed a 2-litre gradient is manufactured from 5 and 70% sucrose. Table tennis balls are used to represent subcellular particles. These all have identical size and shape but they are filled with sucrose solution at different densities. When they are placed on the gradient they rapidly fall to the level of isodensity.

Prepare a 2-litre gradient from 5 and 70% sucrose.

↓

Add table tennis balls containing water, 5, 10, 20, 30, 40, 50, 60 and 70% sucrose (*see* note 1).

- -

Notes and Points to watch

1. It must be possible for the balls to pass each other freely in the gradient. This requires an internal diameter of at least 85 mm.
2. The gradient is remarkably stable so all the balls can be thrown on together. This looks quite spectacular while they sort themselves to their respective density levels.
3. The balls will remain in position for 2–3 weeks should this be required.

MATERIALS

1. A 2-litre gradient maker (MSE ref W101A, 25–28 Buckingham Gate, London, SW1). It is not necessary to use the peristaltic pump which MSE offer with this item as the sucrose drains satisfactorily under gravity if the apparatus is set up in the same way as the student gradient—former detailed below. It is not possible to set up a magnetic stirrer with this item and it is necessary that a mechanical stirrer be lowered into the container from above.
2. The collecting vessel illustrated in Figure 1 was manufactured from standard Pyrex heavy wall tubing (outside diameter 9·5 mm; wall thickness, 5 mm; height 450 mm) (Corning catalogue number 5520/95, R. W. Jennings, Main Street, East Bridgeford, Nottingham). If glass blowing facilities are not available the base should be sealed with a cork or bung.
3. Table tennis balls should be filled to capacity with the appropriate sucrose solution with a syringe. The hole may be sealed with nail varnish or a small piece of tape. Label the balls with the sucrose concentration.

Figure 1

Experiment 2 *(3 h)*

The manufacture of a 100 cm³ sucrose gradient

Set up a gradient-making apparatus as in Figures 2a and 2b (*see* note 1).

↓

Add 50 cm³ of 75 % sucrose solution to tube A with clip C closed (*see* notes 2 and 6).

↓

Open clip C to allow sucrose to run along the connecting tube to tube B to prevent air bubbles forming. As soon as it reaches tube B close clip C (*see* note 3).

↓

Add 50 cm³ of 5 % sucrose solution to tube B with clip D closed (*see* note 6).

↓

Open clip D to allow 5 % sucrose to run along outlet tube 3–5 cm beyond clip D. Close clip D.

↓

Start magnetic stirrer and ensure the outlet tube has a slight downward slope with the Pasteur pipette vertical in the centrifuge tube and extending to within 2 mm of the bottom of the tube (*see* note 6).

↓

Open clips C and D simultaneously. The gradient will now form.

↓

Prepare a second gradient in the same way.

↓   ↓

Determination of the linearity of the gradient

Clamp the first gradient in a vertical position and mark the side of the glass tube in 1 cm steps from the top to the bottom of the gradient.

↓

Lower a 10 cm³ pipette to each mark in turn and draw off the gradient above that mark. If this is done gently the gradient below the mark should not be disturbed (*see* note 4).

↓

Separation of yeast and bacteria on a gradient

Take the second gradient.

↓

Layer 5 cm³ of mixture of bacteria and yeast on top of the gradient (**gently**!) with a pipette.

↓

Obtain the average density of each fraction by mixing it well and weighing 5 cm³ on a balance, weighing to 3 decimal places in a preweighed 10 cm³ beaker. Wash and dry well between each weighing (*see* note 5).

↓

Plot specific gravity against fraction number.

Centrifuge for 30 min at 4°C at 2500 rev. min⁻¹ in an MSE Mistral 6L with a swing out head (this gives 2000 g at the tip of the centrifuge tube). If using a bench centrifuge operate it at maximum speed at room temperature.

↓

Remove centrifuge tube and look for two distinct layers of cells within the gradient. Hold the tube vertically in a clamp and mark off the outside of the tube in 1 cm³ steps from the top of the gradient.

↓

Fractionate the gradient as with the first gradient.

↓

Mix the contents of each tube thoroughly and, using a counting chamber, estimate the concentration of yeast and bacteria in each gradient (*see* notes 7–9).

↓

Plot cell concentration against fraction number on semi-log graph paper.

Notes and Points to Watch

1. Tube A should not be lower than tube B; in fact it helps if it is a few millimetres higher.
2. 75 % sucrose takes some time to dissolve—warming helps. It is very sticky and students need to be careful otherwise it gets all over other apparatus.
3. Sometimes the sucrose will not run in the gradient maker. Obtain a bung to fit the top of tubes A and B and insert a central tube in the bung. Gently blowing into tubes A or B will cause the flow to start.
4. Fractionation is made easier if the pipette is clamped into position so the gradient does not become disturbed.
5. Weigh the lightest (top-most) fractions first and wash and dry the weighing beaker as well as possible with tissues. Alternatively use disposable weighing trays.
6. When separating cell components in an ultracentrifuge very small gradients (capacities below 5 cm³) are used and it would disturb the gradient if the delivery tube had to be withdrawn. To avoid this the sucrose is usually run down the side of the tube with the dense sucrose entering first. This can be demonstrated in this model system by allowing the sucrose to run down the side of the glass centrifuge tube but it is necessary to reverse the sucrose solutions so that 75 % sucrose is in tube B.

(a)

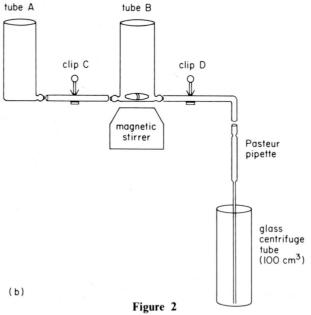

tube A   tube B

clip C   clip D

magnetic
stirrer

Pasteur
pipette

glass
centrifuge
tube
(100 cm³)

(b)

Figure 2

7. Students should be very careful not to get sucrose on the microscope.
8. If using a Helber counting chamber an average of one organism per small square is equivalent to a concentration of 2×10^7 organisms cm^{-3}.
9. It would be possible to estimate the cell concentration in each fraction by some means other than cell counts. For instance the turbidity could be measured at 550 nm using a colorimeter (for details *see* Experiment 6). Alternatively a nitrogen or protein colorimetric determination with Nessler's reagent or the biuret reaction may be used. For further detail *see* Meynell and Meynell (1970).

MATERIALS

Students should work in pairs. Each pair will require the following items:
1. A gradient maker (*see* Figure 2). This comprises: 2 glass vessels made from medium wall Pyrex glass tubing (outside diameter 40 mm; wall thickness 2·3 mm; height 165 mm) (Corning catalogue number 5510/40).
Tube A should have a single outlet at the base.
Tube B should have an inlet and an outlet directly opposite to each other.
Connect as shown using colourless silicone rubber tubing (diameter 5 mm, approximate length 150 mm).
Close off with 2 thumb clips.
Lead the outlet into the centrifuge tube using a Pasteur pipette.
Tube B requires a 2 cm magnetic follower and should stand on a magnetic stirrer.
2. A slide counting chamber (optional: *see* note

10). A suitable counting chamber is a Helber counting chamber with Thoma ruling (Gallenkamp catalogue No. MNK-780-T: Gallenkamp and Co., P. O. Box 290, Technico House, Christopher Street, London EC2P 2ER).
3. A mixture of bacteria and yeast. Any bacterium and yeast can be used. Suitable organisms are *Beneckia natriegens*, or *Escherichia coli* and *Saccharomyces cerevisiae*. The organisms are grown separately to the stationary phase in a rich medium, pelleted and resuspended in saline. The two saline suspensions are mixed and dispensed so that each pair of students receives 5 cm^3.
4. Each pair of students will require access to the following: an analytical balance (e.g. the Oertling R series) a Mistral 6L centrifuge or a bench centrifuge.

SPECIFIC REQUIREMENTS

125 cm^3 5% sucrose
125 cm^3 75% sucrose
2 glass 100 cm^3 centrifuge tubes (straight-sided, round botton (MSE cat. No. 34411–8241)
10 test tubes to hold the fractions

1×10 cm^3 beaker to weigh the fractions (or disposable weighing trays)
1×100 cm^3 measuring cylinder
1×10 cm^3 pipette with bulb
tissues, distilled water
marker pen and ruler

FURTHER INFORMATION

Because both vessels A and B are identical cylinders, the gradient maker should produce a linear gradient. Figure 3 shows that surprisingly linear gradients are in fact obtained in spite of the simplicity of the equipment.

Figure 4 shows a typical student result where a fairly clean separation of the *E. coli* and yeast cells has been achieved. The two bands of cells are clearly visible provided glass centrifuge tubes are used. Two peaks are often obtained for the yeast cells as single cells stabilize higher up the gradient than budding cells.

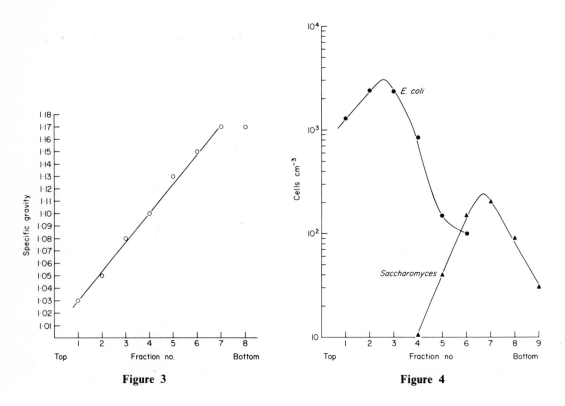

Figure 3   Figure 4

REFERENCES

G. G. Meynell and E. Meynell (1970). "Theory and Practice in Experimental Bacteriology". Cambridge University Press, England.

B. L. Williams and K. Wilson (eds) (1975). "Principles and Techniques of Practical Biochemistry". Edward Arnold, London.

8

Preparation and Analysis of Polar Lipids from *Staphylococcus aureus* and *Escherichia coli*

I. CHOPRA

Department of Microbiology, University of Bristol, Bristol, BS8 1TD, England

> *Level*: All undergraduate years
> *Subject areas*: Microbial chemistry and biochemistry
> *Special features*: Strong visual impact

INTRODUCTION

Polar lipids are a major constituent of bacterial membranes. In Gram-positive bacteria they occur in the cytoplasmic membrane, whereas in Gram-negative species they are distributed both in the inner (cytoplasmic) and outer membranes. Although, in general, bacterial species cannot be identified by their polar lipid composition, *Staphylococcus aureus* and *Escherichia coli* can be distinguished from each other on this basis (Goldfine, 1972). This experiment demonstrates the qualitative differences in polar lipids of *S. aureus* and *E. coli*.

EXPERIMENTAL

Day 1 *(1 h)*

Take 40 ml broth culture of *S. aureus* or *E. coli*.

↓

Transfer to 500 ml separating funnel.

↓

Add 100 ml methanol followed by 50 ml of chloroform and shake for 5–10 min to extract lipid.

↓

Add 50 ml chloroform, shake and add 50 ml water.

↓

Allow phase separation to occur. Collect lower phase into 250 ml quickfit flask.

↓

Concentrate extracts (approximately 1 ml final volume) on a rotary evaporator.

- -

Day 2 (4 h, but not fully occupied)

Each student is provided with a silica gel thin-layer plate (20×20 cm, 0·25 mm layer, without binder) which has been dried overnight at room temperature in a desiccator.

↓

Draw a thin pencil line approx. 3 cm from an edge which is completely coated (but do **not** cut the silica gel layer). Carefully mark 6 origin spots along this line approx. 3 cm apart and at least 2 cm from either edge of the plate.

↓

In addition to the concentrated lipid extracts each student is provided with the following standards: cardiolipin, phosphatidyl glycerol, phosphatidyl ethanolamine and lyso phosphatidyl ethanolamine (all at 5 mg ml^{-1} of chloroform). Apply 10 μl of each standard and 10 μl of extracts to separate origin spots (noting which is which) by applying the tip of the micro-pipette to the plate and allowing a small volume of liquid to form a spot **not more than 2–3 mm in diameter**. Remove the micro-pipette and allow the spot to dry. Reapply, etc. keeping the spot size as small as possible. **The spots at which samples are applied should not damage the surface of the gel.**

↓

The plates should be developed in chloroform–methanol–water (65:25:4). Stop development when the solvent front is 2–3 cm from the top of the plate and mark the position of the solvent front. Development usually takes 3 h.

↓

Allow the solvent to evaporate from the plate and then place the chromatogram in a tank containing iodine crystals. After a few minutes the previously non-coloured polar lipids will develop as yellow spots. This results from iodination of the unsaturated groups in the constituent acyl chains of each polar lipid. Measure the R_F of each spot and then locate phospholipids as described below.

↓

Stand the chromatogram in a glass tray in a **fume cupboard** and spray lightly with molybdenum blue reagent (**caution**, contains strong sulphuric acid) until the adsorbent is moistened uniformly. Neutral lipids and glycolipids do not give a positive reaction, but phospholipids show up immediately as blue spots on a white or light blue–grey background. Their intensity increases on standing, but after several hours the background becomes dark and obscures the spots. Measure the R_F of each spot.

Notes and Points to Watch

1. Students should be warned that the solvents used for lipid extraction are potentially hazardous.
2. The recommended procedure for lipid extraction follows that described by Bligh and Dyer (1959).
3. Occasionally the concentrated lipid extracts are cloudy due to contamination with water or non-lipids. Such contaminated extracts should be purified prior to thin layer chromatography. We have found the procedure described by Radin (1969) utilizing benzene to be most suitable. However, treatment with benzene, if required, should be done by the teacher because of the toxicity of this solvent. Benzene should be removed from the extracts by lyophilization and the samples resuspended in chloroform (1 ml) for use by the students on Day 2. Manipulations involving benzene should always be performed in a fume cupboard.
4. Thin layer plates should be prepared beforehand using conventional apparatus (e.g. that manufactured by Shandon Southern Products, UK). We have found that plates prepared according to the method of Minnikin and Abdolrahimzadeh (1971) give very satisfactory results. Plates are dried overnight at room temperature by placing them in a desiccator.
5. Chromatography tanks should be prepared at least 30 min before immersion of the plates and are lined with filter paper (e.g. Whatman 3MM) wetted in developing solvent.
6. Lipid standards (apart from diglucosyldiglyceride and lysyl phosphatidyl glycerol) are available commercially.
7. If possible, spotting templates (e.g. Shandon) should be provided to minimize possible damage to the gel surface during application of samples.
8. During extraction and concentration of lipid extracts introduction of unnatural lipids should be avoided, e.g. avoid the use of short Pasteur pipettes where the solvent spray inside the pipette can reach the rubber bulb and drain back down. Similarly, the use of Teflon stoppers and sleeves is recommended for the separating funnels and rotary evaporator.

MATERIALS

1. Broth cultures (late log phase) of *Staphylococcus aureus* and *Escherichia coli*.
2. Analar grade chloroform and methanol.
3. Thin layer chromatography plates (20 × 20 cm, 0·25 mm layer) prepared from a slurry of Merck Silica Gel PF$_{254}$ (40 g) in 100 ml of 0·2% aqueous sodium acetate (sufficient to produce 5 plates).
4. Lipid standards dissolved at 5 mg ml^{-1} chloroform. These are best dissolved in chloroform immediately before the start of the work on Day 2.
5. Molybdenum blue reagent for the detection of phospholipids is prepared as follows.

Reagent I. Add 40·11 g of MoO$_3$ to 1 litre of 25 N H$_2$SO$_4$, and boil gently until molybdic anhydride is dissolved.
Reagent II. 1·78 g of powdered molybdenum metal is added to 500 ml of Reagent I, and the mixture is boiled gently for 15 min. The solution is cooled and decanted.
Reagent III. Equal volumes of Reagent I and Reagent II are mixed and the combined solution is mixed with 2 volumes of water. This final Reagent III (molybdenum blue reagent) has a greenish yellow colour and is stable for months.

SPECIFIC REQUIREMENTS

Day 1

40 ml broth cultures of *S. aureus* and *E. coli*

two separating funnels (500 ml) with Teflon stoppers and sleeves

2 × 100 ml methanol; 4 × 50 ml chloroform; distilled water; 100 ml measuring cylinders

2 × 250 ml Quickfit flasks to fit rotary evaporator

rotary evaporator and water-bath

Day 2

1 thin layer chromatography plate

lipid standards

10 μl micro-pipettes

pencil and ruler

chromatography tank containing solvent (chloroform–methanol–water; 65:25:4)

rack for chromatography tank

tank containing iodine crystals

10 ml molybdenum blue reagent in spray gun fume cupboard

FURTHER INFORMATION

The polar lipid compositions of *S. aureus* and *E. coli* are shown in Table 1 (see also Goldfine, 1972). The lipid composition of the two organisms can be deduced (a) from the R_F values of the separated components and (b) from the response to iodine and molybdenum blue. For the *E. coli* extract all

Table 1 Polar lipid compositions of *E. coli* and *S. aureus*. For the structures of these lipids *see* Goldfine (1972).

	Presence of					
Organism	Phosphatidyl ethanolamine	Lyso phosphatidyl ethanolamine	Phosphatidyl glycerol	Cardiolipin	Lysyl phosphatidyl glycerol	Diglucosyl diglyceride
E. coli	+	trace	+	+		
S. aureus			+	+	+	+

E. coli   S. aureus

Figure 1 Separation of *E. coli* and *S. aureus* phospholipids by thin layer chromatography. The phospholipids, listed from the origin in order of increasing R_F values, are follows:.
E. coli, phosphatidyl glycerol, phosphatidyl ethanolamine, cardiolipin.
S. aureus, lysyl phosphatidyl glycerol, phosphatidyl glycerol, cardiolipin.
Phospholipids were detected using the molybdenum blue reagent. The *E. coli* extract contains no detectable lyso phosphatidyl ethanolamine and development with iodine would reveal the staphylococcal diglucosyl diglyceride located between phosphatidyl glycerol and cardiolipin.

the spots staining in iodine vapour should also respond positively in the molybdenum blue test (i.e. only phospholipids found) whereas the *S. aureus* extract should contain a spot (corresponding to diglucosyldiglyceride) that is only detected in iodine vapour. Figure 1 shows typical separation of *E. coli* and *S. aureus* phospholipids and Table 2 the approximate R_F values for these components and the staphylococcal glycolipid.

Identification of polar lipids by the methods used here can only be tentative. Confirmation of the respective identities would require determination of the ratios of the various moieties that comprise the polar lipids (e.g. *see* typical analyses in White and Frerman, 1967).

Table 2 Approximate R_F values for various polar lipids separated by thin layer chromatography.

Lipid	R_F
Cardiolipin	0·81
Phosphatidyl ethanolamine	0·60
Diglucosyl diglyceride	0·60
Phosphatidyl glycerol	0·47
Lyso phosphatidyl ethanolamine	0·36
Lysyl phosphatidyl glycerol	0·20

REFERENCES

E. G. Bligh and W. J. Dyer (1959). A rapid method of total lipid extraction and purification. *Canadian Journal of Biochemistry and Physiology*, **37**, 911–17.
H. Goldfine (1972). Comparative aspects of bacterial lipids. *Advances in Microbial Physiology*, **8**, 1–58.
D. E. Minnikin and H. Abdolrahimzadeh (1971). Thin-layer chromatography of bacterial lipids on sodium acetate-impregnated silica gel. *Journal of Chromatography*, **63**, 452–4.
N. S. Radin (1969). Preparation of lipid extracts. *Methods in Enzymology*, **14**, 245–54.
D. C. White and F. E. Frerman (1967). Extraction, characterization, and cellular localization of the lipids of *Staphylococcus aureus*. *Journal of Bacteriology*, **94**, 1854–67.

Bacterial Corrosion of Iron

M. O. MOSS

*Department of Microbiology, University of Surrey,
Guildford, GU2 5XH, England*

Level: All undergraduate years
Subject areas: Industrial microbiology,
biodeterioriation
Special features: Interaction of microbiology and
inorganic chemistry

INTRODUCTION

Microbial attack on iron objects may occur by several mechanisms (Booth, 1968). (1) As a result of the production of corrosive metabolic products such as acids. (2) By the formation of an oxygen cell, in which the presence of microbial material adhering to a metal surface causes a reduced oxygen concentration in that region. This part of the metal may then become anodic, releasing metal ions into solution, the resulting electrons being conducted away to an aerobic region to be

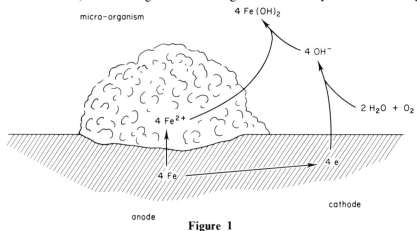

Figure 1

taken up by oxygen (*see* Figure 1). (3) By a process known as cathodic depolarization which is the topic of this experiment.

When iron goes into solution at an anodic area, the electrons released must be conducted away and may be removed at a cathodic area by protons to form nascent hydrogen:

Anode:
$$4 \, Fe \rightarrow 4 \, Fe^{2+} + 8 \, e \qquad (1)$$

$$8 \, H_2O \rightarrow 8 \, OH^- + 8 \, H^+ \qquad (2)$$

Cathode:
$$8 \, H^+ + 8 \, e \rightarrow 8 \, (H). \qquad (3)$$

Once a layer of nascent hydrogen is absorbed onto the surface of the cathodic area of the metal it is said to be polarized and further reaction is inhibited. Depolarization is the process by which this protective film of nascent hydrogen is removed so that further corrosion is possible. Depolarization in a torch battery is carried out by an oxidizing agent such as manganese dioxide. To account for depolarization and continued corrosion under anaerobic conditions, Von Wolzogen Kühr and Van der Vlugt (1934) postulated that the hydrogen may be removed by the hydrogenase activity of sulphate-reducing bacteria such as species of *Desulfovibrio*:

$$SO_4^{2-} + 8 \, (H) \rightarrow S^{2-} + 4 \, H_2O. \qquad (4)$$

The accumulation of excess hydroxyl ions and sulphide ions arising in equations 2 and 4 are removed as ferrous hydroxide and ferrous sulphide at the anodic region of the metal:

Anode:
$$Fe^{2+} + S^{2-} \rightarrow FeS \qquad (5)$$

$$3 \, Fe^{2+} + 6 \, (OH^-) \rightarrow 3 \, Fe(OH)_2. \qquad (6)$$

The overall reaction represented by all these processes, chemical and microbiological, may thus be represented by the following stoichiometry:

$$4 \, Fe + SO_4^{2-} + 4 \, H_2O \rightarrow FeS + 3 \, Fe(OH)_2 + 2(OH^-).$$

Although Booth and Tiller (1968) certainly showed that the hydrogenase-negative sulphate-reducing organism *Desulfotomaculum orientis* was negative in their test for anaerobic corrosion, whereas the hydrogenase positive strains of *Desulfovibrio* were positive, there has been some debate about the relative roles of the hydrogenase activity and of sulphide. In an attempt to resolve this question, Booth *et al.* (1968) used a fumarate utilizing strain of *Desulfovibrio desulfuricans* which grew in the absence of sulphate (and hence in the absence of bacterially produced sulphide). They found that, although corrosion occurred, it was accelerated by the addition of ferrous sulphide. King and Miller (1971) have also suggested that iron sulphide is actively involved in corrosion, although they accept that cathodic depolarization is also important.

The aim of this experiment is to demonstrate the enhanced corrosion of a mild steel surface by *Desulfovibrio desulfuricans*. It is a reproduction of the experiment described by Iverson (1966), in which sulphate is replaced as the electron acceptor of the hydrogenase enzyme by benzyl viologen, a useful redox dye which is colourless when oxidized and violet when reduced.

Although the experiment is carried out in the absence of sulphate it should be remembered that, not only may the formation of sulphide from sulphate have a direct influence on corrosion, but that sulphate reduction does play a very important part in the oxidative metabolism of this interesting group of anaerobic bacteria (LeGall and Postgate, 1973).

EXPERIMENTAL

Day 1 *(30 min)*

Obtain strips of mild steel (approx. 6 × 1 cm), shaped so that two completely flat ends are joined by an arch (*see* Figure 2 and note 1).

↓

Polish the flat surfaces with emery cloth, degrease the metal strips in acetone and sterilize them by dipping into ethanol and flaming.

↓

Place a generous loopful of *Desulfovibrio*, taken from a plate culture (*see* note 2), onto a small area of a plate of benzyl viologen agar and mark the position of the inoculum on the bottom of the Petri dish.

↓

Place a strip of mild steel onto the surface of the agar so that one end covers the inoculum and both ends make good contact with the agar. Place a second strip on an uninoculated plate as a control. Set the experiment up in duplicate.

↓

Incubate the plates for 24 h at 30 °C in an anaerobic jar containing an atmosphere of nitrogen.

Day 2 *(30 min)*

Remove the metal strips and immediately note any differences in colour between the inoculated and uninoculated ends (*see* note 4).

↓

Flood the plate with 10 % aqueous potassium ferricyanide and observe any changes in colour.

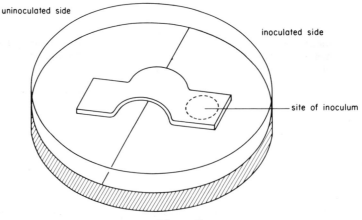

Figure 2

Notes and Points to Watch

1. The metal strips should be carefully shaped for it is essential that both ends make good contact with the agar surface.
2. It is essential to have a large inoculum, for the benzyl viologen plate does not contain any nutrients that would support the growth of the organism.
3. The benzyl viologen plates should be stored under nitrogen until required.
4. The metal surfaces themselves cause some reduction of the benzyl viologen giving a pale violet colour which usually disappears fairly rapidly when the plates are exposed to air.

MATERIALS

1. *Desulfovibrio desulfuricans* NCIB 8312 and/or NCIB 8446.
2. Maintenance medium containing 0·5 g KH_2PO_4, 1·0 g NH_4Cl, 0·05 g $CaCl_2$, 0·02 g $MgCl_2$, 16 g sodium fumarate, 4·2 g sodium lactate and 1 g yeast extract per litre of distilled water.
3. Growth medium (TSA) containing 40 g Tryptone Soya Agar (Oxoid) and 0·5 g $FeSO_4 \cdot 7H_2O$ per litre of distilled water.
4. Buffered benzyl viologen agar containing 1·21 g tris (hydroxymethyl) aminomethane, 0·1 g benzyl viologen (B.D.H.) and 20 g inoagar (Oxoid) per litre of distilled water. This medium is adjusted to pH 7·0 with HCl before

autoclaving at 103·4 kPa (15 lbf in^{-2}) for 15 min.
5. Strips of mild steel (approx. 6 × 1 cm) shaped in the form of an arch joining two flat regions of about 1·5 cm long (*see* Figure 2).
6. Hydrogen (99%), carbon dioxide (1%) for maintenance and growth.
7. Nitrogen for corrosion experiment.
8. Acetone for degreasing metal strips.
9. Industrial methylated spirits for sterilizing metal strips.
10. 10% aqueous potassium ferricyanide.
11. 1% potassium ferrocyanide in 1·0 N HCl. (Optional, *see* further information.)

SPECIFIC REQUIREMENTS

Day 1
A 3-day culture of *D. desulfuricans* grown on TSA at 30 °C under 99% H_2, 1% CO_2
4 plates of benzyl viologen agar
4 mild steel strips, emery cloth and acetone wire loop and alcohol

anaerobic jar and nitrogen

Day 2
10% aqueous potassium ferricyanide
1% potassium ferrocyanide in 1·0 N HCl (optional)

FURTHER INFORMATION

On removing the metal strip it is usual to find that the imprints at both ends are violet because the metal itself is able to reduce benzyl viologen to some extent. However, if a hydrogenase has been active then the region of the inoculum should be more deeply and persistently coloured.

On flooding the plate with aqueous potassium ferricyanide any ferrous ions in solution will be precipitated as ferrous ferricyanide, a deep blue pigment known as Turnbull's blue. The acceptance of electrons by the hydrogenase of the inoculum will make that end of the metal strip cathodic. It should then be at the other end that the anodic solution of iron (i.e. corrosion) take place.

Sometimes there is a brown surface stain surrounding the parts of the metal in contact with the agar. This may be ferric hydroxide arising from the incomplete removal of oxygen during the setting up of the experiment. This possibility can be tested by flooding a duplicate plate with acidified potassium ferrocyanide. As the ferric hydroxide goes into solution in the acid, it is reprecipitated as ferric ferrocyanide (Prussian blue).

A useful account of the isolation and maintenance of sulphate-reducing bacteria is given by Pankhurst (1971).

REFERENCES

G. H. Booth (1968). Microbiological corrosion. *Process Biochemistry*, **3**, December, 17–20.

G. H. Booth and A. K. Tiller (1968). Cathodic characteristics of mild steel in suspensions of sulphate-reducing bacteria. *Corrosion Science*, **8**, 583–600.

G. H. Booth, L. Elford and D. S. Wakerley (1968). Corrosion of mild steel by sulphate-reducing bacteria: an alternative mechanism. *British Corrosion Journal*, **3**, 242–5.

W. P. Iverson (1966). Direct evidence for the cathodic depolarization theory of bacterial corrosion. *Science*, **151**, 986–8.

R. A. King and J. D. A. Miller (1971). Corrosion by the sulphate-reducing bacteria. *Nature, London*, **233**, 491–2.

J. Legall and J. R. Postgate (1973). The physiology of sulphate-reducing bacteria. *Advances in Microbial Physiology*, **10**, 81–133.

E. S. Pankhurst (1971). The isolation and enumeration of sulphate-reducing bacteria. *In* "Isolation of Anaerobes," (D. A. Shapton and R. G. Board, eds), Society of Applied Bacteriology Technical Series, No. 5, pp. 223–40. Academic Press, London and New York.

C. A. H. Von Wolzogen Kühr and L. S. Van der Vlugt (1934). The graphitization of cast iron as an electrochemical process in anaerobic soils. *Water (Den Haag)*, **16**, 147.

10

Precision and Accuracy in the Use of Microbiological Pipettes or, Statistics without Tears

A. C. WARDLAW

Microbiology Department, Glasgow University,
Glasgow, G11 6NU, Scotland

Level: Intermediate undergraduates
Subject areas: Quantitative microbiology; statistics and data processing; design of experiments
Special features: Minimal requirements; provides a framework of very simple exercises on which a large number of statistical ideas and tests can be based

---

INTRODUCTION

The title of this chapter indicates its twin objectives: (1) to check the students' manipulative skill in using ordinary 1 cm^3 graduated pipettes, and (2) to introduce a variety of statistical ideas and tests by getting the students to analyse data which they themselves have generated, and which therefore have an immediacy that is often absent from the numerical problems encountered in statistics books or lectures.

A central aim is to emphasize the crucial interrelationship—often overlooked—between how an experiment or survey is actually done and how it is subsequently evaluated statistically: in particular, the fact that the statistical approach has to be built in **before** the experiment is done and not tacked on as an afterthought, only after the data have been gathered. This leads to the highlighting of the unpalatable, and hence infrequently stated, fact that "statistical respectability" in even a simple experiment may be difficult to achieve, and is often in conflict with the practicalities of doing the benchwork quickly, economically and without mistakes. The student should be encouraged to realize that in many types of quantitative experimental work, there is a definite "Art of Science" in choosing some middle way between the demands of absolute statistical perfection and the limitations imposed by time and resources in the laboratory.

The exercise, in addition to providing information on student accuracy and precision in the use of pipettes, introduces the following statistical ideas and procedures: avoidance of bias in the design and conduct of an experiment; use of random-number tables; the concepts of the **sample** and the **population**; determination of mean, median, range, standard deviation, standard error of the mean and 95 % confidence limits; one-way and two-way analysis of variance; F-test; summary of a large mass of data by histogram, cumulative frequency diagram and Probit plot; use of Rankits; parametric and non-parametric tests; Student t-test and Mann–Whitney U-test.

EXPERIMENTAL

Each student is given three ordinary $1 \, cm^3$ microbiological pipettes (unsterile) and asked to find out how accurately and how consistently he or she can make replicate deliveries of $1 \, cm^3$ of water. Three deliveries are to be made from each pipette into an initially tared weighing vessel, and the amount of water delivered each time measured by the increment in weight, read to $\pm 0.1 \, mg$ on an analytical balance. This is equivalent to a volume of $0.0001 \, ml$ and thus permits a much more accurate measurement of the amount delivered than would be possible by simple volumetric methods.

The exercise sets out to answer the following questions:

1. How **consistent** are the nine replicate deliveries?
2. How **accurate** are they? (not the same as question 1).
3. Are there significant differences in the volumetric capacities of the three pipettes?
4. Is it reasonable to treat the data as being normally distributed?
5. Do students differ significantly in the amounts of water they deliver from the same bits of glassware?

Most students if given only the above instructions would probably sit down at the balance and, having weighed the empty receiving vessel, make three deliveries from each pipette in turn, reweighing after each. However, this is **not** the proper procedure if there are any pretensions to "statistical respectability", especially in answering question 3. To do it in this straightforward way would be rather like taking someone into a shooting alley to fire three shots from each of three rifles: if the third rifle gave a better score than the other two, how could it ever be settled afterwards whether this was due to that rifle being a more accurate or stable weapon, or to the marksman improving with practice? Clearly the shots should be fired from the different rifles in some **predetermined random sequence**, which is not disclosed to the marksman until the end of the test. Therefore the recommended procedure is for the students to work in pairs in which one individual "masterminds" the design and conduct of the experiment, while the other does the actual pipetting and weighing. They then change places.

Day 1 (*20 min per student*)

The "masterminder":

stands out of sight of the pipetter and places the 3 pipettes on sheets of paper labelled A, B and C or better, in notches labelled A, B and C in a pipette rack;

↓

consults a Table of Random Numbers (*see* note 1) and arranges the 9 deliveries in a predetermined random sequence such as A, A, C, B, C, A, B, C, B (*see* note 2). This sequence is not divulged to the other student until later;

↓

wets each pipette, so that the pipetter does not know whether any particular pipette that is handed to him/her has been used before. Possibilities for operator bias should therefore be minimal.

The pipetter:

draws up a blank version of Table 1 (*see* Further Information), leaving col. 1 empty until the weighings are complete;

↓

on an analytical balance, weighs an empty weighing bottle with lid, or a small beaker with aluminium foil cover, and enters the tare weight in the bottom line of Table 1, col. 2;

↓

takes the first pipette from the masterminder, fills it (*see* note 3) with distilled water at 20°C, adjusts the level of the meniscus as in normal usage and delivers the water to the tared vessel. The last drop may or may not be expelled according to whether the pipette is of the "blow-out" or "delivery" variety. The pipette is passed back to the masterminder and the vessel is reweighed (col. 2);

↓

proceeds likewise with pipettes passed from the masterminder in the predetermined random sequence until a total of 9 deliveries has been made and weighed into the same vessel;

↓

fills in the rest of Table 1: i.e. obtains the random sequence in which deliveries had been made (col. 1); the weight of each delivery in grams (col. 3), by subtraction of successive values in col. 2; the weight of each delivery in milligrams (col. 4) and in "working milligrams" (*see* note 4) in col. 5;

↓

the pipetter and masterminder then change places and proceed with a new random sequence, but keeping the identity of pipettes A, B and C the same (*see* note 5).

Day 2 et seq. *(3–5 h)*

This period would be taken up with the calculations made on the weight data, and could be done partly on Day 1, but is probably best spread over about three class periods so that the students can be given explanatory tutorials, at suitable intervals, according to background and need. The various calculations, together with sample results, are given in Further Information.

Notes and Points to Watch

1. Random Number tables are provided in each of the three references cited. It is convenient to allocate digits 1, 2, 3 to pipette A; 4, 5, 6 to B and 7, 8, 9 to C. The code is therefore:

Trial:	A_1	A_2	A_3	B_1	B_2	B_3	C_1	C_2	C_3
Random no.:	1	2	3	4	5	6	7	8	9

Thus for example, the first line of the Random Number table (p. 343) in Campbell (1974) is:

$$2\ 0\ 1\ 7\ 4\ 2\ 2\ 8\ 2\ 3\ 1\ 7\ 5\ 9\ 6$$

which gives the pipette-usage sequence:

2	0	1	7	4	2	2	8	2	3	1	7	5	9	6
A	—	A	C	B	—	—	C	—	A	—	—	B	C	B,

where dashes are used to indicate numbers no longer required because that "trial" has already been obtained or is not wanted. This random sequence is therefore A A C B C A B C B.

2. By chance, the Random Number table may yield unscrambled sequences such as B B B A A A C C C. It is legitimate to discard these, so as to avoid bias, provided one's intention to do so is declared in advance. Note, however, that if a learning effect is suspected or feared, then it would be better to use a balanced sequence such as A C B B A C C B A and to allow for it in the subsequent analysis. Randomization should therefore be regarded as part of the "middle way" that gives **some** protection against misleading results.

3. This exercise was done for many years by mouth-pipetting. However to conform to modern safety rules, the pipettes should be used with a mechanical dispenser such as a Pi-pump or safety bulb.

4. The use of "working milligrams" is purely for arithmetical convenience in the subsequent calculations. By having 2- or 3-digit numbers rather than 4- or 5-digit numbers there is less chance of mistakes. There is no rounding up or down around the decimal point and therefore no loss of accuracy from the original data. All students in the class should subtract the same constant (*see* Table 1).

5. To achieve a higher level of "statistical respectability" it would be better—indeed some statisticians would say, **essential**—to randomize the overall sequence in which the **two** students take their nine readings. This would require a third student masterminding the two performers, so that the 18 weights would be fully randomized, rather than being collected as an uninterrupted sequence of nine from one student and then nine from the other. Alternatively, a randomized block design could be used, with each block having a set of six observations, one from each pipette and from each student, in a predetermined random sequence. The whole experiment would then consist of three such blocks.

MATERIALS

1. Analytical balances, preferably capable of weighing to 0·1 mg.
2. Weighing vessels with lid, or 25 ml beakers with aluminium foil covers, one vessel per balance.
3. Thermometer to measure temperature of water.
4. Good-quality pocket calculators.
5. Statistical tables (Random numbers, t, F, U, Probits and Rankits), e.g. as in Campbell (1974) and Bliss (1967).

SPECIFIC REQUIREMENTS

Day 1
analytical balances
50 ml bottles of distilled water
weighing vessels with lids, 1 per balance
thermometer
1 cm^3 pipettes, sterility unimportant
notched pipette-racks (not essential)

Pi-pumps or Safety dispensing bulbs

Day 2
pocket calculators
statistical tables
graph paper, simple 1 cm, ruled in mm

FURTHER INFORMATION

Initially the nine weight readings (y) from each student are treated as nine replicate observations, ignoring temporarily that they came from three different pipettes. Later they will be retabulated to investigate possible between-pipette differences.

Algebraic symbols are introduced as each item is discussed.

Mean ($\bar{y}$) and Standard Deviation (s)

With a good-quality pocket calculator, the arithmetic mean and standard deviation may be obtained directly from the program in the instrument by entering the y-values (i.e. "working mg") and pressing the appropriate buttons (details not given here). It may, however, be useful to remind, or inform, students of the following points:

The formula for arithmetic mean ($\bar{y}$) is:

$$\bar{y} = \Sigma y/N, \tag{1}$$

where N is the number of observations; here $N = 9$ and the Table 1 data yield $\bar{y} = 10.92$.

The formula for standard deviation (s) is:

$$s = \left(\frac{\Sigma y^2 - (\Sigma y)^2/N}{(N-1)} \right)^{\frac{1}{2}}, \tag{2}$$

application of which gives $s = 5.40$.

It is worthwhile "dissecting" this equation to label its parts for subsequent use in the analysis of variance and elsewhere below:

Σy^2	= crude sum of squares
$(\Sigma y)^2/N$	= correction factor
$\Sigma y^2 - (\Sigma y)^2/N$	= corrected sum of squares
$(N-1)$	= degrees of freedom
$\dfrac{\Sigma y^2 - (\Sigma y)^2/N}{(N-1)}$	= variance.

Implicit in calculating $\bar{y}$ and s is the underlying assumption that the data are from a normal, or approximately normal, distribution. The validity of this assumption will be tested later.

The nine y-values are to be regarded statistically as a random sample from an infinitely large "universe" or "population" of weights which, in theory, could be gathered by getting an infinitely long-lived student to repeat the exercise infinitely often. $\bar{y}$ is an **estimator** of the theoretical underlying "universe" mean (μ), and s is an **estimator** of the underlying "universe" standard deviation (σ). Thus as N, the number of observations, approaches infinity, the value of s approaches σ more and more closely.

The Median and the Range

The **median** is the middle value of the nine observations when they are arranged in rank order, i.e. in this example, the median is 11·2. Non-parametric tests (below) make use of medians rather than means.

The **range** is the difference between the highest and lowest value in the series, i.e. $22·6 - 3·7 = 18·9$. Both the range and the standard deviation are "measures of dispersion". The standard deviation (σ) can be estimated from the range by multiplying the latter by a constant whose value depends on the size of N, e.g. for N = 9, the constant is 0·337 (*see* Snedecor and Cochran, 1967, p. 40), which gives an estimate of $\sigma = 6·40$. An additional exercise is to get each student to tabulate the estimates of σ obtained by equation (2) and by the range-multiplier method. These may be plotted as a correlation diagram with the data from the whole class. It should show a linear relationship.

Standard Error of the Mean (s.e.m.)

On the one hand, s.e.m. is an estimate of the standard deviation of $\bar{y}$, i.e. it measures $\bar{y}$ variability about μ in repeated samples each of size N. On the other hand, s.e.m. measures the uncertainty concerning μ as illustrated by its use in calculating 95% Confidence Limits (below). It is defined by

$$\text{s.e.m.} = s/\sqrt{N}. \tag{3}$$

With the example given:

$$\text{s.e.m.} = 5·40/\sqrt{9} = 1·80.$$

The conceptual difference between s and s.e.m. should be emphasized by discussion of the central ideas of the **sample**, i.e. the nine observations, and the underlying hypothetical "universe". If the sample gets very large, i.e. $N \to \infty$, then s.e.m. $\to 0$, despite s retaining a finite value of, say, around 5·4. That is, the individual replicate deliveries still have a certain inherent amount of scatter, as expressed by s, but despite this, if we have enough observations we can determine the underlying "universe" mean (μ) with an s.e.m. that approaches zero.

95% Confidence Limits (95% CL)

Although the s.e.m. is a useful index of the error, an uncertainty, associated with $\bar{y}$, it can be made still more meaningful in a quantitative sense by conversion into 95% confidence limits (95% CL) defined by:

$$95\% \text{ CL} = \bar{y} \pm t.s/\sqrt{N}, \tag{4}$$

where t is the Student t-statistic for $(N-1)$ degrees of freedom and $P = 0.05$. For any set of $N = 9$ observations, $t = 2.306$ for $P = 0.05$.

Inserting the above values into equation (4) gives 95% CL of 6·8 and 15·1. The idea behind the 95% CL is that they utilize the information contained in the sample to make the prediction that the "true" value of the mean, i.e. the "universe" value (μ), has a 95% probability of lying within the limits calculated. Note that there is an implicit assumption here that the data may legitimately be regarded as random samples from an underlying normal distribution.

Precision and Accuracy

The conceptual difference between these two should be emphasized by taking target-shooting, for example, as an illustration. Precision is concerned with how similar are replicate shots or pipette deliveries, and may be measured by standard deviation or range. Accuracy is concerned with closeness of the shots to the bullseye or "target" value. One can have **precise** shooting which is either **accurate** or **inaccurate** as judged respectively by whether the shots all lie within the bull's-eye or are tightly grouped in one of the outer rings. With pure water at 20°C, the weight of $1\,cm^3$ is 998·2 mg.

It is not particularly meaningful to use the term "accurate" without defining **limits of tolerance**. With ordinary Grade B glass pipettes, the manufacturer's tolerance limits are $\pm 2\%$, or from 978·23 to 1018·16 mg of water at 20°C. In the example provided (Table 1) all nine readings are therefore "accurate". However if we set the standard of accuracy higher (equivalent to asking a marksman to fire at a smaller bull's-eye) to say, tolerance limits of $\pm 0.5\%$, this would decrease the allowed range to 993·21–1003·19, and only seven of the nine deliveries would be "accurate".

Table 1 An example of weight data gathered in the pipetting exercise. The weighings are collected in col. 2 from the bottom upwards.

| Pipette | Balance reading (g) | Weight of water delivered | | |
		grams	milligrams	"working milligrams"[a]
B	35·6466	1·0028	1002·8	12·8
C	34·6438	1·0126	1012·6	22·6
B	33·6312	0·9969	996·9	6·9
A	32·6343	1·0024	1002·4	12·4
C	31·6319	0·9964	996·4	6·4
B	30·6355	1·0001	1000·1	10·1
C	29·6354	1·0022	1002·2	12·2
A	28·6332	0·9937	993·7	3·7
A	27·6395	1·0012	1001·2	11·2
Tare	26·6383	0·0000		

[a] Obtained by subtracting 990, a convenient whole number which permits all the data to be expressed as small positive values. When dealing with data from a whole class of students it is usually necessary to subtract a smaller constant, say 960 or 970, to allow for the odd, highly aberrant, values from the least skilful students in the class.

Summarizing the Class Data as a Histogram

First, prepare a table as illustrated by Table 2, in which a range large enough to encompass all the results is divided into classes of 5 mg width, starting with the class with the lowest readings. To avoid double entry of a value, the extent of each histogram class should be identified, e.g. from 10·0001 to 15·0000 working mg. Therefore a value of 15·0 would unambiguously go into the class whose upper boundary is 15·0000 rather than into the next higher class, which starts at 15·0001. For simplicity, each class can be labelled by its upper boundary. Figure 1(a) shows the histogram obtained from values summarized in Table 2. The other three histograms in the figure are from the same set of data but categorized differently. Figure 1(b) shows the effect of assigning the 126 results to 5 mg-wide classes whose boundaries were shifted slightly, e.g. so that each **began** at 15·0000 and ended at 19·9999 instead of from 15·0001 to 20·0000. Figure 1(c) and 1(d) illustrate the effect of having a narrower (1·0 mg) class-width and wider (45 mg) class-width.

Table 2   Assigning the 126 pipetting results from a group of 14 students into classes of 5 mg width, preparatory to plotting a histogram. The two right-hand columns are for the cumulative-frequency and probit plots which follow.

Upper boundary of class (mg)	Accumulation of individual weights	Total number	% of total	Cumulative frequency	Probit
5	//	2	1·59	1·59	2·86
10	𝟕𝑯𝑳 ///	8	6·35	7·94	3·59
15	𝟕𝑯𝑳 𝟕𝑯𝑳 𝟕𝑯𝑳 𝟕𝑯𝑳	19	15·08	23·02	4·26
20	𝟕𝑯𝑳 𝟕𝑯𝑳 𝟕𝑯𝑳 𝟕𝑯𝑳 𝟕𝑯𝑳 𝟕𝑯𝑳 //	32	25·40	48·62	4·96
25	𝟕𝑯𝑳 𝟕𝑯𝑳 𝟕𝑯𝑳 𝟕𝑯𝑳 𝟕𝑯𝑳 𝟕𝑯𝑳	30	23·81	72·43	5·59
30	𝟕𝑯𝑳 𝟕𝑯𝑳 𝟕𝑯𝑳 𝟕𝑯𝑳	19	15·08	87·51	6·15
35	𝟕𝑯𝑳 𝟕𝑯𝑳 //	12	9·52	97·03	6·88
40	///	3	2·38	99·41	7·51
45	/	1	0·79	100·20[a]	–
Totals		126	100·00		

[a] This would be 100·00 exactly but for rounding-off errors in the previous column.

Cumulative Frequency Diagram

In col. 5 of Table 2, the percentages of col. 4 are successively totalled to give the cumulative frequencies, which are also expressed as percentages of the total. A plot of these figures against weight gives the S-shaped cummulative-frequency distribution curve in Figure 2(a). This in itself is not particularly useful, except as an intermediate stage in transforming the bell-shaped normal distribution curve (to which the histogram in Figure 1(a) approximates) into a straight line by a further transformation to probits.

Probit plot

The probit plot is a mathematical device for transforming the bell-shaped, normal distribution curve into a straight line. It may therefore be used to check whether the 126 pipette deliveries approximate to a normal distribution. This is done by converting the cumulative frequencies in col. 5 of Table 2 into probits, by consulting a Probit table (e.g. *see* Bliss, 1967). The probit values corresponding to each cumulative frequency have been entered in col. 6 of Table 2. Figure 2(b) is a plot of probit against weight with, superimposed, the theoretical straight line calculated independently from the mean and standard deviation of the 126 weights by equations 1 and 2. The line is fitted through the points by making use of the fact that a probit value of 3·0 corresponds to $\bar{y} - 2s$, and a probit value of 7·0 corresponds to $\bar{y} + 2s$. Data that fit a normal distribution exactly would give points lying exactly on the straight line. The data given here are clearly a good approximation to a normal distribution.

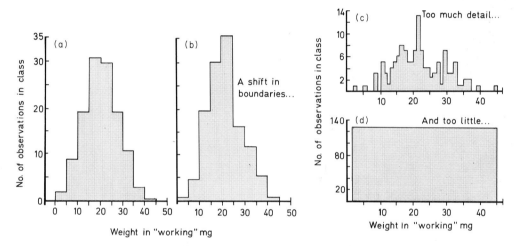

Figure 1 Histograms plotted from the **same** collection of student pipetting results but with the class boundaries defined differently (**a**) and (**b**) (see text), or (**c**) with narrow class-widths or (**d**) very wide class-width.

A more objective procedure for testing for significant departures from normality is given in Snedecor and Cochran (1967, pp. 84–8).

Normality Check by Means of Rankits

The Probit technique is satisfactory for large collections of data, say more than fifty observations. Frequently, however, one needs a rapid (even if only approximate) check on Normality with small-size groups such as the two sets of nine observations obtained by each pair of students. Here the Rankit plot may be useful (*see* Bliss, 1967, p. 108). First arrange the observations (Table 3) in descending rank order starting with the largest value (cols 3 and 5). Then look up the Rankit value (col. 2) corresponding to the Rank no. (col. 1). Graphs are then plotted of Rankit against "working mg" (Figure 3) and the points joined. As with the Probit plot, a straight line is superimposed by independent calculation of the mean ($\bar{y}$) and standard deviation (s) by equations 1 and 2, making use of the fact that a Rankit value of $-1·0$ corresponds to $(\bar{y}-s)$ and a rankit of $+1·0$ corresponds to $(\bar{y}+s)$. If the points show a random scatter around the independently fitted straight line, then one can assume at least provisionally that the data are a random sample from a normal distribution. However, a single Rankit plot based on only nine observations cannot be definitive. If, however, several Rankit plots are available, and if collectively they show no **consistent** pattern of departures from the fitted straight line, then it is reasonable to assume provisionally that the data are approximately normally distributed. The plots in Figure 3 do not reveal obvious departures from normality, but the displacement of student K to higher weight values suggests that this individual consistently delivered larger weights than student J.

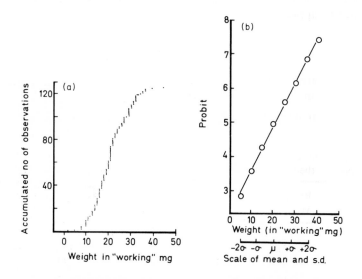

Figure 2 (a) S-shaped cumulative-frequency distribution curve obtained by plotting the student pipetting results summarized in Table 2. (b) Probit plot of the 126 pipetting results arranged in 5 mg classes. The independently positioned (see text) straight line is that expected for a normal distribution with $\mu = 20·67$ and $\sigma = 7·81$.

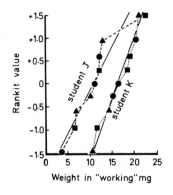

Figure 3 Rankit plots of the pipetting data of two paired students J and K who used the same three pipettes, A, ●; B, ▲; C, ■.

In addition to providing a check on normality, the Rankit diagram may provide a readily visible indication of whether there are likely to be any significant differences between the three pipettes. This is done by observing the extent of mutual overlap of the different types of points along the length of the Rankit plot. Here there is no such evidence.

Application of the Student t-test

The Student t-test may be used to determine whether there is a significant difference in the mean volumes of water delivered by the two students J and K of each pair. Each student will already have calculated the mean and standard deviation of the nine weighings. Thus the following version of the t-test formula may be applied:

$$t = \frac{(|\bar{y}_J - \bar{y}_K|)\sqrt{n}}{(s_J^2 + s_K^2)^{\frac{1}{2}}}, \tag{5}$$

where $\bar{y}_J$, $\bar{y}_K$, s_J and s_K are the means and standard deviations of the working mg from students J and K. From the data in Table 3, these values are:

$$\bar{y}_J = 10\cdot9222, \quad \bar{y}_K = 16\cdot3778 \quad s_J = 5\cdot3997, \quad s_K = 3\cdot7749, \quad n = 9.$$

Table 3   Pipette results from students J and K arranged in descending rank order and tabulated opposite their Rankit values.

Rank no.	Rankit value[a]	Student J weight	Student J pipette	Student K weight	Student K pipette
1	1·485	22·6	C	21·1	B
2	0·932	12·8	B	20·4	A
3	0·575	12·4	A	19·1	C
4	0·275	12·2	C	18·6	C
5	0·000	11·2	A	16·6	A
6	−0·275	10·1	B	15·3	A
7	−0·572	6·9	B	14·2	B
8	−0·932	6·4	C	11·6	C
9	−1·485	3·7	A	10·5	B

[a] From Table of Rankits (Bliss, 1967, p. 498).

This gives $t = 2\cdot484$. The tabulated value of t for $n_J + n_K - 2 = 16$ degrees of freedom and $P = 5\%$ is $2\cdot120$. Therefore there is a significant difference in the mean volumes of water delivered by these two students using the same pipettes.

Note that the t-test should only be used if the data (1) have been obtained by random sampling, (2) are from a normal, or approximately normal, distribution and (3) do not differ significantly in the variances of the two groups. This last mentioned may be checked by the F-test. Here

$$F = \frac{s_J^2}{s_K^2} = \frac{5\cdot3997^2}{3\cdot7749^2} = 2\cdot05$$

which is not significant for 8 and 8 degrees of freedom (tabulated $F = 3\cdot44$ for $P = 0\cdot05$). In addition, this t-test is only appropriate if no differences are assumed to exist between the pipettes. The formula in equation 5 must be changed if the numbers of observations in the two groups are different.

Mann–Whitney U-test

The same data taken for the t-test may also be analysed by the Mann–Whitney U-test if one wishes to avoid the assumption of normality of the data. The following procedure is adapted and condensed from Campbell (1974, p. 59).

Arrange the values in ascending rank order:

Student J: 3·7, 6·4, 6·9, 10·1, 11·2, 12·2, 12·4, 12·8, 22·6.
Student K: 10·5, 11·6, 14·2, 15·3, 16·6, 18·6, 19·1, 20·4, 21·1.

Produce a joint ordering of the ranks of the two groups:

Student J: 1 2 3 4 6 8 9 10 18
Student K: 5 7 11 12 13 14 15 16 17.

Calculate the sum (R) of the ranks for each student

$$R_J = 61 \qquad R_K = 110.$$

Check that the ranking has been done correctly from

$$R_J + R_K = \tfrac{1}{2}(n_J + n_K)(n_J + n_K + 1), \tag{6}$$

where n_J and n_K = no. of observations in each group = 9. Here $61 + 110 = \tfrac{1}{2} \times 18 \times 19$, which checks.

Calculate the U-statistics

$$U_J = n_J . n_K + \tfrac{1}{2}n_K(n_K + 1) - R_K \tag{7}$$
$$= 16$$

$$U_K = n_J . n_K + \tfrac{1}{2}n_J(n_J + 1) - R_J \tag{8}$$
$$= 65.$$

Take the **lower** value of U_J or U_K; here $U_J = 16$ is the lower value. Compare this with the tabulated value of the U-statistic for a 9×9 comparison (*see* Campbell, 1967, p. 347); in this case the tabulated value is 17. If the found U is **less** than the tabulated value, then the difference between the medians of the two groups is significant at the $P = 0.05$ level. These data therefore indicate a barely significant difference between the two students.

One-way Analysis of Variance

Table 4 presents the data from students J and K tabulated in a form suitable for one-way and two-way analysis of variance (*see* Campbell, 1974, Chap. 7, for further information). The former can be done on the data from an individual student as shown in Table 5 for student J. This analysis indicates that there is no significant difference in the mean volumes of water delivered by the three different pipettes used by this student.

Table 4 "Working mg" data (*y*-values) of two students *J* and *K* who worked as a pair in the pipetting experiment. The table also shows group totals, pipette totals and student totals which are needed for the analysis of variance.

Student	Pipette A	Pipette B	Pipette C
J	$\left.\begin{array}{l}3\cdot7\\11\cdot2\\12\cdot4\end{array}\right\}$ 27·3[a]	$\left.\begin{array}{l}6\cdot9\\10\cdot1\\12\cdot8\end{array}\right\}$ 29·8[a]	$\left.\begin{array}{l}22\cdot6\\6\cdot4\\12\cdot2\end{array}\right\}$ 41·2[a]
		overall student *J* total = 98·3 (S)	
K	$\left.\begin{array}{l}16\cdot6\\20\cdot4\\15\cdot3\end{array}\right\}$ 52·3[a]	$\left.\begin{array}{l}14\cdot2\\10\cdot5\\21\cdot1\end{array}\right\}$ 45·8[a]	$\left.\begin{array}{l}19\cdot1\\18\cdot6\\11\cdot6\end{array}\right\}$ 49·3[a]
		overall student *K* total = 147·4 (S)	
Pipette totals	79·6 (Q)	75·6 (Q)	90·5 (Q)

Grand total of all readings: 245·7

[a] Each set of three replicate readings is referred to as a group total (T).

Table 5 One-way analysis of variance of pipette data of student *J*.

Source of variation	Degrees of freedom (d.f.)	Sums of squares	Mean square	Variance ratio (F)	P[a] (%)
Total	8	$\Sigma y^2 - (\Sigma y)^2/N = 233\cdot2556$ (1)	—	—	
Between pipettes	2	$\Sigma T^2/3 - (\Sigma y)^2/N = 98\cdot4089$ (2)	49·204	2·19	> 5
Residual	6	$1 - 2 = 134\cdot847$	22·474	—	

[a] Probability of such an extreme value of *F*, assuming the null hypothesis to be correct. The tabulated values of *F* for d.f. 2 and 6 are respectively 5·14 and 10·92 for *P* = 5% and 1%.

Two-way Analysis of Variance

This may be done on the 18 results from a pair of students, e.g. the whole of the data in Table 4, and provides answers to three questions: (1) are there significant differences between the three pipettes as used jointly by both students? (2) Do the two students differ in the overall amounts of water delivered? (3) Is there significant interaction of pipettes and students? i.e. taking the pipettes one by one, is there a consistent (insignificant interaction) or variable (significant interaction) difference between the individual pipette totals from each student? The analysis of variance in Table 6 shows that there **is** a significant difference between students but not between pipettes and not in regard to interaction.

Table 6 Two-way analysis of variance of pipette data for students J and K.

Source of variation	Degrees of freedom (d.f.)	Sums of squares[b]	Mean square	Variance ratio[b] (F)	P^c
Total	17	$\Sigma y^2 - (\Sigma y)^2/N = 481{\cdot}345$ (1)	—	—	—
Between groups[a]	5	$\Sigma T^2/3 - (\Sigma y)^2/N = 177{\cdot}592$ (2)	—	—	—
Between pipettes	2	$\Sigma Q^2/6 - (\Sigma y)^2/N = 19{\cdot}8233$ (3)	$9{\cdot}9117$	$0{\cdot}39$	> 5
Between students	1	$\Sigma S^2/9 - (\Sigma y)^2/N = 133{\cdot}9339$ (4)	$133{\cdot}9339$	$5{\cdot}29$	< 5
Student's X pipettes	2	$(2) - (3) - (4) = 23{\cdot}8348$	$11{\cdot}9174$	$0{\cdot}47$	> 5
Residual	12	$(1) - (2) = 303{\cdot}753$	$25{\cdot}3127$	—	—

[a] The three replicate readings which each student produced for each pipette constitutes "a group".
[b] T = group totals; Q = pipette totals; S = students totals; the divisor in each formula is determined by the number of observations making up each total.
[c] Probability of such an extreme value of F, assuming the null hypothesis to be correct. The tabulated values of F for d.f. 2 and 12 are $3{\cdot}88$ and $6{\cdot}93$ for $P = 5\%$ and 1% respectively; for d.f. 1 and 12 the values are $4{\cdot}75$ and $9{\cdot}33$.

Other Tests and Variations

Instead of delivering $1\,cm^3$ volumes of water, students could try their skill at pipetting smaller volumes, viscous liquids such as glycerol, or water with a detergent added. They could also use the general procedure outlined here to check the precision and accuracy of other devices for delivering liquids, such as dropping pipettes, calibrated loops and automatic dispensers. A refinement in the weighing operation would be to check for evaporative losses from the vessel while the lid is off.

Among the other statistical tests which may be applied to the data presented above are the Bartlett test for homogeneity of variances (Snedecor and Cochran, 1967, p. 296), the Kolmogorov–Smirnov test for comparing two groups of data while avoiding assumptions of either normality or similarity in shape of the underlying distribution (*see* Campbell, 1974, p. 71) and the Kruskal–Wallis procedure for non-parametric analysis of variance (*see* Campbell, 1974, p. 61).

REFERENCES

C. I. Bliss (1967). "Statistics in Biology", Vol. 1, 558 pp. McGraw-Hill, New York.
R. C. Campbell (1974). "Statistics for Microbiologists", 2nd edn, 385 pp. Cambridge University Press, London.
G. W. Snedecor and W. G. Cochran (1967). "Statistical Methods", 6th edn, 593 pp. Iowa State University Press, Iowa.

Part Two
The Micro-organism Feeds

11

Feeding Strategies in Free-living Ciliated Protozoa

D. J. PATTERSON

*Department of Zoology, University of Bristol,
Bristol, BS8 1UG, England*

> *Level*: All undergraduate years
> *Subject areas*: Nutrition and diversity of ciliates
> *Special features*: Use of living, active and relatively
> large micro-organisms

INTRODUCTION

Ciliates are a common, and occasionally dominant, group of aquatic micro-consumers. They feed on bacteria, detritus, algae, other protozoa or even small metazoa. Food is ingested with the help of a discrete mouth and digestion occurs within membrane-bound vacuoles inside the cell. The mouth may have several components, but common to all mouthed ciliates is the cytostome. This is the specific site on the surface of the cell at which food vacuoles form. Microtubular structures (nemadesmata) internal to the cytostome may help in the ingestion of food or in the movement of food vacuoles away from the cytostome. Cilia lying external to the cytostome may be used to drive food particles towards the cytostome.

Most Kinetofragminophoran ciliates (the most primitive of the three classes of ciliates) consume food as large particles. Food may be other ciliates, algae, detritus or bacterial films. Ingestion typically involves well developed nemadesmata. Predatory ciliates employ extrusible extrusomes to kill, paralyse or capture prey. Originally the cytostome lay at the anterior pole of the cell, but in many species it is found in a lateral or ventral position. Examples of this class are *Discophrya* (Figure 3) and *Homalozoon* (Figure 4).

The most significant evolutionary step within the ciliates has been the development of structures which allow suspended particles, primarily bacteria, to be consumed. These include blocks of cilia called membranelles, each block acting as a functional unit, and a tightly packed row of cilia, the undulating membrane. The second class of ciliates, the Oligohymenophora, are characterized by having three membranelles and one undulating membrane (together they are called a hymenium). Typically, these structures lie in a depression called the buccal cavity at the bottom of which is the cytostome. This group includes *Tetrahymena* (Figure 5), *Paramecium* (Figures 1 and 6) and *Vorticella* (Figure 9a).

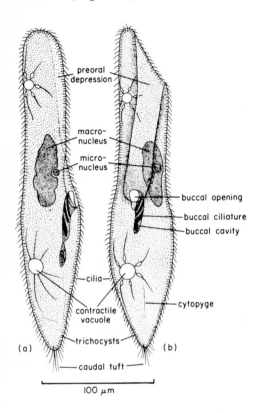

Figure 1 *Paramecium caudatum.* The organism is viewed from its right side (a) and from the ventral surface (b). The buccal cavity is preceded by a scoop-shaped preoral depression. The buccal cavity, with its membranelles of densely packed cilia, lies at the posterior extremity of this depression. Food vacuoles form at the cytostome, at the posterior extremity of the buccal cavity.

The experiment described in this chapter exploits a bactivorous species of *Paramecium* (Figure 1). In this genus, the buccal cavity containing the membranelles is preceded by a sculpting of the body surface, the preoral depression. The body cilia of this depression generate currents of water carrying bacteria to the buccal cavity which opens at the posterior end of the depression. The ciliary structures within the cavity select and concentrate bacteria which are packed into food vacuoles at the cytostome which lies at the base of the buccal cavity. Food vacuoles pinch off from the cytostome and pass into the cytoplasm. The contents of the vacuole become acid and digestion occurs. Inert material and undigested residues are discharged from the cell at a specific site, the cytopyge.

A third class of ciliates, the Polyhymenophora, have many membranelles associated with the mouth. These lie as a band called the adoral zone of membranelles (AZM), extending from the anterior pole of the cell to the cytostome. Most members of this group feed on bacteria and other suspended matter. The group includes *Blepharisma* (Figure 7), *Stentor* (Figure 8) and *Euplotes* (Figure 9b).

This chapter presents a quantitative experiment on the feeding by *Paramecium* on bacteria; and the inhibitory effects of cold and colchicine are investigated. The experiment illustrates the capacity of bactivorous ciliates for bacteria, the functioning of the mouth-parts and the contribution made by microtubular elements. Other ciliates, including some from a natural environment, are described and may be used to illustrate the diversity of feeding strategies within this group of protozoa.

EXPERIMENTAL

Day 1 *(1 h, then a gap of 3½ h during which other work may be done, then 3½ h).*

↓

Prepare paramecia as follows (*see* notes 1 and 2). Use about 25 ml thick culture per pair of students. Filter cultures through thinly teased cotton wool to remove gross debris.

↓

Concentrate cells by low speed centrifugation into one-tenth the original volume (*see* note 3).

↓

Divide concentrated paramecia into equal thirds (*see* note 4).

↓   ↓   ↓

Dilute one-third to 2 ml with Chalkley's fluid, designate as stock (A) (*see* note 5); do not refrigerate.	Dilute one-third to 2 ml with Chalkley's fluid, designate as stock (B) and refrigerate. (*see* note 6).	Dilute one-third to 2 ml with colchicine to give a final colchicine concentration of 10 mM 1^{-1}, designate stock (C) and refrigerate (*see* note 6).

Leave stocks (B) and (C) in refrigerator for appropriate period (*see* note 6).

↓

Students work in pairs. Each student should be provided with:
   (a) 2 ml concentrated untreated paramecia (A);
   (b) 2 ml concentrated cold-treated paramecia (B);
   (c) 2 ml concentrated cold-treated paramecia in 10 mM 1^{-1}. colchicine (C);
   (d) 2 ml stained bacteria (*see* notes 7 and 8).

↓

At time designated zero add 0·5 ml stained bacteria to each of stocks (A), (B) and (C) of paramecia.

↓

Divide each stock of paramecia into 6 or 7 replicate volumes (*see* notes 9 and 10).

↓

After 5 min fix the first set of replicate samples of stocks (A), (B) and (C) by the addition of one drop of 25 % glutaraldehyde solution (*see* note 5).

↓

Allow cells to settle for several minutes and then collect the settled cells with a Pasteur pipette and place on a slide.

↓

Add coverslip and count the number of food vacuoles in each of twenty cells. Express results as the average number of food vacuoles per cell (*see* notes 11 and 12).

↓

Repeat fixing and food-vacuole counting of further replicates from stocks (A), (B) and (C) at 15, 30, 60, 90, 120 (and 180) min (*see* note 13).

↓

Express results graphically as average number of food vacuoles per cell against time (*see* note 14).

As time permits, observe *Homalozoon* or *Discophrya*, (as examples of carnivorous ciliates), *Paramecium bursaria* or *Stentor polymorphus* (*see* note 15) (as examples of symbiotic associations) and *Tetrahymena vorax* or *Blepharisma* sp. (for cannibalism) and material from sewage works (*see* notes 16–19).

When time permits, take a small drop of living paramecia from an unfixed replicate of stock (A). Place on a slide and avoiding compression (*see* note 15) observe using phase contrast microscopy. Note surface ciliature, preoral depression, buccal cavity and the generation of currents carrying stained bacteria towards the mouth. Note also the red (alkaline) and blue (acid) food vacuoles.

↓

Compress the cells by drawing off some fluid from beneath the coverslip using some filter paper or a piece of tissue paper. Observe again. The buccal cavity, the membranelles and food vacuoles should be more obvious. Note also the surface ciliature, contractile vacuoles, trichocysts and the macronucleus.

Notes and Points to Watch

1. It is usually preferable that students do not prepare the paramecia, but are presented with stocks (A), (B) and (C) after the appropriate experimental treatment. The end of the chilling period should correspond with the beginning of the student practical.
2. *Paramecium caudatum*, *P. jenningsi* and *P. aurelia* can be cultured relatively easily and so are suitable species for this practical. As much material as possible should be available (*see* Methods).
3. Centrifuge with bench centrifuge at 1500–2000 rev. min^{-1} for several minutes.

4. The large size of paramecia makes them susceptible to damage if pipetted using conventional volumetric pipettes. The use of inverted pipettes or of pipettes with broken tips is advocated. Recently chilled cells are particularly fragile. Divide the paramecia into lots destined for each pair of students before chilling.

5. Details of solutions are given in methods.

6. It is imperative that a trial experiment be carried out to establish the appropriate extent of the cold treatment. Too little will leave the cells unaffected, too much will kill them. About 210 min at 3–4°C is usually appropriate. The end of the cold period should correspond with the beginning of the student practical.

7. *Klebsiella aerogenes* is suitable, 2 ml of suspension being provided for each pair of students. Other small bacteria are probably equally suitable.

8. In the absence of bacteria, coloured inert particulate material with a particle size of 2 μm or less may be used. These of course will not illustrate the pH changes that occur within food vacuoles. It would be advisable to check the toxicity of alternative "food particles" before using them in a practical.

9. The students will need about 20 small (1–5 ml) containers each. Multidepression dishes (*see* Materials) are recommended as an alternative.

10. Keep the experimental stocks in the dark as much as possible since colchicine is unstable in light.

11. Students should be told to count only food vacuoles containing stained bacteria and to ignore unstained food vacuoles which are due to feeding activities before the beginning of the experiment.

12. Counting is more convenient if coverslips measuring 32×22 mm or larger are used.

13. At the beginning of the practical, students can accumulate fixed organisms faster than they can count food vacuoles. The excess can usually be dealt with in the period between the 30 and 60 min samples.

14. It can be very instructive to produce a set of averaged class results.

15. Compression between glass slide and coverslip can be prevented without having a significant effect on the quality of the phase contrast image by supporting the coverslip on a bridge formed from two fragments of a broken coverslip.

16. Ideally, students should be given a Kinetofragminophoran, an Oligohymenophoran and a Polyhymenophoran ciliate to observe. *Homalozoon, Paramecium bursaria* and *Blepharisma* sp. are recommended.

17. All cultures should be allowed to settle for 10 min or so before dispensing. The ciliates will be found in greatest concentration near the debris on the bottom of the container. It may be necessary to concentrate the organisms with light centrifugation, as used for *Paramecium* (above).

18. Students frequently complain that protozoa move too fast to be observed easily. The best answer to this difficulty is patience, since all treatments which slow down or immobilize the ciliates will also have additional effects. Movement may be slowed or arrested by (a) removing the excess fluid from between the glass slide and coverslip with a piece of filter paper until the coverslip presses gently on the ciliate, (b) mixing one drop of ciliate culture with one drop of 2% (w/v) methyl cellulose, (c) mixing one drop of the ciliate culture with one drop 10 mM l^{-1} nickel chloride or (d) fixing the organisms with a small amount of 25% gluta-raldehyde.

19. Sewage treatment works are normally willing to allow a sample of activated sludge to be collected for teaching purposes. An exploratory telephone call is advisable. The sample should be collected on the day of the practical from the sludge return channel. It should not be taken from a tank which is bulking. Five millilitres is adequate for a class of 20. Ensure that the samples remain well oxygenated (i.e. collect in a large bottle such that the sample forms only a thin layer of fluid).

MATERIALS

1. All organisms, excepting activated sludge, are available from Philip Harris Ltd., Oldmixon, Weston-super-Mare, Avon BS24 9BJ. When ordering, at least two weeks' notice should be given. If paramecia are required, specific mention should be made that they are for the practical outlined in this chapter. Culture methods for all of the organisms are given below. Activated sludge must be obtained from a local sewage works by arrangement.

2. *Culture*. Glassware can be sterilized in dry heat, solutions may be autoclaved and vegetable material sterilized by dry heat for several hours. All cultures are kept at room temperature (15–25°C). Keep the culture fluid depth to less than 1 cm to ensure adequate oxygenation. Crystallizing dishes make appropriate culture vessels.

Blepharisma spp. should be grown in bacteria-reinforced Chalkley's fluid with 2 wheat or barley grains and 2 rice grains per 50 ml of medium. *Chilomonas* may be added to supplement food. Subculture by transferring 5 ml of good culture to fresh medium every 2–3 weeks. Cannibalism occurs in older cultures.

Chilomonas paramecium a colourless flagellate, is a useful food organism for some larger ciliates. Use an infusion of 4 rice grains per 50 ml Chalkley's fluid. Subculture at 3–4 week intervals by transfer of 5 ml rich culture to fresh medium.

Colpidium sp. is a ciliate suitable as food for carnivorous ciliates such as *Discophrya* or *Homalozoon*. It may be grown in an infusion similar to that used for *Paramecium* spp. Subculture similarly.

Discophrya sp. should be maintained in Chalkley's fluid. Twice every week add 5 ml rich *Colpidium* sp. or *Paramecium* sp. culture as food. Subculture at 4–6 week intervals. Silk threads may be added to the culture dish as a substrate upon which larval forms of this ciliate may settle. As the *Discophrya* attach to surfaces, excess fluid may be removed at any time.

Homalozoon vermiculare should be grown in Chalkley's fluid. Feed and subculture as *Discophrya* sp. but without threads.

Paramecium jenningsi, *P. caudatum* and *P. aurelia* are grown in bacteria-rich infusions of vegetable material in the dark and will produce good cultures in 2–3 weeks. To 50 ml Chalkley's solution, add several lengths, 2–3 cm each, of Timothy grass (*Phleum pratense*) and/or several barley or wheat grains. One or two rice grains may be added after 1 week. An alternative is to dry lettuce leaves until crisp and brown, pulverize and boil 1·5 g of the resulting powder with 1 litre of distilled water. The solution is left overnight, filtered and may then be dispensed into smaller containers and autoclaved. This medium is used with or without added vegetable material, diluted to one-third strength with distilled water. The fluid in culture vessels should be kept less than 1 cm deep for densest cultures. Subculture at 2–3 weekly intervals by transferring 5 ml of good culture to fresh medium. In this practical, 25 ml thick culture will be required for every pair of students. Double the required amount should be set up to allow for culture failures, etc.

Paramecium bursaria are maintained as other species of *Paramecium*, but with half-quantities of vegetable material. Keep cultures in bright artificial or natural light, but avoid direct sunlight which might cause the cultures to become overheated.

Stentor polymorphus will grow under the same conditions as *Paramecium bursaria* and 5 ml thick *Chilomonas* culture may be added weekly. Subculture at 4–5 week intervals by transferring as many organisms as possible from an old culture to fresh medium. Maintain in lighted conditions as *Paramecium bursaria*.

Tetrahymena vorax are grown axenically in 1% Bacteriological peptone (BDH Chemicals Ltd., Poole, England), 0·25% yeast extract (Oxoid Ltd., Southwark Bridge Road, London SE1) in distilled water. Adjust pH to 7·0 using NaOH or HCl. Volumes of 15 ml are placed in test tubes which are capped and autoclaved at 103·4 kPa (15 lbf in^{-2}) for 20 min. Subculture at 2–3 week intervals using sterile techniques. Cannibalism occurs in older cultures, but can be stimulated by concentrating the cells by centrifugation.

Activated sludge cannot be cultured. For collection, etc. *see* Notes and Points to Watch.
3. *Solutions.* 2% methyl cellulose (w/v) in distilled water (BDH Chemicals Ltd., Poole, England). Takes a day to dissolve.

10 mM l^{-1} nickel chloride in distilled water.
Colchicine (BDH Chemicals Ltd., Poole, England). A stock solution of 20 mM l^{-1} is usually appropriate. Make up immediately before required in distilled water and keep it in the dark.

25% glutaraldehyde in distilled water (BDH Chemicals Ltd., Poole, England).
Chalkley's fluid: NaCl, 80 mg
                NaHCO$_3$, 4 mg
                KCl, 4 mg
                CaCl$_2$, 4 mg
                CaH$_4$ (PO$_4$)$_2$H$_2$O, 1·6 mg
dissolved in 1 litre of distilled water. Adjust to pH 7·0 with HCl or NaOH as appropriate. A × 1000 concentrated stock solution containing all salts may be prepared but must be well agitated before use.
4. *Bacteria.* Grow *Klebsiella aerogenes* in nutrient broth at 28°C. Harvest by centrifugation at 4°C at 3000 rev. min^{-1}. Suspend in a solution of Congo red vital stain in distilled water (Hopkins and Williams) at a final concentration of 1% (w/v). Stain by placing suspension in a boiling water-bath for 10 min. Remove excess stain by centrifugation; repeat the removal of the supernate and the resuspension in Chalkley's fluid until the supernate is clear. Other small bacterial species would probably work equally well.

SPECIFIC REQUIREMENTS

supply of protozoa and bacteria (*see* notes 2, 7 and 16)
1 compound microscope (preferably with phase contrast facilities) per student; 1 dissecting microscope for every 4 or 5 students should also be available
glass microscope slides
coverslips (*see* note 12)
18 or 21 watch-glasses or small vials or 1 multi-depression dish (*see* note 9) (Model FB-16-24-TC Multidishes marketed by Flow

Laboratories (P.O. Box 17, Second Avenue Industrial Estate, Irvine KA12 8NB) are particularly suitable) one dish per pair of students
10 ml 25% glutaraldehyde solution per pair of students
10 ml 2% methyl cellulose solution per pair of students
10 ml 10 mM l^{-1} NiCl$_2$ per pair of students
Pasteur pipettes with teats

FURTHER INFORMATION

Typical results of an experiment on feeding by *Paramecium* on stained bacteria is given in Figure 2. The results indicate the involvement of microtubules in the uptake of food by *Paramecium* as food vacuole formation is depressed by cold treatment and by colchicine, both of which adversely affect microtubular integrity. Cold treatment causes depolymerization of microtubules, but re-polymerization begins immediately on rewarming. This treatment (Figure 2) causes a transient depression of food vacuole formation. Colchicine prevents re-assembly of depolymerized microtubules. Combination of this drug with the cold treatment (Figure 2) leads to a more pronounced and sustained depression of food vacuole formation.

Microtubules are involved in several aspects of the feeding process and in food vacuole formation. They are integral components of the cilia which direct food to the cytosome, they are involved in passing a newly formed food vacuole from the cytosome into the cytoplasm (as the cytopharynx), they assist in the recycling of old food vacuole membranes back to the mouth for re-use, and they are involved in the expulsion of undigested residue from old food vacuoles (as the cytopyge).

The biology of *Paramecium* is extensively discussed by Wichterman (1953), but a more recent textbook by van Wagtendonk (1974) contains a useful chapter dealing with the ultrastructure of members of the genus. Detailed descriptions of the mouth region and cytoproct of *P. caudatum* are provided by Allen (1974) and Allen and Wolf (1974). The observations made by these authors almost certainly apply to other species in the genus. The biology of microtubules has been reviewed extensively, but two convenient sources are Olmsted and Borisy (1973) and Dustin (1978).

Some notes on the species of ciliates recommended to illustrate non-bactivorous feeding habits are given below.

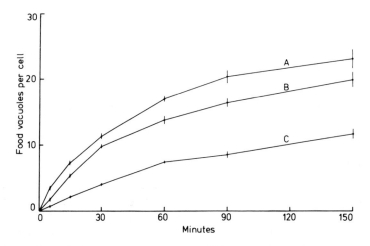

Figure 2   Food vacuole uptake by *Paramecium jenningsi*. (a) Untreated cells, (b) cells chilled at 3 °C for 3½ h immediately preceding time zero, (c) cells chilled at 3 °C for 3½ h in 10 mM l^{-1} colchicine immediately preceding time zero. Vertical lines indicate ±1 standard error.

Discophrya (*Kinetofragminophora*) (*Figure 3*)

Discophrya is a predatory but sessile ciliate. The organism has, in effect, many mouths, each represented by a tentacle. Potential prey organisms (ciliates) are caught by extrusomes located in the knob at the end of each tentacle. Prey cytoplasm is then sucked down the tentacles to form food vacuoles at the base of the arms. The prey organism is not necessarily killed, and may be released after the *Discophrya* has fed. *Discophrya* has a single apical contractile vacuole and a large central (macro-) nucleus. Attachment to the substrate is by a non-contractile secreted stalk. Distribution in the environment is assured by having a ciliated larva which buds from the apical portion of the cell.

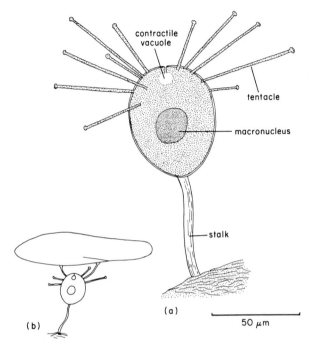

Figure 3 *Discophrya* sp. (a) General organization, (b) feeding on a prey ciliate.

Homalozoon vermiculare (*Kinetofragminophora*) (*Figure 4*)

H. vermiculare is a predatory ciliate. The mouth is located at the anterior spatulate end. Below it lie large numbers of extrusomes that are used to kill the potential prey, namely other ciliates. The organism is ribbon-like and has cilia only on its locomotor surface, the ventral side. There are many contractile vacuoles and the nucleus (macronucleus) looks like a string of sausages. These two features are adaptions to large size.

Feeding can only be satisfactorily observed under a low-power (dissecting) microscope. A ciliate such as *Paramecium* is added in excess. Contact is random. A prey organism impaled with extrusomes will exhibit an escape response before dying. *Homalozoon* appears to be chemo-attracted to the dead prey. The entire front end of the *Homalozoon* opens up to envelop the prey organism.

Tetrahymena vorax (*Oligohymenophora*) (*Figure 5*)

The ciliate *T. vorax* exists in a bacteria-feeding form and as a cannibal, the former being the more characteristic. This form has a small buccal cavity supplied with a poorly developed hymenium which is difficult to see. There is a central (macro-) nucleus and a contractile vacuole in the posterior part of the cell. Food vacuoles contain bacteria or, with axenically grown organisms, are

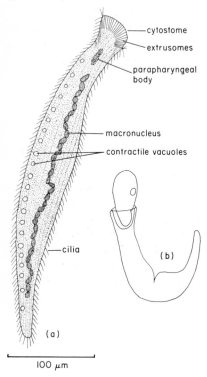

Figure 4 *Homalozoon vermiculare.* (a) General organization, (b) ingesting a prey ciliate.

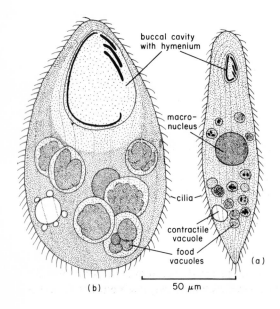

Figure 5 *Tetrahymena vorax.* (a) Microstome form, with a small mouth and food vacuoles containing bacteria, (b) macrostome form, with buccal cavity greatly expanded to fill a large part of the front of the cell. Buccal ciliary organelles, the hymenium, are enlarged. The food vacuoles contain other ciliates.

empty. The cannibalistic form is much larger, with the buccal cavity occupying the anterior third of the cell. Its food vacuoles contain other ciliates.

The cannibalistic form develops when the population of the organism becomes very dense. The old mouth is lost and the enlarged one develops. Dividing cells ultimately revert to bacteria-feeding forms in the absence of large numbers of the same species, or of other suitable prey.

This so-called microstome/macrostome change allows this ciliate to feed on either bacteria or on small ciliates. In the presence of a limited supply of bacterial food, the occurrence of cannibalistic ciliates serves to regulate the size of the ciliate population.

Paramecium bursaria (*Oligohymenophora*) (*Figure 6*)

P. bursaria shares many obvious features of the other *Paramecium* species: the body is evenly ciliated and there is a posterior tuft of longer cilia; there is a dense layer of trichocysts (extrusomes) around the periphery of the cell; there are two contractile vacuoles each with radiating collecting canals; there is a single large central macronucleus; there is a preoral depression in front of the buccal cavity, and three elongate membranelles lie in the buccal cavity. However, the shape of the cell differs from that of *P. caudatum*, *P. jenningsi* and *P. aurelia* and there are also differences in the arrangement of the micronucleus. The most obvious difference is the colour. *Paramecium bursaria* is green because it contains large numbers of a green alga as endosymbionts. Nevertheless, *P. bursaria* remains capable of feeding upon bacteria. It is the only member of this genus to exhibit phototropism.

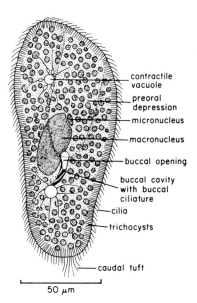

contractile vacuole
preoral depression
micronucleus
macronucleus
buccal opening
buccal cavity with buccal ciliature
cilia
trichocysts
caudal tuft

50 μm

Figure 6 *Paramecium bursaria*. The organism contains many green algal symbionts. It has a functioning buccal cavity preceded by a preoral depression.

Blepharisma americanum (*Polyhymenophora*) (*Figure 7*)

A large pink organism, *B. americanum* is ciliated all over. The mouth lies about half-way down the body and is preceded by a large number of membranelles which are easily distinguished from the cilia of the body. There is also a veil-like undulating membrane. The cytoplasm is vacuolated and the (macro-) nucleus has the appearance of a string of sausages—these features being adaptations to the large size. The single contractile vacuole lies at the posterior end where the cytoproct is also located. The colour is derived from granules, whose function is obscure, lying under the plasma membrane.

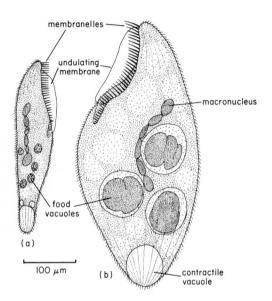

Figure 7 *Blepharisma americanum.* (a) Bacteria-feeding form. The organism is slim and the food vacuoles contain only bacteria or similar small particles. (b) Cannibalistic form. The organism is greatly enlarged, and the food vacuoles contain the remains of other organisms.

Typically these ciliates feed on suspended bacteria. In dense populations they may also become cannibals. The cannibalistic form is larger than the bacteria-feeding form, and often contains dense, red, food vacuoles. The colour is the condensed pigment from an ingested *Blepharisma*.

This species also forms dense cysts, the wall of which has a single plugged opening.

Stentor polymorphus (*Polyhymenophora*) (*Figure 8*)

S. polymorphus is one of the largest ciliates. It is trumpet-shaped when feeding, with the thin posterior end attached to the substrate. The body is evenly covered with cilia, and there is a long band of membranelles which extends around the anterior rim of the organism and leads into a buccal cavity. The (macro-) nucleus takes the form of a string of sausages. This and the long canal extending from the anterior contractile vacuole are adaptations to large size. The organism may release its hold of the substrate and actively swim. When this happens, its form becomes much more rounded.

This ciliate has a green colour due to large numbers of symbiotic unicellular algae. *S. polymorphus* does survive without algae, but it is generally thought that the algae supplement the particulate food which is filtered from the medium by supplying products of photosynthesis to the ciliate.

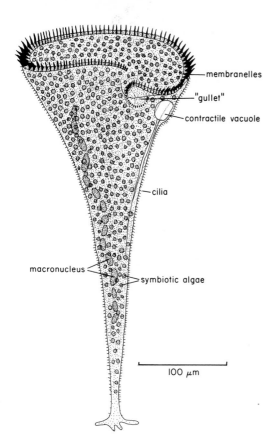

membranelles

"gullet"

contractile vacuole

cilia

macronucleus

symbiotic algae

100 μm

Figure 8 *Stentor polymorphus.* The organism is trumpet-shaped when feeding. A row of membranelles extends from the invaginated funnel-like buccal cavity to form a conspicuous ring around the anterior end of the cell. The cytoplasm is rich in symbiotic green algae.

Activated sludge

The activated sludge process of sewage treatment depends on protozoa removing suspended bacteria from sewage undergoing treatment. This favours the growth of floc-forming bacteria which convert the soluble and suspended organic matter into a readily sedimentable material. Some of the sewage sediment (sludge) is returned to the treatment tanks, and so a selective advantage is given to those organisms intimately associated with the flocs of bacteria which form the sludge. This material is always rich in protozoa, and particularly in attached peritrich ciliates of the genus *Vorticella* (Figure 9a) and related colonial forms (Oligohymenophora) and "walking" hypotrich ciliates such as *Euplotes* (Figure 9b) and *Aspidisca* (Figure 9c) (Polyhymenophora). Predatory or detritus-feeding ciliates such as *Trachelophyllum* (Figure 9d) also occur.

Vorticella attaches to the flocs by a contractile stalk. The only cilia that are normally present are those of the hymenium, which extend as two long rows of cilia from the buccal cavity around the rim of the "bell" of this organism. A posterior band of cilia develops under unfavourable conditions and this serves to propel organisms which have separated from their stalk. Larvae with a posterior band of cilia bud from the body of the sessile organism.

Euplotes has a large number of membranelles which curve around the front end of the cell and terminate in the buccal cavity on the ventral surface of the organism. The locomotor organelles which occur on the ventral surface are cirri. These are blocks of cilia acting in synchrony and which allow the ciliates to "walk" over the flocs formed by bacteria. *Aspidisca* also has cirri, but the membranelles are attached to a small flap projecting from the ventral surface.

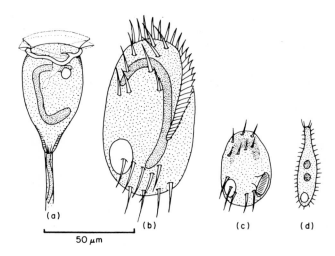

Figure 9   Several common ciliates found in activated sludge plants. (a) *Vorticella* sp., a peritrichous Oligohymenophoran. (b) *Euplotes* sp., and (c) *Aspidisca* sp., hypotrichous Polymenophorans. (d) *Trachelophyllum pusillum*, a small, predatory Kinetofragminophoran.

A suitable general textbook on Protozoa is Sleigh (1973) although it does not contain details of the systematic division of the ciliates as used here. Corliss (1979) reviews the present state of ciliate systematics, while Corliss (1973) is a more convenient though slightly dated source.

Sleigh's book contains no details of the biology of *Homalozoon*, *Blepharisma* or *Tetrahymena vorax*. These deficiencies can be rectified by reference to Weinreb (1955a, b), Giese (1973), Elliott (1973) and Kuhlmann *et al.* (1980).

Finally, Curds (1969) has provided a good guide to the various ciliates encountered in activated sludge.

REFERENCES

R. D. Allen (1974). Food vacuole membrane growth with microtubule-associated transport in *Paramecium*. *The Journal of Cell Biology*, **63**, 904–22.

R. D. Allen and R. W. Wolf (1974). The cytoproct of *Paramecium caudatum*: structure and function, microtubules, and fate of food vacuole membranes. *Journal of Cell Science*, **14**, 611–31.

J. O. Corliss (1973). Taxonomic characterization of the suprafamilial groups in a revision of recently proposed schemes of classification for the Phylum Ciliophora. *Transactions of the American Microscopical Society*, **94**, 224–67.

J. O. Corliss (1979). "The Ciliated Protozoa. Characterization, Classification and Guide to the Literature", 455 pp. Pergamon Press, Oxford.

C. R. Curds (1969). "An Illustrated Key to the British Freshwater Ciliated Protozoa commonly found in Activated Sludge". Water Pollution Research Technical Paper No. 12, 90 pp. HMSO, London.

P. Dustin (1978). "Microtubules", 452 pp. Springer Verlag, Berlin.

A. M. Elliott (1973). "Biology of *Tetrahymena*", 508 pp. Dowden, Hutchinson and Ross, Pennsylvania.

A. C. Giese (1973). *Blepharisma*. "The Biology of a Light-sensitive Protozoan", 366 pp. Stanford University Press, California.

S. Kuhlmann, D. J. Patterson and K. Hausmann (1980). Untersuchungen zu Nahrungserwerb und Nahrungsaufnahme bei *Homalozoon vermiculare*, Stokes 1887 I. *Protistologia*, **16**, 39–55.

J. B. Olmsted and G. G. Borisy (1973). Microtubules. *Annual Review of Biochemistry*, **42**, 507–40.

M. A. Sleigh (1973). "The Biology of the Protozoa". Arnold, London.

W. J. van Wagtendonk, (1974). "*Paramecium*: a Current Survey", 499 pp. Elsevier Scientific, Amsterdam.

S. Weinreb (1955a). *Homalozoon vermiculare* (Stokes): I. Morphology and reproduction. *Journal of Protozoology*, **2**, 59–66.

S. Weinreb (1955b). *Homalozoon vermiculare* (Stokes): II. Parapharyngeal granules and trichites. *Journal of Protozoology*, **2**, 67–70.

R. Wichterman (1953). "The Biology of *Paramecium*", 527 pp. Blakiston, Toronto.

Sugar Transport in Yeast: the Temperature Dependence of Galactose Efflux from Baker's Yeast

R. K. POOLE

Department of Microbiology, Queen Elizabeth College, Campden Hill, London, W8 7AH, England

> *Level*: All undergraduate years
> *Subject areas*: Microbial physiology
> *Special features*: Relatively simple, well-tried experiment on substrate transport across cell membranes

INTRODUCTION

Biological membranes allow the movement of solutes into and out of cells by several distinct mechanisms. One mechanism, facilitated diffusion, involves the binding of a substrate for transport, e.g. a sugar, to a protein carrier in an otherwise predominantly lipid membrane, thus allowing more rapid solvation of the sugar in the membrane and transmembrane mobility. As in simple diffusion, an equilibrium is attained in which the solute concentration is equal on both sides of the membrane. However, unlike simple diffusion, rate and concentration are not simply related. In facilitated diffusion processes, the protein carrier or porter affects the rate but not the equilibrium state of the transmembrane solute diffusion.

Normally, yeast cells cannot metabolize galactose but can be induced to do so; the transport system and the necessary catabolic enzymes are synthesized co-ordinately. In non-induced cells, galactose uptake occurs by facilitated diffusion. Controversy exists as to the mechanism of galactose transport in induced cells (Jennings, 1974).

This experiment is based on an early, but important, paper by Burger *et al.* (1959). Yeast cells are equilibrated with *D*-galactose, then washed and the subsequent rate of sugar efflux determined when the "loaded" cells are resuspended in galactose-free medium. The aim is to demonstrate the movement of a solute across the cell membrane in response to a concentration gradient, and to illustrate the temperature dependence of the process.

EXPERIMENTAL

Day 1 *(4–5 h including 60–70 min interval early in the experiment)*

The experiment is simple to conduct and requires little "last minute" preparation.

Weigh out 60 g ($\pm$ 1 g) of fresh baker's yeast. Resuspend in buffered saline (300–400 ml) and pellet the cells by centrifugation. Acceleration to 6000 rev. min^{-1} in the 6 × 250 ml rotor of an MSE "High Speed 18" centrifuge at room temperature, followed by immediate deceleration, is suitable. Resuspend the cells in the same medium to a final volume of about 150 ml. This procedure may conveniently be done once for the whole class in advance: the quantities given are sufficient for 10 groups of students. Directions below are for each student or group.

↓

Pipette 13 ml of the yeast suspension, 8·5 ml of galactose solution and 8·5 ml of saline into a 100 ml Erlenmeyer flask. Incubate in a water-bath at 28 °C for 60–70 min.

↓

Pour cell suspension from the flask into centrifuge pots that have been precooled by standing in an ice-bath or refrigerator. Using the MSE "High Speed 18" centrifuge fitted with a precooled (4 °C) 8 × 50 ml rotor, harvest the cells by acceleration to 6000 rev. min^{-1}, followed by immediate deceleration. Resuspend the pelleted cells in ice-cold saline and repeat the centrifugation. Wash the cells once more as above and resuspend to a **final volume** of 30 ml in saline. **Keep the suspension cold** (*see* note 1).

↓

Have 4 tubes standing in a water-bath at **either** 18 **or** 28 °C. Half of the class should perform the experiment at each temperature (*see* note 2). Quickly dispense 6 ml of the yeast suspension into each tube as shown in Figure 1.

↓

Estimate the galactose present in each filtrate.

Into a boiling tube (approx 2·5 × 15 cm) pipette 5 ml of reagent A. Add 3 ml of the filtrate for assay and 2 ml distilled water. Mix well, cover with a metal cap or marble, and place in a boiling water bath for 15 min (*see* note 5). Transfer tubes from the hot water-bath to a bath of cold water (e.g. 18 °C). Leave for 5 min, and then proceed immediately with the titration.

↓

Transfer the solution into a conical flask (50–250 ml). Add, while shaking, 2 ml of reagent B, followed by 1 ml of reagent C. These additions need not be precise. Mix well for about 30 s and then titrate with Na$_2$S$_2$O$_3$ using a few drops of sodium starch glycollate as indicator.

↓

Repeat the last two steps using, instead of the filtrate plus water, 5 ml of water (in duplicate) and then 5 ml of solutions containing 0·1, 0·3 and 0·5 mg of galactose ml^{-1}, respectively.

↓

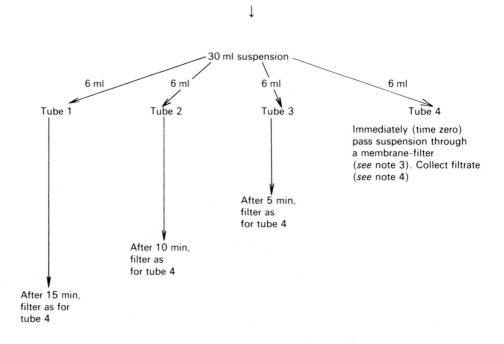

Figure 1 Dispensing and treatment of yeast suspension.

Plot a standard curve of $Na_2S_2O_3$ (ml) against amount of galactose (mg) in standard.

Calculate the sugar content of each filtrate and plot sugar content (mg) against time (min).

Calculate the rate of galactose efflux (mg ml^{-1} h^{-1}). Exchange results with those who performed the experiment at the alternative temperature and calculate Q_{10}, i.e. [rate at $(T+10)°$]. [rate at $T°$]$^{-1}$.

Notes and Points to Watch

1. It is essential that the suspension be kept as cold as possible by standing it in an ice-bath prior to transfer to the tubes at 18 or 28 °C. Efflux is very low at 1 °C and increases considerably with increasing temperature (Burger *et al.*, 1959).
2. Other pairs of temperatures (separated by 10 °C) may be found more convenient.
3. Membrane filters should be replaced after filtering each sample, and care should be taken to see that they are not damaged. Ideally, a completely new filter holder plus filter should be used for each sample; with such high cell concentrations it is easy to contaminate and cloud successive filtrates with cells from the filter or holder.

4. A convenient means of collecting the small volumes of filtrate without contaminating the Buchner flask is to collect the filtrate directly into a boiling tube (Figure 2).
5. During reduction, the heat treatment should be carefully standardized; all tubes should be kept vertical and as free as possible from vibration.

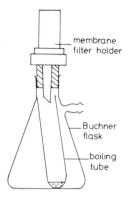

Figure 2   Collection of filtrate for analysis.

MATERIALS

1. Baker's yeast.
2. 0·15 M NaCl.
3. Iodometric reagents:
   (a) $K_3Fe(CN)_6$,   6·6 g (weighed accurately)
   K_2HPO_4,   70 g ⎱ (rough balance
   K_3PO_4,   21 g ⎰ adequate)
   Dissolve in distilled water to a final volume of 1 litre.
   (b) $ZnSO_4.7H_2O$,   50 g
   Conc. H_2SO_4,   70 ml
   Dissolve in distilled water to a final volume of 500 ml.
   (c) KI,   15 g
   Na_2CO_3,   0·05 g

   Dissolve in distilled water to a final volume of 100 ml.
   (d) 0·01 N $Na_2S_2O_3$.
   (e) 0·5 % (w/v) sodium starch glycollate (Fison's Scientific Apparatus, Loughborough, Leics., England). Dissolve by suspending in cold water, then heating and stirring.
4. Standard galactose solutions containing 0·1, 0·3 and 0·5 mg ml^{-1}.
5. Centrifuge with rotors suitable for harvesting cells from 20–400 ml of suspension at temperatures close to 5 °C.

SPECIFIC REQUIREMENTS

Requirements shown are for each student or group, with the exception of the first four items, which should be sufficient for about 10 groups.

baker's yeast (approx. 60 g)
1 litre 0·9 % saline, room temperature
centrifuge pots, e.g. 250 ml capacity, and suitable rotor

1 measuring cylinder (500 ml)
pipettes (non-sterile)
10 ml 20 % (w/v) galactose solution
100 ml Erlenmeyer flask
water-baths or incubators at 28 and 18 °C
precooled centrifuge rotor and tubes, suitable for centrifuging about 30 ml of suspension at 0–4 °C
100 ml 0·9 % saline, at 4 °C

cont.

cont.

measuring cylinder, 50 or 100 ml	10 conical flasks (50–250 ml) or similar for titration
ice bucket and ice	
clock	40 ml reagent B
20 boiling tubes and 10 caps or large marbles	20 ml reagent C
racks for boiling tubes	10–20 ml starch glycollate
membrane filter assembly with Buchner flask and at least 4 filters (e.g. 0·45 μm pore size)	burette (25–50 ml), stand, funnel
	200 ml $Na_2S_2O_3$ solution
100 ml reagent A	10 ml each of standard solutions containing 0·1, 0·3 and 0·5 mg galactose ml^{-1}
distilled water for dilution	
boiling water-bath, containing racks for boiling tubes	

FURTHER INFORMATION

The assay method for determination of reducing sugars used here is modified from that of Fujita and Iwatake (1932). The deproteinization step described by these authors is unnecessary in the present experiment.

Using a similar experimental procedure to that described here, Burger *et al.* (1959) showed that the influx of D-galactose and D-arabinose into cells of *S. cerevisiae* followed an exponential curve, reaching an apparent equilibrium within 60 min. The apparent equilibrium concentration of the monosaccharides was lower than that of the medium; thus, transport against a concentration gradient (active transport) was not observed.

As is true for most chemical reactions, the rate of an enzyme-catalysed reaction increases with temperature. The rate of most enzymatic reactions, however, approximately doubles for each 10 °C rise in temperature ($Q_{10} \triangleq 2$). Values for Q_{10} in excess of 2 were found by Burger *et al.* (1959) for both the influx and efflux of D-galactose. Such values do not of themselves constitute conclusive evidence for an enzyme-catalysed mechanism of transport across the membrane, as opposed to transport by physical forces alone. However, Burger *et al.* (1959) obtained further evidence for the involvement of enzymes by showing that this transport was competitively inhibited by glucose.

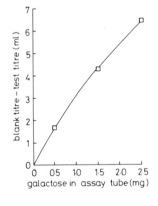

Figure 3 Standard curve for the essay of galactose.

Results obtained by undergraduates in routine practical classes are illustrated below (Figures 3 and 4).

Over the past five years, values of Q_{10} obtained by third year undergraduates have been in the range $2 \cdot 0 - 2 \cdot 6$.

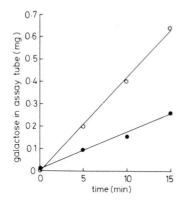

Figure 4   The efflux of galactose from baker's yeast suspended in galactose-free medium at $18\,^{\circ}C$ (●) and $28\,^{\circ}C$ (○).

Rate of efflux at $18\,^{\circ}C = 0 \cdot 35\ \text{mg ml}^{-1}\,\text{h}^{-1}$.
Rate of efflux at $28\,^{\circ}C = 0 \cdot 83\ \text{mg ml}^{-1}\,\text{h}^{-1}$.

$$\therefore Q_{10} = \frac{0 \cdot 83}{0 \cdot 35} = 2 \cdot 5.$$

ACKNOWLEDGEMENT

This experiment is closely based on an earlier version passed to me by Professor S. J. Pirt.

REFERENCES

M. Burger, L. Hejmova and A. Kleinzeller (1959). Transport of some mono- and di-saccharides into yeast cells. *Biochemical Journal*, **71**, 233–242.

A. Fujita and D. Iwatake (1932). Bestimmung des echten Blutzuckers ohne Hefe. *Biochemische Zeitschrift*, **242**, 43–60.

D. H. Jennings (1974). Sugar transport into fungi: an essay. *Transactions of the British Mycological Society*, **62** (1), 1–24.

13

Cation Adsorption by Lichens

G. W. GOODAY

*Department of Microbiology, University of Aberdeen,
Aberdeen, AB9 1AS, Scotland*

> *Level*: All undergraduate years
> *Subject areas*: Microbial ecology, symbiosis, pollution
> *Special features*: Simplicity of technique and of requirements

INTRODUCTION

Lichens bind cations and therefore concentrate such materials as heavy metal pollutants and radioactive fall-out products from the atmosphere. The uptake process is closely similar to that shown by ion-exchange resins. At a particular pH, the binding of added cations involves the release of an equivalent amount of another cations. Thus the relative affinities for different cations can be compared by measuring their ability to replace each other.

In this experiment, the cationic dye methylene blue is used to saturate the binding sites of the lichen, *Peltigera polydactyla*, and any subsequent release of this methylene blue can be measured when the lichen is incubated in solutions of cations.

EXPERIMENTAL

3 h, with two periods of incubation

Dilute the solution of methylene blue provided (5×10^{-3} molar) to give 100 ml of 5×10^{-4} molar stock solution. Construct a calibration curve of absorption at 540 nm against concentration of methylene blue, using a linear series of dilutions of this stock solution.

↓

Wash the lichen (*see* note 1) thoroughly with distilled water.

↓

Cut discs of lichen with a cork borer on moist filter paper to give 5 discs for each cation to be tested.

↓

Immerse each group of 5 discs in 5 ml of the 4×10^{-4} M methylene blue in a universal bottle and shake at 20 °C in a shaking water-bath for 1 h (*see* note 2).

↓

Decant the supernate from the discs and discard. Using a tea-strainer, wash the discs very well in distilled water.

↓

Transfer each group of 5 discs to 5 ml of a 10 mM solution of the metal salt to be tested, or to 5 ml of a distilled water control. Suitable salts are chlorides of lithium, sodium, potassium, calcium, magnesium, copper, cobalt, nickel, aluminium, ferric iron, but any soluble metal chloride can be used.

↓

Shake at 20 °C in the shaking water-bath for 1 h.

↓

Measure the concentration of methylene blue in each supernate. If the salt solution is coloured, it should be used to zero the spectrophotometer.

↓

Calculate the number of moles of methylene blue eluted from 5 discs by each metal salt.

↓

Construct a "league table" of efficiency of the different cations at displacing the methylene blue. Correlate position in the "league table" with properties of the cation, such as valency and atomic number.

- -

Notes and Points to Watch

1. Other lichens can be used. *Peltigera polydactyla* is particularly easy to handle as discs can be cut readily. Crustose and fruticose lichens can be weighed out as equal samples for the experiment.
2. The anatomy of sections of the lichen can be examined, with and without staining with methylene blue, during the incubation periods.

MATERIALS

1. *Peltigera polydactyla* or *P. canina* can be obtained readily in many parts of Great Britain, growing in woodland rides, sand dunes or lawns. For its conservation it should be collected only where abundant, and only in quantities just sufficient for the classwork.

SPECIFIC REQUIREMENTS

cork borer no. 3
moist filter paper in Petri dish
Universal bottles
shaking water-bath at 20 °C with rack for
  Universal bottles
5×10^{-3} M methylene blue (0·94 g per 500 ml
  water)
selection of metal chlorides (5 ml of 10 mM
  solutions)

balance
pipettes
spectrophotometer at 540 nm
cuvettes
forceps
tea-strainer

FURTHER INFORMATION

The binding of metal cations by lichens has particular consequences for man, as lichens can be used to monitor airborne metal pollution, but also lichens can accumulate radioactive metals such as ^{55}Fe and ^{137}Cs, which then enter food chains, e.g. lichen →reindeer →laplander. Both of these aspects are discussed in detail by Richardson (1976).

The results obtained in this experiment are consistent with the lichen acting as a passive cation-exchanger. The most strongly bound cations are trivalent iron and aluminium, then divalent copper, calcium and nickel, and then monovalent potassium, sodium and lithium. This phenomenon is discussed by Farrar (1973). It is probable that the main sites of binding are anionic groups in the fungal walls. Killed lichens, or lichens incubated at 0 °C or in the presence of metabolic poisons still give the same results.

REFERENCES

J. F. Farrar (1973). Lichen physiology: progress and pitfalls. *In* "Air Pollution and Lichens", (B. W. Ferry, M. S. Baddeley and D. L. Hawksworth eds), pp. 238–82. Athlone Press, London.
D. H. S. Richardson (1976). "The Vanishing Lichens". David and Charles, Newton Abbot.

14

Polarographic Measurements of Oxygen Uptake Rates of Intact Yeast Cells

R. K. POOLE

*Department of Microbiology, Queen Elizabeth College,
Campden Hill, London, W8 7AH, England*

Level: All undergraduate years
Subject area: Microbial biochemistry
Special features: Simple experiments on microbial
metabolism using the oxygen
electrode

INTRODUCTION

Although the mitochondria of eukaryotic cells exhibit a variety of different functions, there is little doubt that the most important one is electron transport linked to ATP synthesis (oxidative phosphorylation). Analagous reactions are performed in prokaryotes by components firmly bound to the cytoplasmic membrane. The major components of the respiratory chain in both eukaryotes and prokaryotes are proteins that have prosthetic groups able to undergo oxidation or reduction by the removal or addition of either electrons or hydrogen atoms. Electrons are supplied to this chain of electron-carrying components by the oxidation of metabolites such as malate or succinate. Ultimately, electron transport results in the reduction of a terminal electron acceptor which in aerobic organisms is generally oxygen.

The synthesis of ATP is coupled to electron transfer at three discrete "sites" in the mitochondrial electron transfer chain. These three sites of energy conservation (I, II and III) are also the sites of action of the classical inhibitors of electron transfer, rotenone, antimycin and cyanide, respectively (Figure 1). The ionophores referred to as uncoupling agents allow electron transfer to proceed without concomitant ATP synthesis. Control of respiration rates by the availability of ADP (respiratory or acceptor control) is thus abolished in a system previously tightly coupled, and in the presence of limiting amounts of ADP, addition of such ionophores stimulates the respiration rate.

The aim of this experiment is to demonstrate the oxidation of endogenous and various added substrates by intact cells of the yeast *Schizosaccharomyces pombe*, and to demonstrate the action of the classical respiratory inhibitors and an uncoupling agent.

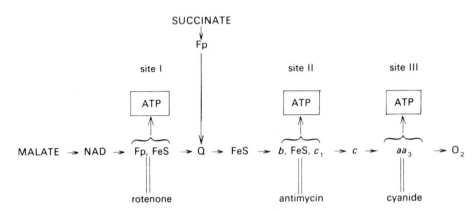

Figure 1 Generalized scheme of the functional organization of the respiratory chain in eukaryotic mitochondria. The points of entry of electrons from malate and succinate are shown, as well as the sites of inhibition of electron transport and the probable sites of ATP synthesis. Cytochromes are indicated by italic lower case letters; Fp represents flavoproteins; FeS indicates iron–sulphur centres; Q is ubiquinone. The stoichiometries of ATP synthesis at each "site" and the positions of FeS in the chain are still uncertain.

EXPERIMENTAL

The procedure for using the oxygen electrode apparatus manufactured by Rank Brothers (High Street, Bottisham, Cambridge CB5 9DA, England) is described, but many other commercial (and home-made) systems should function equally well. It is recommended that the growth protocol is followed closely. The following describes the methods and materials suitable for a few groups of students (say 4) working simultaneously. In most laboratories, the number of students will be limited by the number of electrode assemblies available.

Day 1 *(5 min)*

Inoculate 200 ml medium with cells from an agar slant culture. Incubate with shaking at 30 °C.

Day 2

(a) *approx. 24 h later; 5 min*

Inoculate 200 ml medium with 5 ml of the above starter culture. Incubate as above. This is culture 1.

(b) *late afternoon, see Day 3; 5 min*

Inoculate 200 ml medium with 1 ml of the above starter culture. Incubate as above. This is culture 2.

Day 3 *(4 h or longer)*

Harvest the cells from culture 1 **only** (which will have reached the early stationary phase of growth) by membrane filtration. Pour about 20 ml KCl–Tris buffer over the filter to wash the cells, then transfer the filter (+ cells) to a McCartney bottle. Resuspend the cells in a further 20 ml of buffer by "Whirlimixing". This suspension should suffice for 6–10 groups of students, each performing the experiments below, and so could be done once before the practical class begins.

↓

Pipette 2 ml of air-saturated KCl–Tris buffer into the electrode vessel, maintained at 30° C by water circulation. Switch on the stirrer and then insert the vessel plug. Ensure all air is excluded by displacement through the fine hole in the plug. Allow to equilibrate for 5–10 min and set the recorder pen to maximal deflection (*see* note 1).

↓

Add a few grains of sodium dithionite through the fine hole. On anaerobiosis, check that the recorder pen gives almost zero deflection.

↓

Thoroughly rinse out the vessel with distilled water and repeat the first operation to check calibration with air-saturated buffer again.

↓

Pipette 1·8 ml of air-saturated buffer into the vessel, begin stirring, and then add 0·2 ml of the suspension of washed cells (*see* note 2). Insert the vessel plug, always ensuring that all air is displaced.

↓

Record O_2 tension for 3–5 min or until a constant rate of endogenous O_2 uptake is obtained. Then add substrate, using 10 μl of either 2 M glucose or 2 M ethanol (*see* note 3). Follow respiration until O_2 is exhausted. Be wary of anomalous recorder traces (*see* note 6).

↓

Repeat the experiment, observing the effects of the following inhibitors in turn on the substrate-stimulated respiration rates (**caution**, *see* note 4):
  (a) rotenone;
  (b) antimycin A;
  (c) KCN.
The inhibitors should be added as soon as a constant rate of respiration is observed. The vessel should be carefully cleaned between experiments (*see* note 5).

↓

Repeat the experiment but using 2 ml withdrawn directly from culture 2. (This culture should be in mid-exponential phase about 20 h after inoculation.) As soon as a constant rate of O_2 uptake is attained, add small volumes (e.g. 2 μl) of CCCP **caution**, *see* note 4). As soon as a second constant (faster?) rate is attained, add a further 2 μl. Repeat until O_2 is exhausted.

↓

Repeat, investigating the effects of similar volumes of methanol on respiration rates.

Notes and Points to Watch

1. At $30°C$, the oxygen content of the air-saturated buffer described is 220 nmoles ml^{-1}.
2. Cultures and washed cell suspensions will obviously vary in their rates of O_2 uptake. The volumes suggested are guidelines only.
   The untreated cell culture (culture 2) could be diluted with fresh medium or concentrated accordingly.
3. Other inhibitors or substrates can be used. Suitable alternatives are given by Heslot et al. (1970).
4. Students should be warned of the poisonous nature of all inhibitors and ionophores used. It is recommended that the solutions be prepared by technical staff or the instructor. They should be stored on ice.
5. Rotenone, antimycin and CCCP are dissolved in methanol. To ensure that these compounds are totally cleaned from the vessel between experiments, the vessel should be rinsed with methanol as well as distilled water. A convenient and safe method of rinsing the vessel is to use a suction pump to remove cell suspensions and washings.
6. The most common difficulties are:
   (a) failure to remove all air from the vessel;
   (b) torn membrane, perhaps due to inserting a microsyringe too far;
   (c) inadequate stirring.

MATERIALS

1. *Schizosaccharomyces pombe* strain 972 h$^-$ (agar slant culture).
2. Defined medium, prepared as described in Experiment 26.
3. Oxygen electrode apparatus (e.g. from Rank Brothers, High Street, Bottisham, Cambridge, CB5 9DA, England), consisting of: incubation vessel, magnetic stirrer, battery-operated polarizing voltage circuit.
4. Water circulator (not essential).
5. Potentiometric recorder with input sensitivity 1–10 mV full scale deflection (FSD) (for Rank electrode) or appropriate for the electrode used.

Suitable chart speeds are about 1–2 cm min^{-1}.
6. Micro-syringes (e.g. Hamilton). Most useful sizes are in the range 5–50 μl total capacity.
7. Wash and reaction medium containing (per litre): KCl (7·5 g); Tris base (2·42 g). Bring to pH 7·4 at $30°C$ with 1 M HCL (about 3–3·3 ml).
8. Rotenone, antimycin and carbonylcyanide *m*-chlorophenyl hydrazone (CCCP) (Sigma London Chemical Co Ltd., Fancy Road, Poole, Dorset BH17 7NH, England), each dissolved in methanol at 1 mg ml^{-1}.
9. Solution (0·1 M) of KCN, brought to pH 7–8 with dilute HCl.

SPECIFIC REQUIREMENTS

Day 1
200 ml growth medium in 1 litre Erlenmeyer flask
agar slant culture of S. pombe
orbital or reciprocating shaker (30°C) with capacity for 1 litre flasks

Day 2
culture from Day 1

2 × 200 ml growth medium in 1 litre Erlenmeyer flasks
shaker as above

Day 3
cultures from Day 2
Membrane filters (0·45 μm pore size), at least 50 mm diam. and appropriate filter holder and Buchner flask

cont.

cont.

100 ml KCl–Tris buffer, pH 7·4	ice and ice buckets
McCartney bottle or similar	methanol
pipettes	lens tissue
Rotamixer or Whirlimixer or similar	O₂ electrode membrane ⎫ for fitting membrane
methanol and distilled water wash-bottles	saturated KCl ⎬ and electrolyte
sodium dithionite	scissors ⎭
vacuum pump operated by water supply	

FURTHER INFORMATION

The Clark-type oxygen electrode is a convenient alternative to Warburg manometers for experiments like the one described above. Useful accounts of the instrument and clear diagrams are given by Eastbrook (1967) and Whittaker and Danks (1978).

The first systematic study of the respiratory metabolism of *S. pombe* was made by Heslot *et al.* (1970). Glucose, ethanol, L-lactate and malate are rapidly oxidized by cells grown aerobically on either glucose or glycerol. Glucose-grown cells have a greatly reduced rate of endogenous respiration by comparison with glycerol-grown cells and are thus more suitable for demonstrating substrate-stimulated respiration.

Respiration of intact cells from exponentially growing cultures is strongly inhibited by both antimycin and KCN (Poole *et al.*, 1973). Somewhat higher inhibitor concentrations than those given in that paper are required to inhibit substrate-stimulated rates of respiration by washed cells from stationary phase cultures (Table 1). Rotenone inhibits neither cell growth nor respiration of intact cells or isolated mitochondrial particles (Heslot *et al.*, 1970). A possible correlation between sensitivity to rotenone, specific e.p.r.-detectable signals, and the presence of site I energy

Table 1 Observed respiration rates of *S. pombe* in the absence or presence of substrate and inhibitors of electron transfer. Concentrations shown are final concentrations in the electrode vessel. Inhibition is expressed as a percentage of the substrate-stimulated respiration rate in the absence of inhibitor. The electrode vessel contained approx. 0·3 mg (dry weight) of yeast in each experiment. (Second-year undergraduate results of Ian Salmon.)

Additions	Rate of O₂ uptake [n moles O₂ min⁻¹ (mg dry wt)⁻¹]	Inhibition (%)
Experiment 1		
None	6·1	—
ethanol (20 mM)	17·8	—
rotenone (0·013 mM)	15·3	14
antimycin A (0·009 mM)	5·5	69
Experiment 2		
ethanol (20 mM)	17·5	—
KCN (0·5 mM)	6·7	62
KCN (1 mM)	2·2	87

conservation has been the subject of an extensive and confusing literature (Lloyd, 1974). P/O (or → H⁺/O) ratios have not been determined for *S. pombe* but it is tempting to speculate that resistance to rotenone reflects the absence of site I.

CCCP is a potent uncoupling ionophore. At least a two-fold stimulation of the respiration rates of intact cells may be demonstrated; the optimal concentration in our hands has been about 15 μM. Sensitivity is maximal during mid-exponential growth and declines as the stationary phase is approached (Poole and Salmon, 1978). It is therefore important to control carefully the growth of cells in the experimental culture.

A typical titration of exponential phase cells with CCCP is shown in Figure 2. Addition of small volumes of CCCP solution elicit a marked stimulation of respiration rate which is abolished on further additions (Figure 3). Care is therefore required in establishing a suitable final concentration of the ionophore.

An interesting variation on this experiment is to grow cells on glycerol, rather than glucose, as sole carbon source, and allow the culture to enter the stationary phase of growth (approx. 3–4 days after inoculation). Such cells exhibit substrate – (e.g. glycerol-) stimulated respiration that is largely

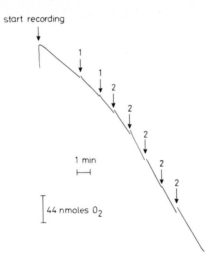

Figure 2 Titration of cells from an exponentially growing culture with CCCP. Time proceeds from left to right and a downward deflection represents O_2 uptake. Figures show the volume (in μl) of a methanolic solution of CCCP (1 mg ml⁻¹) added at each time indicated by the arrows.

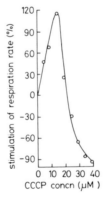

Figure 3 Effects of CCCP on respiration rates of cells in samples removed from an exponentially growing culture. Concentrations shown are final concentrations in the electrode vessel. The measured stimulation by a methanolic solution of CCCP was corrected for the small stimulation elicited by equal volumes of the solvent alone.

insensitive to KCN. A comprehensive treatment of cyanide-insensitive respiration is that of Henry and Nyns (1975).

REFERENCES

R. W. Estabrook (1967). Mitochondrial respiratory control and the polarographic measurement of ADP:O ratios. *Methods in Enzymology*, **10**, 41–7.

M.-F. Henry and E.-J. Nyns (1975). Cyanide-insensitive respiration. An alternative mitochondrial pathway. *Sub-cellular Biochemistry*, **4**, 1–65.

H. Heslot, A. Goffeau and C. Louis (1970). Respiratory metabolism of a "petite negative" yeast *Schizosaccharomyces pombe* 972 h⁻. *Journal of Bacteriology*, **104**, 473–81.

D. Lloyd (1974). "The Mitochondria of Micro-organisms", pp. 82–158. Academic Press, London and New York.

R. K. Poole and I. Salmon (1978). The pool sizes of adenine nucleotides in exponentially-growing, stationary phase and 2′-deoxyadenosine-synchronized cultures of *Schizosaccharomyces pombe* 972 h⁻. *Journal of General Microbiology*, **106**, 153–64.

R. K. Poole, D. Lloyd and R. B. Kemp (1973). Respiratory oscillations and heat evolution in synchronously dividing cultures of the fission yeast *Schizosaccharomyces pombe* 972h⁻. *Journal of General Microbiology*, **77**, 209–20.

P. A. Whittaker and S. M. Danks (1978). "Mitochondria: Structure, Function and Assembly", pp. 68–9. Longman, Harlow, UK.

15

Effect of the Growth Environment on the Assimilation Pathways for Glycerol and Ammonia in *Klebsiella aerogenes*

O. M. NEIJSSEL, S. HUETING and D. W. TEMPEST

Laboratorium voor Microbiologie, Universiteit van Amsterdam, Nieuwe Achtergracht 127, 1018 WS Amsterdam, The Netherlands

> *Level*: Advanced undergraduates
> *Subject area*: Bacterial physiology
> *Special features*: Chemostat culture

INTRODUCTION

In nature the nutrients necessary for cell synthesis are not usually present in concentrations which allow microbial growth at the maximum rate: growth is mostly nutrient-limited. To cope with the scarcity of food, micro-organisms have evolved uptake systems with a high affinity for a wide variety of nutrients such as sugars, amino acids, phosphate, metal ions, etc. (for review *see* Tempest and Neijssel, 1978). On the other hand, if a particular nutrient suddenly is present in excess of the growth requirement, the high-affinity uptake system for that nutrient is generally repressed and replaced by an uptake system of relatively low affinity. There may be two reasons for this behaviour: first, high-affinity uptake systems generally involve the expenditure of extra metabolic energy (either as ATP or as a membrane potential) which can be saved when the nutrient is plentiful; second, when an organism expressing a high-affinity uptake system is suddenly faced with an excess of nutrient, the flux of intermediates in the metabolic pool may be so increased as to pose severe metabolic problems, e.g. "substrate accelerated death" (Calcott and Postgate, 1972).

The experiment described here deals with the effect of changes in the glycerol and ammonia concentration of the medium on the regulation of the assimilation of these two nutrients by *Klebsiella aerogenes*. Glycerol can be assimilated via two pathways (Figure 1): the upper one involves glycerol kinase and then glycerol phosphate dehydrogenase; the lower one uses glycerol dehydrogenase and then dihydroxyacetone (or glyceraldehyde) kinase.

The product of both pathways is glyceraldehyde 3-phosphate. Since glycerol is not taken up via an active transport system, but by facilitated diffusion, the kinetics of the first enzyme of the two pathways will determine the efficiency of uptake. Glycerol kinase has an apparent K_m of about 20 μM for glycerol, whereas the apparent K_m of glycerol dehydrogenase for glycerol is 20–40 mM.

There is an analogous situation for ammonia which, after diffusion through the cell membrane, can be assimilated via glutamate dehydrogenase ($K_m = 10$ mM) or through the glutamine synthetase (GS, $K_m = 0.5$ mM) and glutamate synthase (GOGAT) reactions (Figure 2).

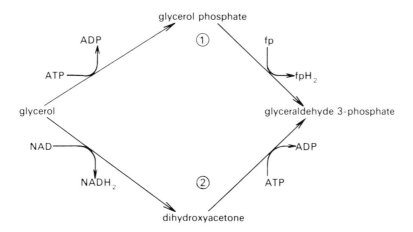

Figure 1   Dual pathways of glycerol assimilation in *Klebsiella aerogenes*. 1 high-affinity pathway; 2 low-affinity pathway.

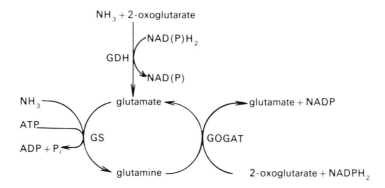

Figure 2   Dual pathways of ammonia assimilation. High-affinity pathway: GS + GOGAT; low-affinity pathway: GDH.

In this experiment, *K. aerogenes* is grown in a chemostat under either glycerol- or ammonia-limiting conditions, and the assimilation pathways for glycerol and ammonia under these conditions is studied.

EXPERIMENTAL

This experiment should be performed with two chemostats, but if only one is available the scheme can be modified according to note 1. If the chemostats are not equipped with a pH-control unit, the medium has to be modified to increase its buffering capacity (*see* note 2). Make sure that the cultures are fully aerobic, and if necessary adjust the amount of biomass in the steady state to avoid oxygen-limitation (*see* note 3).

Day 1 *(2h: preparation and sterilization of media and chemostats + 1h: starting up and inoculation of chemostats)*

The chemostats are autoclaved. From 10 to 20 litres (depending on the size of the medium reservoirs of the chemostats) of medium I and the same amount of medium II are made up. After the media have been autoclaved (*see* note 4) and cooled sufficiently to allow handling, the medium reservoir bottles are connected to the chemostats: medium I with chemostat 1 (glycerol limitation) and medium II with chemostat 2 (ammonia limitation). The growth vessels are filled with medium and the chemostats are checked for proper functioning: pH value 6·8 and temperature 37 °C (*see* note 6). The chemostats are then inoculated with 5–10 ml of an overnight culture of *K. aerogenes* NCTC 418 in nutrient broth. The dilution rate is set at $0·1–0·2$ h^{-1}. Both chemostats must have the same dilution rate.

Day 2 *(1 h: checking chemostats + 1 h: preparation of reagents)*

The chemostats should be checked: determine pH and optical density (*see* note 5), culture purity, dilution rate and dry weight of the culture and adjust if necessary.

↓

Prepare reagents and solutions to be used on Day 4.

Day 3 *(1 h: checking chemostats + 1 h: preparation of reagents)*

Check the chemostats as on Day 2. Prepare the rest of the reagents and solutions to be used on Day 4.

Day 4 *(4h)*

Check the chemostats as on Day 2. If they are in a steady state, samples can be drawn for the experiments.

↓

Take a sample from chemostat 1 containing approximately 50 mg dry weight of cells (sample 1) (*see* note 3). Take a similar sample from chemostat 2 (sample 2). Centrifuge both samples for 15 min at 6000 *g* (4 °C). Wash the samples with phosphate buffer. Centrifuge again for 15 min at 6000 *g* (4 °C).

↓

Suspend the cells of sample 1 and 2 separately in 3 ml of phosphate buffer. Sonicate both suspensions for 3×1 min with an ultrasonic disintegrator (the suspensions must be cooled in an ice bath). Centrifuge the sonicated suspensions for 20 min at 48 000 g at 4 °C. The clear supernates should be used for the following enzyme analyses: glutamate dehydrogenase, glutamate synthase, glycerol kinase and glycerol dehydrogenase. Details of the assays are given in Materials.

- -

Notes and Points to Watch

1. If only one chemostat is available, then one should first grow the organisms carbon-limited (Day 1, 2, 3, 4). The ammonia-limited medium is prepared on the second day and at the end of Day 4 the medium is changed from glycerol limitation (medium I) to ammonia limitation (medium II). On Days 5 and 6 the chemostat is checked as usual and on Day 7 or 8 the second part of the analyses can be done.
2. The buffering capacity of the media can be increased by increasing the phosphate concentration from 10 to 40 mM, and the medium should be adjusted to pH 7·0. In addition, it is also preferable to lower the steady-state dry weight to about 1 mg ml^{-1} (see note 3).
3. The quality of the chemostat determines the amount of biomass which can be maintained in steady-state under conditions of oxygen excess. Medias I and II will give a steady state dry weight of about 4 mg ml^{-1}. If a lower dry weight is required, the concentration of the growth limiting nutrient (either glycerol or ammonia) should be lowered correspondingly; the concentrations of the other nutrients remain the same.
4. In general, an autoclaving time of 30 min is sufficient for 10 litres of medium.
5. The optical density of bacterial suspensions may be measured at various wavelengths so in this experiment wavelength is not critical. At 540 nm, a 25-fold dilution of a *K. aerogenes* culture (initially 4 mg ml^{-1}) has an absorption of approximately 0·5 (1 cm light path).
6. 2 M NaOH can be used as the titrant for the pH control unit. This solution does not need to be sterilized. To suppress foaming of the culture the regular addition of an antifoam solution will be necessary. If an antifoam control unit is not available it is advisable to lower the steady-state dry weight of the culture to about 1 mg ml^{-1} according to note 3.

MATERIALS AND METHODS

1. *Klebsiella aerogenes* NCTC 418 (which is equivalent to NCIB 418 and ATCC 15380).
2. 2 chemostats with a working volume of minimally 350 ml.
3. *Medium I*:

NaH$_2$PO$_4$.2H$_2$O,	1·56 g l^{-1}, 10 mM
KCl,	0·75 g l^{-1}, 10 mM
MgCl$_2$,	0·12 g l^{-1}, 1·25 mM
NH$_4$Cl,	5·35 g l^{-1}, 100 mM
Na$_2$SO$_4$.10H$_2$O,	0·64 g l^{-1}, 2 mM
citric acid^a,	0·42 g l^{-1}, 2 mM
CaCl$_2$,	0·002 g l^{-1}, 0·02 mM
glycerol,	10 g l^{-1}, 109 mM
trace elements solutions,	20 ml l^{-1}

Na$_2$MoO$_4$ (1mM) solution, 0·1 ml l^{-1}.
^{a} Citric acid is added as chelating agent for metal ions.

Medium II: as medium I but

NH$_4$Cl,	1·60 g l^{-1}, 30 mM
glycerol,	30·0 g l^{-1}, 326 mM.

Trace elements solution

ZnO,	2·04 g l^{-1}
FeCl$_3$.6H$_2$O,	27·0 g l^{-1}
MnCl$_2$.4H$_2$O,	10·0 g l^{-1}
CuCl$_2$.2H$_2$O,	0·85 g l^{-1}
CoCl$_2$.6H$_2$O,	2·38 g l^{-1}
H$_3$BO$_3$,	0·31 g l^{-1}
HCl (conc.),	50 ml l^{-1}.

4. Phosphate buffer: 10 mM, pH 7·0.
5. Ultrasonic disintegrator.
6. Refrigerated centrifuge.
7. Double beam spectrophotometer with thermo-statically controlled cell carrier.
8. *Glutamate dehydrogenase assay*:
   Pipette into a cuvette:

imidazole buffer (0·1 M, pH = 7·9),	2·47 ml
2-oxoglutarate (0·2 M, pH = 8·0),	0·2 ml
NADPH (10 mg ml^{-1}),	0·03 ml
EDTA (26 mM),	0·1 ml
ADP (0·1 M),	0·05 ml
cell-free extract,	0·1 ml

   Measure the non-specific oxidation of NADPH against a curvette containing no NADPH for 2 min; then add

ammonium acetate (12·8 M),	0·05 ml

   and follow the oxidation of NADPH.
9. *Glutamate synthase assay*:
   Pipette into a cuvette:

Tris–HCl (50 mM, pH = 8·5),	2·57 ml
2-oxoglutarate (0·2 M, pH = 8·0),	0·2 ml
NADPH (10 mg ml^{-1}),	0·03 ml
cell-free extract,	0·1 ml

   Measure the non-specific oxidation of NADPH against a cuvette containing no NADPH for 2 min; then add

glutamine (120 mM),	0·1 ml

   and follow the oxidation of NADPH.
10. *Glycerol dehydrogenase + glycerol kinase assay*:
   Pipette into a cuvette:

hydrazine/glycine buffer,[a]	2·80 ml
(0·4 M hydrazine, 0·2 M glycine, 2 mM MgCl$_2$, pH = 9·8)	
NAD (10 mg ml^{-1}),	0·10 ml
glycerol phosphate dehydrogenase (600 U/ml),	0·02 ml
MnCl$_2$ (8 mM),	0·03 ml
cell-free extract,	0·10 ml

   Follow the non-specific reduction of NAD against a cuvette containing no NAD for 2 min; then add

glycerol (1 M),	0·30 ml

   Follow the glycerol dehydrogenase reaction (if present); then add

ATP, sodium salt (50 mg ml^{-1}),	0·05 ml

   and follow the glycerol kinase reaction.

   [a] The hydrazine glycine buffer must be set at the right pH only **shortly** before the assay, since the buffer is unstable at high pH values.

SPECIFIC REQUIREMENTS

Day 1
2 complete chemostats
10–20 litres medium I
10–20 litres medium II
pH meter
overnight culture of *K. aerogenes* NCTC 418
sterile pipettes (for inoculation)
stop-watch (dilution rate)

Day 2
4 dry-weight tubes
pipettes
tube rack
pH meter
spectrophotometer
stop-watch
microscope

Day 3
as on Day 2

Day 4
as on Day 2 plus
centrifuge tubes
reagents for enzyme assays (*see* Materials and Methods)
ultrasonic disintegrator
phosphate buffer
cuvettes (minimally 2)
refrigerated centrifuge
double beam spectrophotometer

FURTHER INFORMATION

Because the assay for glutamine synthetase is troublesome, the presence of glutamate synthase is an acceptable indicator of the high-affinity pathway of ammonia assimilation. Sometimes the glutamate synthase assay on the cell-free extract of the carbon-limited cells will give a false positive result due to the presence of glutaminase. The ammonia generated by this reaction will be used by the glutamate dehydrogenase reaction. The presence of glutaminase can be determined easily by the method of Bergmeyer (1974).

The glycerol-limited cells should contain glycerol kinase and glutamate dehydrogenase (and no glutamate synthase and glycerol dehydrogenase), whereas the ammonia-limited cells should contain glutamate synthase and glycerol dehydrogenase and only traces of the other two enzymes.

Another possible way to determine the kinetics of glycerol assimilation is by using washed suspensions (in 100 mM phosphate buffer, pH = 7·0) in a Warburg or Gilson apparatus. The optimal concentration of cells is 2–2·5 mg per vessel and glycerol is added at concentrations that vary between 0·2 and 20 mM. Assuming that the kinetics of the first enzyme determines the rate of glycerol consumption, and that this rate is reflected in the oxygen consumption rate, one can observe the high affinity for glycerol of glycerol-limited cells and the low affinity for glycerol of ammonia-limited cells (*see* Neijssel *et al.*, 1975).

Once the chemostats are in a steady-state the experiments can be repeated every day by different students. The chemostat experiments can be expanded to metabolic studies. Depending on the equipment available, one can determine the differences in $q_{glycerol}$, q_{O_2} and q_{CO_2} between glycerol- and ammonia-limited cultures. It has been reported that ammonia-limited cultures of *K. aerogenes* excrete 2-oxoglutarate (Neijssel and Tempest, 1975). The presence of this metabolite in the culture fluids can be assayed by the method of Bergmeyer and Bernt (1974). A very simple, but non-specific method is with ketone test strips (routinely used in clinical chemistry and commercially available at a low price). Alternatively one can use a spot test for ketones (2,4-dinitrophenylhydrazine-HCl solution).

REFERENCES

H. U. Bergmeyer (1974). Glutaminase. *In* "Methoden der enzymatischen Analyse", (H. U. Bergmeyer, ed.), 3rd edn, Vol. I, pp. 493–4. Verlag Chemie, Weinheim.

H. U. Bergmeyer and E. Bernt (1974). α-Ketoglutarate, UV-spektrophotometrische Bestimmung. *In* "Methoden der enzymatischen Analyse", (H. U. Bergmeyer, ed.), 3rd edn, Vol. II, pp. 1624–7. Verlag Chemie, Weinheim.

P. H. Calcott and J. R. Postgate (1972). On substrate accelerated death of *Klebsiella aerogenes*. *Journal of General Microbiology*, **70**, 115–22.

O. M. Neijssel and D. W. Tempest (1975). The regulation of carbohydrate metabolism in *Klebsiella aerogenes* NCTC 418 organisms, growing in chemostat culture. *Archives of Microbiology*, **106**, 251–8.

O. M. Neijssel, S. Hueting, K. J. Crabbendam and D. W. Tempest (1975). Dual pathways of glycerol assimilation in *Klebsiella aerogenes* NCIB 418. Their regulation and possible functional significance. *Archives of Microbiology*, **104**, 83–7.

D. W. Tempest and O. M. Neijssel (1978). Eco-physiological aspects of microbial growth in aerobic nutrient–limited environments. *Advances in Microbial Ecology*, **2**, 105–53.

16

Uncouplers, Ionophores and Amino Acid Accumulation by Bacteria

I. R. BOOTH and W. A. HAMILTON

Department of Microbiology, Marischal College,
University of Aberdeen, Aberdeen, AB9 1AS, Scotland

> *Level*: Advanced undergraduates
> *Subject areas*: Microbial physiology and growth
> *Special features*: Chemiosmotic principles applied
> to bacterial active transport

INTRODUCTION

The ability of an organism to accumulate intracellularly a substrate in a chemically unmodified form depends upon the availability of metabolic energy and the effective coupling of that energy to the transport process.

One of the principal mechanisms of active transport in bacteria involves energy coupling via proton circulation as envisaged by the chemiosmotic hypothesis (Mitchell, 1966). Respiration-linked proton extrusion from the cell generates gradients of charge (the membrane potential, interior negative) and pH (the pH gradient, interior alkaline), known collectively as the proton-motive force. Within the same membrane are transport proteins which couple the flow of protons to the movement of substrates. Thus as a consequence of proton flow down the electrochemical gradient (generated by respiration) it is possible for substrates to be accumulated against their electrochemical gradient (Figure 1).

An essential feature is the requirement for energy in the form of reducing equivalents which can be respired to bring about proton extrusion and thus generate a proton-motive force. If, however, a compound such as an uncoupling agent is present which rapidly returns the protons (Figure 2) to the milieu of the cell, no proton gradient will be generated and no net accumulation of transport substrate will be observed. Such compounds are called uncouplers because of their ability to separate respiration-driven proton extrusion from the performance of useful work. In this experiment starved *Staphylococcus aureus* cells will be used to demonstrate:

(i) the requirement for energy for the accumulation of isoleucine, and

(ii) the effect of an uncoupling agent, 3′, 4′, 3, 5 tetrachlorosalicylanilide (TCS), on isoleucine uptake.

When *S. aureus* is incubated for 3 h in the absence of a substrate, it depletes its endogenous substrate reserves, including the amino acid pool. Such cells possess little, if any, source of metabolic energy and are unable to accumulate amino acids from the medium to any appreciable extent. If, however, these cells are presented with a substrate, (e.g. glucose), which can be metabolized by an energy-producing reaction sequence, accumulation of an added amino acid may be expected to occur. If, on the other hand, the uncoupling agent TCS is added with the substrate, the metabolic energy produced by catabolism of the substrate will not be coupled to the transport process, and no amino-acid accumulation will be observed.

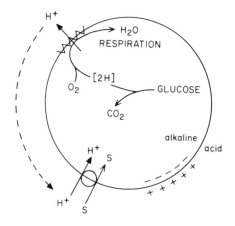

Figure 1 Metabolism leads to the generation of a proton-motive force (inside alkaline and negative) which drives amino acid accumulation via proton symport.

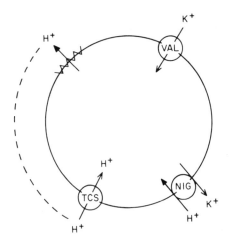

Figure 2 The presence of an uncoupler, TCS, prevents the build up of the proton-motive force by allowing free passage of protons back into the cell. Valinomycin and nigericin reduce $\triangle \psi$ and $\triangle$ pH respectively, by allowing compensatory ion movements.

EXPERIMENTAL

Day 1 ($1\frac{1}{2}$–2 h; see *note 1*)

Transfer 10 ml of starved *S. aureus* cell suspension (approx. 1·05 mg dry weight ml^{-1}) in 33 mM sodium phosphate buffer, pH 7, to a 100 ml conical flask. Incubate for 10 min at 30 °C in a shaking water-bath.

↓

Add 250 μl 0·26 mM [^{14}C]-isoleucine with a microsyringe or automatic pipette. Start clock. Wash out syringe with distilled water (*see* note 2).

↓

After 1 min, remove a 1 ml sample with an automatic pipette and transfer to membrane filter assemby (*see* notes 3–5).

↓

Wash the filtered cells with 3 ml 33 mM sodium phosphate buffer, pH 7. Transfer filter to a scintillation vial, which should be labelled on its shoulder.

↓

Repeat the above procedure with samples taken at 5, 10 and 15 min.

↓

Prepare a second flask **during the first run**, containing cells as before plus 0·25 ml 21 mM glucose. Preincubate for 10 min as above. Measure [^{14}C]-isoleucine uptake as above.

↓

Prepare a third flask **during the second run**, containing cells, glucose and 5 μl TCS (1·05 mg ml^{-1} in acetone). Preincubate for 10 min as above. Measure [^{14}C]-isoleucine uptake as above (*see* note 6).

↓

Dry filters in vials in a warm oven (60–65°C).

↓

Add scintillant and count in liquid scintillation counter.

_ _

Day 2 *(1 h)*

Plot [^{14}C]-isoleucine (ct min^{-1}) mg^{-1} cells against time.

↓

Convert ct min^{-1} to nmoles [^{14}C]-isoleucine using the counting efficiency factor (supplied by demonstrator), the specific activity of the radioactive isoleucine, and the conversion factor $2·2 \times 10^3$ d min^{-1} = 1 nCi.

↓

Calculate the internal concentration of isoleucine (internal volume *S. aureus* = 1·55 μl mg^{-1} dry weight), and compare to the external concentration.

- -

Notes and Points to Watch

1. Students should be warned of the dangers of working with radiochemicals and should be provided with Benchkote, metal trays, radioactive tape and plastic gloves.
2. All radioactive solutions should be handled only with microsyringes and automatic pipettes.
3. Best results have been obtained with Oxoid or Millipore membrane filters, 0·45 μm pore size and 2·5 cm diam.
4. Vacuum filtration is readily achieved by water pumps, but electric pumps can be used if necessary.
5. Filters should be wetted with an appropriate buffer prior to filtration of cells.
6. Carbonyl cyanide-M-chlorophenylhydrazone (CCCP, 25 μM) can be substituted for TCS. 2, 4-Dinitrophenol (DNP) should not be used, since much higher concentrations are required to achieve the same effect.
7. The number of experiments can be increased by substituting ionophores, such as valinomycin (+ K$^+$), or nigericin or both, for TCS.

MATERIALS

1. Sodium phosphate buffer should be made up as described by Dawson *et al.* (1969).
2. [^{14}C]-Isoleucine can be obtained from the Radio-chemical Centre, Amersham, UK and TCS may be obtained from Kodak Chemicals Ltd. CCCP can be obtained from Sigma Chemical Co., UK.
3. Suspension of *S. aureus* (starved for 3 h by incubation with shaking at 30°C in 33 mM sodium phosphate buffer pH 7·0) in 33 mM sodium phosphate buffer pH 7·0 (1·05 mg dry weight ml^{-1}; 1 mg dry weight ml^{-1} = OD$_{650}$ 2·2).

SPECIFIC REQUIREMENTS

Day 1
30 ml starved *S. aureus*
12 × 0·45 μm pore size, 2·5 cm diam. membrane filters
Petri dish in which to wet filters
filtration assembly
Benchkote, metal tray and radioactive tape
vacuum pump
21 mM glucose
0·26 mM [^{14}C]-L-isoleucine
1·05 mg ml^{-1} TCS in acetone
pipettes, 10 ml and 5 ml
25 μl and 250 μl syringe

1 ml automatic pipettes
tweezers
12 scintillation vials
scintillation fluid
50 ml buffer, pH 7·0
30°C shaking water-bath
3 × 100 ml conical flasks
distilled water

Day 2
graph paper
calculators

FURTHER INFORMATION

Figure 3 gives typical results obtained by students. The results clearly demonstrate the requirement for energy for active transport and the effects of uncouplers on amino acid uptake into *S. aureus*.

The experiments may be extended by the use of the ionophores valinomycin (4 μg mg^{-1} cells) and nigericin (0·5 μg mg^{-1} cells). In the presence of external potassium ions (approx. 50 mM = 33 mM potassium phosphate buffer pH 7·0), valinomycin leads to a reduction of the membrane potential. On the other hand, nigericin reduces the pH gradient. Neither of these agents alone fully inhibits isoleucine uptake. When both are added, however, complete cessation of isoleucine uptake results. The net effect of the presence of both antibiotics is that of the uncoupler TCS, i.e. inward proton flux. Further, consideration of these points will be found in Harold's (1974) and Hamilton's (1975) reviews.

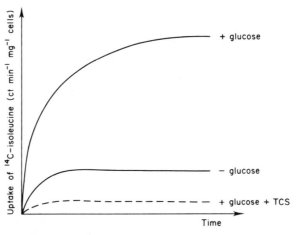

Figure 3   Schematic diagram of expected results.

REFERENCES

R. M. C. Dawson, D. C. Elliott, W. H. Elliott and K. M. Jones (eds) (1969). "Data for Biochemical Research", pp. 475–508. Oxford University Press, Oxford.

W. A. Hamilton (1975). Energy coupling in microbial transport. *Advances in Microbial Physiology*, **12**, 1–53.

F. M. Harold (1974). Chemiosmotic interpretation of active transport in bacteria. *Annals of the New York Academy of Sciences*, **227**, 297–311.

P. Mitchell (1966). Chemiosmotic coupling in oxidative and photosynthetic phosphorylation. *Biological Reviews*, **41**, 445–502.

17

Induction of β-galactosidase in *Escherichia coli*

W. A. HAMILTON

*Department of Microbiology, University of Aberdeen,
Aberdeen, AB9 1AS, Scotland*

> *Level*: Advanced undergraduates
> *Subject areas*: Biochemistry
> *Speical features*: Illustrates major concepts in mole-
> cular biology. The exercise de-
> mands and rewards manipulative
> care and skill

INTRODUCTION

Our knowledge of the expression of the *lac* operon in *Escherichia coli* is a corner-stone of that aspect of biochemistry and microbiology known as molecular biology. Through the identification and study of the phenomena of induction and catabolite repression of β-galactoside permease and β-galactosidase, we have come to recognize **the component parts of the operon** (regulator, promotor, operator, structural genes), **the nature of their interaction** (repression through binding of regulator gene product to operator, induction through binding of inducer to regulator gene product, catabolite repression through effect of lowered levels of cyclic AMP on binding of RNA polymerase to promotor), and **the basic mechanisms of transcription and translation**.

The aim of these experiments is to examine in detail the events during the induction of β-galactosidase in *E. coli*. The experiments are modelled on the work of Nakada and Magasanik (1964) and Pastan and Perlman (1968).

Cells are grown in a defined medium with glycerol as carbon and energy source; under these conditions expression of the *lac* operon is neither induced nor repressed. Synthesis can be induced by the addition of iso-propylthiogalactoside (IPTG), which is not itself catabolized by the cells and does not, therefore, act as a further source of carbon and energy. Such conditions of induction have been described as "gratuitous".

The technique of membrane filtration allows the cells to be separated from the induction assay medium, washed, and resuspended in fresh medium. Transcription and translation occur sequentially and the filtration step allows the separate study of the phases of mRNA and protein syntheses following induction of the *lac* operon, and the analysis of their dependence on the inducer and their sensitivity to inhibitors.

Catabolite repression can be generated in further experiments by the addition of glucose to the system after induction with IPTG. The so-called phases of "transient" and "permanent" repression can be demonstrated, as can the relief from transient repression, by the addition of low concentrations of cyclic AMP.

Assay of the expression of the *lac* operon can be simplified by treatment of the induced cells with toluene. This destroys the membrane permeability barrier and thus the need for β-galactoside permease, which is normally the rate-determining step in whole cells. The activity assay therefore measures only β-galactosidase, the *z*-gene product.

EXPERIMENTAL

These experiments form a series in which a basic concept and associated experimental design are developed through repetition, with slight modifications being introduced at each stage. This reinforces an understanding of the cellular and molecular events being studied, and allows the development of improved manipulative skills.

Each experiment begins in the same way with 1 h incubation period, during which students should make certain that all flasks, tubes, etc. to be used in the experiment are prepared ready for use.

Experiment 1 (3–4h)

Induction of β-galactosidase

Stationary phase culture of *E. coli* ML30 grown in M63 medium with glycerol as sole carbon source.

$$\downarrow$$

Take 1 ml of culture, add to 3 ml 10 % formalin and measure E_{650} in spectrophotometer against a water blank. Obtain the cell density from the relationship, E_{650} 0·6 = 1 mg ml^{-1}.

$$\downarrow$$

Add a sufficient volume of this culture to 100 ml fresh M63–glycerol medium (in a 250 ml sterile conical flask) to give a final cell density of approximately 40 μg ml^{-1}. Incubate at 30° C with shaking for 1 h in an orbital shaker.

$$\downarrow$$

After 1 h withdraw 3 ml of culture and measure E_{650}, **directly and without dilution**. When E_{650} equals approximately 0·2 (between 0·18 and 0·26 is satisfactory) the cells are ready for use. From this point on it is not necessary to use aseptic conditions.

$$\downarrow$$

Transfer 22·5 ml of culture to a 100 ml conical flask and incubate at 30° C in a shaking water-bath.

↓

Add 2·5 ml of 5 mM iso-propylthiogalactoside (IPTG), mix well and immediately withdraw a 1 ml sample and pipette into a test tube, pre-chilled in ice and containing 1 drop of toluene. Shake the tube and return to ice (*see* note 6). This is the 0 min sample. Withdraw 1 ml samples and treat similarly at 1, 2, 3, 4, 5, 6, 7, 8, 9, 10, 12, 14, 16, 18 and 20 min. Retain all samples on ice until ready to assay β-galactosidase (*see* notes 1 and 2).

Experiment 2 (4 h)

Induction of β-galactosidase: translation

Basically the procedure is as for experiment 1.

Again a culture is prepared containing 100 ml cells with $E_{650} = 0·2$. Once growth has reached this point, the culture is removed from the shaker and put on the bench at room temperature and without shaking.

Two separate induction incubations will be carried out with these cells.

Incubation A

Take 22·5 ml culture in 100 ml flask, incubate at 30° C in the shaking water-bath for approximately 5 min and add 2·5 ml 5 mM IPTG. Withdraw 1 ml samples into pre-chilled tubes containing 1 drop of toluene, at 0, 1, 2 and 3 min (*see* note 6).

↓

Immediately after withdrawing the 3 min sample, take 15 ml and filter through **47 mm Oxoid membrane filter** (pore size 0·45 μm) under suction (*see* note 7).

↓

As rapidly as possible wash the filter with 10 ml cold M63–glycerol medium, then with 5 ml M63–glycerol medium prewarmed to 30° C. Transfer filter to a second 100 ml flask containing 15 ml of complete M63–glycerol medium, **also prewarmed to 30° C**. Using a pipette, rapidly wash cells off the filter into the medium.

↓

With a clean pipette withdraw a 1 ml sample into a tube + toluene on ice, and note the time (i.e. the clock has **not** been stopped).

↓

Thereafter, withdraw 1 ml samples every minute until 10 min, and then at 12, 14, 16, 18 and 20 min.

↓

Store all samples on ice until ready for assay.

Incubation B

Exactly as for A except that the medium in which the cells are washed and resuspended in **after** filtering is 13·5 ml M63–glycerol + 1·5 chloramphenicol (200 μg ml^{-1}); i.e. prepare 2 × 15 ml volumes, one for washing and the second for resuspension.

↓

After completion of both incubations, assay separately each batch of tubes for β-galactosidase activity.

Experiment 3 *(4 h)*

Induction of β-galactosidase: transcription

Procedure is basically as for experiment 2, except that we study the effect of 5-fluorouracil as an inhibitor, present both before the filtration step (Incubation C) and after (Incubation D).

Incubation C

Incubate 20 ml culture + 2·5 ml 5-fluorouracil (200 μg ml^{-1}) + thymidine (400 μg ml^{-1}) for 5 min. 2·5 ml 5 mM IPTG is added at 0 min, and incubation and sampling continued for 3 min.

↓

15 ml cells are then filtered, washed and resuspended in 15 ml M63–glycerol. Sampling continued to 20 min as before.

Incubation D

Incubate 22·5 ml of culture + 2·5 ml 5 mM IPTG for 3 min, with 1 ml samples withdrawn at 0, 1, 2 and 3 min.

↓

15 ml cells are then filtered, washed and resuspended in 13·5 ml M63–glycerol + 1·5 ml 5-fluorouracil (200 μg ml^{-1}) + thymidine (400 μg ml^{-1}).

↓

Samples are withdrawn every minute to 10 min, then every second minute to 20 min.

After completion of both incubations, assay separately each batch of tubes for β-galactosidase activity.

Experiment 4 *(4 h)*

Induction of β-galactosidase: catabolite repression

The experiments described above are taken from Nakada and Magasanick (1964), and are designed to demonstrate the separate phases of "induction" and "synthesis". It is suggested that the students should design for themselves an experiment to test the sensitivity of the class strain of

E. coli ML30 to transient and permanent catabolite repression. A suitable model for such an experiment is given in Figure 1 of the paper by Pastan and Perlman (1968). Since the class strain is not genetically identical with strain 3000 used by Pastan and Perlman, two or three groups might use *E. coli* K12 strain 3000.

Before attempting the experiment, the group should discuss its design with the lecturer in charge. To assist in the experimental design a number of points can be made.

(1) Pastan and Perlman use Medium A at $37°$ C supplemented with thiamine and with 0.5% carbon source: we can continue using our M63 at $30°$ C with the carbon source at 0.2% but supplemented with $5\ \mu g\ ml^{-1}$ thiamine ($1\ ml$ of $500\ \mu g\ ml^{-1}$ per $100\ ml$ medium).

(2) Pastan and Perlman use 10^{-3} M IPTG: again we can continue with 4×10^{-4} M.

(3) Pastan and Perlman's starting E_{560} of 0.07, corresponds almost exactly to E_{650} of 0.1 under our conditions, i.e. make initial cell density $20\ \mu g\ ml^{-1}$, and incubate for 1 h at $30°$C until E_{650} is between 0.09 and 0.13.

(4) In assaying β-galactosidase activity, we are already using the method of Pardee, Jacob and Monod.

(5) Remember that the the O.D. of **every** sample must be measured in this experiment: a suggested method would be to withdraw two 1-ml samples at each time, one for enzyme assay, the other to be added to 1 drop of undiluted formalin for reading of E_{650}, using semi-micro cuvettes.

Notes and Points to Watch

1. *β-galactosidase assay.* Place the tubes in a water-bath at $30°$ C for 30 min **prior to** the carrying out the assay of enzymic activity; occasionally mix the contents by shaking gently. After 30 min, add $0.2\ ml$ 13.3 mM *o*-nitrophenylgalactoside (ONPG) to each tube and incubate for **exactly** 10 min. Stop the reaction by adding $0.5\ ml$ 1 M Na_2CO_3.

   Withdraw the solution with a long Pasteur pipette fitted with a rubber bulb and read E_{550} and E_{420} for each sample (using glass semi-micro cells and a water blank).

2. E_{550} gives a measure of light scattering due to cells, emulsion with toluene, etc. and E_{420} gives a measure of light scattering **and** the yellow colour of *o*-nitrophenol (ONP) in alkaline solution.

   $E_{420} - (E_{550} \times 1.65)$ gives us a measure of ONP only.

   Under these conditions of assay 1 m μmol ONP gives:

   $$E_{420} - (E_{550} \times 1.65) = 0.0075.$$

   Calculate m μmol ONP produced $10\ min^{-1}\ ml^{-1}$, and plot against time of incubation in the induction medium.

3. *E. coli* ML30 and K12 3000 are both $i^+ y^+ z^+$.

4. Prior to experiment, strains should be "trained" by several passages in synthetic medium with glycerol as sole carbon and energy source.

5. In order to obtain late logarithmic or early stationary phase cells for class at 0900 hours, culture ($100\ ml$ medium in a $250\ ml$ flask) should be inoculated at 1700 hours the previous evening with $0.1\ ml$ of the previous day's glycerol-grown culture.

6. It is important that the mixing of the cell suspension with toluene prior to assay of β-galactosidase activity be done by **gentle** shaking of the tube. Vigorous shaking can give rise to the formation of an emulsion, the light scattering character of which can adversely affect the

accuracy of determining the colour change due to ONP production. Similarly, care should be taken not to transfer emulsion to spectrophotometer cuvettes for recording of E_{420} and E_{550}.

7. The step of filtration, washing and resuspension of the cells is a major source of difficulty. It is imperative that the whole process be done as rapidly as possible: 2–3 min should be the target.

   (a) Prior to beginning the experiment, 10 ml and 5 ml volumes of washing solution should be dispensed in test-tubes and kept, respectively, on ice or at 30° C; the washing steps can then be carried out rapidly simply by pouring the contents of each test tube in turn onto the filter. Similarly the 15 ml volume of resuspension solution should be dispensed prior to beginning experimental manipulations, and kept at 30° C.

   (b) The suction must be switched off prior to **careful** removal of the filter and transfer to resuspension flask.

   (c) The cells **must** be washed off the filter in fresh medium. This can best be done by drawing up 5 or 6 ml of medium in a 10 ml pipette and ejecting it rapidly across the surface of the membrane filter held against the side of the flask. This should be repeated twice and the filter then pushed to the bottom of the flask; the flask can then be shaken vigorously by hand to make sure any remaining loosely attached cells are freely suspended in medium.

   (d) Oxoid filters (N47/45G) are suitable for this procedure; with certain other makes, the cells become embedded and are difficult to remove from the filter during resuspension.

MATERIALS

1. *Escherichia coli* ML 30 (NCIB 10,000) and K 12 3000 (NCIB 10,584), or other strains with $i^+y^+z^+$ phenotype (*see* note 3).
2. M63 medium with glycerol as carbon source; prepared by mixing 90 ml of solution A and 10 ml of solution B.

Solution A:
   15·1 g KH_2PO_4, 2·2 g $(NH_4)_2SO_4$, 1 ml of 0·006 g ml^{-1} $FeSO_4 \cdot 7H_2O$, 11H_2O, pH to 7·0, autoclaved at 103·4 kPa (15 lbf in^{-2}) for 15 min

Solution B:
   2·0 g glycerol, 0·2 g $MgSO_4 \cdot 7H_2O$, 100 ml H_2O, autoclaved at 103·4 kPa for 15 min.

SPECIFIC REQUIREMENTS

These items are needed for each group of 2–4 students.

Experiment 1
100 ml stationary phase culture of *E. coli* ML30 (*see* notes 4 and 5)
100 ml 10% formalin solution
spectrophotometer, with two 1 cm cuvettes (glass or plastic) and four semi-micro cuvettes (glass)
250 ml conical flask (sterile)
2 × 100 ml M63 medium
access to a shaking incubator (taking 250 ml flasks)
100 ml conical flask

shaking water-bath (taking 100 ml flasks)
25 ml 5mM IPTG
toluene
ice bucket
stop-clock
50 ml 13·3 mM ONPG (4·006 g l^{-1})
100 ml 1 M Na_2CO_3
Pasteur pipettes (long)
pipettes
test tubes and rack

Experiment 2
as per experiment 1 +
47 mm Oxoid filters (0·45 μm pore size)

cont.

cont.

filter assembly mounted in 500 ml or 1 litre Buchner flask electric suction pump (or reliable water pump) forceps for handling filter 4 × 100 ml conical flasks 5 ml 200 μg ml^{-1} chloramphenicol ***Experiment 3*** as per experiment 2 + 5 ml 200 μg ml^{-1} 5-fluorouracil + 400 μg ml^{-1} thymidine	***Experiment 4*** as per experiment 1 + 2 × 250 ml conical flasks (sterile) 3 × 100 ml conical flasks (sterile) undiluted formalin 1 ml 1 M glucose 1 ml 40 mM cyclic AMP (only 0·5 ml required) 5 ml 500 μg ml^{-1} thiamine HCl (filter sterilized)

FURTHER INFORMATION

Alternative procedures are suggested for the fourth day of this practical.

(1) A number of aspects of the experiments given in the manual give rise to technical difficulties, e.g. washing the cells off the filter with anything approaching 100 % recovery of the cells in

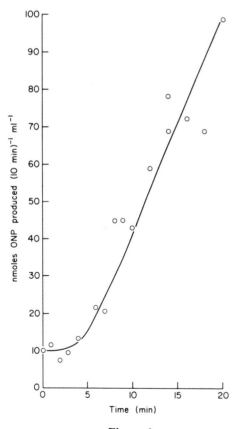

Figure 1

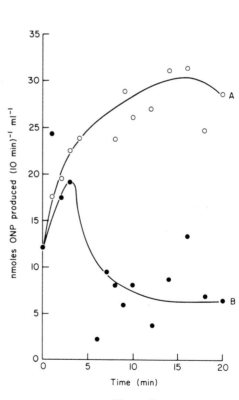

Figure 2

the new suspension, and obtaining satisfactory assays of the very low levels of enzyme activity present. At the first attempt it might not prove possible, therefore, to get wholly convincing results. The fourth day might be occupied then in repeating some of the incubations of the previous three experiments.

(It is possible that some groups might find they can carry out such repeat runs—an essential part of any research project—during Experiments 1–3).

(2) The students might design their own experimental protocol for Experiment 4 to study catabolite repression.

Copies of both papers referred to should be available in the laboratory and the expected or typical results can then be available from the study of these papers (*see* Figures 1–4 for examples). Students should be directed to consider their findings in relation to a number of questions:

Can one analyse the process of induced enzyme synthesis and divide it into two separable phases—induction and synthesis?

Can one describe the molecular events taking place in these two phases?

Does the inducer act at the level of transcription or translation?

Is mRNA stable?

When 5-fluorouracil is used as an inhibitor of mRNA synthesis, why is thymidine also present in the system?

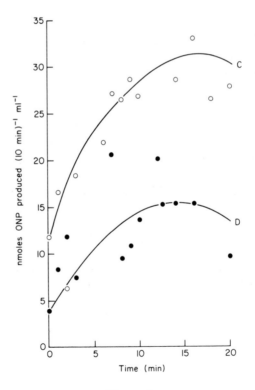

Figure 3

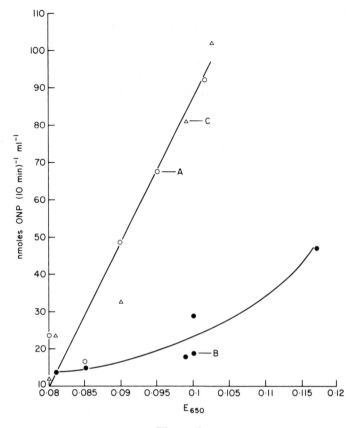

Figure 4

Can one analyse the process of catabolite repression and divide it into two separable phases—transient and permanent?

What is the involvement of cyclic AMP in glucose catabolism and mRNA synthesis?

REFERENCES

D. Nakada and B. Magasanik (1964). The roles of inducer and catabolite repressor in the synthesis of β-galactosidase by *Escherichia coli. Journal of Molecular Biology*, **8**, 105–27.

I. Pastan and R. L. Perlman (1968). The role of the *lac* promotor locus in the regulation of β-galactosidase synthesis by cyclic 3′, 5′-adenosine monophosphate. *Proceedings of the National Academy of Sciences, USA*, **61**, 1336–42.

18

Computer Simulation of a Prey–Predator Model with a Commodore "PET" Micro Computer

P. J. LeB. WILLIAMS

*Department of Oceanography, University of Southampton,
Southampton, SO9 5NH, England*

> *Level*: Advanced undergraduates
> *Subject area*: Microbial population dynamics
> *Special features*: Provides an introduction to inter-
> active modelling. The experiment
> is easy to set up and then needs
> the minimum of supervision. The
> student can easily elaborate on the
> model

INTRODUCTION

Population models are conventionally based upon the LOTKA–VOLTERRA rationalization:

Rate of increase of the biomass of an organism = Biomass production – Biomass removal.

Various removal processes may be put into the model: e.g. respiration, grazing, flushing or export from the environment. From this simple model, complex models may be constructed involving competing species, food chains, etc. The model to be described considers a simple food chain (organic substrate → bacterium → ciliate) in a chemostat.

The program is written in BASIC for a PET Commodore micro computer. The PET is now comparatively inexpensive and used by many university departments and schools. The system is very easy to use and to modify. The exercise can be run on the PET alone, using its visual display unit (VDU) screen. However, there are considerable advantages in having a permanent ("hard") record, and for this purpose either a teletype (TTY) or a PET printer may be used.

The program will solve the growth equations and follow the changes in the population with time. It can also, with small modifications, plot out a phase plane diagram as the experiment progresses. By varying the parameters used to run the model, the following phenomena will be seen: (1) limit-cycle oscillations between prey and predator numbers, (2) stable equilibria and (3) unstable equilibria (*see* Slater (1979) for a useful background to this subject).

The purpose of this exercise is in part to illustrate numerical modelling, but more to introduce the use of models to provoke and answer questions. The student should be encouraged to use and modify the model to examine possible real or hypothetical situations. The "chemostat" for example could be a model for a lake, reservoir or sewage works. It can be used as a closed system (D = 0) to model the oceanic or terrestrial environments. It could be used to explore the effect of a pollutant (e.g. a herbicide) which affects only one species in the model.

EXPERIMENTAL

The model considers the following system:

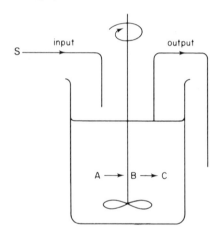

Parameter	Description	Units
S	concentration of substrate in input	mg l^{-1}
A	concentration of substrate in chemostat	mg l^{-1}
B	concentration of bacterium in chemostat	mg l^{-1}
C	concentration of ciliate in chemostat	mg l^{-1}
D	dilution rate of chemostat	h^{-1}
UB	maximum growth rate of bacterium	h^{-1}
KB	growth constant of bacterium	mg l^{-1}
YB	yield constant of bacterium	dimensionless

UC, KC and YC are the equivalent constants for the ciliate. The three kinetic equations are as follows:

$$\frac{dA}{dt} = D.S - D.A - \frac{UB.A.B}{KB+A} \cdot \frac{1}{YB}$$

$$\frac{dB}{dt} = \frac{UB \cdot A \cdot B}{KB + A} - D \cdot B - \frac{UC \cdot B \cdot C}{KC + B} \cdot \frac{1}{YC}$$

$$\frac{dC}{dt} = \frac{UC \cdot B \cdot C}{KC + B} - D \cdot C$$

The program solves these equations with time using a "finite difference" solution, i.e. the rate of change of, for example, A is calculated and then presumed to remain constant for a short period of time and this small change in A (ΔA) is added to A to give a new value of A; the new rate of change A is recalculated and the procedure repeated. Identical solutions are used for B and C. If the time step is too great, ΔA may exceed A and then A will go **negative**; this is quite in order from the mathematical point of view and the program merely enters the world of **negative ciliates** grazing upon **negative bacteria**. If the time step is too small, then computing time becomes unacceptably long. The model allows the time step to be chosen.

COMPUTING EQUIPMENT

Ideally the PET should be used in conjunction with a printing terminal, either a teletype or the Commodore line printer. If this can be arranged, the terminal will be used to produce the graphics, and the VDU screen on the PET will be used to display the actual values of A, B, C and their rates of change; alternatively it can generate a phase plane diagram. If a teletype is not available, then a graphics output can be displayed on the VDU screen.

The program is stored on cassette. Once loaded it can be modified at will and the variants stored if desired. The PET manuals describe how to operate the machine; the program calls for the necessary details.

A listing of the program is provided in the Appendix. The program is written in BASIC, which is a very simple language. Commodore sell auto-teaching programs which students can use to learn the language.

RUNNING THE PROGRAM

Place the tape into the cassette player, rewind and type "LOAD" and RETURN.

↓

Make appropriate changes to the program to give the required display. The default option is a time series display on the VDU screen.

↓

Type "RUN" and RETURN. The program will call for the data as required. Type each number into the keyboard and then enter with the RETURN key. If an error is made it will be necessary to terminate the program (*see* note 1) and start again.

↓

The program will then call for data. It is suggested for the first run that the data in note 2 are used.

This will produce limit cycle oscillations and will require about 20 min to produce a suitable set of curves.

↓

Assuming that the computing facilities consist of the PET and a teletype/line printer terminal, the VDU screen of the former will give the calculated values of A, B, C and their rates of change and the elapsed time of model in hours, or a phase plane diagram, i.e.

ELAPSED TIME

A $\dfrac{dA}{dt}$

B $\dfrac{dB}{dt}$ OR

C $\dfrac{dC}{dt}$

$\log_{10} A$
and
$\log_{10} C$

$\log_{10} B$

Notes and Points to Watch

1. The program may be terminated at any time by pressing the STOP key. If the teletype is in use the key will need to be held down for a few seconds.
2. Use the following data for the first run:

S	= input substrate concentration,	$200\,\text{mg l}^{-1}$
A	= initial chemostat substrate concentration,	$200\,\text{mg l}^{-1}$
B	= initial bacterial concentration in chemostat,	$1{\cdot}0\,\text{mg l}^{-1}$
C	= initial ciliate concentration in chemostat,	$0{\cdot}1\,\text{mg l}^{-1}$
D	= dilution rate of chemostat,	$0{\cdot}1\,\text{h}^{-1}$
UB	= maximum growth rate of bacterium,	$0{\cdot}6\,\text{h}^{-1}$
KB	= uptake constant of bacterium,	$4{\cdot}0\,\text{mg l}^{-1}$
YB	= yield constat of bacterium,	$0{\cdot}45$
UC	= maximum growth rate of ciliate,	$0{\cdot}43\,\text{h}^{-1}$
KC	= uptake constant of ciliate,	$12{\cdot}0\,\text{mg l}^{-1}$
YC	= yield constant of ciliate,	$0{\cdot}54$
P	= printing frequency,	1
T	= number of integration time steps per "model" hour,	10
R	= running time of model in hours,	100

When P = 1, the program will print out every "model" hour, when P = 2 every two "model" hours, etc.; this can be used to condense the output.

By changing statements 199, 349, 360, 570 and 802 as shown below, various combinations of output may be selected. Without alteration, the program is set up to run the first option (the default option).

3. The teletype/line printer will give suitable scaled values of A, B and C and they will be printed as such (Figure 1). The running time of the model is about 10–15 min and can be modified by varying R. T and P will also alter the effective running time.

"GO TO" statements	VDU display	Teletype display
199 GOTO 200 349 GOTO 350 360 GOTO 600	State and rate variables	Time series
199 GOTO 200 349 GOTO 500 570 GOTO 600	Phase plane	Time series
199 GOTO 300 349 GOTO 350 360 GOTO 800	State and rate variables	None
199 GOTO 300 349 GOTO 500 570 GOTO 800 802 GOTO 830	Phase plane	None
199 GOTO 300 349 GOTO 400	Time series	None

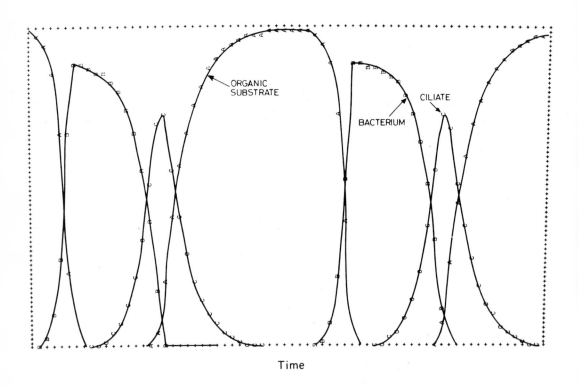

Time

Figure 1   Example of teletype output: data letters are joined up for illustrative purposes.

FURTHER INFORMATION

Results with the data listed in note 2 are presented in Figure 1. The model is simple and is very easy to modify. For example, one can put in two competing species at either trophic levels. The model is based on data provided by Curds (1971), although the notation in the equations has been changed. To vary the model, the following "experiments" are suggested.

1. Reduce the frequency of integration (T) which may send the model into the "negative" world. The program may stop when this "negative" world is entered.
2. Curds's paper provides data which can be used to demonstrate **stable** and **unstable equilibria**. Unstable equilibrium values for A, B and C, calculated from the equations and data given by Curds, are A = 152·752 4; B = 3·636 364 and C = 9·417 534. If approximated to five figures and used with the other data (*see* note 2) they will give an initial equlibrium which eventually breaks down, producing limit-cycle oscillations. This is best demonstrated by using the teletype to produce the time series and the VDU to plot the phase plane of B versus C. The display is improved if a "GOTO 530" statement is placed in line 509. The experiment will need to run for 400 "model" hours.
3. Simple modification may be made to introduce a second competing bacterium in place of the ciliate. Taylor and Williams (1975) give some background of the behaviour of competing organisms in continuous flow systems.
4. Varying arrangements and combinations of substrates, competing bacteria and ciliates may be used. This will involve a greater amount of program modification. Steele (1974) provides examples from the marine food chain.
5. The model may easily be converted from a "chemostat" to a closed system model (D = 0) of, for example, the oceanic planktonic food chain. In this case A is a plant nutrient (e.g. inorganic nitrogen); B, planktonic algae and C, herbivorous zooplankton. For the model to run, it is necessary to recycle plant nutrients (A) and this is logically derived as 1 − YB and 1 − YC. Steele's book (1974) can be used as a source of information and ideas.
6. *See also* Experiment 71, "Use of Mathematical Models in Microbial Ecology".

REFERENCES

C. R. Curds (1971). Computer simulations of microbial population dynamics in the activated-sludge process. *Water Research*, 5, 1049–66.
J. H. Slater (1979). Microbial population and community dynamics, *In* "Microbial Ecology: A Conceptual Approach," (J. M. Lynch and N. J. Poole, eds), pp. 45–63. Blackwell Scientific Publications, Oxford.
J. H. Steele (1974). "The Structure of Marine Ecosystems," 128 pp. Blackwell Scientific Publications, Oxford.
P. A. Taylor and P. J. LeB. Williams (1975). Theoretical studies on the coexistence of competing species under continuous-flow conditions. *Canadian Journal of Microbiology*, 21, 90–8.

APPENDIX

Listing of Program

It is hoped to arrange for the author to copy the program onto cassettes.

Note: Certain symbols associated with the screen and cursor control cannot be accurately represented below. Similarly they are not present in the standard ASCI character set so will not be given in teletype listings. In the listing below these symbols will be represented as follows:

PRINT "HOME" as PRINT "*H*"
PRINT "CLEAR" as PRINT "*C*"
PRINT "CURSOR DOWN" AS PRINT "*D*"
PRINT "CURSOR RIGHT" AS PRINT "R"

```
2 DIM AR$ (70)
3 DIM AS$ (70)
5 PRINT"C"
50 PRINT" TO READ IN CONSTANTS: TYPE IN NUMBER, THEN PRESS RETURN"
100 PRINT"D", "READ IN CONSTANTS", "D"
101 INPUT" INPUT SUBSTRATE CONC="; S
102 INPUT" INITIAL A="; A
103 INPUT" INITIAL B="; B
104 INPUT" INITIAL C="; C
105 INPUT" DILUTION RATE = "; D
106 INPUT" UB="; UB
107 INPUT" KB="; KB
108 INPUT" YB="; YB
109 INPUT" UC="; UC
110 INPUT" KC="; KC
111 INPUT" YC="; YC
112 INPUT" P="; P
113 INPUT" T="; T
114 INPUT" R="; R
150 PRINT"C", "LIST OUT CONSTANTS", "D"
151 PRINT""
152 PRINT"S=", S
153 PRINT"A=", A
154 PRINT"B=", B
155 PRINT"C=", C
156 PRINT"D=", D
157 PRINT"UB=", UB
158 PRINT"KB=", KB
159 PRINT"YB=", YB
160 PRINT"UC=", UC
161 PRINT"KC=", KC
162 PRINT"YC=", YC
163 PRINT"P=", P
164 PRINT"T=", T
165 PRINT"R=", R
170 PRINT"D"
180 PRINT"TO CONTINUE, PRINT '1' THEN PRESS RETURN": INPUT X
185 PRINT"C"
199 GOTO 200
200 REM"LIST OUT CONSTANTS OF TELETYPE", ""
201 OPEN 1, 4
202 CMD 1
205 PRINT""
206 PRINT "LIST OF CONSTANTS"
207 PRINT""
210 PRINT" S=", S
211 PRINT" A=", A
212 PRINT" B=", B
```

```
213 PRINT" C =", C
214 PRINT" D =", D
215 PRINT" UB =", UB
216 PRINT" KB =", KB
217 PRINT" YB =", YB
218 PRINT" UC =", UC
219 PRINT" KC =", KC
220 PRINT" YC =", YC
221 PRINT" P =", P
222 PRINT" T =", T
223 PRINT" R =", R
251 PRINT"D"
252 PRINT "PREY-PREDATOR MODEL"
253 PRINT"D"
254 FOR I = 1 TO 70
255 AS$ (I) = " + "
256 NEXT I
257 FOR I = 1 TO 70
258 PRINT AS$(I);
259 NEXT I
260 CLOSE 1, 4
261 GET F
300 REM"CALCULATION SECTION"
301 REM"SCALE KINETIC PARAMETERS"
305 UB = UB/T
306 UC = UC/T
307 D = D/T
310 REM"FINITE DIFFERENCE SOLUTIONS"
312 N = T
313 M = P
320 FOR J = 1 TO M
321 FOR I = 1 TO N
322 DA = D* S − D* A − ((UB* A* B)/((KB + A)* YB))
323 DB = (UB* A* B)/(KB + A) − D* B − ((UC* B* C)/((KC + B)* YC))
324 DC = (UC* B* C)/(KC + B) − D* C
330 REM"BUDGET EQUATIONS"
331 A = A + DA
332 B = B + DB
333 C = C + DC
335 NEXT I
336 Q = Q + 1
340 NEXT J
349 GOTO 350
350 REM"WRITE OUT STATE & RATE VARIABLES"
351 PRINT"D"; "D"; Q; "D"
352 PRINT A,DA,B,DB,C,DC
360 GOTO 600
400 REM"PRINT OUT TIME SERIES ON VDU"
401 REM"SCALE PARAMETERS"
405 X = INT (A* 40/200)
406 Y = INT (B* 40/100)
407 Z = INT (C* 40/50)
409 REM"CLEAR OUT & LOAD UP STRING"
410 FOR I = 1 TO 40
411 AR$ (I) = " "
412 NEXT I
415 AR$ (1) = " + "
416 AR$ (40) = " + "
417 AR$ (X) = "A"
418 AR$ (Y) = "B"
```

```
419 AR$ (Z) = "C"
420 REM"WRITE OUT STRING ON VDU"
421 FOR I = 1 TO 40
422 PRINT AR$ (I);
423 NEXT I
430 GOTO 800
500 REM"DISPLAY PHASE-PLANE ON VDU"
501 REM"SCALE PARAMETERS ON LOG SCALE"
503 X = LOG (1 + INT (A* 5))
504 Y = LOG (1 + INT (B* 10))
505 Z = LOG (1 + INT (C* 20))
506 X = INT (40* X/(2.303* 3))
507 Y = INT (20* Y/(2.303* 3))
508 Z = INT (40* Z/(2.303* 3))
510 REM"PLOT OUT PHASE-PLANE"
511 PRINT"H"
512 PRINT" × "
513 PRINT"H"
515 FOR I = 1 TO X
516 PRINT"R";
519 NEXT
520 FOR I = 1 TO Y
521 PRINT "D";
522 NEXT
525 PRINT"●"
530 PRINT"H"
531 PRINT"*"
535 FOR I = 1 TO Z
536 PRINT"R";
537 NEXT
540 FOR I = 1 TO Y
541 PRINT"D";
542 NEXT
550 PRINT"○"
561 PRINT"H"
562 PRINT" Q = A VRS. B; W = C VRS. B"
570 GOTO 600
600 REM"PRINT OUT TIME SERIES ON TELETYPE"
601 REM"SCALE PARAMETERS"
605 X = INT (A* 70/200)
606 Y = INT (B* 70/100)
607 Z = INT (C* 70/50)
610 REM"CLEAR OUT & LOAD UP STRING"
611 FOR I = 1 TO 70
612 AR$ (I) = " "
613 NEXT I
615 AR$ (1) = "+"
616 AR$ (70) = "+"
617 AR$ (X) = "A"
618 AR$ (Y) = "B"
619 AR$ (Z) = "C"
620 REM"WRITE OUT STRING ON TELETYPE"
621 OPEN 1,4
622 CMD 1
625 FOR I = 1 TO 70
626 PRINT AR$ (I);
627 NEXT I
628 CLOSE 1
629 GET F
630 GOTO 700
```

```
700 REM"TEST LENGTH OF RUN & TERMINATE WHEN Q=R"
702 IF Q < R THEN 310
703 OPEN 1, 4
704 CMD 1
705 FOR I=1 TO 70
706 PRINT AS$ (I);
707 NEXT
708 PRINT"D"
709 PRINT"END OF PREY-PREDATOR MODEL"
710 PRINT"D"
711 PRINT"MODEL RAN FOR ", Q," HOURS"
720 CLOSE 1, 4
721 GET F
730 STOP
731 END
800 REM"TEST LENGTH OF RUN & TERMINATE WHEN Q=R"
801 IF Q < R THEN 310
803 FOR I=1 TO 40
804 PRINT"+";
805 NEXT
806 PRINT"D"
808 PRINT"END OF PREY-PREDATOR MODEL"
810 PRINT"D"
811 PRINT"MODEL RAN FOR ", Q," HOURS"
830 STOP
831 END
```

19

Catabolite Inactivation of Alcohol Oxidase and Catalase in the Yeast *Hansenula polymorpha*

L. A. HUISMAN

*Department of Microbiology, University of Groningen,
Kerklaan 30, 9751 NN Haren, The Netherlands*

> *Level*: Advanced undergraduates
> *Subject areas*: Microbial physiology

INTRODUCTION

Apart from regulation of enzyme activity on the level of the catalytic process itself, by activation or inhibition, and regulation on the level of enzyme synthesis, by induction or repression, a third type of regulation has been recognized in micro-organisms in recent years, called inactivation. Although the exact mechanism of this process is not yet fully understood, it is possible to discriminate between reversible inactivation, e.g. by covalent modification of the enzyme protein, and irreversible inactivation of the enzyme by (partial) proteolysis (Switzer, 1977). It has been suggested that this kind of regulation might come into play when a rather drastic change in environmental conditions takes place, such as transfer to another source of carbon and energy.

In yeast cells, pregrown on acetate and subsequently cultured on glucose, several enzymes involved in the synthesis of glucose from acetate are inactivated. In view of the analogy with catabolite repression, Holzer (1976) proposed the term "catabolite inactivation" for this phenomenon. Catabolite inactivation also appears to play a role, when methanol-utilizing yeasts are transferred to a medium containing glucose as energy and carbon source (Bormann and Sahm, 1979; Veenhuis *et al.*, 1978). When growing on methanol these organisms contain a special type of organelle, the peroxisome, in which two key enzymes of methanol metabolism (alcohol oxidase and catalase) are located. The first enzyme catalyses the oxidation of methanol to formaldehyde with molecular oxygen, the second one transforms the toxic H_2O_2 produced in the first reaction to H_2O and O_2. The peroxisomes and their enzymes are not present during growth on glucose. Electron-microscopic studies reveal that addition of glucose to methanol-grown *Hansenula polymorpha* results in the breakdown of most of the peroxisomes (Veenhuis *et al.*, 1978). It can also be demonstrated that the activities of alcohol oxidase and catalase concomitantly decrease rapidly, more rapidly than can be accounted for by dilution of the enzymes due to growth of the organism. This decrease of activity is caused by catabolite inactivation.

The aim of this experiment is to demonstrate this phenomenon, by measuring the activities of alcohol oxidase and catalase in extracts of *H. polymorpha* after addition of glucose to a culture pregrown on methanol. At the same time the activities of formic dehydrogenase, a cytoplasmic enzyme involved in methanol metabolism, and of hexokinase, one of the key enzymes of glucose metabolism, are determined to demonstrate that regulation by repression and induction respectively, also are involved in the adaptation of *Hansenula* to another carbon and energy source.

EXPERIMENTAL

The students are supposed to prepare and sterilize all solutions and media themselves. The time required for this is not included in the programme outlined below.

Day 1 *(10 min)*

Inoculate *Hansenula polymorpha* strain CBS 4732 from an agar slant into a 100 ml conical flask with 30 ml sterile growth medium, containing $5 \, g \, l^{-1}$ methanol.

↓

Incubate overnight in a rotary shaker at 37°C and 200 rev. min^{-1}.

Day 2 *(10 min)*

Measure the optical density of the preculture at 663 nm.

↓

Inoculate 800 ml sterile growth medium, containing $5 \, l^{-1}$ methanol in a 2 litre Erlenmeyer flask with enough preculture to obtain an optical density of about 0·1.

↓

Incubate the main culture in a rotary incubator at 37°C and 200 rev. min^{-1} at about 1500 hours.

Day 3 *(short periods throughout the day)*

Measure regularly the optical density of the culture at 663 nm.

↓

When A_{663} has reached a value of about 2·0 add sterile glucose solution to a final concentration of $5 \, g \, l^{-1}$.

↓

Measure the A_{663} of the culture every half-hour. Take samples of 120 ml after 0, 1, 2, 4 and 6 hours.

↓

Centrifuge the samples for 15 min at 4500 g.

↓

Remove the supernate, resuspend the cells in 120 ml demineralized water and centrifuge again.

↓

Repeat the washing procedure once and store the pellets as such in a freezer $(-40°C)$.

- -

Day 4 *(preparation of cell-free extracts and determination of enzyme activities, 5–6 h)*

Resuspend the pellets in 4 ml 50 mM potassium phosphate buffer, pH 7·0, and add 3 g glass beads (0·11 mm diam.).

↓

Sonicate the suspensions 6 times for 45 s, allowing them to cool for 2 min between the sonication periods. Keep the tubes in icewater during the entire procedure.

↓

Centrifuge the suspensions for 20 min at 27 000 g and 0 °C. Keep the supernate on ice (*see* note 2).

Determination of enzyme activities (all determinations are carried out at 37 °C).

1. *Catalase (see note 1)*
The decrease of the hydrogen peroxide concentration can be followed by measuring the absorption of peroxide at 240 nm.

Pipette into a 1 ml quartz cuvette 0·96 ml 50 mM potassium phosphate buffer, pH 7·0, and 20 μl diluted extract (diluted 1:20 with buffer).

↓

Allow the mixture to equilibrate in a thermostated double beam spectrophotometer for a few minutes.

↓

Add 20 μl 3 % H_2O_2 solution and record the change in extinction at 240 nm against a reference cuvette containing water.

↓

Calculate the activity of catalase as ΔE_{240} min^{-1} mg^{-1} of protein.

2. *Methanol oxidase*
The oxygen consumption of the extracts in the presence of added methanol is taken as a measure of enzyme activity: $CH_3OH + O_2 \rightarrow HCHO + H_2O_2$.

Add 20 μl extract and 200 μg catalase to 2·90 ml 50 mM potassium phosphate buffer, pH 7·0, and aerate the suspension until it is saturated with air.

$\downarrow$

Record with an oxygen electrode the oxygen uptake in the absence of substrate for a few minutes.

$\downarrow$

Add 0·1 ml 1 M methanol solution and record the oxygen uptake again.

$\downarrow$

From the slopes of the curves calculate the activity of the enzyme in μmol O_2 consumed per min per mg of protein, assuming that 1 ml reaction mixture saturated with air contains 0·194 μmol O_2 at 37°C (*see* note 3).

3. *Formate dehydrogenase*
The production of NADH, followed spectrophotometrically, is taken as a measure of enzyme activity:

$$HCOOH + NAD^+ \rightarrow NADH + H^+ + CO_2.$$

Pipette into a 1 ml quartz cuvette 680 μl water, 200 μl 0·5 M potassium phosphate buffer, pH 7·0, 50 μl extract and 20 μl 20 mM NAD$^+$.

$\downarrow$

Record the endogenous extinction change against water at 340 nm for a few minutes.

$\downarrow$

Start the reaction by adding 50 μl 2·5 M sodium formate, pH 7·0, and record the extinction change for several minutes.

$\downarrow$

Calculate the specific enzyme activity in nmol substrate used min^{-1} mg^{-1} protein. The molar extinction coefficient of NADH is $6·22.10^3$ cm^{-1} M^{-1}.

4. *Hexokinase*
The activity of this enzyme is determined by measuring the NADP$^+$-dependent oxidation of the reaction product, glucose-6-phosphate, to 6-phospho-gluconate:

$$glucose + ATP \rightarrow glucose\text{-}6\text{-}P + ADP$$
$$glucose\text{-}6\text{-}P + NADP^+ \rightarrow NADPH + H^+ + 6\text{-}P\text{-}gluconate.$$

Pipette into a quartz cuvette 700 μl H_2O, 100 μl 0·5 M Tris-buffer, pH 8·0, 100 μl 50 mM $MgCl_2$, 20 μl 20 mM NADP$^+$ 20 μl 1 M glucose, 20 μl extract and 20 μl glucose-6-P-dehydrogenase (50 u ml^{-1}).

$\downarrow$

Record the extinction change against water at 340 nm in a double beam spectrophotometer.

$\downarrow$

Then add $20 \mu l$ 50 mM trisodium-ATP and follow the reaction for several minutes.

↓

Calculate the specific activity of the enzyme in nmol substrate used per min per mg of protein. For NADPH use the same extinction coefficient as for NADH.

↓

Protein concentrations are determined according to the method of Lowry *et al.* (1951), using bovine serum albumin as a standard.

↓

Make a plot of A_{663} and the four enzyme activities against the time.

↓

Speculate, with the aid of the papers by Bormann and Sahm (1978) and by Veenhuis *et al.* (1978), how in the case of alcohol oxidase the catabolite inactivation is brought about, e.g. by covalent modification (reversible) or by proteolysis (irreversible). Suggest an experimental procedure to discriminate between these two possibilities.

Notes and Points to Watch

1. The catalase activities in all extracts should be determined first on the same day, in view of the instability of this enzyme.
2. When it is not possible to determine all enzyme activities on the same day, store the extracts overnight at $4°C$, not at $-40°C$.
3. In the assay of alcohol oxidase, additional catalase is added to prevent accumulation of hydrogen peroxide. The oxygen produced during the oxidation of peroxide by catalase causes an underestimation of the oxygen consumption by alcohol oxidase. To obtain the specific activity in μmol methanol oxidized per min the oxygen consumption rate has to be multiplied by a factor 2.

MATERIALS

1. *Hansenula polymorpha* strain CBS 4732, sub-cultured every 6 months on malt agar (2% malt extract, 2% agar, distilled water). This strain can be obtained from Dr David Yarrow, Laboratorium voor Microbiologie, Technische Hogeschool, Julianalaan 67A, 2628 BC Delft, The Netherlands.
2. Growth medium, containing 3 g KH_2PO_4, 2 g $(NH_4)_2SO_4$, 0·2 g $MgSO_4·7H_2O$, 0·2 ml trace-element solution (Vishniac and Santer, 1957), 0·5 g yeast extract (source of vitamins) and 5 g glucose or 5 g methanol per litre distilled water, pH 6·0. For convenience, Yeast Nitrogen Base (Difco Laboratories, Detroit, Michigan, USA)

may be used instead of the first five components.
   Methanol and glucose are sterilized separately in concentrated solution (50 or 25%), methanol by filtration and glucose by heat sterilization.
3. Oxygen electrode. The "Biological Oxygen Monitor", type 53 SA, Yellow Springs Instruments (Yellow Springs, Ohio, USA) is very suitable.
4. Catalase (250 mg in 12·5 ml), NAD$^+$, NADP$^+$, ATP, glucose-6-phosphate dehydrogenase (from yeast) can be obtained from Boehringer Mannheim, W. Germany, or from any other supplier of biochemicals.

SPECIFIC REQUIREMENTS

Day 1
agar slant with *Hansenula polymorpha*
several flasks, bottles, beakers, measuring
 cylinders, pipettes (various sizes)
chemicals, spatulas, balance
autoclave, aluminium foil
pH meter, calibration buffers
magnetic stirrer + fleas
rotary shaker

Day 2
sterile pipettes, 5 ml
spectrophotometer
tubes (for diluting culture samples)
tube rack
pipettes, 1 and 5 ml

Day 3
as on Day 2 +
measuring cylinder, 250 ml (for sampling)
low speed centrifuge
freezer

Day 4
sonification unit, glass beads
high speed centrifuge
micropipettes, 100 μl
quartz cuvettes (1 ml, 1 cm light path)
double beam spectrophotometer, thermostated,
 connected to a recorder
oxygen monitor, connected to a recorder
catalase, ATP, NAD$^+$, NADP$^+$, glucose-6-
 phosphate dehydrogenase

FURTHER INFORMATION

The results obtained by the students are in good agreement with data reported in the literature (e.g. Veenhuis *et al.*, 1978). A typical example of results obtained by students is given in Table 1. The

Table 1 Effect of addition of glucose to a methanol-grown culture of *Hansenula polymorpha* on the specific activities of methanol oxidase, catalase, formate dehydrogenase and hexo-kinase. Figures in parentheses are calculated by multiplying specific activities by optical densities.

Specific activities of:	Before glucose addition $t = 0$ h	After glucose addition			
		$t = 1$ h	$t = 2$ h	$t = 4$ h	$t = 6$ h
Methanol oxidase[a]	1·5 (3·0)	0·75 (1·73)	0·22 (0·55)	0·20 (0·60)	0·19 (0·57)
Catalase[b]	70 (140)	20 (46)	6 (15)	5 (15)	5 (15)
Formate dehydrogenase[c]	280 (560)	210 (483)	180 (450)	160 (480)	140 (420)
Hexokinase[c]	150 (300)	170 (391)	282 (705)	700 (2100)	746 (2238)
Optical density at 663 nm	2·0	2·3	2·5	3·0	3·0

[a] μmol O_2 mg^{-1} protein min^{-1}. [b] ΔE_{240} mg^{-1} protein min^{-1}.
[c] nmol substrate used mg^{-1} protein min^{-1}.

specific activities of alcohol oxidase and catalase decrease rapidly upon addition of glucose to the culture. This cannot be explained by redistribution of these enzymes over newly formed cells, since the total amount of enzyme, calculated by multiplying the specific activity by the optical density of the culture, decreases as well. This indicates the "disappearance" of enzyme protein, caused by inactivation. The rate of decrease of the specific activity of formate dehydrogenase, however, does correspond to the growth rate, which one would expect in the case of regulation by repression of enzyme synthesis. The activity of hexokinase, already present before addition of glucose, increases as a result of induction.

REFERENCES

C. Bormann and H. Sahm (1978). Degradation of microbodies in relation to activities of alcohol oxidase and catalase in *Candida boidinii*. *Archives of Microbiology*, **117**, 67–72.

H. Holzer (1976). Catabolite inactivation in yeast. *Trends in Biochemical Science*, **1**, 178–81.

O. H. Lowry, N. J. Rosebrough, A. L. Farr and R. J. Randall (1951). Protein measurement with the folin phenol reagent. *Journal of Biological Chemistry*, **193**, 265–75.

R. L. Switzer (1977). The inactivation of microbial enzymes *in vivo*. *Annual Reviews in Microbiology*, **31**, 135–57.

M. Veenhuis, K. Zwart and W. Harder (1978). Degradation of peroxisomes after transfer of methanol-grown *Hansenula polymorpha* into glucose-containing media. *FEMS Microbiology Letters*, **3**, 21–8.

W. Vishniac and M. Santer (1957). The Thiobacilli. *Bacteriological Reviews* (P. N. Wilson, ed.), **21**, 195–213.

20

Functional Properties of Membrane Vesicles from *Bacillus subtilis* W23

L. A. HUISMAN and W. N. KONINGS

Department of Microbiology, University of Groningen,
Kerklaan 30, 9751 NN Haren, The Netherlands

> *Level*: Advanced undergraduates
> *Subject areas*: Microbial chemistry and
> biochemistry
> *Special features*: Use of radioactive isotope
> technique

INTRODUCTION

Solutes taken up from the environment by a bacterial cell have to pass the cell wall and the cytoplasmic membrane respectively. While the former is permeable to most solutes, the cytoplasmic membrane has a selective function. This holds for transport in both directions. Solutes are taken up from the environment, while important products and intermediates of the intracellular metabolism are prevented from being excreted by the cell.

Transport across the cytoplasmic membrane is mediated by various translocation systems which differ in the mechanism of translocation and/or energization.

(i) Primary transport: transport by enzyme systems which convert chemical or light energy into electrochemical energy. These transport systems comprise among others the electrogenic proton pumps: the electron transfer systems, the Ca^{2+}-, Mg^{2+}-stimulated ATPase and the light-driven proton pump, bacteriorhodopsin.

(ii) Secondary transport: transport systems which are driven by electrochemical gradients. The transport systems are defined as passive when transport occurs without specific interactions with membrane components, and as facilitated when transport is mediated by specific carrier molecules in the membrane.

(iii) Group translocation: the solute is the substrate for a specific enzyme system in the membrane; the enzymatic reaction results in a chemical modification of the solute and a release of the product at the other side of the membrane.

(iv) ATP-driven transport: solute transport is mediated by a specific membrane protein and the energy for translocation is supplied by ATP or phosphate-bond energy directly (in essence this type of transport is a special form of primary transport).

The reader is referred for more detailed information about these transport systems to the reviews by Harold (1977) and Konings and Michels (1980).

In this experiment the coupling between primary transport and facilitated secondary transport is studied. The primary transport system is the respiratory chain. According to the chemiosmotic hypothesis (Mitchell, 1970), electron transport in this respiratory chain results in the translocation of protons from the cytoplasmic side to the medium side of the membrane. As a result of this translocation a proton motive force ($\Delta \tilde{\mu}_{H^+}$) is generated which consists of two components: an electrical potential ($\Delta \psi$, inside negative vs. outside) as a consequence of the translocation of charge, and a transmembrane pH gradient (inside alkaline vs. outside) as a result of the translocation of protons. In equation:

$$\Delta \tilde{\mu}_{H^+} = \Delta \psi - 2{\cdot}3 \frac{RT}{F} \Delta pH (mV),$$

R is the gas constant, T the absolute temperature and F the Faraday constant; $2{\cdot}3$ is the factor for conversion of natural logarithms into decimal logarithms. This proton motive force is the driving force for secondary transport systems such as transport of solute A. This could occur by the following mechanism. A specific carrier protein binds A at the outer surface of the membrane, together with, for instance, one proton. The resulting A–proton carrier complex is positively charged and rotates in the membrane as a result of the proton motive force. The $\Delta \psi$ drives the translocation of charge, the ΔpH the translocation of protons. At the inside the proton and subsequently A dissociate from the carrier. This process can lead to the accumulation of A, because the process will proceed until the inward directed proton motive force equals the outward directed gradient of A.

In this experiment electron transfer in the respiratory chain is studied with several potential electron donors. Besides NAD(P)H, D-lactate and succinate, non-physiological donors also have been found to donate electrons to the electron transfer chain. In the non-physiological donor system ascorbate–phenazine–methosulphate (PMS), ascorbate donates electrons to PMS, which then transfers electrons partially to the terminal part of the electron transport chain, and the majority directly to oxygen. This latter process will not contribute to the generation of a proton motive force.

The electrons from the various physiological donors enter the electron transport chain at different sites. Whether or not a certain substrate functions as electron donor will depend on the presence of an enzyme catalysing the oxidation of that substrate, for instance succinate dehydrogenase or lactate dehydrogenase.

In order to be able to investigate the coupling between electron transfer and facilitated secondary transport, all factors that may have an unwanted influence have to be eliminated. Such factors can be, among others, the cell wall and the cytoplasm. One way to eliminate such factors is to isolate functional membranes, consisting of so-called membrane vesicles (Kaback, 1971; Konings et al., 1973). All components essential for transport processes are present in these preparations but cytoplasmic components are absent.

The aim of this experiment is to isolate membrane vesicles from *Bacillus subtilis* W23, and to establish their functional integrity by investigating the coupling between electron transfer in the respiratory chain and the transport of a radioactively labelled amino acid, L-glutamate.

EXPERIMENTAL

Students are supposed to prepare and, when necessary, sterilize all solutions and growth media themselves. The time required for this is not included in the programme of the experiment outlined below. The experiment can best be performed by two students.

Day 1 *(inoculation of preculture, 10 min)*

Inoculate two volumes of 100 ml sterile growth medium in 500 ml Erlenmeyer flasks from a slant culture of *B. subtilis* W23, or any other strain of *B. subtilis*.

↓

At the end of the day incubate the flasks in a rotary incubator at 37 °C, 200 rev. min^{-1}.

- -

Day 2 *(whole day; start as early as possible)*

With a spectrophotometer measure the optical density at 660 nm of the culture.

↓

Inoculate five 3-litre Erlenmeyer flasks containing 600 ml sterile growth medium each with enough preculture to get an initial $A_{660\,nm}$ of about 0·1.

↓

Incubate the flasks in a rotary incubator at 37 °C (300 rev. min^{-1}) as early as possible (between 0800 and 0900 hours).

↓

Measure the $A_{660\,nm}$ every hour. When the A_{660} reaches 0·9 the cells can be harvested.

↓

Centrifuge the 3 litre culture for 20 min at 13 000 *g* and 4 °C. Discard the supernate.

↓

Resuspend the pellet in 50 ml prewarmed (37 °C) 50 mM potassium phosphate buffer, pH 8·0, using a 25 ml glass syringe fitted with a needle of 10 cm length and 1 mm internal diam.

↓

Start the combined lysis/lysozyme treatment by adding the homogenous cell suspension to 600 ml prewarmed (37 °C) lysis-buffer, consisting of 50 mM potassium phosphate buffer, pH 8·0, supplemented just before use with 150 mg lysozyme, 9 mg RNase, 9 mg DN'ase, and 1 M MgSO$_4$ solution to a final concentration of 10 mM.

↓

Incubate the mixture, stirring gently with a magnetic flea, for 30 min in an incubator at 37 °C. Examine the results of the treatment with a phase contrast microscope.

↓

Add 0·4 M sodium EDTA solution, pH 7, to a final concentration of 15 mM.

↓

Wait for distinct clearing (about 1 min) and add 1 M $MgSO_4$ solution to a final concentration of 20 mM (10 mM was already present).

↓

Centrifuge the membrane vesicles for 45 min at 25 000 g and 4 °C. Discard the supernate.

↓

With the 25 ml glass syringe, resuspend the pellet in 30 ml 0·1 M potassium phosphate buffer, pH 6·6.

↓

Transfer the homogenous suspension to a 50 ml centrifuge tube and centrifuge for 5 min at 500 g and 4 °C to remove whole cells and protoplasts. This step may be omitted when hardly any cells or protoplasts can be detected microscopically.

↓

Centrifuge the supernate (or suspension) in a 50 ml tube for 30 min at 48 000 g and 4 °C. Discard the supernate.

↓

With the 25 ml syringe, resuspend the pellet gently in the smallest possible (maximally 5 ml) volume of 0·1 M potassium phosphate buffer, pH 6·6.

↓

Dispense 1 ml aliquots of the membrane preparation in plastic containers, freeze in liquid nitrogen and store the containers in liquid nitrogen until use. Prior to an experiment the frozen suspension(s) is (are) thawed rapidly in a waterbath at 50 °C.

Day 3 *(activity measurements, 4–5 h)*

1. *Measurement of secondary facilitated transport*

This experiment is done together with an assistant in a laboratory equipped for work with radioactive isotopes. All necessary precautions should be taken.

Pipette the first three components of the following reaction mixture into 10 small test tubes (1·2 × 7·5 cm):

	energized (tubes 1–5)	not energized (tubes 6–10)
0·1 M K-phosphate buffer, pH 6·6	40 µl	40 µl
0·1 M MgSO$_4$	10 µl	10 µl
H$_2$O	15 µl	35 µl
membrane preparation	10 µl	10 µl
0·4 M K-ascorbate, pH 6·6	10 µl	—
1·0 mM phenazine methosulphate (PMS, *see* note 1)	10 µl	—
2 × 10^{-4} M [^{14}C]-labelled L-glutamate	5 µl	5 µl

↓

Incubate the tubes at 25 °C in a water-bath (*see* Figure 1). The contents of the tubes are stirred magnetically and are aerated by a stream of air through a hypodermic needle.

↓

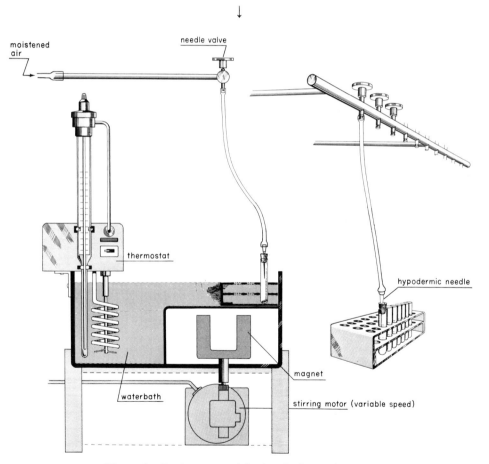

Figure 1 Equipment used for incubating and gassing.

Just before the experiment add membrane suspension, ascorbate and PMS, and start the reaction immediately thereafter by addition of the labelled amino acid (*see* note 2).

↓

Stop the reaction after 0·5, 1, 2, 3 and 4 or 5 min by quickly adding with a dispensor 2 ml 0·1 M LiCl solution to tubes 1 and 6, 2 and 7 and so on. The vesicles are then separated from the reaction mixture by filtration (*see* Figure 2).

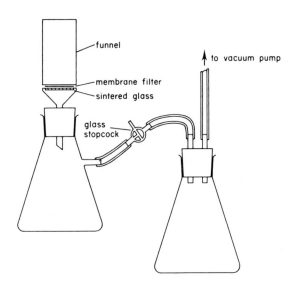

Figure 2 Equipment used for filtration.

↓

Place a membrane filter (2·5–3 cm diam., 0·45 μm pore size) on the filter holder and place the funnel in the correct position.

↓

Pour 2 ml of water in the funnel in order to moisten the filter.

↓

Just before use, open the valve to the vacuum pump, wait till the water has completely passed through the filter and pour the diluted reaction mixture into the funnel.

↓

Rinse the reaction tube with another 2·0 ml 0·1 M LiCl and pour this again into the funnel. In this way most of the non-specifically bound radioactivity is removed from the filter (*see* note 3).

↓

Remove the filter from the suction apparatus.

To determine the remaining amount of non-specifically bound radioactivity, the following procedure is applied in duplicate.

Add all reaction components to a test tube, except for the labelled glutamate.

↓

Dilute the mixture with 2·0 ml 0·1 M LiCl, add the radioactive material and mix well.

↓

Filter immediately, rinse the tube with 2 ml 0·1 M LiCl and filter again. The radioactivity bound to these filters has to be subtracted from all other values.

↓

With a microsyringe dispense 1 μl labelled glutamate on 5 filters. From the radioactivity detected on these filters, the specific radioactivity (number of counts per min per nmol labelled glutamate) can be determined.

↓

Dry all filters in an oven at 110 °C, for 15 min.

↓

Determine the amount of radioactivity on the filters with a gas-flow counter or a scinitillation counter. If the latter procedure is used, the dried filters should be put in vials with 10 ml toluene containing 5 g 2,5-diphenyloxazole (PPO) per litre.

↓

Calculate the uptake of L-glutamate in nmol per mg of protein and plot the amount of radioactivity taken up by the vesicles, against the time. Calculate the concentration gradient of L-glutamate, assuming that the internal volume of the vesicles is 3 μl mg^{-1} of protein.

2. *Activity of membrane-bound enzymes*

The activities of a few membrane bound enzymes are determined by measuring the rate of oxygen uptake by the vesicles in the presence of various electron donors. The enzymes tested are succinate oxidase, lactate oxidase, L-α-glycerol-phosphate oxidase and NADH oxidase.

Pipette in the thermostated reaction vessel (37 °C) of an oxygen monitor a suitable volume of 50 mM potassium phosphate buffer containing 10 mM MgSO$_4$, pH 6·6, and aerate until air saturation.

↓

Put the electrode in the correct position without enclosing air bubbles.

↓

With a micropipette add 30 μl membrane suspension ml^{-1} of phosphate buffer and record the endogenous respiration rate for several minutes.

Then add 30 μl ml^{-1} of phosphate buffer of a 0·3 M substrate solution (sodium succinate, lithium lactate, D, L-α-glycerol-posphate or NADH) and again record the oxygen consumption.

$\downarrow$

Determine the protein concentration of the membrane preparation according to the method of Lowry *et al.* (1951), using bovine serum albumin as a standard.

$\downarrow$

From the slopes of the recorder traces calculate the respiration rates in nmol O_2 per mg of protein per min, after correction for endogenous respiration.

Notes and Points to Watch

1. PMS is light sensitive. Vials, tubes and syringes containing PMS should therefore be wrapped in aluminium foil.
2. During the transport measurements effective mixing is essential. This should be checked regularly during the course of every incubation.
3. The filtration should not take longer then 10 s, otherwise radioactivity may leak out of the vesicles.

MATERIALS

1. *Bacillus subtilis* W23 or another strain of *B. subtilis* subcultured monthly on agar slants.
2. Growth medium: 8 g tryptone, 5 g NaCl, 25 mmol KCl and 0·01 mmol $MnCl_2 \cdot 4H_2O$ per litre distilled water. Sterilize at 120 °C for 20 min.
3. Potassium ascorbate: a solution of ascorbic acid is adjusted to pH 6·6 with KOH.
4. [^{14}C]-L-glutamic acid, 290 mCi mmol^{-1}, from The Radiochemical Centre, Amersham, UK.
5. Membrane filters, pore size 0·45 μm, diameter 2·5–3·0 cm. Suitable is type BA 85, 25 mm diam., cellulose nitrate, Schleicher and Schüll, Dassel, W. Germany.
6. Lithium lactate, BDH Chemicals Ltd, Poole, UK.
7. Oxygen electrode. A suitable model (53 SA) can be obtained from Yellow Springs Instruments Co. Inc., Yellow Springs, Ohio, USA.

SPECIFIC REQUIREMENTS

Day 1
agar slants of *B. subtilis* W23
screw-capped bottles, beakers, measuring cy-
    linders and pipettes, various sizes
chemicals (specified in text) and spatulas
balance
pH meter, calibration buffers
magnetic stirrer + fleas
2 Erlenmeyer flasks, 500 ml + cotton plugs
5 Erlenmeyer flasks, 3 litre + cotton plugs

autoclave, aluminum foil
rotary incubator

Day 2
spectrophotometer
sterile pipettes, 5 ml
tubes, tube rack
centrifuge
glass syringe, 25 ml with needle (specified in text)
Erlenmeyer flask, 1 litre

cont.

cont.

<table>
<tr><td>
measuring cylinder, 1 litre
5 measuring cylinders, 100 ml
magnetic stirrer, flea
lysozyme, DNase, RNase
balance, spatulas

Day 3
water-bath, 50 °C
micro-pipettes (Hamilton syringes, 100 μl, are suitable)
12 small test tubes with magnetic fleas:
The fleas can be made by putting small pieces (4 mm) of a paperclip in a thin glass tube (for instance the thin end of a Pasteur pipette) and
</td><td>
melting the glass at each side of the metal equipment shown in Figures 1 and 2
stop-watch.
dispensor (repeating syringe), 5 ml
20 membrane filters
tweezers
oven
oxygen electrode, reaction vessel(s), water-bath, recorder
micro-pipettes (e.g. Hamilton syringes), 100 μl, for protein determination: 20 pipettes, 1 and 5 ml, 20 test tubes, tube rack, 5 screw-capped bottles, 100 ml, measuring cylinder, 100 ml, bovine serum albumin, spectrophotometer
</td></tr>
</table>

FURTHER INFORMATION

Most students succeed in isolating fairly active membrane vesicles. The protein concentration of the preparations varies between 3 and 10 mg ml^{-1}. Usually the following values are found for the measured activities.

(1) Secondary facilitated transport: after 5 min, about 10 nmoles of L-glutamate mg^{-1} protein have been taken up by the vesicles in the presence of ascorbate and PMS. In the absence of these compounds virtually no uptake can be detected.

(2) Enzyme activities (expressed as nmol O_2 consumed min^{-1} mg^{-1} protein):

succinate oxidase:	50
lactate oxidase:	20–30
D, L-α-glycerol phosphate oxidase:	less than 25
NADH oxidase:	250–1000

REFERENCES

F. M. Harold (1977). "Membrane and Energy Transduction in Bacteria", Current Topics in Bioenergetics, (D. Rao Sanadi, ed.) Vol. 6, pp. 83–149. Academic Press, London and New York.

H. R. Kaback (1971). Bacterial membranes. *In* "Methods in Enzymology", (W. B. Jakoby, ed.), Vol. 22, 99–120. Academic Press, London and New York.

W. N. Konings and P. A. M. Michels (1980). Electron transfer driven transport processes in bacteria. *In* "The Diversity of Bacterial Respiratory Systems", (C. J. Knowles, ed.), Vol. 1. pp. 33–86. CRC Press, Florida.

W. N. Konings, A. Bisschop, M. Veenhuis and C. A. Vermeulen (1973). New procedure for the isolation of membrane vesicles of *Bacillus subtilis* and a microscopical study of their ultrastructure. *Journal of Bacteriology*, **91**, 1456–65.

O. H. Lowry, N. J. Rosebrough, A. L. Farr and R. J. Randall (1951). Protein measurement with the folin phenol reagent. *Journal of Biological Chemistry*, **193**, 265–75.

P. Mitchell (1970). Membranes of cells and organelles: morphology, transport and metabolism. *In* "Organization and Control in Prokaryotic and Eukaryotic Cells". Symposia Society of General Microbiology, Vol. 20, pp. 121–66.

21

Induction of Isocitrate Lyase in *Escherichia coli*

L. A. HUISMAN and T. A. HANSEN

*Department of Microbiology, University of Groningen,
Kerklaan 30, 9751 NN Haren, The Netherlands*

> *Level*: Advanced undergraduates
> *Subject areas*: Bacterial physiology
> *Special features*: Particularly suitable for introducing students to a wide range of instruments, e.g. fermenter, gas chromatograph, oxygen monitor and double beam spectrophotometer

INTRODUCTION

Several micro-organisms, including *Escherichia coli* are able to grow on acetate as their sole source of energy and carbon. Energy can be derived from this substrate by its oxidation to CO_2, with the concomitant production of reducing equivalents (NADH and FADH). Reoxidation of these compounds via the electron transport chain is coupled to the synthesis of ATP (oxidative phosphorylation). The use of acetate as carbon source implies the synthesis of intermediates with more than two carbon atoms for biosynthetic purposes. Since these compounds cannot be derived directly from acetate by carboxylation (except in a few strict anaerobes) obviously there must exist another mechanism for the assimilation of carbon.

It is well established that in many micro-organisms growing on acetate a special bypass of the citric acid cycle, called the glyoxylate cycle, is operative (Kornberg, 1966). This pathway is characterized by two enzymes, isocitrate lyase and malate synthase. The cycle is named after the product of the reaction catalysed by isocitrate lyase:

$$\text{isocitrate} \longrightarrow \text{succinate} + \text{glyoxylate}.$$

Glyoxylate condenses with acetylCoA (formed directly from acetate) to yield malate by the action of malate synthase:

$$glyoxylate + AcCoA \longrightarrow malate + CoASH.$$

Succinate and malate are then converted to oxaloacetate by citric acid cycle enzymes. One of the molecules of oxaloacetate condenses with another AcCoA in the citrate synthase reaction to form citrate, which in turn is converted to isocitrate. The net result of this reaction sequence is then:

$$2 AcCoA \longrightarrow oxaloacetate + 2 CoASH.$$

In this way a C_4-compound is synthesized which can either be used directly for biosynthetic purposes, or can be decarboxylated to phosphoenolpyruvate, the substrate for gluconeogenesis.

In *E. coli* both isocitrate lyase and malate synthase are not constitutively present, but their synthesis is subject to regulation (Kornberg, 1966). When the organism grows on glucose or any other compound that can be broken down via phosphoenolpyruvate, the synthesis of isocitrate lyase and malate synthase is repressed. This appears very clearly from the successive utilization of glucose and acetate, during growth on a mixture of these substrates.

The aim of this experiment is to demonstrate that the activity of isocitrate lyase, measured in cell-free extracts of *E. coli* growing on a mixture of acetate and glucose, is strongly induced only after complete utilization of the glucose.

EXPERIMENTAL

Students should prepare and, where necessary, sterilize all solutions and growth media themselves. The time needed for this is not included in the time schedule given below because it depends very much on the technical skill of the individual students.

Day 1 *(inoculation of preculture (10 min) and preparation and sterilization of the fermenter (3– 4 h, including 2 h for autoclaving))*

Inoculate *E. coli* B from a slant culture into two 100 ml Erlenmeyer flasks containing 30 ml sterile growth medium with 0·4% glucose.

↓

Incubate the flasks for 18 h at 37 °C in a shaker-incubator at 200 rev min^{-1}.

↓

In the meantime prepare a 3–4 litre fermenter which can be thermostated. Use silicone tubing for all connections, and make sure the connections are securely fastened.

↓

Sterilize the fermenter with enough growth medium to obtain a total volume of 3 litres, after addition of the preculture and the separately sterilized phosphate buffer, glucose solution and acetate solution. The final glucose concentration should be 1·0 g l^{-1}.

- -

Day 2 *(main culture, whole day; start as early as possible)*

Set the temperature of the fermenter to 37 °C and add the separately sterilized solutions and both precultures aseptically to the rest of the medium in the fermenter. Pass 1–2 litre of air per min through the culture.

↓

Every hour take a 100 ml sample and measure the optical density at 430 nm. Follow the growth for 8–9 h; the stationary phase will then be reached.

↓

Centrifuge 90 ml of the sample for 15 min at 12 000 g at room temperature.

↓

Store 5 ml of the supernate at −40 °C for determination of glucose and acetate concentrations at a later date.

↓

Resuspend the pellet in 90 ml 50 mM potassium phosphate buffer, pH 7·0, and centrifuge again.

↓

Discard the supernate and resuspend the pellet in the smallest possible known amount of 50 mM potassium phosphate buffer, pH 7·0 (2–3 ml).

↓

Transfer 0·5 ml to a tube and keep on ice; this sample is used immediately for measuring the acetate oxidation capacity of the cells. Store the remainder of the suspension at −40 °C, for later determination of the isocitrate lyase activity.

Measurement of the acetate oxidation capacity
Pipette 3·0 ml 50 mM potassium phosphate buffer, pH 7·0, in a thermostated (37 °C) reaction vessel of an oxygen monitor, and saturate the solution with air.

↓

Add 0·05–0·5 ml (depending on the activity) of the cell suspension and place the oxygen electrode in the vessel.

↓

Record the endogenous respiration for 2–3 min and add 50 μl 1 M sodium acetate, pH 7·0 (*see* note 2).

↓

Record the respiration for another 2–3 min, and calculate the net acetate respiration rate in nmoles of O_2 per hour per unit of optical density. At 37 °C, 1 ml of phosphate buffer, saturated with air, contains 5·145 μl O_2.

- -

Day 3 *(Whole day; preparation of cell-free extracts and determination of isocitrate lyase activity)*

Thaw the cell suspensions, add glass beads (0·11 mm diam.) up to 10 % of the suspension volume, and sonicate the suspensions six times at maximal amplitude for 30 s, allowing them to cool for 2 min between the sonication periods. Keep the suspensions on ice during the entire procedure.

↓

Centrifuge the treated cell suspensions for 10 min at 40 000 g and 4 °C to remove whole cells and large cell debris. Keep the supernates on ice.

↓

Into two 1 ml cuvettes, pipette 0·2 ml reaction mixture, 0·05 ml reduced glutathione (50 mM, pH 6·8), 0·05–0·5 ml cell-free extract (depending on the activity of isocitrate lyase) and make up to 1·0 ml with demineralized water.

↓

To start the reaction, add 0·01 ml trisodium isocitrate solution (0·1 M) to one of the cuvettes and record the extinction at 324 nm at room temperature against the blank cuvette using a double beam spectrophotometer. The reaction mixture consists of 1·0 ml imidazole–HCl buffer (0·5 M, pH 6·8), 1·0 ml MgCl$_2$ (0·1 M), 0·2 ml disodium EDTA (0·1 M), 0·8 ml phenylhydrazine–HCl (0·1 M) and 1·0 ml demineralized water. This mixture is stable for only 2 h. Keep on ice. At 324 nm the formation of glyoxyl-phenylhydrazone from glyoxylate and phenylhydrazine can be followed (*see* note 1).

↓

Calculate the specific activity of isocitrate lyase in nmol per min per mg of protein from the extinction change (the extinction coefficient of phenylhydrazone is $1·7 \times 10^4$ M^{-1} cm^{-1}). Protein concentrations are estimated according to the method of Lowry *et al.* (1951), using bovine serum albumin as a standard.

- -

Day 4 (*several hours; determination of glucose and acetate in the supernates obtained on Day 2*)

1. *Determination of glucose*
Glucose is determined using the GOD-Perid method from Boehringer Mannheim, W. Germany.

Add 0·1 ml of the supernate or 0·1 ml of the glucose standard solution to 5 ml glucose reagent.

↓

Incubate for 25–50 min at room temperature and read the extinctions against a blank, consisting of 0·1 ml distilled water and 5·0 ml glucose reagent.

↓

Calculate the glucose concentration in the sample from the extinctions in the standard and sample tubes.

2. *Determination of acetate*
Acetate is determined gas chromatographically, after extraction with diethylether.

Pipette 1·0 ml of the supernate into a small tube (15 × 70 mm) containing 0·4 g solid NaCl.

↓

Add 0·1 ml concentrated formic acid and 1·0 ml diethylether.

↓

Close the tube with a stopper and mix well, by turning the tube over twenty times.

$\downarrow$

Inject 2 μl from the ether phase in a gas chromatograph, equipped with a flame ionization detector.

$\downarrow$

Use an ether extract of a 20 mM acetate solution as a standard and calculate the acetate concentrations in the supernates from the ratios of the mean heights (mean of 3 injections) of the peaks obtained from the samples and from the standard. Chromatograph settings: column temperature, 125 °C; detector oven temperature, 170 °C; sample inlet temperature, 175 °C; nitrogen flow through the column, 15 ml min^{-1}.

Notes and Points to Watch

1. In our experience, the isocitrate lyase activity can best be measured between 20 and 25 °C. We found repeatedly that at 37 °C the extinction at 324 nm, after a short initial increase, started to decrease rapidly. We do not have an explanation for this.
2. Oxidation of reserve materials by *E. coli* may cause a high endogenous respiration rate. As a result virtually no difference in respiration rate with or without acetate can be observed. In that case, incubate the cells for 1–2 h in 50 ml 50 mM phosphate buffer, pH 7·0, in a shaking incubator at 37 °C, to starve the cells. Centrifuge the suspension and resuspend the pellet in the original volume of phosphate buffer (2–3 ml).

MATERIALS

1. *Escherichia coli* B can be obtained from the ATCC or NCIB.
2. Growth medium: final concentration (g l^{-1}) NH_4Cl (1·0), $MgSO_4 \cdot 7H_2O$ (0·5), KH_2PO_4 (1·6), K_2HPO_4 (2·4), glucose (1·0 or 4·0), sodium acetate (2·0) and trace-element solution (according to Vishniac and Santer, 1957) 1 ml l^{-1}, final pH 7·0. Phosphate buffer (pH 7·0), glucose and acetate are sterilized separately in concentrated solutions, and added aseptically to the rest of the medium. All components are heat-sterilized at 120 °C. All chemicals except $MgSO_4$ are anhydrous.
3. Test combination Glucose, GOD-Perid® method, catalogue no. 124 036, Boehringer Mannheim GmbH, W. Germany.
4. Oxygen electrode. The "Biological Oxygen Monitor", type 53 SA, Yellow spring Instruments (Yellow Springs Ohio, USA) is very suitable.
5. Gas chromatograph with flame ionization detector. Column for gas chromatograph: glass, internal diameter 0·4 cm, length 120 cm. Solid phase: Chromosorb W.AW, 100–200 mesh, stationary phase 10 % SP-1000 and 1 % H_3PO_4. Chrompack, Middelburg, The Netherlands.

SPECIFIC REQUIREMENTS

Day 1
agar slant of *E. coli* B
1 beaker (3 litre), 3 measuring cylinders (100 ml and 1 litre), 3 bottles (100 ml), 5 sterile pip-ettes (5 ml), 2 Erlenmeyer flasks (100 ml) chemicals, spatulas, balance, pH meter, calibration buffers
autoclave, cotton plugs, aluminium foil

cont.

cont.

magnetic stirrer + fleas	high speed centrifuge + centrifuge tubes, 10 ml
fermenter vessel with accessory parts, silicone tubing, scissors, sterilizable plastic tape, tubing clips	double beam spectrophotometer, quartz cuvettes
rotary shaker	20 pipettes, 1 and 5 ml
	20 tubes and tube rack
Day 2	***Day 4***
spectrophotometer	10 tubes, normal and small size, see text
centrifuge, centrifuge bottles, 100 ml	20 pipettes (0·1 ml), 10 pipettes (1 ml), 1 pipette (5 ml)
freezer	stoppers
oxygen electrode	gas chromatograph, stabilized for at least 2 h
	micro-syringe, 10 μl
Day 3	spectrophotometer
glass beads, 0·11 mm diam.	
sonifier + tubes	

FURTHER INFORMATION

Typical student results are given in Figure 1. As can be seen, *E. coli* preferentially utilizes glucose. When the glucose concentration has decreased to about 25% of its original value, acetate

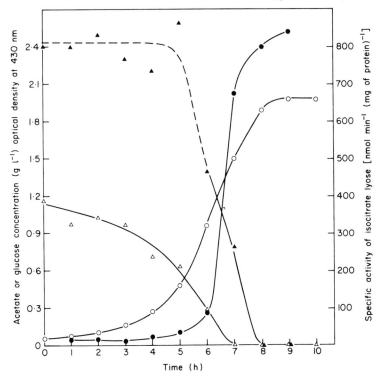

Figure 1 Induction of isocitrate lyase in *E. coli*, growing on a mixture of glucose and acetate. O——O, optical density; △——△, glucose concentration; ▲——▲, acetate concentration; ●——●, isocitrate lyase activity.

consumption begins and isocitrate lyase is induced, resulting in a 60-fold increase in activity after a few hours. The capacity of the organism to oxidize acetate (not shown in the figure) increases by 3 to 4 times during growth on acetate, in this typical experiment from 0.2 to 0.6 μl O_2 per min per unit of optical density.

REFERENCES

H. L. Kornberg (1966). Anaplerotic sequences and their role in metabolism. *In* "Essays in Biochemistry", (P. N. Campbell and G. D. Greville, eds), Vol. 2, pp. 1–31. Academic Press, London and New York.

O. H. Lowry, N. J. Rosebrough, A. L. Farr and R. J. Randall (1951). Protein measurement with the folin phenol reagent. *Journal of Biological Chemistry*, **193**, 265–75.

W. Vishniac and M. Santer (1957). The Thiobacilli. *Bacteriological Reviews* (P. N. Wilson, ed.), **21**, 195–213.

Part Three
The Micro-organism Grows

22

The Evaluation of Bacterial Growth in Liquid Media

L. MULLENGER

Department of Biological Sciences, Lanchester Polytechnic,
Coventry, CVI 5FB, England

> *Level*: Beginning undergraduates
> *Subject area*: Introductory microbiology/microbial physiology
> *Special features*: Safe, simple, rapid, with immediate results

INTRODUCTION

Choice of Organism

A number of organisms has been used in the past for experiments on bacterial growth (*see* the list of suitable bacteria in DES Education pamphlet No. 61). The organism often chosen has been *Escherichia coli*, usually for historical reasons. The choice of the correct organism for this type of practical is particularly important as the demonstration of a microbial growth curve is often used as an elementary introduction to microbiology, with the result that the experiment may be performed by students who have not perfected their aseptic technique. I consider the use of *E. coli* in this situation to be unsatisfactory both for technical and public health reasons. Although previously regarded as a safe organism, *E. coli* is an opportunistic pathogen; it does survive in the gut and has the ability to donate plasmids to other Gram-negative organisms. Because of this, even such laboratory strains as *E. coli* K12 should not be used in the classroom at an elementary level. With this in mind I sought a robust, but harmless, organism suitable for such work and for some years have recommended the use of *Beneckea natriegens* (*Vibrio natriegens*) (Mullenger and Gill, 1973). This is a marine micro-organism originally called *Pseudomonas natriegens* (Payne *et al.*, 1961). This organism has proved to be highly suitable for a wide range of laboratory demonstrations, having all the usual responses to varying growth conditions, antibiotics, etc. that might be used in classroom experiments. There are also distinct technical advantages in the use of this organism. It is unicellular so that it can be assayed by a plate count and, of particular pertinence to classroom use, it demonstrates a very rapid rate of growth in common laboratory media. A mean generation time of 10 min has been obtained with this organism, but under the usual shake culture conditions it is not unusual to obtain a mean generation time of 15 min. *Beneckea* also has the advantage that it demonstrates a very short lag period (see results reported below). Many organisms such as *E. coli* demonstrate lag periods of up to 120 min under similar conditions, followed by a longer mean generation time. Thus the use of *Beneckea* in even the most basic apparatus enables a typical student to obtain a fairly complete growth curve in a three hour laboratory session. These advantages have now led to the distribution and use of this organism in a number of teaching institutions (e.g. North and Saggers, 1975).

The aims of the following experiments are that students should obtain the lag and exponential phases of a normal growth curve for *Beneckea natriegens* from which the mean generation time can be calculated. Subsequent experiments examine the effects of temperature, salt concentration, antibiotics and a lytic enzyme on the growth of this organism. The final experiment demonstrates the diauxie effect.

EXPERIMENTAL

The experiments can be conveniently grouped into four sessions of three hours which have been labelled experiments 1 to 4. It is suggested that all students attempt experiment 1 before attempting any other experiments.

Experiment 1

The determination of the growth curve of Beneckea natriegens *by following the increase in cell mass with a spectrophotometer and calibration by a simultaneous plate count*

Students should work in pairs with one responsible for obtaining absorbance readings and the other responsible for plate counts.

Day 1 *(3 h)*

Take 20 cm^3 overnight culture of *B. natriegens* (*see* note 1).

↓

Determine absorbance at 550 nm (*see* note 2).

↓

Dilute with Beneckea broth to give a final volume of approx. 100 cm^3 with an A_{550} between 0·02 and 0·05 (*see* note 3).

↓

Incubate in a shaking water-bath at 37 °C (*see* note 5).

↓

Start clock.

Student 1

Sample at 10 min intervals from 0 to 120 min → and measure A_{550} (*see* note 4).

↓

Tabulate results and plot on semi-log paper (*see* note 6).

↓

Determine the length of the lag phase and the mean generation time.

Student 2

Take samples at 30, 50, 70 and 90 min (check from the graph that these form part of the exponential growth phase).

↓

Make ten decimal dilutions in saline (1 cm³ culture in 9 cm³ saline)

↓

Make triplicate pour-plates from dilutions in the range 10^{-5} to 10^{-10} (1 cm³ of diluted culture per plate filling it one-third full with molten agar—mix well and leave to set) (*see* note 7).

↓

Incubate at 37 °C overnight.

- -

Day 2 *(1 h)*

Count the most appropriate plates. Calculate the number of viable bacteria cm^{-3} in the original culture and plot to obtain a calibration curve. Determine how many bacteria correspond to one absorbance unit.

- -

Notes and Points to Watch

1. Allow the overnight culture to cool at room temperature for about an hour before use otherwise a lag-phase may not be obtained.
2. The spectrophotometer should be zeroed using the growth medium as a blank — not distilled water.
3. An overnight culture will usually give an A_{550} reading of 0·8. The spectrophotometer scale is reasonably linear at its lower end and to obtain a starting culture with a reading of 0·05 units it will be necessary to make a 1/16 dilution in fresh broth. As approximately 100 cm³ are required, add 6 cm³ of the overnight culture to 96 cm³ broth. Mix well by pouring between two flasks and take the zero-time reading. The actual starting value is not critical.
4. The incubation flask should be removed for the minimum possible time when taking samples to the cuvette. Samples should not be returned to the incubation flask but discarded after reading (except for those to be plated).

5. The optimum growth temperature for *B. natriegens* is 35 °C but this is so close to 37 °C that it does not warrant recalibrating incubators normally set at 37 °C.
6. Spectrophotometer readings should be plotted directly onto 3-cycle semi-log paper (Chartwell 5531). Tabulating sheets (Chartwell 4580) may be found useful.
7. Spread plates may be used instead of pour plates. In this eventuality use sample sizes of $0.1\,cm^3$. Pour plates are preferable for large classes as it saves on pre-class technical-support time.

MATERIALS

1. *Beneckea natriegens* (*Vibrio natriegens*; NCMB 857, National collection of Marine Bacteria, Ministry of Agriculture, Fisheries and Food, Torry Research Station, P.O. Box 31, 135 Abbey Road, Aberdeen, AB9 8DG).

   For general storage this organism should be kept at room temperature on saline nutrient agar slopes (nutrient agar $28\,g\,l^{-1}$, NaCl $20\,g\,l^{-1}$, pH 7.5). Refrigerated cultures soon lose their viability whereas room-temperature cultures remain healthy but do require subculturing at least once a month.

2. The experimental medium (referred to as Beneckea broth or agar) provides very rapid growth and consists of brain–heart infusion broth ($37\,g\,l^{-1}$), NaCl ($20\,g\,l^{-1}$) adjusted to pH 7.5. For use in pour plates add 1.5% ionagar.
3. All the experiments require a simple colorimeter or spectrophotometer (e.g. EEL Spectra) with 1 cm cuvettes (plastic disposable cuvettes are quite satisfactory).

SPECIFIC REQUIREMENTS

These items are needed for each pair of students.

$20\,cm^3$ cooled, overnight culture of *Beneckea natriegens* grown in Beneckea broth
1 shaking water-bath at 37 °C
$2 \times 250\,cm^3$ empty sterile conical flasks for use in incubating and mixing
$4 \times 100\,cm^3$ molten Beneckea agar held at 60 °C in water-bath
$150\,cm^3$ Beneckea broth held at 37 °C for use as a blank and as a diluent
72 sterile Petri dishes
$40 \times 9\,cm^3$ 0.85% NaCl in universal bottles for use in decimal dilutions

$5 \times 10\,cm^3$ pipettes
$60 \times 1\,cm^3$ pipettes (use Volac disposable pipettes)
$1 \times 100\,cm^3$ sterile measuring cylinder
1 spectrophotometer $+ 2 \times 1\,cm$ cuvettes
1 Whirlimixer (for ease of mixing dilution bottles)
1 clock
tissues
pipette bulbs for $10\,cm^3$ and $1\,cm^3$ pipettes
disposal jars of disinfectant for used pipettes and discarded culture samples
wash-bottle of distilled water
37 °C incubator for Petri dishes

FURTHER INFORMATION

The student following the growth of *Beneckea* over a period of $2\frac{1}{2}$ h should obtain a lag period of no more than 20 min followed by rapid exponential growth demonstrating a mean generation time of 17–20 min. An A_{550} of 0.05 corresponds to approx 5×10^6 cells cm^{-3}. The stationary phase is not always reached before the end of the practical session. However, Figure 1 shows the result obtained by a pair of first year students in which the stationary phase was reached.

Experiment 2 (*3h*)

To determine the effects of a variety of antimicrobial agents on the growth of Beneckea natriegens

Students should work in pairs with one being responsible for taking samples and the other responsible for absorbance readings and for result plotting.

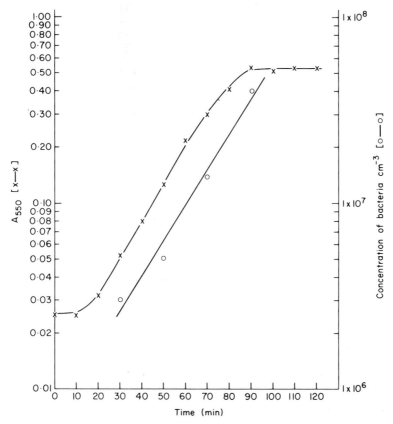

Figure 1   The growth curve for *Beneckea natriegens* incubated at 37 °C in Beneckea broth. $\times$——— $\times$ absorbance readings at 550 nm; O———O viable counts cm^{-3} of culture.

Take 50 cm^3 overnight culture of *B. natriegens*.

$\downarrow$

Determine absorbance at 550 nm.

$\downarrow$

Dilute with Beneckea broth to give approx. 600 cm^3 final volume with an A_{550} between 0·02 and 0·05 (*see* note 1).

$\downarrow$

Dispense into 6 flasks each containing 100 cm^3.

↓

Label flasks A to F.

↓

Start clock and insert flasks at 2-min intervals into a shaking water-bath at 37 °C (flasks G and F go in together at 8 min) (*see* note 2).

↓

At 10 min sample flask A and obtain A_{550} (*see* note 3).
At 12 min sample flask B and obtain A_{550}.
At 14 min sample flask C and obtain A_{550}.
At 16 min sample flask D and obtain A_{550}.
At 18 min sample flask E and F and obtain A_{550}.

Each of these samples represents the results of 10 min of incubation.

↓

Continue taking a sample every 2 min and thereby sampling **each** flask every 10 min until 30 min of exponential growth has been obtained (approx. 50 min from time zero).

↓

Take the next reading due for each flask and then add the appropriate antimicrobial agent, mix well and immediately take a second reading and note the lowered reading due to dilution.

↓

To flask A add 10 cm^3 distilled water (control flask) (*see* note 4).
To flask B add 10 cm^3 lysozyme.
To flask C add 10 cm^3 sodium lauryl sulphate (SLS).
To flask D add 10 cm^3 penicillin solution.
To flask E add 10 cm^3 chloramphenicol solution.
To flask F add 10 cm^3 distilled water and transfer to a low-temperature bath (*see* notes 5, 6 and 7).

↓

Continue incubation at 37 °C taking absorbance readings at 10 min intervals. You may cease when the organisms in the penicillin culture have been destroyed (*see* note 8).

↓

Having obtained the A_{550} value for flask B add 1 drop of SLS to the cuvette and note the effect (*see* note 9).

- -

Notes and Points to Watch

1. The Beneckea broth for use in the experiment does not need to be autoclaved if it is made up just before the practical. This can save a lot of time for large classes.

2. Because it is not possible to sample six flasks simultaneously the sample times are staggered. However, this still results in each flask being sampled once every 10 min. When plotting the results ignore the staggering: thus the second sample from flasks E and F are plotted at time 10 min even though the real time of sampling was 18 min.
3. Use one cuvette for each flask, using Beneckea broth as a blank; not distilled water.
4. All solutions to be added to the flasks should be preheated to 37 °C. Because of the dilution there will be a discontinuity in the growth curve.
5. The low temperature flask B should be placed in an ice-bath the temperature of which should be noted; 4 °C is adequate.
6. Samples from flask B are cold and this results in condensation forming on the cuvette which gives a false reading. The sample should be warmed quickly before the reading is taken. This may conveniently be done by dipping the cuvette into the 37 °C bath for a few seconds. Dry the cuvette with tissue before reading.
7. The low temperature bath is not being shaken. The student should be aware of the effects of this on aeration of the culture and must overcome the sampling problems due to sedimentation.
8. The experiment should be run until a distinct fall in the A_{550} has been followed for at least 30 min.
9. The addition of one drop of SLS to the culture treated with lysozyme will result in a very rapid lysis of the cells. Thus the cuvette should be in the spectrophotometer when the drop of SLS is added. Readings should be noted every 5 **seconds**. Within 30 s the suspension will have lysed so this will not interfere with the 2-min sampling regimen.
10. Students will require two sheets of 3-cycle semi-log graph paper. The results for each flask should be plotted on a separate cycle which means the cycles will overlap. Plot them in alphabetical order with A starting on the second cycle, B on the third, etc. F (low temperature) can be conveniently plotted on the same cycle as A.

MATERIALS

1. *Beneckea natriegens* as in experiment 1.
2. Beneckea broth as in experiment 1.
3. lysozyme $100 \mu g \, cm^{-3}$
   penicillin $200 \mu g \, cm^{-3}$

chloramphenicol $200 \mu g \, cm^{-3}$
sodium lauryl sulphate 2.5% solution,
obtainable from Sigma Chemical Co. Ltd, Fancy Road, Poole, Dorset BH17 7NH.

SPECIFIC REQUIREMENTS

These items are needed by each pair of students.

$50 \, cm^3$ cooled overnight culture of *B. natriegens* grown in Beneckea broth
1 ice-bath with thermometer
1 spectrophotometer $+ 7 \times 1$ cm disposable cuvettes

5 empty, sterile $250 \, cm^3$ conical flasks for incubation
$700 \, cm^3$ Beneckea broth (held at 37 °C)
$100 \, cm^3$ sterile measuring cylinder
1 large empty flask for mixing
$10 \, cm^3$ of each antimicrobial agent (held at 37 °C)

---

cont.

cont.

$2 \times 10 \text{ cm}^3$ sterile distilled water (held at 37 °C) $1 \times 10 \text{ cm}^3$ pipette with bulb (for dispensing overnight culture) 1 Pasteur pipette (for dropping SLS) stop-clock	tissues disposal jars for pipettes and samples In addition, 1 or more shaking water-baths, set at 37 °C, are required by the class.

FURTHER INFORMATION

The optimum temperature for growth of *Beneckea natriegens* is 35 °C (Mullenger and Gill, 1973), in spite of its normal habitat being estuarine mud. Thus the effect of transferring the organism to an ice-bath is an extension of the mean generation time to several hours so that, over the time span of this experiment, growth seems to stop. Students are expected to discuss the physiology of growth in their report. (If incubators, etc. are already calibrated for 37 °C this is quite satisfactory.)

Beneckea natriegens is a Gram-negative organism and lysozyme, or detergents like sodium lauryl sulphate, do not lead to lysis of the cells during the short time span of this experiment. The lysozyme cleaves the linkage between muramic acid and the glucosaminyl residues in the cell wall, whereas the SLS attacks the lipid component. The result of simultaneous attack by both agents is dramatic, leading to an immediate loss in turbidity due to lysis.

Chloramphenicol has an immediate effect upon protein synthesis due to the inhibition of peptide bond formation at the ribosome: thus growth ceases immediately (but would resume upon washing the cells as chloramphenicol is bacteriostatic but not bactericidal).

Penicillin inhibits the synthesis of the murein component of the cell wall. This is an immediate reaction but the observable effect is delayed because the cell wall remains intact. This becomes progressively weaker as the cells grow until eventually the cells start to lyse.

Students are expected to investigate the mechanism of action of each agent by a library search before presenting their report.

The results presented in Figure 2 are an actual class result and it will be seen that the overnight culture had been insufficiently cooled and no lag phase was obtained on this occasion.

Experiment 3 *(3 h)*

To determine the effect of various concentrations of sodium chloride on the growth of Beneckea natriegens

Students should work in pairs with one being responsible for taking samples and the other responsible for absorbance readings and for plotting the results.

Take 50 cm³ overnight culture of *B. natriegens* (*see* notes 1 and 3).

↓

Determine absorbance at 550 nm.

↓

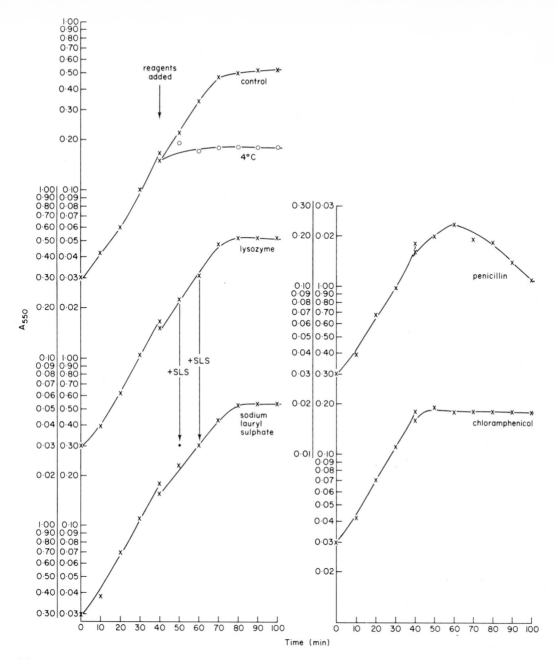

Figure 2 In descending order the figure illustrates the effects of the following procedures on a culture of *Beneckea natriegens* growing exponentially at 37 °C in Beneckea broth:
a temperature shift to 4 °C;
addition of lysozyme (to a final concentration of 10 $\mu g\,cm^{-3}$) with the further addition of 1 drop of a 2·5 % solution of sodium lauryl sulphate (solid descending arrows);
addition of penicillin to a final concentration of 10 $\mu g\,cm^{-3}$;
addition of chloramphenicol to a final concentration of 10 $\mu g\,cm^{-3}$.
For greater clarity the 550 nm absorbance readings have been plotted on overlapping semi-logarithmic cycles.

Prepare 5×100 cm^3 samples with a final A_{550} between 0·02 and 0·05 using the Beneckea broths containing various concentrations of sodium chloride (*see* notes 4 and 5).

↓

Incubate and determine A_{550} at staggered intervals as in Experiment 2.

Notes and Points to Watch

1. This experiment requires an overnight culture of high cell density (above 0·80 on the spectrophotometer scale), otherwise the carry over of sodium chloride will be significant.
2. It is helpful in plotting the graph if the initial cell density increases slightly with increasing sodium chloride concentration so that all the curves remain separate. In this way all the results can be plotted on the same log cycles.
3. It is not necessary to induce a lag phase for this experiment.
4. Brain–heart infusion broth powder already contains 0·5% NaCl.
5. All broths should be preheated to 37 °C.

MATERIALS

1. *Beneckea natriegens* as in experiment 1.

2. Brain–heart infusion broth containing 0, 0·5, 1·0, 1·5 and 2·0% additional NaCl.

SPECIFIC REQUIREMENTS

The items are needed for each pair of students.

50 cm^3 exponentially growing culture of *B. natriegens* in Beneckea broth at high cell density
150 cm^3 Beneckea broth held at 37 °C
150 cm^3 Beneckea broth + 0·5% NaCl held at 37 °C
150 cm^3 Beneckea broth + 1·0% NaCl held at 37 °C
150 cm^3 Beneckea broth + 1·5% NaCl held at 37 °C

150 cm^3 Beneckea broth + 2·0% NaCl held at 37 °C
1 shaking water-bath at 37 °C
1 spectrophotometer + 6 cuvettes (1 cm)
5 empty, sterile 250 cm^3 conical flasks for incubation
100 cm^3 sterile measuring cylinder
1 large empty flask for mixing
1 × 10 cm^3 pipette with pipette bulb
stop-clock
tissues
disposal jar for pipettes and samples

FURTHER INFORMATION

Beneckea natriegens is a halophile requiring at least 0·5% NaCl to prevent cell lysis. However, it grows well on 2% NaCl (Figure 3), which is less than that demanded by many strict halophiles, as might be expected from its natural habitat, estuarine mud, where the salinity can fall considerably.

Beneckea natriegens is unusual in having a specific requirement for sodium chloride, hence the specific epithet *natriegens* (natrium: sodium, egens (egere): to need). Most halophiles will grow in any isotonic medium where the sodium ion might be replaced by K^+, Li^+, Rb^+, or CS^+. *B. natriegens* has a specific, and unusual, requirement for the sodium ion both in the induction of certain oxidative enzymes and for substrate penetration where the sodium ion seemed to be required for the binding of the substrate to a permease or some other penetration mechanism (Payne, 1960).

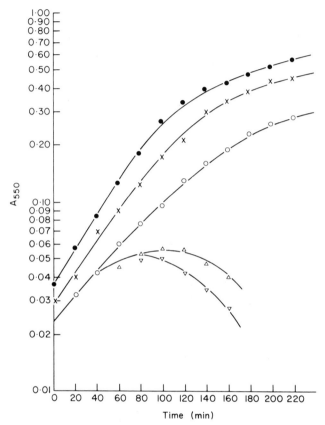

Figure 3   The effect of various concentrations of sodium chloride on the growth of *Beneckea natriegens* at 37 °C. ●——● 2 % NaCl; ×——× 1·5 % NaCl; ○——○ 1 % NaCl; △——△ 0·5 % NaCl ▽——▽ no NaCl (NB. Brain–heart infusion broth contains 0·5 % NaCl which must be added to these figures). Absorbance readings obtained at 550 nm.

Experiment 4 *(3 h)*

The demonstration of diauxic growth

Take 25 cm³ exponential culture of *B. natriegens*.

↓

Spin in bench centrifuge to pellet cells (usually about 10 min at top speed).

↓

Resuspend cells in saline to wash (*see* note 1).

↓

Repeat spin to pellet.

↓

Resuspend in Diauxie minimal medium to an A_{550} between 0·02 and 0·03 (*see* note 2).

↓

Incubate 100 cm^3 at 37 °C, sampling every 5 min.

↓

Plot A_{550} (*see* note 5).

Notes and Points to Watch

1. Time is at a premium in this experiment if it is to be completed in a half-day session. It would help if the class could be presented with the washed culture in which case I suggest two washings in saline.
2. The starting cell concentrations is lower than usual (between A_{550} 0·02 and 0·03). *Beneckea* shows an unusual relationship between inoculum size and mean generation time (Mullenger and Gill, 1973) and the lower starting concentration produces a more rapid rate of growth.
3. If more time is available an increase of the glucose concentration to 0·001% increases the duration of the first exponential rise.
4. The Diauxie minimal medium should be held at 37 °C.
5. Use the minimal medium to zero the spectrophotometer.
6. Use single-cycle semi-log graph paper (Chartwell 5511).

MATERIALS

1. Exponential culture of *B. natriegens* (as in Experiment 1).
2. Diauxie minimal medium:

sodium glucuronate,	2 gm l^{-1}
MgSO$_4$.7H$_2$O,	1·3 g l^{-1}
KCl,	0·75 g l^{-1}
(NH$_4$)$_2$.HPO$_4$,	5·3 g l^{-1}
NaCl,	15 g l^{-1}

Filter and then sterilize by autoclaving or by filtration.
Autoclave 0·5% glucose and 3% xylose separately and add to the above, to a final concentration of 0·0005% glucose and 0·003% xylose.

SPECIFIC REQUIREMENTS

bench centrifuge
25 cm³ exponential culture of *B. natriegens* in
   Beneckea broth
sterile centrifuge tubes (50 cm³ capacity)
30 cm³ sterile saline at 37 °C
150 cm³ Diauxie minimal medium at 37 °C
250 cm³ sterile empty conical flask
1 spectrophotometer plus 2 × 1 cm disposable
   cuvettes

1 shaking water-bath at 37 °C
100 cm³ sterile measuring cylinder
3 × 10 cm³ sterile pipettes + pipette bulbs
stop-clock
tissues
disposal jars for pipettes and samples

FURTHER INFORMATION

Some enzymes in bacteria are constitutive and are, therefore, present all the time. Other enzymes
are inducible and are either absent or present in only very small amounts in the absence of a suitable

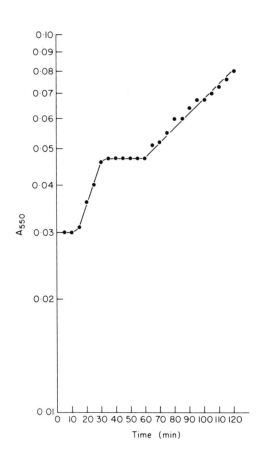

Figure 4 The diauxic growth curve obtained at
37 °C by incubating *Beneckea natriegens* in a minimal medium containing 0·005% glucose and
0·003% xylose. Absorbance reading obtained at
550 nm.

substrate. When presented with a suitable substrate enzyme production proceeds at a very high rate, but there will be an observable lag period whilst the enzyme is being induced. For a full discussion of enzyme induction and repression *see* Lewin (1974).

When *Beneckea* is presented with a choice of two substrates, glucose and xylose, the enzymes required to utilize glucose are preferentially induced. Following a short lag *Beneckea* demonstrates a normal rate of growth (mean generation time of approximately 20 min). This continues until all the glucose has been utilized; xylose then becomes the only available metabolite and the cell culture enters a second lag phase during which the necessary enzymes to deal with this energetically poorer substrate are induced. This is then followed by a diminished rate of growth. The resulting biphasic growth curve is described as a diauxic growth curve (Figure 4).

REFERENCES

D. E. S. Education Pamphlet No. 61 (1977). "The Use of Micro-organisms in Schools". H.M.S.O.

B. M. Lewin (1974). "Bacterial Genomes", Gene Expression, Vol. I. Wiley-Interscience, London.

L. Mullenger and N. R. Gill (1973). "*Vibrio natriegens*: a rapidly growing micro-organism ideally suited for class experiments". *Journal of Biological Education*, **7** (5), 33–9.

L. G. North and B. A. Saggers (1975). Letter to the Editor. *Journal of Biological Education*, **9** (3/4), 181.

W. J. Payne (1960). Effects of sodium and potassium ion on growth and substrate penetration of a marine Pseudomonad. *Journal of Bacteriology*, **80**, 696–700.

W. J. Payne, R. G. Eagon and A. K. Williams (1961). "Some observations on the physiology of *Pseudomonas natriegens*". *Antonie van Leeuwenhoek Journal of Microbiology and Serology*, **27**, 121–8.

23

Hyphal Growth of *Neurospora crassa*

G. W. GOODAY

*Department of Microbiology, University of Aberdeen,
Aberdeen, AB9 1AS, Scotland*

> *Level*: All undergraduate years
> *Subject areas*: Growth; eukaryotic cell structure
> and function
> *Special features*: Direct observation and measure-
> ment of fungus growth and
> the effect of inhibitors

INTRODUCTION

Apical growth of fungal hyphae was described in 1897 by Reinhardt, who observed that particles deposited onto the domed side of an apex did not move forward with the apex, but moved outwards to remain on the lateral wall of the hypha. Autoradiography, with tritiated wall precursors, has confirmed that the wall polysaccharides, such as chitin and glucans, are deposited in a very polarized manner at the hyphal apex. Full discussions of apical growth and of the resultant growth of a colony are given by Burnett (1976) and Trinci (1977).

The growth of individual fungal hyphae can be observed readily by culture on an agar medium underneath a transparent plastic film that is permeable to oxygen. This allows direct microscopic observations without drying or disturbance of the specimen. The aim of the experiment is to demonstrate and measure apical extension of individual fungal hyphae, and to assess the effect of chemicals on this rate of growth.

EXPERIMENTAL

Examine the circular colonies of fungus, which should be about 50 mm in diameter, growing underneath polypropylene film to locate their outermost edges. These will be the sites of the leading hyphal tips. Carefully cut the cultures (membrane, fungus and agar) into 8 sectors, as if cutting a cake, with a very sharp scalpel.

Growth (1½ h)

Mount a sector on a microscope slide and examine microscopically.

↓

With a low power objective, locate a leading hyphal tip. With a medium power objective, examine it to ensure that it is straight, intact, and growing on or very near the surface of the agar (and not down into it).

↓

Line it up along the eyepiece micrometer, and record its position every minute for at least 10 min. Draw a graph of its growth in length (μm) against time (min). Is it growing at a constant rate? If so, express its growth rate as μm min^{-1}. Repeat for two or more leading tips to gauge the variability between different hyphae.

↓

Draw in scale a hypha from its apex back to its older parts, measuring its diameter and locating septa and branches. Measure hyphal growth rates and diameters for side branches and compare them with values from the leading hyphae.

↓

Look for protoplasmic streaming in the hyphae. Determine its direction of flow. Look for streaming near to a septum, and decide whether the septum is a barrier to the movement of material. Using the eyepiece micrometer, measure the speed of streaming of particles in μm min^{-1}. Look for septa being formed. Describe and time this process.

↓

These observations can also be made with an oil immersion objective, with a drop of oil being added directly to the polypropylene film. Use of oil immersion phase-contrast microscopy is especially valuable to examine the cytology of the apex. It should be possible to see the apical body (the "Spitzenkörper", corresponding to the central region of the cluster of apical vesicles seen in electron micrographs) within the domed region, then a group of snake-like mitochondria behind it, and then a group of nuclei (*see* Figure 1).

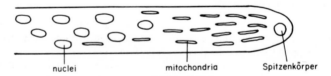

nuclei   mitochondria   Spitzenkörper

Figure 1

- -

Inhibition of growth (1½ h)

Mount a sector of the fungal culture on a slide, and locate a growing apex as above with the medium power objective. Measure its growth and plot against time for 10 min.

↓

Swing aside the objective, very carefully lift the membrane with fine forceps, and cover the growing tip with a drop of the test solution (*see* note 3). Lower the membrane, and measure the growth as before for at least 10 min.

↓

If growth ceases, examine the hyphal tip with the oil immersion phase contrast lens, and compare it with the growing tips that were viewed before.

↓

Re-test the same chemical at ten-fold serial dilutions (diluting with the malt extract solution) until no interruption of growth is observed. In each case, record the time taken for inhibition of growth to occur, and whether any recovery of growth occurs. Express results as the per cent reduction in growth rate against the concentration of inhibitor.

↓

As a control, test the effect of adding the malt extract solution.

Notes and Points to Watch

1. The hyphal tips are very fragile, and are burst easily. This observation, however, has been used to support a model of their dynamic nature during growth (Bartnicki-Garcia, 1973).
2. Any filamentous fungus can be used. *Neurospora crassa* is suggested here because of its ready availability, and its fast growth rate (of the order of 40 μm min^{-1} in these conditions). A wide range of mutants of this fungus is available, some of which, for example the clock mutants, could be used for further experiments using this system.
3. Any potentially antifungal substance or treatment could be tested. Suggestions are copper sulphate (10 mM), cycloheximide (0·2 mM) and 2, 4-dinitrophenol (5 mM), made up in 2% (w/v) malt extract to be isotonic with the growth medium. The effect of osmotic pressure on growth rate, shape and integrity of the apex can be observed by adding a drop of a hypotonic solution such as water or a hypertonic solution such as 1 M glucose to a growing tip.

MATERIALS

1. *Neurospora crassa*, any strain (*see* note 2). This should be inoculated onto the centre of the malt agar plate, between 24 h and 6 h before the class, and overlain with a sterile disc of plastic film. Incubation is between 20 and 37 °C, as convenient to give a circular colony about 50 mm diameter by the start of the class.
2. The malt agar plate is a Petri dish of 2% malt extract, 2% agar, poured as a shallow layer with 5–10 ml molten medium in a 9 cm diameter plate.
3. Plastic film. The experiment works well with a variety of thin plastic films, but they must be clear, flat and readily permeable to oxygen. The

most suitable have been cut from clear polypropylene autoclavable disposal bags. Teflon film (as used for oxygen electrodes) is satisfactory. Cooking film (used for wrapping food in ovens) also works, but is very difficult to handle. The discs, slightly smaller than the size of a Petri dish, should be sterilized, wrapped in aluminium foil by brief autoclaving (5 min at 34·5 kPa (5 lbf in^{-2}) is satisfactory for polypropylene). Whatever film is used, it should be layered onto the agar with care, and air bubbles should be removed by stroking them out with the flat end of a spatula.

SPECIFIC REQUIREMENTS

culture

microscope with calibrated eyepiece micrometer

microscope slides

Pasteur pipettes

2% (w/v) malt extract

growth inhibitors, dissolved in 2% (w/v) malt
  extract

pipettes and tubes for serial dilutions of
  inhibitors

sharp scalpel

fine forceps

FURTHER INFORMATION

Analysis of the growth of individual hyphae shows that in the typical fungal colony the leading hyphae grow faster than primary branches, which in turn grow faster than secondary branches (Trinci, 1977). This would be revealed in a "race" amongst members of the class, where first results often show a bimodal distribution for growth rate, some students having inadvertently chosen primary branches. Leading hyphae are wider than the branches. The cytology and the biochemistry of apical wall deposition are discussed by Grove (1977) and Gooday (1977) respectively.

REFERENCES

S. Bartnicki-Garcia (1973). Fundamentals of hyphal morphogenesis. *Symposium of the Society for General Microbiology*, **23**, 245–67.

J. H. Burnett (1976). "Fundamentals of Mycology", 2nd edn. Arnold, London.

G. W. Gooday (1977). The enzymology of hyphal growth. *In* "The Filamentous Fungi", (J. E. Smith and D. R. Berry, eds), Vol. III, pp. 51–77. Arnold, London.

S. N. Grove (1977). The cytology of hyphal tip growth. *In* "The Filamentous Fungi", (J. E. Smith and D. R. Berry, eds), Vol. III, pp. 28–50. Arnold, London.

A. P. J. Trinci (1977). The duplication cycle and vegetative development in moulds. *In* "The Filamentous Fungi", (J. E. Smith and D. R. Berry, eds), Vol. III, pp. 132–63. Arnold, London.

24

The Effect of Water Activity on Microbes from Different Environments

S. T. WILLIAMS and S. A. LANNING

*Department of Botany, University of Liverpool,
Liverpool, L69 3BX, England*

> *Level*: Advanced undergraduates
> *Subject area*: Microbial ecology
> *Special features*: Demonstrates a range of microbial responses to osmotic stress

INTRODUCTION

The effects of high solute concentrations on microbes may be due to the solutes themselves or to the effects of solutes on water activity (a_w). Everything that dissolves in water decreases the amount of "free water". Water activity, in the thermodynamic sense, provides a measure of "free water" and is defined by the equation;

$$a_w = \frac{P}{P_0} = \frac{n_2}{n_1 + n_2},$$

where P = the vapour pressure of the solution and P_0 that of the pure water; n_1 and n_2 are the number of moles of solute and solvent respectively. Thus as solute concentration increases, a_w decreases; for pure water a_w is 1·0.

Most microbes will tolerate only small decreases in a_w below 1·000. However, as with other environmental factors, there are notable exceptions of microbes which tolerate or even require reduced water activities. These include halophilic bacteria, xerophytic moulds and osmophilic yeasts (Ingram, 1957; Brown, 1976; Kushner, 1978). In some cases the tolerance of reduced a_w is the same irrespective of the solute (e.g. salts or sugars) concerned. In others, the reactions to the various solutes differ.

The aim of this experiment is to compare the a_w requirements, in culture, of microbes originating from different habitats. Reduced a_w is obtained by adding glucose or sodium chloride to the culture media.

EXPERIMENTAL

Day 1 *(2h)*

The experiment starts with two slope-cultures of 4 microbes:

> A: from an animal gut,
> B: from salted fish,
> C: from concentrated fruit juice,
> D: from dried prunes.
> (*See* notes 1 and 2.)

↓

Inoculate each microbe from the slopes into the following media, using one loopful of inoculum per tube (*see* note 3):

	Nutrient broth		
% Glucose (w/v)	a_w	%Nacl (w/v)	a_w
0	0·999	0	0·999
10	0·990	2.5	0·986
20	0·980	5	0·974
40	0·958	10	0·940
60	0·935	20	0·875
		30	0·800

In addition, inoculate microbe B into tubes of nutrient broth containing the above range of NaCl concentrations, but which also each contain 2% (w/v) $MgSO_4 \cdot 7H_2O$ (*see* note 5).

↓

Incubate cultures of A and B at 37° C for 7 days and those of C and D at 25° C for the same period.

- -

Day 2 *(1 h, 7 days later)*

With a nephelometer, measure the turbidities of cultures A and B, using a tube of the appropriate uninoculated medium to set the zero point.

↓

Use the same procedure with microbe C but **shake** the tubes before placing them in the nephelometer (avoid bubbles).

↓

Assess growth of microbe D visually or by nephelometry after maceration with a tissue grinder (*see* note 4).

↓

Present the results to compare the responses of each microbe to reduced a_w in the glucose and NaCl media.

↓

How far do the results reflect the normal habitats of these microbes?

Notes and Points to Watch

1. The microbes used are cultures of *Escherichia coli* (A), *Halobacterium salinarum* (B), *Saccharomyces rouxii* (C) and *Aspergillus repens* (D). Other cultures or isolates from these habitats with similar responses to a_w may be substituted.
2. Results will depend on the strains or species used. Nevertheless the differences in response to a_w should be clear cut.
3. If desired, replicate tubes for each a_w can be inoculated to provide information on variation in growth and as a precaution against loss of results due to contamination.
4. Nephelometry of mould macerates provides a rough, but useful, quantitative comparison.
5. Growth of *Halobacterium* is sometimes slow, so longer than 7 days' incubation may be needed. For optimum growth it requires Mg^{2+} ions.

MATERIALS

1. Strains suitable for this experiment are:
   Escherichia coli NCIB 9132
   Halobacterium salinarum NCMB 764
   Saccharomyces rouxii CBS 411
   Aspergillus repens CBS 119.36.
2. Media used for growth of these strains are as follows:
   Nutrient agar (Oxoid, CM3) for *Escherichia coli* (12 h at 30° C).
   Halobacterium medium consists of:

casamino acids (Difco),	0·75 % (w/v)
yeast extract (Difco),	1·0 %
trisodium citrate,	0·3 %
KCl,	0·2 %
$MgSO_4 \cdot 7H_2O$	2·0 %
Fe^{2+},	10·0 parts 10^{-6}
Mn^{2+},	0·1 parts 10^{-6}
NaCl,	4·3 M

   agar, 2·0 %
   pH adjusted to 7·4.
   Grow *Halobacterium* in this medium at 37° C for 14 days, in a container over saturated NaCl solution.
   The fungal medium contains:

malt extract (Oxoid L39),	3·5 %
glucose,	20·0 %
agar,	1·2 %

   pH adjusted to 7·0.
   Grow *Saccharomyces* and *Aspergillus* in this medium at 25° C for 7 days.
3. Equipment needed:
   nephelometer (e. g. Evans Electroselenium Ltd., Halstead, Essex);
   tissue grinder (e.g. Teflon tissue grinder, 10 ml capacity; Aimer Products, Ltd., Rochester Place, Camden Town, London).

SPECIFIC REQUIREMENTS

Day 1
2 slope cultures of:
   E. coli (A)
   H. salinarum (B)
   S. rouxii (C)
   A. repens (D)
9 tubes of nutrient broth (Oxoid CM1) (1 tube as uninoculated control for all media)

5 tubes each of nutrient broth + 10, 20, 40 and 60 % (w/v) glucose
5 tubes each of nutrient broth + 2·5, 5, 10, 20 and 30 % (w/v) NaCl
2 tubes each of nutrient broth + 2 % (w/v) $MgSO_4 \cdot 7H_2O$ + 2·5, 5, 10, 20 and 30 % (w/v) NaCl (for *Halobacterium*)

FURTHER INFORMATION

Typical Class Results

A typical set of results is given in Table 1. These show the tolerance ranges and optimum growth requirements for each microbe in the presence of increasing concentrations of glucose and salt.

Table 1 Effect of a_w on growth of four microbes from different environments.

		Nephelometer reading (%)			
	a_w	*E. coli* (A)	*H. salinarum* (B)	*S. rouxii* (C)	*A. repens* (D)
% glucose					
0	0·999	82	0	34	12
10	0·990	72	0	100	33
20	0·980	51	0	99	46
40	0·958	0	0	100	49
60	0·935	0	0	100	97
% NaCl					
0	0·999	100	0	47	33
2·5	0·986	85	0	50	30
5	0·974	35	0	55	36
10	0·940	20	0	50	35
20	0·875	0	97	0	40
30	0·800	0	41	0	0

Discussion Points

E. coli behaves like a "typical" microbe with reduced growth as a_w is lowered and none at lowest values. *H. salinarum* is an extremely "halophilic" bacterium. *S. rouxii* is tolerant of high sugar concentrations, but not the highest salt concentrations; it is an "osmophilic" yeast. *A. repens* grows better at higher sugar concentrations and tolerates most NaCl concentrations; it is a "xerophytic" mould.

REFERENCES

A. D. Brown (1976). Microbial water stress *Bacteriological Reviews*, **40**, 803–49.

M. Ingram (1957). Micro-organisms resisting high concentrations of sugars or salt. *Symposium Society of General Microbiology*, **7**, 90–133.

D. J. Kushner (1978). Life in high salt and solute concentrations: halophilic bacteria. *In* "Microbial Life in Extreme Environments", (D. J. Kushner, ed.), pp. 317–68. Academic Press, London and New York.

25

Estimation of the *D*-period in Exponentially Growing Cultures of *Escherichia coli*

R. K. POOLE

*Department of Microbiology, Queen Elizabeth College,
Campden Hill, London, W8 7AH, England*

> *Level*: All undergraduate years
> *Subject areas*: Microbial physiology and
> growth
> *Special features*: Extreme simplicity. Minimal
> facilities required. Illustrates cell
> cycle analysis without
> resort to synchronous cultures

INTRODUCTION

The most widely accepted model for the relationship of DNA replication to the cell cycle in *Escherichia coli* is that of Cooper and Helmstetter (1968). This model is based on the assumption that, in cultures with doubling times of less than 60 min, there are two constants during the cell cycle. One constant, *C* (of 40 min duration), is the time required for a replication fork to proceed from the origin to the terminus of the chromosome. The second, the *D*-period (of 20 min duration), is the interval between replication of the terminus and the succeeding cell division.

In slow-growing cultures, various patterns of DNA synthesis have been observed. Results from several laboratories demonstrated that the lengths of *C* and *D* were dependent on the generation time, with DNA synthesis occurring over most or all of the first two-thirds of the protracted cell cycle, and a *D* period that occupied the final third of the cycle. More recent evidence, reviewed by Kubitschek (1974), suggests that both the *C* and *D* periods are independent of generation time and that DNA synthesis invariably terminates near the end of the cell cycle.

This experiment is based on Kubitschek's (1974) approach that utilizes residual cell division as a simple means of examining the length of the *D*-period. Residual cell division was first observed when growing cultures were exposed to an inhibitor of DNA synthesis, such as nalidixic acid (Clark, 1968). In the presence of the inhibitor, division occurred only in those cells that had completed a round of DNA replication.

Because measurements of bacterial numbers are troublesome to make with a high degree of precision in undergraduate laboratories, culture turbidity, rather than cell numbers, is used to follow growth. However, when nalidixic acid is used as an inhibitor, growth in cell size continues in the absence of division, giving rise to short filaments and confusing analysis of the data. We have, therefore, used chloramphenicol as the inhibitor and found that the results obtained by either cell counting or turbidity measurements are very similar. The requirement for protein synthesis for replication of the terminal segment of the chromosome in *E. coli* was established by Marunouchi and Messer (1973).

The aim of this experiment is to determine, using the method of residual division, the length of the *D* period in exponential cultures of *E. coli* at different growth rates.

EXPERIMENTAL

The doubling time of the organism is varied by using media of different composition. Five suitable media are described later but others can be utilized. It is recommended that each student (or pair of students) works with one of these.

Day 1 *(15 min)*

Inoculate 20 ml of each medium with *E. coli* K12. Incubate with shaking at 37 °C overnight.

- -

Day 2 *(4–8 h)*

The time required from inoculation to completion of the experiment is dependent on growth rates. There will be a gap of about 2–4 h, again depending on growth rate, between steps (a) and (b).

(a) 15 min

Use a small volume of each overnight starter culture to inoculate two further flasks containing 20 ml of the same medium. Use 0·2 ml for medium 1, 0·5 ml for media 2 to 4, and 1·5 ml for medium 5. Shake at about 200 rev. min^{-1} and 37 °C and allow to grow to the exponential phase of growth (*see* note 1).

(b)

Record the absorbance or turbidity of each flask at 10–30 min intervals depending on the growth rate that the medium supports. The optical instrument should be zeroed before each reading with a suitable blank (e.g. distilled water in a test tube or "nephlos" flask). Between readings, plot the growth of each culture on semi-logarithmic graph paper.

↓

When both cultures are in the mid-exponential phase of growth, add chloramphenicol powder

(4 mg) to one flask (*see* note 2). Continue recording and plotting the growth of each culture until the turbidity of the chloramphenicol-treated culture is constant.

↓

Calculate *D* from the expression

$$D = \frac{(N/N_0)T}{0\cdot693} \qquad (see \text{ note 3})$$

where,

N = final number of cells when residual division is complete
N_0 = initial number of cells at time of adding inhibitor
T = generation time.

Notes and Points to Watch

1. Different strains of *E. coli*, with different or no additional nutritional requirements, will probably exhibit growth rates on the media described that are different from those obtained with strain A 1002, and shown in Figure 1. Consequently, the size of inocula required and the period between inoculation and the mid-exponential phase of growth will be different and it is strongly recommended that preliminary, trial experiments are conducted if an alternative strain is used.
2. It is essential that the inhibitor is added sufficiently early so that the control culture continues growing exponentially until residual division in the experimental culture is complete. Examples of satisfactory results are illustrated in Figure 1; *see also* Further Information.
3. When an inhibitor that permits division of only those cells that have completed a round of DNA replication is applied to a culture in the exponential growth phase, the fraction of cells capable of performing division, *R*, is related to the initial number of cells (N_0) at the instant of adding the inhibitor, to the final number of cells (N) when residual division is complete, and to the duration (D) of the period between termination of DNA replication and cell division by

$$\frac{N}{N_0} = 1 + R = 2^{D/T},$$

where *T* is the generation time.
Thus, *D* is given by $((N/N_0)T)/0\cdot693$.
The results shown in Table 1 and Figure 1 demonstrate the validity of replacing measurements of cell numbers (*N* and N_0) by those of turbidity.

MATERIALS

1. *Escherichia coli* K12, strain A1002 (K12 Y, *mel ilv⁻ lac* I⁻ *met* E⁻), an amino acid auxotroph, was used in the experiments described here, solely for reasons of ready availability in our own laboratory (*see* note 1).

2. Defined medium, each litre containing, the following:

K_2HPO_4,	4 g
KH_2PO_4,	1 g
NH_4Cl,	1 g

CaCl$_2$.2H$_2$O,   0·01 g
K$_2$SO$_4$,   2·6 g
trace element solution, 10 ml

The trace element solution is prepared by dissolving Na$_2$EDTA (5 g) in approx. 700 ml of water, adjusting the pH to between 7 and 8 with HCl and adding separately the following: FeCl$_3$. 6H$_2$O, 0·5 g; ZnO, 0·05 g; CuCl$_2$. 2H$_2$O, 0·01 g; CoNO$_3$. 6H$_2$O, 0·01 g; H$_3$BO$_3$, 0·01 g; (NH$_4$)$_6$Mo$_7$O$_{24}$. 4H$_2$O, 0·01 g. The pH should be maintained between 7 and 8 during additions. Make up to 1 litre and store at 0–5 °C.

Dissolve the medium components together, dispense in 20 ml portions to 250 ml "nephlos" flasks and autoclave at 103·4 kpa (15 lbf in^{-2}) for 10 min. When cool, add aseptically the following: a filter-sterilized solution of iso-leucine, valine and methionine, each at 10 mg ml^{-1}, 1 ml; 1 M MgCl$_2$ · 6H$_2$O, 0·02 ml; 10% solution of vitamin-free casamino acids (Difco), 0·2 ml (where indicated); 0·25 ml of one of the following solutions, each at 40% (w/v), glucose, glycerol, sodium succinate, sodium acetate (trihydrate).

The five media used contain the following carbon sources and amino-acid supplements:

Medium	Carbon source	Casamino acids
1	Glucose	+
2	Glucose	−
3	Sodium succinate	−
4	Glycerol	−
5	Sodium acetate	−

3. Colorimeter, nephelometer, or spectrophotometer suitable for "nephlos" flasks in use. The results illustrated here were obtained with an EEL Portable Colorimeter (Evans Electroselenium Ltd., Halstead, Essex, England) fitted with a green 404 filter. Ideally the instrument should be in a 37 °C constant temperature room that also houses the shaker.
4. "Nephlos" flasks (see Experiment 26, for further details).
5. D-Chloramphenicol, e.g. from Sigma (London) Chemical Company Ltd. or Chloromycetin Powder from Parke-Davis.

SPECIFIC REQUIREMENTS

These items are for each student (or pair of students) working with one medium.

Day 1
20 ml medium in 250 ml Erlenmeyer (or "neph-los") flask
stock culture of E. coli K12
shaker operating at 37 °C

Day 2
2 × 20 ml portions of medium in "nephlos" flasks
sterile pipettes (1–2 ml) for inoculation
shaker operating at 37 °C
chloramphenicol powder (4 mg)
semi-logarithmic graph paper
colorimeter, nephelometer or spectro-photometer

FURTHER INFORMATION

Kubitschek (1974) applied the method of residual division to E. coli B/r, E. coli K12 (AB1157) and Salmonella typhimurium. His results for E. coli K12, incorporated in Table 1, showed the mean D period measured at three growth rates to vary between 18·6 and 28·4 min. The mean D period for E. coli B/r was found to be similar (26·1 min) but was much shorter in S. typhimurium (13·3 min). It was concluded that for each organism the length of the D period was independent of growth rate.

Results obtained in our laboratory are shown in Table 1 and Figure 1. They differ from those of Kubitschek in suggesting that the length of the D period is related to the doubling time, T. It must

Table 1   Values of T, residual division (N/N_0) and D from cultures of *E. coli* K12 exposed to chloramphenicol (200 μg ml^{-1}). Data in the bottom three lines are from Kubitschek (1974).

Medium	T (min)	From turbidity measurements		From cell counts	
		N/N_0	D (min)	N/N_0	D (min)
1	30	1·27	10·3	1·20	7·8
2	42	1·20	11·1	1·17	9·3
3	48	1·18	11·6	1·18	11·3
4	66	1·22	18·7	1·27	22·5
5	160	1·18	35·8	1·17	33·4
—	79	—	—	—	28·4 ± 16·3
—	125	—	—	—	18·6 ± 7·5
—	327	—	—	—	24·2 ± 5·8

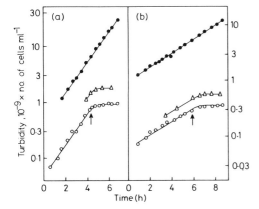

Figure 1   Kinetics of cell growth after adding chloramphenicol to exponentially-growing cultures of *E. coli* K 12, strain A 1002. In (a) the carbon source was glycerol (doubling time 66 min); in (b) it was sodium acetate (160 min). At the times indicated by the arrows, *D*-chloramphenicol (200 μg ml^{-1}) was added to one flask of each medium (open symbols) whilst a second control flask (closed symbols) received no inhibitor. Culture growth was monitored either by turbidity (●, ○) or by counting cells in small samples of culture using a Coulter Counter Model Z_{BI} (△). Values of turbidity of the control flasks have been multiplied by 10 for clarity. Values of T, N/N_0 and D from these experiments are presented in Table 1.

be admitted, however, that there is considerable scatter in the measured values. There is continuing disagreement in the literature regarding the patterns of chromosome replication in the cell cycle of *E. coli*, especially in slow-growing cultures. One explanation for this is that different strains have been used in various laboratories; some support for this viewpoint has been presented by Helmstetter and Pierucci (1976) who used three sub-strains of *E. coli* B/r and found two different replication patterns. A recent compilation of relevant literature is given by Kubitschek and Newman (1978). Of greater importance in the present report, however, is that measurements of culture turbidity, which are easy to make in the undergraduate laboratory, gave values for the D periods very similar to those obtained from cell counts. In agreement with the findings of Kubitschek (1974), we have found cell size distributions and mean cell volumes to remain almost unchanged in samples taken from the plateau regions of these cultures. Thus, culture turbidity is a valid estimate of cell numbers in this situation.

We have tried extending this simple technique to the inhibitor of DNA synthesis, nalidixic acid. Although residual division can be measured by cell counting and used to estimate the D period, the measure of culture turbidity gives misleading results due to continuing cell growth (filamentation) in the absence of division.

REFERENCES

D. J. Clark (1968). The regulation of DNA replication and cell division in *E. coli* B/r. *Cold Spring Harbor Symposia on Quantitative Biology*, **33**, 823–36.

S. Cooper and C. E. Helmstetter (1968). Chromosome replication and the division cycle in *Escherichia coli* B/r. *Journal of Molecular Biology*, **31**, 519–40.

C. E. Helmstetter and O. Pierucci (1976). DNA synthesis during the division cycle of three substrains of *Escherichia coli* B/r. *Journal of Molecular Biology*, **102**, 477–86.

H. E. Kubitschek (1974). Estimation of the *D* period from residual division after exposure of exponential phase bacteria to chloramphenicol. *Molecular and General Genetics*, **135**, 123–30.

H. E. Kubitschek and C. N. Newman (1978). Chromosome replication during the division cycle in slowly-growing steady-state cultures of three *Escherichia coli* B/r strains. *Journal of Bacteriology*, **136**, 179–90.

T. Marunouchi and W. Messer (1973). Replication of a specific terminal chromosome segment in *Escherichia coli* which is required for cell division. *Journal of Molecular Biology*, **78**, 211–28.

26

Preparation and Analysis of Synchronous Cultures of a Fission Yeast

R. K. POOLE

*Department of Microbiology, Queen Elizabeth College,
Campden Hill, London, W8 7AH, England*

> *Level*: All undergraduate years
> *Subject area*: Microbial physiology and growth
> *Special features*: Use of density gradients; illustrates
> the important distinction between
> culture turbidity and cell numbers
> in monitoring culture growth

INTRODUCTION

Schizosaccharomyces pombe is a fission yeast, having a simple cell cycle in which cell division occurs by binary fission. Some principal "landmarks" in the cell cycle are shown in Figure 1. The cycle is deemed to start with the division of a binucleate "mother" cell into two approximately equal daughters. Growth in cell volume is almost linear throughout the first three-quarters of the cycle. The attainment of a doubling in size at 0·75 of the cycle is coincident with nuclear division. Shortly after, a cell plate is laid down between the two nuclei indicating the site of subsequent septum formation. Throughout the cycle, cell diameter remains almost constant so that a measure of cell length is a good measure of cell volume and thus the age of the cell in the cycle. The S-phase (period of DNA synthesis) is centred at cell division.

The most powerful technique for analysing these events and others in the cell division cycle involves the preparation of synchronous cultures, i.e. cultures in which all the cells progress through the cycle and divide more or less synchronously. One method for preparing such cultures is by size-selection. This involves the removal, from an exponentially growing (asynchronous) culture, of a small population of cells, fairly homogeneous with respect to size and, therefore, age in the cell cycle. When resuspended in fresh medium this population constitutes a synchronous culture. The method described here utilizes a rate (velocity) sedimentation in a linear sucrose gradient and is a modification of the method of Mitchison and Vincent (1965).

The aims of this experiment are to familiarize the student with a centrifugal density gradient method for separating cells and to investigate the relationship between cell numbers and culture turbidity. It also provides valuable practice with microscopy and turbidimetric measurements of growth.

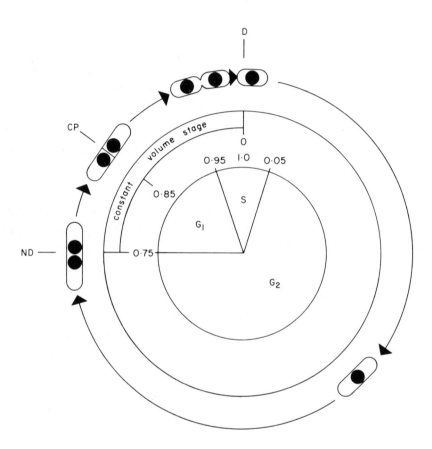

Figure 1   The cell cycle of *S. pombe*. Figures refer to that fraction of the cycle elapsed since the previous cell division (D). ND denotes the timing of nuclear division and CP that of formation of the cell plate. Data from various sources (*see* Mitchison, 1970; Mitchison and Creanor, 1971).

EXPERIMENTAL

Day 1 (*5 min; mid-day*)

Inoculate 200 ml of medium (in a 1 litre Erlenmeyer flask) with the cells from an agar slant culture. Incubate with shaking at 30 °C.

Day 2 *(10 min)*

Two 1 litre flasks, each containing 200 ml of medium, are required per group of students. Inoculate each with 1·0 ml of culture from Day 1.

Day 3 *(6 h)*

The following procedure is to be adopted by each group of students (perhaps 2 or 3 per group).

Clamp a 50 ml centrifuge tube vertically and pipette 6 ml of the densest sucrose solution (40 %, w/w) into the bottom. Do not allow the sucrose to touch the walls of the tube.

↓

Using a 10 ml hypodermic syringe fitted with a long needle bent into a U-shape, carefully layer 6 ml of the next sucrose solution (34 %, w/w). Keep the upward-pointing needle tip level with the top of the applied band (*see* note 1).

↓

Layer successively lighter solutions until all six 6 ml layers have been applied. (When done carefully, the boundaries between the layers will be clearly visible at first, but will disappear as the steps diffuse to form a linear gradient.)

↓

Leave undisturbed while the cells are harvested.

Use an exponentially growing culture about 20 h after inoculation when the absorbance at 600 nm (1 cm cuvette) is about 1·0 (*see* note 2).

↓

Pass one culture (200 ml) through a membrane filter and holder by reduced pressure.

↓

Using forceps, remove the filter with cells and transfer to a McCartney bottle.

↓

Insert a new filter in the filtration apparatus and harvest cells from the second flask in the same way, transferring the second filter to a second McCartney bottle.

↓

Add 2 ml of culture filtrate to one bottle, "Whirlimix" to resuspend the cells and then transfer the suspension to the second bottle and resuspend the cells in the same way. Mix well.

Use a 5 ml syringe and U-shaped needle to layer the remaining suspension on top of the density gradient. Use the syringe calibrations to estimate volume applied.

Pipette out 0·5 ml of this suspension into a test tube containing one small drop of formalin and retain. Later, dilute about 400-fold with distilled water. Determine the cell count and cell plate index and describe the cell morphology as outlined below.

↓

Using a glass rod, immediately mix the suspension into the top of the gradient to give a band 2–3 times broader than that applied (*see* note 3).

↓

Carefully transfer the tube to a balance and prepare a tube containing only water (or sucrose) of equal weight.

↓

Centrifuge both tubes in a swinging-bucket rotor of a bench centrifuge (*see* note 4) until the trailing edge of the band of cells has moved between one-third and one-half the tube length (typically 3 min at half-maximal centrifuge speed).

↓

Remove the tube and return it to the clamp. Use a fresh syringe and a long straight needle. Immerse the tip of the needle to a point about 2 mm below the upper edge of the band of cells and, holding the syringe stationary, withdraw about 0·5 ml of the gradient (*see* notes 5 and 6).

↓

Inoculate immediately into a nephlos flask (*see* note 7) containing 20 ml of fresh medium at 30 °C. Start a clock, then incubate the culture on a shaker at the same temperature.

↓

At about 20 min intervals, perform the following procedure.

(*a*) Record the culture turbidity using a spectrophotometer, colorimeter or nephelometer appropriate to the nephlos flask (*see* note 7). Zero the instrument before each reading with a water blank.

(*b*) Using a 1 ml pipette, remove a sample (0·5 ml) and transfer to a small test tube containing one small drop (about 25 μl) of formalin. "Whirlimix" well and count the number of cells in a drop of this suspension using a microscope counting chamber (*see* note 8).

Take another drop of the formalin-fixed sample, mount on a plain slide under a coverslip and examine the cells using a high power (perhaps oil immersion) objective. Make drawings of cell morphology and, in randomly selected fields, determine the number of cells that exhibit cell plates. Express as a percentage of the total number of cells present (the cell plate index).

_____

Notes and Points to Watch

1. The U-shaped syringe needle allows gradient layers and the sample to be **floated on** existing layers rather than **squirted into** lower layers. Nevertheless, care is required.
2. It is essential that cultures used are in the mid or late exponential phases of growth. As the stationary phase is approached, cells tend to accumulate in certain cell cycle phases and the size distribution is not that of an exponential culture.
3. Gentle mixing of the loaded cells with the top of the gradient column tends to produce an inverse gradient of cells and reduces local instability of the sample zone. (For a survey of fractionation methodology, *see* Lloyd and Poole, 1979.) Instability is manifest as "streaming"—the bulk movement of cells, often in clumps, into the gradient before centrifugation. In this experiment, in which only the cells which sediment the slowest are selected, the lack of resolution that results from streaming is generally unimportant, but the yield of small cells may be drastically reduced.
4. The duration and speed of centrifugation is critical. We use an MSE "Super Minor" (round) bench centrifuge at room temperature. The 4-place swinging bucket rotor provides an average radius (measured from the axis of rotation to the mid-point of the bucket in the horizontal position) of about 11 cm. Suitable tubes are MSE clear polycarbonate tubes (ref. no. 5912 6 0872) of outside dimensions 11×2.9 cm. They are nominally of 50 ml capacity, but 36 ml of gradient plus 2 ml of sample is an appropriate volume to use to avoid spillage. The conditions described previously, 3 min centrifugation at half maximal speed (measured on an arbitrarily calibrated dial), may need to be modified depending on room temperature, geometry of the swinging bucket rotor and other factors. It is essential for the instructor to conduct a trial run to establish suitable conditions.
5. This organism rapidly ferments sucrose. Care should be taken that the cells are selected as soon as possible after centrifugation and before bubbles of CO_2 disturb the distribution of cells in the gradient.
6. By far the most common cause of disappointing results is the removal from the gradient of too large (and thus too heterogeneous) a sample. Only about 5 % of the cells loaded should be selected.
7. We use the term "nephlos" flask to describe an Erlenmeyer flask to which has been attached a test tube side-arm (Ellinghausen, 1959). The tube can be selected to be suitable for use in the colorimeter, nephelometer or spectrophotometer available. The results shown here were obtained with a Pye-Unicam SP600 spectrophotometer (with carriage for tubes, rather than cuvettes) and nephlos flasks manufactured from 250 ml Erlenmeyer flasks fitted with 15×1.2 cm Pyrex test tubes as side-arms. Bellco Nephelo Culture Flasks are obtainable from Arnold R. Horwell Ltd., 2 Grangeway, Kilburn High Road, London NW6 2BP.
8. A counting chamber with improved Neubauer rulings is recommended (Hawksley, Lancing, Sussex). It is important to distinguish between cells that have divided but not separated and those that have not divided. An arbitrary, but useful, rule is to count as two any cell that shows marked constrictions in its equatorial plane. Mitchison (1970) offers valuable advice in this and other aspects of handling *S. pombe*.

MATERIALS

1. *Schizosaccharomyces pombe* 972 h⁻ (ATCC 26189).
2. Bench centrifuge with swinging-bucket rotor for tubes of about 50 ml capacity.
3. Spectrophotometer, colorimeter or nephelometer with "nephlos" flasks.
4. Defined medium, prepared as described below (modified from Mitchison, 1970).
   For 1 litre of medium:
   (a) Dissolve 5 g $(NH_4)_2SO_4$ in 770 ml distilled water.
   (b) Add 100 ml of a salts solution that contains the following:

Na acetate.$3H_2O$,	17·6 g	in 1 litre
KCl,	10 g	(do not
$MgCl_2.6H_2O$,	10·8 g	autoclave;
$NaH_2PO_4.2H_2O$,	0·132 g	store at
$CaCl_2.2H_2O$,	0·132 g	4 °C)

   (c) Add 10 ml of a trace elements solution that contains the following:

H_3BO_3,	50 mg
$MnSO_4.4H_2O$,	52·8 mg
$ZnSO_4.7H_2O$,	40 mg
$FeCl_3.6H_2O$,	20 mg
$H_2MoO_4.H_2O$,	16 mg
KI,	10 mg
$CuSO_4.5H_2O$,	4 mg
citric acid,	100 mg

   in 1 litre (do not autoclave; store at 4 °C)
   (d) Autoclave the above (a)–(c) together at 103·4 kPa (15 lbf in⁻²) for 15 min. When cooled, add aseptically the following sterile solutions:
   (e) 100 ml of a 10 % (w/v) solution of glucose.
   (f) 10 ml of a 4 % (w/v) solution of $NaH_2PO_4.2H_2O$.
   (g) 10 ml of a vitamin solution that contains the following:

inositol	1·0 g
nicotinic acid	1·0 g
calcium	
pantothenate	0·10 g
biotin	1·0 mg

   in 1 litre
   Dissolve the biotin separately in 1 ml 50 % (v/v) ethanol and then add to the solution of the other 3 components. Sterilize by membrane filtration and store at 0–4 °C.
   The final pH is about 5·5 and does not require adjusting.

SPECIFIC REQUIREMENTS

Day 1
200 ml growth medium in 1 litre Erlenmeyer flask
agar slant culture of *S. pombe*
orbital or reciprocating shaker (30 °C) with capacity for 1 litre flasks

Day 2
culture from Day 1
per group of students: 2 × 200 ml growth medium in 1 litre Erlenmeyer flasks
shaker as for Day 1

Day 3
Items are for each student group.
400 ml of exponentially-growing culture of *S. pombe* from Day 2
10 ml each of the following aqueous sucrose solutions:
10 %, 16 %, 22 %, 28 %, 34 % and 40 % (all w/w), which need not be sterile
retort stand and clamp

5 and 10 ml hypodermic syringes with long needles (19 gauge), one needle to be bent into a "U"-shape at the bottom
2 × 50 ml centrifuge tubes (*see* note 4)
1 membrane filter assembly with 2 membrane filters (0·45 µm pore size, at least 50 mm diam.)
500 ml Buchner flask for filter assembly
2 wide-necked McCartney bottles
forceps
Whirlimixer, Rotamixer or similar
pipettes, including 1 ml sterile pipettes
formalin, either in dropper bottles or with dispensing glass rod
"nephlos" flask containing a volume of medium similar to that of the side-arm (e.g. 20 ml)
16–20 test tubes
tube rack
distilled water
counting chambers (*see* note 8)
clock

FURTHER INFORMATION

This technique has been applied to a wide range of cell types with satisfactory results (Mitchison, 1971). Typical class results are shown in Figure 2. Selection methods such as this are generally agreed to reflect most accurately the time sequence of events occurring in the cell cycle of an "unperturbed" culture. An advantage of the method is the ease with which control experiments can be performed to study the effects of exposure to potentially perturbing conditions such as the presence of gradient media, centrifugation and perhaps prolonged anaerobiosis. Such controls involve mixing the entire contents of the gradient after centrifugation, and inoculating into fresh medium. Division synchrony has not been reported in cultures of this type (Kubitschek, 1968), but recent disturbing results indicate that, for *S. pombe*, fluctuations in enzyme activity (Mitchison, 1977) or adenylate pools (Poole and Salmon, 1978) occur, even in the absence of division synchrony, after gradient selection (Mitchison, 1977) or merely short centrifugation and resuspension in growth medium (Poole and Salmon, 1978). Suitable controls could readily be incorporated into the experiment described here (Figure 2).

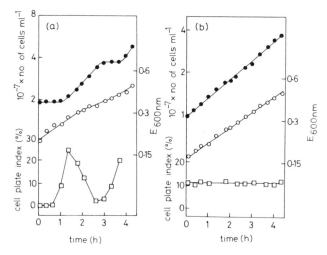

Figure 2 Growth of *S. pombe* after density gradient centrifugation. Cells (about 1.6×10^{10}) harvested from an exponential culture were layered on a sucrose gradient and centrifuged as described in the text. About 1 ml (containing 3 % of the loaded cells) from the top of the separated band was removed and inoculated into 20 ml of fresh medium (a). The remaining gradient and cells were mixed well and a portion (0.5 ml) inoculated into a second flask of medium (b). Cell numbers (●), culture turbidity (○) and cell plate index (□) are shown.

Students should be encouraged to consider the following questions in assessing their data:

What is the importance/unimportance of aseptic technique during the experiment?

How perfect is the "synchronous" increase in cell numbers? What are the reasons for imperfections?

What is the relationship between cell numbers and turbidity? Can turbidity be used to monitor changes in cell numbers? If not, of what is it a measure?

What is the cell plate index in the original culture and how does this compare with the values obtained during synchronous growth?

REFERENCES

H. C. Ellinghausen (1959). Nephelometry and a nephelo-culture flask used in measuring growth of leptospires. *American Journal of Veterinary Research*, **20**, 1072–6.

H. E. Kubitschek (1968). Linear cell growth in *Escherichia coli*. *Biophysical Journal*, **8**, 792–804.

D. Lloyd and R. K. Poole (1979) Subcellular fractionation: isolation and characterization of organelles. *In* "Techniques in the Life Sciences, Biochemistry", vol. B2/1, "Techniques in Metabolic Research" (H. L. Kornberg, ed.). Elsevier/North Holland Biomedical Press, B202/1–B202/27.

J. M. Mitchison (1970). Physiological and cytological methods for *Schizosaccharomyces pombe*. *In* "Methods in Cell Physiology", (D. M. Prescott, ed.), Vol. 4, pp. 131–65. Academic Press, London and New York.

J. M. Mitchison (1971). "The Biology of the Cell Cycle", pp. 25–57. Cambridge University Press.

J. M. Mitchison (1977). Enzyme synthesis during the cell cycle. *In* "Cell Differentiation in Micro-organisms, Plants and Animals", (L. Nover and K. Mothes, eds.), pp. 377–401, North Holland, Amsterdam.

J. M. Mitchison and J. Creanor (1971). Further measurements of DNA synthesis and enzyme potential during cell cycle of fission yeast *Schizosaccharomyces pombe*. *Experimental Cell Research*, **69**, 244–7.

J. M. Mitchison and W. S. Vincent (1965). Preparation of synchronous cell cultures by sedimentation. *Nature, London*, **205**, 987–9.

R. K. Poole and I. Salmon (1978). The pool sizes of adenine nucleotides in exponentially-growing, stationary phase and 2′-deoxyadenosine-synchronized cultures of *Schizosaccharomyces pombe* 972h$^-$. *Journal of General Microbiology*, **106**, 153–64.

Amoeba-to-Flagellate Transformation by *Naegleria gruberi*

ELIZABETH U. CANNING

Department of Zoology, Imperial College, London, England

and

A. W. BARK

Department of Applied Biology, Chelsea College, London, England

> *Level*: Beginning undergraduates
> *Subject area*: Cell biology, general microbiology
> *Special features*: Good, reliable model for studying phenotypic change

INTRODUCTION

Amoebae may be found in any habitat where there is moisture and organic matter. Most of them feed on bacteria and, wherever bacterial populations develop, so too will populations of amoebae. Many amoebae form cysts, which enable them to survive desiccation and this, coupled with their rapid rate of growth and division, allows them to colonize habitats that are only transitorily damp, such as soil and even bare rocks or tree trunks. Cultivated soil typically contains between 10^4 and 10^6 amoebae per gram.

Soil amoebae are much smaller than the familiar *Amoeba proteus*, most of them measuring only 10–40 µm. The four commonest genera are *Naegleria*, *Acanthamoeba*, *Hartmannella* and *Vahlkampfia*. Under optimum conditions these have a generation time of 2–3 h. *Naegleria* is one of several types of soil organism which can exist in an amoeboid or a flagellated phase but is dominantly amoeboid whereas others may be dominantly flagellated.

Naegleria gruberi (Figure 1) is an excellent experimental organism for studying cell transformation as it is easy to cultivate and handle and "performs" reliably. Its normal form is a small monopodial amoeba, about 25 µm in length. Under adverse conditions, such as drought or starvation, it will round up and secrete a tough double-walled cyst around itself. The cyst has a number of plugged pores and, when favourable conditions return, the amoeba will emerge through one of them.

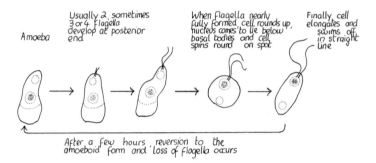

Figure 1   Flagellate transformation in naegleria.

In response to dilution of the medium in which they are living, the amoeboid forms will transform into flagellates. The flagellates are ovoid, with two anterior flagella, and measure about 17 μm with a length:breadth ratio of about 2. They are good swimmers, progressing at 90–140 μm s^{-1} at room temperature, many times faster than the amoebae, and may therefore be useful in dispersal (Fulton, 1977). The flagellate phase is only transitory, no feeding or division take place and the organism reverts to the amoeboid form, usually within a few hours.

The aim of this experiment is (a) to illustrate the potential for dramatic phenotypic change in cells in a system where the changes occur so rapidly that they may be observed directly, and (b) to indicate the close taxonomic relationship of amoebae (Sarcodina) and flagellates (Mastigophora). **Warning**—another species of *Naegleria*, *N. fowleri* causes primary amoebic meningo-encephalitis, a rare but fatal disease of man. It is morphologically and physiologically distinct from *N. gruberi*. Of particular importance is the tolerance of *N. fowleri* of temperatures higher than those reached in fever, from which a patient would survive. *N. fowleri* gains access to the brain and spinal cord by the olfactory route after entry of contaminated water into the nostrils, e.g. when swimming. Although *N. fowleri* is very rare in soil, it is recommended that the following experiment be carried out with established cultures of *N. gruberi* (*see* Materials), rather than by attempting to isolate *N. gruberi* from soil.

EXPERIMENTAL

The use of exponential phase cultures is essential since amoebae encyst rapidly on reaching the stationary phase. The experiment could be conducted over one or two days, depending on whether the exponential phase cultures of *N. gruberi* are provided for the class or set up by the students. If the work for Day 2 is set up in the morning, the reversion (flagellate to amoeba transformation) may be observed later in the day.

Day 1 *(30 min)*

A group of 4 or 5 students is provided with a spread plate culture of *Escherichia coli* on nutrient agar.

Harvest the *E. coli* by adding 2–3 ml of sterile water to the culture and bring all the cells into suspension by thoroughly rubbing the agar surface with a loop (*see* note 2).

↓

Using a sterile pipette dispense **all** the *E. coli* suspension between 5 amoeba saline agar plates and distribute the bacteria evenly using a sterile loop or glass spreader. This gives a suitable density of bacteria for the amoebae (*see note* 1).

↓

Add 2–3 ml of sterile water to a stock plate culture of *N. gruberi* and make a suspension of cells by thoroughly rubbing the agar surface with a loop. (*see* note 2).

↓

Inoculate each amoeba saline plate with 3 or 4 drops of *N. gruberi* suspension and distribute evenly (*see* note 2).

↓

Incubate these plates at 30°C for 24 h.

Day 2 *(10 min to set up; 5 min to examine preparations after 1 h and every 15 min for another hour; a further 5 min towards the end of the day).*

Students now work alone.

Add 2 or 3 ml of distilled water to one of the plates of *N. gruberi* set up on Day 1 and thoroughly rub the surface with a loop to suspend the amoebae.

↓

Transfer the amoeba-suspension to several cavity slides and place the slides in a moist chamber to prevent evaporation. Do not use coverslips at this stage as transformation may be inhibited. Note the time.

↓

To observe the amoebae, cover one of the slides with a coverslip and examine immediately at × 400, preferably with a phase contrast objective. The nucleus, contractile vacuole, ectoplasm, pseudo-podium and monopodial movement should be obvious. Discard this slide after examination.

↓

Scan the other slides at × 100 after 1 h and at intervals of 15 min thereafter, replacing them in the moist chamber between examinations. No coverslips are necessary when examinations are made at low power. The first flagellates normally appear 1–2 h after dilution of the medium (on rare occasions this may take 3–4 h).

↓

Add a coverslip to a slide in which a few flagellates have been observed and examine at higher magnifications with phase contrast (if possible). In addition to actively swimming flagellates the majority of amoebae should exhibit various stages of flagellar growth as in Figure 2. Note the position of the nucleus and the polarity of the organism as indicated by the flagella and pseudopodium.

↓

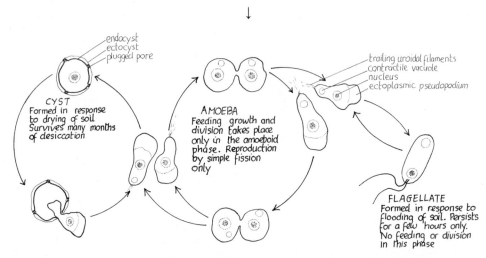

Figure 2   Life cycle of *Naegleria*

Estimate and record the relative percentages of flagellates and amoebae present at 15 min intervals, after the first flagellates are seen.

↓

After some hours reversion from the flagellates to the amoebae will automatically take place by resorption of the flagella and resumption of amoeboid movement. The flagellates may be induced to undergo amoeboid movement at any time by the application of pressure. For instance, the pressure caused by the coverslip as a slide preparation evaporates will cause the cells to become amoeboid while retaining their flagella. Release of the pressure by adding more water to the slide will allow the cells to recommence swimming.

Day 3 *(optional, 30 min–1 h)*

Any *N. gruberi* cultures not used on Day 2 will now contain cysts. Excystment of the amoebae can be induced readily. Make a suspension of cysts in distilled water. Examine under a coverslip, if necessary sealing the edges with vaseline to prevent evaporation. Examine, preferably with phase contrast, at × 400 magnification to observe the emergence of the amoebae through the pores. Excystment usually occurs within 30 min.

Notes and Points to Watch

1. It is essential for the success of the experiment to have a good growth of amoebae in the exponential phase of culture. The instructions for Day 1 are important.
2. The cultures of amoebae should be set up on Day 1 using aseptic technique.

MATERIALS

1. *Naegleria gruberi* culture (The Culture Centre of Algae and Protozoa, 36 Storey's Way, Cambridge, CB3 ODT, England). Stock cultures can be maintained on amoeba saline agar (*see* note 3 below) with a lawn of *E. coli*. These plates are made up and inoculated as specified for Day 1 (Experimental). *N. gruberi* cysts remain viable for several months, whether wet or dry, so that subculturing is only necessary every 3 months or so. Once a culture has been purchased the organism is easy to maintain. The use of agar slopes in screw-cap bottles, rather than agar plates provides protection against contamination over a long period.

2. *Escherichia coli* K12 (or any other non-pathogenic strain of *E. coli*) grown on nutrient agar.

3. Amoeba-saline plates, containing agar, KH_2PO_4, Na_2HPO_4, NaCl, $CaCl_2$, $MgSO_4$, distilled water. Some agars do not support reliable transformation. Lab M Agar No. 1 is recommended (London Analytical and Bacteriological Media Ltd., Adelaide House, London Bridge, London EC4R 9DR, England). In view of the small weights of chemicals involved, it is simpler to make up concentrated stock solutions of the individual reagents and use as follows:

   2·0 ml of 6·8% KH_2PO_4
   10 ml of 1·42% Na_2HPO_4
   10 ml of 1·2% NaCl
   1·0 ml of 0·4% $CaCl_2.2H_2O$
   1 ml of 0·4% $MgSO_4.7H_2O$
   agar 15 g
   distilled H_2O to 1 litre.

4. Incubator at 30°C.

5. Moist chamber. A moist chamber consists of a large Petri dish with 2 or 3 sheets of wet filter paper in the bottom. A U-shaped glass rod is placed on the filter paper and slides are laid across the rod. Stored in this way, a few drops of water on a cavity slide will not evaporate for several days.

SPECIFIC REQUIREMENTS

Day 1
Per group of 4 or 5 students:
1 spread plate culture of *E. coli* on nutrient agar
   (24 h at 37°C)
1 stock plate culture of *N. gruberi*
5 amoeba-saline agar plates
5 sterile Pasteur pipettes
glass spreader and alcohol
per class:
1 incubator at 30°C

Day 2
Per student:
1 culture of *N. gruberi* set up on Day 1

distilled water
6 cavity slides
coverslips
Pasteur pipettes
moist chamber

Day 3
cultures of *N. gruberi* set up on Day 1
distilled water
microscope slides (not cavity slides)
coverslips
Pasteur pipettes
vaseline

FURTHER INFORMATION

Fulton (1977) has reviewed the events of amoeba to flagellate transformation in *Naegleria*. A combination of several stimuli, some acting synergistically, were found to initiate transformation of axenically grown amoebae in liquid medium. These were: lowering of elecrolyte concentration, lowering of temperature, increasing mechanical agitation and lowering the concentration of a low molecular weight "factor" which could be extracted from yeast. The stimuli controlling transformation of amoebae grown on plates with bacteria were more difficult to determine, but it is unlikely that the stimulus is a simple flow of water and ions, as has been inferred from the occurrence of transformation when amoebae are transferred from an agar to an aqueous solution.

When transformation is induced, there is a period of about 50 min when the amoebae are active, but no morphological changes are obvious. Two basal bodies, which are not present in amoebae, are then organized just beneath the plasmalemma and the nucleus moves to a fixed position beneath the basal bodies. The flagellar axonemes are assembled on the basal bodies and, at 60 min, the external flagella are visible. These grow to their full length by about 90 min. A periodically banded structure, the rhizoplast, the function of which is not clear, is also assembled as the flagella are formed: the rhizoplast passes from the basal bodies alongside the nucleus and terminates free in the cytoplasm.

During formation of the flagella the amoebae round up and begin to spin. When the flagella are fully formed, the cells elongate and swim actively. It is noteworthy that a change of polarity takes place at transformation: during reversion, as the flagella are resorbed, pseudopodia become active at the opposite end to the flagella. Publications showing electron micrographs of the amoebae and the flagellates of *Naegleria* on display in the laboratory would greatly enhance the benefits of the practical work (e.g. Schuster, 1963; Dingle and Fulton, 1966).

Representatives of quite diverse groups of unicellular organisms, while being dominantly flagellates, have amoeboid tendencies. Many, which lack a definite pellicle or cell wall, including some photosynthetic species, produce pseudopodia and feed in an amoeboid fashion by food cup formation, while retaining their flagella. Others may lose their flagella and metamorphose completely. Conversely, predominantly amoeboid organisms not related to *Naegleria*, e.g. the Myxomycetes or Mycetozoa, can also exist as flagellates. Other amoeboid groups, e.g. Foraminifera, have flagellated gametes. Yet others are permanently flagellated amoeboid cells. There is thus no clear-cut distinction between flagellates and amoebae and for this reason, modern taxonomic schemes in animal biology classify the two types of organism together as a single phylum, the Sarcomastigophora.

REFERENCES

A. D. Dingle and C. Fulton (1966). Development of the flagellar apparatus of *Naegleria*. *Journal of Cell Biology*, **31**, 43–54.

C. Fulton (1977). Cell differentiation in *Naegleria gruberi*. *Annual Review of Microbiology*, **31**, 597–629.

F. Schuster (1963). An electron microscope study of the amoeba-flagellate *Naegleria gruberi* (Schardinger) I. The amoeboid and flagellate stages. *Journal of Protozoology*, **10**, 297–313.

28

Endospore Formation in *Bacillus subtilis*

J. G. COOTE

Department of Microbiology, University of Glasgow,
Garscube Estate, Bearsden, Glasgow, G61 1QH, Scotland

> *Level*: Advanced undergraduates
> *Subject area*: Bacterial physiology
> *Special features*: Illustrates the sequence of events
> during bacterial sporulation

INTRODUCTION

Many micro-organisms have the ability to undergo physiological and morphological changes which result in the formation of dormant structures (usually as a response to starvation conditions). These structures include the bacterial endospores which are readily recognized microscopically by their intracellular location and high refractility. The development of the endospore involves the formation of a new type of cell within the existing vegetative cell. As this new cell possesses markedly different structural and biochemical characteristics, the process may be regarded as a primitive form of cellular differentiation.

Formation of the spore occurs by a time-ordered sequence of events brought about to a considerable extent by the sequential activation of sporulation-specific genes which remain repressed during vegetative growth. As a result, biochemical properties characteristic of the developing spore appear at regular intervals during the process and, by assaying some of these properties, the temporal sequence of events during sporulation can be monitored. In this experiment an increase in two enzymic activities, *exo*-protease and alkaline phosphatase, will be followed during sporulation together with an increase in resistance to heat and a surface-active agent, *n*-octanol. Resistance to heat and such agents are characteristic properties of spores and reflect the structural and metabolic alterations which occur during their maturation.

The aim of this experiment is to follow endospore formation in *Bacillus subtilis* 168 by monitoring the sequential appearance of various marker events characteristic of the process. Bacteria are allowed to grow exponentially in a rich broth medium and sporulation is initiated when the cells enter the stationary phase of growth. At this point the medium will have become depleted of an essential carbon or nitrogen source and the cells, faced with starvation of an essential growth requirement, will channel their metabolism towards spore formation rather than growth.

EXPERIMENTAL

B. subtilis 168 is grown overnight on nutrient agar before inoculation into broth medium on the day of the experiment (*see* note 1). Growth of the organism and the subsequent formation of spores will take the whole of a normal working day even at an incubation temperature of 42°C. The organism should be inoculated no later than 0830 to 0900 hours in order to finish the experiment by 1630 to 1700 hours. In addition, the experiment should be conducted by a group of 3 to 4 students to spread the work load throughout the first day.

Day 1 (*regular intervals all day*)

Take 60 ml culture of *B. subtilis* shaken at 42°C (*see* note 2).

↓

At 30 min intervals record the extinction at 600 nm of a portion of the culture (*see* note 3). Plot the readings on semi-log graph paper as they are obtained.

↓

When it is clear that the cells are growing exponentially, note the E_{600} at this time and take samples for assay of *exo*-protease, alkaline phosphatase, octanol-resistance and heat-resistance (*see* note 4). The assays during the growth phase provide control basal activities for comparison with samples taken during the stationary phase.

↓

Eventually, exponential growth will cease and the cells will enter the stationary phase. Note the time that this occurs, as this time (*t*-zero) is taken as the start of spore formation (*see* note 5). Continue to measure the E_{600} of the culture at hourly intervals for the next 4 h.

↓

Take samples for assay at the following times after *t*-zero:

at $\frac{1}{2}$, 1, 1$\frac{1}{2}$, 2 and 3 h for *exo*-protease activity;

↓

at 1, 2, 3 and 4 h for alkaline phosphatase activity;

↓

at 2$\frac{1}{2}$ h for a total cell viable count (*see* note 6);

↓

at 3, 4 and 4$\frac{1}{2}$ h (or 5 if there is time) for octanol-resistance and heat-resistance;

↓

at intervals after 3 h for refractile spores within the cells.

Assay Procedures

Exo-*protease* (see *note* 7)   The substrate for the enzyme, Hide powder azure, is an insoluble complex. The action of the protease releases a soluble blue dye which can be monitored in the spectrophotometer.

Pipette a 1·0 ml cell sample into a tube containing Hide powder azure. Add 2·5 ml imidazole buffer (pH 7·2) and 0·05 ml calcium acetate solution. Whirlimix to disperse the substrate and incubate at 37°C for 15 min. **Whirlimix every 2 or 3 min to keep the substrate in suspension**. Stop the reaction by putting the tube in ice for 10 min. Keep all samples, taken at different times, standing in ice until all have been collected. Prepare a blank in the same way from Hide powder azure plus 1·0 ml H_2O, 2·5 ml imidazole buffer and 0·05 ml calcium acetate solution. At the end of the day, centrifuge all the samples for 10 min and decant the supernates into separate test tubes. Read the extinction at 595 nm of the samples against the blank in the spectrophotometer.

Alkaline phosphatase (see *note* 7)   The substrate for the enzyme, *p*-nitrophenyl phosphate, is colourless. The action of the phosphatase releases *p*-nitrophenol which is yellow and can be measured spectrophotometrically.

To a 1·0 ml cell sample add one drop of toluene from a Pasteur pipette. Whirlimix for 60 s. Add 1·0 ml *p*-nitrophenylphosphate (PNPP) in diethanolamine buffer (pH 9·4) and incubate at 37°C for 30 min. Stop the reaction with 1·5 ml 2 N NaOH. Mix and store all samples, taken at the different times, at room temperature. Prepare a blank in the same way from 1·0 ml H_2O, 1·0 nl PNPP in diethanolamine buffer and 1·5 ml NaOH. At the end of the day centrifuge all the samples for 10 min and decant the supernates into separate tubes. Read the extinction at 410 nm of all the samples against the blank in the spectrophotometer.

Refractile spores   Examine the cells in the microscope under phase-contrast and oil immersion. Put one drop of the culture onto a microscope slide. Put on the cover slip and put one drop of lens oil onto the coverslip. Mature endospores have a low water content and appear phase-bright against the dark background of the mother cell when viewed by phase-contrast.

Octanol-resistance   Pipette a cell sample (0·1 ml) into 0·9 ml sterile H_2O and add 0·01 ml of *n*-octanol. Whirlimix for 60 s. A 10^{-4} dilution is the most convenient single dilution to use. Add 0·1 ml of the cell suspension treated with *n*-octanol to 9·9 ml of sterile H_2O, Whirlimix thoroughly and spread 0·1 ml onto a nutrient agar plate. Incubate the plate at 37°C overnight. The number of colonies on the plate after incubation will equal the number of octanol-resistant cells per ml of the original culture $\times 10^{-4}$.

Heat-resistance   Pipette a cell sample (0·1 ml) into 0·9 ml sterile H_2O and heat at 80–85°C for 15 min in a water-bath. Whirlimix the sample at the end of this time and dilute and spread samples on nutrient agar plates as described for octanol-resistance.

Total viable count   Immediately dilute the cell sample (0·1 ml) plus sterile H_2O (0·9 ml) to 10^{-6} dilution and spread on a nutrient agar plate.

Day 2 *(60 min)*

Count the colonies on the nutrient agar plates from the viable count, octanol and heat-resistance determinations. Calculate *exo*-protease and alkaline phosphatase activities, at each time interval, as a change in extinction at 595 or 410 nm min^{-1} ml^{-1} of culture (*see* note 8). The rate at which the activity of each enzyme increases during sporulation can simply be determined by plotting enzyme activity against time of sampling. The basal activity found during growth should be subtracted from each value obtained during sporulation. Refractility, octanol and heat-resistance values can be expressed as a percentage of the total cell population.

To demonstrate more clearly the sequential appearance of the marker events during sporulation, all values should be plotted against time (h) after *t*-zero on a single graph.

Notes and Points to Watch

1. Inoculation procedure. The **evening before** the experiment, *B. subtilis* is streaked onto nutrient agar and incubated overnight at 30°C. Growth from one plate is removed with a sterile loop to 2·0 ml Penassay broth. After Whirlimixing, 10 drops of the suspension ($\sim$ 0·5 ml) from a sterile Pasteur pipette can be used to inoculate 60 ml Penassay broth, plus additions (*see* Media).
2. The bacteria should be cultured with vigorous aeration at 42°C throughout the experiment, in order to maintain aerobic conditions and to allow growth and sporulation to occur as rapidly as possible. A shaking water-bath should be used in preference to a gyratory incubator, where constant lifting of the lid will tend to lower the temperature of incubation. Cultures should be sampled without removing flasks from the water-bath and aeration should be resumed immediately after a sample has been taken.
3. Estimation of growth. If micro-cuvettes are available, small volumes can be removed from the culture, the E_{600} noted against a water or medium blank and the bacteria subsequently washed from the cuvette into disinfectant. If only large cuvettes are available, the E_{600} of the sample should be noted as quickly as possible and the cells then returned to the main culture. This will prevent the loss in culture volume which will occur if many large samples are removed. The lack of sterile conditions involved in this procedure is unimportant as a high cell density was used for the original inoculum.
4. Culture samples (0·1 ml) are pipetted into 0·9 ml sterile distilled water for either octanol or heat-resistance determinations. Care must be taken to ensure that no bacteria are deposited on the lip of the tube when this is done as these cells will escape the subsequent octanol or heat treatment and give rise to a falsely increased level of survival.

    Theoretically no spores should be present in the growing population of cells. Resistant colonies are sometimes found, however, and these arise from spores added to the broth medium from the overnight growth on agar which is used as the inoculum (*see* note 1). A few spores may fail to germinate in the broth and will give rise to resistant colonies.
5. After a short lag-period, growth of *B. subtilis* in the broth medium at 42°C will be rapid, with a doubling time of less than 30 min. The bacteria will reach an $E_{600}^{1\,cm}$ of between 1·0 and 2·0 before growth ceases. This will take 2–2$\frac{1}{2}$ h from the time of inoculation. The transition from the growth phase to the stationary phase tends to be gradual rather than abrupt, and usually the point at which the cells enter the stationary phase (*t*-zéro or the start of sporulation) is only clear after subsequent optical density readings remain constant. It is recommended therefore that E_{600} readings are taken at frequent intervals when it is obvious that exponential growth is beginning to slow down.

6. It is only necessary to determine the total cell population at one point during the stationary phase as there will be no increase in cell numbers during the sporulation process. The total viable count should be determined at 10^{-6} dilution.

7. For the *exo*-protease and phosphatase assays it is only necessary to prepare **one** blank. A blank should not be prepared each time the assay is performed.

   The *exo*-protease is an extracellular enzyme and a whole cell suspension can therefore be assayed directly. Alkaline phosphatase, however, can only be readily detected if the cells are first made permeable by treatment with toluene.

   The Hide powder azure used in the protease assay is insoluble and tends to settle to the bottom of the tube. It is important that the tube is shaken at regular intervals throughout the assay in order to keep the substrate suspended.

   In the phosphatase assay, it is sometimes found that after centrifugation the cell debris precipitated by the NaOH does not pellet satisfactorily. Some material may remain on the surface. This is probably caused by the presence of the toluene. If this occurs the supernate has to be withdrawn with a Pasteur pipette. To minimize this effect it is important to use only a small drop of toluene to render the cells permeable.

   If there is insufficient time on Day 1 to read the optical densities for the protease and phosphatase assays the tubes can safely be kept at 4°C until the following day.

8. A more accurate estimate of the specific activities of *exo*-protease and alkaline phosphatase can be obtained by standardizing the activities at each time interval to a constant E_{600} value of 1·0. In this way the difference in cell numbers between samples taken during growth and those taken in the stationary phase will be eliminated.

MATERIALS

1. *Organism. Bacillus subtilis* strain 168. The strain used in this laboratory carries the *trp-C2* mutation and requires added L-tryptophan for growth. Other strains of *B. subtilis* could probably be used to obtain similar results.

2. *Media.* Penassay broth (Difco antibiotic medium 3). Before inoculation, add 1·0 ml of the following per 100 ml of medium: 2·5% $MgSO_4.7H_2O$, 2·5% $Ca(NO_3)_2$, 0·05% $MnCl_2.4H_2O$, 0·05% $FeSO_4.7H_2O$, (0·2% L-tryptophan if the auxotrophic strain is used). Nutrient agar (Oxoid CM3). All media are autoclaved at 103·4 kPa (15 lbf in^{-2}) for 15 min.

3. *Apparatus.* Shaking water-bath which will carry 250 ml flasks and maintain a temperature of 42°C. Static water-baths, one maintained at 37°C and the other at 85°C. Incubator capable of being maintained at either 30 or 37°C. Spectrophotometer and cuvettes suitable for determination of optical densities at 410, 595 and 600 nm. Whirlimixer. Low speed bench centrifuge.

4. *Reagents.* Hide powder azure (Calbiochem), imidazole, L-tryptophan, *p*-nitrophenylphosphate (Sigma), calcium acetate, diethanolamine, *n*-octanol, toluene (Fisons). All other chemicals are obtained from B. D. H.

SPECIFIC REQUIREMENTS

These items are needed for each group of students.

Day 1

60 ml broth medium plus additions (*see* Media)

in a 250 ml flask inoculated with *B. subtilis* grown overnight on nutrient agar (*see* note 1)

9 nutrient agar plates

7 small tubes containing approx. 20 mg Hide powder azure

cont.

cont.

9 small tubes containing 0·9 ml sterile distilled water	ice
10 large tubes containing 9·9 ml sterile distilled water	semi-log and 1 cm graph paper
20 ml 0·1 M calcium acetate,	phase-contrast microscope, slides, coverslips and oil
20 ml 0·1% *p*-nitrophenylphosphate in M diethanolamine–HCl (pH 9·4)	distilled water wash-bottle
20 ml 2 N NaOH	container of disinfectant
20 ml 0·1 M imidazole–HCl (pH 7·2)	small and large tubes with racks
10 ml *n*-octanol	sterile pipettes
10 ml toluene	glass spreader and alcohol
	sterile Pasteur pipettes

FURTHER INFORMATION

The appearance of *exo*-protease activity is one of the earliest events associated with sporulation (for reviews *see* Mandelstam, 1969; Piggot and Coote, 1976). An increase in activity should be detected within 30 min from *t*-zero, the point of initiation of spore formation. The activity will

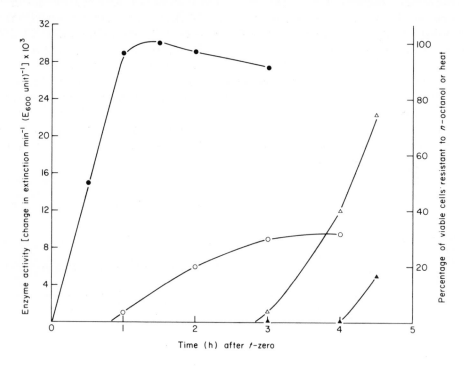

Figure 1 Sequential appearance of four marker events during endospore formation in *B. subtilis* 168 at 42 °C. The time at which the cells cease exponential growth and enter the stationary phase (*t*-zero) is taken as the point of initiation of sporulation. ●, *exo*-protease activity; ○, alkaline phosphatase activity; △, resistance to *n*-octanol; ▲ resistance to heat, 85 °C for 15 min.

increase over the next $1-1\frac{1}{2}$ h and then should remain constant. This is a reflection of increased synthesis of the enzyme during the initial stages of sporulation which then ceases as the process moves into its later stages.

Alkaline phosphatase activity is associated with a later stage in the sporulation process, formation of the spore septum. It should therefore begin to increase in activity at a later time than the *exo*-protease, about 60 min after *t*-zero, and continue to increase for $1\frac{1}{2}-2$ h before remaining constant. The increase in phosphatase activity is not as marked as that of the protease. It should be noted that the function, if any, of these two enzymes during sporulation is not clear. The phosphatase activity increases during sporulation even in the presence of inorganic phosphate which represses the enzyme during growth.

The most characteristic events of sporulation, development of refractility, of organic-solvent resistance and of heat-resistance are associated with the formation of the cortex and coat layers. As the developing prespore forms the cortex so it becomes white when viewed by phase-contrast. This can usually be observed $3-3\frac{1}{2}$ h after *t*-zero. Octanol-resistance is associated with the laying down of the coat layers around the cortex and usually begins to occur around $3\frac{1}{2}$ h after *t*-zero. Finally, full maturation of the endospore, when it becomes phase-bright, is accompanied by resistance to high temperatures. This is the last property developed by the endospores and begins to occur around 4 h after *t*-zero.

The experiment attempts to illustrate the sequential nature of this primitive differentiation process by showing the time-dependent appearance of each marker event. Mutants unable to sporulate have been studied extensively (*see* Waites *et al.*, 1970; Piggot and Coote, 1976). They illustrate quite clearly the dependent sequence of events in sporulation. A mutant blocked at a particular stage in the process will not show any characteristics associated with later stages, a feature which demonstrates that the occurrence of later events depends on the successful completion of earlier events. The experiment outlined here can therefore be developed by replacing the wild-type strain of *B. subtilis* with one or more asporogenous mutants. In this case a mutant blocked at the very earliest stage in the process will not show protease or phosphatase activities, nor will it develop refractility or resistance. A mutant blocked at a later point, however, may develop protease and phosphatase activities, but no refractility and resistance.

REFERENCES

J. Mandelstam (1969). Regulation of bacterial spore formation. *Symposia Society for General Microbiology*, **19**, 377–402.

P. J. Piggot and J. G. Coote (1976). Genetic aspects of bacterial endospore formation. *Bacteriological Reviews*, **40**, 908–62.

W. M. Waites *et al.* (1970). Sporulation in *Bacillus subtilis*. Correlation of biochemical events with morphological changes in asporogenous mutants. *Biochemical Journal*, **118**, 667–76.

Requirements for the Germination and Outgrowth of Spores of *Bacillus subtilis*

J. G. COOTE

*Department of Microbiology, University of Glasgow,
Garscube Estate, Bearsden, Glasgow, G61 1QH, Scotland*

> *Level*: All undergraduate years
> *Subject area*: Bacterial physiology
> *Special features*: Correlation of spectrophoto-
> metric, microscopical and
> physiological observations

INTRODUCTION

The transition from the dormant bacterial spore to the actively growing vegetative cell is conveniently divided into the two processes of germination and outgrowth. Germination, once initiated, is irreversible and is essentially a degradative process whereby the spore loses its characteristic properties. It is accompanied by rehydration and swelling of the spore protoplast and this, coupled with a loss of refractility and fragmentation of spore peptidoglycan, promotes a drop in optical density of the spore suspension which can be monitored. Outgrowth follows germination and involves the metabolic changes necessary to promote active cell-division. A new germ cell emerges from the spore and as it grows and divides this will promote an increase in optical density.

The aim of this experiment is to follow the germination and outgrowth of *Bacillus subtilis* spores in a rich broth medium and to observe the effect of chloramphenicol, an inhibitor of protein synthesis, on these processes. In a separate experiment the minimal requirements for germination and outgrowth will be determined. Both processes can be monitored by noting the change in optical density of a spore suspension.

EXPERIMENTAL

As a preliminary to this experiment, spores of *B. subtilis* 168 have to be prepared (part (a)). This will take about 3 h. Parts (b) and (c) of the experiment can be done separately by a single student, but it is assumed that a group of 2 to 3 students will perform the exercise. In this case both parts (b) and (c) can be run concurrently over a period of about 3 h.

Day 1 *(part a, 3 h; parts b and c, 3 h)*

(a) *Preparation of spores*

Take 300 ml culture of *B. subtilis* which has been incubated overnight at 37 °C (*see* note 1).

↓

Centrifuge at full speed on a bench centrifuge for 15 min, discard the supernate, drain the tubes and take up the pellets in 30 ml 0.1 M KH_2PO_4; 0.01 M $MgCl_2$ (pH 7.8) containing lysozyme (500 μg ml^{-1}). Incubate at 37 °C for 45–60 min until the cells are lysed and free spores released.

↓

Centrifuge the spore suspension at full speed for 15 min, discard the supernate, drain the tubes and resuspend the pellets in a total volume of 30 ml of sterile distilled H_2O.

↓

Repeat the washing procedure twice more, take up the pellets after 3 washings in an equal volume of sterile distilled water and heat them at 80 °C for 20 min. This procedure kills any remaining vegetative cells.

↓

Centrifuge the suspension at full speed for 15 min, discard the supernate and take up the spores in a total of 10 ml of sterile distilled H_2O. Distribute the spore suspension into sterile bijou bottles and store at -20 °C unless used immediately (*see* note 2).

(b) *Germination and outgrowth of spores in a rich medium*

Flask 1	*Flask 2*
10 ml broth in a 100 ml flask	10 ml broth in a 100 ml flask
$+0.2$ ml sterile distilled H_2O.	$+0.2$ ml 0.5% chloramphenicol.

↓　　　　　　　　　　↓

Add 10 drops of spore suspension to each flask and incubate at 30 °C in a shaking water-bath (*see* note 3).

↓　　　　　　　　　　↓

Monitor the E_{600} of the suspensions and examine samples under phase-contrast at 0, 15, 30, 45, 60 min intervals thereafter (*see* note 4). Continue to observe the E_{600} and appearance of the spores for about 3 h.

(c) *Requirements for germination and outgrowth*

Five 100 ml flasks with minimal salts medium plus additions (*see* note 5).

Three 100 ml flasks with distilled water plus additions (*see* note 5).

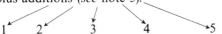

1　2　3　4　5

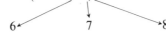

6　7　8

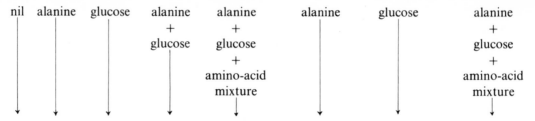

Add 10 drops of spore suspension to each of the 8 flasks and incubate at 37 °C in a shaking water-bath.

↓

Monitor the E_{600} of each of the 8 suspensions and examine samples under phase-contrast at 0, 15, 30, 45, 60 min, and thereafter every 30 min for the next 2 to 3 h (*see* note 4).

Notes and Points to Watch

1. The inoculating suspension is prepared the **day before** the experiment by streaking the organism onto nutrient agar, allowing it to grow at 37 °C for 6–8 h, and then removing the growth from one 9 cm plate with a sterile loop into 1·0 ml Penassay broth. Penassay broth (300 ml containing additions of salts; *see* Media) in a 2 litre flask is then inoculated with 0·5 ml of the cell suspension and incubated with vigorous aeration at 37 °C overnight.
2. The spore preparation procedure will give enough spore suspension for at least 20 inoculations into germination media. If necessary the spore suspension can be prepared in advance of the class and parts (b) and (c) can then be started immediately.
3. The suspended spores, after preparation and final resuspension in water, tend to settle. It is important that the preparation is shaken well before drops are dispensed into germination media. The addition of 10 drops of spore suspension from a Pasteur pipette to 10 ml of germination medium is an arbitrary figure. The quantity of spores obtained will vary from one preparation to another. Sufficient spores should be added to the germination medium to give a starting E_{600}^{1cm} of between 0·7 and 1·0.

   Aerobic conditions should be maintained throughout each experiment. A shaking water-bath should be used in preference to a gyratory incubator where constant lifting of the lid will tend to lower the temperature of incubation. Germination and outgrowth occur more rapidly in the broth medium (part (b)) than in the minimal medium (part (c)). For this reason it is more convenient to follow the processes in the broth medium at 30 °C and in the minimal medium at 37 °C.
4. If micro-cuvettes are available, small volumes can be removed from the suspension, the E_{600} noted against a water or medium blank and the suspension subsequently washed from the cuvette into disinfectant. If only large cuvettes are available, the E_{600} of the samples should be noted as quickly as possible and the sample then returned to the main suspension. This will prevent the loss in volume which will occur if many large samples are removed.

   Suspensions should be sampled without removing flasks from the water-bath and aeration should be resumed immediately after a sample has been taken.
5. If a number of individual students or groups of students are performing this experiment, it would be more convenient for each student or group to do only two of the eight germination

media. In a large class this would also allow the option of introducing further combinations of metabolic germinants and/or inhibitors.

Use 10 ml amounts of minimal salts medium or distilled water and add, as required, 0·5 ml 10% D-glucose, 0·1 ml 0·5% L-alanine and 0·1 ml 0·2% amino-acid mixture (*see* Specific Requirements).

MATERIALS

1. *Organism. Bacillus subtilis* strain 168. The strain used in this laboratory carries the *trpC2* mutation and requires added L-tryptophan for growth. Other strains of *B. subtilis* could probably be used to obtain similar results.
2. *Media.* Penassay broth (Difco antibiotic medium 3). This medium is used with no further additions for germination and outgrowth experiments (parts (b) and (c)). It is also used to prepare the spores (part (a)), but in this case, before inoculation, add 1·0 ml of the following per 100 ml of medium: 2·5% $MgSO_4.7H_2O$, 2·5% $Ca(NO_3)_2$, 0·05% $MnCl_2.4H_2O$, 0·05% $FeSO_4.7H_2O$, (0·2% L-tryptophan, if the auxotrophic strain is used).

Minimal salts medium has the following composition (gl^{-1}); KH_2PO_4, 13·6; $(NH_4)_2SO_4$, 2·0; adjusted to pH 7·2 with 40% KOH. Before use, the following are added per 100 ml of medium; 0·8 ml 2·5% $MgSO_4.7H_2O$, 1·0 ml 0·05% $FeSO_4.7H_2O$, 1·0 ml 0·05% $MnCl_2.4H_2O$; (1·0 ml 0·2% L-tryptophan, if the auxotrophic strain is used).

Nutrient agar (Oxoid CM3). All media are autoclaved at 103·4 kPa (15 lbf in^{-2}) for 15 min.
3. *Apparatus.* Orbital incubator which will carry a 2 litre flask and maintain a temperature of 37 °C. Shaking water-bath which will carry 100 ml flasks and maintain a temperature of 30 or 37 °C. Static water-baths, one maintained at 37 °C and the other at 80 °C. Spectrophotometer and cuvettes suitable for determination of optical density at 600 nm. Whirlimixer. Low speed bench centrifuge.
4. *Reagents.* Chloramphenicol, lysozyme, L-alanine, L-tryptophan, L-aspartate, L-asparagine, L-glutamate, L-glutamine, L-arginine (Sigma). All other chemicals are obtained from B.D.H.

SPECIFIC REQUIREMENTS

These items are sufficient for 6 students or groups.

overnight culture of *B. subtilis* (*see* note 1)
250 ml 0·1 M KH_2PO_4; 0·01 M $MgCl_2$ (pH 7·8). Lysozyme (500 $\mu g\ ml^{-1}$) should be added on the day of use
150 ml sterile Penassay broth
250 ml sterile minimal salts medium
300 ml sterile distilled water
10 ml 0·5% chloramphenicol made up in 1 : 1 H_2O : absolute alcohol
10 ml solutions of 0·5% L-alanine, 10% D-glucose and a solution containing L-aspartate, L-asparagine, L-glutamate, L-glutamine and L-arginine, all at 0·2%, sterilized by filtration
1 cm graph paper
phase-contrast microscope, slides, coverslips and oil
distilled water wash-bottle
container of disinfectant
sterile pipettes
sterile Pasteur pipettes
sterile small bijou bottles
sterile 100 ml flasks

FURTHER INFORMATION

Germination of bacterial endospores is an irreversible process which occurs when the spores are exposed to specific metabolites (usually amino acids, sugars or purine ribosides). The resistant, dormant spore is transformed into a sensitive, metabolically active cell and all germination events can be supported by endogenous resources within the spore (for reviews *see* Gould, 1969; Keynan, 1973, 1978). The process does not require the synthesis of new macromolecules and so will occur in the presence of chloramphenicol, an inhibitor of protein synthesis. During germination the cortex peptidoglycan is degraded; the spore consequently rehydrates, swells and loses its refractile properties. This leads to a fall in optical density of a spore suspension and although this is a rather late event in germination it is the most convenient to measure.

For each medium the E_{600} readings can be plotted against time and an assessment made, firstly, as to whether germination has occurred (a drop in E_{600} and phase-darkening of the spores) and, secondly, whether this is subsequently accompanied by outgrowth (a rise in E_{600} coupled with elongation and formation of new vegetative cells).

In the first experiment spores are germinated in a broth medium which contains a rich supply of metabolic germinants. The E_{600} should fall steadily for 60 min as the spores germinate and become phase-dark. The E_{600} will cease to fall when the majority of spores have germinated and by about 90 min it will begin to rise again. The increase in E_{600} is the period of outgrowth and is

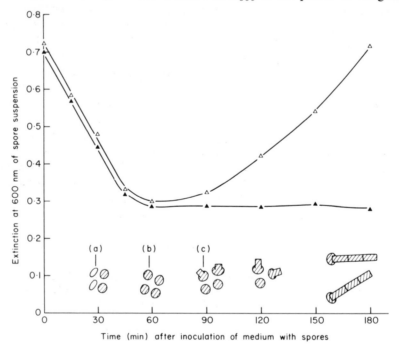

Figure 1 Change in appearance and extinction at 600 nm of spores of *B. subtilis* 168 incubated in a rich broth medium at 30 °C. ▲, medium with added chloramphenicol ($100 \mu g \, ml^{-1}$); △, medium with no additions. As the spores germinate, their appearance under phase-contrast changes from oval and phase-bright (a), to round and phase-dark (b). During outgrowth a new cell emerges from the spore (c). Outgrowth does not occur in the medium with added chloramphenicol.

accompanied by emergence of a new vegetative cell from the spore. By 120 min, division of the newly emerged cell will have occurred and, after 3 h, chains of cells should be seen with the spore case often still attached to one end of the chain. Germination should occur normally in the spore suspension treated with chloramphenicol, but outgrowth will be inhibited. Consequently, there should be a normal fall in E_{600} accompanied by phase-darkening of the spores, but thereafter the E_{600} will remain constant and no emergence of vegetative cells will occur. Outgrowth is characterized by renewed RNA and protein synthesis, as during this phase proteins and structures characteristic of the vegetative cell are synthesized. The process is thus susceptible to inhibition by chloramphenicol.

The second experiment is designed to test the efficiency of an amino-acid, L-alanine, and a sugar, D-glucose, to initiate germination when used either singly or in combination. In addition, inorganic ions have been shown to stimulate germination in some *Bacillus* species (Gould, 1969). Resuspension in salts medium alone should have no effect on the spores. L-alanine alone acts as a good germinant, whether added in water or salts medium. Glucose is a poor germinant in either medium, but gives a slightly faster rate of germination in combination with L-alanine than L-alanine alone. L-alanine generally promotes phase-darkening in about 80 % of the spores examined, while only 20 % show this characteristic property of germination with D-glucose.

Outgrowth in *B. subtilis* 168 should not occur unless certain additional amino-acids, besides L-alanine, are supplied (Strange and Hunter, 1969). This is probably due to the absence of the necessary biosynthetic enzymes in the germinated spore. Outgrowth, an increase in E_{600} and emergence of new vegetative cells, normally occurs only in the salts medium containing L-alanine, D-glucose and the 5 additional amino acids (*see* above). Outgrowth requires a ready supply of carbon, nitrogen, sulphur and phosphorus for synthetic processes and does not occur to any significant extent when the salts in the medium described above are replaced by water.

The trigger reaction which sets the process of germination in train has not been defined, but it could well involve allosteric alteration of a receptor molecule rather than metabolism of the germinant (reviewed by Keynan, 1978). Non-metabolizable analogues of glucose (e.g. 2-deoxyglucose) or inhibitors of electron transport (e.g. sodium cyanide, sodium azide or 2-*n*-heptyl-4-hydroxyquinoline-*N*-oxide) and ATP production (e.g. *N, N'*-di-cyclohexylcarbodiimide) allow normal germination in *B. megaterium* (Dills and Vary, 1978; Shay and Vary, 1978), although ATP production is, of course, necessary for outgrowth. The experiments described above could therefore be extended to include an investigation of the effects of other compounds besides chloramphenicol on germination and outgrowth in *B. subtilis*.

REFERENCES

S. S. Dills and J. C. Vary (1978). An evaluation of respiration chain-associated functions during initiation of germination of *Bacillus megaterium* spores. *Biochimica et Biophysica Acta*, **541**, 301–11.

G. W. Gould (1969). *In* "The Bacterial Spore," (G. W. Gould and A. Hurst, eds), pp. 397–444. Academic Press, London and New York.

A. Keynan (1973). The transformation of bacterial endospores into vegetative cells. *Symposia of the Society for General Microbiology*, **23**, 85–123.

A. Keynan (1978). Spore structure and its relations to resistance, dormancy and germination. *In* "Spores VII," (G. Chambliss and J. C. Vary, eds), pp. 43–53. American Society for Microbiology, Washington D.C.

L. K. Shay and J. C. Vary (1978). Biochemical studies of glucose-induced germination of *Bacillus megaterium*. *Biochimica et Biophysica Acta*, **538**, 284–92.

R. E. Strange and J. R. Hunter (1969). *In* "The Bacterial Spore", (G. W. Gould and A. Hurst, eds), pp. 445–83. Academic Press, London and New York.

Determination of the Molar Growth Yield of *Beneckea natriegens* in Batch and Continuous Culture

A. G. SCHAFFER

Department of Science, Bristol Polytechnic, Frenchay, Bristol, BS16 1QY, England

> *Level*: Advanced undergraduates
> *Subject area*: Bacterial physiology
> *Special features*: Theory and practice of batch and
> continuous cultures

INTRODUCTION

The molar growth yield (Y) of a micro-organism is defined as the yield of cells (μg dry weight) per μmol of the growth-limiting substrate. Usually the limiting substrate is the energy source, e.g. glucose. The relationship between yield and the limiting substrate in both batch and continuous (chemostat) culture is given by:

$$X = Y(S_R - S),$$

where X is the yield of cells (μg dry weight ml^{-1}) during growth; and

Y is the molar growth yield (μg dry weight μmol^{-1} substrate).

In **batch culture**, S_R and S are the initial and final concentrations (μmol ml^{-1}) of the substrate, respectively.

In **continuous culture**, S_R and S are the concentrations of the substrate in the reservoir medium and in the culture, respectively.

In this experiment, *Beneckea natriegens* is grown aerobically in a defined medium with glucose as the sole source of carbon and energy. The initial period of growth is in batch culture and, when the glucose is exhausted, conditions for continuous culture are established. During the experiment, the turbidity (and hence dry weight) of the culture and the glucose in the medium are determined. The value of Y, during logarithmic growth in the batch culture, is determined from the plot of dry weight

---

against glucose used. During continuous culture, the residual glucose is less than can be detected and thus Y, which is equal to X/S_R can be re-determined.

The primary aim of the experiment is to determine value(s) for the molar growth yield but the exercise also illustrates some of the more important principles used in the analysis of batch and continuous culture of bacteria.

EXPERIMENTAL

The work on Day 2 should be distributed among a group of students, four probably being optimum (*see* note 3).

Day 1 *(10 min)*

Grow an inoculum of *B. natriegens* at 20 °C with aeration in 100 ml of defined Beneckea broth (DBB) containing 5 mM glucose.

Day 2 *(6 h including 1 h to set up apparatus; see note 2)*

Set up the culture apparatus as in Figure 1. Note especially the facility for continuous monitoring of the turbidity of the culture (*see* notes 1, 2 and 5).

Batch culture
The volume of medium (about 700 ml) must be known and includes the medium in the circulating tubes. Set the temperature to 35 °C and the aeration rate to 4 litres min^{-1}; the stirrer speed should be about 250 rev. min^{-1}. Pump the medium through the flow-through cell at about 1 litre h^{-1}. Set full scale deflection on the recorder to correspond to 100 % transmission (zero absorbance) at 550 nm on the colorimeter. Then set zero on the recorder to correspond to 0 % transmission on the colorimeter.

↓

Add sufficient of the inoculum culture (about 30 ml) to give an initial turbidity of 90 % transmission. Remove a sample (8 ml) of the culture from tube (b) for a glucose determination (*see* note 4). Start the recorder at a chart speed of 0.5 or 1 cm min^{-1}.

↓

Add the calculated volume of 1.0 M glucose to give a concentration of 2 mM. The time of addition is zero time; after 2 min, when the culture should be fully mixed (this can be checked in a preliminary run with a dye) take a second sample (8 ml) from tube (b) for the determination of the actual initial glucose concentration.

↓

Take samples (8 ml) at 5 min intervals. About 20 samples are needed; this removes about 200 ml from the vessel. Note the chart recorder reading at each sampling time and convert the readings to absorbance values and/or to cell dry weight per ml from a calibration curve (*see* note 5). Plot log absorbance (or log dry weight) against time of incubation so as to determine (i) the specific growth rate (μ) of the organism and (ii) when growth has ceased. The value of μ, which is assumed to be μ_{max}, is needed if the experimenter is to make a decision about a suitable dilution rate for the continuous culture.

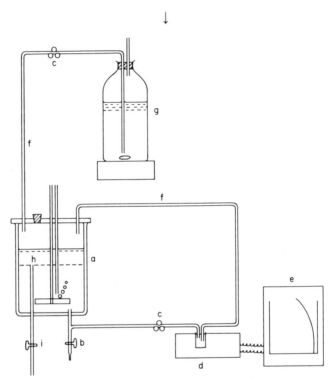

Figure 1 Culture apparatus.
(a) Culture vessel of 1 litre capacity with 700 ml of medium, stirrer and sterile air input. The temperature control system is not shown. (b) Sampling point for use during batch culture operation. (c) Peristaltic pump. (d) Spectrophotometer with flow through cell and output to recorder. (e) Recorder. (f) Silicone rubber tubing. (g) Reservoir medium for continuous culture; magnetic stirrer with thermostat. (h) Overflow for continuous culture set for 500 ml of medium. (i) Outflow and sampling point for use during continuous culture.

Continuous culture
To establish a continuous culture at the final cell density achieved in the batch culture, a flow of medium containing glucose at the concentration supplied at the **start** of the batch culture (plus an addition for the cells added in the inoculum; *see* note 6) is required. To the reservoir of medium DBB add the calculated volume of 1·0 M glucose to give the correct final concentration of glucose which should be about 2·2 mM. Check the concentration of glucose in the reservoir.

↓

Open the tap in outflow tube (i) to reduce the culture volume to the overflow level (h). If the level is below the overflow, medium with glucose should be pumped into the vessel and the culture allowed to grow to a steady population before the continuous culture is started. Start the continuous culture by pumping medium from (g) at a dilution rate less than μ_{max} for the organism (*see* note 7). The continuous culture is in operation as soon as the culture emerges from the outflow. The turbidity measurement should remain constant unless there is a mismatch between the glucose concentration in the reservoir and the cell population; in which case, the turbidity will increase, or decrease, slowly. Check the flow rate by measuring the output from the vessel for 10 or 20 min. Take two or more samples (8 ml) for glucose assay from the outflow at 10 min intervals. These samples should show no detectable glucose.

$$\downarrow$$

Calculate the values of the molar growth yield (*see* note 8). Estimate a value for K_s (*see* note 9).

Notes and Points to Watch

1. The rapid growth of *B. natriegens*, the high salt medium, the short duration of the experiment and the method of collecting samples mean that there are few microbiological hazards in the experiment. However it is necessary to flush the sampling tube with growing culture and, as this needs to be done rapidly, it could lead to contamination of the laboratory.
2. Sterilization of the apparatus can be performed *in situ* with a water wash, hypochlorite (1 % v/v "Chloros" for 30 min), thiosulphate (1 % sodium thiosulphate for 15 min) and a final sterile distilled water wash.
3. Students gain most from the experiment if they perform all the operations themselves. Since this requires familiarity with the equipment it is best if a preliminary experiment is done with batch cultures in which 1, 2 and 5 mM glucose are added, but without the glucose assays.
4. *Determination of glucose*
   The method is a modification of Trinder's (1969) glucose oxidase procedure. The modification is designed to achieve maximum sensitivity which is about $0.001 \, \mu mol \, ml^{-1}$.
   Standard curve for glucose. From a 0.1 M solution of glucose prepare standards of 250, 100 and 10 μM glucose by serial dilution in DBB. Use these to make 4 ml volumes of DBB containing 1 to 200 or 250 μM glucose. To these, and a blank of 4 ml DBB, add 1 ml of HCl–phenol reagent; mix the solutions. Transfer 2 ml of these standard solutions to fresh tubes and add 2 ml of glucose oxidase (GO) reagent (**caution—this reagent contains azide**). Mix the solutions and incubate for 15 min at 37 °C. Vortex the tubes at intervals to ensure aeration is adequate. Read the coloured solutions at 515 nm against the blank.
   Glucose in culture samples. Prepare a series of 10 ml graduated centrifuge tubes by adding 2 ml of HCl–phenol reagent to each. Flush the static culture from the delivery tube (b) and run 8 ml of culture into the centrifuge tube. The HCl–phenol kills the cells. Mix the contents rapidly and centrifuge the mixtures in a bench centrifuge to remove the cells; retain the supernate for the glucose determination. The volume of supernate is sufficient for replicate assays if desired.
   To 2 ml of the supernate add 2 ml of the GO reagent and proceed as for the standard solutions of glucose. Where the glucose concentration is beyond the upper limit of the method, a sample of the supernate must be diluted with a DBB : HCl–phenol (4 : 1) mixture and the assay repeated. Note that the phenol participates in the colour development.
5. The accuracy of the Y value determination depends on an accurate dry weight–turbidity calibration curve. The geometry of the flow-through cell has a major effect on the instrument

response to turbid suspensions. Thus the calibration curve must be prepared for the instrument and cell used in the experiment.

The flow-through cell can cause other problems. The internal surfaces must be free of adhering organisms. Any particles which remain in the cell can cause considerable "noise" on the recorder. Some "noise" seems to be inevitable when an ordinary peristaltic pump is employed. This is attributed to the small clumps of cells, characteristic of *B. natriegens*, scattering the light unequally during the different phases of the pumping cycle.

6. Under the conditions of the experiment the cell population increases about 10-fold; thus about 10 % of the final population was present at the outset and did not require the glucose provided at the start for cell growth. If the glucose in the continuous culture reservoir medium is to support the population achieved at the end of the batch culture, its concentration needs to be about 10 % higher than that used in the batch culture. The actual concentration of glucose to be used must be calculated during the experiment.

7. The dilution rate (D, h^{-1}) is the flow rate $(l\,h^{-1})$ per culture volume (l). It is suggested that D is set at half μ_{max} or about $0.8\,h^{-1}$. To determine the dilution rate it is necessary to know the culture volume and have a calibration curve relating pump setting to flow rate.

8. Plot X (μg dry weight of cells per ml) against glucose used (μmol ml^{-1}). The slope of the linear regression during logarithmic growth is Y in batch culture. The value of Y, during continuous culture, is calculated from X/S_R; its accuracy depends on both the dry weight calibration curve and accurate knowledge of the concentration of glucose in the reservoir medium.

9. It is also possible to calculate an **upper** limit for K_s, the constant defined as the concentration of the substrate (glucose) which limits the specific growth rate (μ) to half its maximum value, using the equation derived in 1942 by Monod (Tempest, 1970):

$$D = \mu = \frac{\mu_{max}S}{K_s + S},$$

where S is the concentration of glucose in the vessel (i.e. a concentration **less** than can be detected by the method employed).

MATERIALS

1. *Beneckea natriegens* (NCMB 857), maintained on nutrient agar + 2 % sodium chloride at room temperature or frozen on glass beads in nutrient broth + 15 % glycerol.

2. Culture apparatus as shown in Figure 1. The components which have been used are:
   fermenter: Type LHE 1/1000 (L. H. Engineering Co. Ltd., Stoke Poges, Bucks) with 1 litre continuous culture vessel. Other makes can be used.
   colorimeter: Corning EEL, 253, but other makes can be used.
   flow through cell: 1 cm cell with internal bubbler and silicone rubber connections. Corning EEL.
   recorder: Servoscribe, Potentiometric. Other makes can be used.

peristaltic pumps: Type MHRE/7, Watson–Marlow Ltd., Falmouth, Cornwall.
tubing: silicone rubber, 9 mm internal diam.

3. Defined Beneckea broth (Niven *et al.*, 1977), containing (laboratory grade reagents): tris (hydroxymethyl) aminomethane/HCl buffer, pH 7·5 (50 mM), NaCl (400 mM), KCl (10 mM), K_2HPO_4 (3·3 mM), $MgSO_4$ (1 mM), $CaCl_2$ (1 mM), NH_4Cl (25 mM), $FeSO_4$ (0·1 mM) and disodium ethylenediamine tetraacetic acid (0·1 mM). The magnesium, calcium and iron salts must be dissolved separately; even so, a slight precipitate may develop after sterilization (121 °C for 15 min). Glucose is sterilized and added separately.

4. Reagents for the assay of glucose.

Glucose oxidase (GO) reagent: dissolve 3 g Na$_2$HPO$_4$ in 220 ml of deionized water; add 5 ml of glucose oxidase (Fermcozyme, 653AM), 5 mg of peroxidase (60 u mg^{-1}), 300 mg of sodium azide and 100 mg of 4-aminophenazone. The solution keeps for at least 4 weeks at 4 °C. The enzymes are obtainable from most biochemical suppliers.

HCl–phenol solution: dissolve 5 g phenol in 1 litre of 0·1 M HCl. Alternatively, ready-to-use glucose assay kits are available from most biochemical suppliers.

5. Dry weight–turbidity calibration curve. The cells from 1 litre of a logarithmic phase culture in DBB + 10 mM glucose are sufficient for the measurement of dry weight and the construction of a turbidity correction curve (to correct for the deviation from Beer's law shown by turbid suspensions). The cells should be resuspended in DBB without glucose (*see* note 5).

6. Peristaltic pump: flow rate calibration curve.

SPECIFIC REQUIREMENTS

Inoculum culture of *B. natriegens* (100 ml) grown overnight at 20 °C, shaken, in DBB + 5 mM glucose. This yields a culture depleted of glucose but able to initiate logarithmic growth at 35 °C within 30 min
culture apparatus
supply of DBB (at least 3 litres in 100 ml and 500 ml batches)
glucose (1·0 M)
dry weight–turbidity calibration curve

for the determination of glucose
glucose standard (0·1 M)
GO reagent
HCl–phenol reagent
volumetric flasks (50 ml)
bulb pipettes (2 ml, 5 ml)
graduated pipettes (1 ml, 5 ml)
centrifuge tubes with 10 ml graduation
test tubes and racks
low speed centrifuge
water-bath (37 °C)
spectrophotometer

FURTHER INFORMATION

Values for the molar growth yield on glucose (Y_G) in defined media under aerobic conditions have been determined for bacteria and yeasts (Stouthamer, 1969; Payne, 1970). The values range from 48 to 120 μg μmol^{-1}.

The results of a class experiment are shown in Figures 2 and 3. The Y_G from these data are 57 g μmol^{-1} for logarithmic growth ($\mu = 1·53$ h^{-1}) in batch culture and 70 μg μmol^{-1} in continuous culture ($D = 0·8$ h^{-1}). The difference between these values is more likely to be due to experimental error than to real differences in the metabolism of the cells.

An unexpected phenomenon is observed when glucose is added to starved cells. As is shown in Figure 2, on addition of glucose the turbidity declines by about 10% within 2 min; the culture returns to its original turbidity during the next 10 or 15 min. During the last 10 min of the batch culture, the turbidity continues to increase although there is no detectable glucose remaining in the medium; if glucose is added when the turbidity has reached a stationary value, it then declines again by 10%. This change also occurs when the continuous culture is started. The response to glucose suggests that there is an error of about 10% in the dry weight–turbidity calibration curve we have used.

There are several ways in which the experiment can be extended. For example, comparisons can be made of growth and Y_G under aerobic and anaerobic conditions; or during continuous culture, the dilution rate could be increased to a value approaching μ_{max}, when glucose would become detectable in the medium.

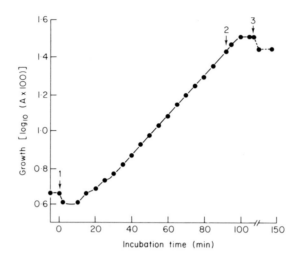

Figure 2   The growth of *Beneckea natriegens* in the defined medium with limiting glucose. Growth is expressed as $\log_{10}$ corrected absorbance ($\times 100$) as determined from the continuous recording of percentage transmission and a turbidity correction curve. 1: glucose (2·04 mM) added; 2: glucose exhausted; 3: continuous culture started.

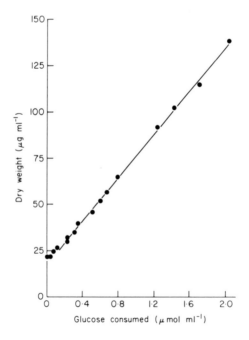

Figure 3   The determination of molar growth yield for glucose in a batch culture of *Beneckea natriegens* during the logarithmic phase of growth $Y_G = 57\,\mu\mathrm{g}\,\mu\mathrm{mol}^{-1}$.

REFERENCES

D. F. Niven, P. A. Collins and C. J. Knowles (1977). Adenylate energy charge during batch culture of *Beneckea natriegens. Journal of General Microbiology*, **98**, 95–108.

W. J. Payne (1970). Energy yields and growth of heterotrophs. *Annual Review of Microbiology*, **24**, 17–52.

A. H. Stouthamer (1969). Determination and significance of molar growth yields. *In* "Methods in Microbiology" (J. R. Norris and D. W. Ribbons, eds), Vol. 1, pp. 629–63. Academic Press, London and New York.

D. W. Tempest (1970). The continuous cultivation of micro-organisms I: Theory of the chemostat. *In* "Methods in Microbiology", (J. R. Norris and D. W. Ribbons, eds), Vol. 2, pp. 259–76. Academic Press, London and New York.

P. Trinder (1969). Determination of blood glucose using 4-aminophenazone as oxygen acceptor. *Journal of Clinical Pathology*, **22**, 246.

31

The Growth of *Pseudomonas aeruginosa* in Acetamide-limited Chemostat Culture: the Influence of Dilution rate on Amidase Activity

J. H. SLATER

Department of Environmental Sciences, University of Warwick, Coventry, CV4 7AL, England

---

Level: Advanced undergraduates
Subject area: Microbial physiology and growth

INTRODUCTION

Pseudomonas aeruginosa synthesizes an enzyme, aliphatic amidase (or acylamide aminohydrolase), during growth on a number of aliphatic amides, such as acetamide and propionamide. Acetamide induces the synthesis of this catabolic enzyme and permits the organism to grow on acetamide as the carbon and energy source and/or nitrogen source. The enzyme catalyses the following reaction:

$$CH_3CONH_2 + H_2O \rightarrow CH_3CHOO^- + NH_4^+.$$

The enzyme also catalyses a non-physiological reaction in which acethydroxymate is formed by the amidase-catalysed condensation of hydroxylamine with acetate thus:

$$CH_3CONH_2 + NH_2OH + H^+ \rightarrow CH_3CONHOH + NH_4^+$$

and acethydroxymate reacts with ferric ions to yield a coloured complex, providing a convenient colorimetric assay (Brammar and Clark, 1964).

The amidase is subject to catabolite repression in the presence of a number of other growth substrates, including succinate. Furthermore, it is possible to demonstrate the balance between induction and repression in chemostat grown cultures. Thus the major aim of this experiment is to examine the specific activity of the enzyme amidase in various steady state cultures grown at varying dilution rates in an acetamide-limited chemostat (Clarke *et al.*, 1968; Clarke and Lilly, 1969). In addition this experiment has the aim of introducing students to continuous-flow culture methods and an examination of the behaviour of micro-organisms growing under these conditions.

EXPERIMENTAL

Pseudomonas aeruginosa is grown in a defined, acetamide-limited medium as described in the materials section, in a chemostat continuous-flow culture system with a working culture volume of 1·0 litre. Amidase activity is measured in steady state cultures (absorbance measurements remaining constant for three times the culture doubling time) at dilution rates, $D\,(h^{-1})$: 0·1, 0·2, 0·3, 0·35, 0·45 and 0·6. Other dilution rates may be selected if necessary to give an adequate enzyme activity against dilution rate profile.

This practical has to be spread over a number of days since different steady state cultures are required, unless a number of chemostats operating at different dilution rates are available. Each steady state culture is treated in a similar fashion.

Amidase assay (60 min)

Using a bench centrifuge, harvest the cells from 20 ml of a steady state culture of *P. aeruginosa* from the chemostat.

↓

Discard the supernate culture fluid and re-suspend the pellet of organisms in 10·0 ml (20·0 ml, depending on enzyme activity) 0·1 M Tris–HCl buffer, pH 7·2.

↓

Store the cell suspension on ice until required for use.

↓

Measure the absorbance, at 600 nm, of the cell suspension

↓

Equilibrate 2·0 ml of the amidase assay mixture (*see* Materials) in acid washed test tubes at 37 °C for 5·0 min.

↓

At time zero add 2·0 ml of the cell suspension and incubate at 37 °C for 10 min (or an appropriate incubation time depending on the activity of the enzyme) (*see* note 1). Remember to include a reagent control containing 2·0 ml

Acethydroxymate standard curve (45 min)

Prepare a standard curve by heating 5·0 ml 0·4 M acetamide with 10·0 ml 5 M hydroxylamine–HCl in 4·5 M NaOH in a boiling water-bath for 15 min.

↓

Cool and neutralize with concentrated HCl and make up to 10·0 ml with distilled water in a volumetric flask. 1·0 ml of this solution contains 5 μmol acethydroxymate.

↓

Set up a standard curve over the range 0–5 mol acethydroxymate per assay tube, remembering to make the volume up to 4·0 ml with distilled water.

↓

Add 4·0 ml $FeCl_3$/HCl reagent and mix thoroughly.

↓

Read the absorbance at 500 nm in the spectrophotometer.

distilled water in place of the cell suspension
(*see* note 2).

↓

Terminate the enzyme's reaction by adding
4·0 ml 6% (w/v) $FeCl_3$ in 2% (w/v) HCl
reagent and mix thoroughly.

↓

Read the absorbance at 500 nm in the spectrophotometer (*see* note 3).

↓

Convert the enzyme assay readings to μmol acethydroxymate produced. Calculate the specific
activity in terms of μmol acetyhydroxmate produced per unit time (min) per unit cell suspension
(absorbance).

↓

Plot the amidase specific activity against the organisms' growth rate (= dilution rate) (*see* note 4).

- -

Notes and Points to Watch

1. Students should be encouraged to determine the optimum enzyme assay conditions with respect
   to the volume of cell suspension added and the incubation time.
2. With little increase in effort an enzyme assay against time may be set up.
3. By including a protein determination of the cell suspension, the enzyme specific activity may be
   recorded in terms of activity per unit time per unit mg protein.
4. Further information on the growth of *P. aeruginosa* in chemostat culture may also be obtained
   readily by measuring the biomass concentration (by measuring the culture dry weight) and the
   acetamide concentration (which may be measured by modifying the standard curve procedure).

MATERIALS

1. *Pseudomonas aeruginosa* strain 8602,
2. Any suitable chemostat continuous-flow culture
   system,
3. The mineral salts component of the growth
   medium in distilled water contains $(g\,l^{-1})$;
   K_2HPO_4, 12·5; KH_2PO_4, 3·8; $(NH_4)_2SO_4$, 1·0;
   $MgSO_4.7H_2O$, 0·1; 5·0 ml trace element
   solution.

   The trace element solution contains $(mg\,l^{-1})$;
   $FeSO_4(NH_4)_2SO_4.6H_2O$, 116; HBO_3, 232;
   $CoSO_4.7H_2O$, 95·6; $CuSO_4.5H_2O$, 8·0;
   $MnSO_4.4H_2O$, 8·0; $(NH_4)6Mo_7O_{24}.4H_2O$,
   22·0; $ZnSO_4.7H_2O$, 174.

After autoclaving the mineral salts medium,
filter sterilized acetamide solution is added to
give a final concentration of 1·25 g acetamide l^{-1}
(Brammar and Clarke, 1964).

Prepare the *amidase assay mixture* as follows:
1 vol. 0·4 M acetamide +
1 vol. 2·0 M hydroxylamine–HCl, pH 7·0 +
2 vol. 0·1 M Tris–HCl buffer, pH 7·2.
Make up sufficient assay mixture for the total
number of assays to be performed, and store on
ice until required for the assays.

SPECIFIC REQUIREMENTS

These items are needed for each assay.

50 ml 0·4 M acetamide
50 ml 2·0 M hydroxylamine–HCl, pH 7·0
  (N.B. this must be freshly neutralized just
  before use)
100 ml 0·1 M Tris–HCl buffer, pH 7·2
200 ml 6·0% (w/v) FeCl$_3$ in 2·0% (w/v) HCl
50 ml 5 M hydroxylamine–HCl in 4·5 M NaOH

20 acid-washed test tubes in test tube rack
selection of 1·0, 2·0, and 5·0 ml non-sterile
  pipettes
ice in ice-bucket
spectrophotometer and cuvettes
100 ml volumetric flask
bench/low speed centrifuge and centrifuge tubes
amidase assay mixture

FURTHER INFORMATION

Typically the amidase specific activity shows a peak of activity at a medium dilution rate with lower activities at higher and lower dilution rates (Figure 1). Clarke *et al.* (1968) showed that the low specific activities at low dilution rates were due to very low unused acetamide concentrations which, since this compound acts as an inducer, failed to induce high levels of amidase activity. At high dilution rates in faster growing organisms full induction was possible, but the lower levels were due to the effects of catabolite repression.

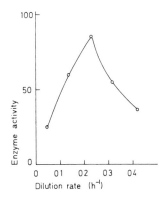

Figure 1

REFERENCES

W. J. Brammar and P. H. Clarke (1964). Induction and repression of *Pseudomonas aeruginosa* amidase. *Journal of General Microbiology*, **37**, 307–19.

P. H. Clarke, M. A. Houldsworth and M. D. Lilly (1968). Catabolite repression and the induction of amidase synthesis by *Pseudomonas aeruginosa* 8602 in continuous culture. *Journal of General Microbiology*, **51**, 225.

P. H. Clarke and M. D. Lilly (1969). The regulation of enzyme synthesis during growth. *In* "Microbial Growth", 19th Symposium of the Society for General Microbiology (P. M. Meadow and S. J. Pirt, eds), pp. 113–59.

32

Microbiological Assay of a Vitamin—Nicotinic Acid

B. J. B. WOOD

Department of Applied Microbiology, University of Strathclyde, Glasgow, G1 1XW, Scotland

and the late

F. W. NORRIS

formerly *Department of Biochemistry, University of Birmingham, Birmingham, England*

> *Level*: Advanced undergraduates
> *Subject areas*: Bioassay, statistics
> *Special features*: Biological application of statistics

INTRODUCTION

Although they are now being superseded to some extent by chemical and radio-immune assays, microbiological assays still offer a valuable combination of simplicity and specificity in clinical, nutritional and research work. As an exercise in the undergraduate teaching laboratory, they possess the additional advantage of being a good test of care and precision in procedures and calculations.

Bioassays for vitamins and other growth factors depend for success on meeting six criteria:

(1) The test organism must have an absolute requirement for the nutrient in question, i.e. be **auxotrophic** for it.

(2) The growth of the organism must bear a linear relationship to the concentration of the nutrient, or a relationship which can be made linear (e.g. by taking logarithms) (Norris, 1955).

(3) Extraction procedures must convert the nutrient into a standard form. Some organisms may respond differently to different chemical forms of nutrients. For example with vitamin B12, the use of different test organisms determines which forms of the vitamin are assayed (*see*, for example, Provasoli and Carlucci, 1974; Hashami, 1973).

(4) The basal medium must contain none, or only trace amounts, of the nutrient being assayed.

(5) The basal medium must contain an adequate amount of all other nutrients needed by the test organism.

(6) The means by which growth is measured must bear a linear relationship to the amount of growth which has taken place. When lactic acid bacteria are used as assay organisms, homofermentative strains must be used if growth is to be assessed by measuring the amount of acid which has been produced, since the ratios of fermentation products from heterofermentative strains might be affected by their nutritional status.

When assays of this type are being carried out in a manufacturer's or analyst's laboratory, it is essential that tests be made from time to time to check for any interference with the assay by inhibitory substances present in the extract. This type of test is time-consuming, and therefore inappropriate for most undergraduate classes; the vitamin, test organism and procedure which have been selected for the present exercise minimize the danger of such interference, but in professional work it should be checked routinely.

The aim of this experiment is to analyse a range of foods for their nicotinic acid content by measuring the production of lactic acid by *Lactobacillus plantarum*. For consistency, the term "nicotinic acid" has been used throughout, but "niacin" is now widespread and refers to the same chemical compound.

EXPERIMENTAL

Ideally a group of students should work together assaying a range of foods between them (*see* note 1). For simplicity the procedure for examining ground brewer's malt (nicotinic acid content about $100 \, \mu g \, g^{-1}$) is described. The amounts taken and dilutions used should be adjusted accordingly when assaying other materials. Careful preparation of glassware, etc. is vital to the success of the experiment (*see* notes 2 and 3).

Day 1 *(1 h)*

Weigh a sample of the food (*see* notes 4 and 5) in a **dry** beaker.

↓

Add 1 M hydrochloric acid (50 ml).

↓

Cover the beaker with a watch glass.

↓

Autoclave for 15 min at 121 °C.

↓

Remove from the autoclave.

↓

Cool and store in a refrigerator overnight.

- -

Day 2 *(3h)*

To the autoclaved, cooled mixture, add 2 ml of 2·5 M sodium acetate solution, then correct to approximately pH 4·5 with 1 M sodium hydroxide, using bromocresol green indicator solution or a narrow range indicator paper to judge the end-point (*see* note 6).

↓

Make up to 500 ml in a volumetric flask.

↓

Mix thoroughly and filter a portion of the liquid through a **dry** filter paper.

↓

Take 25·0 ml of the clear filtrate, adjust to pH 6·8 and make to final dilution, e.g. 1/10.

↓

For one assay, 30 test tubes fitted with glass or aluminium caps are required. Each dose-level of standard vitamin and sample is set up in triplicate thus:

Tubes	1–3	4–6	7–9	10–12	13–15	16–18	19–21	22–24	25–27	28–30
Standard vitamin	0	0·5	1·0	1·5	2·0	3·0	—	—	—	— ml
Sample ext.	—	—	—	—	—	—	1·0	2·0	3·0	4·0 ml
Water	5	4·5	4·0	3·5	3·0	2·0	4·0	3·0	2·0	1·0 ml
Medium	5	5	5	5	5	5	5	5	5	5

(double-strength) (*see* notes 7 and 8).

↓

Cap the tubes and autoclave them for 10 min at 121 °C (*see* notes 9 and 10).

↓

Prepare the inoculum of *Lactobacillus plantarum* strain NCIB 6376 by transferring a loopful of stock culture into 10 ml of sterile Bacto Micro-Inoculum broth contained in a 1 oz. screw-capped "Universal" bottle or a capped or plugged centrifuge tube.

↓

Incubate for 15–24 h at 30 °C (*see* note 11).

- -

Day 3 *(1h)*

Centrifuge the cell crop of *L. plantarum*, decant the spent medium, resuspend the cell pellet in sterile 0·85% sodium chloride solution (10 ml). Re-centrifuge.

↓

Repeat the washing procedure.

↓

Suspend the twice-washed pellet in sterile saline.

↓

Add one drop of the suspension to a 10 ml portion of sterile saline. Inoculate each assay tube with one drop of this diluted cell suspension. Mix each inoculated tube by rotating it between the hands. Randomize the tubes (*see* note 12) if required, and place in the incubator.

↓

Incubate for 72 h at 30°C. Mix the contents of each tube at least once every 24 h.

_____

Day 4 *(1½ h)*

At the end of the incubation period (*see* note 13) take each tube in turn, wash its contents into a conical flask and make up to about 60 ml with distilled water. Add bromothymol blue indicator solution (*see* note 14).

↓

Titrate with 0·1 M sodium hydroxide to pH 6·8 (*see* notes 15, 16 and 17).

↓

Record the results in the manner shown.

_____

Notes and Points to Watch

1. Many breakfast cereals and some other foods list a minimum niacin content on the packet. "Ovaltine" used to do this but has now discontinued the practice; at the time of writing, the preparation still assays at about 0·23 mg g^{-1}. An interesting exercise is to compare branded foods with similar ones sold on "own brand" labels by supermarkets, etc. which do not list vitamin content on the packet.

2. All water used in preparing media, etc. must be glass-distilled; de-ionized water is not satisfactory.

3. For the greatest precision, glassware should be cleaned with chromic acid. Since use of this dangerous preparation in laboratories should be kept to a minimum, it will normally be sufficient to clean the glassware stringently with Decon, or another similar proprietary cleanser. It is suggested that freshly prepared cleaning solution be used for vitamin assay glassware, and that solutions previously used on other glassware not be used. It is convenient to set aside a set of glassware reserved for the vitamin assay.

4. Take samples of food, etc. to give about 0·05 μg nicotinic acid ml^{-1} after final dilution. Thus 1 g of ground malt, which contains about 100 μg nicotinic acid, is diluted finally to the equivalent of 2000 ml. Sample size and dilution of other materials will need to be adjusted accordingly.

5. Breakfast cereals, malt, "Ovaltine", etc. are very hygroscopic, and weighing must be completed as rapidly as possible if accuracy is to be maintained. Therefore, weigh to within a band $\pm 10\%$ of the calculated "target" weight, recording the actual weight to four decimal places.

6. Adjust the pH of half the solution to about the desired value (it does not matter if you slightly

"overshoot"), noting the amount of alkali used. Then add the rest of the solution and add most of the alkali needed, mix and add the alkali drop-wise until the desired pH is obtained. A pH meter may seem more appropriate but adequate precision is obtained using pH paper or an indicator solution and permits several students to be working at the same time, whereas to specify a pH meter may cause unnecessary delays.

7. The tubes are conveniently filled from burettes; the double-strength medium could be added with an automatic pipette if available. The materials should be added in the order indicated, as accurately as possible, and the contents of each tube must be carefully mixed.

8. The full assay procedure described here has several advantages. However, if the amounts of materials to be taken for assay are carefully prescribed and the students' workings are checked, economies of glassware and materials can be achieved by using a five-point assay. In the latter case, only the first, second, fifth, seventh and ninth sets of tubes will be required.

9. Since the tubes are about two-thirds full, there is a danger of spurting and loss of contents if the autoclave loses pressure too rapidly, leading to boiling of the tubes' contents at the end of the autoclaving cycle. It is a useful precaution to autoclave a set of tubes filled with 10 ml each of water as a check on this problem, before commencing a series of experiments.

10. The tubes can be cooled after autoclaving and inoculated at once. In this case the preparation of the inoculum should be started earlier. It is desirable that a 24 h culture be used to inoculate the experimental tubes and since more than one day may well elapse between two "Days" in the experimental description imposed here, the experimental design may require some modification to allow for the preparation of the inoculum.

11. Growth of the *Lactobacillus* will be indicated by the development of a silky turbidity.

12. For the most exact work it is essential that the order in which the tubes are inoculated, and their position in the incubator or water-bath, should be completely at random. Any orderly arrangement may lead to conditioned results; however, for undergraduate class work it is often easier to accept this risk, for the sake of smoother and more expeditious working.

13. It is quite acceptable for the growth measurement to be delayed for one or even two more days, but for the most precise work you may prefer to transfer the tubes to a refrigerator after the 72 h incubation has been completed.

14. The amount of indicator solution which should be added will obviously be influenced by the intensity of colour of the medium and will need to be found experimentally for each set of assays.

15. It is suggested that one of the "blank" tubes is titrated at the pH meter, or with the aid of narrow-range pH paper, to pH 6·8, then used as a comparison for the remaining tubes. If the turbidity caused by the bacterial cells causes difficulty in estimating the end-point at the higher vitamin concentrations, it is helpful to add a pinch of filter-aid to the reference flask.

16. As described, the programme calls for titration of the samples. Turbidimetry or nephelometry can be used if preferred; with optically matched tubes, these methods can be performed directly on the tubes used for incubation. We have found that titration is easy, economical and as it requires no special equipment it can be carried out by several workers at the same time. It also permits the use of ordinary, inexpensive glassware throughout, rather than cuvettes or optically matched tubes.

17. *Lactobacillus plantarum* represents a minimal health hazard, but as a matter of good hygienic practice, flasks after titration should be emptied into Lysol, Domestos or other suitable disinfectant, pending disposal to the drains. An alternative is to collect the wastes and autoclave them.

MATERIALS

1. *Lactobacillus plantarum* strain NCIB 6376. This is the organism referred to as *L. arabinosus* 17/5 in the older literature.
2. Nicotinic acid assay medium, e.g. Oxoid No. CM.211 made up according to the manufacturer's directions.
3. Materials for assay, e.g. finely ground pale ale malt; "Ovaltine"; breakfast cereals which report a niacin or nicotinic acid content on the package.
4. Stock nicotinic acid solution. About 50 mg is weighed out accurately and made up to 50·0 ml with water. The working solution is prepared by diluting the stock solution 1:10000, giving a solution containing 0·1 μg nicotinic acid ml^{-1}.

SPECIFIC REQUIREMENTS

These items are needed for each student.

Day 1
materials for assay
molar hydrochloric acid
0·1 M sodium hydroxide
a 25 ml bulb pipette
1 ml and 10 ml pipettes
nicotinic acid solution
100 ml volumetric flasks
100 ml beaker, plus watch glass to cover it

Day 2
L. plantarum
nicotinic acid assay medium
30 test tubes plus caps
test tube racks
burettes (1 per student assuming that they work in groups of 4)
2·5 M sodium acetate solution
molar sodium hydroxide solution
Bromocresol green indicator, **or** narrow range
pH paper to cover pHs 4·5 and 6·8, **or** a pH meter suitable for student use
100 ml volumetric flask
500 ml volumetric flask
filter paper (Whatman no. 1)
filter funnel
25 ml bulb pipette
1 ml and 10 ml pipettes

Day 3
sterile saline, 10 ml portions
centrifuge
sterile Pasteur pipettes

Day 4
burette
3 conical flasks (250 ml capacity)
bromothymol blue indicator
0·1 M sodium hydroxide.
bucket containing disinfectant for discarding titration residues

---

FURTHER INFORMATION

A typical set of data is presented in Table 1. By recording the results in this format the subsequent calculations are made easier. The results were those obtained by a student and the standard solution had been made up at twice the correct strength resulting in the standard curve departing from linearity at the higher doses (Figure 1); therefore only the results between the 0 and 1·0 ml dose levels were used in subsequent calculations. The sample points all lie on a straight line and were therefore all used for calculation. All points on the sample plot lie below the highest point on the linear portion of the standard plot; experienced workers prefer to discard points on the sample plot which lie above the highest point on the plot of the standard. The sample and standard plots show reasonable convergence at the origin.

Table 1   Protocol and results of a microbiological assay of nicotinic acid.

1. *Description of sample*
   Weight of sample taken: 1·0165 g
   This was dissolved in 500 ml and then diluted 1/10
   1 ml of sample extract dilution as used in the assay = 0·2033 mg of sample

2. *Stock nicotinic acid solution:* 0·2064 g $(100 \text{ ml})^{-1}$
   Dilution: 1:10 000
   Working standard solution: 0·2064 μg ml^{-1}

3. *Titrations*

Vol. of standard, ml	0	0·5	1·0	1·5	2·0	3·0
Titres, ml (i)	2·3	5·85	8·8	11·0	12·45	14·35
Titres, ml (ii)	2·3	5·65	9·0	11·0	12·15	14·25
Titres, ml (iii)	2·0	5·6	8·8	10·8	12·0	14·4
Mean titre, ml	2·2	5·7	8·87	10·93	12·2	14·33

Vol. of sample extract, ml			1	2	3	4
Titres, ml (i)			3·3	4·5	5·8	6·8
Titres, ml (ii)			3·25	4·5	5·95	6·9
Titres, ml (iii)			3·2	4·5	5·65	6·8
Mean titre, ml			3·25	4·5	5·8	6·83

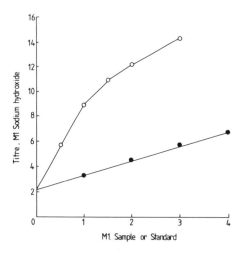

Figure 1   Niacin assay; plot of experimental results: O——O, standard; ●——●, sample.

A graph of the results should always be prepared and examined as above and for less precise work the results for the sample can be read off against a line of best fit drawn by eye for the standard. For more precise work a statistical computation must be made. Many methods are available and the choice is a matter for individual preference. We use a procedure involving evaluating the individual points of the sample results in terms of the calculated line of best fit for the standard and also a slope ratio calculation. This shows that even closely related methods will give slightly different results with a typical set of data.

Calculation of a Line of Best Fit

Take the average for each set of 3 titres, discarding any result which obviously differs from its two companions.

The line of best fit is then found by use of the standard equation of a straight line, $y = mx + c$, as

$$T = a + bV, \tag{1}$$

where

T = titre (mean of the 3 observed values)
V = volume of vitamin solution
a = intercept on the "response" axis, i.e. the blank value
b = slope of the line.

Further, let

ΣV = sum of the volumes of vitamin solution
ΣVT = sum of the products of V and T at each dose level
ΣV^2 = sum of the squares of V at each dose level
ΣT = sum of the mean titres at each dose level
$\Sigma^2 V$ = square of the sum of volumes of vitamin solution
N = the number of dosage levels actually used in the computation (remembering that some may have to be discarded because of obvious error.

Using the results in Table 1, a line of best fit for the standard may be calculated as follows:

$$T = a + bV$$

$$a = \frac{(\Sigma V \times \Sigma VT) - (\Sigma V^2 \times \Sigma T)}{\Sigma^2 V - N\Sigma V^2}$$

$$= \frac{(1 \cdot 5 \times 11 \cdot 72) - (1 \cdot 25 \times 16 \cdot 77)}{2 \cdot 25 - (3 \times 1 \cdot 25)}$$

$$= 2 \cdot 255$$

$$b = \frac{(\Sigma V \times \Sigma T) - (N\Sigma VT)}{\Sigma^2 V - N\Sigma V^2}$$

$$= \frac{(1 \cdot 5 \times 16 \cdot 77) - (3 \times 11 \cdot 72)}{2 \cdot 25 - (3 \times 1 \cdot 25)}$$

$$= 6 \cdot 67,$$

whence,

$$T = 2 \cdot 255 + 6 \cdot 67 V$$

i.e.

$$V = \frac{T - 2 \cdot 255}{6 \cdot 67}.$$

The values for the mean sample titres are each substituted for T in this equation, to give values for V in each case, and V for 1 ml is calculated for every level as described. These calculations are shown in Table 2.

Table 2   Steps in the calculation of the nicotinic acid content of the sample.

Sample (ml)	Mean titre (T) (ml)	V^a	V for 1 ml	μg nicotinic acid (g sample)$^{-1}$
1	3·25	0·1492	0·1492	151·5
2	4·5	0·3366	0·1683	170·9
3	5·8	0·5315	0·1772	179·9
4	6·83	0·6859	0·1715	174·1
			Mean	169·1
			s.d.	12·3

^{a} Calculated by substituting each T value into $V = (T - 2·255)/6·67$ which describes the dose–response line of the standard.

Since the standard solution contained 0·2064 μg nicotinic acid ml^{-1}, and the sample extract dilution contained 0·2033 mg dry sample ml^{-1} the values of V for 1 ml are therefore multiplied by

$$0·2064 \times \frac{1000}{0·2033} = 1015·25$$

to give the calculated contents of nicotinic acid in μg g^{-1} sample; these results are also listed in Table 2. From these results, the mean figure for the nicotinic acid content of the malt was 169·1 μg nicotinic acid (g malt)$^{-1}$. This result was rather higher than expected but not unacceptably so. The results exhibit rather a wide scatter (standard deviation 12·3), but this is not unreasonable for a student's first attempt at an assay.

For the calculation using the slope ratio method, a line of best fit is calculated for the sample plot, remembering to use the blank value as one of the points on the plot (this is frequently overlooked, so strong is the association of the blank with the standard).

$$a = \frac{(10 \times 56·97) - (30 \times 22·58)}{100 - (5 \times 30)}$$

$$= 2·154$$

$$b = \frac{(10 \times 22·58) - (5 \times 56·97)}{100 - (5 \times 30)}$$

$$= 1·181,$$

whence

$$T = 2·154 + 1·181\, V.$$

It will be observed that the calculated blank of 2·15 ml is in reasonable agreement with that calculated for the standard (2·26 ml) and the value of 2·2 ml actually found.

The vitamin content of the sample is then calculated using the equation

$$\frac{(\text{slope for sample}) \times (\text{vol. of extract}) \times (\text{std. vitamin ml}^{-1})}{(\text{slope for standard}) \times (\text{mass of sample})}.$$

Inserting numerical values this becomes

$$\frac{1{\cdot}181 \times 5000 \times 0{\cdot}2064}{6{\cdot}67 \times 1{\cdot}0165}$$
$$= 179{\cdot}8 \ \mu\mathrm{g}\,\mathrm{g}^{-1},$$

which compares reasonably well with the previously calculated value of $169{\cdot}1\ \mu$g vitamin (g malt)$^{-1}$.

Clearly it is of advantage to put these calculations onto a computer, but such is the power of modern calculators that they will fit easily onto quite inexpensive programmable pocket calculators. For example, the Texas Instruments Programmable 57 model can handle the program because of its useful combination of pre-programmed statistical functions and the nesting facility within its 50 program-step capacity. It is recommended that students be encouraged to try preparing and running a program.

ACKNOWLEDGEMENT

Thanks are due to Mrs M. Provan and Mrs J. Winter for stoically enduring massive revisions of this manuscript.

REFERENCES

M.-U.-H. Hashami (1973). "Assay of Vitamin in Pharmacuetical Preparations", 512 pp. John Wiley and Sons, Chichester and New York

F. W. Norris (1955). The microbiological assay of vitamins. "The Extra Pharmacopoeia", Vol. II, pp. 528–38.

L. Provasoli and A. F. Carlucci (1974). Vitamins and growth regulators, *In* "Algal Physiology and Biochemistry", (W. D. P. Stewart, ed.), pp. 749–51. Blackwell Scientific, Oxford.

Manufacturers of media produce useful and informative literature relating to their products for these assays. The Difco and Oxoid Companies' pamphlets on microbiological assay are particularly recommended.

Note added in Press

Dr F. W. Norris died while this chapter was being prepared. I had discussed it with him in general terms, but he did not comment in detail upon the chapter as presented here. He taught me the methods, and the application of statistical procedures to the evaluation of the results is his also, but I alone am responsible for (or guilty of) errors of fact or infelicities of style, which I know that he would have detected and excised with his typical combination of ruthlessness and good humour.

Brian J. B. Wood

Induction of an Anaerobic Electron Transport Chain in *Escherichia coli*

L. A. HUISMAN and W. N. KONINGS

Department of Microbiology, University of Groningen,
Kerklann 30, 9751 NN Haren, The Netherlands

> *Level*: Advanced undergraduates
> *Subject areas*: Microbial chemistry and
> biochemistry
> *Special features*: Involves preparation and use of
> membrane vesicles

INTRODUCTION

Escherichia coli is a facultative anaerobe, which obtains energy by substrate-level phosphorylation and electron transfer under both aerobic and anaerobic conditions. Aerobically, a respiratory chain is operative. Under anaerobic conditions, in the absence of exogenous terminal electron acceptors, a fumarate reductase system is present in which fumarate functions as electron acceptor and is converted to succinate. In the absence of fumarate in the growth medium this system has a low activity. Supplementation of the growth medium with fumarate results in an increase of the activity.

Another anaerobic electron transfer system that can be formed in *E. coli* is the nitrate respiration system. This system is induced during anaerobic growth in the presence of nitrate (for review *see* Konings and Boonstra, 1977). Nitrate functions in this system as terminal electron acceptor and is reduced to nitrite. All electron transfer systems are tightly incorporated in the cytoplasmic membrane. The nitrate respiratory chain resembles in a number of aspects the respiratory chain (in which oxygen functions as terminal electron acceptor).

Anaerobic growth of *E. coli* in the presence of nitrate can occur at the expense of a variety of energy sources. The electron donor for nitrate respiration is most likely formate. Formate is oxidized to CO_2 and H_2O by a membrane-bound formate dehydrogenase, a protein with a molecular weight of about 400 000 daltons, which contains molybdenum, selenium, a *b*-type cytochrome and iron–sulphur centres. The electrons derived from formate are subsequently transferred via ubiquinone to nitrate reductase, an enzyme catalysing the reduction of nitrate to nitrite. Nitrate reductase contains molybdenum, non-heme iron and a cytochrome of the b-type, cyt $b_{556}^{NO_3^-}$. The components of this anaerobic respiratory chain are located in the cytoplasmic membrane in such a way that the flow of electrons through the chain results in the generation of a proton motive force (p.m.f.) across this membrane (Boonstra and Konings, 1977). According to the chemiosmotic coupling theory (Mitchell, 1970), this p.m.f. is the driving force for energy dependent processes, like ATP synthesis and transport of solutes against a concentration gradient.

The aim of this experiment is to measure the activities of two enzymes of the nitrate respiration system of $E.$ $coli$, formate dehydrogenase and nitrate reductase. Since the nitrate respiration system is membrane-bound, the activities of both enzymes can be measured in isolated membrane preparations (so-called membrane vesicles). In this way interference of membrane-bound activites with activities of the cytoplasm can be avoided (Kaback, 1971).

EXPERIMENTAL

Students are supposed to prepare and, if necessary, sterilize all solutions and growth media themselves. The composition and sterilization of growth media is described in the Materials section, the concentrations of all other solutions are specified in the text. The time needed for preparation of the solutions is not included in the programme outlined for this experiment.

The experiment can best be performed by students working in pairs.

Day 1 (inoculation of preculture, 10 min)

Inoculate a loopful of cell material of $E.$ $coli$ ML 308–225 from a slant culture in a 100 ml Erlenmeyer flask containing 40 ml basal medium supplemented with 1 % (w/v) sodium succinate.

↓

Incubate the flask in a rotary incubator at 37 °C and 200 rev. min^{-1} for 14–18 h.

- -

Day 2 (inoculation of main culture, 10 min)

Pipette 5 ml of the preculture in a 3 litre Erlenmeyer flask, entirely filled with sterile basal medium supplemented with 0·5 % glucose, 50 mM KNO_3, 1 μM Na_2SeO_3 and 1 μM Na_2MoO_4.

↓

Close the Erlenmeyer flask with a rubber stopper which should be secured with tape.

↓

Incubate the flask at 37 °C, with gentle stirring for 16–20 h (use a magnetic flea).

- -

Day 3 (isolation of membrane vesicles, 4–5 h)

Near the end of the incubation period make regular measurements of the optical density at 660 nm of the culture. When A_{660} reaches about 0·7 the cells should be harvested.

↓

Add chloramphenicol (final concentration 50 μg ml^{-1}) to stop growth.

↓

Centrifuge the culture in preweighed centrifuge bottles for 20 min at 13 000 g.

↓

Discard the supernate and determine the wet weight of the cells by weighing the bottles. Usually a cell yield of about 6 g wet weight is obtained from a 3 litre culture.

↓

Resuspend the pellets in prewarmed (37 °C) 30 mM Tris–HCl buffer, pH 8·0, containing 20 % (w/v) sucrose. Use 80 ml buffer per gram wet weight.

↓

To each ml of suspension add 50 μg chloramphenicol, 200 μg lysozyme and 0·4 M potassium-EDTA to a final concentration of 10mM.

↓

Incubate the suspension at 37 °C and follow spheroplast formation with a phase contrast microscope. Continue the incubation until almost all cells are transformed into spheroplasts; this will take about 30 min.

↓

Centrifuge the suspension for 30 min at 13 000 g.

↓

Discard the supernate and suspend the spheroplast pellet gently in 10 ml solution consisting of 50 mM potassium phosphate, pH 6·6, 20 % sucrose and 2 mM MgSO$_4$. The pellet should be suspended with the aid of a 25 ml syringe fitted with a needle of 10 cm length and an internal diameter of 1 mm.

↓

Dilute the spheroplast suspension to 300 ml in a solution consisting of 10 mM potassium phosphate buffer, pH 6·6, 10 mM MgSO$_4$, 10 mg DNase and 10 mg RNase. The spheroplasts will lyse in this buffer by the difference between the internal and external osmotic pressures. DNase and RNase will hydrolyse the nucleic acid and thereby decrease the viscosity of the solution. The membrane fragments will reseal and will form closed membrane vesicles.

↓

Incubate the mixture for 15 min at 37 °C.

↓

Add potassium–EDTA, pH 7, to a final concentration of 10 mM, in order to release membrane bound DNA and RNA fragments.

↓

Incubate again for 15 min, then add $MgSO_4$ to a final concentration of 15 mM (10 mM was already present). Mg^{2+} is required for DNase activity. DNase will be reactivated and degrades DNA fragments.

$$\downarrow$$

Incubate again for 15 min and centrifuge the suspension at low speed (1 h at 800 g), in order to remove remaining spheroplasts and whole cells, and discard the pellet.

$$\downarrow$$

Centrifuge the supernate containing the membrane vesicles for 1 h at 48000 g.

$$\downarrow$$

Discard the supernate and resuspend the vesicles in 50 mM potassium phosphate buffer, pH 6·6, to obtain a protein concentration of about 5 mg ml^{-1}. Usually about 10 ml of phosphate buffer is required. Keep the membrane preparation on ice or freeze it rapidly in liquid nitrogen. Prior to the experiment the frozen suspension is thawed rapidly in a water-bath at 50 °C.

$$\downarrow$$

Determine the protein concentration of the preparation by the method of Lowry *et al.* (1951), using bovine serum albumin as a standard.

- -

Day 4 *(enzyme assays, 3–4 h)*

1. *Formate dehydrogenase*
The rate of oxidation of formate is determined by measuring the reduction rate of an artificial electron acceptor, the redox dye 2,6-dichlorophenolindophenol (DCPIP). The specific light absorption of this compound at 600 nm decreases upon reduction.

Into an anaerobic cuvette (1 cm light path, 1 ml volume, *see* Figure 1) pipette:

0·1 M potassium phosphate buffer, pH 6·6,	0·5 ml
0·1 M $MgSO_4$,	0·1 ml
H_2O,	0·36 ml

$$\downarrow$$

Bubble nitrogen through the solution for 5 min and remove the hypodermic needles.

$$\downarrow$$

With a micro-syringe add through the rubber seal:

10 mM DCPIP,	10 μl
33 mM phenazine methosulphate (PMS, *see* note 1),	10 μl
membrane preparation,	10 μl

$$\downarrow$$

Record the endogenous DCPIP reduction rate for some minutes with a recording spectrophotometer at 600 nm and 37 °C.

$$\downarrow$$

Add 10 μl 1 M potassium formate to start the reaction, and again record the reduction rate of DCPIP.

↓

Calculate from the slopes of the curves the formate dehydrogenase activity in nmol formate oxidized per min per mg of protein. The extinction coefficient of oxidized DCPIP is 21 mM^{-1} cm^{-1} at 600 nm.

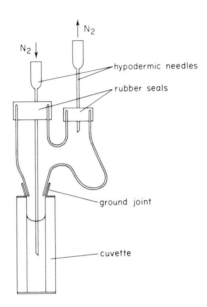

Figure 1   Anaerobic cuvette.

2. Nitrate reductase

The activity of this enzyme is determined by measuring the oxidation rate of reduced benzyl viologen. This redox dye donates electrons to nitrate reductase, which in turn reduces NO_3^- to NO_2^-. The oxidation of benzyl viologen causes an increase in its absorption at 600 nm.

Pipette into an anaerobic cuvette (1 ml, 1 cm light path):

0·1 M potassium phosphate buffer, pH 6·6,	0·5 ml
0·1 M MgSO$_4$,	0·1 ml
166 mM benzyl viologen (reduced) (*see* note 1),	10 μl

↓

Bubble nitrogen through the solution with a hypodermic needle for 5 min. Then add:

0·2 M sodium dithionite,	10 μl
membrane preparation,	10 μl

↓

Record the endogenous oxidation of benzyl viologen with a spectrophotometer at 600 nm and 37 °C.

↓

Start the reaction by adding 10 μl 1 M potassium nitrate and record again the oxidation of benzyl viologen.

↓

Calculate the nitrate reductase activity in nmol nitrate reduced per min per mg of protein, taking into account that reduced benzyl viologen donates two electrons to nitrate reductase. The extinction coefficient of oxidized benzyl viologen is 7·4 mM^{-1} cm^{-1} at 600 nm.

3. *Coupled activities of nitrate reductase and formate dehydrogenase*

The functioning of the complete nitrate respiration chain is tested by measuring the production of nitrite from nitrate in the presence of formate as electron donor.

Into 5 sample and 5 reference tubes (small test tubes, 1·2 × 7·5 cm) pipette the following reaction mixture:

	sample tubes 1–5	reference tubes 1–5
0·1 M potassium phosphate buffer, pH 6·6,	40 μl	40 μl
0·1 M MgSO$_4$,	10 μl	10 μl
0·1 M sodium formate,	10 μl	— μl
H$_2$O,	25 μl	35 μl
membrane preparation,	5 μl	5 μl

↓

Incubate the tubes in a water-bath at 37 °C. The solutions are stirred with small magnetic fleas. A suitable experimental set up is given in Experiment 20.

↓

Using a hypodermic needle, blow water-saturated nitrogen into the tubes, just above the surface of the liquid, for 5 min.

↓

Start the reaction by adding 10 μl 0·1 M potassium nitrate to all tubes.

↓

Stop the reactions in both sample and reference tubes by adding 0·4 ml ice-cold water to both tubes 1 after 1 min, tubes 2 after 2 min, and so on until tubes 5 after 5 min. Each time, immediately add 0·5 ml of a 1 % (w/v) sulphanilamide solution in 2·5 N HCl.

↓

Incubate for 15 min at 0 °C (on ice).

↓

Add 0·5 ml 0·02 % (w/v) N-1-naphtylethylene diamine. Incubate for 30 min at 30 °C.

↓

Add 2·0 ml water and measure the extinction at 540 nm with a spectrophotometer.

↓

Calculate the nitrate concentration from a calibration curve as outlined below.

↓

Weigh accurately about 70 mg $NaNO_2$, and dissolve in 1000 ml distilled water.

↓

Dilute 3 ml of this stock solution into 100 ml distilled water.

↓

From this diluted solution pipette 0·1, 0·2, 0·3, 0·4 and 0·5 ml into test tubes.

↓

Add distilled water to a total volume of 0·5 ml.

↓

Proceed from here as described above, by adding 0·5 ml 1 % sulphanilamide in 2·5 N HCl, and so on. Extinctions are measured against a reference containing 0·5 ml distilled water instead of nitrite.

↓

Plot the extinction values against the $NaNO_2$ concentrations.

↓

Calculate the coupled activities of formate dehydrogenase and nitrate reductase in nmol NO_2^- produced per mg protein per min.

- -

Notes and Points to Watch

1. PMS and benzyl viologen are light sensitive. Vials, tubes and syringes containing these compounds should therefore be protected from the light by wrapping them in aluminum foil.

MATERIALS

1. *Escherichia coli* strain ML 225-308, subcultured monthly on agar slants. This strain is available from the author.
2. Basal medium: K_2HPO_4 (7·0 g l^{-1}), KH_2PO_4 (3·0 g l^{-1}), trisodiumcitrate. $3H_2O$ (0·5 g l^{-1}), $MgSO_4 . 7H_2O$ (0·1 g l^{-1}), yeast extract (1 g l^{-1}) in distilled water; pH is corrected to 7·0 with KOH. All components except the phosphate buffer are heat-sterilized together in 80 % of the final volume. The phosphate buffer is sterilized separately in a 10-fold concentrated solution, 10 % of the final volume, and added aseptically to the basal medium after cooling. Supplements, such as succinate or a mixture of glucose, KNO_3,

Na_2SeO_3 and Na_2MoO_4 are also sterilized separately in 10-fold concentrated solutions, in 10% of the final volume of the complete growth medium. When necessary, agar is added up to 1·5%. All sterilizations are carried out in an autoclave at 120 °C, for 20 min.

3. Chloramphenicol, Boehringer Mannheim, W. Germany.
4. DCPIP, sodium salt, Fluka A. G., Buchs, Switzerland.

5. Benzyl viologen, BDH Chemicals Ltd., Poole, England.
6. Lysozyme, Merck, Darmstadt, W. Germany.
7. RNase and DNase, Miles Laboratories Ltd., Elkhart, Indiana, USA.
8. PMS, Sigma Chemical Co., St. Louis, Missouri, USA.
9. Anaerobic cuvettes, Hellma GmbH, Müllheim/Baden, W. Germany.

SPECIFIC REQUIREMENTS

Day 1
agarslants of *E. coli* ML 225–308
several screw-capped bottles, beakers, calibrated cylinders, pipettes (various sizes)
2 sterile pipettes, 5 ml
chemicals, spatulas
(analytical) balance
Erlenmeyer flask, 100 ml with cotton plug
Erlenmeyer flask, 3 litre with rubber stopper and flea
pH meter, calibration buffer(s)
magnetic stirrer + fleas
autoclave, aluminum foil
incubator shaker, 37 °C

Day 2
sterile pipette, 5 ml
tape
magnetic stirrer

Day 3
chloramphenicol, spatula
centrifuge
balance
0·4 M potassium-EDTA. Dissolve a suitable quantity of EDTA (free acid) in 90% of the final volume of distilled water. Add KOH pellets until pH is 7·0. Make up to final volume

incubator or water-bath, 37 °C
calibrated cylinder (500 ml)
lysozyme, DNase, RNase, spatulas
syringe, 25 ml with 10 cm needle, 1 mm internal diam.
for protein determinations (including calibration curve): 20 pipettes (1 ml); 20 test tubes, tube rack, 5 screw-capped bottles (100 ml), calibrated cylinder (100 ml), bovine serum albumin, spectrophotometer, reagents, cuvettes

Day 4
5 micro-syringes, 100 μl
2 anaerobic cuvettes, 1 μl
supply of oxygen-free nitrogen, pressure reduction valve, PVC tubing, hypodermic needles
10 small test tubes, (12 × 75 mm)
equipment for anaerobic incubation (see text)
stop-watch
10 magnetic fleas. These can be made as follows: put a 4 mm length of paperclip in the thin end of a Pasteur pipette, and melt the glass on each side of the metal piece
analytical balance
volumetric flasks, 1 litre, 100 ml
5 pipettes, 1 ml
5 pippetes, 5 ml
12 test tubes, tube rack
spectrophotometer

FURTHER INFORMATION

Most students succeed in preparing active membrane vesicles, usually showing the following enzyme activities:

(1) formate dehydrogenase: 100–200 nmol formate oxidized
                               per min per mg of protein

(2) nitrate reductase: 1000–2000 nmol nitrate reduced
                                per min per mg of protein
(3) coupled activities: 200–500 nmol nitrite produced
                                per mg of protein after 5 min.

The functional integrity of the anaerobic electron transport chain is clearly demonstrated by the coupling between the activities of formate dehydrogenase and nitrate reductase. From the fact that the coupled activity can be measured in isolated mambrane vesicles it can be concluded that the nitrate respiration system, like the aerobic respiratory chain, is incorporated in the cytoplasmic membrane (*see also* Experiment 20).

REFERENCES

J. Boonstra and W. N. Konings (1977). Generation of an electrochemical proton gradient by nitrate respiration in membrane vesicles from anaerobically grown *Escherichia coli*. *European Journal of Biochemistry*, **78**, 361–8.

H. R. Kaback (1971). Bacterial membranes. *In* "Methods in Enzymology" (W. B. Jakoby, ed.), Vol. 22, pp. 99–120. Academic Press, London and New York.

W. N. Konings and J. Boonstra (1977). "Anaerobic Electron Transfer and Active Transport in Bacteria", (F. Bronner and A. Kleinzeller, eds), Current Topics in Membranes and Transport, Vol. 9, pp. 177–231. Academic Press, London and New York.

O. H. Lowry, N. J. Rosebrough, A. L. Farr and R. J. Randall (1951). Protein measurement with the folin phenol reagent. *Journal of Biological Chemistry*, **193**, 265–75.

P. Mitchell (1970). Membranes of cells and organelles: morphology, transport and metabolism. *In* "Organization and Control in Prokaryotic and Eukaryotic Cells". Symposia Society of General Microbiology, Vol. 20, pp. 121–66.

34

Isolation and Identification of the Reserve Material of *Bacillus megaterium*

L. A. HUISMAN

Department of Microbiology, University of Groningen,
Kerklaan 30, 9751 NN Haren, The Netherlands

Level: Advanced undergraduates
Subject areas: Bacterial physiology
Special features: Use of a fermenter for culturing
bacteria

INTRODUCTION

Many micro-organisms form reserve materials under appropriate culture conditions. The physiological significance of these compounds is that they can serve as source of carbon and/or energy, when no exogenous substrates are available. In this way the organisms are able to survive under unfavourable environmental conditions for a prolonged period.

A reserve material which is characteristic for bacteria is poly-β-hydroxybutyrate (PHB). It is present in the cell in the form of granular inclusions, which can be discerned either by staining with a lipophilic dye (e.g. Sudan black) or by phase contrast microscopy, because of their strong light refracting capacity.

The synthesis of PHB proceeds best in an environment with a high carbon/nitrogen ratio (*see* MacRae and Wilkinson, 1958). After complete utilization of the nitrogen source, growth is no longer possible and the energy derived from the excess of the carbon and energy source can be used for synthesis of reserve material. PHB is synthesized from acetylCoA; two molecules of AcCoA combine to form acetoacetylCoA, which is reduced by NADH to β-hydroxybutyrylCoA. About 60 of these moieties are polymerized to give one molecule of PHB, CoASH being liberated.

The fact that the reserve material is stored as a polymer has two important advantages: first, it is osmotically inert in this form, and secondly, the acidic groups of the hydroxybutyrate are neutralized by esterification (Stanier *et al.*, 1976). In this way, large amounts of reserve materials can be stored intracellularly without causing damage to the cell. In some organisms the PHB content may account for 60–80 % of the cellular dry weight (Stanier *et al.*, 1976).

The aim of this experiment is to isolate the reserve material from the aerobic, spore-forming bacterium, *Bacillus megaterium*, and to identify it as PHB.

EXPERIMENTAL

The students are supposed to prepare and, if necessary, sterilize, all solution or media themselves. The time needed for this is not included in the time schedule given below because it depends very much on the technical skill of the individual students. The experiment can best be performed by two students together.

Day 1 *(inoculation of preculture (10 min) and preparation and sterilization of the fermenter (3– 4 h, including 2 h autoclaving))*

Inoculate a loopful of cell material of *B. megaterium* from a slant culture into two 250 ml conical flasks with 60 ml growth medium each.

↓

Incubate the flasks at 30 °C for 18–24 h in a rotary incubator at 200 rev. min^{-1}.

↓

In the meantime prepare a 2 litre fermenter vessel, which can be thermostated. Use silicone tubing for all connections, and make sure the connections are tightened securely.

↓

Sterilize the vessel with enough growth medium to obtain a total volume of 2 litres after addition of the separately sterilized phosphate buffer and glucose solutions, and the preculture (2 × 60 ml).

↓

Also sterilize a tube with 4 ml antifoam solution, consisting of 2 drops of polypropylene per ml water. Addition of this solution is only needed in case of heavy foaming of the culture.

- -

Day 2 *(main culture, whole day; start as early as possible)*

Set the temperature of the fermenter to 30 °C and add the separately sterilized solutions and both precultures to the rest of the medium in the vessel. Pass 1–2 litres of air min^{-1} through the culture.

↓

Every 2 h take a sample of about 10 ml. Measure the optical density of the culture at 430 nm with a colorimeter (when O.D. > 0·5, the sample has to be diluted with water). Take one drop for the Sudan black staining (*see* below).

↓

Centrifuge 5 ml sample for 15 min at 4000 g in a screw-capped tube. The supernate is used for the determination of glucose, the pellet for the PHB assay (*see* below). The pellets may be stored in the freezer for PHB determination at a later date. When the glucose concentration in the culture is almost zero, the cells can be harvested.

↓

Take a 40 ml sample for measuring the percentage dry weight of the culture (*see* below). Centrifuge the rest of the culture for 20 min at 5000 g. Discard the supernate and resuspend the pellet in 400 ml 0·9 % NaCl solution.

↓

Centrifuge again, and pour off the supernate carefully.

↓

Transfer the cell material with a small volume of distilled water into a 500 ml round bottomed flask

↓

Add 250 ml 10 % NaOCl solution and incubate the mixture at 55 °C in a water-bath. Check the progress of the lysis regularly with a microscope.

↓

When lysis is complete (45–60 min) and 250 ml distilled water and transfer the solution to a dialysis tube. Dialyse for 20 hs against running tapwater.

Staining of PHB according to Burdon (1946)
Place one drop of the bacterial suspension on a thoroughly cleaned, fat-free slide and let it dry.

↓

Fix the preparation by moving the slide quickly several times through a small flame.

↓

Place a few drops of Sudan black B solution (0·3 % in 70 % ethanol) on the fixed preparation.

↓

After 10 min remove the dye carefully with a piece of filter paper (without rubbing) and immerse the slide in xylol until decolouration is complete.

↓

Carefully dry the slide again and apply the counter stain, 0·5 % safranine in water.

↓

After 10 s, rinse the slide with running water and dry again.

↓

Examine the preparation with an oil-immersion lens. The PHB can be seen as very dark granules in pink cells.

Determination of glucose
Glucose is assayed by the glucose–oxidase method (GOD-Perid method, Boehringer, Mannheim, W. Germany) or any other suitable method.

Determination of poly-β-hydroxybutyrate (*see* note 2)
This is done according to Law and Slepecky (1961), as cited by Herbert *et al.* (1971).

Suspend the pellet (from 5 ml culture) in 1·0 ml 10% NaOCl solution and incubate 1 h at 37 °C in a screw-capped tube.
↓

Add 4 ml distilled water, mix well, and centrifuge for 15 min at 4000 *g* (*see* note 1).
↓

Discard the supernate, resuspend the pellet in 5·0 ml acetone and centrifuge again for 15 min at 4000 *g*.
↓

Resuspend the pellet in 3·0 ml icecold chloroform, and incubate for 2 min at 75 °C in a water-bath with the cap on the tube.
↓

Pipette a suitable amount of the extract (containing not more than 50 µg PHB) into a screw-cap tube (cleaned at 100 °C with concentrated sulphuric acid) and evaporate the chloroform in a boiling water-bath.
↓

Add 10 ml concentrated sulphuric acid (**caution**), and incubate for 10 min in a boiling water-bath (tubes closed).
↓

Measure the extinction at 235 and 280 nm in quartz cuvettes against concentrated H_2SO_4 (**caution**). The PHB concentration in the sample can be calculated from the following equation:

$$\frac{E_{235} - E_{280}}{1·56 \times 10^4} \times 86 \times 1000 \times F = [\text{PHB}] \ (\text{mg} \, l^{-1})$$

$1·56 \times 10^4$ is the extinction coefficient of crotonic acid, 86 its molecular weight, and F is a factor to convert the cuvette concentration to the sample concentration.

Dry weight measurement
Centrifuge a 40 ml culture sample for 20 min at 5000 *g*.
↓

Resuspend the pellet in 30 ml distilled water and centrifuge again. Repeat this procedure once.
↓

Transfer the sediment with the smallest possible amount of distilled water to a weighing bottle (heated until constant weight at 120 °C and cooled in a desiccator).

↓

Dry till constant weight at 120 °C and determine the net weight of the sediment. Calculate from this the dry weight content of the culture in $mg \, ml^{-1}$.

- -

Day 4 *(extraction of PHB, $1\frac{1}{2}$ h + 20 min, separated by 4 h refluxing) (*see note 2)*

Centrifuge the dialysed material for 20 min at 50 000 g (*see* note 1).

↓

Transfer the sediment with the smallest possible amount of distilled water to a 500 ml round bottomed evaporating flask.

↓

Connect the flask to a vacuum film evaporator and dry the sediment completely.

↓

Then add 200 ml chloroform, put a coil condenser on the flask and extract the PHB by refluxing for 4 h. Use a heating mantle as heat source.

↓

Filter the chloroform-insoluble fraction using fine filter paper.

↓

Add 300 ml ice-cold ether and place the mixture overnight in the freezer at −40 °C. Under these circumstances the PHB will sediment as a flocculent white precipitate.

- -

Day 5 *(identification, 2 h + 2 h, separated by $2\frac{1}{2}$ h hydrolysis of PHB) (*see note 2)*

Centrifuge the ether suspension for 30 min at 50 000 g and discard the supernate (*see* note 1).

↓

Resuspend the PHB in ice-cold ether and centrifuge again in a preweighed centrifuge bottle.

↓

Pour off the ether and measure the net weight of the PHB.

↓

Add 2 ml 2 N NaOH solution to about $\frac{1}{4}$ of the PHB and heat the suspension for $2\frac{1}{2}$ h in a boiling water-bath, to hydrolyse the polymer into the monomers.

↓

Allow to cool to room temperature and correct to pH 2 with 5 N HCl.

↓

Extract the 3-hydroxybutyric acid by shaking three times with 2 ml ether.

↓

Remove the ether layers with a pipette and collect them in a tube.

↓

Evaporate the ether by passing air through the solution, and add 0·5 ml distilled water to the residue.

↓

The PHB-monomer is identified by thin layer chromatography, using 3-hydroxybutyric acid as a standard, as follows.

↓

Apply 4 samples, 20 and 50 μl of a standard 3-hydroxybutyric acid solution (10 mg ml^{-1}) and 20 and 50 μl of the PHB hydrolysate, respectively, at equal distances from each other on a line drawn at a distance of 2 cm from one of the sides of a 20 × 20 cm plate coated with a thin layer of cellulose. Dry the spots with a hair-dryer (see note 3).

↓

Prepare the solvent by mixing 1 part of glacial acetic acid with 4 parts n-butanol and 5 parts distilled water, in a separating funnel.

↓

Allow the plate to equilibrate with the waterphase of this solvent in a small chromatography chamber for at least 5 h.

Day 6 (10 min + 20 min, separated by 3½ h developing and 1 h drying)

Develop the plate in the organic part of the solvent until the front of the solvent has almost reached the upper end of the plate (about 3½ h).

↓

Remove the plate from the tank and let it dry at 80 °C for 1 h.

↓

Spray the plate with a solution of 5 g aniline and 5 g glucose in 100 ml 50 % ethanol, and heat it for 10 min at 120 °C.

↓

Calculate the R_F-values of the spots, and compare the values obtained from the standard solution with those obtained from the PHB-hydrolysate.

↓

Make a graph of the time course of the optical density at 430 nm and of the glucose and PHB concentrations of the culture and calculate the PHB content of the cells at the time of harvesting in mg PHB per mg cellular dry weight.

Notes and Points to Watch

1. Use only polypropylene and glass (for low speed runs) centrifuge bottles and tubes for all centrifugations. Polycarbonate is damaged by most organic solvents. For the PHB determination, normal glass centrifuge tubes can be used instead of screw-capped tubes. When screw-capped tubes are used the centrifugal force should not exceed $4000\,g$.
2. The procedures for PHB determination, extraction and identification involve work with highly inflammable solvents and concentrated acids. Students should be warned to handle these compounds very carefully.
3. The sample spots applied to the thin layer plate should be kept as small as possible without damaging the cellulose layer. Instead of a thin layer plate, a piece of Whatman no. 1 chromatography paper may be used.

MATERIALS

1. Any strain of *Bacillus megaterium*, subcultured monthly on growth medium, solidified with 1.5% agar.
2. Growth medium. Final concentrations $g\,l^{-1}$): Na_2HPO_4 (6·0), KH_2PO_4 (3·0), $NaCl$ (3·0), NH_4Cl (1·0), Na_2SO_4 (0·1), $MgCl_2 \cdot 6H_2O$ (0·1), $MnCl_2 \cdot 4H_2O$ (0·01), casamino acids (0·1), glucose (15·0), trace elements solution according to Vishniac and Santer (1957), $1\,ml\,l^{-1}$; final pH 7·2. All chemicals except $MgCl_2$ and $MnCl_2$ are anhydrous. Phosphate and glucose are sterilized separately as concentrated solutions. All components are heat-sterilized in an autoclave at 120 °C.
3. Test combination glucose, GOD-Perid® method, catalogue no. 124 036, Boehringer Mannheim GmbH, W. Germany.
4. 10% $NaOCl$ solution. Commercially available bleach (containing 10% active chlorine) is very suitable.
5. 3-hydroxybutyric acid, Fluka AG, Buchs, Switzerland.
6. Thin layer plate: TLC aluminium sheets, cellulose layer thickness 0·1 mm, article no. 5552, Merck, Darmstadt, W. Germany.
7. Chromatography chamber: Shandon Scientific Co. Ltd., London, UK.

SPECIFIC REQUIREMENTS

These items are needed for each pair of students.

Day 1
agar slant of *B. megaterium*
1 beaker (3 litre), 1 measuring cylinder (1 litre), 2 measuring cylinders (100 ml), 2 screw-capped bottles (250 ml), 2 Erlenmeyer flasks (250 ml) with cotton plugs, 2 sterile pipettes (5 ml), 1 tube with cotton plug
chemicals, spatulas, balance
pH meter, calibration buffers

autoclave
magnetic stirrer + fleas
fermenter with accessory parts, silicone tubing, scissors, sterilizable PVC-tape, tubing clips
sterile pipettes, 5 and 10 ml
incubator shaker

Day 2
colorimeter with 430 nm filter or spectrophotometer
centrifuge + bottles (polypropylene)

cont.

cont.

glucose reagent
test tubes, tube rack
10 pipettes (1, 5 and 10 ml)
screw-cap tubes (can also be used as centrifuge
  tubes at speeds less than 4000 rev. min^{-1})
water-bath
quartz cuvettes (3 ml, 1 cm light path)
UV-spectrophotometer
measuring cylinder, 500 ml
round-bottomed flask, 500 ml
dialysis tubing, measuring cylinder(s), 3 litre
weighing bottle
oven
desiccator

Day 4
centrifuge (high speed), polypropylene bottles
round-bottomed flask (500 ml) with ground
  joint
vacuum film evaporator
glass measuring cylinder, 500 ml

coil condenser (ground joint), with stand and
  clamps
heating mantle
fine filter paper, funnel
freezer

Day 5
centrifuge
5 pipettes (5 ml)
balance, spatula
5 tubes, tube rack
2 micro-pipettes (20 and 50 μl)
TLC plate
chromatography chamber with trough
hairdryer
separating funnel
glass measuring cylinders

Day 6
oven
spray

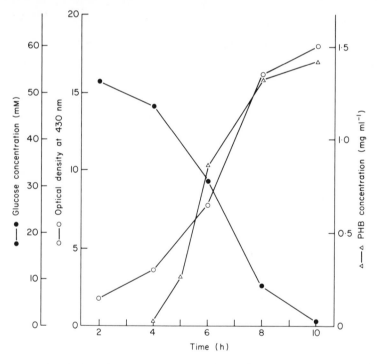

Figure 1 Time course of optical density, glucose concentration and PHB concentration in a batch culture of
B. megaterium. O—O, optical density; ●—● glucose concentration; △—△, PHB concentration.

FURTHER INFORMATION

The growth of *B. megaterium* is completed usually after 8–10 h, at an optical density above 10·0. At this stage the cells contain high amounts of PHB as judged by microscopic observation after staining, or by the measured PHB concentration. Typical student results are given in Figure 1. At the time of harvesting the culture usually contains about 3–4 mg cellular dry weight ml^{-1}, and 0·7 − 1·3 mg PHB ml^{-1}, the PHB percentage varying between 25 and 35%.

The R_F-values of the standard 3-hydroxybutyric acid and the PHB hydrolysate vary between 0·7 and 0·8. In some cases the samples are separated into two spots, characterized by R_F-values of 0·7–0·8 and 0·8–0·9 respectively. We do not have an explanation for this phenomenon.

REFERENCES

K. L. Burdon (1946). Fatty material in bacteria and fungi revealed by staining dried, fixed slide preparations. *Journal of Bacteriology*, **52**, 665–78.

D. Herbert, P. J. Phipps and R. E. Strange (1971). Chemical analysis of microbial cells. *In* "Methods in Microbiology" (J. R. Norris and D. W. Ribbons, eds), Vol. 5B, pp. 209–344. Academic Press, London and New York.

J. H. Law and R. A. Slepecky (1961). Assay of poly-β-hydroxybutyric acid. *Journal of Bacteriology*, **82**, 33–6.

R. M. Macrae and J. F. Wilkinson (1958). Poly-β-hydroxybutyrate metabolism in washed suspensions of *Bacillus cereus* and *Bacillus megaterium*. *Journal of General Microbiology*, **19**, 210–22.

R. Y. Stanier, E. A. Adelberg and J. L. Ingraham (1976). "General Microbiology", 4th edn, p. 353. Macmillan, London.

W. Vishniac and M. Santer (1957). The thiobacilli. *Bacteriological Reviews* (P. N. Wilson, ed.), **21**, 195–213.

Part Four
The Micro-organism Excretes

35

Determination of Ethanol Content of Alcoholic Beverages by Gas Chromatography

D. B. DRUCKER

Department of Bacteriology and Virology, University of Manchester, Stopford Building, Oxford Road, Manchester, M13 9PT, England

Level: Advanced undergraduates
Subject area: Applied microbiology; microbial biochemistry
Special features: Microbiology applied to a domestic article. Data can be used for statistical analysis

INTRODUCTION

The manufacture of beer, cider, wines, reinforced wines and spirits by fermentation of carbohydrates represents an important aspect of applied microbiology. All industrial syntheses require quality control, and the production of alcoholic beverages is no exception. Products are assessed in terms of colour, aroma, taste and chemical composition. Sherry is a useful example: it is a reinforced wine, i.e. a wine of boosted ethanol content. The ethanol content varies from approximately 15 to 21 % v/v depending on whether the sherry is dry or sweet, respectively. The ethanol content can be measured by specific gravity (which decreases as sugar is converted into ethanol during fermentation), by enzyme assay or by gas chromatography. The last method provides very rapid analyses, and accurate and precise data. It also detects any undesirable fermentation products, e.g. amyl alcohol or lactic acid, the formation of which causes "off-flavours".

Gas chromatography is similar in principle to paper chromatography. In the gas chromatograph (Figure 1) the solute partitions between a stationary phase and a mobile phase. In gas chromatography the mobile phase is a gas, usually nitrogen (but **not** a liquid). A highly volatile solute tends to travel quickly through the column, swept along by carrier gas; conversely, a relatively non-volatile solute is localized in the stationary phase, travelling only slowly through the column. There are various types of detector. All work on the principle that eluate molecules alter the electric current flowing through the detector which is amplified and fed to a chart recorder, which records a trace or chromatogram. Under standard conditions each compound has a characteristic retention time, i.e. a time, in seconds, from the time of injection to the top of a peak. The area (or height) of a peak is related to the concentration of solute passing through the detector.

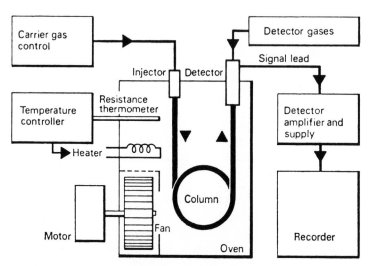

Figure 1   Schematic representation of a gas chromatograph. (Courtesy of Pye-Unicam Ltd.)

The aims of this experiment are to illustrate the use of gas chromatography as a tool for quantitative chemical analysis and to demonstrate the different alcohol contents of dry and sweet sherry. Sherry is a particularly suitable beverage for this kind of experiment. Its high ethanol content means samples can be stored without much risk of conversion to vinegar and analysed without confusion due to peaks of alkanals (aldehydes) and other compounds present in much smaller amounts than ethanol.

This practical can be used as a means of illustrating good experimental design: before commencing, the class can be asked to design a suitable experiment to measure the ethanol content of a sherry sample. A brief seminar will reveal whether they have realized the need to (a) dilute the sherry until a more dilute solution is obtained, (b) make a series of ethanol standard solutions of known strength, (c) include an internal standard (propan-1-ol), (d) inject pure samples of ethanol and propanol to establish their positions on the chromatogram, and (e) inject each sample several times to obtain statistically valid data, viz. means with standard errors.

EXPERIMENTAL

3 h + 1 h for data handling

Make a series of dilutions of absolute ethanol in water to give concentrations ranging from 0·05 to 0·5% w/v. Also dilute sherry 1/100 with distilled water.

↓

Add propan-1-ol to all samples as an internal standard (*see* note 2).

↓

Inject $1.0\,\mu l$ samples into a 5 foot glass column packed with Chromosorb 101 (*see* note 1), operated at 190°C with a carrier gas flow rate of 45 ml min^{-1}. The gas chromatograph should have a flame ionization detector.

↓

Repeat the analysis of each sample several times (*see* note 3) and include pure samples of ethanol and propan-1-ol.

- -

Notes and Points to Watch

1. A Chromosorb 101 porous polymer is far superior to a column of support coated with a polymer stationary phase for it can tolerate injections of water or dilute sherry. Also, it can be operated at relatively high temperatures, providing extremely rapid analysis time (2–3 min per sample). The columns can easily be emptied and repacked.
2. An internal standard is essential if errors in measurement of small injection volumes are to be avoided. All samples should contain exactly the same volumes of *n*-propanol; 0.2% w/v is suitable.
3. The use of a gas chromatograph including delicate microsyringes, should be carefully **demonstrated** to the student. The syringe should be thoroughly rinsed out between injections.

MATERIALS

1. Gas chromatograph with flame ionization detector and 5-foot column of Chromosorb 101

(Phase Separations Ltd., Queensferry, Wales) for three groups of students.

SPECIFIC REQUIREMENTS

pipettes (non-sterile)
Universal bottles (non-sterile)
sherry samples (small!)

samples of absolute ethanol and *n*-propanol
distilled water
5 μl micro-syringe

FURTHER INFORMATION

The method described is based on the Carlsson (1973) technique for analysis of fermentation end-products of anaerobic micro-organisms (Holdeman and Moore, 1972). Further information on gas chromatography techniques in microbiology are described elsewhere (Drucker, 1981). Absolute retention times vary from one laboratory to another, but the relative retention of propan-1-ol:ethanol should be approximately 1·6 at 190 °C (Drucker, 1976).

There should be a linear relationship between the ethanol peak height, relative to propan-1-ol, and the ethanol concentration (Figures 2 and 3).

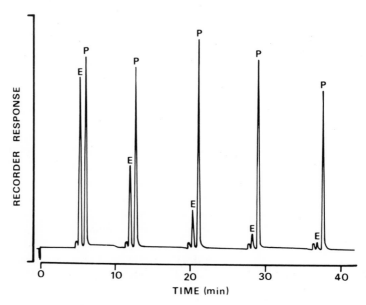

Figure 2   Results obtained when different amounts of ethanol are injected into a gas chromatograph. E, ethanol peak; P, peak of propan-1-ol standard.

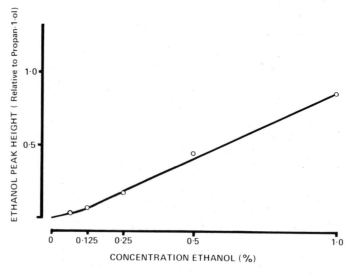

Figure 3   The linear relationship between the height of ethanol peak, relative to peak height of propan-1-ol standard, and the amount of ethanol injected into the gas chromatograph.

REFERENCES

J. Carlsson (1973). Simplified gas chromatographic procedure for identification of bacterial products. *Applied Microbiology* **25**, 287–9.

D. B. Drucker (1976). Gas–liquid chromatographic chemotaxonomy. *In* "Methods in Microbiology", (J. R. Norris, ed.), Vol. 9, pp. 51–125. Academic Press, London and New York.

D. B. Drucker (1981). "Microbiological Applications of Gas Chromatography". Cambridge University Press, Cambridge.

L. V. Holdeman and W. E. C. Moore (1972). Anaerobe Laboratory Manual. V. P. I. Anaerobe Laboratory, Blacksburg, Va.

Assessment of Gas-evolving and Microbiological Quality of Yeast Samples

I. DAVIES and C. J. GRIFFITH

Department of Science, South Glamorgan Institute of Higher Education, Colchester Avenue Centre, Cardiff, Wales

> *Level*: All undergraduate years
> *Subject areas*: Food microbiology
> *Special features*: Simplicity of apparatus

INTRODUCTION

Strains of the yeast *Saccharomyces cereviseae* find use both in the baking and brewing industry. Of importance to the baker is that the yeast metabolizes the glucose present to produce sufficient carbon dioxide to raise/lift the dough.

$$C_6H_{12}O_6 + 6O_2 \rightarrow 6CO_2 + 6H_2O. \tag{1}$$

In brewing, a true fermentation is preferred where the desirable end products are alcohol and carbon dioxide.

$$C_6H_{12}O_6 \rightarrow 2CO_2 + 2C_2H_5OH. \tag{2}$$

It can be seen from the two equations that the volume of carbon dioxide produced is greater in the presence of oxygen and the amount of alcohol produced greater in the absence of oxygen. In the breadmaking process the biochemical pathway of glucose breakdown does not strictly follow the route outlined in equation (1); there will always be some alcohol produced as well.

In the conventional breadmaking process, a dough of flour, water, salt, etc. is allowed to react with *S. cerevisiae*, originally overnight, now for about 3 h. The dough is worked, allowed to rise, then reworked, placed into tins, allowed to rise for a fruther 50 min (final proof) and baked. This is now being replaced by the high energy system (Chorleywood Bread Process) employing mechanical working of the dough in combination with increased amounts of yeast and the addition of fast-acting oxidizing agents. This latter process has several advantages over the original process, but it does demand efficient production of carbon dioxide by the yeast. Samples of yeast used by a baker must have a gas evolving power of at least 40 cm^3 g^{-1} h^{-1}. Gassing power below this rate would not produce enough gas for the short proving stages used in Chorleywood-type processes. It is therefore of interest to the baker to know the gas-evolving power of a sample of yeast and any factors which influence this. One such factor is the age and hence viability of yeast samples, for their gas-evolving powers deteriorate with age. This experiment is designed to show under aerobic conditions the relationship of gas-evolving power to viability of various samples of yeast and to show the effect of anaerobiosis on yeast activity. Another feature of the yeast sample, which is assessed in this experiment, is its degree of bacterial contamination. Of particular interest is contamination with spores of *Bacillus subtilis* for this may lead to a type of bread spoilage known as "rope".

EXPERIMENTAL

To illustrate the relationship of viability to gassing power, "aged" samples of yeast are required. These may be commercial samples of yeast stored in a refrigerator for up to 3 months.

Day 1 (7 h)

1. *Testing of gas evolution*

Set up the apparatus described in Figure 1.

↓

Weigh out 10 g of fresh commercial yeast and distribute evenly in the sugar solution in the round bottomed flask.

↓

Immediately replace the rubber bung in the neck of the flask ensuring that there are no air leaks (*see* note 2). Carbon dioxide evolved passes through the flexible plastic tubing and is collected in the inverted burette for up to 1 h (*see* note 1).

↓

Record the volume of carbon dioxide evolved every 10 min.

↓

After 1 h filter the contents of the flask (*see* note 4). Rapidly freeze the filtrate and store at −18°C.

↓

Record temperature and barometric pressure and convert volume of gas evolved to those evolved at standard temperature and pressure (S.T.P.)

↓

Plot a graph of gas evolved against time.

↓

Repeat experiment using samples of yeast 3, 5 and 8 weeks old.

2. *Microscopical assessment*
Add a knife point of yeast to approximately $10 \, cm^3$ of $\frac{1}{4}$ strength Ringer's solution and rinse thoroughly.

$\downarrow$

Place one drop of the suspension on a glass slide, cover with a coverslip and examine under high power and immersion objectives of a light microscope.

$\downarrow$

Note the size of the cells and the size of the cell vacuole (should be at least $\frac{1}{3}$ of the size of the cell).

3. *Quantitative microbiological assessment*
Aseptically weigh out 1 g of yeast into $9 \, cm^3$ of sterile $\frac{1}{4}$ strength Ringer's solution; serially dilute samples as far as 10^{-10}.

$\downarrow$

Aseptically transfer $1 \, cm^3$ aliquot of dilutions 10^{-6} to 10^{-10} into five labelled Petri dishes.

$\downarrow$

Pour approximately $10 \, cm^3$ of cooled molten wort agar (50° C) into each Petri dish.

$\downarrow$

Distribute agar (*see* note 3), allow to set and incubate at 25° C. This gives the viable yeast count.

$\downarrow$

Aseptically transfer $1 \, cm^3$ aliquot of dilutions 10^{-1} to 10^{-3} inclusive into three labelled Petri dishes and prepare pour plates using Actidione agar (*see* note 3). Incubate at 37° C for 3 days. This enumerates the viable bacteria.

$\downarrow$

Heat dilutions 10^{-1} and 10^{-2} in a water-bath at 80° C for 10 min.

$\downarrow$

Prepare pour plates using plate count agar (*see* note 3) and incubate at 37° C for 2 days. This gives the endospore count.

Day 2 *(3½ h)*
Set up gas evolution apparatus as for Day 1 but boil the sugar solution in the flask for 10 min.

$\downarrow$

Allow to cool while flushing nitrogen gas through the apparatus until the temperature is 30° C.

$\downarrow$

Add 10 g fresh yeast and record the volume of carbon dioxide evolved over a 1 h period.

↓

Filter the sugar solution remaining in flask (*see* note 4) and freeze. Convert the volume of gas evolved to those evolved at S.T.P. Compare the graph using anaerobic conditions to that obtained under aerobic conditions.

Day 3 *(4 h)*

1. *Endospore count*
Count the colonies which have grown and record the results as the number of spores per gram of yeast. Confirm that the colonies that have developed are from sporing organisms by checking with the spore stain (malachite green, safranin).

2. *Estimation of alcohol in flask filtrates*
Defrost samples of filtrates and prepare a series of dilutions of each filtrate from 10^0 to 10^{-3} (*see* note 5).

↓

Estimate the amount of alcohol present using alcohol dehydrogenase (*see* note 6) (kit containing instructions and reagents, available from Boehringer Mannheim GmbH).

Day 4 *(45 min)*

Viable bacterial count
Count the colonies and express the result as the number of bacteria per gram of yeast sample.

Viable yeast count
Count the colonies and express the result as the number of yeasts per gram of yeast sample. Relate to gas-evolving capabilities.

Notes and Points to Watch

1. The inverted burette can be filled by means of a vacuum line connected to the delivery point of the burette and the desired level of brine controlled by means of the tap.
2. If an air leak around the rubber bung is encountered smear with petroleum grease.
3. Ensure that dilutions for plate counts are thoroughly mixed, e.g. use a vortex mixer.
4. Use filtration equipment readily available. It is not necessary to have the filtrate free of bacteria. In fact, if too fine a pore size is used difficulties in filtration may occur due to blockage.
5. For estimation of ethanol the ethanol content of the cuvette should range between 1 and 15 μg. The sample must therefore be sufficiently diluted in order to obtain an ethanol **sample** concentration between 10 and 150 mg l^{-1}.
6. Ethanol is oxidized in the presence of the enzyme alcohol dehydrogenase (ADH) by nicotinamide-adenine dinucleotide (NAD) to acetaldehyde:

$$\text{Ethanol} + \text{NAD} \xrightarrow{+\text{ADH}} \text{acetaldehyde} + \text{NADH} + \text{H}^+.$$

The equilibrium of this reaction lies on the side of ethanol and NAD. It can, however, be completely displaced to the right by an excess of NAD, alkaline conditions and by trapping the acetaldehyde with semicarbazide. The amount of NADH formed in the above reaction is stoichiometric with the amount of ethanol. NADH is determined by means of its absorption at 340 nm.

MATERIALS

1. Commercial samples of *Saccharomyces cerevisiae*, including "aged" samples.
2. Sugar solution made up in a 250 ml wide-necked round-bottomed flask.

Sucrose,	10 g
KH_2PO_4,	0·8 g
$NH_4H_2PO_4$,	0·4 g
$MgSO_4$,	0·1 g
$CaSO_4$,	0·8 g

distilled water 160 cm³.
3. Gas evolution test apparatus (*see* Figure 1).
4. Plate count agar, actidione agar, wort agar.
5. Solutions for the estimation of ethanol (available as a kit from Boehringer Mannheim GmbH).

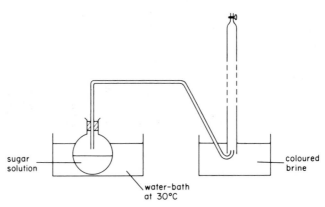

sugar solution

coloured brine

water-bath at 30°C

Figure 1 Place the flask in a water-bath at 30° C and connect by means of a flexible plastic tubing to an inversely calibrated 400 cm³ capacity burette inverted in a 23 % solution of sodium chloride (coloured with a few drops of methyl red).

SPECIFIC REQUIREMENTS

Day 1
gas-evolution test apparatus (Figure 1) (one per group)
four round-bottomed flasks and 800 ml of sugar solution (per group)
thermometer
barometer
tubes for dilutions, tube racks
sterile ¼ strength Ringer's solution (15 ml per group)
sterile pipettes

micrometer eyepiece and calibrated slide
vortex mixer
sterile Petri dishes (10 per group)
molten plate count, actidione and wort agar in water-bath at 50° C (50 ml of each per group)
500 ml beaker for use as an 80° C water-bath
filtration apparatus

Day 2
gas-evolution test apparatus
filtration equipment

cont.

cont.

Day 3   spectrophotometer
colony counting equipment
spore stains   **Day 4**
solutions for estimation of ethanol   Colony counting equipment
pipettes

FURTHER INFORMATION

Results vary depending upon the actual strains of yeast involved and their exact storage conditions, but typical sets of results would show the following trends. Under anaerobic conditions production of alcohol increases and the production of carbon dioxide decreases, by at least 50%, compared with aerobic conditions. Ten-gram samples of fresh yeast should be capable of producing 400 cm^3 of carbon dioxide within 35–40 min (fresh samples frozen and defrosted immediately before the experiment take slightly longer; 35–45 min). Such yeast samples give viable counts in the order of 10^{10} yeast g^{-1}, 10^3 bacteria g^{-1} and less than 10^2 spores g^{-1}. Samples of yeast refrigerated for 4–5 weeks are likely to take 50–55 min to produce requisite carbon dioxide levels and demonstrate the upper limit of usefulness for refrigerated storage. Samples refrigerated for longer than 5 weeks are unlikely to be suitable for use in modern breadmaking processes.

The Production of Organic Acid from Glucose by *Lactobacillus plantarum*

R. K. POOLE

Department of Microbiology, Queen Elizabeth College,
University of London, Campden Hill, London, W8 7AH, England

Level: Beginning undergraduates
Subject areas: Microbial chemistry and
biochemistry
Special features: A simple quantitative experiment
in bacterial metabolism

INTRODUCTION

Lactic acid was identified as the major acid in sour milk as long ago as 1780. The identification and association of lactic acid-producing organisms with fermentation was established by Bondeau, Pasteur, Schultze and Lister. Since the 1880s, when the first commercial fermentations were initiated, the production of lactic acid by fermentation has become an important industry. All members of the genera *Streptococcus*, *Pediococcus*, *Microbacterium*, a large number of *Lactobacillus* species (including *L. casei* and *L. plantarum*) and certain *Rhizopus* species ferment glucose predominantly to lactic acid. Trace amounts of volatile acids, ethanol, fumarate and CO_2 are also produced. The term "homofermentative" is used to describe these organisms and to distinguish them from "heterofermentative" types in which other products are formed in major amounts (for a review *see* Wood, 1961).

In the lactic acid bacteria glycolysis is coupled to reduction of pyruvate to lactate. This is in contrast to sugar fermentation by yeast, where the pyruvate is decarboxylated and reduced to ethanol. The earliest evidence of Embden-Meyerhof intermediates in the lactic acid bacteria was that of Stone and Werkman (1937) who reported that *L. plantarum* accumulated phosphoglyceric acid when glucose was fermented in the presence of acetaldehyde or other hydrogen acceptors and fluoride.

The following equation describes simply the chemical and energetic transformations involved.

$$C_6H_{12}O_6 + 2ADP + 2P_i \rightarrow 2CH_3CHOHCOOH + H_2O + 2ATP.$$

glucose   lactate

In this experiment, the production of acid from glucose by *L. plantarum* is followed by periodic titration of the suspension with standard alkali. Acid production may also be measured in the presence of a specific inhibitor of glycolysis (fluoride) and an alkylating agent (iodoacetamide).

EXPERIMENTAL

6 Days Before the Experiment (10 min)

Inoculate a small batch of medium (e.g. 100 ml) with a freeze-dried culture of *L. plantarum*. Incubate without shaking at 30 °C.

----

3 Days Before the Experiment (10 min)

Distribute the above starter culture amongst five 540-ml batches of medium. Incubate without shaking at 30 °C.

The Day of the Experiment (1–2 h, exclusive of harvesting of cells)

Harvest cells by centrifugation (e.g. at 10 000 g for 10 min) at room temperature. Wash once with distilled water, centrifuge again and resuspend the cells in distilled water. Each student or group will require about 30 ml of suspension. These could be prepared in advance of the class.

↓

Into a boiling tube, pipette 5 ml *L. plantarum* suspension and 1 ml water. Add a few drops of bromocresol purple solution and titrate with 0·01 M NaOH until the indicator is just perceptibly purple. This is the initial colour that should be matched in any subsequent titrations (*see* note 1).

↓

Note the burette reading. Add 1 ml of 0·05 M glucose (50 µmol glucose). Note the time of glucose addition. At suitable intervals (e.g. 5 min) add NaOH from the burette to keep the pH constant. Note the burette reading and time when the mixture is neutralized at each titration (*see* note 2).

↓

When fermentation has ceased add a further 1 ml of glucose and continue the experiment.

↓

Repeat the entire experiment but replace the water with 1 ml of inhibitor solution, i.e. reaction mixture before addition of glucose is: 5 ml *L. plantarum* suspension, 1 ml fluoride or iodoacetamide, and bromocresol purple.

↓

Plot graphs showing the time-course of acid production.

$$1 \text{ ml } 0.01 \text{ M NaOH} = 10 \, \mu\text{Eq acid} = 5 \, \mu\text{mol glucose.}$$

- -

Notes and Points to Watch

1. To ensure reproduction of the same end point at each neutralization, a colour standard is useful. A convenient standard is a corked tube with a little Kieselguhr and indicator, which has been adjusted to give a colour similar to that in the experimental tube before addition of glucose. Alternatively, add 6 ml water to 2 ml of cell suspension and a few drops of bromocresol purple **but add no glucose**. Titrate to the initial colour (purple) with NaOH. A second standard, lacking both glucose and NaOH, and thus yellow, can be useful in assessing the progress of titration.
2. Each titration should be carried out as quickly as possible so that there is minimal production of acid while the NaOH is being added.

MATERIALS

1. Freeze-dried cultures of *Lactobacillus plantarum*. A suitable strain is NCIB 6376.
2. *Lactobacillus* medium, containing in 3 litres:

Bacteriological peptone (Oxoid),	30 g
"Lab Lemco" powder (Oxoid),	30 g
Yeast extract powder (Oxoid),	15 g
Glucose,	60 g
Sodium acetate trihydrate,	15 g
Tween 80 (Hopkin and Williams),	3 ml
KH_2PO_4,	6 g
ammonium citrate,	6 g
$MgSO_4.7H_2O$,	0.6 g
$MnSO_4.4H_2O$,	0.15 g

Check that the pH is 6–6.4. Dispense 540 ml into each of five serum bottles (glass bottles with screw caps and rubber gaskets, total capacity about 620 ml). Dispense the remainder into small screw-cap bottles, almost filling each bottle. Autoclave for 15 min at 103.4 kPa (15 lbf in^{-2}).

SPECIFIC REQUIREMENTS

These items are needed for about 10 students or 10 groups of students.

First inoculation
freeze-dried *L. plantarum*
small batches of medium described above
30 °C incubator

Second inoculation
"starter" culture from 3 days previously
5 × 540 ml batches of medium described above
30 °C incubator

Day of experiment
culture from 3 days previously

centrifuge suitable for harvesting approx. 3 litres of culture at room temperature
distilled water
boiling tubes, about 50
pipettes (5 ml and 1 ml), about 50 of each, non-sterile
racks for boiling tubes
bromocresol purple indicator solution
0.01 M NaOH, 1 litre
0.05 M glucose, 100 ml
0.1 M NaF, 50 ml
0.1 M iodoacetamide, 50 ml
corked tubes
Kieselguhr, about 30 g

FURTHER INFORMATION

Results obtained by a second year undergraduate are illustrated in Figure 1. In the absence of inhibitors, acid production proceeds rapidly, with a half-time of about 2 min. The amount of acid produced following each of two sequential additions of glucose is in excellent agreement with that expected (100 μEq) from total conversion of 50 μmol glucose to lactate.

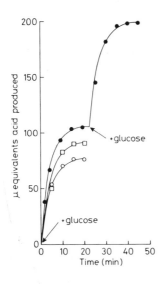

Figure 1 Production of acid from glucose by *L. plantarum*. Glucose (50 μmol) was added at the points indicated by arrows and the acid produced measured in the absence of inhibitors (●) or in the presence of 0·017 M NaF (□) or 0·017 M iodoacetamide (○). Results were kindly provided by Dr L Richards.

The conversion of 2-phosphoglycerate to phosphoenol pyruvate is catalysed by enolase. It has an absolute requirement for a divalent cation (Mg^{2+} or Mn^{2+}) which makes a complex with the enzyme before the substrate is bound. The enzyme is strongly inhibited by fluoride, particularly if phosphate is present, the inhibitory species being the phosphofluoridate ion which forms a complex with Mg^{2+}. In Figure 1, fluoride is shown to inhibit both the rate and extent of acid production. It should be pointed out that the concentration of fluoride used here (0·017 M) is considerably lower than those widely used (0·2–0·5 M) to cause accumulation of phosphoglycerate (Stone and Werkman, 1937). Iodoacetamide, an alkylating reagent, irreversibly inhibits enzyme activity by modifying the SH component of cysteine residues (Stryer, 1975) and is seen to be a more potent inhibitor than fluoride under these experimental conditions.

ACKNOWLEDGEMENTS

This experiment has an excellent pedigree. Suffice it to say that Dr C. F. Thurston (this department), Prof. D. P. Kelly (Warwick), Prof. P. Syrett (Swansea), and Mr H. Tristram (University College, London), in that order, have each given their permission for this experiment to be included here.

REFERENCES

R. W. Stone and C. H. Werkman (1937). The occurrence of phosphoglyceric acid in the bacterial dissimilation of glucose. *Biochemical Journal*, **31**, 1516–23.

L. Stryer (1975). "Biochemistry," pp. 129–30, 277–304. W. H. Freeman, San Francisco.

W. A. Wood (1961). Fermentation of carbohydrates and related compounds. *In* "The Bacteria," (I. C. Gunsalus and R. Y. Stanier, eds), Vol. II. pp. 59–149, Academic Press, London and New York.

Isolation and Characterization of Mutants of *Serratia marcescens* with Altered Pigmentation

S. B. PRIMROSE

*Department of Biological Sciences, University of Warwick,
Coventry, CV4 7AL, England*

> *Level*: All undergraduate years
> *Subject areas*: Bacterial genetics
> *Special features*: Strong visual impact

INTRODUCTION

The pioneering work of Beadle and Tatum (1941) established that genes exert their function through enzymes and that mutation of a gene leads to an alteration in the structure of the corresponding enzyme, usually resulting in loss of its function. By the isolation and analysis of many such "biochemical" mutants it was found that most of the simple monomers such as amino acids, purines and pyrimidines are synthesized by a series of small steps, each step under the control of an enzyme and forming a biochemical or metabolic pathway. Thus the biosynthesis of compound D would be carried out by means of the three enzymes a, b and c, controlled respectively by genes α, β and γ, acting successively to transform precursor A through the intermediates B and C to the end-product D.

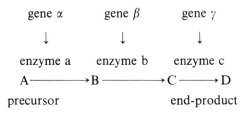

A groups of mutants with a common phenotype (e.g. an inability to synthesize end-product D) can be subdivided on the basis of cross-feeding (syntrophism) which relies on the accumulation and

---

diffusion of biosynthetic intermediates by mutants. Thus, if the compound accumulated and diffusing from one mutant is one which comes after the block of a second mutant, the cells of the latter will be able to utilize it for growth. Accordingly, a mutant which is blocked at the terminal biosynthetic step should be able to support the growth of mutants blocked at any one of the preceding steps. By cross-feeding experiments, mutants can therefore be arranged in the sequential order of steps of the metabolic pathway.

The aim of this experiment is to isolate colour mutants of *Serratia marcescens* defective in the biosynthesis of prodigiosin and to elaborate a biosynthetic pathway based upon syntrophic (cross-feeding) interactions between various mutants.

EXPERIMENTAL

Methods for the induction of colour mutants by either ultra-violet light or nitrosoguanidine (NTG) are given below but there is no reason why other mutagens could not be used. It should be noted that mutants with the same colour phenotype may be blocked at different steps in prodigiosin biosynthesis (*see* Tables 1 and 2). In order to identify as many different mutants as possible, students should be encouraged to exchange mutants.

Day 1 *(1½h)*

Take 10 ml culture of *S. marcescens*.

↓ | ↓

Centrifuge to pellet cells. Resuspend the cells in 10 ml saline.	Centrifuge to pellet cells. Resuspend the cells in 10 ml fresh broth.

↓ | ↓

Transfer the cell suspension to a glass Petri dish with magnetic flea. Begin stirring.	Add NTG to a final concentration of $30 \mu g\, ml^{-1}$. Incubate at $37\,°C$ for 15 min (*see* note 3).

↓ | ↓

Remove the lid of the Petri dish and irradiate with short-wave UV light (*see* note 2). Remove samples at 0, 30, 60 and 90 s.	Centrifuge, wash twice in saline, resuspend in 10 ml broth and incubate at $37\,°C$ for 60 min.

↓ | ↓

Make 5 decimal dilutions of each sample, e.g. 0·1 ml sample → 0·9 ml saline, and spread 0·1 ml of each on PG medium. Incubate plates (*see* note 5).	Make 5 decimal dilutions of the broth culture and spread 0·1 ml of each dilution on PG medium. Incubate plates (*see* note 5).

Day 2 *(30 min)*

Examine plates for colour mutants many of which may be present in sectored colonies. Pick representative colour mutants and purify by re-streaking on PG medium or PG medium plus spectinomycin (*see* note 6).

Day 3 *(30 min)*

Set up cross-feeding tests as shown below, checking for the excretion of both soluble and volatile intermediates.

To check for excretion of soluble intermediates make a 5 cm long streak of a mutant across PG medium. Make equivalent streaks from each of 3 other mutants perpendicular to the first but separated from it by 3 mm (*see* Figure 1). Incubate for 1–2 days (*see* note 5).

To check for excretion of volatile intermediates, streak one mutant across PG medium and a selection of other mutants across a second plate of PG medium. Discard the Petri-dish lids and tape the 2 plates together to form a box (*see* Figure 1). Incubate for 1–2 days (*see* note 5).

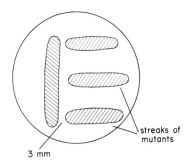

streaks of mutants

3 mm

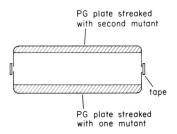

PG plate streaked with second mutant

tape

PG plate streaked with one mutant

Figure 1

Day 4 or 5 *(10 min)*

Examine plates for cross-feeding resulting in restoration of red pigmentation to mutants.

Notes and Points to Watch

1. Students should be warned not to look directly at sources of UV irradiation or at reflections from surfaces.
2. The output from different UV lamps varies greatly. In this experiment the object is to achieve $0 \cdot 1 \%$ survival of the *S. marcescens* and irradiation for 90 s with $400 \,\mu\text{W cm}^{-2}$ will be satisfactory. If the output from your lamp is not known you should construct a killing curve to determine at what distance to place the lamp from the culture in order to achieve $0 \cdot 1 \%$ survival in 60–90 s. Note also that cells should be irradiated in buffer or saline as organic compounds absorb UV light and consequently exert a protective effect.
3. Chemical mutagens, especially NTG, should be treated with great respect and **never** mouth-pipetted. It is best if the instructor prepares a stock-solution beforehand rather than allowing

the students to weigh out the material themselves. NTG is heat labile and decomposes in solution. Consequently, it should be freshly prepared and filter-sterilized. Dispose of NTG containing solutions by flushing down a sink with plenty of running water.

4. Pigment formation is inhibited by many chemicals, e.g. phosphates, so it is best to use saline rather than buffers for making all dilutions. The PG medium used in this experiment gives better pigmentation than the common laboratory media.

5. Pigment formation does **not** occur at 37 °C. Temperatures of 25–30 °C are optimal and pigment formation is enhanced in the presence of light. If a suitable incubator is not available the plates can be incubated at room temperature, preferably near a window.

6. It is possible that contaminants could be mistaken for non-pigmented mutants of S. marcescens. Consequently we use a mutant strain which is spectinomycin-resistant and check all "suspect" isolates for resistance to spectinomycin. However, this is not essential and other identification methods can be used if desired.

7. Some strains of S. marcescens yield colour mutants more readily than others. Strain STM 38 is particularly good in this respect.

8. Other mutagens can be used. Suitable procedures are given by Miller (1972).

MATERIALS

1. *Serratia marcescens* strain STM 38 which is resistant to 25 μg ml^{-1} spectinomycin.
2. PG medium containing 5 g peptone, 10 ml glycerol and 20 g agar l^{-1} of distilled water.
3. Source of short wavelength UV light. Suitable portable lamps are manufactured by Ultra-Violet Products Ltd., Winchester, Hants.
4. Nitrosoguanidine (*N*-methyl-*N'*-nitro-*N*-nitro-so-guanidine; Sigma Chemical Co.).

5. Spectinomycin (optional, *see* note 6) is dissolved in water, filter-sterilized and added to media to a final concentration of 25 μg ml^{-1}. Spectinomycin for injection is marketed by the Upjohn Co. as "Trobicin" and is best obtained from your local hospital pharmacy.

SPECIFIC REQUIREMENTS

Day 1
10 ml broth culture (18 h, 37 °C) of S. marcescens (in a container which will fit centrifuge, e.g. Universal bottle)
100 ml sterile saline
low speed centrifuge
sterile glass Petri dish containing sterile magnetic flea (UV irradiation only)
magnetic stirrer (UV irradiation only)
20 plates PG medium
nitrosoguanidine solution (1 mg ml^{-1}) **caution!**
pipettes

glass spreader and alcohol
tubes for dilutions
tube rack

Day 2
5 plates PG medium (this is average requirement; more may be required if many mutants isolated)

Day 3
5 plates PG medium (approximately).

FURTHER INFORMATION

Morrison (1966) isolated several hundred colour mutants of *S. marcescens* and examined them for syntrophic pigmentation interactions. The mutants that gave clear feeding patterns could all be placed in ten distinct classes. The interaction patterns defining six of these classes are shown in Table 1 and the location of the mutant blocks on the pathway of prodigiosin biosynthesis is shown in Figure 2. The characteristic colour of mutants of each class is shown in Table 2.

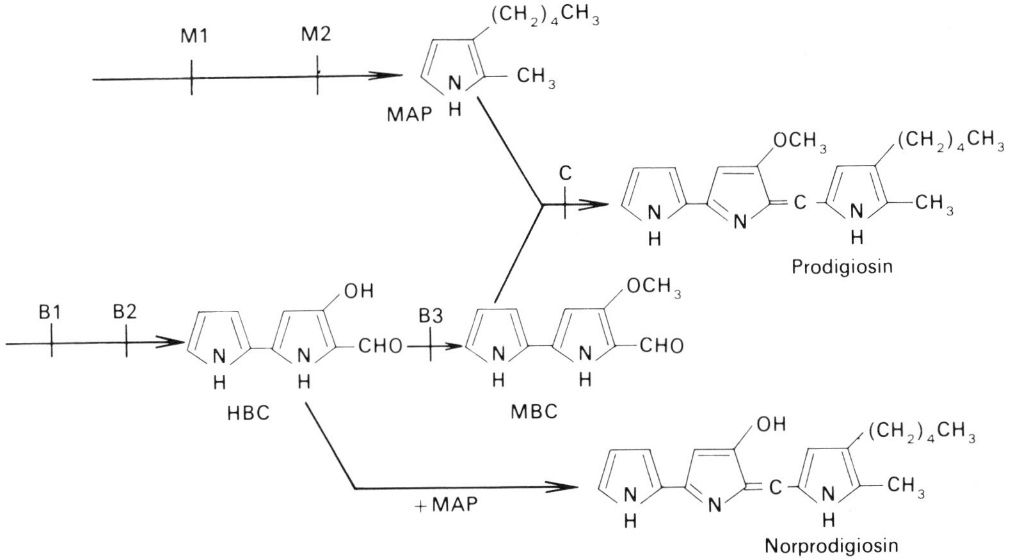

Figure 2 Location of mutant blocks on the pathway of prodigiosin biosynthesis. Prodigiosin is the usual red pigment and norprodigiosin the organge pigment formed by mutants of class **B3**. MAP, 2-methyl-3-*n*-amyl pyrole; MBC, 4-methyoxy-2,2′-bipyrole-5-carboxyaldehyde; HB, 4-hydroxy-2′-bipyrole-5-carboxyaldehyde.

Table 1 Syntrophic pigmentation pattern among certain colour mutant classes. Red pigmentation in the acceptor strain is induced by a volatile product (V) or a soluble product (S);—, no induced pigmentation. (Adapted from Morrison, 1966.)

		Donor class					
		C	B1	B2	B3	M1	M2
Acceptor	C		—	—	—	—	—
class	B1	S		S	S	S	S
	B2	S	—		S	S	S
	B3	S	—	—		S	S
	M1	V	V	V	V		V
	M2	V	V	V	V	—	

Table 2 Colour characteristics of pigmentation mutants. With some mutant classes the exact colour depends on the strain of *S. marcescens.* (Adapted from Morrison, 1966.)

Class	Colour
C	Pink
B1	White or yellow
B2	Pink or white
B3	Orange
M1	Light pink
M2	Pink or white

Morrison (1966) also described four other classes of mutant whose interpretation is more complex. Class I mutants behaved neither as donors nor acceptors and may be control mutants. Class X mutants appear to have blocks in both MAP and MBC synthesis. Class Y and class Z mutants are unusual in that red pigmentation can be induced by co-cultivation with strains of *Escherichia coli* as well as other mutants of *S. marcescens.*

Our experience has been that pink and white mutants are readily obtainable but that only 10–20% of the class isolate orange mutants. Most students manage to isolate 2–3 mutants and a few isolate as many as 15 or 20. The majority of the mutants isolated by the students cross-feed suitable acceptor strains but a few will not and these probably belong to classes I, X, Y or Z as outlined above.

REFERENCES

G. W. Beadle and E. L. Tatum (1941). Genetic control of biochemical reactions in Neurospora. *Proceedings of the National Academy of Sciences, USA,* **27**, 499–506.

J. H. Miller (1972). "Experiments in Molecular Genetics", 466 pp. Cold Spring Harbor Laboratory.

D. A. Morrison (1966). Prodigiosin synthesis in mutants of *Serratia marcescens. Journal of Bacteriology,* **91**, 1599–1604.

39

Distribution of *β*-lactamase in Cultures of *Staphylococcus aureus* and *Escherichia coli*

I. CHOPRA and M. H. RICHMOND

*Department of Microbiology, University of Bristol,
Bristol, BS8 1TD, England*

> *Level*: All undergraduate years
> *Subject areas*: Microbial chemistry and
> biochemistry

INTRODUCTION

A wide variety of Gram-positive and Gram-negative bacteria produce *β*-lactamases. These enzymes hydrolyse the *β*-lactam bond of penicillins and cephalosporins, rendering these compounds biologically inactive. With penicillins as substrates the corresponding penicilloic acid is formed:

Penicillin

Pencilloic acid

$R = \langle \bigcirc \rangle - CH_2$ for benzyl penicillin

With the cephalosporins an analogous reaction has been suggested except that the hydrolysis is followed by a series of further changes, some of which have not been elucidated in detail.

β-lactamases were originally divided into two classes depending upon their ability to hydrolyse certain substrates: the "penicillinases" hydrolyse benzyl-penicillin and related compounds at a high rate, while the "cephalosporinases" only hydrolyse cephalosporin C and its derivatives at high rate. This classification has since proved misleading, as the majority of enzymes possess a broad spectrum of activity, hydrolysing both families of substrate at roughly similar rates.

The aim of this experiment is to examine the distribution of β-lactamases in cultures of penicillin-resistant *Staphylococcus aureus* 8325 pI524 and *Escherichia coli* K-12 JC3272 RP1. The β-lactamase assay depends on the reaction of iodine with penicilloic acids, but not penicillins. Thus, by estimating iodine uptake it is possible to measure the amount of penicilloic acid formed and hence the β-lactamase activity. In this procedure the enzyme is mixed with penicillin for a standard period, an excess of standard iodine is added and any remaining unreacted is estimated by back-titration with standard sodium thiosulphate. Addition of iodine stops the reaction between enzyme and substrate.

EXPERIMENTAL

Day 1 *(2h)*
Assay the β-lactamase content of each whole culture (*see* note 3).

↓

While this assay is proceeding, centrifuge 10 ml of each culture in a universal bottle. Assay the resulting supernate for β-lactamase (*see* note 3).

↓

Take samples of the original cultures and subject them to ultrasonic disintegration (*see* note 4). Keep 1 ml of the original cultures for serial dilution (viable count, *see* below).

↓

Assay an appropriate sample of each sonicated culture for β-lactamase activity (*see* note 3).

↓

Transfer the remainder of the sonicated cultures to Universal bottles and centrifuge for 10 min. Assay the supernate for β-lactamase activity.

↓

Plate out 0·1 ml of 10^{-3}, 10^{-4}, 10^{-5}, 10^{-6} and 10^{-7} dilutions of the original and sonicated cultures for estimation of the viable count. For *S. aureus* use nutrient agar containing 5 % (w/v) NaCl and for *E. coli* use MacConkey's agar (*see* note 5).

↓

Incubate plates overnight at 37 °C.

Day 2 *(20 min)*

Count the colonies on the plates set up on Day 1 and calculate the viable bacteria per ml of the original and sonicated samples.

Notes and Points to Watch

1. The synthesis of most plasmid-specified β-lactamases in enteric bacteria is constitutive (including that mediated by RP1), whereas plasmid-coded β-lactamases in *S. aureus* are frequently inducible. To avoid the necessity for induction of the staphylococcal enzyme a derivative of plasmid pI524 constitutive for the synthesis of the enzyme has been chosen.

2. Since plasmids can be lost from bacteria on prolonged subculture in the absence of antibiotic selection pressure, it is essential for the organizers of the practical to check the marker pattern of the strains and select plasmid-positive cells for use in the experiment. Bacteria containing RP1 can conveniently be selected by growth on nutrient agar containing $20\,\mu g$ ampicillin ml^{-1}. Because of the extracellular nature of the staphylococcal enzyme which may cause excessive degradation of penicillin on solid media, plasmid-positive staphylococci are best selected with nutrient agar containing $0\cdot1$ mM cadmium chloride (resistance to cadmium is also coded by pI524).

3. To assay a sample for β-lactamase activity proceed as follows.

    (a) Pipette $5\cdot0$ ml of benzyl penicillin ($2\cdot5$ mg ml^{-1} in $0\cdot1$ M phosphate buffer, pH $5\cdot9$) into each of two Universals. One will serve for the enzyme estimation, the other as the control. Prewarm both at $30\,°C$. The control is required to estimate binding of iodine by organic compounds.

    (b) At time zero, add a suitable volume of enzyme to bottle no. 1, mix and incubate at $30\,°C$.

    (c) After a suitable period (usually 10 or 50 min, *see* Further Information) add $5\cdot0$ ml of standard iodine ($0\cdot0166$ N) to both the test and control bottles. Immediately afterwards add the same volume of enzyme to the control as was added at time zero to the test.

    (d) Allow the iodine to react with the penicilloic acid for 5 min and then titrate the contents of both bottles with standard sodium thiosulphate ($0\cdot0166$ N), dealing with the experimental sample first. Add a few drops of 1% starch solution near the end-point. The method is most satisfactory when the titration difference is 1–2 ml. To obtain this the volume assayed and the period of incubation can be varied accordingly.

    (e) Using the following relationship, calculate the β-lactamase activities in terms of μmoles penicillin degraded per hour per ml of bacteria or per ml of supernate.

$$\text{Enzyme activity} = \text{titration difference} \times \frac{60}{t} \times \frac{1}{V} \times S$$

(in μmol penicillin degraded h^{-1} ml^{-1}), where t is the incubation period (min), V is the volume of enzyme assayed (ml), and S is the stoichiometric factor ($= 2$ for benzyl penicillin).

4. Sonication is best performed in a room away from the main laboratory, or else in a machine equipped with adequate sound insulation. We subject 15 ml samples in Universals to three 20 s periods at 4 A using a Dawe Soniprobe Type 1130A. Samples should be cooled (2–4 °C) throughout sonication and given 1 min cooling intervals between exposures. During sonication "cavitation unloading", i.e. the growth of large air bubbles on the surface of the probe should be

avoided. These bubbles inhibit transmission of energy into the liquid and cause decreased cell disruption. "Cavitation unloading" is detected by a sudden change in the amplitude and frequency of the sound coming from the cavitating liquid.

5. The solid media recommended for determination of viable counts are designed to be partially selective for *S. aureus* and *E. coli*. This minimizes the need for aseptic techniques during sonication.

MATERIALS

1. *Escherichia coli* K-12 JC3272 RP1 and *Staphylococcus aureus* 8325 pI 524(i⁻) (*see* notes 1 and 2) grown aerobically in batch culture (from 1% inocula) to mid-logarithmic phase (4–5 × 10⁸ bacteria ml⁻¹). Nutrient medium should be used. These strains are available from Dr I. Chopra.

2. Nutrient agar plates containing 5% (w/v) sodium chloride. MacConkey agar plates.

3. Ultrasonic disintegrator.

4. Potassium phosphate buffer (0·1 M, pH 5.9). Benzyl penicillin (Glaxo laboratories) (2·5 mg ml⁻¹ phosphate buffer). Sodium thiosulphate (0·0166N). Starch solution (1%). Iodine (0·0166N) (I_2 + KI) is made by dissolving 4·2 g iodine and 20 g potassium iodide in 1 litre 2M sodium acetate buffer, pH 4. Acetate buffer (pH 4) contains (per litre): 85 ml glacial acetic acid, 43 g sodium acetate (anhydrous) made up to 1000 ml with distilled water.

SPECIFIC REQUIREMENTS

These items are needed for each student, where appropriate.

Day 1
40 ml broth cultures (mid-log phase) of *S. aureus* and *E. coli*
ultrasonic disintegrator and ice bath
10–15 containers (e.g. Universal bottles) to fit bench centrifuge and for use in titration
low speed centrifuge
100 ml penicillin solution
100 ml iodine solution
200 ml sodium thiosulphate solution
10 ml starch solution

30 °C water-bath
burette
pipettes
glass spreader and alcohol
10 plates of MacConkey's agar
10 plates of salt agar
Universals for dilution (viable count) containing sterile saline
Universal racks
37 °C incubator
clock

Day 2
colony counter

FURTHER INFORMATION

The staphylococcal β-lactamase specified by pI524 is predominantly extracellular (Richmond, 1968) and therefore high activity is found in the culture supernate after centrifugation. Sonication in fact has little effect on the viability of staphylococci, due in part to their shape and the rigidity of their cell walls. However, even if the staphylococci were broken by another method, e.g. solid-shear in a Hughes press, little increase in enzyme activity would be noted because of the small amount of intracellular enzyme. The apparent viable count of the sonicated staphylococci should in fact increase 2–3 fold due to disruption of clumps.

Many β-lactamases in *E. coli* are periplasmic (Richmond and Sykes, 1973) and therefore liberated by sonication which effectively disintegrates *E. coli*. This applies to the RP1-coded enzyme, as activity is low in the untreated culture, but increases after sonication. Sonication of *E. coli*, as described, results in 95–98 % killing.

Approximate values of β-lactamase activity for the cell suspensions and extracts are shown in Table 1.

Table 1. Typical β-lactamase activities in various *S. aureus* and *E. coli* cell suspensions and extracts. Bacteria were grown in Difco nutrient broth.

Organism	Sample	Assay conditions (ml)	Assay conditions (min)	Approximate titration difference (ml)	Approximate activity ml^{-1}
S. aureus	Untreated culture	1	10	1·4	17
S. aureus	Culture supernate	1	10	1·2	14
S. aureus	Sonicated culture	1	10	1·4	17
S. aureus	Supernate from sonication	1	10	1·2	17
E. coli	Untreated culture	1	50	1·0	2
E. coli	Culture supernate	1	50	0·5	1
E. coli	Sonicated culture	0·5	10	2·0	48
E. coli	Supernate from sonication	0·5	10	2·0	48

REFERENCES

M. H. Richmond (1968). The plasmids of *Staphylococcus aureus* and their relation to other extrachromosomal elements in bacteria. *Advances in Microbial Physiology*, **2**, 43–88.

M. H. Richmond and R. B. Sykes (1973). The β-lactamases of Gram-negative bacteria and their possible physiological role. *Advances in Microbial Physiology*, **9**, 31–88.

Production of Alcohol and Conversion to Vinegar

B. J. B. WOOD and M. C. ALLAN

Department of Applied Microbiology, University of Strathclyde, Glasgow, G1 1XW, Scotland

> *Level*: Advanced undergraduates
> *Subject areas*: Industrial microbiology
> *Special feature*: Two-stage fermentation; scope for kinetic and yield studies

INTRODUCTION

There are three main processes for the microbial conversion of dilute ethanolic solutions to vinegar:
  (a) the Orleans process, in which acetic acid bacteria grow very slowly as a film over the surface of still liquid. This is said to produce the finest culinary vinegar;
  (b) the "Quick" vinegar process, introduced in the last century, which is the basis for the present experiment;
  (c) deep fermentation processes which have had commercial success only recently.

Despite the attractions of modern, deep fermentations, the "Quick" vinegar process retains considerable commercial importance. The basic process is as follows.

Malt extract is prepared as in beer brewing, except that no hops are used. The extract (wort) is fermented anaerobically with yeast to convert the sugars in the wort to alcohol. After removing the yeast, the alcoholic solution is then converted to vinegar by passing it down a tower containing solid material impregnated with appropriate strains of acetic acid bacteria, while air is simultaneously blown up the tower. The solid support material has traditionally been shavings or twigs of wood low in resins, etc. such as beech and birch, although "wood wool" is also used today.

After percolating down the tower, the liquid is pumped back to the top, then passed down the tower again. This recycling is continued until acetification is complete. The vinegar is then drained off, and a fresh batch of alcoholic liquor is charged into the tower. After six or seven batches have been processed, the acetic acid bacteria have accumulated in the tower to such an extent that the

---

tower starts to clog, and it must then be cleaned out and re-inoculated. As might be expected, the speed of acetification increases from batch to batch, as the population of acetic acid bacteria builds up. The present experiment tries to repeat on a small scale the production of vinegar by the "Quick" process.

EXPERIMENTAL

A basic experiment is described here, but many variations are possible. In the authors' classes, students work in groups of 4, and this experiment is conveniently operated in parallel with other experiments since much of the time is taken in routine daily sampling.

Day 1 *(1½ h)*

Prepare a 5 litre batch of either the synthetic medium or the malt extract medium in a 10 litre flask or bottle and sterilize it by autoclaving.

Acetobacter inoculum
Prepare 2×200 ml batches of malt extract medium in flasks and sterilize by autoclaving.

Day 2 *(10 min)*

After overnight cooling, inoculate with a tea-spoonful of dried brewer's or baker's yeast, or pressed yeast and mix until evenly dispersed.

After overnight cooling, inoculate with acetic acid bacteria *(see* note 3).

↓

↓

Aseptically remove 50 ml samples into 4 tared, sterilized flasks and reweigh them *(see* note 2).

Incubate for at least 7 days at 25–30 °C.

↓

Incubate both the large bottle and the 4 small bottles at 30 °C.

Days 3–7 *(15 min each day)*
Shake all the yeast culture to expel as much dissolved CO_2 as possible.

↓

Weigh the small samples *(see* note 2).

↓

Continue the fermentation (normally for 4–5 days) with daily shaking until no further weight loss occurs. Plot a graph showing the loss of weight.

Days 8–14 (see *notes 4–9*)
Carefully decant off as much as possible of the
supernate liquid from the yeast deposit. The
yeast remaining in suspension must then be
removed, using a continuous centrifuge of the
Sharples or Alfa-Laval type, batch centrifugation or filtration.

↓

Place the clear liquid into the vinegar tower (*see* Figure 1) and add the inoculum of acetic acid
bacteria if required (*see* note 3).

↓

Take samples at least once a day and analyse them for total acid. Take 10 frequent samples for
volatile acid and alcohol analyses (*see* note 9).

At Completion of the Run

If the tower is to be used to acetify another batch of alcoholic liquid it should be left operating until
required for the next batch: in this way it is maintained in good working condition, with a healthy
film of bacteria over the tower packing material (*see* note 10).

ANALYTICAL PROCEDURES (*see* note 1)

1. *Alcohol: precise method*
   Ethanol is the only volatile, oxidizable compound present in significant amounts in the system
   and is estimated by distillation followed by oxidation under acidic conditions with dichromate
   (*see* note 11).
      The following method, which is based on that described by Maxon and Johnson (1945), uses
   back-titration of the excess dichromate remaining after oxidation of the ethanol to determine
   the amount of ethanol in the sample (*see* note 12).
      Put a suitable aliquot of ethanolic wash into a small distillation flask and make up to about
   15 ml with distilled water.
      Distil about 12–13 ml into 10·0 ml of dichromate reagent contained in a boiling tube (*see*
   note 13).
      Thoroughly mix the contents of the boiling tube.
      Immerse the boiling tube in a bath of boiling water for 30 min.
      Cool the tube and contents.
      Transfer the contents to a conical flask.
      Add solid potassium iodide, sufficient to effect solution of all the iodine which is liberated.
      Titrate the mixture with 0·1 N thiosulphate solution (*see* note 14).
      When most of the iodine colour has been discharged, stop titrating.
      Add a knife-point of starch iodine indicator (*see* note 15).
      Continue titrating.

Stop when the colour of the iodine-indicator complex is fully discharged (*see* note 16).
Note the burette reading.

Calculate (*see* note 18) the amount of alcohol from the difference between the sample titration and a blank (*see* note 17).

2. *Total acid*

   Place 10 ml of wash from the vinegar tower in a flask.
   Add phenolphthalein indicator (*see* note 19).
   Mix.
   Titrate with 0·1 N sodium hydroxide.
   Stop when the first pink colour appears.
   Note the burette reading.
   Calculate the amount of acid present.
   Record as milli-equivalents of acid (*see* note 23).

3. *Volatile acid (modified after Fred et al., 1919)*

   Place a sample (10·0 ml) of wash from the vinegar tower into the Markham or other semi-micro steam distillation apparatus (Markham, 1942).
   Add syrupy phosphoric acid (2 ml) (*see* note 20).
   Distil for about 15 min (*see* note 21).
   Titrate the distillate exactly as described for the determination of total acid (*see* note 22).
   Calculate the amount of acetic acid present in the sample (*see* note 23); 1 ml of 0·1 N acetic acid contains 6 mg of the acid.

Notes and Points to Watch

1. Because of the nature of our classes we find it very instructive to have the students prepare and standardize all reagents themselves, and prepare calibration curves. Since this is very time-consuming it should only be undertaken if ample time (student and staff) is available.
2. The loss of weight which occurs as sugar is converted into carbon dioxide and ethanol provides a rough check on the progress of the fermentation. Shake the bottles with care so that no liquid is lost due to frothing over. The fermentation normally takes 4–5 days. Remember to agitate the main batch of medium each day as well. Plot a graph of weight change against time. The weight-loss method of following the alcoholic fermentation is only approximate, and weighing to two decimal places is quite adequate.
3. If a tower is being used for the first time it should be sterilized by autoclaving (together with accessory tubing, reservoirs, etc.) and then inoculated with acetic acid bacteria. A mixture of *Acetomonas oxydans* and *Acetobacter rancens* gives good results, but there are other possibilities (*see* Further Information). The inoculum may conveniently be prepared by growing the organism(s) in shallow layers of malt extract medium.
4. The receiver vessel for the vinegar tower **must** be large enough to contain the whole charge of liquid, in case of pump failure.
5. Pumps and other electrical apparatus must be deployed around the tower in such a way as to minimize risks of water and electricity coming into contact with each other.

6. Silicone rubber tubing passing through a peristaltic pump must be of the correct diameter and wall-thickness, and be firmly secured, all in accordance with the instructions supplied with the pump. The tubing should be moved every 2 or 3 days so that a different section is being acted upon by the pump; this minimizes the dangers of wear and of fatigue-cracking of the tubing.

7. Silicone rubber tubing and bungs should be used. Ordinary rubber swells and deteriorates very rapidly under these conditions, but silicone rubber lasts well and soon repays the higher initial expense.

The tower shown in the diagram may be replaced by one with glass or stainless steel end-pieces if desired. The body of the tower should be of glass to allow it to be seen in operation.

8. Despite what might be expected, it is uncommon for contamination to be a problem. For instance, the continuous centrifuge used to clarify the alcoholic wash is unlikely to remove all the yeast, yet yeast growth in the tower is very rare indeed. Gram-stained smears of samples

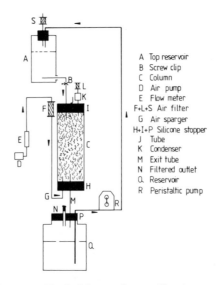

A   Top reservoir
B   Screw clip
C   Column
D   Air pump
E   Flow meter
F+L+S Air filter
G   Air sparger
H+I+P Silicone stopper
J   Tube
K   Condenser
M   Exit tube
N   Filtered outlet
Q   Reservoir
R   Peristaltic pump

Figure 1 Typical layout for acetification tower.

Operation of the tower is as follows: the supernate from the yeast fermentation stage is placed into the top reservoir (A) of the vinegar tower, and the inoculum of acetic acid bacteria added (if required). The screw-clip B regulates the flow of liquid to the column C which is packed with wood shavings or other suitable inert support material. Air is blown through the column from pump D, through flow-meter E and filter F to entrance tube G passing through bung H and the bottom of the column. If possible, bungs H and I should be of silicone rubber and all tubing should be of a good grade silicone rubber.

Air escapes from the top of the column through tube J, and, if desired, may pass through a water-jacketed reflux condenser K (although this is not essential) before emerging to the atmosphere via filter L. If filter L is kept dry by heating it (e.g. with a low voltage electric current passed through heating tape, or with a hot water jacket), operation of the acetification tower will be improved.

Meanwhile, the liquid percolates through the tower and exits via tube M. The possibility of tube M being blocked by packing material will be much reduced, and visual inspection will be greatly aided, if the bottom 5 cm or so of the tower is packed with Raschig rings. From reservoir Q, the liquid is pumped back to the top of reservoir A, via pump R; we use a peristaltic pump for this purpose. Operation of the tower is improved if the contents of the reservoir vessel (A) are mixed once or twice a day, to disperse any sludge of acetic acid bacteria which sometimes accumulates.

taken from the acetifier should nevertheless be made periodically and examined for contaminants.

9. Samples should be taken at least once a day, and more frequently if acetification is rapid. Every sample should be titrated for total acid; other analyses (volatile acid, alcohol) should be performed on at least 3 occasions (beginning, middle, end of run). Depending upon the precise configuration of the tower, it may be more convenient to sample from reservoir A (preferably just after mixing its contents) or from receiver Q, after stopping pump R for a few minutes (*see* Figure 1).

10. (a) When convenient, the liquid in the apparatus is accumulated in reservoir A by closing the clamp and leaving the pump running until all surplus liquid has drained from the tower and been pumped up to the reservoir. The vinegar is then removed from reservoir A, and the fresh batch of alcoholic medium placed therein, and operation of the tower restarted as quickly as possible.

   If more convenient, the vinegar may be accumulated in receiver Q, by stopping pump R, then removed from there after reservoir A has emptied and all surplus liquid has drained from the tower.

   When the apparatus has been charged with freshly fermented malt extract this should be cycled rapidly through the tower for half an hour or so, in order for it to equilibrate with the residual vinegar, and only then should the first sample be taken for analysis.

   (b) If the apparatus is to be cleaned and re-sterilized, the vinegar should be removed as described above, then the apparatus disassembled, sterilized and cleaned as for initial use.

   If the tower is to be re-used, then back-washing with water, draining overnight and autoclaving is an adequate treatment.

11. Alternative procedures include the Cavett flask (Kent-Jones and Taylor, 1954) and the Kozelka and Hynes (1941) method. The acid dichromate method forms the principle of the police force's breathalyser test. Ethanol may also be measured by gas chromatography and enzymatic methods.

12. The range of the method is very wide, depending on the amount and concentration of the dichromate solution used. Using 10 ml of the reagent specified permits accurate determinations in the range from 2 to 20 mg ethanol.

13. Avoid charring the contents of the distillation flask as the volatile products thus formed could react with the dichromate reagent giving false results. Distillation is best carried out using a small flame directed onto the flask, without using wire gauze.

14. Previously standardized against potassium iodate or periodate solution (Reagent 4c).

15. The indicator must be added late in the titration since otherwise the iodine molecules become too firmly embedded in the starch molecules and the end-point of the titration is difficult to observe.

16. The end-point of this titration is rather difficult to observe, and practice is recommended before tackling a sample from the distillation. Because the oxygen of the air liberates iodine from iodide, the titration must be carried out immediately and quickly, standardizing the time of each titration as closely as possible.

17. The blank is performed in exactly the same way as the sample titration except that alcohol is absent.

18. Where 1·0 ml of 0·1 N thiosulphate is equivalent to 1·15 mg ethanol.

19. The strength and quantity of the phenolphthalein solution required will be determined by the

nature of the substrate, a darker material needing more indicator in order to obtain a reproducible end-point. Since carbon dioxide discharges the red colour of phenolphthalein, titrations should be completed as quickly as possible, and new alcoholic wash should be thoroughly aspirated to remove all carbon dioxide before titration commences.

20. Syrupy phosphoric acid is used because it is not volatile under the conditions being used (unlike hydrochloric acid, for example) and is not an oxidizing agent (sulphuric acid might oxidize organic matter under these conditions).

21. Use the procedures and precautions appropriate to the apparatus being employed. Substrates such as malt extract have a marked tendency to froth in the early stages of the distillation, and care must be exercised to prevent carry-over of froth. It is preferable to check that the distillation procedure is satisfactory by carrying out distillations on standard solutions of a salt of acetic acid.

22. Since the distillate is colourless only a small amount of phenolphthalein indicator will be needed.

23. Since only acetic acid is present in the distillate prepared, in the volatile acid procedure the results can be reported as acetic acid. The total acid procedure will measure other acids formed in the fermentation (e.g. sugar acids such as gluconic acid) and must therefore be recorded as milli-equivalents.

MATERIALS (*see* note 1)

1. *Media*
   (a) *Synthetic medium*:

sucrose,	500 g
potassium dihydrogen phosphate,	2·75 g
potassium chloride,	2·125 g
calcium chloride dihydrate,	0·625 g
hydrated magnesium sulphate,	0·625 g
ferric chloride ($6H_2O$),	0·0125 g
hydrated manganese sulphate,	0·0125 g
ammonium sulphate,	1·0 g
yeast extract,	5·0 g
citric acid/sodium citrate buffer, pH 5·0,[a]	250 ml
water to	5 litres

   [a] prepared by mixing:

0·1 M citric acid solution,	102·5 ml
0·1 M tri-sodium citrate solution,	147·5 ml

   (b) *Malt extract medium*:

liquid malt extract,	700 g

   (use unhopped concentrate from a home-brewers suppliers) or (much more expensive):

spray-dried malt extract,	550 g
water to,	5 litres.

2. A vessel capable of holding 5 litres of medium and with sufficient head-space to allow for frothing.

3. 5 × 100 ml bottles or flasks, capped or plugged, and sterilized.

4. *Analytical reagents*:
   (a) Potassium dichromate (0·2 N) solution in 5 N sulphuric acid.
   (b) Sodium thiosulphate (0·1 N).
   (c) Potassium iodate or periodate (Analytical Reagent grade dried to constant mass). Primary standard for thiosulphate solution.
   (d) Potassium iodide crystals.
   (e) Sulphuric acid (2 N).
   (f) Starch iodine indicator.
   (g) Sodium hydroxide (0·1 N).
   (h) Phenolphthalein solution.
   (i) Syrupy phosphoric acid (Analytical Reagent grade).

5. Markham still (Markham, 1942) or equivalent small-scale steam-distillation apparatus.

6. Apparatus for distillation of alcohol, comprising a 50 ml round bottomed flask with a B19 ground glass neck; still head, B19; Liebig condenser, B19 (as short as is practicable); receiver adaptor.

7. Boiling tubes, approximately 60 ml capacity.

8. Burettes, 25 or 50 ml graduated in 1/10 ml.

9. Bulb pipettes, 10 ml capacity.

10. Acetification tower: *see* Figure 1. A list of parts is not appended, since the detailed construction will depend on the materials available to the laboratory in question.

11. Acetic acid bacteria: *Acetomonas oxydans* and *Acetobacter rancens*: departmental strains.

SPECIFIC REQUIREMENTS

Day 1
materials and vessels for synthetic medium or
  malt extract medium

Day 2
yeast
acetic acid bacteria
tared, sterilized flasks

Days 3–7
balance

Days 8–14
centrifuge for large volume harvest

vinegar tower
Markham still
phosphoric acid
phenolphthalein
0·1 N NaOH
burettes
dichromate reagent
boiling tubes
KI
0·1 N thiosulphate
starch (iodine indicator)

FURTHER INFORMATION

In a particular experiment using malt extract medium; dried baker's yeast inoculum for the alcoholic fermentation; a newly packed tower inoculated with a mixture of *Acetobacter aceti* and *Gluconobacter oxydans* subspecies *suboxydans* (formerly *A. suboxydans*); alcoholic fermentation conducted at 30 °C; vinegar tower operated at ambient temperature (18–20 °C); a class obtained the results shown in Tables 1 and 2. In Table 1 the 5 bottles of subsamples in the yeast fermentation behaved in a fairly consistent manner, suggesting that no spillage had occurred. The total weight loss from 258·7 g of liquid was 8·6 g. Since 180 g hexose would give rise to 88 g carbon dioxide and 92 g ethanol, an 8·6 g weight loss corresponds to the fermentation of 9·0 g ethanol in 250 ml medium or 36·0 g ethanol per litre medium.

Table 1 Weights and weight losses of bottles containing subsamples of the inoculated medium for alcoholic fermentation. These bottles were incubated at 30 °C alongside the main bulk of the fermentation, and were well shaken to remove as much dissolved carbon dioxide as possible before each weighing.

	Weight (g)				Weight loss (g) after incubation for (days)			
Bottle no.	Tare	Bottle + Medium	Medium	Bottle + medium after 1 day	1	2	3	6
1	86·9	138·7	51·8	137·6	1·1	1·4	1·6	1·8
2	92·1	143·8	51·7	142·7	1·1	1·4	1·5	1·7
3	103·1	155·0	51·9	153·8	1·2	1·5	1·6	1·7
4	104·7	156·4	51·7	155·4	1·0	1·4	1·5	1·7
5	107·1	158·7	51·6	157·6	1·1	1·5	1·5	1·7

Table 2 Progress of the vinegar tower. Volatile acid is expressed in grams of acetic acid per litre. The tower was operated at ambient temperature with a charge of 4·5 litres alcoholic medium inoculated with 2 × 100 ml cultures of acetic acid bacteria grown on malt wort medium.

Day	0	1	2	3	4	5	6	7	8
Total acid, mEq l^{-1}	28·4	34·2	65·6	134·4	214·0	—	416·8	376·8	285·2
Volatile acid, g l^{-1}	0·10	—	—	—	—	—	22·44[a]	—	—
Alcohol, g l^{-1}	32·7	32·1	30·0	23·9	18·5	—	9·2	8·3	7·5

[a] Equal to 373·3 mEq l^{-1}.

Table 2 shows that the progress of acetification, as judged by the daily estimation of total acid, was satisfactory. Note that the ommission of a sample on Day 5 makes the peak level of acid difficult to ascertain. The results show the eventual consumption of acetic acid by the bacteria, called "over-oxidation". Volatile acid analyses **always** give lower results than do the corresponding total acid analyses, because of acids other than acetic.

Yields of alcohol on sugar supplied (where the latter figure is known) and of acid on alcohol, should be calculated. At the highest observed acid concentration, there were 22·44 g acetic acid present per litre of wash, and 9·2 g ethanol per litre. Now 22·44 g acetic acid are equivalent to $(22·44 \times 46)/60 = 17·20$ g ethanol. Hence a total of $9·2 + 17·2 = 26·4$ g of ethanol can be accounted for, leaving 6·3 g of the original 32·7 g ethanol unaccounted for; the bulk of this was presumably lost through evaporation.

Numerous other analyses can be carried out depending on the time available and the requirements of the course; an important example is an analysis for reducing sugars.

Many variations on this basic experiment are possible. Thus, the effect of temperature could be investigated, an increase improving the rate of reaction (up to a point) but also increasing the evaporative losses. Other studies might include using different media, examining gas transfer rates, various aspect ratios (height:diameter ratios) for the acetifier; plastics, etc. as replacements for wood shavings; the effect of different acetic acid bacteria; identification of other fermentation products; attempts at deep fermentation in stirred tank reactors; analysis for carbon dioxide in the gas effluent; estimates of heat production in the fermentation. The simple experiment described takes about two weeks to run, but it could easily provide the basis for an investigation occupying several students for a term. The choice of media for the experiment offers considerable scope, the two suggested here, malt extract and a synthetic medium, both work well but show interesting differences in rates of fermentation.

The experiment may be given further interest by using raw materials such as over-ripe bananas, cane molasses, grape juice, etc. which serve as the raw materials for small local vinegar industries in the developing world.

Perhaps surprisingly, a reasonably good quality vinegar can be produced by this simple experiment.

The conversion of alcohol to vinegar is practically unique to the acetic acid bacteria and these bacteria carry out many other unusual fermentations, regulated according to the Bertrand–Hudson Rules (Rose, 1976). The formation of gluconic acids (Kulka and Walker, 1954) is an interesting class experiment and has diagnostic value in identifying acetic acid bacteria. Other conversions include D-sorbitol to L-sorbose and glycerol to di-hydroxyacetone. In vinegar production, *Acetobacter* spp. are preferred to *Acetomonas* or *Gluconobacter* (because they tend to be more metabolically active) despite being able to further oxidize acetic acid to water and carbon dioxide when the supply of ethanol runs low.

ACKNOWLEDGEMENT

The original version of the above experiment was introduced to our undergraduate classes in collaboration with the late Dr. J. Meyrath (formerly Professor of Hochschule für Bodenkultur, Institut für Angewandte Mikrobiologie, Vienna, Austria).

We also wish to thank Mr J. McElroy for assistance each year in assembling and running the apparatus, and Mrs M. Provan and Mrs J. Winter for their patience over the numerous retypings of the manuscript.

REFERENCES

R. B. Fred, W. H. Peterson and A. Davenport (1919). Acid fermentation of xylose. *Journal of Biological Chemistry*, **39**, 347–83.

D. W. Kent-Jones and G. L. Taylor (1954). Determination of alcohol in blood and urine. *Analyst*, **79**, 121–36.

F. L. Kozelka and C. H. Hynes (1941). Determination of alcohol for Medico-Legal purposes. *Industrial and Engineering Chemistry, Analytical Edition*, **13**, 905–7. (Abstracted in *Analyst* (1942) **67**, 174–5.)

D. Kulka and T. K. Walker (1954). The ketogenic activities of *Acetobacter* species in a glucose medium. *Archives of Biochemistry and Biophysics*, **50**, 169–79.

R. Markham (1942). A steam distillation apparatus suitable for micro-kjeldahl analysis. *Biochemical Journal* **36**, 790–1.

W. D. Maxon and M. J. Johnson (1945). Aeration studies on propagation of baker's yeast. *Industrial and Engineering Chemistry*, **45**, 2554–64.

A. H. Rose (1976). "Chemical Microbiology", 3rd edn, p. 204. Butterworths, London.

41

Environment and Variation in Streptococcal Carbohydrate Metabolism

Department of Science, South Glamorgan Institute of Higher Education, Colchester Avenue, Cardiff, Wales

---

Level: Advanced undergraduates
Subject areas: Bacterial metabolism
Special features: Introduction to the concept that bacteria are variable and their properties depend upon the environment

---

INTRODUCTION

Streptococci are usually described as being homofermentative, i.e. converting glucose almost exclusively to lactic acid, even though it has been realized that the products of their carbohydrate metabolism are dependent, to some extent, on growth conditions (Friedmann, 1939). All of the early work on streptococcal metabolism was carried out in batch culture, where it is difficult to control closely all the environmental growth conditions. By using a chemostat it is possible to control the environment precisely and the effect of single changes in growth conditions on the nature of streptococcal fermentation end-products can be observed.

The aim of the experiment is to show that changing the growth rate, growth limiting substrate and pH alter the fermentation end-products of *Streptococcus mutans* JC2 and thus illustrate that, depending upon environment, the organism may be homofermentative or heterofermentative.

EXPERIMENTAL

The first part of the experiment is to allow *S. mutans* JC2 to establish steady state conditions under carbon limitation at a stated temperature, pH and growth rate, and then sample the culture vessel

---

© 1982, The Society for General Microbiology
Those wishing to photocopy this experiment must follow the instructions given on page v.

and analyse the fermentation end products. After changing to nitrogen limitation, and allowing steady state conditions to be re-established, the culture vessel is sampled and the fermentation end-products analysed. After returning to carbon limitation the effect on fermentation end-products of single changes in growth rate and pH are investigated.

Day 1 *(1 h + 10 min separated by 7 h gap)*

Check sterility and pH of culture medium 1 in assembled chemostat.

↓

Inoculate with 10 ml of an 18 h batch culture of *S. mutans* JC2 grown in medium 1.

↓

After 7 h switch on the pump supplying medium to the culture vessel so that the dilution rate = 0·13 h^{-1}. Remember it is dilution rate which is important as culture volume may vary (suggested culture volume 150–400 ml). Note time.

- -

Days 2 and 3 *(1¼ h) Note times*

Sample 5 ml of the culture, check for purity on blood agar and Mitis/Salivarius agar *(see* notes 1 and 2). Incubate aerobically and anaerobically.

↓

Perform Gram stain and read optical density of sample at 420 nm against medium blank *(see* note 4).

↓

Determine dry weight of bacteria produced. Use the supernate from the dry weight determination for glucose determination by glucose oxidase method (Kit from Boehringer Corporation) *(see* note 3).

- -

Day 4 *(4 h) Note time*

Approximately 72 h after the start of the medium supply steady state conditions will have been established. (Check constant optical density and level of glucose in medium.) Take 20 ml of culture from the fermentor vessel and use as follows.

- (a) Two 5 ml portions for dry weight determination. Duplicate 5 ml samples of culture are centrifuged at 4000 rev. min^{-1} for 30 min, washed once in 10 ml distilled water, re-centrifuged and then dried for 2 days at 110 °C.
- (b) 7 ml for optical density measurement and streaking onto blood agar and Mitis/Salivarius agar.
- (c) Three 1 ml samples for gas–liquid chromatography (but *see* note 6). Take 1 ml of steady state chemostat culture and mix with 1 g of cation exchange resin (AG 50W – X4) 200–400 mesh suspended in water to a volume of 2 ml.

↓

Filter the suspension through glass wool in a Pasteur pipette.

↓

Inject 0·5 μl of the effluent into a gas chromatograph with a hydrogen flame detector.

↓

Standard curves relating peak areas to concentration should be obtained by injecting 0·5 μl portions of standard solutions of the following fatty acids and alcohols in 0·1 M sulphuric acid: acetic acid, pyruvic acid, lactic acid, succinic acid, isovalearic acid and ethyl alcohol (*see* note 8).

↓

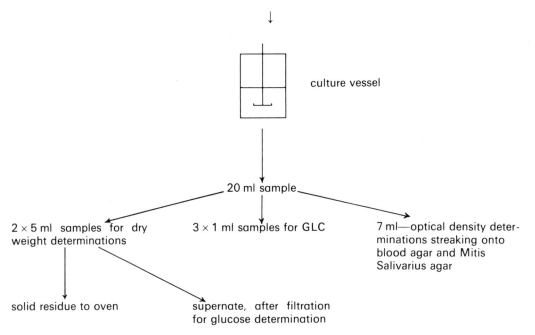

culture vessel

20 ml sample

2 × 5 ml samples for dry weight determinations

3 × 1 ml samples for GLC

7 ml—optical density determinations streaking onto blood agar and Mitis Salivarius agar

solid residue to oven

supernate, after filtration for glucose determination

Repeat with sample of uninoculated medium.

↓

Change supply medium to the chemostat to medium 2.

- -

Days 5 and 6 *(1¼ h) Note time*

As for Day 2.

- -

Day 7 *(4 h) Note time*

As for Day 4 but revert back to medium 1 and alter dilution rate to 0·3 h^{-1}.

- -

Day 8 *(1¼ h) Note time*

As for Day 2.

Day 9 *(4 h) Note time*

As for Day 4 but revert back to $D = 0.13 \, \mathrm{h}^{-1}$ and alter pH to 6·0.

Days 10 and 11 *(1¼ h) Note times*

As for Day 2.

Day 12 *(5 h) Note time*

As for Day 4. Halt experiment. Sterilize and dismantle chemostat.

Day 13 *(20 min)*

Examine plates from Day 12. Determine dry weights.

Notes and Points to Watch
1. The Gram stain is useful in checking the purity of the culture, but also note if there is any difference in the average chain length of the organism. Optical density measurements are affected by chain length, but it should be possible to plot a graph of optical density against dry weight.
2. Streaking on blood agar and Mitis/Salivarius agar not only serves as an indication of contamination by outside organisms, but also if mutants with different colonial morphology are growing in the culture vessel.
3. Levels of glucose remaining in used culture media can be used to calculate the % conversion of glucose to lactic acid.
4. Dilution of samples may be necessary for optical density determinations. Use a medium blank to zero the instrument.
5. A sample of uninoculated medium should also be analysed for acetic acid, lactic acid, etc.
6. If preferred, samples from the culture vessel may be deep frozen and fermentation end-products analysed at the end of the experiment.
7. If no GLC is available then the concentrations of lactic acid can be estimated colorimetrically, e.g. using the method of Nanni and Baldini (1964). Alternatively both lactic and acetic acid can be estimated enzymatically (kits obtained from Boehringer Corp. GmbH).
8. This method does not detect formic acid.

MATERIALS

1. Organism: *Streptococcus mutans* JC2 (or other strain) or *S. sanguis.*
2. Chemostat: with stirring, pH, temperature control.

3. Culture media. Two culture media are used in the chemostat, medium 1 and medium 2. Both are made up from the following stock solutions.
   (a) T.Y.C. solution made up from 10 g tryptose, 5 g yeast extract, 2·5 g casamino acids in 900 ml de-ionized distilled water.
   (b) Salt solution containing per millilitre: $MgSO_4 . 7H_2O$, 40 mg; NaCl, 2 mg; $FeSO_4 . 7H_2O$, 2 mg; $MgSO_4 . 7H_2O$, 2 mg.
   (c) Glucose solution 20% w/v.
   (d) 1 M potassium dihydrogen orthophosphate solution.

   Medium 1 is glucose-limiting and contains 26 ml of glucose and 900 ml of T.Y.C. in addition to 5 ml of stock salt solution and 10 ml of phosphate solution in 1 litre.

   Medium 2 is nitrogen-limiting and contains 50 ml of glucose, 75 ml of T.Y.C. 5 ml salt solution 10 ml phosphate solution in 1 litre.

   Mitis/Salivarius agar is obtainable from Difco.

4. Culture conditions: atmosphere of 95% CO_2, 5% H_2 sterilized by filtration at a flow rate of 50 ml min^{-1}, pH 7·0 (unless otherwise stated), temperature 37 °C, dilution rate 0·13 h^{-1} (unless otherwise stated).

5. Gas chromatograph with hydrogen flame detector (but see note 7). The chromatograph should be equipped with a 1·8 m × 0·63 cm o.d. (0·20 cm i.d) coiled glass column packed with chromosorb 101. The temperature of the injection port is 200 °C and that of the detector 240 °C; the carrier gas is nitrogen. Samples are run isothermally at 200 °C.

6. Glucose oxidase kits for estimation of glucose. These can be obtained from most biochemical supply houses.

7. Cation exchange resin AG 50W-X4 (200–400 mesh). This can be obtained from most chemical supply houses.

SPECIFIC REQUIREMENTS

Day 1
chemostat and supply of culture medium 1
10 ml of 18 h batch culture of *S. mutans* JC2 grown in medium 1
pipettes
pH meter

Days 2–12 inclusive
2 blood agar plates
2 Mitis/Salivarius agar plates
pipettes
spectrophotometer
Gram stains
microscope slides

5 ml sterile sample bottles
bacterial filters 0·45 μm
low speed centrifuge + weighed centrifuge tubes
oven at 110 °C
reagents for estimation of glucose

Days 4, 7, 9, 12
sample bottles capable of holding 20 ml
GLC equipped as described previously
glass wool
Pasteur pipettes, 0·5 μl
syringe for GLC
cation exchange resin

FURTHER INFORMATION

Typical results should show (*see* Figure 1) that a change in growth limitation from glucose to nitrogen reduces bacterial yield and causes a change in fermentation end products from ethanol, acetic acid (and formic acid if estimated) to lactic acid. Similarly, under glucose limitation, increasing growth rate and lowering the pH should cause an increase in the amount of glucose converted to lactic acid.

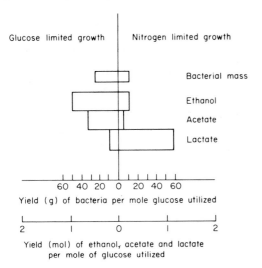

Glucose limited growth   Nitrogen limited growth

Bacterial mass

Ethanol

Acetate

Lactate

60 40 20 0 20 40 60

Yield (g) of bacteria per mole glucose utilized

2   1   0   1   2

Yield (mol) of ethanol, acetate and lactate
per mole of glucose utilized

Figure 1 The yield of fermentation products and bacterial mass in glucose limited and nitrogen limited culture of *S. mutans* grown in a chemostat, pH 7·0, $D = 0·13\,h^{-1}$ at 37 °C.

Figure 2 Proposed explanation for regulation of fermentation end-products.

Proposed scheme for the regulation of glucose metabolism of *S. mutans* JC2 during (a) continuous growth under glucose limitation and (b) nitrogen limitation in the presence of an excess of glucose.

G-6-P, glucose-6-phosphate; FDP, fructose-1, 6-dephosphate; GAL-3-P, D-glyceraldehyde-3-phosphate; PEP, phosphoenolpyruvate; PK, pyruvate kinase; LDH, lactate dehydrogenase; PFL, pyruvate formate-lyase. $\oplus$: Activation; $\ominus$: Inhibition. Capitals: high concentration of intracellular intermediates in the cell; small capitals: low concentration of intracellular intermediates in the cell.

When *S. mutans* JC2 is growing in the presence of an excess of glucose with the nitrogen source-limiting growth, the intracellular concentrations of fructose-1,6-diphosphate and of D-glyceraldehyde-3-phosphate are high. Fructose 1, 6-diphosphate activates lactate dehydrogenase, D-glyceraldehyde-3-phosphate inhibits pyruvate formate-lyase and *S. mutans* JC 2 produces almost exclusively lactate. When *S. mutans* JC 2 is growing under glucose limitation, the cellular level of pyruvate formate-lyase will increase and the intracellular concentrations of fructose-1, 6-diphosphate and of D-glyceraldehyde-3-phosphate will be low. This condition stops the activation of lactate dehydrogenase, releases any inhibition of pyruvate formate-lyase and *S. mutans* JC 2 produces mainly formate ethanol and acetate.

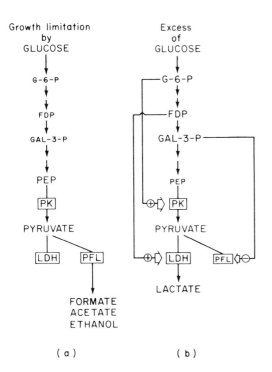

Growth limitation
by
GLUCOSE

Excess
of
GLUCOSE

G-6-P

FDP

GAL-3-P

PEP

PK

PYRUVATE

LDH   PFL

FORMATE
ACETATE
ETHANOL

(a)

G-6-P

FDP

GAL-3-P

PEP

⊕ PK

PYRUVATE

⊕ LDH   PFL ⊖

LACTATE

(b)

It would seem that the nature of the fermentation end-products is regulated by the environment via certain glycolytic intermediates (Carlsson and Griffith, 1974; Yamada and Carlsson 1976; de Vries *et al.*, 1970) (*see* Figure 2). The experiment could be extended to investigate the effects of other growth limiting substrates using defined media (described by Carlsson and Griffith, 1974).

REFERENCES

J. Carlsson and C. J. Griffith (1974). Fermentation products and bacterial yields in glucose limited and nitrogen limited cultures of streptococci. *Archives of Oral Biology*, **19**, 1105–9.

T. E. Friedmann (1939). The carbohydrate metabolism of streptococci. *Journal of Biological Chemistry*, **130**, 757–61.

G. Nanni and I. Baldini (1964). A micro method for the estimation of lactic acid. *Italian Journal of Biochemistry*, **13**, 135–7.

W. De Vries, W. M. C. Kapteijn, E. G. van der Beek and A. H. Stouthamer (1970). Molar growth yields and fermentation balances of Lactobacillus casei L 3 in batch cultures and continuous cultures. *Journal of General Microbiology*, **63**, 333–45.

Y. Yamada and J. Carlsson (1976). The role of pyrovate formate-lyase in glucose metabolism of *Streptococcus mutans*. *In* "Microbial Aspects of Dental Caries", (Stiles, Loesche and O'Brien, eds). A special supplement to *Microbiology Abstracts* Vol. III, pp. 809–19.

42

Movement of Metabolites from Alga to Fungus in a Lichen

G. W. GOODAY

Department of Microbiology, University of Aberdeen, Aberdeen, AB9 1AS, Scotland

> *Level*: Intermediate/Advanced undergraduates
> *Subject areas*: Symbiosis; algal and fungal physiology
> *Special features*: Identification of a metabolite by isotope dilution technique

INTRODUCTION

A major factor in the lichen symbiosis is the movement of photosynthate from alga to fungus. This movement takes place predominantly as one particular carbohydrate which may differ in different lichens. The nature of this algal metabolite can be elucidated by the use of the "inhibition technique" developed by Smith and colleagues (*see* review by Smith, 1975). Radioactive carbon dioxide is fed during photosynthesis, and the lichen is incubated in a medium containing a non-radioactive putative algal metabolite. If radioactivity is released from the lichen to the medium, then this inhibition of the transport from alga to fungus is assumed to show competition between the naturally translocated metabolite and the large excess of this metabolite that has been added to the bathing solution. This can be checked by testing if the radioactive metabolite released is chemically the same as the compound in the bathing medium. If so, the translocated metabolite has been identified.

In all lichens the translocated metabolite produced specifically by the alga during photosynthesis is rapidly converted to another chemical when received by the fungus.

The aim of the experiment is to demonstrate photosynthetic fixation of carbon dioxide by the lichen, to identify the translocated metabolite, and to estimate the proportion of photosynthate that is translocated. The lichen of choice is *Peltigera polydactyla*, in which the alga is a blue-green *Nostoc* species, but other lichens can be used.

EXPERIMENTAL

(2½ h with a 1 h incubation period)

In a fume cupboard, make the following additions to four Universal bottles, each of which contains 1 ml of water with $2\mu Ci$ (micro-Curies) of $NaH^{14}CO_3$: bottle A, 1 ml water; bottle B, 1 ml 2% glucose; bottle C, 1 ml 2% carbohydrate (*see* note 4); bottle D (the "dark control"), 1 ml 2% glucose.

↓

Cut discs from the washed thallus of *Peltigera* or other lichen (*see* note 3) with a cork borer no. 3, to give 4 samples of 5 discs, each on moist filter paper.

↓

Add 5 discs of lichen, green side uppermost, to each of the Universal bottles, A, B, C, D. Immediately wrap bottle D completely with foil to exclude all light. Incubate in the light. Occasionally shake the bottles gently.

↓

After incubation for 1 h, process the discs and the incubation medium from each bottle, in a fume cupboard.

↓ ↓

Pipette 200 μl incubation medium onto a glass fibre disc in a scintillation vial (labelled at the top).	Lift out each set of lichen discs with forceps into a Universal bottle containing 2 ml 80% methanol.
↓	↓
Add 100 μl 5% acetic acid to eliminate any remaining $NaH^{14}CO_3$.	Shake in a water-bath at 45°C for 30 min.
↓	↓
Dry completely on a hot plate.	Pipette 20 μl onto a glass fibre disc in a scintillation vial (labelled at the top).
↓	↓
Add 4 ml scintillation fluid.	Add 100 μl 5% acetic acid.
↓	↓
Count in a scintillation counter set for ^{14}C.	Dry completely.
	↓
	Add 4 ml scintillation fluid.
	↓
	Count in a scintillation counter set for ^{14}C.
↓	↓

After counting, calculate for tubes A, B, C, (1) ^{14}C released to medium for each of the 5 discs, expressed as ct min^{-1} h^{-1}; (2) total ^{14}C fixed as measured by that extracted with methanol plus that released to the medium; (3) the percentage that is released. The results from tube D should be used to assess whether the experiment is dealing with a photosynthetic phenomenon. A comparison between percentage ^{14}C released in tubes A, B, C demonstrates the result and specificity of the "inhibition technique".

Notes and Points to Watch

1. This experiment uses radioactivity. Adequate facilities for handling NaH^{14}CO$_3$ must be provided on the bench and in the fume cupboard. Students must be well versed in the correct procedures. Samples must be disposed of properly.
2. The glass fibre discs should be translucent after the addition of scintillation fluid. If not, they have been inadequately dried.
3. Other lichens can be used. *Peltigera* is particularly easy to handle as discs can be cut readily, and it gives a clear result in a short time. Crustose and fruticose lichens can be weighed out as equal samples for the experiment, but a longer incubation time may be needed (Hill and Smith, 1972). If a lichen contains a blue–green symbiont the translocated metabolite is glucose; if a green symbiont it is a polyol: sorbitol, ribitol or erythritol (Smith, 1975).
4. A range of carbohydrates can be used in bottle C, with a different compound being used for each member of the class. 2-Deoxyglucose, a glucose analogue, and the saponin digitonin (at 0·01 %) both cause release of the ^{14}glucose to the medium, by mechanisms discussed by Chambers *et al.* (1976).
5. The identity of the radioactive metabolite in the incubation medium B can be investigated by running samples, with appropriate standards, in a thin layer or paper chromatographic system suitable for separating carbohydrates (cf. Hill and Smith, 1972), and locating the radioactive spots by radioautography or end-window scanning.
6. The anatomy and cytology of the lichen can be examined during the incubation period.

MATERIALS

1. *Peltigera polydactyla* or *P. canina* can be obtained readily in many parts of Great Britain, growing on woodland paths, sand dunes or lawns. For its conservation it should be collected only where abundant, and only in quantities just sufficient for the classwork.

2. A scintillation fluid, compatible with counting ^{14}C in the available counter. A common recipe is 5 g 2, 5-diphenyloxazole (PPO) and 0·1 g 1, 4-di-2-(5-phenyloxazolyl)-benzene (POPOP) in 1 litre toluene.

SPECIFIC REQUIREMENTS

4 Universal bottles, each tightly capped and containing freshly dispensed NaH^{4}CO$_3$ (Radiochemical Centre, Amersham, UK) at 2 μCi in 1 ml water

4 empty Universal bottles with caps
8 scintillation vials
automatic pipettes for 200 μl, 100 μl, 20 μl
1 ml pipettes

---

cont.

cont.

2·5 cm glass fibre discs (Whatman GF/C)	cork borer no. 3
shaking water-bath at 45°C with rack for Universal bottles	moist filter paper in Petri dish
hot plate or source of heat to dry discs in vials	range of carbohydrates and facilities to make up 2% solutions
radioactive requisites: gloves, trays, labels, disposal facilities, fume cupboard	80% (v/v) methanol
	5% (v/v) acetic acid
light source for photosynthesis	2% (w/v) glucose
forceps	scintillation fluid
aluminium foil	

FURTHER INFORMATION

The original explanation for the results obtained with the inhibition technique with *P. polydactyla* is that the ^{14}C–glucose in the medium swamps the uptake of the ^{14}C–glucose by the fungus:

(a) Translocation uninhibited (i.e. tube A)

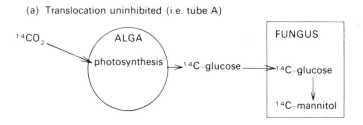

(b) Translocation inhibited (i.e. tube B)

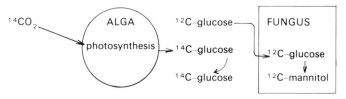

Hill and Smith (1972) have questioned whether this is too simplistic a model, and suggest that there might be more direct effects on the algal metabolism by the exogenous glucose; such as exchange and competition between ^{12}C and ^{14}C carbohydrates. Nevertheless the technique has proved very valuable as a means of identifying the mobile carbohydrates, and has been used for many lichens.

With *P. polydactyla*, this experiment usually gives a release of up to 50% of the total counted ^{14}C–photosynthate in the presence of exogenous glucose, compared with less than 1% in its absence, or in the presence of a non-competing carbohydrate such as galactose or mannitol. One fate of ^{14}C–photosynthate that is not accounted for by this procedure is the insoluble residue of the lichen. This is ignored for the calculation since (1) in the time of the experiment little ^{14}C will have accumulated in insoluble products and (2) the only efficient way of assessing it by liquid scintillation counting would be to burn it to CO_2, requiring facilities out of reach of most undergraduate laboratories.

The experiment could be taken further by analysing (1) the ^{14}C released to the medium, which in tube B would be almost totally glucose, and (2) the ^{14}C in the methanol extract, which would be chiefly the fungal metabolite, mannitol.

REFERENCES

S. Chambers, M. Morris and D. C. Smith (1976). Lichen physiology XV. The effect of digitonin and other treatments on biotrophic transport of glucose from alga to fungus in *Peltigera polydactyla*. *New Phytologist*, **76**, 485–500.

D. J. Hill and D. C. Smith (1972). Lichen physiology XII. The "inhibition technique". *New Phytologist*, **71**, 15–30.

D. C. Smith (1975). Symbiosis and the biology of lichenised fungi. *In* "Symbiosis". Symposia of the Society for Experimental Biology, Vol. 29, pp. 373–405.

Part Five
The Micro-organism Dies

Differential Bacteriolytic Response Associated with the Addition Order of Lysozyme and Salt

J. DOUGLAS and Z. MOUSSAVI-JAHED

Department of Applied Biology, Brunel University,
Uxbridge, UB8 3PH, England

> *Level*: Beginning undergraduates
> *Subject areas*: Microbial chemistry and
> biochemistry
> *Special features*: Unexpected results severely test
> powers of observation

INTRODUCTION

Lysozyme is an enzyme which dissolves peptidoglycan, thereby causing bacterial cells to disintegrate. It was discovered by Alexander Fleming in nasal mucus and is present in many tissues and secretions. Tears are a rich source of it, as is egg-white. Chang and Carr (1971) showed that its activity had a curious "pH-cation profile" that did not permit optima to be determined. Metcalf and Deibel (1969) reported that the order of addition of lysozyme and salt had a profound effect on the lysis of *Streptococcus faecium*.

The object of the simple experiment described here is to show, using *Micrococcus lysodeikticus*, that the sequence of mixing of lysozyme, salt and bacteria greatly influences the extent and rapidity of the turbidity changes in the suspension. This exercise can serve as an introduction to the nephelometer, to lysozyme, or both, or as a test of the students' general awareness and understanding. Note that turbidities can also be compared satisfactorily by eye and a nephelometer is not absolutely necessary.

EXPERIMENTAL

(1 h, see note 1)

The same amounts of the same reagents (*see* note 3) are mixed in tubes 1–3 in different sequences, as detailed below. The turbidity of each mixture is compared at each stage with that of an appropriate control consisting of bacteria (lysozyme substrate) only, correspondingly diluted with water. Control A is made from 5 ml substrate + 5 ml water and control B from 5 ml substrate + 10 ml water.

Tube 1	*Tube 2*	*Tube 3*
Mix 5 ml 0·3 mg ml^{-1} lysozyme substrate with 5 ml 4 M NaCl and compare turbidity with control A (*see* note 4).	Mix 5 ml 0·1 mg ml^{-1} lysozyme with 5 ml 4 M NaCl and compare turbidity with control A (*see* note 4).	Mix 5 ml 0·1 mg ml^{-1} lysozyme with 5 ml 0·3 mg ml^{-1} lysozyme substrate and compare turbidity with control A (*see* note 4).
↓	↓	↓
Wait 10 min (*see* note 2) and again compare turbidity with control A.	Wait 10 min (*see* note 2) and again compare turbidity with control A.	Wait 10 min (*see* note 2) and again compare turbidity with control A.
↓	↓	↓
Add 5 ml 0·1 mg ml^{-1} lysozyme and compare turbidity with control B.	Add 5 ml 0·3 mg ml^{-1} lysozyme substrate and compare turbidity with control B.	Add 5 ml 4 M NaCl and compare turbidity with control B.

Notes and Points to Watch

1. One hour is ample; much time is saved by running the tubes concurrently, but they must be started at intervals of **at least 2 min** to allow observation of any immediate reactions.
2. The substrate tends to settle out. Remix the suspension frequently.
3. Tubes and pipettes may be re-used for repetitions after **thorough** rinsing with distilled water.
4. Temperature is not critical; the experiment may be performed in a water-bath at 20 °C or on the bench.

MATERIALS

1. Lysozyme (crystalline, from egg white).

2. Lysozyme substrate (Difco). This is UV-killed *Micrococcus lysodeikticus*.

SPECIFIC REQUIREMENTS

Day 1
lysozyme, 0·1 mg ml^{-1} in distilled water, 50 ml
substrate, 0·3 mg ml^{-1} in distilled water, 50 ml
4 M NaCl in distilled water, 50 ml
five 6″ × $\frac{3}{4}$″ test tubes, in a rack

four 5 ml pipettes
distilled water, about 300 ml in a beaker
nephelometer (optional)
(NB none of the above needs to be sterilized)

FURTHER INFORMATION

The origins of this experiment can be found in an obscure paper by Nakamura (1923) who discovered that suspensions of bacteria treated with lysozyme in distilled water remained turbid, but cleared immediately on addition of either NaOH or HCl and erroneously ascribed this "Nakamura effect", as it came to be called, to pH changes, having omitted to test the effect of adding NaCl after the lysozyme. Although the hydrolysis of peptidoglycan by lysozyme is now so thoroughly documented (*see* Blake *et al.*, 1967) as to need no further treatment here, the consequent turbidity changes in bacterial suspensions remain enigmatic. It will be seen (Table 1) that when lysozyme and substrate are mixed in salt-free solution the turbidity actually increases, indicating that some sort of interaction does occur (analysis of free reducing groups would confirm that peptidoglycan is being destroyed). Then, on adding salt, almost complete clearing is instantaneous; but if salt is added first, either to substrate or to the lysozyme, subsequent clearing is slow. The initial increase in turbidity brought about by lysozyme alone could be due to change in refractive index of some cell component(s), to the formation of additional reflective surfaces within the cell, to the liberation of colloidal material or to any combination of these. The sudden clearing on addition of NaCl is too great to be accounted for by the increased refractive index of the medium. Optical and electron microscope studies on Gram-positive and Gram-negative cells at all stages of treatment (Boassen, 1937; Grula and Hartsell, 1957) reveal dramatic changes but shed little light on the turbidity problem. A satisfactory solution probably requires the application of physical techniques at the macromolecular level.

Table 1 Turbidity changes in relation to the sequence of addition of lysozyme and salt to a suspension of *M. lysodeikticus.*

	Turbidity (on a subjective scale, 1–5, control = 4)		
Tube	When the first two reagents are mixed	After 10 min	When the third reagent is added
1	4	4	4, decreasing slowly
2	0	0	4, decreasing slowly
3	4	5	$< \frac{1}{2}$

Specimen results are shown in Table 1. Students are given the blank table to complete. In tube 3, students usually fail to observe the initial rise in turbidity because they are expecting a decrease; similarly they may not believe their eyes when salt is added. On repetition (the reagents are not expensive) the truth usually dawns.

REFERENCES

C. C. F. Blake, N. Johnson, G. A. Mair, A. C. T. North, D. C. Phillips and V. R. Sarma (1967). Crystallographic studies of the activity of hen egg-white lysozyme. *Proceedings of the Royal Society, Series B,* **167**, 378–88.

F. H. Boassen (1937). On the bacteriolysis by lysozyme. *Journal of Immunology*, **34**, 281–93.

K. Y. Chang and C. W. Carr (1971). Studies on the structure and function of lysozyme. I. The effect of pH and cation concentration on activity. *Biochimica et Biophysica Acta*, **229**, 496–503.

E. A. Grula and S. E. Hartsell (1957). Lysozyme in the bacteriolysis of Gram-negative bacteria. I. Morphological changes during use of Nakamura's technique. *Canadian Journal of Microbiology*, **3**, 13–21.

R. H. Metcalf and R. H. Deibel (1969). Differential lytic response of enterococci associated with addition order of lysozyme and anions. *Journal of Bacteriology*, **99**, 674–80.

O. Nakamura (1923). Ueber Lysozymwirkungen. *Zeitschrift für Immunitätsforschungen*, **38**, 425–49.

44

Assay for the Presence of Antibiotic Residues in Milk

J. F. LOWE

*Department of Microbiology, School of Agriculture,
West Mains Road, Edinburgh, EH9 3JG, Scotland*

Level: Beginning undergraduates
Subject areas: Milk microbiology; antibiotic assay
Special features: Results obtained in 3 h

INTRODUCTION

Cows may excrete antibiotics in milk for up to 72 h after treatment. This can cause a number of problems, the most important of which, in economic terms, is that antibiotic residues may kill or inhibit starter cultures when the milk is to be used for cheese or yoghurt manufacture.

Various methods have been developed for assaying antibiotic residues in milk. The experiment described here is based on that recommended by the Scottish Milk Marketing Board and reveals the presence of an antibiotic by a zone of inhibition around an antibiotic assay disc.

EXPERIMENTAL

Day 1 *(3 h including incubation time)*

	Aseptically remove antibiotic assay disc from vial.
	↓
Take 1 ml culture of *Bacillus calidolactis*.	Dip into sample of milk and allow to drain (*see* note 3).
↓	↓

Add to 5 ml of Tryptone glucose yeast agar.

↓

Pour into a Petri dish and allow to solidify (*see* note 1).

↓

Repeat until all squares of the grid have been filled (*see* note 2).

↓

Incubate at 55 °C for 2·5 h.

↓

Read results (a milk is regarded as being + ve for residue if the zone of inhibition is larger than that of the standard).

Place disc on agar using the grid as a guide.

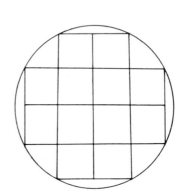

Notes and Points to Watch

1. In order for as large a zone of inhibition as possible to develop in the short incubation period, the plate is poured very thinly. It is advisable to warm the dish so that the 6 ml of culture in agar spreads evenly over its base.
2. The grid allows for 12 discs to be put onto the agar. It is probably best to restrict the number of milk samples tested to 3, which, with the Standard (0·02 iu penicillin ml^{-1}), gives three replicates of each.
3. It is vitally important that the antibiotic assay discs are drained of excess milk. This is best done by touching against the edge of the container until the milk is seen to have drained. If this is not done, it is likely that any zones of inhibition will run into each other.

MATERIALS

1. Liquid culture of *Bacillus calidolactis* strain NIZOC 953.
2. Five ml quantities of tryptone glucose yeast agar (Plate Count Agar, Oxoid). This **must** be adjusted to pH 8·0.
3. Incubator set at 55 °C.

SPECIFIC REQUIREMENTS

Petri dishes
pipettes
antibiotic assay filter paper discs (6 mm diam.), about 15 per student

forceps and alcohol for flaming
samples of milk for testing and a Standard containing 0·02 iu penicillin ml^{-1} (the Scottish Standard)

FURTHER INFORMATION

There are one or two problems which may arise in this assay. Firstly, *B. calidolactis* is an organism that needs careful handling and is best grown in tryptone glucose yeast extract broth (pH 8·0) for a few days at 55 °C, subculturing frequently, to ensure that rapid growth occurs once the assay is set up. It is also wise to keep the organism at 55 °C until the last possible moment so that it remains in the log phase.

Sometimes students have difficulty in reading the assay after 2·5 h. In this case, it is quite easy to continue the incubation for a little longer. If, as a regular feature, the experiment is done with a longer incubation period, it might be wise to pour slightly thicker plates, i.e. to use more than 5 ml of agar.

The last problem is one of obtaining milk containing antibiotic. In theory, this should be difficult and I have "doctored" samples, but it is not uncommon to find antibiotics in milk, mainly because milk for human consumption is not often checked. Also, in a large class there are frequently students with farming connections who may be able to help in providing samples.

REFERENCE

R. J. M. Crawford and J. H. Galloway (1964). Testing milk for antibiotic residues. *Dairy Industries*, **29**, 256–62.

The Effect of Antiseptics on a Test Organism on Human Skin

S. G. B. AMYES

Department of Bacteriology, The Medical School, University of Edinburgh, Teviot Place, Edinburgh, EH8 9AG, Scotland

and

J. T. SMITH

Microbiology Section, Department of Pharmaceutics, The School of Pharmacy, University of London, 29–39 Brunswick Square, London, WC1N 1AX, England

---

Level: All undergraduate years
Subject areas: Medical microbiology, hygiene
Special features: Strong visual impact; relevance to everyday life

---

INTRODUCTION

The action of chemical disinfectants is never instantaneous and is often affected by a variety of factors. Thus, it is essential to ensure that the correct disinfectant procedure is used for each specific situation. Antiseptics are chemical disinfectants that can be applied to certain areas of the human body without causing serious side effects. For this reason they are always milder than the chemical disinfectants used for hospital or domestic control of micro-organisms on inanimate objects. They act more slowly and are never able to sterilize the area to which they are applied.

Almost everyone is aware of the considerable number of antiseptics that are available from pharmacists and we are all exposed to various claims about the power of these products from both television and press advertisements. However, much of the advertising often fails to state under what conditions the antiseptic is most potent, which organisms it is able to deal with and the length of time needed for the desired result.

The aim of this experiment is to test the efficacy of a variety of commonly available antiseptic preparations against a test organism and to show that not all antiseptics are equally effective. The experiment is also designed to show that the length of time that the antiseptic is in contact with the organism is important.

EXPERIMENTAL

This experiment is most successfully performed on the ventral surface of the forearm. It is important to use skin surfaces that do not possess many surface hairs so as to minimize contamination from the resident skin bacteria. The whole experiment cannot be performed on one forearm.

Day 1 *(1 h)*

Swab, for 20 s (*see* note 1), the ventral surface of one forearm of each of four members of the group, with 70 % alcohol on a piece of sterile, absorbent cotton wool to eliminate, as far as possible, the normal skin flora.

↓

When the alcohol has dried thoroughly, mark out six squares (with sides of 2 cm) along the cleansed areas of each of the four forearms with a felt-tip pen (*see* note 4).

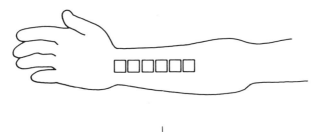

↓

Swab each of the marked areas with a sterile swab moistened with an overnight broth culture of *Serratia marcescens* and allow to dry.

↓

Using a separate swab for each antiseptic, swab each of five of the marked squares on forearm 1 with a "swabful" of the antiseptics numbered 1–5, leaving the sixth area untreated as a control. Repeat this procedure with the five antiseptics numbered 6–10 on forearm 2, and use the sixth square again as a control (*see* note 5).

↓

Using separate swabs again repeat the procedure with all ten antiseptics on forearms 3 and 4, each with its own control square. Allow the antiseptics to act on the four forearms and keep the test areas facing downwards and clear of any bench surfaces to prevent contamination.

↓

After 5 min, moisten a sterile swab in sterile nutrient broth and swab one test area on forearm 1. Spread the contents of the swab over the surface of half a nutrient agar plate. Repeat this with the five remaining squares of forearm 1 and all the squares of forearm 2 using a new swab for each area and a new half of a nutrient agar plate (*see* note 6).

↓

After 15 min, repeat this procedure with the twelve squares of forearms 3 and 4.

↓

At the end of the experiment, swab the arm with 70 % alcohol, allow to dry and then wash with soap and water.

↓

Incubate all 12 nutrient agar plates at 28 °C for 24 h (*see* note 2).

Day 2 *(30 min)*

Examine the plates and note the amount of growth of the red pigmented colonies. It is convenient to consider the growth on the four control squares, from the four forearms, as equivalent to "+ + +" and the growth on the remaining plates as "−", "+", "+ +" or "+ + +".

Notes and Points to Watch

1. *S. marcescens* is the organism most suitable for this experiment as it produces pigmented colonies which are easily distinguishable from those of any residual skin bacteria. *S. marcescens* should be used with some care as, in rare cases, it may be pathogenic (Traub, 1978). Non-pigmented *S. marcescens* have been shown to have a possible pathogenic role in immunologically compromised people (Clayton and von Graevenitz, 1966). Other bacteria may be used as long as they are not pathogenic.
2. The formation of pigment with *S. marcescens* does not occur at 37 °C. It is best achieved if the plates are incubated at 25–30 °C exposed to light. If a suitable incubator is not available, the plates may be incubated at room temperature near a window.
3. Tincture of iodine may give hypersensitivity reactions with some people. If any person is known to give an adverse reaction to any of these antiseptics let him or her do the swabbing of the arms.
4. An alcohol-based felt-tip pen is the best means of marking out the squares on the forearm. If this is not available a ball-point pen may be used but this may be more difficult to remove afterwards.
5. A control must be performed on each test forearm because a small proportion of the population have skin which is bactericidal for some organisms. The result of each square should be compared with the control on the same forearm.
6. When using a swab to remove a test sample from a square, ensure that the side of the swab that takes up the sample is the same that is spread on the surface of the nutrient agar plate.

MATERIALS

1. *S. marcescens* strain NCTC 1377.
2. Ten antiseptics diluted according to the manufacturers' instructions: suitable preparations which are easily obtained from pharmacies are listed in Table 1.

Table 1 Suitable antiseptics and dilutions.

Antiseptic	Type of antiseptic	Dilution
1. Hibitane	Surface active agent	1 in 100
2. Roccal	Surface active agent	1 in 10
3. Milton	Halogen	1 in 10
4. TCP	Two halogens and phenol	1 in 3
5. 70 % Alcohol	Alcohol	None
6. Savlon	Two surface active agents	1 in 10
7. Cetavlon	Surface active agent	1 in 10
8. Listerine	Phenol and others	None
9. Dettol	Phenol	1 in 20
10. Tincture of iodine (*see* note 3)	Halogen plus alcohol	None

---

SPECIFIC REQUIREMENTS

Day 1
10 ml broth culture (18 h, 37 °C) of *S. marcescens* in nutrient broth
10 antiseptic solutions
cotton wool (sterile, absorbent)
70 % alcohol for swabbing

alcohol-based felt-tip pen or ball-point pen (*see* note 4)
45 disposable medical swabs
10 ml sterile nutrient broth
12 nutrient agar plates
clock or timer

FURTHER INFORMATION

The experiment, as detailed here, uses well-known specific antiseptics against the test organism. The results show that the action of antiseptics, in general, is not instantaneous and in some cases very little reduction in the number of micro-organisms is observed even after 15 min exposure. The results will vary from class to class and can be partly dependent on the forearms used. However, typical results obtained by students are given in Table 2.

The results show the efficacy of various types of antiseptics against a specific organism under limited test conditions and may not reflect the full potential of each antiseptic. In order to gain a more complete view of the efficiency of the antiseptics, other organisms may be used. However, in a practical class it is essential that these organisms be non-pathogenic. The experiment can be

Table 2   The growth of *S. marcescens* after exposure to various antiseptics on the human skin.

Antiseptics	Time (min)	
	5	15
	forearm 1	forearm 3
Control	+ + +	+ + +
1. Hibitane	+ + +	+ +
2. Roccal	+ +	+
3. Milton	+	−
4. TCP	+ + +	+ + +
5. 70% Alcohol	+ +	+
	forearm 2	forearm 4
Control	+ + +	+ + +
6. Savlon	+ + +	+ + +
7. Cetavlon	+ +	+ +
8. Listerine	+ + +	+ +
9. Dettol	+ +	+
10. Tincture of iodine	+	−

+ + +: confluent growth
  + +: many colonies, not confluent
+: few colonies—less than 100
−: no growth

modified by leaving the antiseptics in contact with the organism over a wider range of time or by using a larger number of antiseptics.

Our experience has shown that it is best to divide the whole class into four groups and let each group test five antiseptics plus a control and pool all the results at the end. This reduces the materials required with only a small loss of comparability.

REFERENCES

E. Clayton and A. von Graevenitz (1966). Non-pigmented *Serratia marcescens*. *Journal of the American Medical Association*, **197**, 1059–64.

W. H. Traub (1978). *In* "Bacteriocin Typing of Clinical Isolates of *Serratia marcescens*," T. Bergan and J. N. Norris, eds), Methods in Microbiology, Vol. 11, Chap. VI, pp. 223–42. Academic Press, London and New York.

46

Finger Imprint Test of the Skin-substantivity of Germicides in Soaps

B. J. B. WOOD

Department of Applied Microbiology, University of Strathclyde, Glasgow, G1 1XW, Scotland

---

Level: Advanced undergraduates
Subject area: Hygiene, medical microbiology
Special features: Microbiological assessment of an everyday material, namely soap

---

INTRODUCTION

In general, the antimicrobial activities of soaps and similar preparations are of two kinds:
  (i) **transitory**, i.e. effective while the preparation is being used, but removed as the hands, etc. are rinsed; the fatty acids of soaps belong to this category, being appreciably germicidal, but easily rinsed away; and
 (ii) **skin-substantive**, i.e. adhering more or less firmly to the skin, and retaining appreciable germicidal action on the skin after washing and drying. Some toilet soaps which claim to protect the user against body-odour contain skin-substantive germicides, as do the specialist grades of preparations such as "Zalpon".
  In order to exhibit skin-substantive properties, a germicide will often be lipid-soluble, and the reader may recall the controversy caused by claims that certain agents used in preparations for infant hygiene might cause damage through absorption into the nervous system. These claims were countered by claims that for measurable uptake to occur, a body would need to be treated with the preparations in question at a frequency and level which would never be used in real life. The debate is an illuminating example of the manner in which sincere people can reach very different conclusions from the same data.
  The present experiment uses a simple finger-imprinting method onto agar plates seeded with *Staphylococcus aureus*, to compare the skin-substantive properties of a range of toilet soaps, etc.

EXPERIMENTAL

Day 1 *(2½ h)*

Prepare test solutions by making 10% w/v aqueous suspensions of the soaps, etc. to be tested (*see* note 1). Prepare a solution of Lux toilet soap or other germicide-free formulation as a control.

↓

Since there will be variation in the responses to the same soap when used by different individuals, it is recommended that at least 15–20 volunteers be used if 4 or 5 soaps are being compared. At least two individuals should supervise the experiment.

↓

All volunteer subjects first wash their hands with a control soap which contains no germicide; Lux toilet soap (*see* note 2) is satisfactory.

↓

Use the following procedure:

apply soap,	15 s
lather,	30 s
rinse (under running water),	15 s.

↓

Dry the hands, preferably on paper towels from a newly opened package.

↓

Immerse the fingers of one hand in the solution of test soap and those of the other hand in the solution of the control soap. Immersion is for 30 s. Keeping the hands separate, rinse each for 15 s under cold running water. Then dry the hands on paper towels and press the fingers lightly on the surface of agar plates seeded with an overnight culture of *S. aureus* (*see* note 3). After 30 s remove the hands and replace the lid on the plate, which is then incubated for 24 h at 37 °C. All subjects must wash their hands again after completing the test.

- -

Day 2 *(30 min)*

Examine the plates and grade them according to the type and amount of inhibition, which is a measure of the amount and/or effectiveness of the germicide remaining in the skin. Score the plates as follows:

4: a sharply outlined area of no growth;
3: a clear area of no growth with a hazy outline;
2: area of growth inhibition but with slight growth throughout;
1: area of growth inhibition just barely perceptible;
0: no inhibition; contact area indistinguishable from its surroundings.

The control plates should give a reading of zero. If a control plate shows inhibition, this must be due to other materials being present on the subject's hands, and the result from that subject's test plate must be disregarded.

- -

Notes and Points to Watch

1. It is strongly recommended that the students be encouraged to supply the samples for testing. In addition, a preparation which is known to give a strong positive test (such as germicidal Zalpon) should be included. In this way student interest is maximized, but the supervisor can be reasonably certain of having one clear positive reaction.

   About 100 ml of each soap solution will be sufficient in most cases.

2. Volunteers should be invited to report any recent exposure of their hands to substances which might have an effect on the test, and asked what toilet soap they are currently using. This information should be used in interpreting the results of the test, particularly if anomalous results are obtained. They should also be told that the test involves placing their fingers on a plate seeded with *S. aureus*. It is recognized that some people would consider the use of a potential pathogen to be highly undesirable. Its use should be confined to classes small enough, or otherwise so organized, that direct supervision by the competent microbiologist is possible. Thorough handwashing as soon as the test is completed is mandatory for all personnel. A further precaution could be swabbing the hands with 70% alcohol. Another possibility is to replace *S. aureus* by *S. epidermidis* or perhaps *Sarcina* sp.

3. It is essential that the layer of seeded agar be as thin as possible for clear-cut results to be obtained. For normal classes it is suggested that the plates be prepared for the class, but for advanced classes the students could prepare their own plates and try additional experiments, e.g. investigating the effect of the thickness of the agar layer, or trying different test organisms.

MATERIALS

1. A 24-h broth culture of *Staphylococcus aureus*, recommended strain NCTC 6571. To each 100 ml of prepared plate count agar medium, at 50 °C, 1 ml of 18 h, 37 °C broth culture is added. After thorough mixing, the seeded medium is immediately dispensed into sterile Petri dishes. The minimum volume is used which will give a thin, even layer of medium, the thinner the better.

SPECIFIC REQUIREMENTS

Day 1

Petri dishes prepared with seeded agar as above, one plate prepared for each subject recruited for the experiment (each plate can be used for 2 sets of fingerprints)

balance suitable for weighing to 0·1 g

test soaps (see text)

bar(s) of germicide-free soap, e.g. Lux toilet soap (Unilever, Port Sunlight, Cheshire)

paper towels (we have not experienced problems with contamination from these, but if necessary they can be autoclaved before use)

sufficient beakers for test and control soap solutions

knife for cutting soap samples

FURTHER INFORMATION

Assessment of the plates is fairly easy, but it is a help for inexperienced personnel if they work in pairs. Inevitably there is a range of responses to a particular soap, and it is necessary to take this into consideration in assessing the results. Typically (*see* Table 1), specialist germicidal preparations such as Zalpon will give a grade 4 response. Currently, Lifebuoy (Lever Bros.) and similar preparations will give a grade 2 or 3 response. It must be emphasized that soaps which give no zone of inhibition may contain antimicrobials which are not skin-substantive, and this test alone would not be grounds for action under the Trade Descriptions Act! More advanced classes might enjoy devising ways to test soaps for this other type of germicidal action. Other developments of the basic exercise could include testing different concentrations of soaps, the effect of hard water, the difference between using soap solutions according to the protocol, and using the soaps to wash the hands in the usual way. A range of organisms could be tried in the test, possibly including isolates from subjects' skin, although the possibility that these may include pathogens must be taken into account in designing the experiment. For further information *see* Vinson, *et al.* (1961).

Table 1   Finger imprint responses to various soaps (class results). All subjects except one showed zero inhibition from the control soap, and the exception was discarded from the results compiled here.

A zero response was expected from Simple soap and from the plain white soap procured from the university hand-basins in lavatories. Some scatter in the results obtained from germicidal soaps is normal, but the rather wide scatter observed here probably reflects imperfections in experimental procedure.

Soap	Numbers showing response score				
	4	3	2	1	0
Simple	0	0	0	1	4
Lifebuoy	1	2	2	1	0
Zalpon	3	1	1	0	0
Plain white	0	0	0	0	5

ACKNOWLEDGEMENT

I should like to thank friends and former colleagues at Unilever Ltd., Colworth House Research Centre, for the helpful and interesting discussions on this test and on its suitability for student practical classes.

REFERENCE

L. J. Vinson, E. L. Ambye, A. G. Bennett, W. C. Schneider and J. T. Travers (1961). *In vitro* tests for measuring antibacterial activity of toilet soap and detergent bars. *Journal of Pharmaceutical Sciences*, **50** (10), 827–830.

47

The Bacteriostatic and Bactericidal Action of Antibiotics

L. J. DOUGLAS and J. H. FREER

*Department of Microbiology, University of Glasgow,
Glasgow, G11 6NU, Scotland*

> *Level*: Advanced undergraduates
> *Subject areas*: Microbial physiology and growth
> *Special features*: Illustrates important principles in
> chemotherapy

INTRODUCTION

Chemotherapeutic agents, a group of substances which includes antibiotics, may be defined as chemicals that, at concentrations tolerated by the host, can interfere directly with the proliferation of pathogenic micro-organisms. Thus the essential feature is one of selective toxicity. Some drugs are bacteriostatic, the inhibition of growth being reversed when the drug is removed; others are bactericidal, exerting an irreversible lethal effect. The distinction between bactericidal and bacteriostatic action is not, however, absolute since at relatively low concentrations some normally bactericidal agents appear to be bacteriostatic.

The aim of this experiment is to define the type of antibacterial action of chloramphenicol and rifampicin.

EXPERIMENTAL

Overnight shake culture of *Escherichia coli* B in SK medium.

↓

Subculture 5 ml samples into flasks C and R, each containing 50 ml fresh SK medium prewarmed to 37 °C (*see* note 1).

↓ ↓

Flask C

Incubate at 37 °C with shaking until E_{600}^{1cm} is about 0·4 (approx. 60 min).

↓

Add 2 mg chloramphenicol in 1 ml ethanol (*see* note 2).

↓

Continue to read the E_{600} every 10 min for 30 min (*see* note 3).

↓

Centrifuge at 15 000 *g* for 15 min at 37 °C. Wash the cells twice with 20 ml fresh SK medium prewarmed to 37 °C (*see* note 4). Resuspend the final pellet (*see* note 5) to 50 ml in a clean 250 ml shake flask.

↓

Continue incubation at 37 °C, with shaking, and read E_{600} every 15 min for a further 60–90 min.

Flask R

Incubate at 37 °C with shaking until E_{600}^{1cm} is about 0·4 (approx. 60 min).

↓

Add 1 mg rifampicin in 1 ml methanol (*see* note 2).

↓

Continue to read the E_{600} every 10 min for 30 min (*see* note 3).

↓

Centrifuge at 15 000 *g* for 15 min at 37 °C. Wash the cells twice with 20 ml fresh SK medium prewarmed to 37 °C (*see* note 4). Resuspend the final pellet (*see* note 5) to 50 ml in a clean 250 ml shake flask.

↓

Continue incubation at 37 °C, with shaking, and read E_{600} every 15 min for a further 60–90 min.

Notes and Points to Watch

1. It is crucial for the success of this experiment that the bacteria remain in the exponential phase of growth throughout. Periods during which flasks are removed from the shaker for additions or for sampling should therefore be kept to an absolute minimum. Incubation of cultures is best done in bench-top shaking water-baths, preferably with one water-bath per group of students. Sterile medium may be prewarmed in these baths before the start of the class.
2. Because of their low water solubilities, the antibiotic solutions should be discharged directly into the medium without contacting the sides of the flask. For speed, safety and convenience, each antibiotic solution should be added with a disposable 1 ml syringe.
3. Because of the relatively high cell densities used and the short duration of the experiment, aseptic technique during antibiotic addition and cell sampling is not essential. Samples removed for optical density measurements may be read in unsterile cuvettes and returned to the culture flasks.
4. The initial centrifugation and subsequent washing of inhibited cells is best done at 37 °C to avoid any unnecessary temperature stress.
5. Resuspension of cell pellets can be conveniently achieved by progressive addition of small volumes of fresh SK medium interspersed with vortex mixing. Centrifuge tubes with leak-proof push-on caps are best used for this purpose.

MATERIALS

1. *Escherichia coli* strain B.
2. SK medium (Snyder and Koch, 1966) containing (per litre): KH_2PO_4, 0·78 g; K_2HPO_4, 2·3 g; $(NH_4)_2SO_4$, 1·0 g; $MgSO_4·7H_2O$, 0·1 g; $Na_3C_6H_5O_7·2H_2O$ (sodium citrate), 0·6 g.

Sterile 10 % (w/v) D-glucose solution is added to give a final concentration of 0·2% (w/v). The medium has a pH value of 7·3.

3. Rifampicin and chloramphenicol (Sigma Chemical Co.).

SPECIFIC REQUIREMENTS

50 ml overnight shake culture of *E. coli* B in SK medium
100 ml sterile SK medium
2 × 50 ml SK medium in 250 ml shake flasks
2 shake flasks (250 ml)
rifampicin (1 mg ml^{-1} in methanol, 1 ml, freshly made)
chloramphenicol (2 mg ml^{-1} in ethanol, 1 ml, freshly made)
2 glass micro-cuvettes
1 can 1 ml pipettes (or disposable syringes)
1 can sterile 5 ml pipettes

2 pipette fillers
1 vortex mixer
temperature-controlled centrifuge with a rotor that takes 50 ml tubes and is capable of 15 000 g
plastic centrifuge tubes (50 ml) with push-on caps (4)
2 × 1 ml plastic syringes
shaking water-bath, with platform for 250 ml flasks, at 37 °C
spectrophotometer to read at 600 nm

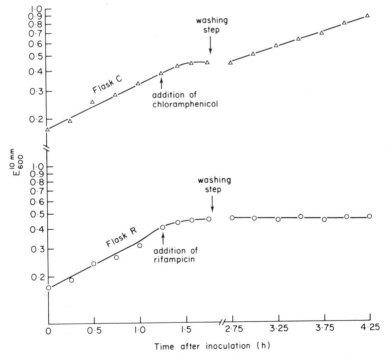

Figure 1   Growth of *E. coli* B as affected by chloramphenicol and rifampicin.

FURTHER INFORMATION

A typical set of class results for this experiment is shown in Figure 1. The washing of inhibited cells in fresh medium alleviates growth-inhibition due to chloramphenicol, but not that caused by rifampicin. This reflects the different binding characteristics of the two drugs: chloramphenicol binds reversibly to the 50s ribosomal subunit while rifampicin binds very tightly, although not covalently, to RNA polymerase (Franklin and Snow, 1981).

REFERENCES

I. S. Snyder and N. A. Koch (1966). Production and characteristics of hemolysins of *Escherichia coli. Journal of Bacteriology*, **91**, 763–7.

T. J. Franklin and G. A. Snow (1981). "Biochemistry of Antimicrobial Action", 3rd edn, 217 pp. Chapman and Hall, London.

48

Heat Resistance of *Bacillus cereus* Spores

J. F. LOWE

Department of Microbiology, School of Agriculture,
West Mains Road, Edinburgh, EH9 3JG, Scotland

> *Level*: Beginning undergraduates
> *Subject area*: Sterilization
> *Special features*: Effective demonstration of death
> logistics

INTRODUCTION

Effective thermal processing of foods depends on a knowledge of the heat-resistance of the potential contaminating micro-organisms. Heat resistance is usually defined as the *D*-value, which is the time required at a particular temperature to reduce the numbers of a specified micro-organism by 90% (Jay, 1978). *D*-values decrease with increase in temperature.

A further figure, the *z*-value, is used to predict the changes in *D*-value with change in temperature.

The aim of this experiment is to work out a series of *D*-values for *Bacillus cereus* spores and from these determine the *z*-value.

EXPERIMENTAL

Day 1 *(1 h)*

Spore suspension of *B. cereus*.

↓

Add to preheated sterile water in vessels at 85, 90 and 95°C to give a final concentration of 10^5 spores ml^{-1}.

↓

At the intervals shown in Table 1 take 1 ml samples from the vessels.

↓

Make 3 dilutions (to give final 10^{-1}, 10^{-2} and 10^{-3}) in sterile water.   Melt plate-count agar and cool.

↓   ↓

Put 1 ml of each dilution into the bottoms of Petri dishes.   Pour onto the 1 ml dilutions in the Petri dishes.

Mix well and allow to solidify.

↓

Incubate at 30°C overnight.

Table 1   Suggested temperatures and heating times.

Temperature (°C)	Heating time (min)					
85	0	5	10	20	30	40
90		5	10	15	20	30
95		2	5	7	10	15

- -

Day 2 *(1 h)*

Count all plates and calculate the numbers of survivors at each temperature/time combination.

↓

Obtain the D-value for B. *cereus* at each temperature by plotting, on the same semi-logarithmic graph paper, temperature on the arithmetic scale against numbers of survivors on the logarithmic scale. (The D-value is the time taken for the graph to drop one log-cycle.)

↓

Obtain the z-value by plotting temperature (arithmetic scale) against the three D-values (logarithmic scale), again on semi-logarithmic graph paper. (The z-value is the temperature difference associated with a change in the D-value of one log-cycle.)

- -

Notes and Points to Watch

1. This experiment has been written as for a single student. However, it works very well as a class experiment. Here, the students should be arranged in pairs, each of which should be allocated at least two temperature/time combinations, preferably not too close together. It is important in

allocating the work to ensure that a sufficient number of replicates is obtained. Students can then determine the D-values by plotting the mean of the class results put on the blackboard.

2. As originally designed, the experiment used closed vessels, fitted with stirrers, which were held in water-baths. Though this works well, it is equally satisfactory to use ordinary flasks as long as they are mixed well when the spore inoculum is added.

3. It is probably advisable for the demonstrators to add the spore inoculum to the flasks at a signal from the lecturer in charge.

4. It is important to check the temperature of each water-bath and to use that temperature in determining the D- and z-values (Day 2). This is particularly true of the 85°C-temperature, which has a slow killing effect.

5. A zero time sample is taken from the 85°C flask (see table 1) immediately after adding the spores and mixing.

6. The one safety point that has been a problem is the risk of students scalding themselves when steam is released from beneath the water-bath lid. The use of a layer of polypropylene spheres on the water overcomes the problem and helps keep the temperature of the bath stable.

MATERIALS

1. A spore suspension of *Bacillus cereus* strain T (West German Culture Collection, DSM 626), approximately 10^7 spores ml^{-1}. The spores are obtained by growing the organism for 2 days at 30°C on Oxoid potato agar with 0·5% yeast extract and 0·5% glucose. The spores are scraped off the medium, washed and can be stored in the freezer.

2. Water-baths set at 85, 90 and 95°C.

SPECIFIC REQUIREMENTS

B. cereus spore suspension
3 × 9 ml amounts of sterile water for each count
3 Petri dishes for each count
3 × 1 ml pipettes for each count
30 ml plate count agar (Oxoid) for each count

flasks, containing 100 ml sterile water, in water-baths at 85, 90 and 95°C
semi-logarithmic graph paper (two- to five-cycle)

FURTHER INFORMATION

This experiment has continually worked well in our classroom (see Figures 1 and 2). If the original spore count is approximately 10^7 ml^{-1}, the addition of 1 ml with thorough mixing to the flasks in the water-baths gives an initial experimental count of approximately 10^5 ml^{-1}.

Because this is an approximation, it is difficult to specify how many cycles of semi-logarithmic graph paper are required and usually I have a range available for plotting the D-value. For the z-value, two-cycle paper suffices.

The specific requirements have been given in quantities per count. In practice, it is at least desirable to duplicate counts and, therefore, the experiment requires 96 Petri dishes, 96 pipettes and so on. It would be worthwhile increasing the replication, but that is probably determined by class size.

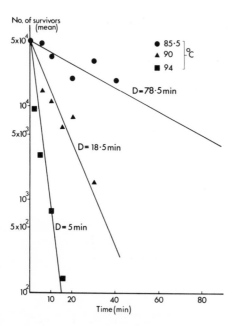

Figure 1 Plot of time (min) against log of survivors obtained in a class experiment. (The D-values are obtained by deriving the time for each curve to pass through one log cycle.)

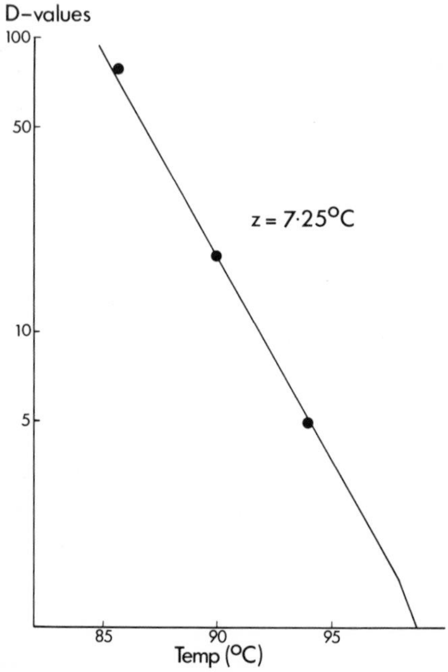

Figure 2 Plot of temperature ($^\circ$C) against D-values (log scale) obtained in same class experiment. (The z-value is obtained by deriving the temperature difference for the curve to pass through one log cycle).

REFERENCE

J. M. Jay (1978). "Modern Food Microbiology", 2nd edn, chap. 12. Van Nostrand, New York.

Inactivation of Bacteriophages by Ultraviolet Light

M. IOSSON

Microbiology Department, University of Reading,
London Road, Reading, RGI 5AQ, England

> *Level*: Beginning undergraduates
> *Subject areas*: Phage genetics
> *Special features*: Introduction to inactivation
> kinetics and repair mechanisms

INTRODUCTION

When ultraviolet (UV) light of wavelength 254 nm (the peak output of a normal mercury vapour lamp) irradiates a suspension of bacteriophage, the prime effect is on the nucleic acid. The bases in the nucleic acid have different absorption maxima in the wavelength range 248–268 nm. The energy absorbed may be dispersed in different ways, a very common effect being the cross-linking of adjacent pyrimidine bases to form dimers. Such structures cannot be replicated by the enzyme systems of the host bacterium. The effect of pyrimidine dimer-formation is thus to block normal replication of phage nucleic acid (Setlow *et al.*, 1963), which, unless repaired, is lethal for the phage.

Repair mechanisms in bacteria may reverse the effects of radiation on the phage nucleic acid. One such repair mechanism, termed **photoreactivation** (Dulbecco, 1950) cleaves pyrimidine dimers in the presence of visible light (Setlow and Setlow, 1963) to restore the original pyrimidine bases. This system operates on both RNA and DNA. In the dark (as in a dark incubator) the system is inoperative, but repair of damage may be effected by the "excision-repair" system (sometimes known as "dark-repair"). Specific nucleases excise the damage, leaving a single-stranded gap to be filled in by DNA polymerase I and DNA ligase, using the complementary DNA strand as a template (Hanawalt and Haynes, 1967). Excision-repair can thus operate only on double-stranded DNA: any attempt to excise damage in a single-stranded DNA leads to a strand break and loss of integrity of the molecule. For a single-stranded DNA, **a single-hit is sufficient to inactivate the virus under conditions of excision repair**. RNA cannot be repaired by this system. For double-stranded DNA, damage is lethal only if the two strands are damaged at nearly the same point, so that when excision repair is effected, a double-strand break results. Double-stranded DNA viruses should be much more resistent to UV-irradiation than single-stranded viruses.

The kinetics of inactivation of a virus (bacteriophage) may thus give information on the nature of its nucleic acid. Virus particles containing single-stranded nucleic acids will be more sensitive to UV irradiation than those with double-stranded molecules **of the same length**. (It should be obvious that the bigger the "target", the more chance there will be of obtaining a UV "hit" upon it.) The practical effect of this influence of "target size" is that the degree of strandedness of the nucleic acid of the phage cannot be estimated directly from the UV inactivation kinetics. It is possible, however, to extrapolate from the data to estimate the degree of strandedness.

The aims of this experiment are to determine the UV inactivation kinetics of different bacteriophages and to estimate the strandedness of their nucleic acid.

EXPERIMENTAL

Four different bacteriophages (*see* note 3) are used; all preparations are diluted in phosphate-buffered saline (PBS) to approximately 2×10^6 plaque-forming units per ml before use (*see* note 2). The low initial titre avoids any complication due to multiplicity reactivation at high inactivation rates (*see* Experiment 50).

Day 1 *(1 h)*

Take 10 ml phage suspension.

↓

Transfer to a sterile glass Petri dish (*see* note 6).

↓

Remove a 0·1 ml sample and dilute in broth for Time zero sample.

↓

Remove the lid and irradiate with short wave (254 nm) UV light (*see* note 1). Remove 0·1 ml samples at 20, 40, 60, 90, 120, 150 and 180 s.

↓

Dilute samples by serial 10-fold and/or 100-fold (or intermediate) dilutions in broth, according to the dilution schedule shown in Table 1.

↓

Titrate the various dilutions using the agar overlay method (*see* note 4). Use 4 plates (i.e. 4 × 0·1 ml) for each dilution tested, to increase the statistical accuracy of the titrations.

↓

As soon as each agar overlay has set immediately place the plate in a 37°C incubator to minimize any photoreactivation.

Table 1

Time of irradiation	Phage			
(s)	λ cI	MI_5	fi	T4rII
0	10^{-3}	10^{-3}	10^{-3}	10^{-3}
20	10^{-3}	10^{-3}	10^{-3}	10^{-3}
40	10^{-3}	10^{-3}	10^{-3}	10^{-3}
60	10^{-3}	10^{-3}	10^{-3}	5×10^{-3}
90	10^{-3}	10^{-3}	10^{-2}	5×10^{-2}
120	10^{-3}	10^{-3}	10^{-2}	10^{-1}
150	10^{-2}	10^{-2}	10^{-2}	10^{-1}
180	10^{-2}	10^{-2}	10^{-1}	nil

Day 2

Count plaques on all plates. Plot the results in the form of a graph to relate the irradiation time to the surviving fraction ($\log_{10} P_t/P_0$) of the phage population, where P_t is the phage titre at time t and P_0 is the phage titre at time 0. If the slope of the graph is extrapolated back to the ordinate, the value of the intercept, $\log_{10} S$ gives an estimate of the degree of strandedness of the nucleic acid, S (*see* Atwood and Norman, 1949).

Notes and Points to watch

1. (a) Students should not look directly at the UV lamp, or at reflections from it since UV light can damage the cornea and cause blindness. Students should be advised to wear the appropriate safety glasses.
   (b) During the irradiation, the dish should be gently rocked backwards and forwards to maintain a steady circulation. Irradiation may be stopped at any time (e.g. for sampling) by replacing the lid of the Petri dish (since glass absorbs short wavelength UV light, the lid acts as an efficient "shutter" for the UV exposure).
   (c) The output from different UV lamps varies greatly. The lamp should be placed at such a distance that gives approximately 1% survival of T4rII phage at 2 min irradiation (approximately $140\,\mu W\,cm^{-2}$ at 30 cm).
   (d) Note that phage suspensions should be prepared in inorganic buffers, since organic compounds in broth absorb UV light and greatly reduce the effective dose.
2. Phage suspensions are prepared by confluent lysis: approximately 10^4 p.f.u. are mixed with indicator bacteria in molten soft agar and poured onto the surface of a nutrient agar plate. After the top layer has set, the plates are incubated (inverted) overnight at 37°C. Plates are then harvested: 5 ml of nutrient broth is pipetted on the surface of each plate and allowed to stand for 30 min. The top layer is then broken up by the use of a sterile glass spreader, and the mixture of agar and liquid transferred to a centrifuge tube. Centrifugation should be for a time and speed sufficient to produce a clear supernate and a hard-packed pellet. Decant off the supernate and store in the cold. Storage over chloroform will prevent bacteria from growing, but should not be

used with filamentous phages since they are chloroform-sensitive. Typical yields from confluent lysates are in the region of 10^{10}–10^{11} p.f.u. ml^{-1}, for tailed phages; 10^{11}–10^{12} p.f.u. ml^{-1} for RNA phages and 10^{12}–10^{13} p.f.u. ml^{-1} for filamentous phages.

3. The strains of phage used in this experiment are λcI_{60}, T4rIIa$_{164}$, and two strains isolated by the author and named "fi" and "MI$_5$". The first two phages are used as examples of double-stranded DNA phages of different target sizes; any conveniently available strain of λ or T4 could be used. The other two phages are used as examples of single-stranded DNA and RNA phages respectively; suitable alternatives would be the filamentous phages fd, M13 and f1 for "fi", and MS2, R17 or Qβ for "MI$_5$".

4. Phage titration is normally performed by the agar overlay technique (Adams, 1959). Small tubes or bottles containing 2·5–3·5 ml of 0·6% agar in nutrient broth ("soft agar") are melted, then stored at 46°C until required. Into each bottle of molten soft agar is inoculated a volume (0·1–0·4 ml) of an overnight suspension of indicator bacteria (see note 5). Normally, 0·1 ml is sufficient to give good plaques, but larger volumes give clearer plaques with male-specific bacteriophages.

   A known volume of diluted phage suspension (usually, 0·1 ml), containing 100–200 p.f.u. of phage, is then added and the two are mixed. The contents are then poured over the surface of a deep (25–30 ml) nutrient agar base plate, and allowed to set. Incubation at 37°C overnight is normally sufficient to allow plaque development.

5. *Escherichia coli* strain K12F$^+$ is suitable as indicator for all of the phages used in this experiment, although we have found that the strain K12(λ^+)F$^+$ is better for filamentous phage, and strain C600 is better for λ phage.

6. A glass Petri dish must be used. Plastic dishes are resistant to wetting and volumes as small as 10 ml would not be evenly dispersed.

MATERIALS

1. *Escherichia coli* strain K12F$^+$ (strains K12(λ^+)F$^+$ and C600 optional: see note 5). These strains are available from the ATCC or NCIB.
2. Phage suspension diluted in PBS to contain approximately 2×10^6 p.f.u. ml^{-1}.
3. Short wavelength UV light, e.g. CAMAG dual wavelength UV light manufactured by CAMAG A.G., Switzerland. Set to a suitable height (see note 1(c)) and allow 10 min for the output to stabilize before commencing irradiation.
4. Nutrient broth medium contains 10 g NaCl, 10 g peptone (Oxoid) and 8 g "Lab-Lemco" powder (Oxoid) per litre, with pH adjusted to 6·9. For plates, add 10 g (Difco) agar per litre; for soft agars, add 6 g agar per litre. Plates should contain 25–30 ml per plate; soft agars, 2·5–3·5 ml per bottle.
5. Bacteriophages λ, T4, fd, R17, etc. may be obtained from the NCIB: National Collection of Industrial Bacteria, Torry Research Station, PO Box 31, Aberdeen, AB9 8DG, Scotland.

SPECIFIC REQUIREMENTS

Day 1
10 ml, overnight broth culture of *E. coli* indicator strain in Universal or McCartney bottle
32 nutrient agar plates per phage tested

32 soft agars (per phage), melted and kept molten at 46°C
sterile 0·1 ml (or 50-dropper), 1 ml and 10 ml pipettes

cont.

cont.

pipette teats of appropriate sizes	phage suspensions
nutrient broth as diluent	UV light source
sterile 1 oz. McCartney bottles or capped tubes for dilutions (capped tubes will require the use of a tube rack)	safety glasses
	Day 2
1 sterile glass Petri dish per phage suspension tested	light boxes and tally counters (or other plaque-counting equipment)

FURTHER INFORMATION

In the absence of photoreactivation, single-stranded phages (DNA or RNA) give typical "single hit" inactivation kinetics. The slope of the curve depends on the "target size". Thus RNA phages which

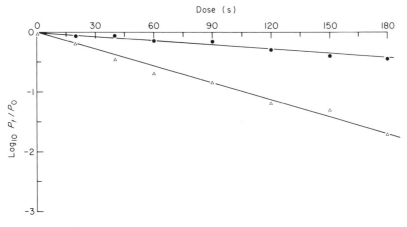

Figure 1 Inactivation of single-stranded phages by UV irradiation. ●, RNA phage; △, filamentous DNA phage.

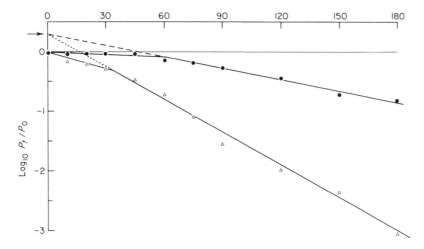

Figure 2 Inactivation of double-stranded DNA phages by UV irradiation. ●, λ phage; △, T4rII phage.

are icosahedral, 25 nm in diameter ($8160\,\mathrm{nm^3}$), are much less sensitive than filamentous DNA phages which are approx. 1000 nm × 8 nm ($50\,265\,\mathrm{nm^3}$) (Figure 1). The nucleic acid for each phage is of approximately the same molecular weight ($1{\cdot}1 \times 10^6$ for RNA phage; $1{\cdot}9 \times 10^6$ for filamentous phage), the difference in molecular weight being insufficient to explain the difference in UV sensitivity.

For λ phage and T4rII phage, differently shaped curves are observed (Figure 2). The "shoulder" in the inactivation curve can be attributed to the time required to accumulate sufficient "hits" in the double-stranded DNA before "inactivating hits" occur. Lambda phage has a smaller DNA (mol. wt 31×10^6) than T4 phage (approx. 120×10^6) and requires a longer "accumulation period" before inactivation starts. When inactivation commences, the rate is 2 % of that of the larger phage. For both phage, if the "decay portion" of the graph is extrapolated back to the ordinate, the value of the intercept is approximately $0{\cdot}3$. The degree of strandedness, S, is therefore given by the antilog of the intercept value ($\log_{10} S$), i.e. for λ and T4 phage, the degree of strandedness is antilog $0{\cdot}3$ or 2.

More time points may be necessary to accurately determine the duration of the "accumulation period". The linearity of the "decay portion" of the graph holds for at least a one thousand-fold inactivation.

REFERENCES

M. H. Adams (1959). Appendix. *In* "Bacteriophages". Wiley/Interscience, New York and London.

K. C. Atwood and A. Norman (1949). On the interpretation of multi-hit survival curves. *Proceedings of the National Academy of Sciences of the United States of America*, **35**, 696–709.

R. Dulbecco, (1950). Experiments of photoreactivation of bacteriophages inactivated with ultraviolet light. *Journal of Bacteriology*, **59**, 329–47.

P. C. Hanawalt and R. H. Haynes (1967). The repair of DNA. *Scientific American*, **216** (6), 36–43.

J. K. Setlow and R. B. Setlow (1963). Nature of the photoreactivable ultraviolet lesion in DNA. *Nature*, **197**, 560–2.

R. B. Setlow, P. A. Swenson and W. L. Carrier (1963). Thymine dimers and inhibition of DNA synthesis by UV irradiation of cells. *Science*, **142**, 1464–5.

50

Multiplicity Reactivation of Bacteriophage T4

D. A. RITCHIE

Department of Genetics, University of Liverpool,
Liverpool, L69 3BX, England

> *Level*: Advanced undergraduates
> *Subject areas*: Microbial genetics and virology
> *Special features*: Demonstrates the principles of
> inactivation kinetics and the use
> of survival curves in
> microbiology

INTRODUCTION

A variety of physical and chemical agents are able to destroy the infectivity of phage particles. This is seen as a loss of plaque-forming ability. Ultraviolet light irradiation (UVL) has been used extensively to study various properties of micro-organisms, in particular those mechanisms concerned with the repair and reactivation of damage to DNA (Stent, 1963; Hayes, 1968).

The rate of UVL inactivation of phage particles is exponential, i.e. the fraction of phage particles inactivated is identical for successive identical doses of irradiation. This is seen as a straight line which passes through the origin when the logarithm of the phage titre (i.e. the logarithm of the surviving fraction) is plotted against the dose of UVL. This indicates that each particle is inactivated by a single quantum of UVL; i.e. inactivation is a "one-hit" process. The slope of the survival curve is a measure of the sensitivity of the phage to the inactivating agent. For UVL, the target for inactivation is the phage DNA which becomes chemically modified at discrete sites by the dimerization of adjacent pyrimidine bases. This prevents both the expression and replication of the DNA molecule.

When the DNA genomes of two or more inactive phage particles are present in the same bacterial cell they may co-operate to produce viable progeny. This type of reactivation is known as **multiplicity reactivation** and occurs very efficiently with phage T4 (Luria and Dulbecco, 1949). Multiplicity reactivation is a process whereby undamaged DNA molecules are formed by the

---

recombination of undamaged regions from two or more DNA molecules each bearing sites of lethal damage (Stent, 1963; Hayes, 1968).

This experiment demonstrates both the inactivation kinetics of phage T4 following UVL irradiation and the phenomenon of multiplicity reactivation. The general methods used are suitable for many types of inactivating agent.

EXPERIMENTAL

The experimental basis of this experiment is straightforward—a stock of T4 phage particles is irradiated with several doses of UVL and samples from each dose are used to infect cells, (a) under conditions of single infection, and (b) under conditions of multiple infection. The curve derived from (a) gives the phage survival curve and that from (b) gives the multiplicity reactivation survival curve which should show a much increased survival rate.

Day 1 *(1½ h)*

Make the following preparations before starting the experiment:
  (a) Label 4 small tubes 0, 1, 2 and 3.
  (b) Label 4 small tubes S0, S1, S2, S3 (S = single infection).
  (c) Label 4 small tubes M0, M1, M2, M3 (M = multiple infection).
  (d) Prepare 25 dilution tubes of buffer using the large tubes; 13 tubes containing 10 ml each and 12 tubes containing 4·5 ml each.
  (e) Label 15 NA plates 1–15.

↓

Transfer the 10 ml of phage T4 stock to a sterile Petri dish and keep the lid in place at all times except during UVL irradiation and sampling. Remove 1·0 ml to tube 0—this is the unirradiated control sample which provides the initial viable phage titre. Expose the phage stock to UVL for three successive periods of irradiation determined to give accumulated surviving fractions of 10, 1 and 0·1 % (*see* notes 2 and 5). After each UVL dose transfer 1·0 ml samples to tubes 1, 2 and 3 respectively.

↓

Transfer 0·5 ml samples from tubes 0, 1, 2 and 3 to tubes M0, M1, M2 and M3 respectively.

↓

Dilute the samples remaining in tubes 0, 1, 2 and 3 serially to give an overall 10^{-3} dilution (0·1 ml into 10 ml followed by 0·5 ml into 4·5 ml). Transfer 0·5 ml of the 10^{-3} dilutions to tubes S0, S1, S2 and S3 respectively.

↓

Put tubes S0–S3 and M0–M3 into the 37 °C water-bath and infect by adding bacteria as indicated.

Time (min)	Action
0	To S0 add 0·5 ml bacteria (*see* notes 6 and 7).
1·5	To S1 add 0·5 ml bacteria.
3·0	To S2 add 0·5 ml bacteria.
4·5	To S3 add 0·5 ml bacteria.
6·0	To M0 add 0·5 ml bacteria.
7·5	To M1 add 0·5 ml bacteria.
9·0	To M2 add 0.5 ml bacteria.
10·5	To M3 add 0·5 ml bacteria.
12·0	Dilute S0 serially 0·1 ml into 10 ml (10^{-2}) followed by 0·5 ml into 4·5 ml (10^{-3}). Plate 0·2 ml from the 10^{-2} dilution on plate 1 and 0·2 ml from the 10^{-3} dilution on plate 2 (*see* note 1).
13·5	Dilute S1 serially 0·5 ml into 4·5 ml (10^{-1}) followed by 0·5 ml into 4·5 ml (10^{-2}). Plate 0·2 ml from the 10^{-1} dilution on plate 3 and 0·2 ml from the 10^{-2} dilution on plate 4 (*see* note 1).
15·0	Dilute S2 0·5 ml into 4·5 ml (10^{-1}). Plate 0·2 ml from the undiluted sample on plate 5 and 0·2 ml from the 10^{-1} dilution on plate 6 (*see* note 1).
16·5	Plate 0·2 ml from the undiluted S3 tube on plate 7 (*see* note 1).
18·0	Dilute M0 serially 0·1 ml into 10 ml (10^{-2}) followed by 0·1 ml into 10 ml (10^{-4}) followed by 0·5 ml into 4·5 ml (10^{-5}). Plate 0·2 ml from the 10^{-4} dilution on plate 8 and 0·2 ml from the 10^{-5} dilution on plate 9 (*see* note 1).
19·5	Dilute M1 and plate samples as for M0 using plates 10 and 11 (*see* note 1).
21·0	Dilute M2 and plate samples as for M0 using plates 12 and 13 (*see* note 1).
22·5	Dilute M3 and plate samples as for M0 using plates 14 and 15 (*see* note 1).

↓

When the top layers have solidified incubate the plates overnight at 37°C (but *see* note 4).

Day 2 *(1 h)*

Count the plaques on each plate and calculate the titres of plaque-forming units (infectious centres) in samples S0–S3 and M0–M3.

↓

Construct survival curves for the single infection (S series) and multiple infection (M series) experiments. This is best done by plotting the logarithm of the fraction surviving each dose against the UVL dose (time of irradiation). For calculating the surviving fraction, the unirradiated infections (S0 and M0) are given the value 1.

↓

Compare the slopes of the two curves.

Notes and Points to Watch

1. As a rule the plaque-forming titre of a phage stock is assayed by the agar top layer or overlay procedure. For this method a 0·2 ml volume of indicator bacteria is pipetted into a small tube containing 2·5 ml of molten top agar which is kept molten by standing in a water-bath at 46 °C. The appropriate volume of phage sample (0·2 ml for this experiment) is pipetted into the agar–bacteria mixture and the contents of the tube is poured over the surface of an agar plate. The plate is gently rocked to distribute the molten agar over the entire surface. The plate is left undisturbed on a flat and level bench for 10 min to allow the top agar layer to solidify.

2. Bare skin and particularly eyes should not be exposed to UVL; thin plastic gloves and goggles should be worn during the irradiation stages.

3. Because UVL damage can be repaired by a visible light-induced process called photoreactivation, this experiment should be done in dim light. Alternatively the assay plates can be covered.

4. Following incubation the plates may be stored for a week, or possibly a little longer, at 4 °C.

5. Because the output from different UVL lamps varies considerably it will be necessary to construct an inactivation curve for phage T4 in order to determine the correct exposure times.

6. The culture of host bacteria contains KCN to inhibit phage development during adsorption. This allows the dilution and plating stages to be completed in a relatively leisurely manner. Without the use of KCN the infected cells would lyse at about 25 min after infection and so make analysis of the results very difficult to impossible.

7. KCN is a **poison** and should not be mouth-pipetted. As an alternative to KCN, chloramphenicol at a final concentration of $20 \mu g \, ml^{-1}$ may be used to inhibit cell metabolism.

MATERIALS

1. *Escherichia coli* strain B (available from ATCC or NCIB) for use as the host bacterium and as the plating indicator.
2. Phage T4 wild type (available from ATCC or NCIB).
3. Nutrient broth for growing bacterial cultures containing: 5 g bacto peptone, 8 g nutrient broth, 5 g NaCl and 1 g glucose per litre of distilled water.

4. Dilution buffer containing: 7 g Na_2HPO_4 (anhydrous), 3 g KH_2PO_4 (anhydrous), 5 g NaCl, 0·12 g $MgSO_4$, 0·01 g $CaCl_2$ and 0·01 g gelatin per litre of distilled water.
5. Ultraviolet lamp with a maximum emission at 260 nm.
6. KCN 0·1 M (**caution**).
7. 37 and 46 °C water-baths.

SPECIFIC REQUIREMENTS

Day 1

5 ml broth culture of exponentially growing *E. coli* B at 2×10^8 cells ml^{-1} containing 0·002 M KCN (**poison**): the host culture

5 ml broth culture of stationary-phase *E. coli* B for use as indicator bacteria

10 ml of phage suspension at 10^9 plaque-forming units per ml.

sterile Petri dish (glass or plastic)

15 nutrient agar plates

50 ml molten top agar

200 ml bottle of dilution buffer

small tubes, about 4″ × ½″ diam.

large tubes, about 6″ × ¾″ diam. for dilutions

tube racks

pipettes

FURTHER INFORMATION

The single infection inactivation curve should show an exponential loss of infectivity at approximately 1 log for each of the 3 doses of UVL. The multiple infection curve should show much higher survival levels indicating almost complete rescue of UVL damage by the reactivation process. Typically the survival rate should not fall below 50 % at a dose in which the singly infected cells show about 10^{-3} survivors (Figure 1).

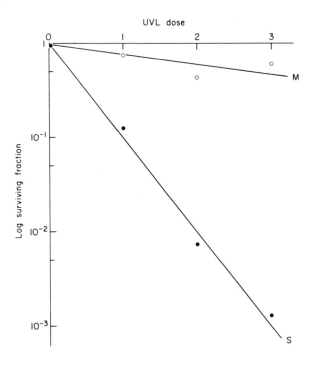

Figure 1 Typical UVL inactivation curves for the survival of the plaque-forming ability of phage T4 when infecting bacteria under condition of single (S) and multiple (M) infection.

REFERENCES

W. Hayes (1968). "The Genetics of Bacteria and their Viruses", 2nd edn, pp. 522–30. Blackwell Scientific Publications, Oxford.

S. E. Luria and R. Dulbecco (1949). Genetic recombinations leading to production of active bacteriophages from ultraviolet inactivated bacteriophage particles. *Genetics*, **34**, 93–125.

G. S. Stent (1963). "Molecular Biology of Bacterial Viruses", pp. 277–305. W. H. Freeman, San Francisco.

51

Cytological Study of the Effects of Penicillin and Mitomycin

R. W. A. PARK

Microbiology Department, University of Reading,
London Road, Reading, RG1 5AQ, England

Level: All undergraduate years
Subject areas: Physiology and growth
Special features: Cytological effects of inhibitors

INTRODUCTION

The mode of action of an antibacterial agent can often be deduced from simple cytological observations. For example, microscopical studies on cells treated with penicillin first indicated that the cell wall is its site of action (Gardner, 1940; Duguid, 1946).

High concentrations of penicillin interfere with formation of rigid murein, supposedly by inhibiting transpeptidation. Thus growing bacteria treated with penicillin lyse in hypotonic media because the defective murein lacks sufficient mechanical rigidity. In a medium containing magnesium ions to stabilize membranes and a large concentration of solute, the cells become spherical but do not burst. The spherical structures thus formed are spheroplasts; they lack a rigid wall component but, unlike protoplasts, still possess other wall components. Low concentrations of penicillin act more specifically, interfering with formation of cell cross-walls in some way not understood. Under such circumstances the bacteria get longer, lysing only when they have increased their length many times.

Abnormal elongation of bacteria can also occur when division is inhibited by many other agents (Hughes, 1956). One such agent is mitomycin C, which inhibits replication of DNA. In the absence of new rounds of DNA replication cell division normally does not proceed. However, increases in cell components other than DNA occur for the equivalent of several generation times to give elongated organisms with only one chromosome.

The aims of this exercise are: (1) to show how simple cytological observations can give useful information about the action of some inhibitors; and (2) to encourage an awareness of the relationship between structure and physiology.

EXPERIMENTAL

(5–6 h with breaks)

Six 100 ml conical flasks aseptically receive the additions indicated in Table 1 and are then incubated in a shaking water-bath at 37°C.

Table 1 The amount of various components to be added to 100 ml conical flasks for incubation in shaking water-baths at 37°C.

Component	Quantity (ml) of each component in flask no.					
	1	2	3	4	5	6
Double strength nutrient broth at 37°C	4·0	4·0	4·0	4·0	4·0	4·0
2 M sucrose at 37°C	0	0	2·0	2·0	0	0
0·1 M $MgSO_4.7H_2O$ (*see* note 1)	0	0·9	0	0·9	0	0
Pencillin solution (25 000 iu ml^{-1})	0	0·1	0	0·1	0	0
Pencillin solution (1250 iu ml^{-1})	0	0	0	0	0·1	0
Mitomycin C solution (0·05 mg ml^{-1}) (*see* note 5)	0	0	0	0	0	0·1
Demineralized H_2O at 37°C	4·0	3·0	2·0	1·0	3·9	3·9
Escherichia coli culture, actively growing at 37°C	2·0	2·0	2·0	2·0	2·0	2·0

↓

After incubation at 37° for 1½–2 h, the turbidity in flasks 1–4 should be compared and the contents examined by microscopy. Add 0·2 ml formalin to each of flasks 1–4, and allow to stand for 5 min (*see* note 5). Then pour about 2 ml into separate centrifuge tubes and centrifuge to pellet organisms and debris. Pour off the supernate and resuspend the precipitate in 0·1 ml demineralized H_2O. Make smears of the material, dry, stain with dilute carbol fuchsin for 30 s, rinse, dry and examine microscopically with an oil immersion objective.

↓

After incubation for 2–4 h, examine the bacteria in flasks 5 and 6. Add formalin and centrifuge as before but **do not resuspend the pellet in water**. Stain by the acid-Giemsa method to show DNA. Also make a simple smear using water and stain with dilute carbol fuchsin to detect gross morphology, noting dimensions.

Staining of DNA by the acid-Giemsa method
Make a dense smear on a coverslip of organisms from a pellet obtained by centrifuging a formalin fixed suspension. Do not add water.

↓

Allow the smear to dry in air for 10–15 min.

↓

Put the coverslip in 1 M HCl at 55°C and maintain at 55°C in a water-bath for 7–10 min.

↓

Tip the HCl from the container and rinse the coverslip thoroughly by running tap-water into the container for 30 s.

↓

Put about 3 ml dilute Giemsa stain on a slide on a staining rack and float the coverslip on the stain with the organisms downwards (*see* note 3). Leave to stain for 30 min.

↓

Remove the coverslip, rinse, and then place, organisms downwards, on a drop of water on a slide (*see* note 2). Take care to avoid air bubbles.

↓

Blot the preparation and then seal at the edges of the coverslip with vaseline applied with the edge of a warm slide. Carefully examine under oil immersion (*see* note 4).

Notes and Points to Watch

1. Magnesium sulphate is added to stabilize unprotected membranes; it is required only in flasks 2 and 4.
2. Preparations stained to show DNA are mounted in water to improve the contrast.
3. Floating the coverslip on the Giemsa stain ensures that the precipitate which forms cannot fall on the preparation to give a misleading appearance.
4. The staining technique for DNA may not be successful at the first attempt. If it is not it should be repeated, the time being altered. It may help to improve technique if you remember that organisms which stain uniformly blue have been under-hydrolysed, those which stain uniformly pink have been over-hydrolysed, and those which do not appear coloured have probably been treated with old Giemsa solution. When the organisms are stained correctly the DNA appears blue in otherwise pink cells. Fixation for visualizing DNA using, for example, osmic acid vapour (Norris and Swain, 1971) is preferable but too hazardous for student classes.
5. **Safety**
   (a) Mitomycin C is possibly a carcinogen.
   (b) Formalin and formalin-treated cultures must not be brought into contact with hypochlorite disinfectants as a volatile carcinogen is produced.

MATERIALS

1. *Escherichia coli.* Strain B is satisfactory but any penicillin-sensitive strain of *E. coli* should work.
2. Penicillin (sodium benzyl), 500 000 iu vial (Glaxo, Greenford, Middlesex or Sigma Chemical Co.)
3. Mitomycin C: 2 mg vial (Dales Pharmaceutical, Steeton, Keighley or Sigma Chemical Co.)
4. Giemsa stain R66 (G. T. Gurr, Searle Scientific, High Wycombe, Bucks) diluted on the day of the class by adding 1 part of stain to 9 parts of Burroughs-Wellcome pH 7·0 buffer.
5. Nutrient broth: Peptone (Oxoid), 5·0 g; Lab Lemco powder (Oxoid), 4·0 g; NaCl, 5·0 g; in demineralized water, 1 litre. Double strength nutrient broth has twice these concentrations.

SPECIFIC REQUIREMENTS

These items are needed for each set of flasks.

actively growing culture of *E. coli* in nutrient
  broth (The stock strain is subcultured daily
  for 3 days in nutrient broth at 37°C. On the
  afternoon before the class 0·5 ml culture is
  inoculated into 20 ml nutrient broth in a
  100 ml conical flask. This is then put in a
  switching incubator at 5°C set to switch to
  37°C 5 h before the class. 2 h before the class
  the flask is transferred to a shaking
  incubator.)
penicillin solution (25 000 iu ml^{-1}) (On the day
  of the class add 2 ml sterile H$_2$O to a vial
  containing 500 000 iu penicillin. Withdraw
  1 ml and add to 9 ml sterile H$_2$O.)
penicillin solution (1250 iu ml^{-1}) (Add 1 ml of
  25 000 iu ml^{-1} solution to 19 ml sterile H$_2$O.)
mitomycin C solution (0·05 mg ml^{-1}) (On the
  day of the class add 2 ml sterile H$_2$O to a vial
  containing 2 mg mitomycin C. Withdraw 1 ml
  and add to 19 ml sterile H$_2$O.)
5 × 10 ml sterile graduated pipettes

7 × 1 ml sterile graduated pipettes
6 × 100 ml sterile conical flasks
6 ml 2 M sucrose (34 g made up to 50 ml; sterilized by autoclaving)
4 ml 0·1 M MgSO$_4$·7H$_2$O (1·2 g made up to
  50 ml; sterilized by autoclaving)
30 ml sterile demineralized H$_2$O
formalin (Shake 20 ml formaldehyde solution
  with MgCO$_3$ and allow to settle. Transfer
  10 ml of the supernate to 90 ml H$_2$O. Store
  above 2°C.)
shaking water-bath at 37°C to take 100 ml
  conical flasks
water-bath at 55°C for acid hydrolysis for
  staining for DNA
20 ml diluted Giemsa stain
containers, e.g. boiling tubes, to hold coverslips
  during acid hydrolysis and racks to hold these
  in the 55°C water-bath
1 M HCl
bench centrifuge and tubes
30 ml double strength nutrient broth

FURTHER INFORMATION

The morphology of cells in flasks 1 and 3 should be normal. In flask 2, debris from lysed cells should be apparent. Flask 4 should contain large spherical structures (spheroplasts) which are protected by the sucrose from lysis. The time for incubation of flasks 5 and 6 is not critical; the longer the incubation with penicillin, and to some extent with mitomycin, the longer the resultant organisms. However, as the cultures age the long forms tend to disintegrate. Students may assess the number of doubling times by comparing mean lengths of the long forms in flask 5 with the length of normal cells in flask 1. DNA staining of bacteria from flask 5 should show regular distribution of DNA throughout the long forms, indicating that inhibition of cell division has occurred at some stage subsequent to DNA replication. Organisms in flask 6 elongate less than those in flask 5. DNA stains reveal only one chromosome per organism, clearly much less per unit length than in either untreated or penicillin treated organisms.

ACKNOWLEDGEMENT

The part of the exercise using a large concentration of penicillin is based on a suggestion of Lichstein and Oginsky (1965).

REFERENCES

J. P. Duguid (1946). The sensitivity of bacteria to the action of penicillin. *Edinburgh Medical Journal*, **53**, 401.
A. D. Gardner (1940). Morphological effects of penicillin on bacteria. *Nature, London*, **146**, 837.
W. H. Hughes (1956). The structure and development of the induced long forms of bacteria. *In* Symposium no. 6, of the Society for General Microbiology, (E.T.C. Spooner and B.A.D. Stocker, eds), pp. 341–60.
H. C. Lichstein and E. L. Oginsky (1965). "Experimental Microbial Physiology", p. 12. Freeman, London.
J. R. Norris and H. Swain (1971). Staining bacteria. *In* "Methods in Microbiology", (J. R. Norris and D. W. Ribbons, eds), Vol. 5A, pp. 105–34. Academic Press, London and New York.

52

The Kelsey–Sykes Capacity Use–Dilution Test for Disinfectants

L. B. QUESNEL

Department of Bacteriology and Virology, University of Manchester, Manchester, M13 9PL, England

> *Level*: All undergraduate years
> *Subject areas*: Antibacterial agents, disinfection
> *Special features*: Demonstration of an important but frequently misunderstood concept

INTRODUCTION

A wide variety of disinfectants is used in laboratories, hospitals, canteens, etc. to reduce and if possible eliminate contaminant organisms. Such an objective will not be achieved if the effective concentration at the site of action is too low, or not maintained for an adequate time. Alternatively, concentrations applied at excessive levels are wasteful and may give unpleasant odours and flavours to food or be irritant to the skin, etc. In current practice a use–dilution test is performed in order to obtain an indication of the best concentration.

Historically, tests for preservative (hence germicidal) action can be traced back much earlier than the knowledge of the causes of disease and putrefaction. Hugo (1978) has traced the development of some early tests devised to evaluate the antimicrobial activity of various preservatives. As early as 1750, Pringle used sea-salt as a standard and ascribed coefficients to other salts in a way which foreshadowed the determination of phenol coefficients, used for most of this century for the evaluation of germicidal power.

Rideal and Walker (1903) described a procedure for comparing other disinfectants with phenol. It was subsequently modified to take account of the variable reaction of different antibacterials in the presence of organic matter. In the Chick and Martin (1908) modification, standard amounts of faeces were added as organic "soil" to simulate possible in-use conditions. Later, standard yeast suspension was incorporated in place of faeces and this procedure, proposed by Garrod (1934), became the British Standard Specification No. 808 (1938).

Since phenol coefficients do not indicate *per se* the concentration of a disinfectant to be used in practice, Kelsey *et al.* (1965) first proposed a capacity use–dilution test which would give a direct indication of in-use concentrations to be recommended under clean and dirty conditions. A similar procedure published later by Kelsey and Sykes (1969) became a generally accepted test for disinfectant evaluation. After five years' experience of this test yet another modification designed to increase reproducibility and simplicity was published by Kelsey and Maurer (1974). This test, described in simplified form below, is suitable for all types of disinfectants.

The aim of this experiment is to find the use dilution for one or more disinfectants.

EXPERIMENTAL

Four organisms are recommended for use in the official test, viz.: *Pseudomonas aeruginosa* NCTC 6749, *Proteus vulgaris* NCTC 4635, *Echerichia coli* NCTC 8196, *Staphylococcus aureus* NCTC 4163, and the most resistant of these to the given disinfectant is used for the test. This could be done by the instructor prior to the class (*see* note 1) or a different organism could be given to different groups of students.

Day 1 *(1½ h)*

"Clean" conditions test
Centrifuge 10 ml inoculum culture in bench centrifuge. Resuspend cells in 10 ml standard hard water by agitation on a "Rotamixer" (*see* note 2).

"Dirty" conditions test
Centrifuge and resuspend 10 ml inoculum culture as for "clean" conditions.

Discard 4 ml of suspension; add 4 ml of standard 5% w/v yeast suspension (i.e. 2% final yeast concentration). "Rotamix" (*see* note 2).

Make decimal dilutions of suspensions in ¼ strength Ringer's solution, and do a Miles and Misra (1938) count (*see* note 3).

Label 3 Universal bottles A, B and C. Into each put 3 ml of appropriately diluted disinfectant (*see* note 4). (Separate sets for "clean" and "dirty" tests.)

At time zero add 1 ml bacterial suspension to disinfectant A; shake to mix.

At time zero add 1 ml bacterial/yeast suspension to disinfectant A; shake to mix.

At time 1 min add 1 ml of bacterial suspension to disinfectant solution B; shake to mix.

At time 1 min add 1 ml of bacterial/yeast suspension to disinfectant solution B; shake to mix.

↓

↓

At time 5 min add 1 ml of bacterial suspension to disinfectant solution C; shake to mix.

At time 5 min add 1 ml of bacterial/yeast suspension to disinfectant solution C; shake to mix.

↓

↓

At time 8 min, transfer 1 drop (by calibrated Pasteur pipette) of 0·02 ml to each of 5 tubes of recovery medium series A1 (*see* note 5). Return unused sample to bottle A (*see* note 6).

At time 8 min transfer 1 drop (0·02 ml) by calibrated Pasteur pipette to each of 5 tubes of recovery medium series A'1 (*see* note 5). Return unused sample to bottle A (*see* note 6).

↓

↓

Proceed with further additions and transfers according to the schedule of times given in Table 1.

↓

Incubate all recovery tubes at $32 \pm 1°$C for 48 h (*see* note 7).

Table 1   Kelsey–Sykes test timetable.

	Disinfectant concentration				
A		**B**		**C**	
Time (min)	Action	Time (min)	Action	Time (min)	Action
0	Inoc. 1 ml to A	1	Inoc. 1 ml to B	5	Inoc. 1 ml to C
8	Drops to broths A1	9	Drops to broths B1	13	Drops to broths C1
10	Inoc. 1 ml to A	11	Inoc. 1 ml to B	15	Inoc. 1 ml to C
18	Drops to broths A2	19	Drops to broths B2	23	Drops to broths C2
20	Inoc. 1 ml to A	21	Inoc. 1 ml to B	25	Inoc. 1 ml to C
28	Drops to broths A3	29	Drops to broths B3	33	Drops to broths C3

- -

Day 2 *(15 min)*

Score growth in recovery tubes as + or −.

- -

Notes and Points to Watch

1. For the official test it is recommended that the most resistant organism to the particular agent under test be chosen. To do this each of the four organisms is grown in Wright and Mundy (1960) dextrose broth, inoculated in 6 ml aliquots and subcultured each day at $32 \pm 1°$C (*see*

note 7) for at least 5, but not more than 14 days (not essential for classroom experiment) and minimum inhibitory concentrations (MIC) determined.

To perform the MIC test, doubling dilutions of 5 ml aliquots of disinfectant solution in nutrient broth, or Wright and Mundy dextrose broth, contained in "Universal" bottles are inoculated with 1 drop (0·02 ml) each of a 10^{-1} dilution of the 24 h culture of the test organism in Wright and Mundy broth.

The inoculated series are incubated at $32 \pm 1°C$ for 72 h and the organism with the highest MIC is selected for the use–dilution test with a particular disinfectant.

2. For the test, a 5th to 14th day, 24-h-old culture in 10 ml Wright and Mundy broth grown at $32 \pm 1°C$, is used. After incubation the culture is spun down in a bench centrifuge at 3000 rev. min^{-1} for 15 min, the supernate removed and the pellet resuspended in 10 ml of standard hard water. Because of pellicle formation, when *Pseudomonas aeruginosa* is used the culture must be filtered through a sterile Whatman No. 4 filter paper before centrifugation. In the official test the suspension is shaken with sterile glass beads.

3. A method of counting, other than the Miles and Misra (1938) method, may be used. The number of viable organisms must be between 10^8 and 10^{10} ml^{-1}.

4. Three concentrations of test disinfectant are made in sterile standard hard water. Concentration B is the expected use–dilution concentration; A is half that of B and C is 50% higher than B. Dilution must be made on the same day as the test.

5. Care should be taken when transferring drops of test mixture to recovery media to add only 1 drop per tube and to ensure that the drop falls into the medium and does not lodge high on the inner wall of the tube. All the tubes should be shaken afterwards. The test specifies 10 ml medium per tube; for class purposes 3 ml per tube may be used.

6. A single Pasteur pipette may be used per disinfectant if it is thoroughly "washed out" in the mixture each time inoculum addition is made.

7. Although $32 \pm 1°C$ is recommended, 37°C can be used to save on incubator use.

MATERIALS

1. Wright and Mundy (1960) broth is a chemically defined liquid medium, obtainable dehydrated from Difco Limited as "Bacto Synthetic Broth, AOAC Code Number 0352".

    The medium is sterilized at 121°C for 20 min and sterile 10% glucose solution added to a final concentration of 0·1%. The pH should be 7·1 $\pm 0·1$ at 25°C.

2. Standard hard water (World Health Organization Standard: 342 parts 10^{-6} hardness).
    $CaCl_2$, anhydrous, 0·304 g
    $MgCl_2.6H_2O$,   0·139 g

Dissolve and make up the volume to 1 litre; sterilize by autoclaving at 121°C for 15 min.

3. Standard yeast suspension for the official test is made according to British Standard 808: 1938, but using standard hard water as diluent. For class purposes suspend 5 g baker's or brewer's dry yeast in 95 ml standard hard water and autoclave at 69 kPa (10 lbf in^{-2}) for 20 min. Check for sterility before use.

4. Recovery broth: Oxoid nutrient broth—Oxoid CM 67—containing 3% w/v Tween 80 (TB culture grade is suitable).

SPECIFIC REQUIREMENTS

These items are needed for each test, i.e. 3 concentrations of a single disinfectant under either "clean" or "dirty" conditions; and are as specified for official test.

incubator(s) set at 32°C (but *see* note 7)
10 ml culture in Universal bottle of test organism grown for 24 h in Wright and Mundy broth at $32 \pm 1°C$ (*see* note 7) (Culture should be between 5th and 14th daily subculture.)
low speed centrifuge
Pasteur pipettes, sterile
sterile filter and Whatman No. 4 filter paper; sterile Universal to receive filtrate (If using *Pseudomonas aeruginosa*, *see* note 2.)
sterile glass beads (optional, *see* note 2)

4 ml sterile standard yeast suspension
70 ml sterile $\frac{1}{4}$ strength Ringer's solution, sterile test tubes and pipettes for dilution series, calibrated Pasteur pipettes and nutrient agar plates for viable count
45 test tubes containing 10 ml sterile recovery broth each; test tube racks
solutions of disinfectant to make 3 ml aliquots; Universal bottles to contain same
rack to hold test disinfectant bottles
calibrated, sterile 50 dropper Pasteur pipettes; at least 1 separate pipette per disinfectant concentration
pen or pencil for labelling tubes, plates, etc.

FURTHER INFORMATION

Typical results of a Kelsey–Sykes test are given in Table 2. The use dilution to be recommended is the lowest initial concentration which shows no growth in at least 2 out of 5 of the recovery broths in recovery series 1 and 2 (columns 1 and 2, Table 2). This concentration must pass the test on 3 separate days using freshly prepared inoculum and freshly prepared disinfectant dilution on each occasion. It should be noted that it is the initial (made up) concentration of disinfectant that passes the test. The fact that this is diluted significantly in the course of the test provides a useful safety margin.

Use dilutions of this kind are now commonly quoted for disinfectants on sale in Great Britain. Some of the published values of commonly available disinfectants are given in Table 3.

Table 2 Typical results of a Kelsey–Sykes test of Clearsol disinfectant using *Pseudomonas aeruginosa*.

Concn.	% v/v	Viable organisms ml^{-1}	Recovery broths 1	Recovery broths 2	Recovery broths 3	Result
"Clean"						
A	0·3125	$1·2 \times 10^9$	+ + + + +	+ + + + +	+ + + + +	fail
B	0·625	$1·2 \times 10^9$	− − − − −	− − − − −	− − + + +	pass
C	0·9375	$1·2 \times 10^9$	− − − − −	− − − − −	− − − − −	pass
"Dirty"						
A	0·5	$7·2 \times 10^8$	− − − − −	+ + + + +	+ + + + +	fail
B	1·0	$7·2 \times 10^8$	− − − − −	− − − − −	− − − − −	pass
C	1·5	$7·2 \times 10^8$	− − − − −	− − − − −	− − − − −	pass

Table 3   Some use–dilutions of commonly available disinfectant compounds.

Disinfectant	Type	Use–dilution % "clean"	"dirty"
Chloros			
Domestos	Hypochlorite yielding chlorine	0·2	1·0
Clearsol	Clear soluble phenolic plus detergent	0·625	1·0
Hycolin	Combination of synthetic phenols	0·5	1·5
Izal	"White fluid" emulsion of coal tar acids	0·5	1·0
Printol	Clear soluble phenolic plus detergent	1·25	2·0
Resiguard	Picloxydine digluconate + benzalkonium chloride with detergent	0·625	2·5
Savlon HG	Chlorhexidine + cetrimide	0·5	2·5
Stericol	Clear soluble phenolic with detergent	1·0	2·0
Sudol	Clear soluble phenolic	0·5	1·0

Considerable variety may be introduced in the experiment by using different organisms, using different types of organic soil, e.g. serum, milk, and at different concentrations. Different disinfectants have different sporicidal properties and the experiment may be used to compare, for example, sporicidal ability of hypochlorites and phenolics, using standard spore suspensions. The time for which an organism is in contact with disinfectant is a crucial parameter in determining killing ability. The timetable may be rescaled to shorten or lengthen times as required.

Some possible causes of failures are:

(1) inaccurate measurement of concentrations;

(2) prolonged use of solution and prolonged storage of diluted solution before use;

(3) presence of inactivating material, e.g. too many organisms, or organic matter, incompatible detergent, hard water, soap, plastics, cotton, cork, nylon, wood and rubber.

It should be noted that the quoted "official use–dilutions" are obtained by performing the test according to the detailed official method. Use dilutions obtained by any modification of the test cannot be compared with those using the published (1974) test method.

REFERENCES AND FURTHER READING

British Standard 808 (1938) confirmed (1960). "Modified Technique of the Chick–Martin test for Disinfectants". British Standards Institute, London.

H. Chick and C. J. Martin (1908). The principles involved in the standardisation of disinfectants and the influence of organic matter on the germicidal value. *Journal of Hygiene, Cambridge*, **8**, 698.

L. P. Garrod (1934). A study of the Chick-Martin test for disinfectants. *Journal of Hygiene, Cambridge*, **34**, 322.

W. B. Hugo (1978). Early studies in the evolution of disinfectants. *Journal of Antimicrobial Chemotherapy*, **4**, 489–94.

J. C. Kelsey and I. M. Maurer (1974). An improved (1974) Kelsey–Sykes test for disinfectants. *Pharmaceutical Journal*, **207**, 528–30.

J. C. Kelsey, M. M. Beeby and C. W. Whitehouse (1965). A capacity use-dilution test for disinfectants. *Monthly Bulletin, P.H.L.S.*, **24**, June, 152–60.

J. C. Kelsey and G. Sykes (1969). A new test for the assessment of disinfectants with particular reference to their use in hospitals. *Pharmaceutical Journal*, **202**, 607–9.

A. A. Miles and S. S. Misra (1938). The estimation of the bactericidal power of the blood. *Journal of Hygiene*, **38**, 732.

J. Pringle (1750). Experiments on substances resisting putrefaction. *Philosophical Transactions of the Royal Society*, **46**, 408–17.

S. Rideal and J. T. A. Walker (1903). Standardization of disinfectants. *Journal of the Royal Sanitary Institute, London*.

E. S. Wright and R. A. Mundy (1960). Defined medium for phenol coefficient tests with *Salmonella typhosa* and *Staphylococcus aureus*. *Journal of Bacteriology*, **80**, 279–80.

Note added in proof   Useful discussions of the advantages and disadvantages of this and other methods of evaluating disinfectants can be found in "Disinfectants: their Use and Evaluation of Effectiveness", edited by C. H. Collins, M. C. Allwood, Sally F. Bloomfield and A. Fox, Society for Applied Bacteriology Technical Series No. 16. Academic Press, London and New York (1981).

53

The *In vitro* Synergy between Combinations of Antimicrobial Agents

E. G. BEVERIDGE and I. BOYD

Department of Pharmaceutics, Sunderland Polytechnic, Sunderland, SR1 3SD, England

> *Level*: All undergraduate years
> *Subject areas*: Antimicrobial agents
> *Special features*: Strong visual impact

INTRODUCTION

The addition of two or more antimicrobial agents to a microbial population sensitive to each of the individual compounds may have various outcomes. The overall inhibitory effect may be insignificantly different from that of the sum of the individual agents (indifference), it may show enhancement over that of the sum of the individual agents (synergy), or it may be considerably lower than that anticipated from the individual responses (antagonism). It is difficult to predict which of these possibilities will occur and the phenomenon of synergy, at least, is often confined to quite narrowly limited drug ratios (*see* Jawetz, 1968; Garrod *et al.*, 1973; Philip, 1974; Williams, 1979).

Reports of synergy *in vitro* are relatively common while reports of antagonism between antimicrobial agents tend to be scarce. For example, combinations of trimethoprim and sulphonamides are markedly synergistic to a variety of micro-organisms (*see* Finland and Kass, 1974), while polymyxins and sulphonamides show synergy towards some *Proteus* species (Herman, 1961). *In vitro* antagonism between naladixic acid and nitrofurantoin has been reported for some *Proteus* isolates (Garrod *et al.*, 1973), and *Staphylococcus aureus* sensitivity to lincomycin is antagonized by erythromycin (Barber and Waterworth, 1964).

Despite the widespread use of combined antibiotic treatment for infections there is usually little evidence of proven synergy *in vivo*. One of the few established examples is the combined use of trimethoprim and sulphamethoxazole ("Bactrim" and "Septrin") for treatment of a wide variety of infections (*see* Finland and Kass, 1974). However, claims for the success of these products have been criticized (Williams, 1979). A system which produces an apparent marked synergistic effect is currently undergoing encouraging clinical trials and involves the use of clavulanic acid and β-lactam antibiotics such as the penicillins. Clavulanic acid is a potent inhibitor of many β-lactamases, including the plasmid-mediated β-lactamases of *Escherichia coli*, *Klebsiella aerogenes* and *Enterobacter cloacae*, while the chromosomally mediated β-lactamases of *E. coli* and the cephalosporinase type of β-lactamase found in *Pseudomonas aeruginosa* are less well inhibited (Cole, 1979; Ball *et al.*, 1980) (*see* note 5).

The technique used in this experiment involves placing filter paper strips soaked in solutions of different antimicrobial agents onto seeded agar plates such that they overlap, and, after incubation, examining the nature of any areas of growth inhibition surrounding the intersections. Ideal signs of drug interactions would appear as shown in Figure 1.

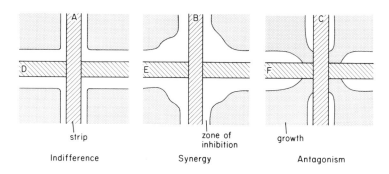

Figure 1   The effect on microbial growth of combinations of different antimicrobial agents using the paper strip diffusion technique.

The aims of this experiment are:
(1) to demonstrate the synergistic action of a sulphonamide and a polymyxin (colistin) towards *Proteus vulgaris* but not towards *E. coli*;
(2) to demonstrate the synergistic action of trimethoprim and a sulphonamide towards *E. coli* and *S. aureus*;
(3) to demonstrate that *p*-aminobenzoic acid (PABA), a microbial metabolite of the pathway interfered with by sulphonamides, destroys the synergistic action of the sulphonamide–trimethoprim combination.
(4) to indicate other possible uses for the basic technique.

EXPERIMENTAL

Day 1 *(1½h)*

Take 10 ml overnight nutrient broth cultures of *S. aureus*, *E. coli* and *Pr. vulgaris*.

↓

Agitate cultures well with vortex mixer. Pipette 0·2 ml *S. aureus* onto one MH agar plate (*see* note 1) and one MH + PABA agar plate. Spread cultures evenly over the whole surface of the plates with sterile glass spreaders. Prepare a similar pair of plates with *Pr. vulgaris* and two pairs of plates with *E. coli*. Leave the plates for 30 min to allow the liquid to be absorbed into the gel (*see* note 2).

↓

From the concentrated antimicrobial solutions provided, aseptically prepare aqueous dilutions (10 ml) in sterile test tubes to the levels indicated in Figure 2.

↓

Rinse two pairs of blunt ended forceps, dry thoroughly and flame. Using the cooled forceps dip one filter paper strip into one of the antimicrobial solutions, leave for 10 s and remove, drawing the strip against the side of the tube to remove excess liquid. With both pairs of forceps carefully place the soaked strip onto the agar surface allowing the centre of the strip to touch first and lowering the ends onto the agar ensuring that the whole of the strip is in good contact with the surface (*see* note 3).

↓

Repeat the previous operation with the different antimicrobial solutions, building up the chequerboard designs exactly as indicated in Figure 2 (*see* note 3).

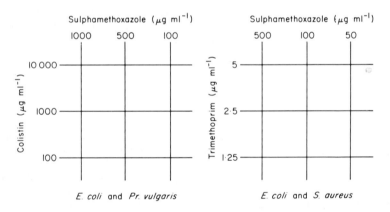

Figure 2 Recommended pattern for placing paper strips onto seeded MH and MH–PABA plates.

↓

Incubate the plates at 37 °C for 18 h (agar side uppermost—the strips will not fall off).

- -

Day 2 *(30 min)*

Examine MH agar plates for signs of synergy by comparing the extent of inhibition of microbial growth surrounding the intersections of the paper strips with any inhibitions at a distance from the intersections. Note whether any observed synergy is related to the relative concentrations of the two antimicrobial agents. Examine the MH + PABA plates and note whether the presence of PABA in the medium influences the results (*see* note 4).

_____

Note and Points to Watch

1. Mueller Hinton agar (MH) was specifically designed for sensitivity testing involving sulphonamides because many routine laboratory media contain sulphonamide antagonists such as PABA.
2. We have obtained the best results with surface seeded plates. The seeding should be done carefully to obtain complete and uniform distribution of the inoculum. A glass spreader gives better results than an inoculating loop.
3. Placement of the strips onto the agar surface requires care to ensure good even contact, particularly at the intersection of the strips. Care must be taken to avoid the cross-contamination of the test solutions by inadequate washing of the forceps. Avoid dripping solution onto the plate. **Do not attempt to move the strips once they are on the plate**.
4. The size of the areas of inhibition are often interpreted clinically as a measure of relative sensitivity of micro-organisms to various antimicrobial agents. However, quantification of sensitivity by this technique requires rigid experimental control. Liquid dilution experiments may give a better indication of possible clinical response.
5. Clavulanic acid in combination with amoxycillin is now available for therapy (Augmentin, Beecham Research Laboratories). Sensitivity test discs (Mast Diamed Ltd.) demonstrate good synergy compared with amoxycillin using agar plates surface-seeded with 0·2 ml of a 10^{-3} dilution of overnight broth cultures of β-lactamase producing *S. aureus*, *Pr. vulgaris* or *K. aerogenes* isolates.

MATERIALS

1. The extent of synergy in this experiment varies with different type culture strains and fresh isolates. Consistent results have been obtained with *Escherichia coli* NCTC 5933, *Staphylococcus aureus* NCTC 4163 and *Proteus vulgaris* NCTC 4175.
2. Mueller Hinton agar (MH) is obtainable from Oxoid, Lab M and Difco, the latter being used routinely here with an extra 0·5 % w/v agar added. MH + PABA agar is prepared by adding 100 μg ml^{-1} *p*-aminobenzoic acid. Both media can be autoclaved at 115°C for 30 min. Good results are obtained if 40 ml aliquots of the sterile agar are poured into 10 cm square Petri dishes (Sterilin) and the gel allowed to dehydrate for 2–3 days at room temperature before use.
3. Whatman No. 1 filter paper cut into 8 mm × 9 cm strips has been found to be the most suitable for this technique. The strips should be free from printing ink and "biro". They may be sterilized by heating in a metal capped test tube at 150 °C for 1 h (dry heat) or by autoclaving at

115 °C for 30 min. If sterilized by dry heat the strips should be discarded if they appear more than faintly discoloured otherwise unusual results may be obtained. Strips which have been autoclaved should be dried before use.
4. Suitable spreaders can be made from 4 mm diam. glass rods bent to an L shape, the shorter arm being 4 cm long with its end smoothed in a Bunsen flame.
5. Colistin (polymyxin E, colistin sulphate or, less satisfactory, colistin methanesulphonate), trimethoprim, and sulphamethoxazole can be obtained readily in a pure form through your local retail chemist. As a requirement of the Medicines Act (1968) it is probably necessary to present a written request signed by your head of department stating that the materials are required for education or research. Except for sulphamethoxazole these chemicals and *p*-aminobenzoic acid can be obtained from Sigma. Sulphacetamide (Sigma) can replace sulphamethoxazole but with less pronounced demonstrations of synergy.

SPECIFIC REQUIREMENTS

Day 1

10 ml nutrient both cultures (18 h, 37 °C) each of
E. coli, S. aureus and *Pr. vulgaris*

4 × 10 cm square Petri dishes containing 40 ml
MH agar

4 × 10 cm square Petri dishes containing 40 ml
MH + PABA agar (prepared and allowed to
dry at room temperature for 2–3 days before
use)

4 × 20 ml metal capped test tubes each containing 15, 8 mm × 9 cm filter paper strips

2 pairs of blunt-nosed forceps

10 ml colistin sulphate solution, 10 mg ml^{-1}

10 ml trimethoprim solution, 5 μg ml^{-1}

10 ml sulphamethoxazole solution, 1 mg ml^{-1}
(These solutions can be prepared on the previous day with sterile water and left at room
temperature in 20 ml sterile test tubes or
sterilized by filtration and supplied in 20 ml
capped test tubes.)

10 sterile capped test tubes for dilutions

500 ml sterile water for dilutions and rinsing
sterile pipettes, 1 ml, 5 ml, 10 ml

vortex mixer ("Whirlimixer", etc.)

5 sterile glass spreaders, individually wrapped

100 ml beaker to aid rinsing of forceps

test tube rack

FURTHER INFORMATION

It has been our experience with this technique that a variety of *Proteus* isolates are virtually
insensitive to either colistin or sulphamethoxazole individually, but that dramatic synergy occurs
with the higher drug levels over a wide range of drug ratios. *E. coli* isolates are frequently sensitive
to both drugs but show little or weak signs of synergy. This synergy appears to be associated with
Proteus species, a feature also noted by Herman (1961). However, Quesnel and Handley (1974) also
reported synergy with *Pseudomonas* species. Explanations for this synergistic action may be related
to the ability of polymyxins to disrupt the outer membrane of Gram-negative bacteria (LaPorte *et
al.*, 1977) allowing penetration of sulphonamides into the cell (Herman, 1961). Alternatively, it has
been suggested that sulphonamides expose or sensitize the lipopolysaccharides of the outer
membrane to polymyxin action followed by its more ready penetration and disruption of the
cytoplasmic membrane (Handley *et al.*, 1974).

The combination of sulphonamides and trimethoprim demonstrates good synergy towards
a wide variety of micro-organisms (*see* Finland and Kass, 1974). The ratio for optimal synergy *in*

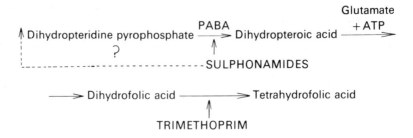

Figure 3   The probable sites of action of sulphonamides and trimethoprim upon the synthesis of folic and
folinic acids (*see* Gale *et al.*, 1972).

vitro seems to be trimethoprim:sulphonamide about 1:20. Commercial formulations contain a 1:6 ratio for this gives relative blood levels *in vivo* of 1:20. Sulphonamides and trimethroprim act at separate points during the synthesis of folic and folinic acids required for pyrimidine synthesis (*see* Figure 3).

The inclusion of the natural metabolite *p*-aminobenzoic acid in the MH agar overcomes the action of the sulphonamide destroying the colistin–sulphamethoxazole and tri-methoprim–sulphamethoxazole synergy. The activity of trimethoprim is not disrupted since it acts at a point beyond the incorporation of PABA into the pathway. Polymyxins act against various cytoplasmic membrane-related functions and are also unaffected by PABA (*see* Gale *et al.*, 1972).

We use this technique for routine qualitative screening of combinations of antimicrobial agents, particularly antimicrobial preservatives and disinfectants for pharmaceutical and food use. Success depends on at least one of the agents diffusing relatively easily into the agar gel. Since there is occasionally a narrow limit of drug ratios for synergy it is necessary to study a wide range of drug ratios and concentrations before deciding whether or not interactions occur. The technique has been particularly useful for examining possible inactivators of antimicrobial agents and the possible interference of medicinal or food ingredients with possible preservatives. For water-insoluble ingredients a suitable solvent is necessary when impregnating the paper strip, the solvent being allowed to evaporate before placing the strip on to the agar gel. A good demonstration of drug inactivation can be made with plates seeded with *S. aureus*, and strips soaked in 500, 100, 50 and 25 μg ml^{-1} benzalkonium chloride solution in combination with strips soaked in 10, 5 and 1 % w/v of the inactivator Tween 80.

REFERENCES

A. P. Ball, P. G. Davey, A. M. Geddes, I. D. Farrell and G. R. Brookes (1980). Clavulanic acid and amocycellia, a clinical, bacteriological and pharmacological study. *Lancet*, (1) 620–3.
M. Barber and P. M. Waterworth (1964). Antibacterial activity of lincomycins and pristinamycin: a comparison with erythromycin. *British Medical Journal*, **2**, 603–6.
M. Cole (1979). Inhibiters of antibiotic-inactivating enzymes. *In* "Antibiotic Interactions" (J. D. Williams, ed.), pp. 99–131. Academic Press, London and New York.
M. Finland and E. H. Kass (eds) (1974). "Trimethoprim–Sulfamethoxazole Microbiological, Pharmacological and Clinical Considerations". University of Chicago Press, Chicago.
E. F. Gale, E. Cundliffe, P. E. Reynolds, M. H. Richmond and M. J. Waring (1972). "The Microbiological Basis of Antimicrobial Action", pp. 11, 33, 137–40. John Wiley and Sons, London.
L. P. Garrod, H. P. Lambert and F. O'Grady (1973). "Antibiotic and Chemotherapy", 4th edn, pp. 282–6. Churchill Livingstone, London.
P. S. Handley, L. B. Quesnel and M. M. Sturgiss (1974). Ultrastructural changes produced in *Proteus vulgaris* by a synergistic combination of colisin and sulphadiazine. *Microbios*, **10**, 211–23.
L. G. Herman (1961). Antibiotic sensitivity testing using pretreated plates. III. Inhibition of *Proteus* species with a combination of sulphathiazole and colistimethate sodium. "Antimicrobial Agents and Chemotherapy (1961)", pp. 943–52. American Society for Microbiology, Detroit.
E. Jawetz (1968). The use of combinations of antimicrobial drugs. *In* "Annual Review of Pharmacology", (H. W. Elliott, W. C. Cutting and R. H. Drersbach, eds), Vol. 8, pp. 151–70. Annual Review Inc., Pao Alto.
D. C. LaPorte, K. S. Rosenthal and D. S. Storm (1977). Inhibition of *Escherichia coli* growth and respiration by polymyxin B covalently attached to agarose beads. *Biochemistry*, **16**, 1642–8.
J. R. Philip (1974). Drug interactions in the management of infections. *In* "Clinical Effects of Interactions between Drugs", (L. E. Cluff and J. C. Petrie, eds), pp. 175–92. Exerpta Medica, New York.
L. B. Quesnel and P. S. Handley (1974). Synergism between polymyxins, polysorbate and antimetabolites, of folic acid synthesis, and a paper disc technique for routine testing for synergism. *Microbios*, **10**, 19–210.
J. D. Williams (1979). "Antibiotic Interactions". Academic Press, London and New York.

54

The Effect of Lipophilicity on Antimicrobial Activity

I. BOYD and E. G. BEVERIDGE

*Department of Pharmaceutics, Sunderland Polytechnic,
Sunderland, SR1 3SD, England*

Level: All undergraduate years
Subject areas: Antimicrobial activity: pharmaceutical
microbiology
Special features: Structure–function relationships
of a series of antimicrobial
agents; statistical analysis
of data

INTRODUCTION

Albert (1973) briefly surveyed how structure–activity relationships have been used to interpret changes in drug activity from changes in the structure of the active molecules. Quantitative structure–activity relationships have been developed where numerical values have been assigned to molecular constituents (structures) in such a way that changes in number can be correlated with biological activity. Physico-chemical measurements on a series of compounds may be used to quantify the term "structure", which has been used variously to mean shapes, dimensions, physical properties or chemical affinities of molecules.

Ferguson (1939) noted that in a homologous series of compounds the biological activity of the compounds followed a geometric progression as did those physical properties which were examples of distribution phenomena, e.g. partition coefficient, vapour pressure and surface tension. On this basis he suggested that the good correlation between distribution phenomena (lipophilicity) and biological activity was due to the large effect of distribution in determining toxic concentrations.

Hansch and Fujita (1964) also recognized the importance of the physico-chemical properties of a molecule in determining its biological activity and in particular controlling its movement to the site of biological action within the cell. They also developed the comparative substituent π which could be used as a calculated partition parameter in investigating structure–activity relationships. Lien *et al.* (1968) produced equations correlating log partition coefficients with log biological activity for many series of compounds. They found that simple linear regressions of the form

$$\log \frac{1}{\text{biological activity}} = a + b \log(\text{partition coefficient})$$

gave good correlations for many antibacterial agents.

Hansch and Fujita (1964) also pointed out that multiple correlations involving a partition parameter and its squared term were parabolic relationships capable of indicating the properties of a compound with optimal activity. Additionally they used other parameters to develop multiple correlations—a procedure which Hansch subsequently used to incorporate factors correlating the steric or electronic properties of molecules. The whole subject is reviewed by Tute (1971) who also outlines the statistical methods used.

Partition coefficients are not easily measured with accuracy and in any case are very time-consuming measurements. However, Boyce and Millborrow (1965) suggested the use of R_M values as a measure of partitioning ability, and these values are readily obtained from reverse-phase thin lay chromatography (TLC) measurements.

The aim of this experiment is to develop a simple linear regression which expresses the relationship between the antibacterial activity, defined as the minimum inhibitory concentration (MIC), of a series of esters of gallic acid and their R_M values. This relationship is then used to predict the MIC of octyl gallate.

EXPERIMENTAL

Day 1 *(3h)*

Part A

Prepare solutions of the nine esters of gallic acid by dissolving about 10 mg of each ester in 1 ml ethanol—precise concentrations are not required (*see* note 1).

↓

Spot 1 µl volumes of each solution onto TLC plates. Air dry (*see* note 2).

↓

Place TLC plate in a tank with 100 ml 40:60 acetone:water solvent, and allow solvent to ascend plate for about 15 cm.(During this period of about 90 min carry out the MIC determinations as in Part B.)

↓

Remove plates from the tank, mark the position of the solvent front, and air dry. Spray with 0·05 % ferrous sulphate solution to detect the position of the spots (*see* note 3).

↓

Air dry and then measure the R_f value for each compound and calculate the corresponding R_M

values from the formula

$$R_M = \log_{10} (\frac{1}{R_f} - 1).$$

Part B

Prepare peptone–water solutions of each ester (excluding octyl gallate) at concentrations (in μmol ml^{-1}) about ten times that of the expected MIC (*see* Table 2). Sterilize the solutions by filtration.

↓

Dilute 2·5 ml of ester solution with 2·5 ml peptone–water. Mix.

↓

Dilute 2·5 ml of ester solution with 2·5 ml peptone–water, mix, and repeat the dilution sequence until a 6-tube series of 2-fold dilutions has been prepared.

↓

Inoculate each tube with 0·05 ml of 1:1000 dilution of *E. coli* overnight broth culture. Mix.

↓

Incubate overnight at 37 °C.

- -

Day 2 (1h)

Read the MIC results, noting the presence or absence of growth in each tube. Establish the approximate MIC (in μmol ml^{-1}) for each ester (the lowest concentration of ester which prevents growth).

↓

Prepare, in peptone–water, solutions containing the minimum inhibitory concentration of each ester (as determined above). Sterilize the solutions by filtration (*see* note 4).

↓

Take 25·0 ml of one ester solution. Aseptically transfer 2·5 ml to a sterile tube.

↓

Aseptically add 2·5 ml peptone–water to the remaining 22·5 ml of solution. Mix and aseptically transfer 2·5 ml of this dilution to another sterile tube.

↓

Repeat until eight dilutions of the ester have been prepared (*see* note 5).

↓

Repeat the last three stages for each ester solution.

↓

Inoculate each tube with 0·05 ml of 1:1000 dilution of *E. coli* overnight broth culture. Mix.

↓

Incubate at 37 °C overnight.

Day 3 *(1h)*

Note the presence or absence of growth in each tube and establish a precise MIC (in μmol ml^{-1}) for each ester. Calculate the $\log_{10}$ (100/MIC) value for each ester.

↓

Tabulate the data and the data from other students (as in Table 1), so collecting replicate $\log_{10}$ (100/MIC) and R_M values for each ester. Plot each paired $\log_{10}$ (100/MIC), R_M point on a scatter diagram, with $\log_{10}$ (100/MIC) along the *y*-axis and R_M along the *x*-axis. Calculate the correlation coefficient (*r*), the percentage of data explained by the correlation ($r^2 \times 100\%$), and the linear regression line (*see* notes 6 and 7).

↓

Superimpose the calculated regression line on the scatter diagram (*see* Figure 1). Observe and comment on how well the line fits the data. Substituting the R_M value for octyl gallate in the regression equation will allow an estimated value for the MIC of octyl gallate to be calculated. Compare the estimated MIC for octyl gallate with the solubility of this ester (*see* Table 2), and draw conclusions.

Notes and Points to Watch

1. The experiment is designed for four students working together. Each student prepares initial solutions of two of the eight esters for both the TLC and MIC experiments, but every student prepares a complete TLC plate and determines the MIC for each compound.
2. An electric hair-drier may be used to speed the drying of TLC plates.
3. Solutions of other iron salts may be used, since they all give blue colours with the gallate esters.
4. Solutions of the gallate esters deteriorate on standing, especially if aerated, and should be prepared freshly each day.
5. The procedure described in stages 3–5 of Day 2 results in concentrations changing by 10% from one tube to the next.
6. The statistics mentioned above may be computed on scientific calculators with statistical facilities (e.g. Texas Instruments SR 51 calculator) or by using standard programs from computer centres. Alternatively, the statistics may be computed by hand.
7. The following example gives specimen data (tabulated in Table 1 and plotted as a scatter diagram in Figure 1) and illustrates the calculations required.

Table 1 The pooled results from two students.

Ester	(y) $\log_{10} 100/\text{MIC}$	(x) R_M
methyl	1·22	−0·91
	1·27	−0·79
ethyl	1·30	−0·75
	1·35	−0·66
n-propyl	1·52	−0·50
	1·57	−0·45
iso-propyl	1·46	−0·53
	1·49	−0·45
n-butyl	1·74	−0·29
	1·80	−0·18
iso-butyl	1·60	−0·33
	1·66	−0·25
n-amyl	1·92	−0·07
	1·96	+0·03
iso-amyl	1·85	−0·10
	1·89	+0·02

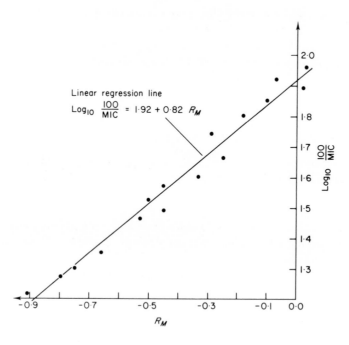

Figure 1 Scatter diagram of the $\log_{10}$ (100/MIC), R_M values for esters of gallic acid tabulated in Table 1, and the superimposed calculated regression line.

The statistics may be calculated manually as follows:

Number of pairs of datum points, N, = 16

$$\Sigma y = 25 \cdot 60 \qquad\qquad \Sigma x = -6 \cdot 21$$

$$\bar{y} = \frac{\Sigma y}{N} = 1 \cdot 6 \qquad\qquad \bar{x} = \frac{\Sigma x}{N} = -0 \cdot 388$$

$$\Sigma(y^2) = 41 \cdot 8466 \qquad\qquad \Sigma(x^2) = 3 \cdot 69$$

$$\Sigma(xy) = -8 \cdot 8847.$$

Correlation coefficient

$$r = \frac{\Sigma(xy) - (\Sigma(x)\Sigma(y)/N)}{\sqrt{\left[\left(\Sigma(x^2) - (\Sigma x)^2/N\right)\left(\Sigma(y^2) - (\Sigma y)^2/N\right)\right]}}$$

$$= \frac{-8 \cdot 8847 - \left(-6 \cdot 21 \times \dfrac{25 \cdot 6}{16}\right)}{\sqrt{\left[\left(3 \cdot 69 - \dfrac{-6 \cdot 21^2}{16}\right)\left(41 \cdot 8466 - \dfrac{25 \cdot 6^2}{16}\right)\right]}}$$

$$= \frac{1 \cdot 0513}{\sqrt{[1 \cdot 2797 \times 0 \cdot 8866]}} = 0 \cdot 987.$$

Percentage of data explained

$$r^2 \times 100\% = 97 \cdot 4\%.$$

Regression equation

$$b = \frac{\Sigma(xy) - (\Sigma(x)\Sigma(y)/N)}{\Sigma(x^2) - \left((\Sigma x)^2/N\right)} = \frac{-8 \cdot 8847 - \left(-6 \cdot 21 \times \dfrac{25 \cdot 6}{16}\right)}{3 \cdot 69 - \dfrac{6 \cdot 21^2}{16}}$$

$$= \frac{1 \cdot 0513}{1 \cdot 2797} = 0 \cdot 82.$$

$$a = \bar{y} - b\bar{x}$$

$$= 1 \cdot 6 - 0 \cdot 82(-0 \cdot 388) = 1 \cdot 92.$$

The regression equation is

$$\log_{10} \frac{100}{\text{MIC}} = 1 \cdot 92 + 0 \cdot 82\, R_M.$$

MATERIALS

1. Esters of gallic acid are available from Nipa Laboratories Ltd., Pontypridd, Glamorgan, Wales.
2. Liquid paraffin loaded TLC plates. 20 cm by 20 cm TLC plates are coated with a 300 μm layer of silica gel and dried at 105 °C for 15 min. They are then placed in a TLC tank containing a solution of liquid paraffin (5 ml in 100 ml of n-hexane) and the solvent allowed to ascend the plates completely. The plates are air dried (**care**—inflammable vapour) and are then ready for use. They may be stored for at least 1 week if kept over silica gel.
3. Peptone–water: 5 g peptone, 5 g sodium chloride in a 1 litre glass distilled water.
4. *Escherichia coli* NCTC 5933. Other strains and other organisms could be used providing their resistance to the gallate esters is not too great.

SPECIFIC REQUIREMENTS

These items are needed for each four students.

Day 1
about 1 g of each of the following esters: methyl, ethyl, n-propyl, iso-propyl, n-butyl, iso-butyl, n-amyl, iso-amyl gallate and octyl gallate
25 ml ethanol
8 small containers with good closures for the alcoholic solutions of the esters
4 TLC plates
8×1 μl capillaries and holders
2 TLC tanks
200 ml 40:60 acetone–water solution
hair-driers
50 ml 0·05% ferrous sulphate solution in an atomizer
8×25 ml glass stoppered measuring cylinders
balance
4×200 ml peptone water
200 capped sterile test tubes
1:1000 dilution of overnight *E. coli* broth culture in peptone–water
test tube racks
sterile pipettes, graduated and Pasteur
incubator at 37 °C

Day 2
about 1 g of each ester
8×25 ml glass stoppered measuring cylinders
balance
4×200 ml peptone water
350 capped sterile test tubes
1:1000 dilution of overnight *E. coli* broth culture in peptone–water
test tube racks
sterile pipettes, graduated and Pasteur
incubator at 37 °C

Day 3
access to calculators/computer terminal

FURTHER INFORMATION

The experiment which has been described is a simple example of the development of a quantitative structure–activity relationship. Table 2 gives some necessary solubility data and a typical set of student measurements of R_f values and MICs. These give rise to the following equation:

$$\log_{10} \frac{100}{\text{MIC}} = 1\cdot94 + 0\cdot84\, R_M$$

with a correlation coefficient $(r) = 0\cdot987$, indicating the % data explained to be 97·4%.

Computation of the MIC for octyl gallate from this equation gives a value of 0·44 μmol ml^{-1} which is in excess of the compound's solubility (Table 2), suggesting that saturated solutions of this

Table 2 Solubility data and typical student results.

	Solubility of the gallate esters at 35°C			MIC
	(μmol ml^{-1})	(%)	R_f	(μmol ml^{-1})
methyl gallate	92	1·7	0·89	6·0
ethyl gallate	140	2·7	0·85	5·0
n-propyl gallate	32	0·7	0·76	3·0
iso-propyl gallate	320	6·7	0·77	3·5
n-butyl gallate	16	0·35	0·66	1·8
iso-butyl gallate	5	0·11	0·68	2·5
n-amyl gallate	7	0·18	0·54	1·2
iso-amyl gallate	10	0·24	0·56	1·4
n-octyl gallate	0·07	0·002	0·24	*

ester will not inhibit growth. Experiments confirm this lack of activity; for a fuller discussion see Boyd and Beveridge (1979).

The experiment may be developed in several ways, for example:

(1) by using different species of test organism;

(2) by using other series of test compounds, e.g. phenols and halogenated phenols, hydroxy-benzoic acid esters, penicillins;

(3) using other measures of antimicrobial activity, e.g. minimum bactericidal concentration, concentration to give 50% inhibition of growth, concentration to give 90% kill, phenol coefficients;

(4) by using calculated π values as measures of partitioning ability;

(5) by incorporating steric and/or electronic parameters in equations.

In some of these cases multiple regression analysis is needed, and, whilst this is impracticable on a calculator or by hand, most computer centres have readily available statistical packages with this facility.

Other approaches to quantitative structure–activity relationships have recently been compared with the Hansch approach by Hall and Kier (1978). James (1974) usefully reviews a variety of structure–activity techniques.

REFERENCES

A. Albert (1973). "Selective Toxicity", 5th edn, Chap. 7, Appendices II and III. Chapman and Hall, London.

C. B. C. Boyce and B. V. Millborrow (1965). A simple assessment of partition data for correlating structure and biological activity using thin-layer chromatography. *Nature, London*, **208**, 537–9.

I. Boyd and E. G. Beveridge (1979). The relationship between the antibacterial activity towards *E. coli* NCTC 5933 and the physico-chemical properties of some esters of 3, 4, 5-trihydroxybenzoic acid (Gallic acid). *Microbios*, **24**, 173–84.

J. Ferguson (1939). The use of chemical potentials as indices of toxicity. *Proceedings of the Royal Society, Series B*, **127**, 387–404.

L. H. Hall and L. B. Kier (1978). A comparative analysis of molecular connectivity, Hansch, Free–Wilson and Darc–Pelco methods in the SAR of halogenated phenols. *European Journal of Medicinal Chemistry–Chimica Therapeutica*, **13**, 89–92.

G. Hansch and T. Fujita (1964). ρ-σ-π analysis. A method for the correlation of biological activity and chemical structure. *Journal of the American Chemical Society*, **86**, 1616–26.

K. James (1974). Linear free energy relationships and biological action. *In* "Progress in Medicinal Chemistry", (G. P. Ellis and E. B. West, eds), Vol. 10. North Holland, London.

E. J. Lien, C. Hansch and Susan M. Anderson (1968). Structure–activity correlations for antibacterial agents on Gram-positive and Gram-negative cells. *Journal of Medicinal Chemistry*, **11**, 430–41.

M. S. Tute (1971). Principles and practice of Hansch analysis. *In* "Advances in Drug Research", (N. J. Harper and Alma B. Simmonds, eds), Vol. 6. Academic Press, London and New York.

55

Estimation of Nisin in Foodstuffs

J. F. LOWE

Department of Microbiology, School of Agriculture,
West Mains Road, Edinburgh, EH9 3JG, Scotland

> *Level*: Beginning undergraduates
> *Subject areas*: Food microbiology; antibiotic assay
> *Special features*: Illustrates problems associated
> with adsorption of a biologically
> active substance to inert
> materials.

INTRODUCTION

Nisin is one of the antibiotics allowed in foodstuffs. It consists of more than one polypeptide and has a molecular weight of about 7500. It is only slightly soluble in neutral solutions but much more soluble at lower pH values. It is particularly effective against spore-formers and is consequently used in the canning of some foodstuffs (Tramer, 1966; Hall, 1966).

This assay is an example of a plate diffusion method. There are three main problems to be overcome. Firstly, nisin diffuses only slowly through agar but can be speeded up by addition of the detergent Tween 20.

Secondly, nisin adheres strongly to the proteins in foods and thus any method for its assay must bring it into the free state. Here an acid treatment is used.

Finally, a standard curve must be prepared using part of the foodstuff to be assayed as a source of diluent. It is therefore necessary to ensure that any nisin in that part is destroyed. This is done by treatment with alkali.

EXPERIMENTAL

Day 1 *(2h)*

Homogenize (*see* note 1) about 25 g food (in water if necessary).

↓

Make a 20 % w/v suspension in 0·02 N HCl, adjust to pH 2·0 with 5 N HCl, boil for 5 min and centrifuge at 670 g for 5 min.

↓

Divide the supernate into two fractions.

↓   ↓

Fraction I   *Fraction II*
Adjust to pH 11·0 with 5 N NaOH, hold at   Dilute 1:1 and 1:3 with remainder* of
63 °C for 30 min, cool and reacidify to pH 2·0   fraction 1.
with 5 N HCl (*see* note 2).

↓

Divide into 4 small parts (and a remainder*)
and use these to dilute a stock nisin solution to
give known standards of 20, 15, 10 and 5
μg ml^{-1}.

↓   ↓

Soak some antibiotic assay discs in the various known (from fraction I) and unknown (from fraction II) samples. Remove excess sample and place discs randomly over plates of assay agar. The assay should be done in triplicate.

↓

Incubate plates at 30 °C overnight.

- -

Day 2 *(30 min)*

Measure the diameters of the zones of inhibition produced. A standard straight line should be obtained by plotting the logarithm of the nisin concentration (from fraction I) against zone diameter. From the standard curve, the amounts of nisin in the original materials can be determined.

- -

Notes and Points to Watch

The only item of concern in this experiment is that of safety and there are two things to watch.

1. Homogenization of the foodstuff: this can be done in a macerator but, if this is glass, there is always the risk of it shattering. These procedures are best performed by staff. Preferably, a Stomacher should be used.
2. Release and destruction of nisin: this is achieved by using warm strong acids and alkalis. Care must be taken at all times, and particularly when boiling the 5 N HCl.

MATERIALS

1. *Micrococcus flavus* NCIB 8166.
2. Stock nisin (from Koch-Light) solution: a stock solution of 1000 μg ml^{-1} in 0·2 N HCl is required. It may be further diluted with 0·02 N HCl, if necessary.
3. Suitable foods: try tinned peas, soups, tomatoes and processed cheeses.
4. Assay medium:

Peptone,	10 g	Yeast extract,	1·5 g
Beef extract,	3 g	Sugar,	1 g
NaCl,	3 g	Agar,	15 g

Distilled water to 1 litre.

The assay medium is melted and cooled to about 50 °C. Tween 20, warmed to 50 °C, is added to give a final concentration of 1 %.

A suspension of the test organism is made by adding 7 ml of physiological saline to a slope and scraping off the growth. This is often too thick and can be further diluted 1 in 10 before adding 2 ml to 100 ml of the molten medium and thoroughly mixing. The inoculated medium is poured to a depth of 3–4 mm in sterile Petri dishes and allowed to solidify.

SPECIFIC REQUIREMENTS

Day 1
assay medium
HCl (5 and 0·2 N) and NaOH (5 N)
antibiotic assay discs
forceps and alcohol for flaming
pipettes

various test tubes
water-bath(s) at 63 °C

Day 2
graph paper

FURTHER INFORMATION

Nisin is a food additive that does not have to be declared on labels and it is therefore difficult to be sure whether is present in a particular food. It may be present in processed cheese and various canned foods and at the time of writing is certainly present in Hartley's peas.

As a test of the procedure, nisin can be added to the foodstuff to be assayed and the procedure followed through. Should you wish to do this, it is recommended that sufficient nisin in 0·02 N HCl is added to the sample to give a concentration of 100 μg g^{-1}, immediately after the food is homogenized.

Tramer and Fowler (1964) recommend the use of wells in the agar rather than antibiotic assay discs. Wells are not used here mainly because it is time-consuming to produce them. However, they would allow the testing of more material and avoid the problem of students tending to overload the antibiotic assay discs.

REFERENCES

R. H. Hall (1966). Nisin and food preservation. *Process Biochemistry* **1**, 461–4.
J. Tramer (1966). Nisin in food preservation. *Chemistry and Industry*, 446–50.
J. Tramer and G. G. Fowler (1964). Estimation of nisin in foods. *Journal of the Science of Food and Agriculture*, **15**, 522–8.

56

Inhibition of Bacterial Growth by Antibiotics: Effect of Rifampicin on RNA Synthesis

L. J. DOUGLAS and J. H. FREER

*Department of Microbiology, University of Glasgow,
Glasgow, G11 6NU, Scotland*

> *Level*: Intermediate/Advanced undergraduates
> *Subject areas*: Microbial physiology and growth
> *Special features*: Introduction to radioisotope
> techniques

INTRODUCTION

Antibiotics were originally defined by Waksman (1945) as "chemical substances produced by micro-organisms which possess the ability to kill or to inhibit the growth of bacteria and other micro-organisms". Most of these compounds act by inhibiting the formation of a particular type of macromolecule in the microbial cell. For example, penicillins and cephalosporins inhibit peptidoglycan synthesis; chloramphenicol and the tetracyclines interfere with protein biosynthesis; rifampicin and actinomycin D prevent nucleic acid synthesis. Some antibiotics have an irreversible action while the action of others, when used in low concentration, is reversible.

One of the most rapid and convenient techniques for following macromolecular synthesis in micro-organisms involves supplying growing cells with an appropriate radioactively labelled precursor which is directly incorporated into the polymer in question. For example the kinetics of RNA synthesis can be determined by growing cells in a medium containing $[^3H]$-uridine, and measuring uptake of label into acid-insoluble material.

The aim of this experiment is to assess the effect of rifampicin on RNA synthesis in *Escherichia coli* by this technique.

EXPERIMENTAL

(3–4 h)

Overnight shake culture of *E. coli* B in SK medium.

↓

Subculture 5 ml samples into flasks **R** and **U**, each containing 50 ml fresh **SK** medium prewarmed to 37°C (*see* note 1).

↓ ↓

Flask R | *Flask U*

Incubate at 37°C with shaking, until E_{600}^{1cm} is about 0·5 (60–90 min). Meanwhile prepare Universal containers (*see* note 2).

Incubate at 37°C with shaking, until E_{600}^{1cm} is about 0·5 (60–90 min). Meanwhile prepare Universal containers (*see* note 2).

↓ ↓

Add, with a disposable syringe, 1 ml of [^{3}H]-uridine solution to each flask, at zero time (*see* note 3).

↓ ↓

After 1 min and 2 min further incubation, remove 1 ml samples to TCA in Universal containers labelled R_1 and R_2 (*see* notes 4–6) Mix well on vortex mixer and store on ice.

After 1 min and 2 min further incubation, remove 1 ml samples to TCA in Universal containers labelled U_1 and U_2 (*see* notes 4–6) Mix well on vortex mixer and store on ice.

↓ ↓

At 2·5 min, add 0·5 ml of rifampicin solution (*see* note 7).

At 2·5 min, add 0·5 ml of methanol.

↓ ↓

Continue incubation; remove samples to TCA in Universal containers at 3, 5, 10 and 15 min. Mix well and store on ice for at least 10 min.

↓ ↓

Pass each sample through a membrane filter. Wash filters twice with ice-cold 10% TCA.

↓ ↓

Transfer each membrane with forceps to a scintillation vial, labelled on the cap. Place vial under an infrared lamp or in a 37°C incubator until membrane is dry. Add 10 ml of scintillant and count in scintillation counter.

- -

Notes and Points to Watch

1. Incubation of cultures is best done in bench-top shaking water-baths, preferably with one water-bath per group of students. Sterile medium may be prewarmed in these baths before the start of the class.

2. During the initial stage of the experiment (i.e. before the cultures have reached an E_{600}^{1cm} value of 0·5) a series of twelve Universal containers should be prepared, each containing 5 ml of 10% trichloroacetic acid (TCA). The containers should be clearly labelled R_1, R_2, R_3, R_5, R_{10}, R_{15}, and U_1, U_2, U_3, U_5, U_{10}, and U_{15}, and stored on ice. Trichloroacetic acid is corrosive. Skin contact or inhalation of vapour should therefore be avoided.

3. All transfers of radioactive material should be done over trays lined with absorbent paper or on sheets of "benchkote" or similar product. **No** mouth-pipetting should be allowed; syringes or hand-operated pipettes should always be used together with disposable gloves. Adequate provision should be made for the disposal of all contaminated material.

4. The success of this experiment depends upon the cells remaining in the exponential phase of growth throughout. Samples should therefore be taken as rapidly as is consistent with safety. Periods when flasks are removed from the shaker for sampling should be kept to an absolute minimum.

5. Because of the relatively high cell densities used and the short duration of the experiment, aseptic technique is not necessary during antibiotic addition and cell sampling.

6. Cell samples should be expelled from pipette or syringe directly into the TCA, without contacting the sides of the Universal container.

7. Because of its low water solubility, the antibiotic solution should be discharged directly into the medium without contacting the sides of the flask. For speed, safety and convenience, this is best done with a disposable 1 ml syringe.

8. The samples containing rifampicin are subject to colour quenching. The counts obtained for these samples should be multiplied by a factor of 1·15 to correct for this.

MATERIALS

1. *Escherichia coli* strain B.
2. SK medium (Snyder and Koch, 1966) containing (per litre): KH_2PO_4, 0·78 g; K_2HPO_4, 2·3 g; $(NH_4)_2SO_4$, 1·0 g; $MgSO_4.7H_2O$, 0·1 g; $Na_3C_6H_5O_7.2H_2O$ (sodium citrate), 0·6 g. Sterile 10% (w/v) D-glucose solution is added to give a final concentration of 0·2% (w/v). The medium has a pH value of 7·3.
3. Rifampicin (Sigma Chemical Co.)
4. [5-^{3}H]-Uridine (25–30 Ci mmol^{-1}; The Radiochemical Centre, Amersham, England).
5. Plastic filter units (50 ml; Gelman Hawksley Ltd., Northampton, England), complete with 250 ml Buchner flasks; filter membranes (25 mm diam.; 0·45 μm pore size, Oxoid N25/45 UP or equivalent).
6. Toluene-based scintillant containing: toluene, 1 litre; 2,5-diphenyloxazole, 4 g; 1, 4-bis-(4-methyl-5-phenyloxazol-2-yl) benzene, 0·1 g.

SPECIFIC REQUIREMENTS

50 ml overnight shake culture of *E. coli* B in SK medium

2 × 50 ml sterile SK medium in 250 ml shake flasks

[5-^{3}H]-uridine solution containing 25 μCi [5-^{3}H]-uridine plus 1.25 mg unlabelled carrier per ml of solution (2 ml)

rifampicin solution, 2 mg ml^{-1} in methanol (0·5 ml; freshly made)

0·5 ml methanol

200 ml 10% (w/v) trichloroacetic acid

1 filter unit

12 filter membranes

12 glass scintillation vials with caps

12 Universal bottles with plastic caps

1 can 1 ml blowout pipettes (or disposable syringes)

4 × 1 ml plastic syringes

1 can sterile 5 ml pipettes

2 pipette fillers

shaking water-bath at 37°C with 250 ml flask platform

cont.

cont.

1 vortex mixer	large plastic tank for glassware decontamin-
spectrophotometer set at 600 nm	ation in "Alconox"
2 small vacuum pumps with traps and tubing	disposable gloves
1 large enamel working tray or sheet	radioactive label tape
"Benchkote"	2 glass micro-cuvettes
ice bucket	toluene-based scintillant
2 pairs flat-ended forceps	scintillation counter
1 box tissues	infrared lamp

FURTHER INFORMATION

Rifampicin is a semi-synthetic member of the group of antibiotics known as rifamycins produced by *Streptomyces mediterranei*. It strongly inhibits RNA synthesis in sensitive bacteria, and also in cell-free extracts, by binding to DNA-dependent RNA polymerase. Rifampicin forms a tight 1:1 complex with the β subunit of the enzyme and thereby interferes with the initiation sequence at some point after the attachment of the polymerase to the DNA (Franklin and Snow, 1981).

The effect of rifampicin on RNA synthesis in *E. coli* as demonstrated in this experiment is shown in Figure 1.

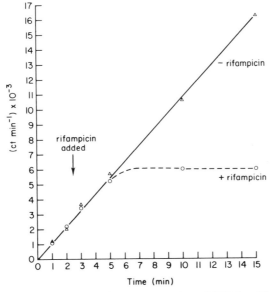

Figure 1 Effect of rifampicin on the incorporation of [³H-]-uridine by *E. coli* B.

REFERENCES

I. S. Snyder and N. A. Koch (1966). Production and characteristics of hemolysins of *Escherichia coli*. *Journal of Bacteriology*, **91**, 763–7.

T. J. Franklin and G. A. Snow (1981). "Biochemistry of Antimicrobial Action", 3rd edn, 217 pp. Chapman and Hall, London.

S. A. Waksman (1945). "Microbial Antagonisms and Antibiotic Substances". The Commonwealth Fund, New York.

Inhibition of Bacterial Growth by Antibiotics: Effect of Chloramphenicol on Protein Synthesis

L. J. DOUGLAS and J. H. FREER

Department of Microbiology, University of Glasgow, Glasgow, G11 6NU, Scotland

> *Level*: Intermediate/Advanced undergraduates
> *Subject areas*: Microbial physiology and growth
> *Special features*: Introduction to radioisotope techniques

INTRODUCTION

The overall procedure is similar to the rifampicin experiment except that here, protein synthesis is monitored by following the incorporation of the precursor [^{3}H]-methionine into protein (acid-precipitable material) and the effect of chloramphenicol is studied.

EXPERIMENTAL

(3–4 h)

Overnight shake culture of *E. coli* B in SK medium.

↓

Subculture 5 ml samples into flasks C and M, each containing 50 ml fresh SK medium prewarmed to 37°C (*see* note 1).

↓   ↓

Flask C
Incubate at 37°C, with shaking, until $E_{600}^{1\,cm}$ is about 0·5 (60–90 min). Meanwhile prepare

Flask M
Incubate at 37°C, with shaking, until $E_{600}^{1\,cm}$ is about 0·5 (60–90 min). Meanwhile prepare

Universal containers (*see* note 2).   Universal containers (*see* note 2).

↓   ↓

Add, with a disposable syringe, 1 ml of [^{3}H]-methionine solution to each flask at zero time (*see* note 3).

↓   ↓

After 1 min further incubation, remove 1 ml sample to TCA in Universal container labelled C_1 (*see* notes 4–6). Mix well on vortex mixer and store on ice.   After 1 min further incubation, remove 1 ml sample to TCA in Universal container labelled M_1 (*see* notes 4–6). Mix well on vortex mixer and store on ice.

↓   ↓

At 3–5 min (note time exactly) add 0·25 ml of chloramphenicol solution (*see* note 7).   At 3–5 min (note time exactly) add 0·25 ml of ethanol.

↓   ↓

Continue incubation; remove samples to TCA in Universal containers at 5, 10, 15, 20 and 30 min. Mix well and store on ice for at least 10 min.

↓   ↓

Pass each sample through a membrane filter. Wash filters twice with ice-cold 10 % TCA.

↓   ↓

Transfer each membrane with forceps to a scintillation vial, labelled on the cap. Place vial under an infrared lamp or in a 37°C incubator until membrane is dry. Add 10 ml of scintillant and count in scintillation counter.

Notes and Points to Watch

1. Incubation of cultures is best done in bench-top shaking water-baths, preferably with one water-bath per group of students. Sterile medium may be prewarmed in these baths before the start of the class.
2. During the initial stage of the experiment (i.e. before the cultures have reached an $E_{600}^{1\,cm}$ value of 0·5) a series of twelve Universal containers should be prepared, each containing 5 ml of 10 % trichloroacetic acid (TCA). The containers should be clearly labelled $C_1, C_5, C_{10}, C_{15}, C_{20}, C_{30}$, and $M_1, M_5, M_{10}, M_{15}, M_{20}$ and M_{30}, and stored on ice. Trichloroacetic acid is corrosive. Skin contact or inhalation of vapour should therefore be avoided.
3. All transfers of radioactive material should be done over trays lined with absorbent paper or on sheets of "benchkote" or similar product. **No** mouth-pipetting should be allowed; syringes or hand-operated pipettes should always be used together with disposable gloves. Adequate provision should be made for the disposal of all contaminated material.
4. The success of this experiment depends upon the cells remaining in the exponential phase of growth throughout. Samples should therefore be taken as rapidly as is consistent with safety. Periods when flasks are removed from the shaker for sampling should be kept to an absolute minimum.

5. Because of the relatively high cell densities used and the short duration of the experiment, aseptic technique is not necessary during antibiotic addition and cell sampling.
6. Cell samples should be expelled from pipette or syringe directly into the TCA, without contacting the sides of the Universal container.
7. Because of its low water solubility, the antibiotic solution should be discharged directly into the medium without contacting the sides of the flask. For speed, safety and convenience, this is best done with a disposable 1 ml syringe.

MATERIALS

1. *Escherichia coli* strain B.
2. SK medium (Snyder and Koch, 1966) containing (per litre): KH_2PO_4, 0.78 g; K_2HPO_4, 2.3 g; $(NH_4)_2SO_4$, 1.0 g; $MgSO_4.7H_2O$, 0.1 g; $Na_3C_6H_5O_7.2H_2O$ (sodium citrate), 0.6 g. Sterile 10 % (w/v) D-glucose solution is added to give a final concentration of 0.2 % (w/v). The medium has a pH value of 7.3.
3. Chloramphenicol (Sigma Chemical Co.).
4. L-[2(n)-^{3}H] Methionine (2–10 Ci mmol^{-1}; The Radiochemical Centre, Amersham, England).
5. Plastic filter units (50 ml; Gelman Hawksley Ltd., Northampton, England), complete with 250 ml Buchner flasks; filter membranes (25 mm diam.; 0.45 μm pore size, Oxoid N25/45 UP or equivalent).
6. Toluene-based scintillant containing: toluene, 1 litre; 2,5-diphenyloxazole, 4 g; 1, 4-bis-(4-methyl-5-phenyloxazol-2-yl) benzene, 0.1 g.

SPECIFIC REQUIREMENTS

50 ml overnight shake culture of *E. coli* B in SK medium
2 × 50 ml sterile SK medium in 250 ml shake flasks
[^{3}H]-methionine solution containing 25 μ Ci ml^{-1} plus 1 mg unlabelled methionine ml^{-1} (2 ml)
chloramphenicol solution, 10 mg ml^{-1} in **ethanol** (0.25 ml; freshly made)
0.25 ml ethanol
200 ml 10 % (w/v) trichloroacetic acid
1 filter unit
12 filter membranes
12 glass scintillation vials with caps
12 Universal containers with plastic caps
1 can 1 ml blowout pipettes (or disposable syringes)
1 can sterile 5 ml pipettes
4 × 1 ml plastic syringes
2 pipette fillers

shaking water-bath at 37°C with 250 ml flask platform
1 vortex mixer
spectrophotometer set at 600 nm
2 small vacuum pumps with traps and tubing
1 large enamel working tray or sheet of "benchkote"
ice bucket
2 pairs flat-ended forceps
1 box tissues
disposable gloves
plastic bags for radioactive waste
large plastic tank for decontaminating glassware in "Alconox"
radioactive label tape
2 glass micro-cuvettes
toluene-based scintillant
scintillation counter
infrared lamp

·FURTHER INFORMATION

Chloramphenicol, one of the oldest antibiotics useful in chemotherapy, is the only one manufactured by chemical synthesis. It is a broad-spectrum antibiotic active against the majority of Gram-negative bacteria.

The mechanism of chloramphenicol action is inhibition of protein synthesis in bacteria; it is active only against 70s ribosomes and is completely inactive against 80s ribosomes. It binds to the 50s subunit in a 1 : 1 ratio and inhibits the enzyme peptidyl transferase, thus preventing peptide bond formation (Franklin and Snow, 1981).

The effect of chloramphenicol on protein synthesis in *E. coli* as demonstrated in this experiment is shown in Figure 1.

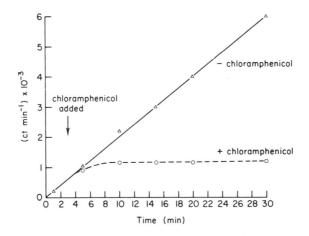

Figure 1 Effect of chloramphenicol on the incorporation of [^{3}H]-methionine by *E. coli* B.

REFERENCES

I. S. Snyder and N. A. Koch (1966). Production and characteristics of hemolysins of *Escherichia coli*. *Journal of Bacteriology*, **91**, 763–7.

T. J. Franklin and G. A. Snow (1981). "Biochemistry of Antimicrobial Action", 3rd edn, 217 pp. Chapman and Hall, London.

58

Four-point Parallel-line Assay of Penicillin

A. C. WARDLAW

Microbiology Department, Glasgow University, Glasgow, G11 6NU, Scotland

---

Level: Intermediate undergraduates

Subject areas: Antibiotic assay; microbiological statistics

Special features: Minimal requirements; randomized block design; analysis of variance and calculation of relative potency and 95% confidence limits

---

INTRODUCTION

The antibiotic activity of a penicillin solution may conveniently be assayed by measuring the zone diameter of growth inhibition around wells filled with dilutions of the solution in nutrient agar plates seeded with *Bacillus subtilis*. Zone diameter is linearly related to the logarithm of the antibiotic concentration over a wide range, e.g. from 0·25 to 8 u ml^{-1}. Having once established this linear relationship, it is then necessary to set up only two known concentrations of penicillin in order to fix the position of the straight dose–response line on subsequent occasions. The 4-point parallel-line assay described here involves measuring the responses produced by two dilutions of Standard benzyl penicillin run simultaneously with two dilutions of the Unknown solution of benzyl penicillin whose potency is to be assayed. The results should yield two parallel straight lines (*see* Fig. 5, in Further Information) whose horizontal displacement is a measure of the relative potency of the Unknown with respect to the Standard.

The aim of this exercise is to provide experience in setting up a simple biological assay, with proper randomization, and then to make a full statistical analysis of the data, leading to an estimate of the potency of the Unknown, with 95% confidence limits.

EXPERIMENTAL

The logistical requirements of this exercise are so simple that even in a large class, each student may do it individually.

Day 1 *(1¼ h, but with a 30 min gap in the middle)*

4 Universal containers with 20 ml molten nutrient agar, in a water-bath at 45 °C.

↓

To each container add 0·2 ml of an appropriately diluted *B. subtilis* spore suspension (*see* note 1) also at 45 °C. By gentle rolling, mix the inoculum evenly throughout the medium. Avoid forming bubbles.

↓

Pour the contents of each container into a Petri dish on a level area of bench and allow to solidify.

↓

Place the plate on the pattern (Figure 1) and punch holes in the gel with a sterilized (*see* note 2) 8 mm (outside diameter) cork borer. Suck out the agar plugs with a Pasteur pipette attached to a vacuum pump and trap (Figure 2).

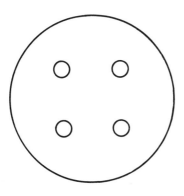

Figure 1 Pattern for punching holes in penicillin assay plates. For use, this should be scaled upto the actual size of the Petri dish.

↓

Place the plates on the special lid-up trays (Figure 3) and put them in a 37 °C incubator (*see* note 3) to dry for 30 min. The lid-up tray raises the lid of the Petri dish by about 4 mm and thus promotes drying.

↓

Meanwhile prepare a 1/4 dilution (1 ml + 3 ml) of the Standard $2\,\mathrm{u\,ml^{-1}}$ penicillin and a 1/4 dilution of the Unknown (*see* note 4), both in phosphate buffered saline (pbs).

↓

Each plate provides a single randomized block (*see* note 5) in the assay. On the base of each plate label the wells S_L, S_H, U_L, and U_H to represent low and high concentrations of Standard and

Unknown respectively. To avoid bias, fill the wells in each plate in a predetermined random sequence (*see* note 5). Using sterile 0·1 or 0·2 ml pipettes, 0·08 ml of each of the 4 solutions is delivered to its appropriate well, taking each plate in turn until completed.

↓

Replace the filled plates on the lid-up trays in a random orientation and incubate for 18–24 h. Without deterioration, they may be refrigerated for up to 1 week or until the time of the next class.

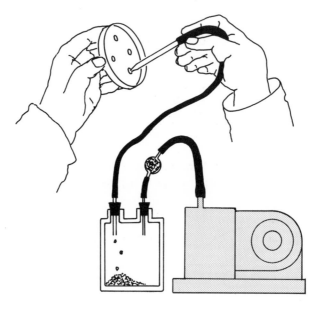

Figure 2   Removal of agar plugs with Pasteur pipette, trap and vacuum pump.

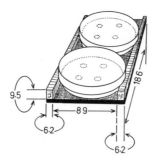

Figure 3   "Lid-up" tray for penicillin assay. The base is made of 3·7 mm hardboard and the edges of 6·2 × 9·5 mm wooden strip, glued and nailed on. The other dimensions given are also in mm.

Day 2   *(20 min benchwork + 1 h for graph plotting and calculations)*

To measure the diameters of the zones of inhibition, place the plates, without lids, in the centre of the illuminated table of an overhead projector and, with a transparent ruler, measure the images projected on a wall or screen about 2 m away (Figure 4). It is best with each zone to measure 2 diameters at right angles and take the average.

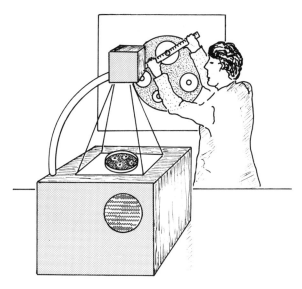

Figure 4 Measurement of zones of inhibition with the aid of an overhead projector.

- -

Notes and Points to Watch

1. The appropriate amount of spore suspension for use in the assay plates is determined before the day of the class by making 10-fold dilutions and adding 0·2 ml of each to melted agar, pouring plates and setting up wells with 2 u ml^{-1} penicillin. A dilution of spore suspension is chosen to give barely confluent growth in the medium but with sharply defined zones of inhibition.
2. One cork borer is enough for a dozen students. It may be sterilized by brief flaming in the Bunsen and allowing it to cool in a sterile tube. It should not be heated red-hot as this damages the metal and is unnecessary.
3. One of the few critical features of this exercise is the drying of the culture plates both before adding the penicillin and during the overnight incubation. If not done correctly, the *B. subtilis* is liable to overgrow the initially formed zones of inhibition and give fuzzy-edged zones or even to obscure them altogether. A walk-in incubator with well-spaced shelves is best because of the good ventilation afforded. However, if a water-jacketed incubator is used, the trays of plates should not be stacked too densely otherwise drying will be inhibited.
4. The Standard penicillin solution should be prepared from benzyl penicillin powder on the day of the class and kept on ice. The Unknown may conveniently be a known dilution of the 2 u ml^{-1} Standard, e.g. 1·5 u ml^{-1} prepared in bulk before the class by adding 10 ml phosphate-buffered saline (pbs) to 30 ml of 2 u ml^{-1} penicillin. The known, "true" potency of the Unknown need not

374 THE MICRO-ORGANISM DIES

be disclosed to the students until they have completed their assays and calculations. It is useful to see how close the student estimates correspond to the true potency.

5. The layout of the randomized block may conveniently be done from a Table of Random Numbers (e.g. Campbell, 1974, p. 343). The dilutions S_L, S_H, U_L and U_H may respectively be assigned the digits 0 and 1; 2 and 3; 4 and 5; 6 and 7. Thus the first row of random digits in Campbell is 2 0 1 7 4 2 2 8 2 3 1 7 5 9 6 6 3 8 6 1 0 2 1 0 . . . Taking these in this order, and compiling the digits for each plate until completed, gives the sequence for filling the wells shown in Table 1. It is best if each student is provided with 4 pipettes for dispensing the 4 dilutions. They can be individually labelled S_L, S_H, U_L and U_H with a glass-writing pen near to the blunt end and kept in the tubes containing the 4 dilutions except when being used.

Table 1 Example of randomized block.

Plate	Sequence of filling the wells			
A	S_H	S_L	U_H	U_L
B	U_L	S_H	S_L	U_H
C	U_H	U_L	S_H	U_H
D	U_H	S_L	S_H	U_L

Note that in determining the random sequence, it is only necessary to fix the first three fillings, since the fourth is then automatically the dose not so far used. Thus the sequence of filling the wells based on the first row of random digits in Campbell (1974, p. 343) and the code given in note 5 is:

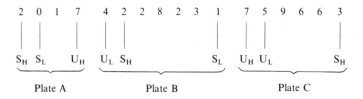

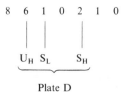

MATERIALS

1. Lab Lemco agar (Oxoid CM 17). Dissolve the powder in distilled water to give $25\,\mathrm{g\,l}^{-1}$, dispense in 150 ml amounts in Roux bottles, autoclave at 121 °C for 15 min and allow the agar to solidify with the bottles laid flat. Five bottles would be ample for a class of 50 students.

2. *Bacillus subtilis* (NCTC 3610) spore suspension.
Inoculate the surface of each of 5 Roux bottles of
Lab Lemco agar with 2 ml of an overnight
nutrient broth culture of the organism. Incubate
for 7 days at 30 °C and check microscopically for
abundant spore formation. Harvest the spores in
a total volume of about 30 ml of sterile saline by
serial transfer of wash fluid from one bottle to
the next and with use of glass beads to dislodge
the growth. The spore suspension may be kept
for several weeks at 4 °C. Before the suspension
is used in the assay, a suitable dilution must be
determined as described in note 1.

3. Benzyl penicillin powder (*see* note 4).
4. Phosphate-buffered saline (pbs): Dulbecco "A"
prepared by dissolving tablets (Oxoid) in distilled water and sterilized by autoclaving.
Alternatively the buffer may be made by dissolving 8 g NaCl, 0·2 g KCl, 1·15 g Na_2HPO_4 and
0·2 g KH_2PO_4 in distilled water to a final
volume of 1 litre. Check that the pH is 7·3 and
sterilize by autoclaving for 15 min at 121 °C.
5. Lid-up trays (*see* Figure 3).
6. Vacuum pump and trap (*see* Figure 2).
7. Pattern for locating wells in dishes (*see* Figure 1).
8. Overhead projector.

SPECIFIC REQUIREMENTS

Day 1
four Universal containers with 20 ml nutrient
agar, melted and put in a water-bath at 45 °C
B. subtilis (NCTC 3610) spore suspension
suitably diluted (1·5 ml) (*see* note 1)
phosphate-buffered saline
penicillin G, 2 u ml⁻¹ in pbs, for the Standard,
and either this or another concentration such
as 1·5 u ml⁻¹ for the Unknown
Petri dishes
12 × 100 mm test tubes
tube rack
0·1 or 0·2 ml and 1·0 ml pipettes
hole-punching pattern (*see* Figure 1)

37 °C incubator, preferably of the walk-in type
45 °C water-bath with thermometer
8 mm (outside diameter) cork borer
lid-up trays
vacuum pump, trap and connector to Pasteur
pipettes
table of random numbers (e.g. Campbell, 1974,
p. 343)

Day 2
overhead projector
transparent metric ruler (30 cm)
good-quality pocket calculator

FURTHER INFORMATION

This section is mainly devoted to working through the calculations on a typical set of results as
provided in Table 2.

Graphical Evaluation

To obtain a graphical estimate of the potency of the Unknown, plot the mean zone diameters
against $\log_{10}$ [concentration of antibiotic] and draw two parallel lines (Figure 5). It is convenient to
use semi-logarithmic paper and to express the concentration, on the logarithmic axis, in terms of
"ml of undiluted Standard or Unknown per ml of solution put in the wells". This allows the
concentration axis to be labelled with 1·0 ml to represent the high doses and 0·25 ml for the low
doses (which were 1/4 dilutions, or 0·25 ml per 1·0 ml) of the high doses. Although seemingly an
awkward way to express concentration, this method has the advantage of subsequently allowing
the figure for relative potency to be read off the graph directly.

Table 2   Sample results for a 4-point parallel-line assay of penicillin.

Plate	Zone diameter (arbitrary units of length)				Plate total (Q)
	S_L	S_H	U_L	U_H	
A	83	101	79	99	362
B	85	103	81	97	366
C	82	95	80	93	350
D	95	101	79	98	373
Dose total (T)	345 (T_1)	400 (T_2)	319 (T_3)	387 (T_4)	1451
Dose mean ($\bar{T}$)	86·25	100	79·75	96·75	

For the analysis of variance the following additional totals are needed:
$\Sigma S = T_1 + T_2 = 745$ (i.e. sum of responses of Standard);
$\Sigma U = T_3 + T_4 = 706$ (i.e. sum of responses of Unknown);
$\Sigma L = T_1 + T_3 = 664$ (i.e. sum of low-dose responses);
$\Sigma H = T_2 + T_4 = 787$ (i.e. sum of high-dose responses);
$\Sigma y = T_1 + T_2 + T_3 + T_4 = 1451$ (i.e. sum of all the responses).

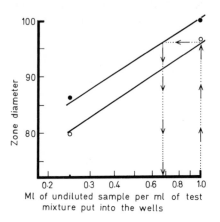

Figure 5   Dose-response lines plotted from the mean zone diameters in Table 2. The lines have been drawn by eye so as to give the best parallel fit to the experimental points. Note that this is not the same as simply connecting up the points, which often would not yield parallel lines. ● standard; ○ unknown

Relative potency (R) is defined as:

$$R = \frac{\text{vol. of Standard}}{\text{vol. of Unknown}} \text{ giving the same biological effect,}$$

i.e. any particular diameter of growth-inhibition zone which may be taken as the end-point. Since the lines are (or should be) parallel, the horizontal distance between them should be constant, and it therefore does not matter what particular zone diameter is chosen. However, it is easiest simply to extrapolate upwards (as shown in Figure 5) from the 1·0 ml mark on the abscissa to the Unknown line, then horizontally across to the Standard line and back down to the abscissa; because the point of final interception on the abscissa is the relative potency value, e.g. in this example, $1.0\,\text{ml}$ of U has the same biological activity as $0.66\,\text{ml}$ of S. The relative potency of U is therefore 0.66. Since the Standard is known to be $2\,\text{u ml}^{-1}$, the Unknown is $2 \times 0.66 = 1.32\,\text{u ml}^{-1}$.

The graphical method for estimating potency is quick and simple and should always be done as a first inspection of the results. However, it is deficient in two important respects: it fails to provide any estimate of limits of error and it also lacks any objective criteria for deciding the point at which departures from parallelism of the dose–response lines should be judged significant and the results therefore discarded. These deficiencies are remedied in the algebraic evaluation which follows.

Algebraic Evaluation

This is done in several connected stages:

analysis of variance;
estimation of mean slope $(\bar{b})$ of the two dose–response lines;
estimation of relative potency (R) of Unknown with respect to Standard;
estimation of the standard error of the $\log_{10}$ relative potency $(S_{\log R})$;
calculation of 95 % confidence limits of R.

For reasons of space, the calculations are given only in summary form and without any attempt being made to derive or explain the formulae.

Analysis of variance

The formulae and results of the analysis of variance are set out in Table 3. The probability (P) values in the right hand column of the table show that there was no significant variation between

Table 3 Analysis of variance of the results given in Table 2.

Source of variation	Degrees of freedom	Sums of squares[a]		Mean[b] square	Variance ratio (F)	P[c] (%)
Total	15	$\Sigma y^2 - CF = 1217.4375$	1	—	—	—
Plates	3	$\dfrac{\Sigma Q^2}{4} - CF = 69.6875$	2	23.23	2.16	> 5
Doses	3	$\dfrac{\Sigma T^2}{4} - CF = 1051.1875$	3	—	—	—
Slope	1	$\dfrac{(\Sigma L)^2 + (\Sigma H)^2}{8} - CF = 945.5625$	4	945.56	88.13	≪ 1
Preparations	1	$\dfrac{(\Sigma U)^2 + (\Sigma S)^2}{8} - CF = 95.0625$	5	95.06	8.86	< 5 > 1
Parallelism	1	$3 - 4 - 5 = 10.5625$		10.56	0.98	> 5
Residual	9	$1 - 2 - 3 = 96.5625$		10.7292	—	—

[a] CF = correction factor = $(\Sigma y)^2/N = 1451^2/16 = 131\,587.5625$.
[b] The square root of the last entry (10.7292) in this column gives the standard deviation, $s = 3.2755$.
[c] Probability of such an extreme value assuming the null hypothesis to be correct. The tabulated values of F for d.f. 3, 9 are 3.86 and 6.99 for $P = 5\%$ and 1 %, and for d.f. 1, 9 are 5.12 and 10.56 respectively.

the replicate plates, that the assay had a highly significant slope, that Standard and Unknown differed significantly in potency and that there was no significant deviation from parallelism in the two dose–response line. If the last P-value **had** been significant, the assay would be invalid and the calculations stopped here. The residual mean square (last entry in col. 4) is the error variance, the square root of which (s) = 3·2755 is used later in the calculations.

Mean slope ($\bar{b}$)

The mean slope is defined by:

$$\bar{b} = \frac{\Sigma H - \Sigma L}{\frac{1}{2} N \cdot \log_{10} D},$$

where ΣH and ΣL are defined in the footnote to Table 2, N = total no. of observations in the whole assay = 16, and D = the dilution interval or ratio of concentrations between low and high dose, which here equals 4. Therefore $\log_{10} D = 0.6021$.

$$\bar{b} = \frac{787 - 644}{\frac{1}{2} \times 16 \times 0.6021}$$

$$= 25.5356.$$

Relative potency (R)

Relative potency is defined by:

$$\log_{10} R = \frac{\Sigma U - \Sigma S}{\frac{1}{2} N \cdot \bar{b}},$$

where ΣU and ΣS are defined in the footnote to Table 2. This gives

$$\log_{10} R = \frac{706 - 745}{8 \times 25.5356} = -0.1909$$

which may also be expressed as $\bar{1}.8091$. Taking the antilogarithm gives $R = 0.64$, which is close to the value 0·66 obtained from the graph. Thus the estimated value in u ml^{-1} of the Unknown = 2 × 0·64 = 1·28, since the Standard was 2 u ml^{-1}.

Standard error of log R ($S_{\log R}$)

The standard error of log R may be estimated either by an exact (but rather long and complicated) formula (*see* Finney (1971) or Hewitt (1977)) or by the approximate formula which is likely to be good enough for present purposes:

$$S_{\log R} = \frac{2s}{\bar{b}} \sqrt{\left\{ \frac{1}{N} \left[1 + \left(\frac{\log_{10} R}{\log_{10} D} \right)^2 \right] \right\}}$$

$$= \frac{2 \times 3.2755}{25.5356} \sqrt{\left\{ \frac{1}{16} \left[1 + \left(\frac{-0.1909}{0.6021} \right)^2 \right] \right\}}$$

$$= 0.06728.$$

95% confidence limits (95% CL)

95% confidence limits are defined by

$$95\% \text{ CL} = \text{antilog}_{10}\,(\log_{10} R \pm t \cdot S_{\log R}),$$

where t is the value of the Student t-statistic for $P = 0.05$ and the no. of degrees of freedom associated with the standard deviation (s); in this example $d.f. = 9$, and $t = 2.262$. This gives

$$95\% \text{ CL} = \text{antilog}\,(-0.1909 \pm 2.262 \times 0.06728)$$

$$= \text{antilog}\,(-0.3431) \text{ and antilog}\,(-0.0387)$$

$$= 0.4538 \text{ and } 0.9147,$$

which may be rounded to 0.45 and 0.91. Multiplying these values by the potency of the Standard $(2\,\text{u ml}^{-1})$ gives confidence limits of the Unknown of 0.90 and $1.82\,\text{u ml}^{-1}$. These easily embrace the true potency of $1.50\,\text{u ml}^{-1}$, although they may seem rather wide to those readers who are not accustomed to calculating confidence limits.

REFERENCES

R. C. Campbell (1974). "Statistics for Biologists", 2nd edn, 385 pp. Cambridge University Press, Cambridge.
D. J. Finney (1971). "Statistical Method in Biological Assay", 2nd edn, 668 pp. Charles Griffin.
W. Hewitt (1977). "Microbiological Assay", 284 pp. Academic Press, London and New York.

Part Six
Micro-organisms in the World around Us

59

Flattened Capillary Tube Technique for Isolating Algae from Natural Aquatic Environments

ELIZABETH G. CUTTER

Department of Botany, University of Manchester, Manchester, M13 9PL, England

> *Level*: Intermediate undergraduates
> *Subject areas*: Phycology; ecology
> *Special features*: Preservation of motility and interaction between organisms

INTRODUCTION

This technique enables students to sample the algal flora (and the microscopic fauna) of ponds or streams by placing special flattened capillary tubes (Figure 1a) called microslide-tubes, in the water for a time, and then removing them and observing the sampled algae directly under the microscope without disturbance. The method is based on the notes by John O. Harris in a booklet provided with the microslide-tubes used in the exercise, and in turn derives from the work of the Russian biologists Perfilev and Gabe.

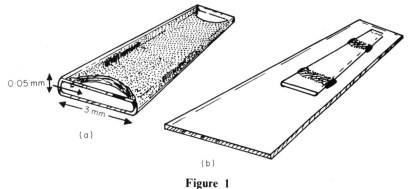

0.05 mm

3 mm

(a)

(b)

Figure 1

EXPERIMENTAL

Day 1 *(15 min, depending upon the location of the aquatic habitat)*

Attach microslide-tubes (3·0 × 0·05 × 50·0 mm) singly or in pairs to ordinary microscope slides using Araldite in 2 places (Figure 1b; *see* note 1).

↓

Fasten the microscope slides, with attached microslide-tubes, to garden canes or place in open-ended slide carriers (*see* note 2).

↓

Immerse the slides in a pond or stream.

Days 7–14

Remove and examine microscopically (*see* notes 3–5 and Figure 2).

Notes and Points to Watch

1. Microslide-tubes may be fixed to the slides in advance.
2. The microslides may be filled with Bold's medium (Stein, 1973) which will rise in the tubes by capillary action, but this is not necessary; they may simply be allowed to fill with pond water.
3. If left exposed to the air the water in the microslide-tube dries up in a few hours. The ends can be sealed with paraffin wax or other impermeable material, but a further difficulty is that animals such as *Paramoecium*, and especially tardigrades contained within the microslide-tubes, consume the algae. This can be turned to advantage in that the process can actually be observed under the microscope, but if the microslide-tubes are left too long the flora becomes markedly diminished.
4. If microslide-tubes have to be transported some distance from the sampling site to the laboratory it is helpful to place the slides in Coplin staining jars of water.
5. Before observation, both slides of the microscope slide and the upper surface of the microslide-tubes should be carefully wiped with tissues while still wet, taking care not to withdraw the liquid from the ends of the microslide-tube during this process.

MATERIALS

1. Microslide-tubes (0·3 mm viewing path) fixed to microscope slides with Araldite. Microslide-tubes are available from Camlab Ltd., Nuffield Road, Cambridge, CB4 1TH, England.
2. Holders for slides.

3. Garden canes or some other means of sinking slide-holders into the pond and marking their position.
4. Coplin jars for transporting slides.

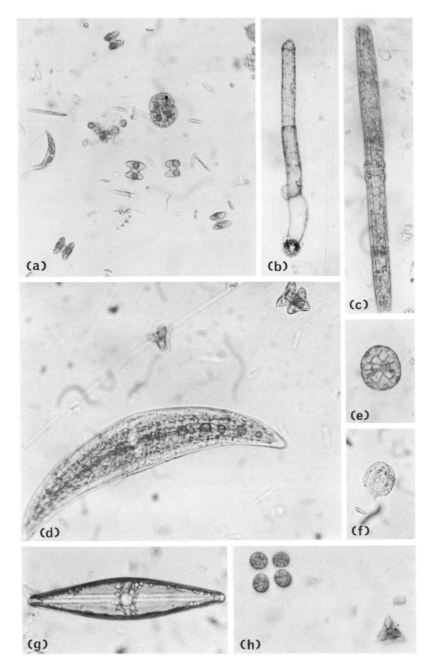

Figure 2 Samples of algae in microslide-tubes from a pond.

(a) General view, including *Cosmarium, Staurastrum*, diatoms, etc. × 200. (b) Young filament of *Oedogonium*. × 130. (c) *Pleurotaenium*. × 200. (d) *Closterium* and two views of *Staurastrum*. × 300. (e) Young developing *Coleochaete*. × 200. (f) *Phacus*. × 200. (g) Pennate diatom. × 300. (h) *Gonium* and *Staurastrum*. × 300.

SPECIFIC REQUIREMENTS

Day 1
materials as above

Days 7–14
medical wipes or other tissues

FURTHER INFORMATION

Samples of algae collected in microslide-tubes immersed in a pond are shown in Figure 2.

REFERENCE

J. R. Stein (1973). "Handbook of Phycological Methods". 448 pp. Cambridge University Press, Cambridge.

60

Some Factors Influencing Dispersal and Adsorption of Microbial Cells

I. Adsorption of bacterial and fungal spores to sand and ion-exchange resins

S. T. WILLIAMS and S. A. LANNING

Department of Botany, University of Liverpool,
Liverpool, L69 3BX, England

Level: Advanced undergraduates
Subject area: Microbial ecology
Special features: Introduces concepts often overlooked in the dispersal of
microbial cells

INTRODUCTION

Many factors are involved in the passive dispersal of microbial cells in soil (e.g. water or air movement, animal carriage and ingestion). However, the nature of the cell surface may markedly influence its reaction to any of these factors.

Most microbial cells carry a net charge, often negative, due to the ionization of specific chemical groups (e.g. $-OH$, $-COOH$) at the cell surface and/or the preferential adsorption of ions from the ambient solution. Most soil mineral and organic particles also carry a net charge, usually negative, and their charged sites also interact with cations in the soil solution. Therefore cations in the soil solution can influence the degree of attraction between microbial cells and soil particles possibly resulting in adsorption of cells, thus limiting their dispersal by water movement.

The aim of this experiment is to determine the net charge on streptomycete spores using ion exchange resins and to study the effect of mono-, di- and trivalent cations on the adsorption of the spores to soil particles. Although the experiment is described using streptomycete spores, *Bacillus* spores or fungal spores may be substituted (*see* note 4).

EXPERIMENTAL

Day 1 (3 h)

1. *Adsorption of spores to ion exchange resins*

The experiment requires a dense aqueous spore suspension of a *Streptomyces* sp. (*see* notes 2 and 4) and 2 columns of resin: De Acidite FF (IP), a strong anion exchange resin, and Zeo-Karb 225, a strong cation exchange resin.

↓

Pipette 5 ml of the very turbid spore suspension on to the top of each column, allow it to pass through and collect the eluate. Add the suspension slowly and carefully to avoid dislodging the resin.

↓

Compare the opacity of the eluate by eye, colorimetry and/or nephelometry (*see* note 1).

↓

Transfer some resin from the top of each column into separate test tubes containing about 5 ml distilled water. Shake for 1 min, allow the resin to settle and decant off the supernate. This removes most unadsorbed spores. Place a few washed resin beads onto a microscope slide and examine them with a light microscope (× 40 objective) to determine which resin holds most spores (*see* note 5).

2. *Adsorption of spores to soil treated with cations*

This requires 3 columns of sterile sand. One has been treated with sodium chloride, one with ferric chloride, and one with distilled water. Slowly add 5 ml of the faintly turbid suspension of the streptomycete to the top of each column. Wash through slowly with 200 ml sterile water per column and collect each eluate in a sterile flask.

↓

Prepare ten-fold dilutions (10^{-1} to 10^{-6}) in distilled water from each eluate and from the original suspension. From each dilution series, plate the 10^{-2}, 10^{-4} and 10^{-6} dilutions. Pipette 1 ml samples from these dilutions into plates, preparing 3 plates for each dilution in each series, i.e. $4 \times 3 \times 3 = 36$ plates. Add molten ($45°C$) nutrient agar to each plate, mix well and incubate at $25°C$ for 5–7 days.

Day 2 *(30 min, 5–7 days later)*

Count the number of colonies at the most appropriate dilutions for the 3 eluates and the original spore suspension.

↓

Calculate the number of spores per ml in the original suspension, in each eluate and remaining in the sand columns.

Notes and Points to Watch

1. Contamination of eluates with resin beads can occur if columns are badly prepared or student technique is poor. This influences the opacity of the eluates.
2. The strain of *Streptomyces* used should be one, such as strain F1, which produces sufficiently wettable spores. Some are very hydrophobic.
3. It is important to remove excess chlorides after treatment before preparation of the columns (*see* Materials). This avoids salt concentrations which could be toxic to spores. Other salts (e.g. $CaCl_2$) may be used.
4. Other bacterial (e.g. *Bacillus*) or fungal spores (e.g. *Penicillium*) may also be used. These also have a net negative charge at pH levels around neutrality (Douglas, 1957; Fisher and Richmond, 1969).
5. If available, a scanning electron microscope can be used for this examination. Resin beads are allowed to air-dry on a specimen stub, and are then coated with metal prior to examination (*see* Figures 1 and 2).

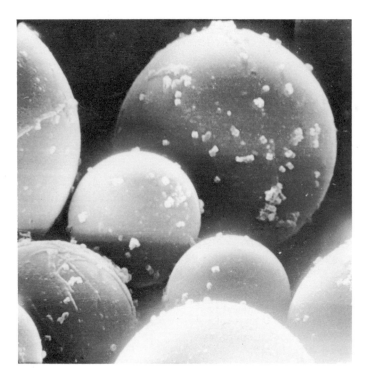

Figure 1 Scanning electron micrographs of beads of the anion-exchange resin De Acidite FF after passage of a suspension of *Streptomyces* spores (× 10 000).

Figure 2 As above, but with Zeokarb 225 resin. Note the infrequency of adhering spores.

MATERIALS

1. *Streptomyces* sp. culture which produces sufficiently wettable spores, e.g. strain F1 (Liverpool soil isolate) (*see* note 2). Most grey-spored strains are suitable. Spores are readily produced after 7 days at 25 °C on oatmeal agar (boil 56 g oatmeal for 30 min, filter, make up the filtrate to 1 litre and add agar). Cultures are prepared on plates of oatmeal agar.
2. Ion exchange resins (BDH): De Acidite FF(IP) and Zeo-Karb 225 are used. Each is placed over a plug of glass wool in a glass tube (diameter about 1·5 cm) to make a column about 4·0 cm in height.
3. Three columns of sterile sand or a sandy soil (to obtain rapid passage of water and spores). Suitable columns can be made in glass tubes (approx. 3·0 cm diam.) with a constricted base. A plug of glass wool is inserted at the constricted end. These are filled with sand to a height of about 12 cm and autoclaved. Before the class, 200 ml sterile water is passed through one column; 100 ml of a sterile 1·0 M solution of sodium chloride is passed through the second and 100 ml of sterile 1·0 M ferric chloride through the third. After 1 h the treated columns are washed free of excess salts by passing 400 ml of sterile water through each (*see* note 3).

SPECIFIC REQUIREMENTS

Day 1
2 columns of resin
3 sand columns in clamp stands
3 sterile 250 ml flasks
2 sterile 5 or 10 ml pipettes
4 × 200 ml sterile water
24 tubes with 9 ml sterile water
40 sterile 1 ml pipettes (blow-out)
36 sterile Petri dishes
36 vials each with 15 ml molten nutrient agar
  (Oxoid CM3) (45 °C)
3 test tubes in rack

15 ml very turbid *Streptomyces* spore suspension in sterile water (growth from 2 plate cultures)
20 ml faintly turbid *Streptomyces* spore suspension
microscope slides
scanning electron microscope (if available)
2 Pasteur pipettes
colorimeter or nephelometer

Day 2
plate counter

FURTHER INFORMATION

Typical class results are shown in Tables 1 and 2. The resin results clearly show that the spores have a net negative charge. It should be emphasized that this is not a fixed property but varies with pH and other factors. Also some microbial cells, e.g. certain fungal spores, are uncharged. The

Table 1 Interactions of *Streptomyces* spores with ion exchange resin.

	Appearance of filtrate	% light transmission of filtrate	Microscopic examination of beads
Original suspension	Turbid	80	—
Zeo-Karb 225	Turbid	85	Few cells adsorbed
De Acidite FF	Clear	100	Many cells adsorbed

Table 2 Interactions of *Streptomyces* spores with sand treated with cations.

	No. spores ml^{-1}		
	In filtrate	Retained in sand	% retained in sand
Original suspension	40×10^5	—	—
Untreated sand	30×10^5	10×10^5	25
Sand treated with sodium chloride	28×10^5	12×10^5	30
Sand treated with ferric chloride	2×10^5	38×10^5	95

relevance of charged cell wall components such as amino acids should also be discussed (*see* Douglas *et al.*, 1970; Fisher and Richmond, 1969).

The sand columns show that some spores are retained in all cases, many probably by a purely physical filtration effect. Monovalent sodium ions have little or no influence on retention as they merely neutralize negatively charged sites in soil. Trivalent ferric ions increase adsorption significantly by providing excess positive charges on soil particles (*see* Zvyagintsev, 1962; Ruddick and Williams, 1972).

REFERENCES

H. W. Douglas (1957). Electrophoretic studies on spores and vegatative cells of certain strains of *B. megaterium*, *B. subtilis* and *B. cereus*. *Journal of Applied Bacteriology*, **20**, 390.

H. W. Douglas, S. M. Ruddick and S. T. Williams (1970). A study of the electrokinetic properties of some actinomycete spores. *Journal of General Microbiology*, **63**, 289–95.

D. J. Fisher and D. V. Richmond (1969). The electrokinetic properties of some fungal spores. *Journal of General Microbiology*, **57**, 51–60.

S. M. Ruddick and S. T. Williams (1972). Studies on the ecology of actinomycetes in soil. V. Some factors influencing the dispersal and adsorption of spores in soil. *Soil Biology and Biochemistry*, **4**, 93–103.

D. G. Zvyagintsev (1962). Adsorption of micro-organisms by soil particles. *Soviet Soil Science* (*English trans.*), 1962, 140–4.

61

Some Factors Influencing Dispersal and Adsorption of Microbial Cells

II. Dispersal of hydrophilic and hydrophobic cells in air and water

S. T. WILLIAMS and S. A. LANNING

*Department of Botany, University of Liverpool,
Liverpool, L69 3BX, England*

Level: Advanced undergraduates
Subject area: Microbial ecology
Special features: Simple but informative about the
influence of cell surface properties
on their dispersal

INTRODUCTION

Many factors are involved in the passive dispersal of microbial cells (e.g. water movement, air movement or animal carriage). However, the nature of the microbial cell surface can markedly influence its reaction to these factors. One important aspect of the cell surface is its reaction to water. Many bacterial cells are hydrophilic and readily form suspensions in water, while many actinomycete and fungal spores formed on aerial hyphae are extremely hydrophobic. These differences may influence their dispersal not only in water but also in air (Davies, 1961; Jarvis, 1962).

The aim of these experiments is to compare the reactions of hydrophilic cells (*Bacillus*) and hydrophobic spores (*Penicillium*) to water, and to study the relevance of these reactions to aerial dispersal. The reactions of cell surfaces to liquids of different surface tensions are compared (i.e. to define hydrophilic and hydrophobic surfaces). The consequent effects of differences in surface reaction to water on dispersal by water droplets are then studied.

EXPERIMENTAL

Day 1 *(Experiment 1, 3 h; Experiment 2, 1 h)*

1. *Reaction of* Bacillus *cells and* Penicillium *spores to liquids of different surface tension.*

The experiment requires plate cultures of *Bacillus subtilis* and *Penicillium spinulosum*. Also needed are water-acetone mixtures with the following surface tensions:

Water	$73 \cdot 05$ dyn cm^{-1}
10 % (v/v) acetone	$48 \cdot 90$ dyn cm^{-1}
20 % (v/v) acetone	$41 \cdot 10$ dyn cm^{-1}
50 % (v/v) acetone	$30 \cdot 40$ dyn cm^{-1}
60 % (v/v) acetone	$28 \cdot 00$ dyn cm^{-1}
70 % (v/v) acetone	$27 \cdot 00$ dyn cm^{-1}
95 %(v/v) acetone	$24 \cdot 00$ dyn cm^{-1}
100 % (v/v) acetone	$23 \cdot 00$ dyn cm^{-1}

Use the following procedure with each microbe and each mixture.

Pipette 5 ml liquid in to a test tube.

↓

Add 5 loopfuls of growth from a culture and mix well by hand or with an automatic mixer for 2 min.

↓

Immediately after mixing, add some of the suspension to a colorimeter tube.

↓

Leave the tube in the colorimeter for 2 min, then take a reading of percentage light transmission.

↓

Note the appearance of the tube contents.

↓

Tabulate light transmission values against surface tension, for each microbe.

2. *Dispersal of* Bacillus *cells and* Penicillium *spores by water droplets*

Place a plate culture of *Bacillus* in a fume cupboard (switched off) or on the bench of an inoculation room. Side draughts should be avoided or, if slight, their direction noted (*see* note 1).

↓

Place 18 plates of nutrient agar around the plate of *Bacillus*, arranging them in 6 equidistant radial arms. The plates in each arm touch adjacent plates. Label the plates to indicate their **exact** position.

↓

Fill a sterile syringe with sterile water. Open the plate of *Bacillus* and remove the lids from all the plates of nutrient agar. From a height of 40 cm allow six drops of water to fall onto the centre of the growth in the *Bacillus* plate. Observe the reaction of the culture to the impact of the water drops.

↓

Replace the lids on all plates after 2 min.

↓

Repeat this procedure using a culture of *Penicillium* and 18 fresh plates of nutrient agar.

↓

Incubate all plates at 25 °C for 5 days.

Day 2 *(1 h, 5 days later)*

1. *Dispersal of* Bacillus *cells and* Penicillium *spores by water droplets*

(a) Count the number of colonies of *Bacillus* and *Penicillium* on the nutrient agar plates (NB a few other aerial contaminants may be seen).
(b) Summarize **all** information obtained on the reactions of the two microbes to water and account for any differences in dispersal patterns observed.

3. *Reaction of* Bacillus *cells and* Penicillium *spores to water droplets*

Place an inoculating needle in a clamp stand at a height of 15 cm above the bench.

↓

Cut out a block of agar with growth from a plate of *Bacillus* (about 1 cm²) and fix it onto the needle so that the growth face is outermost and held vertically (*see* Figure 1).

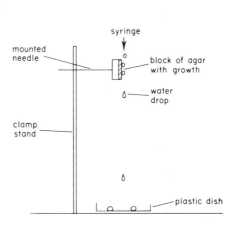

Figure 1   Collection of drops from cultures.

↓

Allow water from a syringe to pass slowly over the growth and collect it in the base of a **plastic** Petri dish. The water will form drops on the hydrophobic plastic surface.

↓

Repeat the procedure using *Penicillium*.

↓

Examine the droplets microscopically to determine the distribution of cells, i.e. are they mainly **within** or **on the surface** of the drop? (*see* note 3).

Notes and Points to Watch

1. The *Bacillus* dispersal test must be done **before** the *Penicillium* to avoid cross-contamination of plates. A **known small** side draught can add to the interpretation of these results.
2. Use of a pigmented bacterium, e.g. an orange–brown *Bacillus subtilis* (e.g. strain Liverpool B2) helps to distinguish specific colonies from other bacterial contaminants.
3. Some students find it necessary to add a small drop of stain (e.g. crystal violet) to the water droplets when observing distribution of *Bacillus*. If drops roll over the plastic dish spores of *Penicillium* are stripped off from the surface of the drops onto the plastic because the two hydrophobic surfaces attract one another.
4. Other bacteria and heavily sporing fungi may be substituted.

MATERIALS

1. Plate culture of *Bacillus subtilis* (strain Liverpool B2) (*see* note 2) should be inoculated onto nutrient agar (Oxoid CM3) to cover the whole of the central region of the dish and incubated at 25 °C for 48 h.

2. *Penicillium spinulosum* (strain Liverpool F27/15) spores well on Czapek-Dox Agar (Oxoid CM 97) plus 0·6 % (w/w) yeast extract after 7 days at 25 °C (*see* note 4).

SPECIFIC REQUIREMENTS

Day 1
3 pipettes, 5 or 10 ml
2 Pasteur pipettes
test tube mixer
colorimeter
wash-bottle + distilled water
water–acetone mixtures (non-sterile), 20 ml
  each:
  a. water
  b. 10 % (v/v) acetone
  c. 20 % (v/v) acetone
  d. 50 % (v/v) acetone
  e. 60 % (v/v) acetone
  f. 70 % (v/v) aceone

g. 95 % (v/v) acetone
h. 100 % (v/v) acetone
16 test tubes + rack
2 plastic Petri dishes
5 plate cultures *Bacillus subtilis*
5 plate cultures *Pencillium spinulosum*
36 plates nutrient agar
2 sterile syringes
50 ml sterile water
clamp stand

Day 5
plate counter

FURTHER INFORMATION

Experiment 1

Typical class results for experiment 1 are given in Table 1. The results with water–acetone mixtures show that *Penicillium* forms optically denser suspensions as the surface tension decreases below that of water, while *Bacillus* acts in the reverse manner. Their hydrophobic and hydrophilic surface properties are thus clearly defined. The hydrophobic *Penicillium* spores have a low "critical surface tension" (point at which they form an even suspension) of **about** 24·20, while that of *Bacillus* is high, between 48·90–73·05.

Table 1   Percentage light transmission of spores and cells mixed in liquids of different surface tension.

Organism	Per cent light transmission of microbial suspension in liquid of surface tension (dyn cm^{-1})							
	73·05 water	48·90	41·10	30·40	28·00	27·00	24·28	23·00 acetone
Penicillium spinulosum	74·20	65·50	54·00	44·50	36·00	34·00	16·00	17·00
Bacillus subtilis	26·00	22·00	36·00	41·00	36·00	86·50	95·00	91·00

Experiment 2

When drops of water land on the *Penicillium* plate, a cloud of released spores should be noticed. Typical class results for the distribution patterns of the two microbes are given in Figure 2. The patterns differ in that (a) more *Penicillium* spores are released, (b) the distribution of *Penicillium* spores is often uneven due to air movements, and (c) the distribution of *Bacillus* cells is more limited radially and tends to be more even (i.e. less affected by air currents).

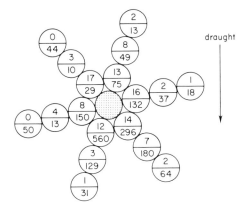

Figure 2 Dispersal of *Bacillus* cells (top numbers) and *Pencillium* spores (bottom numbers) by water droplets (no. of colonies per plate).

Experiment 3

The reactions to water droplets differ (Table 2). While *Bacillus* cells enter the droplets, *Penicillium* spores remain at the surface and are easily removed.

Table 2   Distribution of spores and cells in water droplets.

Organism	Distribution in droplet	Reaction to moving droplet
Pencillium spinulosum	Mostly on surface of droplet	Many spores deposited onto plastic surface
Bacillus subtilis	Mostly within droplet	Few cells removed from droplet

Discussion Points

The observations indicate that the hydrophilic *Bacillus* is primarily dispersed by splash **within** droplets, while the hydrophobic *Penicillium* spores are primarily physically **detached** by drops but do not enter them. They are then subsequently dispersed by air movements. A discussion on possible causes of hydrophobicity of surfaces is worthwhile. Both chemical (e.g. $-CH_3$ groups) and physical (e.g. sculptured surface) factors can be considered. Possible effects of acetone other than those of surface tension (e.g. as a solvent) can also be raised.

REFERENCES

R. R. Davies (1961). Wettability and the capture, carriage and deposition of particles by raindrops. *Nature, London*, **191**, 616–7.
W. R. Jarvis (1962). Splash dispersal of spores of *Botrytis Cinera* Pers. *Nature, London*, **193**, 599.

62

Some Factors Influencing Dispersal and Adsorption of Microbial Cells

III. Effect of temperature and water activity on swarming of *Proteus*

S. T. WILLIAMS and S. A. LANNING

*Department of Botany, University of Liverpool,
Liverpool, L69 3BX, England*

Level: Advanced undergraduates
Subject area: Microbial ecology
Special features: Simple demonstration of rapid
movement by motile bacteria

INTRODUCTION

Many microbes are capable of active movement in their natural habitats, although distances covered may be small compared to those in passive dispersal. Bacteria such as *Proteus* can spread rapidly over solid surfaces in a thin film of moisture. They do so by a cyclical process involving changes in cell shape, motility and division. Various nutritional and environmental factors can influence swarming.

This experiment is designed to study the influence of temperature and water activity on the movement of *Proteus*.

EXPERIMENTAL

Day 1 *(1 h)*

This experiment requires incubators at a range of temperatures and plates of media containing different concentrations of glucose to give a range of water activities (a_w). It is suitable for small groups of students.

Examine a sample of a nutrient broth culture of *Proteus vulgaris* in a hanging drop to check that the bacteria are motile. Do not proceed unless they are.

↓

With a Pasteur pipette, without splashing, place a **small** drop of the culture onto the centre of 15 plates of nutrient agar. Label and incubate these in the upright position at 37, 30, 25, 18 and 4 °C (3 replicate plates at each temperature) for 18–24 h.

↓

Similarly inoculate 3 replicate nutrient agar plates adjusted to the following water activities (a_w): 0·999, 0·990, 0·988, 0·980, 0·974, 0·940. Incubate these at 30 °C for 18–24 h.

Day 2 *(1½h)*

Determine the mean radial extent of swarming at each temperature and at each a_w. Also note any patterns in the colonies.

↓

Select a plate showing good swarming. Take an impression of the growth by laying a slide **gently** onto the plate, ensuring that the edge of the colony is included. Without slippage, lift off the slide with forceps. Heat-fix and stain with dilute carbol fuschin for 2 min. Wash off excess stain and blot dry.

Examine microscopically and note any differences in cell morphology and density across the colony.

↓

Present the results to show the effects of temperature and a_w on swarming and comment on them. Suggest how swarming might occur.

Notes and Points to Watch

1. All plates should be poured with cooled medium (45–55 °C) to reduce condensation. They should then be dried, open at 37 °C for 1 h and used within 24 h. If these precautions are not taken, erratic results may be obtained.
2. The broth culture of *Proteus* must be fresh and show good motility.

MATERIALS

1. 12 h culture of *Proteus vulgaris* (NCIB 4175) in nutrient broth (Oxoid CM1) (*see* note 2).
2. Plates of nutrient agar (Oxoid CM3) for a_w test (*see* note 1):

	Approximate a_w
a. Nutrient agar	0·999
b. Nutrient agar + 10% (w/v) glucose	0·990
c. Nutrient agar + 15% (w/v) glucose	0·988
d. Nutrient agar + 20% (w/v) glucose	0·980
e. Nutrient agar + 30% (w/v) glucose	0·974
f. Nutrient agar + 50% (w/v) glucose	0·940

SPECIFIC REQUIREMENTS

These items are needed for each group of students.

Day 1

1 broth culture (12 h) of *Pr. vulgaris* (NCIB 4175)
2 sterile Pasteur pipettes
slides, coverslips and plasticene
15 plates of nutrient agar (for temperature experiment)
3 plates of nutrient agar a_w 0·999
3 plates of nutrient agar a_w 0·990

3 plates of nutrient agar a_w 0·988
3 plates of nutrient agar a_w 0·980
3 plates of nutrient agar a_w 0·974
3 plates of nutrient agar a_w 0·940

Day 2

wash-bottle + distilled water
blotting paper
slides (grease-free)
carbol fuschin, dilute (BDH)

FURTHER INFORMATION

Typical class results for the effect of temperature on swarming are given in Table 1. Temperature effects are usually clear-cut, with optimum swarming at 30 °C.

Table 1 Effect of temperature on swarming by *Pr. vulgaris.*

Temperature (°C)	Radial extent of swarming (mm)
4	0
18	1·3
25	7·7
30	16·3
37	7·7

Typical class results for the effect of a_w on swarming are given in Table 2. Results obtained tend to be more variable than those with temperature differences. Nevertheless they should demonstrate the need for sufficient "free water" for swarming.

Table 2 Effect of a_w on swarming by *Pr. vulgaris.*

a_w	Radial extent of swarming (mm)
0·999	15·0
0·990	6·0
0·988	1·0
0·980	0
0·974	0
0·940	0

Study of the colony radial patterns indicate the variation in cell density resulting from the cyclical stepwise progression of swarming. Cells at the colony edge ("swarmers") are about ten times (about 30 μm) the length of those near the centre.

The events occurring in swarming of *Proteus* are discussed by Jones and Park (1967a, b).

REFERENCES

H. E. Jones and R. W. A. Park (1967a). The short forms and long forms of *Proteus*. *Journal of General Microbiology*, **47**, 359–67.

H. E. Jones and R. W. A. Park (1967b). The influence of medium composition on the growth and swarming of *Proteus*. *Journal of General Microbiology*, **47**, 369–78.

63

Size Fractionation of Heterotrophic Micro-organisms from Aquatic Environments

P. J. leB. WILLIAMS

Department of Oceanography, University of Southampton, Southampton, SO9 5NH, England

> *Level*: Advanced undergraduates
> *Subject areas*: Microbial ecology
> *Special features*: The heterotrophic microbial
> activity in a water sample is
> demonstrated in a simple
> and direct manner

INTRODUCTION

Heterotrophic micro-organisms (bacteria, protozoa and fungi) play an important role in the decomposition of organic material in aquatic environments. The material may arise from endogenous sources, e.g. phytoplankton photosynthesis in the sea, or from exogenous sources such as organic pollutants. The process of decomposition (or mineralization) of the organic material is often known as **self-purification**. In an essentially closed system, e.g. the oceans or a lake, this is part of the natural cycle of the production and decomposition of organic material, i.e. photosynthesis and respiration. Figure 1 gives a scheme of the cycle in the planktonic community of a lake or ocean.

Whereas photosynthesis in planktonic environments is usually dominated by the algae, all living organisms participate in the respiration–decomposition part of the cycle. The contribution made by micro-organisms is often uncertain and is a matter of argument. Stephens (1967), for example, has demonstrated that a variety of marine animals, e.g. sponges, annelids, echinoderms, etc. may take up dissolved organic components, whereas other authors (*see* e.g. Williams, 1970; Azam and Hodson, 1977) have argued that this is mainly the prerogative of bacteria and other micro-organisms.

The aim of this experiment is to determine the relative contribution of micro-organisms of different sizes to the turnover of intermediates of the biochemical cycle, for which purpose uniformly ^{14}C-labelled glucose is used as a model compound. An analogous experiment on the size distribution of the photosynthetic population may be done with ^{14}C-labelled bicarbonate in place of the ^{14}C-glucose.

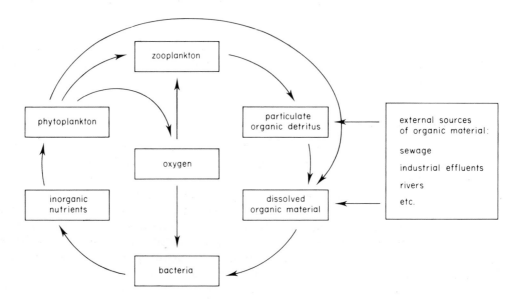

Figure 1 The planktonic biochemical cycle.

EXPERIMENTAL

Day 1 *(20 min + 60 min, separated by 2–6 h: see note 1)*

Take two 500 ml water samples (*see* note 2) from each test site in the river, lake, reservoir, etc. or from the sea.

↓

To one of the 500 ml samples from each site, add (*see* note 3) 0·1 μCi ^{14}C–glucose (*see* notes 4 and 5), mix and incubate for 2–6 h at the natural temperature and in the dark. Keep the other sample, without the isotope, to provide washing water for later use.

↓

At the end of the incubation, mix the contents of the bottle. Take 50 ml samples with a measuring cylinder and pass each through a membrane filter (*see* note 6) of different pore size, e.g. 0·22, 0·45, 1·0, 3·0, 5·0 and 10·0 μm.

↓

Wash each filter with 10 ml of the water sample to which no isotope has been added, to remove traces of unincorporated ^{14}C.

↓

Attach the filters to 10×15 cm index cards by means of staples $120°$ apart and allow them to dry.

↓

Count the radioactivity on the filters with either an end-window counter or liquid scintillation counter. Also count a known amount of the original isotope solution, e.g. 1·0 ml containing 0·1 μCi, for calibration (*see* note 7).

Notes and Points to Watch

1. The incubation time will depend upon time-tabling convenience and other factors, such as the origin of the sample. Overnight incubations are to be avoided for in many instances nearly 100% uptake may occur (which is unsatisfactory from the point of view of interpretation); additionally, it is more convincing if a large amount of isotope is taken up within a short period.
2. Ideally, the water sample(s) should be collected on the day of the experiment. Samples from different areas can usefully be compared, e.g. from the polluted and unpolluted parts of a river or from different parts of a lake or reservoir or different depths in the sea. To minimize sedimentation during the incubation it is best to strain off particles greater than about 100 μm by passage through a 100 μm plankton net attached to a filter funnel.
3. Unless the students have been trained in the use of isotopes, the addition of the ^{14}C–glucose to the sample should be made by a demonstrator.
4. As a alternative to uniformly labelled ^{14}C-glucose, ^{3}H–glucose may be used. With this compound the location of the labelling is unimportant. However, ^{3}H requires a liquid scintillation counter.
5. The labelled glucose should be filter-sterilized, distributed in suitable aliquots and stored frozen until the day of the experiment. Alternatively, the isotope solution may be made in 0·01 N HCl to prevent growth of contaminants and, without filtration, kept frozen until needed. It is not necessary to neutralize the acidity before use because most water samples have enough buffering capacity to cope with the small quantity of acid added with the isotope.
6. Membrane filters are expensive. In the UK, Oxoid filters are the cheapest, but they do not dry flat and are not very satisfactory for end-window counting. Nuclepore filters (available in the UK from Sterilin) are best, but also expensive.
7. The Koch-Light fluor Unisolv 1 is satisfactory for counting both the material collected on the filters and the stock solution of the isotope. For end-window counting, which may have advantages for class work, the stock solutions of the isotope should be dried onto a planchette over a boiling water-bath.

MATERIALS

1. Uniformly ^{14}C-labelled *D*-glucose (Radio-chemical Centre, Amersham, England; either CFB.96 or CFB.2 are suitable). It should be dissolved in distilled water to a concentration of 0·1 μCi ml^{-1} (*see* notes 4 and 5).
2. Membrane filters (47·5 or 50 mm diam.): a

selection of different pore sizes, e.g. 0·22, 0·45, 1·0, 3·0, 5·0 and 10·0 μm. Suitable membrane filters are available from Millipore, Sartorius, Oxoid and Nuclepore. The costs vary considerably from manufacturer to manufacturer (*see* note 6).

3. Filter holders, receiving vessels and suction pump.
4. Radioactivity counter, and either planchettes or scintillation vials and scintillant.

SPECIFIC REQUIREMENTS

isotope solution
water sample(s)
500 ml glass bottle(s) for incubation, sterile
water-bath (only needed if the water sample plus isotope is not incubated *in situ*)
filter holder and vacuum pump (1 set per student or group)

membrane filters of various sizes as suggested above
stapler and 10 × 15 cm filing cards
1 ml pipette
equipment for counting radioactivity

FURTHER INFORMATION

The relative amounts of radioactivity should be calculated on the filters of various pore sizes. Normally the activity in the samples is dominated by organisms of small size, but this varies with the source of the sample. From the counts on the filter, and from that determined for the isotope added, it is possible to calculate the fraction of the substrate utilized during the incubation time. The fraction utilized per day should be calculated and if possible compared with the results of samples from polluted and non-polluted environments.

I have used this technique with samples of offshore water (from the Mediterranean, N. Atlantic and English Channel) (Williams, 1970). Table 1 gives the relative size distribution of heterotrophic uptake of ^{14}C-labelled glucose and amino acids.

Similar results were obtained by Azam and Hodson (1977) using an improved filter (Nuclepore). Wheeler *et al.* (1977) have criticized the technique on the grounds that damage to large organisms could occur during filtrations and one may be measuring the radioactivity incorporated into small fragments from large organisms. Accordingly they have argued that the filtration should be done

Table 1

Size fractions (μm)	Relative uptake rate (%)
> 0·2 } > 0·45	100
> 1·2	60
> 3·0	32
> 5·0 } > 8·0	20

prior to the addition of the isotope and, in doing so, have obtained a different pattern in their results. With an advanced class of students this can be raised as a topic for discussion and the two approaches used and the results compared.

This experiment can be extended to compare the relative size distributions of the heterotrophic and autotrophic (photosynthetic) populations in an environment. The autotrophs can be studied with $NaH^{14}CO_3$ in place of ^{14}C–glucose, but the sample must be incubated in the light. Details are given in Experiment 69. Alternatively, or additionally, the chlorophyll collected on filters of differing pore sizes may be used to give an indication of the size structure of the photosynthetic population. Details of chlorophyll analysis will be found in Strickland and Parsons (1972).

REFERENCES

F. Azam and R. E. Hodson (1977). Size distribution and activity of marine microheterotrophs. *Limnology and Oceanography*, **22**, 492–501.

G. Stephens (1967). Dissolved organic material as a nutritional source for marine and estuarine invertebrates. *In* "Estuaries", (G. H. Lauff, ed.), pp. 367–73. American Association for the Advancement of Science, Washington D.C. 757 pp.

J. D. H. Strickland and T. R. Parsons (1972). "A Practical Handbook of Seawater Analysis", 310 pp. Fisheries Research Board of Canada, Bulletin 167, Ottawa.

P. Wheeler, B. North, M. Littler and G. Stephens (1977). Uptake of glycine by natural phytoplankton communities. *Limnology and Oceanography*, **22**, 900–10.

P. J. leB. Williams (1970). Heterotrophic utilization of dissolved organic compounds in the sea. *Journal of the Marine Biological Association, UK*, **50**, 859–70.

64

Experiments with Photosynthetic Bacteria

C. S. DOW

*Biological Sciences, University of Warwick,
Coventry, CV4 7AL, England*

Level: All undergraduate years
Subject areas: Bacterial photosynthesis, microbial
ecology
Special features: Isolation and characterization of
photosynthetic bacteria and
their pigments. Good visual
impact. Minimal requirements

INTRODUCTION

The photosynthetic prokaryotes have been divided into three distinct, well-defined groups: the cyanobacteria (formerly blue–green algae), the purple bacteria and the green bacteria (Table 1). Although both green and purple bacteria carry out anoxygenic photosynthesis and contain bacteriochlorophyll *a* as one of several pigments, they constitute two remarkably different groups of phototrophs with respect to their major cytological properties. For example, with the exception of *Chloroflexis*, all green bacteria are non-motile whereas the vast majority of purple bacteria are motile, exhibiting both photo- and chemotactic responses to environmental stimulii. The photopigments of the purple bacteria are located on intracytoplasmic unit membrane systems which are continuous with the cytoplasmic membrane, whereas in the green bacteria they are predominantly in chlorobium vesicles. However, all phototrophic bacteria regulate their photopigment content in response to light intensity, the specific concentrations reaching a maximum value in dim light under anaerobic conditions. The extent of this regulation is limited in the strictly anaerobic green and purple non-sulphur bacteria (Rhodospirillaceae) by either high light under anaerobic conditions or oxygen in strongly aerated cultures.

The diversity of structural and physiological organization is greatest in the Rhodospirillaceae. Since they have the capacity to thrive as anaerobic phototrophs and as facultatively micro-

---

Table 1 Structure of the photosynthetic apparatus and the mechanisms of photosynthesis

	Purple bacteria	Green bacteria	Cyanobacteria	Chloroplasts
Location of the photosynthetic apparatus	Extensions of the cytoplasmic membrane	Chlorobium vesicles	Thylacoids and phycobilisomes	Thylacoids
Photosystem I	+	+	+	+
Photosystem II	−	−	+	+
reductants used for CO_2 assimilation	H_2S, H_2 Organic compounds	H_2S, H_2	H_2O	H_2O
Principal photosynthetic carbon source	CO_2 Organic compounds	CO_2 Organic compounds	CO_2	CO_2

aerophilic to aerobic chemo-organotrophs they make the ideal choice of phototroph for undergraduate practicals.

The experiments described in this chapter are:

1. Setting up a Winogradsky column.
2. Enriching and isolating members of the Rhodospirillaceae.
3. Extraction and analysis of photopigments.

EXPERIMENT 1
Ecology of Phototrophs: the Winogradsky Column

The "Winogradsky column" beautifully illustrates the coexistence and interdependence of different ecological niches in the same habitat and it resembles, to a large extent, a natural freshwater habitat.

EXPERIMENTAL

Day 1 (*30 min*)

Take 20–25 g of surface sediment from a freshwater pond (*see* note 1) and mix it with 0·5 g $CaSO_4$ and some organic material. Use either decaying vegetation or, more conveniently, 2 sheets of finely shredded 7 cm diam. Whatman no. 1 filter paper.

↓

Put the mixture into a 500 ml glass measuring cylinder, or similar glass tube, and fill with fresh water (*see* note 2).

↓

Expose the column to natural daylight at room temperature (*see* notes 3 and 4).

Days 2–30: *when convenient (10–60 min; see note 5)*

Visually examine the column. Using long form Pasteur pipettes remove samples from the microbiologically active zones and examine these microscopically.

Notes and Points to Watch

1. The inoculum is of prime importance. Ideally this should come from a freshwater pond (almost stagnant). Take only the top 10–20 mm of sediment.
2. Do **not** use tap-water to fill the column.
3. Do **not** seal the cylinders or substitute bottles for them—there may be sufficient gas pressure generated to shatter the container.
4. Do not disturb the column unnecessarily.
5. The major drawback with this experiment is the time factor involved. This can be alleviated by setting up columns at intervals some weeks prior to the practical class.

MATERIALS

1. A suitable inoculum (*see* note 1).
2. Glass cylinders approximating to the dimensions of a 500 ml measuring cylinder.
3. Phase contrast microscope with × 100 oil immersion objective.

SPECIFIC REQUIREMENTS

These items are needed for each student or student group.

500 ml measuring cylinder or similar sized glass cylinder

Day 1
0.5 g CaSO₄
20–25 g surface sediment
2 sheets Whatman no. 1 filter paper

Days 2–30
long form Pasteur pipettes
phase contrast microscope, slides, coverslips and immersion oil

---

FURTHER INFORMATION

The microbial progression in the column is outlined below. However, to a large extent, the time taken for the described sequence will be dependent on the inoculum.

4–8 days' incubation

The organic material in the sediment is dissimilated by heterotrophs and the end-products of their fermentations are used by bacteria which use sulphate as an electron donor for anaerobic fermentation, i.e. *Desulfovibrio* will predominate and cause blackening of the sediment. The resulting H_2S is subsequently utilized by sulphur bacteria of different kinds.

In the parts of the cylinder where O_2 as well as H_2S are present, *Beggiatoa* and *Thiothrix* may develop. More often, however, motile rods of the genus *Thiobacillus* (energy derived from the oxidation of H_2S) form a whitish veil at some distance above the sediment surface. If this fails to develop it may be induced by pouring out some of the water from the cylinder and replacing with fresh water. This stirs up the mud and the water column becomes rich in H_2S. When the sediment clears, a thin veil (1–3 mm) develops and consists of actively motile *Thiobacilli*. This illustrates beautifully how localized the optimum growth conditions may be.

12–28 days' incubation

Photoautotrophic purple and green bacteria develop using H_2S as the electron donor in photosynthesis and CO_2, derived from fermentations in the sediment, as the main carbon source. These phototrophs tend to stick to the wall of the vessel. The green sulphur bacteria (chlorobiaceae) form the lowermost layer of phototrophs in the sediment underneath the purple sulphur bacteria. They do so for the following reasons: the cells are non-motile, they are obligately phototrophic and sulphide-dependent, the electron donor cannot be stored inside the cell in the form of elemental sulphur, photosynthesis reaches light saturation at lower light intensities (700 lux) than in the purple sulphur bacteria (1000–2000 lux). The uppermost limit of growth of the green sulphur bacteria, therefore, coincides with the level of permanent sulphide production that is not exhausted at the time of maximum photosynthesis. In contrast, the motile and sulphur-storing purple bacteria may adjust themselves to the diurnal changes in sulphide concentration.

There are many more niches in this habitat, but details of the majority are unknown. It is nevertheless well worthwhile dismantling the column carefully and examining each layer of the sediment microscopically, when the diversity of microbial forms will become apparent. The photosynthetic bands are particularly rewarding since they display cells in which elemental sulphur has been deposited (Figure 1).

Finally, it should be mentioned that two physical parameters greatly influence the types of microbial populations that develop, i.e. the wavelength of the light source and the temperature. The latter is the easier to manipulate. At temperatures above 33 °C, sulphate reduction is restricted, i.e. there is a much reduced level of H_2S production and the organic end-products of the fermentative processes occuring in the sediment accumulate. This results in the purple non-sulphur bacteria (Rhodospirillaceae) developing since they can utilize organic substrates as electron donors for photosynthesis and they tolerate micro-aerophilic conditions.

EXPERIMENT 2
Enrichment and Isolation of Members of the Rhodospirillaceae

INTRODUCTION

The cellular organization of the purple bacteria reaches its greatest structural and metabolic diversity in the purple non-sulphur bacteria (Rhodospirillaceae), e.g. *Rhodospirillum rubrum*,

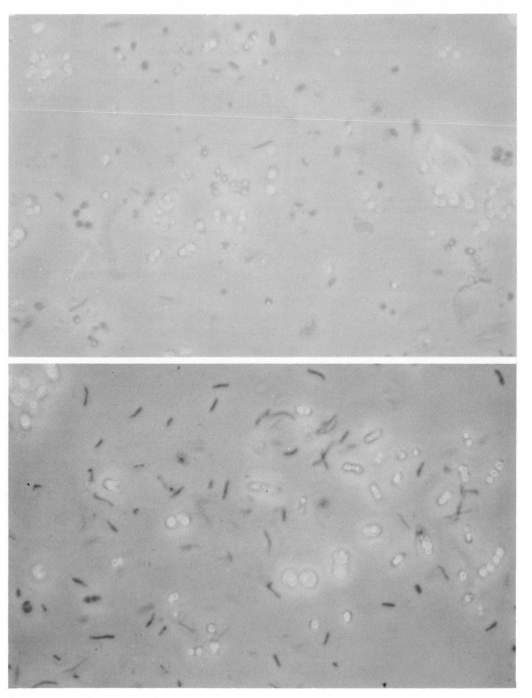

Figre 1 Phase contrast photomicrographs of samples taken from the purple and green zones of the Winogradsky column. Note the microbial diversity and the abundance of elemental S inclusions.

Rhodopseudomonas palustris and *Rhodomicrobium vannielii*. The capacity of most species to grow not only phototrophically in anaerobic culture, but also chemo-organotrophically when cultured aerobically in the dark facilitates purification. The distinctive morphologies facilitate identification.

EXPERIMENTAL

Day 1 *(20 min)*

Put approximately 0·5 g sediment (*see* note 1) into a 20 ml bottle (McCartney bottles are ideal) and fill it with malate yeast extract medium. Cap the bottle and incubate it at 30 °C with an incident light intensity of 1000–2000 lux (*see* notes 2 and 3).

Days 7–10 *(30–45 min)*

Look for red–brown pigmentation in the enrichment cultures. This signifies the growth of phototrophs. Examine the enrichments microscopically and tentatively identify the organisms (*see* Figure 3 and Table 2 in Further Information).

To obtain pure cultures make 6 serial ten-fold dilutions of the enrichments. Treat these dilutions as outlined below.

↓	↓	↓
Agar deeps	*Chemo-organotrophic growth*	*Phototrophic growth*
Add 0·1 ml of appropriate dilution to 10 ml molten malate yeast extract agar. Mix gently (*see* note 4) and allow to set. Cap with 2–3 ml agar and incubate at 30 °C under tungsten lamps (*see* note 2).	Spread 0·1 ml of appropriate dilutions onto malate yeast extract plates. Incubate at 30 °C **aerobically** in the **dark**.	Prepare spread plates of the dilution series. Place in glass or perspex anaerobic jars. Anaerobiosis is achieved by using a Gas Pak H_2–CO_2 generating system. Silica gel **must** be included to minimize condensation. Incubate at 30 °C in the light (*see* note 2).

Days 15–16 *(15 min)*	**Days 17–18** *(15 min)*	**Days 17–20** *(15 min)*
Use the red–brown pigmented colonies to inoculate liquid medium.	Colonies of phototrophic bacteria are easily recognized as many have pigmented centres. Purify by streak plating and incubating as above.	Purify pigmented colonies by streak plating.

Notes and Points to Watch

1. The inoculum should be sediment from a eutrophic pond or a ditch high in vegetative matter.
2. An incident light intensity of 1000–2000 lux is supplied by tungsten lamps (60 W) at a distance of approximately 1 m from the cultures. Care must be taken to ensure that the cultures do not become overheated and if possible a fan should be used to dissipate the unwanted heat.
3. Cultures to be grown phototrophically in liquid medium must be under micro-aerophilic/anaerobic conditions. This can be achieved either by gassing with oxygen-free nitrogen or by filling the bottle almost completely.

   In all instances, take care to ensure when using sealed vessels that the gas pressure does not become excessive.
4. To avoid introducing air bubbles, mixing is best done by rolling the tubes between the hands.
5. Preferential selection of some photosynthetic bacteria is possible by altering the pH of the medium, e.g.

pH 5·0–6·0	selects	*Rhodopseudomonas acidophila*
	and	*Rhodomicrobium vannielii*,
whereas pH7·0	selects	*Rhodospirillum rubrum*
	and	*Rhodopseudomonas palustris.*

MATERIALS

1. Malate yeast extract medium contains

NH_4Cl,	0·5 g
$MgSO_4 \cdot 7H_2O$,	0·4 g
$CaCl_2 \cdot 2H_2O$,	0·05 g
NaCl,	0·4 g
sodium hydrogen malate,	1·5 g
yeast extract,	0·1 g

per litre of distilled water.

The pH is adjusted with KOH prior to autoclaving. After sterilization, 5 ml phosphate buffer of the appropriate pH (*see* note 5) is added aseptically to each 100 ml of medium.

2. Phosphate buffers are prepared from mixtures of 0·1 M $Na_2HPO_4 \cdot 12H_2O$ (35·85 g litre^{-1}) and 0·1 M $NaH_2PO_4 \cdot 2H_2O$ (15·6 g litre^{-1}) in proportions which give the desired pH (*see* note 5).
3. Suitable inocula (*see* note 1).
4. Glass or perspex anaerobic jars and Gas Pak H_2–CO_2 generation systems can be obtained from Becton, Dickinson UK Ltd; York House, Empire Way, Wembley, Middlesex.
5. Phase contrast microscope with × 100 oil immersion objective.

SPECIFIC REQUIREMENTS

Day 1
sediment (*see* note 1)
McCartney bottles
150 ml malate yeast extract
tungsten lamps

Days 7–10
phase contrast microscope
microscope slides, coverslips and immersion oil
Pasteur pipettes
dilution tubes

cont.

cont.

0·1, 1·0, 10 ml pipettes
200 ml phosphate buffer
Agar deeps:
   molten malate yeast extract agar
   sterile test tubes
   tube rack
   water-bath at 50°C
Chemo-organotrophic growth:
   malate yeast extract plates
   glass spreader
   alcohol
Phototrophic growth:
   malate yeast extract plates

glass spreader
alcohol
glass or perspex anaerobic jar
silica gel
Gas-Pak

Days 15–20
malate yeast extract medium in 20 ml
   McCartney bottles
malate yeast extract plates

FURTHER INFORMATION

The purple non-sulphur bacteria are able to assimilate a wide variety of organic compounds, e.g. acetate, pyruvate, the dicarboxylic acids, as well as methanol, ethanol, sugars and sugar alcohols. Complex organic nitrogen sources (yeast extract) give increased growth rates of many species particularly since many have a requirement for one or several vitamins. This is in good agreement with their photo-organotrophic nature and their presence in habitats dominated by the active breakdown of organic matter. However, their inability to break down organic macromolecules, e.g. starch, cellulose, lignin and proteins means that, in natural habitats, they are dependent on the preceding activity of chemo-organotrophic bacteria capable of such degradation. This in part explains why blooms of Rhodospirillaceae are never encountered in nature and why they usually are associated with chemo-organotrophs. The most suitable inocula for enrichments, therefore, comes from situations where the active breakdown of plant residues results in anaerobic conditions and the release of low molecular weight breakdown products. Such environmental conditions are present in eutrophic (nutrient rich) ponds and ditches.

The data presented in Figure 2 and Table 2 should permit the identification of all isolates.

EXPERIMENT 3
Analysis of Photopigments

INTRODUCTION

The spectrum of light absorbed by phototrophic prokaryotes in the natural environment is such that competition for a limited resource is minimized, i.e. cyanobacteria absorb light of a wavelength below 690 nm, the green bacteria between 700 and 760 nm and the purple bacteria above 800 nm.

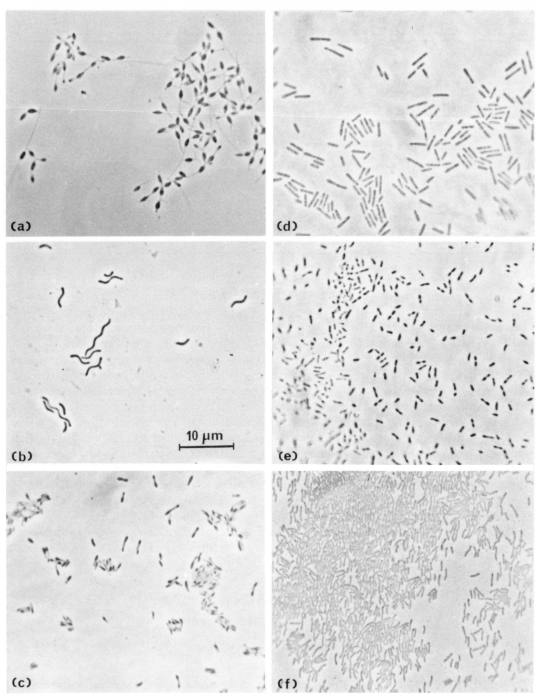

Figure 2 Photomicrographs of species of Rhodospirillaceae. (a) *Rhodomicrobium vannielii*. (b) *Rhodospirillum rubrum*. (c) *Rhodopseudomonas palustris*. (d) *Rh. sphaeroides*. (e) *R. blastica*. (f) *R. gelatinosa*.

Table 2 Properties of the Rhodospirillaceae.

Species	Cell shape (μm)	Motility	Membrane system	Colour anaerobic	Aerobic, micro-aerophilic growth in the dark	Growth factors
Rhodocyclus purpureus	half circle/circle 0·6/2·7–5	—	tubes	purple–violet	m	B_{12} biotin *paba*[a]
Rhodomicrobium vannielii	polymorphic-ovoid and stalked 1·0/2–2·8	swarm cell	lamellae	orange–brown	m	none
Rhodopseudomonas acidophila	dimorphic-rod 1·0/2–5	swarm cell	lamellae	purple–red or orange–brown	m,ae	none
R. capsulata	rod/sphere 0·5–1·2/2–2·5	+	vesicles	yellow to brown	ae	thiamine ± biotin
R. gelatinosa	rod 0·5/1–2	+	tubes	yellow brown to pinkish	ae	thiamine biotin
R. globiformis	sphere	+	vesicles	purple–red	m	biotin *paba*
R. palustris	dimorphic rod 0·6–0·9/1·2–2	swarm cell	lamellae	red–brown	ae	*paba* ± biotin
R. sphaeroides	sphere/ovoid 0·7/2–2·5	+	vesicles	green–brown to brown	ae	biotin thiamine niacin
R. sulphidophila	rod/sphere 0·6–0·9/0·9–2·0	+	vesicles	yellow–brown to red	ae	biotin + thiamine + niacin + *paba*
R. sulfoviridis	rod 0·5–0·9/1·2–2	+	lamellae	olive–green	m	biotin + pyridoxine + *paba*
R. viridis	dimorphic	swarm cell	lamellae	green	m	*paba* + biotin
Rhodospirillum fulvum	spiral 0·5–0·7/3·5	+	stacks	brown	m	*paba*

cont.

Table 2 *cont.*

Species	Cell shape (μm)	Motility	Membrane system	Colour anaerobic	Aerobic, micro-aerophilic growth in the dark	Growth factors
R. molischianum	spiral 0·7–0·1/5–8	+	stacks	brown	m	amino acids
R. photometricum	spiral 1·2–1·5/ 7–10	+	stacks	brown	m	yeast extract
R. rubrum	spiral 0·8–1·0/ 7–10	+	vesicles	red	ae	biotin
R. tenue	spiral 0·3–0·5/3–6	+	tubes	purple– violet or brown– orange	ae	none
Rhodopseudo- monas blastica[a]	dimorphic	−	lamellae	orange– brown	ae	nicotinic acid thiamine

[a] *p*aba: *p* amino benzoic acid.
[b] Eckersley and Dow (1980). Journal of *General Microbiology*. *Rhodopseudomonas blastica* sp. nov.: a member of the Rhodospirillaceae.

The absorption characteristics of the various bacteriochlorophylls are given in Table 3 (*see* Further Information).

The great variety of carotenoids possessed by the phototrophs presumably function as secondary light-harvesting pigments and to prevent photo-oxidation within the cells. They have been ordered into five major groupings (these groupings, the major components and the representative species are given in Table 4, in Further Information).

The main aim of this experiment is to introduce the student to the photobiology of micro-organisms via the isolation and preliminary characterization of the chlorophylls and carotenoids.

EXPERIMENTAL

Day 1 *(part 1, 60 min; part 2, 30 min)*

1. *Absorption spectrum of intact cells*

Centrifuge 20 ml of an exponentially growing phototrophic culture and resuspend the pellet in 3 ml saturated sucrose.

↓

Obtain the absorption spectrum by measuring the absorbance at regular intervals from 300 to 1000 nm using saturated sucrose as the blank (*see* note 1).

↓

The principal bacteriochlorophyll present can be identified from the absorption spectrum (*see* Table 3 and Figure 3 in Further Information).

2. *Extraction and analysis of photopigments*

Take 500 ml of phototrophic culture in the late exponential phase of growth (*see* note 2).

↓

Harvest by centrifuging at 5000 g for 10 min.

↓

Resuspend the pellet in 20 ml of acetone:methanol (7:2 v/v) at 4 °C and leave for a minimum of 4 h but preferably overnight. **Keep the solution in the dark**.

Day 2 *(2–2½ h)*

2. *Extraction and analysis of photopigments*

Centrifuge the acetone/methanol extract at 5000 g for 20 min to remove the cellular debris.

↓

Put the supernate into a separating funnel and add an equal volume of dimethyl ether (**caution**!).

↓

Shake the solution vigorously for 10 min and then allow the two phases to separate (*see* note 3).

↓

Discard the aqueous phase and retain the ether phase which contains the carotenoids. **Keep this fraction in the dark at 4 °C.**

↓

To facilitate thin-layer chromatography (TLC) reduce the volume of the extract by blowing off excess solvent with oxygen-free N_2 (*see* note 4).

↓

Using a capillary tube, spot 10–20 μl of sample onto a pre-dried TLC plate (*see* note 5). The spot should be kept as small as possible.

↓

Put 5–10 ml of the chosen solvent system (*see* note 6) into a 250 ml beaker with a glass Petri dish lid and allow the vapour phase to equilibrate.

↓

Place the TLC plate into the beaker (*see* note 7).

↓

Allow the solvent front to run 3/4 of the way up the TLC plate.

↓

Remove the chromatogram and make a tracing of the resolved carotenoids. Also note the colour of each carotenoid (*see* notes 8 and 9).

_____

Notes and Points to Watch

1. A recording spectrophotometer having the required wavelength span is ideal. If this is not available readings should be taken every 20 nm, closer when considering the carotenoid absorption in the 300–450 nm range.
2. The culture to be used for pigment extraction should be grown under low light, i.e. 1000 lux to enhance the photopigment content per cell.
3. If two phases do not readily appear, 5 ml of distilled water should be added to aid separation.
4. **The solvents are highly inflammable, ensure there are no naked flames in the laboratory.**
5. Ideally, pre-coated plates should be used. Alternatively any similar sized glass plate can be coated with silica gel to a thickness of 0·25–0·4 mm. Although giving reduced resolution, glass microscope slides may be substituted.
6. The following solvent systems are recommended:
    1. ethyl acetate:petroleum ether 1:10 (v/v);
    2. hexane:diethyl ether:glacial acetic 35:15:5 (v/v);
    3. acetone:petroleum ether 1:9 (v/v).
7. The solvent system should only be deep enough to wet the bottom of the TLC plate and on no account should it reach the sample.
8. Tracings of the resolved carotenoids must be carried out immediately as some spots fade very rapidly.
9. If required the R_F values can be calculated and compared with a convenient standard, e.g. β carotene.

MATERIALS

1. Spectrophotometer (*see* note 1).
2. Bench top centrifuge.
3. Oxygen-free N_2.
4. Capillary tubes.

5. Pre-dried TLC plates (*see* note 5), e.g. silica gel 60 TLC plates as supplied by BDH Chemicals Ltd., Poole, England.

SPECIFIC REQUIREMENTS

Day 1
20 ml exponential phototrophic culture
bench top centrifuge
10 ml saturated sucrose
2 glass cuvettes
spectrophotometer (*see* note 1)
500 ml late exponential phototrophic culture
  (*see* note 2)
20 ml acetone: methanol (7:2 v/v)

Day 2
bench top centrifuge
separating funnel
oxygen-free N_2
capillary tubes
TLC plates
solvent systems (see note 6)
250 ml beaker with glass Petri dish lid

FURTHER INFORMATION

The bacteriochlorophylls of the green and purple bacteria may be identified from the absorption spectra of living cells (Table 3 and Figure 3).

Table 3 Characteristic absorption maxima and bacteriochlorophyll designation of living cells of the purple and green bacteria.

Designation	Characteristic absorption maxima of living cells (nm)			
Bacteriochlorophyll *a*	375	590	800–810	830–890
b	400	605	835–850	1015–1035
c	335	460	745–760	812
d	325	450	725–745	805
e	345	450–460	715–725	805

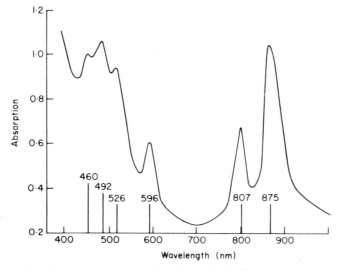

Figure 3 Absorption spectrum of viable cells of *Rhodomicrobium vannielii*.

The purple bacteria contain either bacteriochlorophyll *a* or bacteriochlorophyll *b* as the light-harvesting and photochemical reaction centres. The green bacteria contain a major and minor form. The major form *c*, *d*, or *e* serves as the light harvesting pigment, whereas the photochemical reaction centre is always *a*.

The carotenoids of the purple and green photosynthetic bacteria fall into five major groups (Table 4), each of which is comprised of certain major and characteristic carotenoids.

Table 4 Carotenoid groups of the phototrophic green and purple bacteria.

Group	Designation	Major components	Species
1	Normal spirilloxanthin series	lycopene, rhodopin, spirilloxanthin	*Rhodospirillum rubrum* *Rhodomicrobium vannielii*
2	Rhodopinal branch of spirilloxanthin series	lycopenal, rhodopinal	*Rhodospirillum tenue*
3	Alternative spirilloxanthin series	hydroxyneurosporene, spheroidene, spheroidenane	*Rhodopseudomonas sphaeroides* *R. capsulata* *R. gelatinosa*
4	Okenone series	Okenone	*Chromatium* *Rhodopseudomonas globiformis*
5	Isorenieratene series	β carotene, isorenieratene, γ carotene, chlorobactene	*Chlorobium*

REFERENCE

R. K. Clayton and W. R. Sistrom (1978). "The Photosynthetic Bacteria". Plenum Press, New York.

65

The Isolation and Characterization of Bacteriophages from River Water

S. B. PRIMROSE, N. D. SEELEY and K. B. LOGAN

*Department of Biological Sciences, University of Warwick,
Coventry, CV4 7AL, England*

> *Level*: All undergraduate years
> *Subject area*: Virology

INTRODUCTION

Nearly every species of bacterium investigatd so far has been found to serve as a host to one or more viruses (bacteriophages). Since there are very few fundamental differences between the viruses infecting plants, animals or bacteria (*see* Primrose and Dimmock, 1980, for discussion of this point) the study of bacteriophages offers an easy introduction to basic virology. The aims of the experiments described here are (1) to isolate coliphages from river water, (2) to prepare high titre isolates, (3) to determine, by electron microscopy, the morphology of the phages and (4) to determine their host range.

It is sometimes possible to isolate coliphages by direct plating of water samples, but most workers prefer to start with an enrichment step. There are two disadvantages to enrichment. First, separate enrichments have to be set up with each bacterial host strain. Second, even when more than one phage for a particular host strain is present in the initial inoculum, usually only one phage is preferentially enriched. An alternative procedure is to concentrate the phages from a large volume of water (1–100 litres). The advantage of phage concentrates is that they can be tested on a wide range of bacteria. When only a single host strain is used it is possible to obtain numerous different phage isolates from a single concentrate. If the same concentrate is plated on a range of different bacteria it is possible to isolate a wide range of phages suitable for phage typing.

A number of methods for concentrating bacteriophages have been described and these have been assessed by Seeley and Primrose (1979). For class purposes the best method is adsorption to hydroxyapatite since it is rapid, gives good recoveries (> 50 %) and requires no special apparatus (Primrose and Day, 1977).

Viruses, whether bacterial, plant or animal, carry a net negative charge and bind readily to the positively charged calcium ions in hydroxyapatite, which is a particular crystalline form of calcium phosphate. By subsequently applying high molarity phosphate buffers to the hydroxyapatite it is possible to elute the bound virus. Elution by phosphate appears to be due to a specific competition between phosphate ions of the buffer and carboxylate ions in the protein coat of the virus, for adsorbing sites on the hydroxyapatite, and not simply to an increase in ionic strength.

EXPERIMENTAL

Day 1 *(1 h, see note 1)*

Clarify 1 litre of river water by centrifugation at 5000 *g* for 10–15 min. Carefully decant the supernate fluid and filter it through a 1 cm bed of hydroxyapatite in a 10–12 cm diam. Buchner filter (*see* note 2).

↓

Discard the filtrate. Using a spatula scrape the hydroxyapatite out of the Buchner funnel, transfer to a Universal bottle and resuspend it in an equal volume of 0·8 M sodium phosphate, pH 7·2.

↓

Shake the suspension vigorously for 15–30 s. Pellet the hydroxyapatite by centrifugation at low speed for 5 min. Keep the supernate fluid containing virus particals, eluted off the hydroxyapatite, and discard the pellet (*see* note 3).

↓

Assay the virus concentrate, and a 100-fold dilution of the concentrate, for bacteriophages as follows.

↓

Dispense 2·5 ml soft agar into sterile test tubes held at 45 °C (*see* note 4). Add 0·2 ml of an overnight broth culture of *Escherichia coli* or other coliform (*see* note 5). Add 0·1 ml of virus-containing sample. Immediately mix the contents of the tube either by vortexing or by vigorously rolling the tube between the palms of the hands. Quickly, i.e. before the agar sets, pour the contents of the tube over the surface of a dry nutrient agar plate. Allow the agar to set for 10 min (*see* note 6).

↓

Incubate the plates overnight at 37 °C. Also inoculate 10 ml of nutrient broth with 0·1 ml of the host bacterium used for the assay (the "indicator" bacteria) and incubate, either statically or shaken, overnight at 37 °C (*see* note 7).

- -

Day 2 *(20 min)*

Examine the plates for bacteriophage plaques. Estimate the number of different plaque morphologies which are visible. Count the number of plaques on the plates and determine the number of phage particles per ml of the original water sample that were capable of attacking the particular indicator organisms used. For this purpose assume 100% recovery of the bacteriophages.

↓

Select a well-isolated plaque, preferably a large, clear one, and stab it cleanly through the middle with a sterile wire or Pasteur pipette. Shake the wire stab (or Pasteur pipette) in 1 ml sterile nutrient broth. Make 3 × 10-fold broth dilutions of this phage preparation. Assay 0·1 ml portions of each dilution for bacteriophages by the method described on Day 1. Use as indicator bacteria the culture inoculated on Day 1.

↓

Incubate the plates overnight at 37 °C. Also inoculate 10 ml of nutrient broth with 0·1 ml of indicator bacteria and incubate, either statically or shaken, at 37 °C (*see* note 7).

- -

Day 3 *(5 min + 3 h, separated by 2 h)*

Inoculate 50 ml phage broth with 2 ml of the overnight culture of indicator bacteria. Incubate with shaking at 37 °C for 2 h.

↓

"Punch out" a phage plaque with a sterile Pasteur pipette and transfer it to the 50 ml culture of indicator bacteria. Incubate at 37 °C with vigorous shaking. Examine at 20–30 min intervals for lysis.

↓

When lysis is complete, centrifuge the culture at low speed to pellet cell debris and intact cells. Decant the supernate fluid, add 0·5 ml chloroform to it to kill any residual cells, and store at 4 °C. This sample constitutes a high titre preparation of the selected bacteriophage. Its titre should be determined by making 5 serial 100-fold dilutions and assaying, as before, 0·1 ml portions of the last three dilutions.

↓

With a cotton swab, make a single streak of the high titre phage preparation across the surface of a nutrient agar plate. Then, using an inoculating loop, make streaks of a variety of coliforms (*see* note 5) at right angles to the phage streak (*see* Figure 1). Incubate the plates at 37 °C overnight.

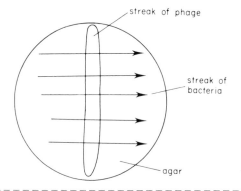

streak of phage

streak of bacteria

agar

Figure 1 Method for testing the sensitivity of different bacteria to a new phage isolate.

- -

Day 4 (*1 h, but* see *note 8*)

Determine the titre of the liquid lysate prepared on Day 3. Record the sensitivity of the various coliforms to the selected bacteriophage. Compare the phage host range with that found by other members of the class with their bacteriophages. Use the pooled data to construct a phage typing scheme.

↓

Prepare a sample of the bacteriophage for examination in the electron microscope. The starting material is the high titre preparation made on Day 3. If a suitable lysate was not obtained a plaque can be used.

↓   ↓

If starting with a bacteriophage plaque (*see* note 8):

↓

carefully remove the capillary part of a Pasteur pipette and use it to stab a well-isolated plaque. Avoid touching the bacterial lawn.

↓

Transfer the agar plug to a 0·75 ml plastic micro-centrifuge tube at 4 °C for 1 h to allow phage to elute from plug.

↓

Remove 5 μl of buffer from the tube and transfer to a Formvar coated electron microscope grid. Leave for 1 min. Return tube to 4 °C (*see* note 10).

↓

If starting with a high titre lysate:

put a drop of high titre phage suspension onto a Formvar coated grid. Leave for 1 min.

↓

Suck off the liquid from the grid by gently touching it with a piece of filter paper. **Immediately** add 4 μl of fresh 1% phosphotungstic acid, pH 7·0 (*see* note 9). Immediately suck off the phosphotungstic acid. Allow the grid to air dry and examine in the electron microscope.

Notes and Points to Watch

1. The 60 min required on Day 1 includes the time required for filtration of the river water. For the whole class only one sample has to be filtered and if this were done beforehand by the instructor, only 20 min would be required.

2. To prevent the hydroxyapatite washing through the pores in the Buchner funnel it is essential to layer it on top of a sheet of coarse filter paper.

3. Any animal viruses present in the river water will be concentrated along with the bacteriophages. All concentrates should be handled carefully and certainly should **not** be mouth pipetted.

4. Water-baths set at 45 °C are suitable. A more convenient alternative is a "Dri-Block" consisting of an electrically heated aluminium block with holes drilled in it to take test tubes. Suitable models can be obtained from Grant Instruments Ltd., Barrington, Cambridge, England. The ideal size of test tube is 100×12 mm.

5. In order to select a wide range of bacteriophages it is best if each student is given a different strain of *E. coli* or other coliform. The sensitivity of all the different strains to each phage isolate can then be determined on Day 3.

6. There are four common reasons for failures with the plaque assay method. First, if too much time is spent in mixing the phage and bacteria with the agar, the agar will set before it is poured over the nutrient agar in the Petri dish. Second, freshly prepared nutrient agar plates are too moist to give a satisfactory plaque assay with the result that all the plaques run together. The plates can be dried by incubation at 37 °C for 18 h with their lids on or 30 min with their lids off. Plates prepared 3–8 days beforehand are just as good. Third, if the nutrient agar plates are poured on a surface which is not level, or if the top agar is added when the plates are not level, all the top agar will run to one half of the dish. This gives results which are difficult to interpret. Fourth, if the bacterial host cells are added to the top agar and left at 45 °C for periods greater than 10–15 min, poor lawns will result due to death of a high proportion of the bacteria.

7. If more than one indicator organism is being used by the class it is essential that each student maintain his/her own culture. If another host is inadvertently used on Days 2 or 3 there is a chance that it will not be susceptible to the phage isolated on Day 1.

8. If a plaque is the source of phage for electron microscopy an additional hour will be required.

9. It is essential that the phosphotungstic acid is freshly prepared and adjusted to pH 7 with potassium hydroxide. Sodium hydroxide must **not** be used.

10. If, after electron microscopy, a particular phage is required for further study the material left in the tube can be used for a plaque assay and hence generate new stocks of phage.

MATERIALS

1. Selection of *Escherichia coli, Klebsiella, Proteus, Aeromonas* and *Pseudomonas* strains.

2. Phage broth is nutrient broth containing $CaCl_2$ (0.22 g l^{-1}).

3. Phage buffer contains per litre, 7 g Na_2HPO_4 (anhydrous), 3 g KH_2PO_4 (anhydrous), 5 g NaCl, 10 ml $CaCl_2$ (0.01 M) and 10 ml $MgSO_4 \cdot 7H_2O$ (0.1 M).

4. Access to electron microscope.

5. Hydroxyapatite can be obtained from Bio Rad Laboratories Ltd., Porter's Wood, Valley Road, St. Albans, Hertfordshire, England.

6. Soft agar is nutrient broth containing 8 g agar l^{-1}.

SPECIFIC REQUIREMENTS

These items are needed for each group of students unless otherwise stated.

Day 1
1 litre river water (per class)
Buchner filter funnel and flask (per class)
filter paper to fit Buchner funnel (per class)
hydroxyapatite (per class)
spatula (per class)
Universal bottle, or other container to fit bench centrifuge (per class)
0·8 M phosphate buffer, pH 7·2, approx. 20 ml (per class)
45 °C water-bath with test tube rack (per class)
37 °C incubator (per class)
2 sterile test tubes, 100 × 12 mm
1 or 2 sterile test tubes (for dilution of virus concentrate)
rack for dilution tubes
soft agar, 5 ml
overnight bacterial culture, 1 ml
2 × 10 ml nutrient broth
2 nutrient agar plates
sterile pipettes
pipetting aids (see note 3)

Day 2
45 °C water-bath with test tube rack (per class)
37 °C incubator (per class)
4 sterile test tubes, 100 × 12 mm

4 dilution tubes
rack for dilution tubes
2 × 10 ml nutrient broth
10 ml soft agar
4 nutrient agar plates
sterile pipettes

Day 3
37 °C incubator (per class)
45 °C water-bath with test tube rack (per class)
50 ml phage broth
chloroform, 1 ml
5 sterile dilution tubes
rack for dilution tubes
5 sterile tubes, 100 × 12 mm
15 ml soft agar
6 nutrient agar plates
1 sterile cotton swab
sterile pipettes
50 ml nutrient broth

Day 4
Formvar-coated grids (allow 3 or 4 per student group)
filter paper
1 % phosphotungstic acid, pH 7·0 (1 ml freshly prepared solution: see note 9)
Pasteur pipettes
micro-centrifuge tube

FURTHER INFORMATION

Numerous plaques should be evident on most of the plates of the undiluted virus concentrate. It is rare not to get at least a few plaques for any particular strain. On many of the plates more than one plaque type should be present. After plaque purification most of the phage isolates should exhibit only a single plaque morphology. Sometimes the plaque morphology does not breed true, but more often than not this is due to failure on Day 2 to select a well-isolated plaque. The preparation of high titre liquid lysates is more of an art than a science and on Day 3 probably no more than 50 % of the cultures will lyse. However, high titre liquid lysates are not essential for the preparation of phage for electron microscopy and an alternative method is given. Most, if not all, of the phages isolated will be of the tailed type, but the length and morphology of the tail will differ considerably between isolates.

It is possible to attempt a one-step growth experiment with any of the phage isolates, but this is not recommended unless you have a small class of advanced students. Not only does such an experiment consume an extensive amount of sterile glassware, particularly pipettes, but it is prone to failure when more than 5 or 6 groups of students are involved. The most likely explanation for this is that so many samples have to be taken at the right time, diluted, and assayed for phage that errors are almost unavoidable unless the students fully understand what they are doing. A more practical alternative is to ask the students to present a detailed flow sheet of the method they would use to determine the kinetics of phage multiplication and then provide them with raw data for analysis.

An interesting variation on the phage isolation procedure is to assay the virus concentrate at 20 and 37 °C using the same bacterial host strain, and to calculate the ratio

$$\frac{\text{no. of plaques at } 20\,^{\circ}\text{C}}{\text{no. of plaques at } 37\,^{\circ}\text{C}}.$$

For relatively unpolluted rivers the ratio will be greater than 1 and for polluted rivers the ratio will be much smaller than 1. The degree of faecal pollution can be confirmed by conventional bacteriological analysis. For further details of the significance of these results, Seeley and Primrose (1980) should be consulted.

REFERENCES

S. B. Primrose and M. Day (1977). Rapid concentration of bacteriophages from aquatic habitats. *Journal of Applied Bacteriology*, **42**, 417–21.

S. B. Primrose and N. J. Dimmock (1980). "An Introduction to Modern Virology", 2nd edn. Blackwell Scientific Publications, Oxford.

N. D. Seeley and S. B. Primrose (1979). Concentration of bacteriophages from natural waters. *Journal of Applied Bacteriology*, **40**, 103–16.

N. D. Seeley and S. B. Primrose (1980). The effect of temperature on the ecology of aquatic bacteriophages. *Journal of General Virology*, **46**, 87–95.

66

Enzyme Activities in Soil

R. G. BURNS

*Biological Laboratory, University of Kent,
Canterbury, Kent, CT2 7NJ, England*

> *Level*: All undergraduate years
> *Subject areas*: Microbial ecology, soil biochemistry
> *Special features*: Development of manual dexterity
> and basic biochemical skills

INTRODUCTION

The breakdown of plant, animal and microbial debris in soil is a vital stage in the cycling of elements in the biosphere. Viable micro-organisms play a major role in this breakdown, frequently by excreting enzymes which attack high molecular weight exogenous substrates thus rendering them suitable for uptake and metabolism. In addition, many enzymes located in the cytoplasm or periplasm of bacteria retain a proportion of their activity long after their release from dead and leaking cells. Whatever their origins, many of these extracellular catalysts are extremely resistant to denaturation and degradation, probably because they have formed an intimate association with the recalcitrant humic colloids. The net result is that soils have an indigenous, steady-state level of enzyme activity that is independent of immediate microbial proliferation (Skujins, 1976; Ladd, 1978).

The contribution of an "accumulated" enzyme to the total turnover of a particular substrate is difficult to assess and yet it has been described (e.g. in the case of urea hydrolysis) as the most important single factor (Kiss *et al.*, 1975; Burns, 1977).

The aims of this experiment are two-fold: to measure the activities of two important soil enzymes in the absence of microbial growth; and to discover if there is any obvious relationship between microbial numbers and activity.

EXPERIMENTAL

A titrimetric method for measuring urease (E.C.3.4.1.5) and a spectrophotometric method for measuring phosphatases (E.C.3.1.3.1. and 3.1.3.2) are given below. The urease assay relies upon the hydrolysis of its natural substrate, urea, and the absorbance of the resulting NH_3 in boric acid. The use of naturally occurring organic phosphorus substrates for phosphatase assays involves lengthy incubation periods and concomitant problems with microbial proliferation. It is therefore common to use a rapidly hydrolysed synthetic organic substrate (e.g. *p*-nitrophenylphosphate) and to measure its product (*p*-nitrophenol).

The experimenter should be aware that the measurements of enzyme activity relate to the almost ideal conditions of the *in vitro* test rather than to substrate turnover *in situ*. In the latter instance, conditions may be far from optimal (i.e. the substrate may be discontinuous in time and space, and the temperature, pH, level of hydration, etc. may be unsuitable).

Differences in soil type—especially differences in texture (i.e. % sand, silt, clay) and organic matter content—will be associated with different activities. Different soils may also contain different numbers of micro-organisms (total count) and different numbers of the relevant enzyme-producing micro-organisms (e.g. total ureolytic microbes). By examining various soil types one can get an indication of the correlation (if any) between microbial numbers and activity; and furthermore, if enzyme measurements are useful indicators of soil microbial biomass.

Urease Assay

Day 1 *(2 h, including 1 h incubation period)*

Weigh 1 g soil directly into the outer well of a Conway diffusion dish (CDD) (*see* note 1).

↓

Add 3 ml tris–maleate (t–m) buffer (0·5 M, pH 7·0) to the outer well of the CDD.

↓

Pipette 3 ml boric acid-indicator solution into the inner well of the CDD.

↓

Initiate the reaction by adding 1 ml urea (6 M in t–m buffer) (*see* note 2) to the outer well of the CDD. Stir **gently** and replace the lid (*see* note 3).

↓

Incubate at room temperature (c. 20 °C) for 60 min.

↓

Terminate the reaction by adding 0·5 ml $AgSO_4$ (10 mM). Stir **gently**.

↓

Release ammonia by adding 1 ml K_2CO_3 (3 M).

↓

Replace the lid and allow 16–20 h for absorption of NH_3 by boric acid.

Day 2 *(30 min, 25 h later)*

> Back titrate directly into the CDD using HCl (0·2 M, 0·02 M) *(see* note 4).
>
> ↓
>
> Express the activity *(see* note 5) as μmol NH_3 evolved g^{-1} dry soil h^{-1}.

Ureolytic Micro-organisms

Day 1 *(1 h)*

Prepare a series of soil dilutions in sterile distilled water (suggested 10^{-5}, 10^{-6}, 10^{-7}, 10^{-8}).

> ↓
>
> Prepare 5 spread plates *(see* note 6) of each dilution.
>
> ↓
>
> Incubate at 25 °C for 48–72 h.

Day 2 *(1 h, 3 or 4 days later)*

> Count colonies *(see* note 7): choose a dilution with >30 and <300.
>
> ↓

Replicate each plate of the chosen dilution onto Christensen's urea agar (5 plates in all).

> ↓
>
> Incubate at 25 °C for 48 h.

Day 3 *(30 min, 4 or 5 days later)*

Count number of colonies producing a purple halo *(see* note 8).

Phosphatase Assay

Day 1 *(3 h)*

> Weigh 1 g soil into a Universal bottle *(see* note 9).
>
> ↓
>
> Add 4 ml of 0·5 M buffers of pH 4·0, 6·0 or 8·0 *(see* note 10).
>
> ↓
>
> Add 1 ml 100 mM *p*-nitrophenylphosphate.
>
> ↓

Incubate on a shaker, in the dark, at 25 °C for 60 min (*see* note 11).

↓

Terminate the reaction by adding 1 ml 0·5 M CaCl₂ and 4 ml 0·5 M NaOH.

↓

Remove the soil by centrifugation (*c.* 200 g, 10 min).

↓

Dilute the supernate (*see* note 12) and measure the absorbance at 400 nm (*see* note 13). Express the activity as μmol *p*-nitrophenol produced g⁻¹ dry soil h⁻¹.

Phospholytic Micro-organisms

Day 1 *(1 h)*

See the procedure given for ureolytic micro-organisms.

Day 2 *(1½ h including 1 h incubation period, 3 or 4 days later)*

Count the number of colonies: choose a dilution with > 30 and < 300.

↓

Flood the plates with 2 ml 1 M *p*-nitrophenylphosphate.

↓

Incubate at 25 °C for 60 min.

↓

Add 0·5 ml 0·5 M NaOH.

↓

Count the number of colonies producing a yellow halo, indicative of *p*-nitrophenol production.

Notes and Points to Watch.

1. Use 4 treatments and 2 controls (4 ml buffer, no urea).
2. Freshly prepared and brought to room temperature before addition.
3. Note the exact time of initiation: 30 s between each replicate will give adequate intervals to terminate the reaction later.
4. Use a CDD containing fresh boric acid-indicator for comparison of end-point. Stand dishes on a white background (e.g. a tile) when titrating. Start with 0·2 M HCl, complete with 0·02 M HCl.

5. Activity is: titre in ml of 0·02 M HCl × 20.
6. Use nutrient agar for bacteria; Czapek Dox agar for fungi.
7. Express as numbers of micro-organisms per gram of air-dried soil.
8. Urea hydrolysis raises the pH which changes the phenol red indicator from yellow to purple.
9. Use 4 treatments and 2 controls for each pH value.
10. The use of three buffers (acetate, tris–maleate, tris–HCl) will reveal acid and alkaline phosphatases. It will also supply a partial pH-activity curve.
11. Add substrate to the controls **after** the 60 min incubation period and **before** the termination step.
12. The dilution factor will obviously vary with activity but will probably be in the range of 1 in 5 to 1 in 20.
13. Relate this to the p-nitrophenol calibration curve which covers the range from 0·01 to 0·1 μmol p-nitrophenol ml^{-1}.

MATERIALS

Urease Assay
1. Soil: hand crumbled, air-dried, sieved (< 2 mm), collected from the surface 10 cm. Store in screw-top jars at room temperature. Choose at least three soils different in texture and organic content (e.g. deciduous forest, evergreen forest, arable land under cultivation, permanent pasture). These characteristics are determined by sedimentation and dry combustion, the standard techniques being available in many soil textbooks (e.g. Black, 1965).
2. Tris–maleate buffer (pH 7·0; 0·5 M): 0·5 M Trizma base + 0·5 M maleic acid adjusted to pH 7·0 with 0·5 M NaOH.
3. Boric acid-indicator solution: boric acid 2% w/v, indicator 2% v/v (indicator composition: bromocresol green 0·084% w/v, methyl red 0·16% w/v, decon 0·005% v/v.
4. Urea.
5. Silver sulphate.
6. Conway diffusion dishes (No. 1 Gallenkamp), 6 per assay.

Ureolytic Micro-organisms
1. Nutrient agar.
2. Czapek Dox agar (Oxoid).
3. Christensens urea agar (urea agar base (Oxoid) + 40% w/v urea).

4. Sterile distilled water (100 ml quantities) for dilutions.
5. Velvet pads.

Phosphatase Assay
1. Soil (as for the urease assay).
2. Acetate buffer (pH 4·0; 0·5 M): 0·5 M sodium acetate adjusted to pH 4·0 with acetic acid.
3. Tris–maleate buffer (pH 6·0; 0·5 M): 0·5 M Trizma base + 0·5 M maleic acid adjusted to pH 6·0 with 0·5 M NaOH.
4. Tris–HCl buffer (pH 8·0; 0·5 M): 0·5 M Trizma base adjusted to pH 8·0 with 0·5 M HCl.
5. p-Nitrophenyphosphate (p-NPP) (100 mM).
6. p-Nitrophenol (p-NP) (0·1 μmol stock solution).
7. Calcium chloride (0·5 M).
8. Sodium hydroxide (0·5 M).

Phospholytic Micro-organisms
1. Nutrient agar.
2. Czapek Dox agar (Oxoid).
3. Sterile distilled water (100 ml) quantities for dilutions.
4. p-Nitrophenylphosphate (1 M).
5. Sodium hydroxide (0·5 M).

SPECIFIC REQUIREMENTS

Urease Assay

Day 1
balance
spatula
pipettes (1 ml, 5 ml)
stirring rod
magnetic stirrer and flea

Day 2
2 × 1 ml burettes
white tile

Ureolytic Micro-organisms

Day 1
tubes for soil dilutions
tube rack
20 plates-nutrient agar (for bacteria); Czapek
    Dox agar (for fungi)
glass spreader and alcohol

Day 2
colony counter
5 plates: Christensen's urea agar
replica plating equipment

Day 3
colony counter

Phosphatase Assay

Day 1
balance
spatula
Universal bottles, 6 per soil assayed
pipettes, 1 ml, 5 ml
p-NPP (100 mM), 6 ml per experiment
shaker
low speed centrifuge
cuvettes, 1 cm
spectrophotometer (e.g. Pye Unicam SP600)
tubes for p-NP dilutions (calibration curve)
test tube rack

Phospholytic Micro-organisms

Day 1
tubes for soil dilutions
tube rack
20 plates nutrient agar (for bacteria); Czapek
    Dox (for fungi)
glass spreader and alcohol

Day 2
colony counter

FURTHER INFORMATION

Table 1 lists the properties of urease and phosphatase in a silt loam soil (sand 16 %, silt 64 %, clay 20 %, organic matter 6·4 %, pH 5·4). It is interesting to note that the optimum pH values for activity are not the same as the soil pH, suggesting that the enzymes are working inefficiently *in situ*. However, bulk soil pH measurements may be misleading as it has been shown that H^+ accumulation at soil colloid surfaces may cause a micro-environment of pH 2–3 units different from that measured in the traditional way (Burns, 1979).

In this soil there is only a "neutral" phosphatase with little evidence of acid or alkaline peaks.

Further characterization of these soil enzymes can be achieved by studying their activity in relation to temperature, and variable substrate concentrations. The latter experiments (see standard enzyme textbooks for experimental details) will reveal the kinetic constants: K_m, the Michaelis constant (concentration of substrate required to produce half maximal activity) and V_{max}

Table 1   Some properties of urease and phosphatase in silt loam soil (Pettit *et al.*, 1976, 1977).

	Urease	Phosphatase
Activity	$11 \cdot 3^a$	$1 \cdot 8^b$
Optimum pH	7·0	6·7
Michaelis constant: K_m (mM)	78·2	0·33
Maximum rate: V_{max}	$8 \cdot 9^a$	$1 \cdot 53^b$
% of total count (bacteria + fungi) which are either urease + ve or phosphatase + ve	60	90

[a] μmol $NH_3 \, g^{-1}$ dry soil h^{-1}.
[b] μmol *p*-nitrophenol g^{-1} dry soil h^{-1}.

(theoretical maximum velocity). These values must be interpreted with caution as they represent substrate interactions with immobilized (insoluble) enzymes in a heterogenous system (McLaren and Packer, 1970). They are not, therefore, strictly comparable with reactions *in vitro* where both substrate and enzyme are free in solution.

The relationship of accumulated enzyme activity to the appropriate microbial population may have been revealed by experiments with different soils. An alternative method is to induce an increase in the entire microbial population (or a component of it) and, over a period of time, compare it with the corresponding changes in enzyme activity. Then, by extrapolating the microbial population to zero, the activity that is independent of immediate microbial growth is shown (Figure 1). The changes in microbial numbers can be brought about by adding easily metabolized substrates to soil (e.g. glucose).

Other enzyme activities frequently measured in soil include arylsulphatase, β 1,4-glucosidase, amylases and proteases (Roberge, 1978).

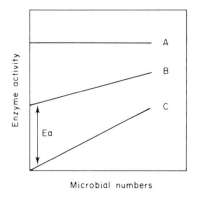

Figure 1   Relating enzyme activity to induced changes in the soil microbial population. A: all activity due to accumulated enzyme; B: Ea is the proportion of total activity that is due to accumulated enzyme; C: no accumulated enzyme activity.

REFERENCES

C. A. Black (1965). "Methods of Soil Analysis", 770 pp. American Society of Agronomy Incorporated, Madison, Wisconsin.

R. G. Burns (1977). Soil enzymology. *Science Progress*, **64**, 281–91.

R. G. Burns (1979). Interaction of microorganisms, their substrates and their products with soil surfaces. *In* "Adhesion of Microorganisms to Surfaces", (D. C. Ellwood, J. Melling and P. R. Rutter, eds), pp. 109–38. Academic Press, London and New York.

S. Kiss, M. Dragan-Bularda and D. Radulescu (1975). Biological significance of enzymes accumulated in soil. *Advances in Agronomy*, **27**, 25–87.

J. N. Ladd (1978). Origin and range of enzymes in soil. *In* "Soil Enzymes", (R. G. Burns, ed.), pp. 51–96. Academic Press, London and New York.

A. D. McLaren and L. Packer (1970). Some aspects of enzyme reactions in heterogenous systems. *Advances in Enzymology*, **33**, 245–308.

N. M. Pettit, A. R. J. Smith, R. B. Freedman and R. G. Burns (1976). Soil urease: activity, stability and kinetic properties. *Soil Biology and Biochemistry*, **8**, 479–84.

N. M. Pettit, L. J. Gregory, R. B. Freedman and R. G. Burns (1977). Differential stabilities of soil enzymes: assay and properties of phosphatase and arylsulphatase. *Biochimica et Biophysica Acta*, **485**, 357–66.

M. R. Roberge (1978). Methodology of soil enzyme measurement and extraction. *In* "Soil Enzymes", (R. G. Burns, ed.), pp. 341–70, Academic Press, London and New York.

J. Skujins (1976). Extracellular enzymes in soil. *CRC Critical Reviews in Microbiology*, **4**, 383–421.

67

Nitrogen Transformations in a Soil Column

R. G. BURNS

Biological Laboratory, University of Kent,
Canterbury, Kent, CT2 7NJ, England

> *Level*: All undergraduate years
> *Subject areas*: Soil biochemistry and microbiology
> *Special features*: An experiment founded upon a classic soil biochemistry technique. Suitable as a group experiment extending over several weeks

INTRODUCTION

The transformation of organic nitrogen to soluble inorganic ions is a microbial process essential to the nutrition of green plants. As a consequence, a great deal of effort has gone into understanding the microbiology and biochemistry of ammonification and nitrification (Walker, 1975). Ammonification (or mineralization) is the conversion of nitrogen from its organic form (e.g. urea) to ammonia (I) and is effected by a wide variety of micro-organisms. Nitrification ((II) and (III)), the oxidation of ammonia via nitrite to nitrate is, in contrast, a rather specialized process performed primarily by two aerobic chemolithotrophs: *Nitrosomonas* (II) and *Nitrobacter* (III).

I $\qquad CO(NH_2)_2 + H_2O \rightarrow 2NH_3 + CO_2$

II $\qquad NH_4^+ + 1\frac{1}{2}O_2 \rightarrow NO_2^- + H_2O + 2H^+ \quad \Delta F \text{ (kcal): } 65$

III $\qquad NO_2^- + \frac{1}{2}O_2 \rightarrow NO_3^- \quad \Delta F \text{ (kcal): } 18.$

The objective of this experiment is to follow the chemical sequence (ammonia → nitrite → nitrate) in a column of soil perfused with an easily assimilated organic nitrogen source (e.g. urea). Perfusion column techniques have been a significant help in our understanding of the nitrogen cycle in soil (*see* for example: Lees and Quastel, 1946; Macura and Kunc, 1965).

EXPERIMENTAL

The spectrophotometric measurement of ammonia, nitrite and nitrate in the perfusate may not accurately reflect events within the column. This is due to a number of factors. For example, a proportion of the positively charged ammonium ions will be adsorbed to the predominantly negatively charged soil colloids (i.e. clays, humic matter) and will not, therefore, appear in the perfusate. Nitrite and nitrate, in contrast, will tend to leach through the column quite freely. Theoretically, the adsorption of ammonia will occur up to the adsorption capacity of the soil (which itself varies with soil type). In practice, however, adsorption will vary with time according to shifts in pH, and the number and valency of exchangeable ions. In addition, assimilation or immobilization will result in nitrogen being tied up in microbial biomass which is itself retained in the column since microbes tend not to leach through soil. Other factors include denitrification (Delwiche and Bryan, 1976) in anaerobic microsites which will reduce the level of nitrate appearing in the perfusate.

All these factors will contribute to the variable rates of transformation within the soil column. Thus four treatment replicates are advised and two control replicates.

Day 1 *(2 h)*

Loosely pack each of 6 columns (*see* note 1) with a mixture of 20 g air-dried soil aggregates (3–5 mm diam.) (*see* note 2) and 10 g glass beads (2 mm). Set up each column as shown in Figure 1.

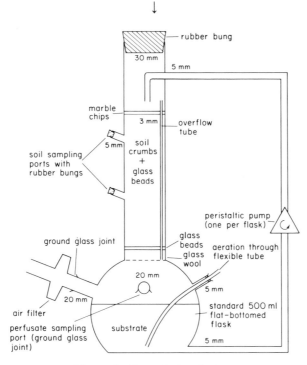

Figure 1   Re-perfusion flask.

Add 300 ml sterile distilled H_2O through the perfusate sampling port (PSP). Perfuse for 24 h to wash out any soluble organic matter which would otherwise interfere with subsequent colorimetric measurements. The perfusion rate should be adjusted to give a regular flow through the column yet not cause waterlogging (*see* note 3). Also bear in mind that nitrification is an obligately aerobic process.

Day 2 *(2 h, 24 h later)*

Turn off the pump and allow each column to drain for 30 min. Pipette off the water and replace it with 300 ml of substrate solution (0·05 M urea or 0·05 M glycine in sterile distilled H_2O). Perfuse the controls with fresh sterile distilled H_2O.

↓

Perfuse for 5 min and then remove 10 ml perfusate; this is the time zero reading.

↓

Adjust all flasks to perfuse at approximately equal rates and check frequently during the experiment.

↓

Sample on as many days as possible (*see* note 4) and at the same time of day. Store the capped (and labelled!) sample tubes in a refrigerator and assay them together at convenient times (suggest once per week).

Days 3, 4, 5, 6 *(4 h, at 7, 14, 21 and 28 days)*

Assay the weekly batch of samples. If there were three sampling times per week this will mean: 3×6 (4 treatments + 2 controls) $\times 3$ (NH_3, NO_2, NO_3) = 54 assays + concentrations used for calibration curves.

Nitrogen Assays (*see* note 5)

(a) *Ammonium*

Pipette a 1 ml sample into a test tube (*see* note 6).

↓

Add 1 ml gum acacia (2%), 1 ml Nessler's reagent and 7 ml distilled H_2O.

↓

Mix well and allow to stand for 30 min.

↓

Measure O.D. at 420 nm against a reagent blank (substitute 1 ml H_2O for sample).

↓

(Prepare calibration curve using 1–30 μg of nitrogen per ml as $(NH_4)_2SO_4$.)

(b) *Nitrite*

Pipette a 1 ml sample into a test tube (*see* note 6).

↓

Add 1 ml modified Griess–Ilosvay's reagent and 8 ml distilled H_2O.

↓

Mix well and allow to stand for 30 min.

↓

Measure O.D. at 540 nm against a reagent blank (substitute 1 ml H_2O for sample).

↓

(Prepare calibration curve using 0·1–2 μg of nitrogen per ml as $NaNO_2$.)

(c) *Nitrate (Sims and Jackson, 1971)*

Pipette 3 ml sample into test tube (*see* note 6).

↓

Add 7 ml of chromotropic acid reagent.

↓

Mix well, cool rapidly to < 40 °C, and allow to stand in a water-bath held at 40 °C for 20 min.

↓

Measure O.D. at 430 nm against a reagent blank (substitute 3 ml H_2O for sample).

↓

(Prepare calibration curve using 0·5–15 μg of nitrogen per ml as KNO_3.)

Notes and Points to Watch

1. Use 4 treatment columns and 2 controls (perfuse with sterile distilled H_2O).
2. As the column will be continuously perfused for 3–4 weeks, it is advisable to choose a soil with stable aggregates. These are most often found in soils under permanent (> 10 year) pasture and will not easily disintegrate when shaken in water.
3. Somewhere in the range 20–50 drops min^{-1} is suitable. Occasionally temporary flooding of the soil column may occur. Replication will enable you to detect this and discard the result if necessary.
4. At least 3 times per week.
5. Take great care with the reagents—several are corrosive; some (e.g. 1-napthylethylenediamine di HCl) are suspected carcinogens.

6. You will probably need to dilute your sample to bring it within the range of the calibration curve. Experience will tell you what dilution factor is appropriate.

MATERIALS

1. Soil: hand crumbled, air-dried, collect from the surface 10 cm, separate 3–5 mm aggregates.
2. Marble chips.
3. Perfusion columns: adapted from standard 500 ml flat-bottomed flasks (*see* Figure 1).
4. Peristaltic pumps (Schuco Scientific Ltd., London), 1 per perfusion flask.

Ammonium Assay
Nessler's reagent (Fisons).

Nitrite Assay
Griess–Ilosvay's reagent (wodified)
Solution A: 5 g sulphanilic acid in 500 ml 2 M HCl. Store in a refrigerator.
Solution B: 1 g   *N*-(1-napthylethylenedia-mine dihydrochloride) (Koch-Light) in 599 ml deionized H_2O. Store at room temperature in the dark.

Mix solution A with solution B in a 1:1 ratio just prior to use.

Nitrate Assay
Stock solution: 1·84 g chromotropic acid disodium salt (Aldrich Chem. Co.) in 1 litre conc. H_2SO_4 (reagent grade) $= 0·1\%$ (w/w).
Working solution: 100 ml of stock solution, 10 ml conc. HCl brought to 1 litre with conc. $H_2SO_4 = 0·01\%$ (w/w).

SPECIFIC REQUIREMENTS

Day 1
perfusion columns (6 per experiment)
glass beads (2 mm)
peristaltic pumps (6)
air filters
glass wool
rubber bungs

Day 2
test tubes
test tube rack
pipettes (10 ml)
urea or glycine solution, 2 l, 0·05 M

Days 3, 4, 5, 6
pipettes (5 ml, 1 ml)
test tubes

test tube rack
cuvettes (1 cm)
spectrophotometer
ice
$(NH_4)_2SO_4$ (stock solution $= 1000\,\mu g$ ammonium–nitrogen ml^{-1}): 2·357 g in 500 ml deionized H_2O. Store at 4 °C.
KNO_3 (stock solution $= 1000\,\mu g$ (nitrate nitrogen ml^{-1}): 3·607 g in 500 ml deionized H_2O. Store at 4 °C.
$NaNO_2$ (stock solution $= 1000\,\mu g$ (nitrite–nitrogen ml^{-1}): 2·643 g in 500 ml deionized H_2O. Store at 4 °C.

FURTHER INFORMATION

Table 1 presents sample data from a re-perfusion experiment. It can be seen that a definite sequence of microbial metabolites appears in the perfusate: firstly ammonia, followed by nitrite and finally

Table 1 Urea ammonification and nitrification in a perfusion flask containing silt-loam soil aggregates. Concentrations of NH_4–N, NO_2–N and NO_3–N are expressed as μg of N per ml of perfusate. Control data are the average of two replicates, urea data are the average of four replicates.

Day	NH₄–N		NO₂–N		NO₃–N	
	Control	Urea + glucose	Control	Urea + glucose	Control	Urea + glucose
0	5	25	0·26	0·31	7	9
3	11	80	0·50	0·48	10	7
6	12	158	0·51	0·62	9	8
9	23	190	0·42	1·02	16	28
12	20	240	0·37	1·56	20	41
15	42	308	0·24	1·82	24	51
18	36	320	0·43	1·65	27	94
21	30	390	0·30	1·29	26	120
24	19	300	0·36	1·40	33	154
27	16	210	0·38	0·96	38	190

nitrate. Nitrite levels are almost always very low as this ion is rapidly utilized by the *Nitrobacter* population. A graphical representation of the data will sometimes illustrate the dynamics of ammonification and nitrification in a more dramatic way.

Remember that what is being measured in the perfusate is only a crude guide to the transformations occurring within the soil column itself. The student should be aware of these shortcomings whilst performing the basic experiment, and be prepared to investigate ways of improving his or her interpretation of the perfusate data.

Once the basic pattern of ammonification and nitrification has been recorded a number of additional experiments are suggested. For example, soil may be extracted through the two sample ports and dilution plate counts made to record changes in the relevant micro-organisms (i.e. ureolytic, *Nitrosomonas* and *Nitrobacter*; *see* Prosser and Cox, 1982). More advanced kinetic studies may be initiated by extracting small quantities of the perfusate from the sample ports and relating the changing concentration of NH_4, NO_2 and NO_3 to time and distance down the column of soil (*see* McLaren, 1978, for a discussion of nitrification kinetics in soil).

Other experiments could involve the addition of various pesticides to the perfusate to discover if they are likely to cause any disruption to the nitrogen cycle. Indeed a high proportion of current research on the nitrogen cycle is concerned with screening pesticides (both new and old) for detrimental side effects (Johnen and Drew, 1977).

In addition, a number of chemicals have been examined for their ability to retard the ammonification of urea fertilizers. This normally rapid hydrolysis is regarded as a disadvantage because gradual release of ammonia (and subsequently nitrate production) is more advantageous for crops, as well as being economically and ecologically sound. Recall that anionic nitrate is readily leached from soil. One group of chemicals that have attracted attention as possible urease inhibitors is the quinones (Bremner and Mulvaney, 1978). It is suggested that the following quinones are added to the perfusate at 50 parts 10^{-6} and their effect on ammonia production evaluated: *p*-benzoquinone, methyl-*p*-benzoquinone and tetrahydroxy-*p*-benzoquinone. Catechol, phenol, formaldehyde and phenylmercuric acetate may also be worth examining in this context.

REFERENCES

J. M. Bremner and R. L. Mulvaney (1978). Urease activity in soils. *In* "Soil Enzymes", (R. G. Burns, ed.), pp. 149–96. Academic Press, London and New York.

C. C. Delwiche and B. A. Bryan (1976). Denitrification. *Annual Review of Microbiology*, **30**, 241–62.

B. J. Johnen and E. A. Drew (1977). Ecological effects of pesticides on soil microorganisms. *Soil Science*, **123**, 319–24.

H. Lees and J. H. Quastel (1946). Kinetics of, and effects of poisons on, soil nitrification, as studied by soil perfusion. *Biochemistry Journal*, **40**, 803–14.

A. D. McLaren (1978). Kinetics and consecutive reactions of soil enzymes. *In* "Soil Enzymes", (R. G. Burns, ed.), pp. 96–116. Academic Press, London and New York.

J. Macura and F. Kunc (1965). Continuous flow method in soil microbiology V. Nitrification. *Folia Microbiologica*, **10**, 125–35.

J. I. Prosser and D. J. Cox (1982). Nitrification. *In* "Experimental Microbial Ecology", (R. G. Burns and J. H. Slater, eds), pp. 178–93. Blackwell Scientific Publications, Oxford.

J. R. Sims and G. D. Jackson (1971). Rapid analysis of soil nitrate with chromotrophic acid. *Soil Science Society of America Proceedings*, **35**, 603–6.

N. Walker (1975). Nitrification and nitrifying bacteria. *In* "Soil Microbiology", (N. Walker, ed.), pp. 133–46. Butterworths, London.

68

Determination of the Concentration of Particulate Organic Material in Aquatic Environments

P. J. LeB. WILLIAMS

*Department of Oceanography, University of Southampton,
Southampton, SO9 5NH, England*

> *Level*: Advanced undergraduates
> *Subject area*: Microbial ecology
> *Special features*: A chemical procedure suitable for
> a microbiological field course
> or to measure the growth
> of a culture

INTRODUCTION

In clean aquatic habitats the amount of particulate organic material may be related to the amount of plankton, but where there has been human influence it may act as an indication of the degree of pollution. Rough estimates of the amount of particulate material may be derived from measurements of light transmission, but chemical determination is generally better. For chemical analysis, the particulate material may be collected on a non-organic filter (e.g. glass fibre mat) and the organic content determined either by measuring the CO_2 produced during high temperature combustion or by measuring the consumption of dichromate during wet oxidation in strong sulphuric acid. The latter procedure is suitable for field work and is described here. Two precautions are necessary: (1) the glass-fibre filter must be combusted before use to remove all organic matter; (2) after filtration, halides must be removed from the filters by washing with Na_2SO_4 solution, because they react with dichromate and would interfere with the estimation.

The object of this experiment is to measure the concentration of particulate organic matter in a sample of sea-water. The method has been adapted from those described by El Wakeel and Riley (1957), Marshall and Orr (1964) and Strickland and Parsons (1972).

EXPERIMENTAL

Caution   This experiment involves the use of hot concentrated sulphuric acid, therefore great care should be taken (*see* note 1).

Day 1 (*15 min*)

Separate ten 4·25 cm diam. Whatman GF/C filters, wrap them loosely in aluminium foil (Figure 1) and heat them overnight at 550 °C in a muffle oven (*see* note 2).

_____

Day 2 (*3 h*)

With clean metal forceps, place a filter on a filtration apparatus and filter a suitable volume (*see* note 3) of the sea-water sample.

$$\downarrow$$

Wash the filter with 30 ml of 0·6 M Na_2SO_4 (*see* note 4). Remove the top of the filter holder and wash the periphery of the filter with Na_2SO_4. Suck the filter dry (*see* note 5).

$$\downarrow$$

After all the samples have been filtered, fold the filters into quarters and place them in **clean** (*see* note 6) 25 × 150 mm heavy-walled test tubes. Run duplicate **control** filters, prepared by wetting the filter with a small volume (i.e. 1 ml) of the water sample and washing, as above. To the tubes containing the filters, add exactly 5·0 ml of 0·05 N potassium dichromate with a volumetric pipette (*see* note 7).

$$\downarrow$$

In addition to the above, run duplicates of a **standard** consisting of the dichromate solution and 1 ml containing 1·9 mg sucrose and two **blanks** consisting of the dichromate solution alone.

$$\downarrow$$

Cover the tubes with a small beaker or a loose glass stopper and heat in **boiling** water for 15 min (**caution**: hot dichromate; *see* note 1).

$$\downarrow$$

After heating, cool the tubes and cautiously add 10 ml distilled water to each. Any sample that turns bright green has probably been fully reduced by the dichromate and may not be worth titrating. This can be readily checked with the indicator. Mix, and transfer the contents of each tube to a 250 ml conical flask. Wash each tube with a further 10 ml distilled water and add to the flask. Then add 1 drop of ferrous-phenanthroline indicator and titrate the digest and washings with 0·05 N ferrous ammonium sulphate from a burette until a pink colour **just** persists. The onset of the end point can be recognized by the formation of a temporary pink coloration around the drops of the titrant (*see* note 8). The burette should be read to 0·02 ml.

_____

Notes and Points to Watch

1. Make sure an eye-wash bottle is at hand and full.
2. For muffling, the filters are best separated by a zig-zag of aluminium foil and then wrapped in an envelope of foil (Figure 1). This enables them to be stored without contamination by atmospheric dust.

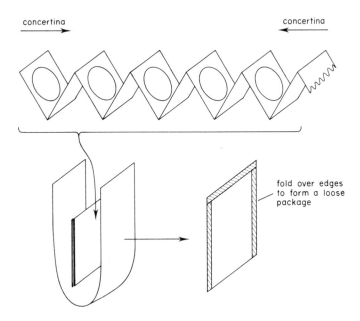

concertina →   ← concertina

fold over edges to form a loose package

Figure 1   A suitable method of wrapping filters.

3. The required volume is difficult to prescribe since it will depend upon locality and season; as an initial guide, try volumes of 100, 250 and 1000 ml. These should be tested in duplicate.
4. The Na_2SO_4 is best provided in a wash-bottle.
5. The filters may be stored for up to one day in clean glass Petri dishes. For longer periods they should either be frozen or stored in a desiccator.
6. The glassware that comes into contact with the dichromate (i.e. the boiling tubes and pipettes) should be cleaned in chromic acid. The boiling tubes, alternatively, may be cleaned by heating them overnight in a muffle oven at 550 °C. The mouth of the boiling tube should be covered with aluminium foil to prevent contamination by dust after cleaning. The volumetric pipettes must **not** be cleaned in this manner for it will alter their calibration.
7. The dichromate/sulphuric acid mixture may be added with a pipette with a safety bulb, although these usually have a very short life when used with strong acid. A simple and inexpensive alternative is a 10 ml disposable syringe fixed on to the pipette with rubber tubing; with practise this can be made to function very satisfactorily.
8. It is advisable to get the students to practise the titration first on trial samples of dichromate. The end point is very exact and easy to recognize.

MATERIALS

1. 0·05 N dichromate: dissolve 2·45 g $K_2Cr_2O_7$ (analytical reagent grade) in a small volume of distilled water and make up to 1 litre with analytical reagent grade sulphuric acid.
2. 0·05 N ferrous ammonium sulphate: dissolve 19·7 g of analytical reagent grade $Fe(NH_4)_2 \cdot (SO_4)_2 \cdot 6H_2O$ in 900 ml distilled water, add 20 ml H_2SO_4 and make up to 1 litre with distilled water.
3. Ferrous-phenanthroline indicator: dissolve 0·34 g o-phenanthroline monohydrate in 25 ml of 0·7 % (w/v) ferrous sulphate solution.
4. Sucrose standard: dissolve exactly 190 mg sucrose (analytical reagent grade) in 100 ml distilled water.
5. 0·6 M sodium sulphate: dissolve 85·2 g anhydrous Na_2SO_4 (or 193 g of the decahydrate) in 1 litre distilled water.
6. Whatman glass-fibre filters grade GF/C, 4·25 cm diam.

SPECIFIC REQUIREMENTS

Day 1
aluminium foil
GF/C filters, 4·25 cm diam.; 10 per student group
muffle furnace, set at 550 °C

Day 2
fine-point metal forceps
filter holder and pump
sodium sulphate in wash-bottle
glass Petri dishes
sample of sea-water for analysis of particulate organic material
25 × 100 mm heavy-wall boiling tubes; 6 per group each with loose glass stopper (see note 6) or small beaker
test tube rack or wooden block drilled with holes

5 ml volumetric pipette (see notes 6 and 7)
1 ml volumetric pipette
10 ml graduated pipette or measuring cylinder
500 ml beaker, tripod and Bunsen or hot plate
wash-bottle containing distilled water
10 ml burette, ideally with a PTFE tap
250 ml conical flask
0·05 N dichromate (approximately 50 ml per group)
0·05 N ferrous ammonium sulphate (approximately 100 ml per group)
sucrose standard, 1·9 mg ml^{-1}, approximately 10 ml per group
ferrous phenanthroline indicator, approximately 10 ml in a dropper bottle.

FURTHER INFORMATION

The results are worked out as follows:
   let A = titre of **blank** in ml,
   let B = titre of **standard** in ml,
   then (A − B) is equivalent to 1900 μg sucrose.
This in turn is equivalent to 1900 × 144/342 = 800 μg C.
   For the **sample** titration,
   let C = titre of **control** in ml,
   let D = titre of **sample** in ml,
   let V = volume of sea-water filtered, in ml.

The concentration of particulate organic material (μg C l^{-1}) in the sample is:

$$800 \times \frac{1000}{V} \times \frac{(C-D)}{(A-B)}.$$

It is possible to determine the residual dichromate spectrophotometrically (Marshall and Orr, 1964), although this turns out to be less convenient than the titrimetric procedure.

A disadvantage of the dichromate oxidation procedure is that certain organic compounds, e.g. uric acid and many plastics, resist oxidation. However, with natural particulate materials the difference in results between the dichromate procedure and high temperature combustion is unlikely to be greater than 10% (El Wakeel and Riley, 1957). On the other hand, the dichromate method has the advantage over the high temperature combustion method in that inorganic carbonates do not interfere.

The standardization with sucrose, used in the present procedure, is arbitrary and strictly the method provides a result in oxidation equivalents. This has an advantage in ecological work on energy relationships if calories are being used as the common unit, for there is a close relationship between equivalents and calories: 1 mEq $\simeq$ 27 g calories.

For unpolluted lakes, rivers and coastal sea-water, particulate organic carbon concentrations in the range 50–500 μg C l^{-1} may be expected. Estuaries, productive lakes and rock pools will give values from 500–2000 μg C l^{-1}. Still higher values (1000–10 000 μg C l^{-1}) may be found with polluted and naturally turbid waters.

REFERENCES

S. K. El Wakeel and J. P. Riley (1957). The determination of organic carbon in marine muds. *Journal du Conseil Permanent International pour l'Exploration de la Mer*, **22**, 180–3.

S. M. Marshall and A. P. Orr (1964). Carbohydrate and organic matter in suspension in Loch Striven during 1962. *Journal of the Marine Biological Association U.K.*, **44**, 285–92.

J. D. H. Strickland and T. R. Parsons (1972). "A Practical Handbook of Seawater Analysis". Fisheries Research Board of Canada, Bulletin 167. Ottawa. 310 pp.

Measurement of Primary Planktonic Production in Aquatic Environments by $^{14}CO_2$ Fixation

P. J. LeB. WILLIAMS

Department of Oceanography, University of Southampton,
Southampton, SO9 5NH, England

> *Level*: Advanced undergraduates
> *Subject areas*: Microbial ecology
> *Special features*: A simple, sensitive and widely used method of measuring plankton photosynthesis

INTRODUCTION

Animal life in aquatic environments is dependent directly or indirectly on the plants, and by far the most important of these in the sea and in lakes are the microscopic algae of the plankton, notably the diatoms and the flagellates. From an ecological point of view, the planktonic algae act as **primary producers**, i.e. they convert inorganic substances (carbon dioxide, nitrate, ammonia, etc.) into organic material by the process of photosynthesis. This can be summarized as:

$$CO_2 + H_2O + \text{light energy} \longrightarrow [CH_2O] + O_2.$$

Steemann-Nielsen (1952) introduced a radio-carbon technique with ^{14}C for measuring the rate of photosynthesis by phytoplankton. The method entails labelling the bicarbonate of sea-water with a known trace amount of ^{14}C and following its uptake over a period of time up to 24 h. The main virtues of the ^{14}C technique are its simplicity, sensitivity and suitability for measuring the rate of CO_2-fixation under the natural environmental conditions of temperature, light and nutrients. The experiment described here, which uses sea-water, is based on Strickland (1960), Vollenweider (1969) and Strickland and Parsons (1972). Its object is to measure the primary productivity of plankton in a sample of sea-water.

EXPERIMENTAL (*see* **note 1**)

(30 min + 2 h with a gap of 2–4 h in between)

Collect a sample of sea-water containing phytoplankton and dispense 250 ml portions into 250 ml bottles.

   ↓   ↓   ↓

Light bottle(s)	*Dark bottle(s)*	*Zero time bottle* (see *note 2*)
Add $5\mu Ci$ $NaH^{14}CO_3$ (*see* notes 3 and 4), mix, stopper and suspend each bottle in the sea at the appropriate depth.	Add $5\mu Ci$ $NaH^{14}CO_3$, mix, stopper and suspend in the sea along with the light bottles.	Add $5\mu Ci$ $NaH^{14}CO_3$, mix, stopper and filter immediately through a $0.45\mu m$ filter. Save the filter and the filtrate.

   ↓   ↓

After 2–4 h exposure in the sea, filter the sample through a $0.45\,\mu m$ membrane filter (*see* note 5). Save the filter and the filtrate

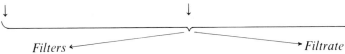

Filters ←   → *Filtrate*

Wash each membrane filter with a small volume (10–50 ml) of membrane-filtered sea-water from the same environment.	Acidify a 15 ml portion of each filtrate with 1.0 ml of 50% H_3PO_4 and decarbonate it by bubbling air through it (*see* note 6) for at least 1 h).
Pick up the filter with forceps, staple it to a 10 × 15 cm filing card, place in HCl vapour (*see* note 7) for 5 min to remove any residual $NaH^{14}CO_3$ and allow to dry.	Add 5 ml of the decarbonated solution to 15 ml Unisolv 1, or equivalent fluor capable of accepting a large volume of water.
Remove the filter from the card, transfer it to a scintillation vial, add sufficient scintillation fluid to cover the filter and count the radioactivity in a liquid scintillation counter (*see* note 8).	Count the radioactivity in a liquid scintillation counter (*see* note 8).
Add an accurately known amount of a ^{14}C-containing standard and recount to determine the quenching correction (*see* note 9).	Add an accurately known amount of a ^{14}C-containing standard and recount to determine the quenching correction (*see* note 9).

Notes and Points to Watch

1. The schedule does not specify the radiochemical procedure and precautions, for these will vary from laboratory to laboratory. In order to instil some awareness into the students, it is often very effective before the start of the experiment to set up a sample containing 1 ml of the ^{14}C–bicarbonate and count it in the liquid scintillation counter, ideally one with an analog display.

   The bench tops where isotope is handled should be covered with Benchkote or similar material.

2. The zero time control will often turn out to be a rather important factor in determining whether or not the procedure has worked properly, and the temptation to leave it out should be resisted. Often it is inconvenient to filter at the sampling site, and a satisfactory and simple alternative is to delay the addition of the isotope until the class returns to the laboratory: the "zero time" sample can be stored until the end of the incubation period, the isotope added then and the sample processed at the same time as the light and dark samples. A further alternative is to add formaldehyde to a final concentration of 4 % immediately after collection and before adding the isotope.

3. The isotope is conveniently added to the sample with a 1 ml disposable hypodermic syringe fitted with a 5 cm, 19 gauge needle, preferably blunt. The bottle is closed and incubated preferably *in situ* in the sea or lake or, alternatively, in an illuminated incubator. Four hours should be sufficient with spring or summer samples.

4. The radioactivity of the $NaH^{14}CO_3$ should be determined directly at the time of use since loss by exchange to the atmosphere may occur during preparation and storage.

5. Although Millipore filters are often used, Oxoid filters are a great deal cheaper in the UK. Use the minimum vacuum necessary for the filtration.

6. A suitable arrangement is shown in Figure 1. Note that air, because it contains CO_2, is preferable

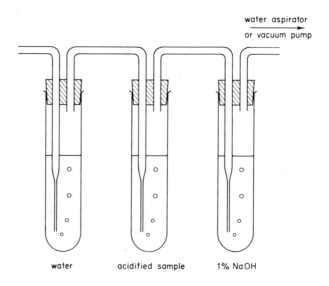

water aspirator or vacuum pump

water   acidified sample   1% NaOH

Figure 1   Apparatus for removing $^{14}CO_2$ from filtrate.

to nitrogen or oxygen for the removal of $^{14}CO_2$. If a vacuum pump is used in place of a water aspirator, a trap should be placed between the pump and the NaOH solution. The output from the pump should be led to a fume cupboard or through a window.

The purpose of the decarbonation is to remove all $^{14}CO_2$ from the filtrate so that any radioactivity that is left can be ascribed to excretion of organic compounds by the algae. High values for "excretion" in the zero time control indicate incomplete removal of $^{14}CO_2$ (*see* Sharp, 1977).

7. This is conveniently done by placing a 100 ml beaker containing 50 ml conc. HCl in a desiccator and placing the filters, stapled to the cards, in the desiccator for 5–10 min.

8. With some fluors it may be necessary to allow chemiluminescence to decay before the sample is counted.

9. The simplest and most convenient method is to add a known volume, e.g. 0·1 ml, of the stock solution of $NaH^{14}CO_3$ whose radioactivity is known, to one or to each of the vials and to recount.

MATERIALS

1. Radioactive bicarbonate solution: The $NaH^{14}CO_3$ is made up at a concentration of 5 μCi ml^{-1} in distilled water containing 0·3 g anhydrous Na_2CO_3 and one pellet of sodium hydroxide per litre. If it is to be used with sea-water samples, 5 % (w/v) NaCl should be added. On standing in open containers it will lose its $^{14}CO_2$ by isotope exchange with the atmosphere. The best procedure is to make up a large stock (say 1 litre, containing 5 mCi) and to dispense 3·5–4 ml volumes into glass ampoules, which are subsequently flame-sealed and sterilized by autoclaving. When autoclaving, it is wise to place the ampoules in a beaker of water, otherwise if one ampoule explodes it will set off a chain reaction and leave you with an autoclave containing 5 mCi of $^{14}CO_2$. After autoclaving, the beaker containing the ampoules is removed whilst still warm and a strong solution of methylene blue added. On cooling, any incompletely sealed ampoule will fill up with water and can be readily identified. This is an easy way of finding a badly sealed ampoule. Each ampoule will contain just sufficient solution for one set of samples (light, dark and zero time).

2. Bottles: each group will need three bottles to incubate their samples. The bottles (clear glass,

soda or Pyrex) in which the samples are incubated should not be too large (not larger than about 250 ml), otherwise there may be too much sample to filter and the filter will become clogged part way through. For lake water samples, where silica may be a limiting nutrient, hard glass or clear plastic bottles should be used. In other situations soda glass will be adequate.

3. Disposable 1 ml plastic hypodermic syringe with a 5 cm, 19 gauge, preferably blunt, needle.

4. Each group will require 6 liquid scintillation vials, 100 ml of fluor, 3 membrane filters (Oxoid, Sartorius or Millipore) water aspirator or vacuum pump and bubbling train (*see* Figure 1), filter holder, pair of forceps.

5. Have available in the laboratory: stapler, 10 × 15 cm filing cards and, if possible, an automatic dispenser for the fluor, set at 5 ml.

6. The experiment is written on the assumption that a liquid scintillation counter is available. If not, an end-window counter can be used although it can not be used to measure the excretion products. It will also be necessary to guess the efficiency of the machine and presume the radioactive content of the $NaH^{14}CO_3$, unless time is available to calibrate with $Ba^{14}CO_3$.

SPECIFIC REQUIREMENTS

bottles, at least 3 per group
isotope in ampoule, glass knife or file
syringe for isotope addition (*see* note 3)
filter holder, vacuum pump
0·45 μm, 47 mm diam. membrane filters
filtered water
forceps
10 × 15 cm filing cards and stapler
HCl in desiccator

50% orthophosphoric acid with 1 ml pipette
25 ml measuring cylinder (to measure out
   filtrate)
bubbling apparatus (*see* note 6)
6 liquid scintillation vials
fluor in automatic dispenser (*c.* 100 ml per
   group)
100 μl pipette
liquid scintillation counter

FURTHER INFORMATION

This experiment is written on the assumption that a sea-water sample is used, but sufficient information is given to allow the calculations to be made for estuarine and freshwater samples. Samples from fast-flowing streams and rivers, which may contain little planktonic material, should be avoided.

The experiment could be used to examine the effect of anthropogenic compounds (e.g. DDT, herbicides, crude oil, etc.) on algal photosynthesis.

Evaluation of Results

For the radioactivity trapped on the filters, the fundamental equation used in primary productivity studies is based on the isotope-dilution principle

$$\frac{^{14}C \text{ fixed (a)}}{^{14}C \text{ present (b)}} = \frac{^{12}C \text{ fixed (c)}}{^{12}C \text{ present (d)}} \times \frac{1}{1·05} \text{(e)},$$

where:
- (a) is measured from the radioactivity collected on the filters (after correcting for quenching and for background);
- (b) is the amount of radioactive $NaH^{14}CO_3$ added to each bottle;
- (c) is the primary productivity value, which is being sought;
- (d) is the total carbonate concentration in the sample;
- (e) is a correction made to allow for $^{12}C/^{14}C$ isotope discrimination during photosynthesis.

Let us consider the above in more detail.
- (a) ^{14}C-fixed is taken as the difference between the ct min^{-1} (counts per minute) of the filters from light and dark bottles at a particular depth, e.g. if the light bottle gave 5000 ct min^{-1} and the dark bottle 400 ct min^{-1} the ^{14}C-fixed would be 4600 ct min^{-1}.
- (b) ^{14}C-present is based on the count of the $NaH^{14}CO_3$ stock solution and should take account of the volume added to each bottle.
- (c) ^{12}C-fixed should be expressed as μgC fixed l^{-1}h^{-1}, or mgC fixed m^{-3}h^{-1}.
- (d) ^{12}C-present, the total carbonate content, is essentially a conservative property in the case of sea-water, and may be deduced from a measurement of salinity; for the present purposes if

the samples were taken from sea-water then we will assume the carbonate content to be equal to 25 mgC l^{-1}. If estuarine or freshwater samples were used, then the total carbonate content will have to be determined from a measurement of alkalinity (details of this may be derived from Strickland and Parsons, 1972, or Mackereth *et al.*, 1978).

(e) Correction for isotope discrimination is conventional, but considering the potential errors in the ^{14}C technique it is probably merely a nicety.

From the radioactivity in the decarbonated filtrates, the rate of production of organic "excretion" products may be calculated by the same equation. It is essential to allow for the fact that only part of the sample (i.e. 5 ml) has been counted, so the corrected ct min^{-1} must be multiplied by $V/5$, where V is the volume of the incubated sample in ml.

Photosynthetic production may be expressed per unit volume (m^{-3} or l^{-1}) or area (usually m^{-2}); per hour, day or year. For physiological studies, rates are typically expressed as mgC l^{-1} h^{-1} and such rates may range from less than 1 μgC fixed l^{-1} h^{-1} for oligotrophic lakes and tropical oceans to greater than 500 μgC l^{-1} h^{-1} for coastal upwelling areas, eutrophic lakes and estuaries. The rates are often standardized by dividing by the chlorophyll *a* content of the sample, and under optimum conditions of light and inorganic nutrients this will give a value (usually termed the productivity index) in the region of 1·5 mgC fixed mg^{-1} Chl *a*. h^{-1}. For ecological work, especially trophic studies dealing with the flow of organic material and energy through the food web, rates expressed per year or day are generally more useful. In the ecological sense a "day" is a 24 h period with a diel light cycle, thus daily rates of primary production are not simply hourly rates multiplied by 24 or the time from dawn to dusk. A similar consideration applies to the annual rates. Daily rates of production vary from less than 10 μgC fixed l^{-1} day^{-1} to greater than 1000 μgC l^{-1} day^{-1}. In many aquatic systems, especially deep ones, light ultimately limits the rate of photosynthetic production of the environment, and as a consequence it is common to express the productivity as the rate per unit area of aquatic surface. Self-shading by the algal population limits the maximum productivity of aquatic systems to rates in the region of 5–10 gC fixed m^{-2} day^{-1}. Such rates are usually only realized for periods of a few days, unless there is some mechanism of recycling or replenishing the inorganic nutrients (nitrate, ammonia, phosphate) used by the planktonic algae.

Estimates of annual marine production show smaller variations than production rates: ranging from 50 gC fixed m^{-2} $year^{-1}$, for tropical areas, to 300 gC fixed m^{-2} $year^{-1}$ in upwelling areas. A median figure for the oceans as a whole is often taken to be 100 gC m^{-2} $year^{-1}$, and estimates of global oceanic production usually lie in the range $2-4 \times 10^{16}$ gC $year^{-1}$, i.e. $2-4 \times 10^{10}$ tonnes C $year^{-1}$. For comparison, estimates for production by land plants are usually in the region of 5×10^{10} gC $year^{-1}$, i.e. roughly twice that of the oceans. Parsons *et al.* (1977) give an excellent account of planktonic processes in the sea.

REFERENCES

F. J. H. Mackereth, J. Heron and J. F. Talling (1978). "Water Analysis: Some Revised Methods for Limnologists", 120 pp. Freshwater Biological Association Scientific Publication 36. Ambleside.

T. R. Parsons, M. Takahashi and B. Hargrave (1977). "Biological Oceanographic Processes", 332 pp. Pergamon, Oxford.

J. H. Sharp (1977). Excretion of organic matter by marine phytoplankton: Do healthy cells do it? *Limnology and Oceanography*, **22**, 381–99.

E. Steemann-Nielson (1952). The use of radioactive carbon (C^{14}) for measuring organic production in the sea. *Journal du Conseil Permanent International pour l'Exploration de la Mer*, **18**, 117–40.

J. D. H. Strickland (1960). "Measuring the Production of Marine Phytoplankton", 172 pp. Fisheries Research Board of Canada, Bulletin 122.

J. D. H. Strickland and T. R. Parsons (1972). "A Practical Handbook of Seawater Analysis", 310 pp. Fisheries Research Board of Canada, Bulletin 167, Ottawa.

R. A. Vollenweider (1969). "A Manual of Methods for Measuring Primary Production of Aquatic Environments", 213 pp. Blackwell Scientific, Oxford.

70

Experiments on the Gas Vacuoles of Planktonic Cyanobacteria

A. E. WALSBY

Department of Botany, University of Bristol,
Woodland Road, Bristol, BS8 1UG, England

> *Level*: All undergraduate years
> *Subject area*: Physiology of cyanobacteria
> *Special features*: Strong visual impact; novelty; application of physiology to ecology

INTRODUCTION

Gas vacuoles are structures which provide many planktonic cyanobacteria (blue–green algae) and bacteria with buoyancy. They appear as bright, refractile granules which can be distinguished from other cellular inclusions by their remarkable property of disappearing when subjected to a moderate pressure. The gas vacuoles are made up of stacks of submicroscopic, hollow structures known as gas vesicles (Bowen and Jensen 1965). The gas vesicles are cylindrical, with cone-shaped ends. They vary in length, growing up to about 1 μm long, but they have a constant diameter, usually 70 nm. They are rigid, brittle structures which can be made to collapse flat under pressure and this explains the disappearance of the gas vacuoles.

The gas vesicles are assembled from a single type of protein (Jones and Jost, 1970), which is arranged in ribs running at right angles to the structure. The protein provides the rigidity of the structure and has a hydrophobic surface, facing the gas space, which prevents liquid water from entering. The gas vesicle is permeable to gases, however, so that the vacuole gas is usually air at a pressure of one atmosphere (Walsby, 1969). Neither the gas nor the protein have any special metabolic function.

The density of the gas vesicles is about one-tenth of that of water and they are therefore very effective in providing buoyancy. The fact that gas vacuoles are restricted to planktonic micro-organisms suggests that the provision of buoyancy is their only function (Walsby, 1972). Some

---

planktonic cyanobacteria are able to regulate their buoyancy by controlling their degree of gas-vacuolation. Under low light intensities they form abundant gas vesicles and become buoyant, while under high intensities they collapse a proportion of their vesicles and sink. In this way they are able to regulate their vertical position with respect to light intensity in lakes. The collapse of gas vesicles at high intensities is brought about by a photosynthesis-dependent rise in cell turgor pressure. The pressure-sensitive gas vesicles can be used to measure this pressure in the cells (Walsby, 1971). For general reviews *see* Walsby (1972, 1978).

The aims of the following seven experiments are to demonstrate the properties of gas vacuoles in planktonic cyanobacteria, to evaluate their role in buoyancy, to isolate and purify the gas vesicles of which they are constituted, and to show how they can be used in the determination of turgor pressure in prokaryotic cells. The five experiments in part A require no special apparatus. The two in part B need a pressure nephelometer, the apparatus described by Walsby (1973). It can be improvised from equipment commonly available in microbiology departments, with a few additional pressure connections.

EXPERIMENTAL

A. Experiments needing no Special Equipment

1. The hammer, cork and bottle experiment (5 min + inspection after 1 h)
This experiment was first described by Klebahn in 1895. Here it is done first with *Anabaena flos-aquae* and then with *Microcystis aeruginosa*.

Fill a McCartney bottle (*see* note 5) to the brim with a suspension of *A. flos-aquae* (*see* note 4). Insert a rubber stopper (*see* note 4), squeezing it out-of-round as you do so, to allow the displaced suspension to escape.

↓

Wrap the bottle in a cloth and then strike the stopper sharply with a hammer.

↓

Note the immediate change in appearance of the suspension as the light-scattering gas vesicles are collapsed by the sudden rise in pressure.

↓

Leave the bottle (and an untreated control) to stand for 1 h and note the *Anabaena* filaments sinking out of suspension (and floating up in the control bottle).

↓

Examine a sample from each bottle under the microscope (preferably phase contrast) using the × 40 objective. Draw a few cells of each, illustrating the gas vacuoles and the change in appearance on pressurizing.

↓

Repeat the hammer, cork and bottle experiment with a suspension of *M. aeruginosa* (*see* note 2) gathered from a waterbloom. Note that the colonies of this unicellular organism float more rapidly before pressure, and sink more rapidly after pressure, than *Anabaena* filaments.

_____

2. Proportion of gas vacuoles required to provide buoyancy (30 min)

Dilute the *Microcystis* suspension with water in a 50 ml beaker. Pick out several (about 10) of the largest colonies individually with a Pasteur pipette and transfer to a second beaker of water.

↓

Select the colony which floats to the surface most quickly and eject it from the pipette at the bottom of a 10 ml measuring cylinder filled to the lip with water (placed in a 1 litre beaker of water to act as a thermostatic water-bath).

↓

As the colony floats up, measure its floating speed by timing it as it crosses the 3 ml and then the 10 ml mark on the measuring cylinder.

↓

Measure the distance between these marks and calculate the speed in mm min^{-1}.

↓

Repeat the measurement twice and calculate the mean speed.

↓

Transfer the colony to a McCartney bottle of water and apply pressure as described in experiment 1. Transfer the colony to the meniscus of the water in the measuring cylinder; let the colony fall out of the end of the pipette rather than expelling it in a rush of water.

↓

Time the colony sinking between the 10 and 3 ml marks and calculate the mean sinking speed (after replication).

↓

Calculate the proportion of gas vacuoles needed to make the colony neutrally buoyant in water (i.e. the same density as water).

_____

3. Collapse of gas vacuoles by centrifugation (1 h)

Place a suspension of the *Microcystis* colonies in a beaker and leave undisturbed for 10 min. Then draw off the floating colonies with a Pasteur pipette.

↓

Set up duplicates of the three centrifuge tubes shown in the diagram (*see* note 6).

↓

Centrifuge each tube opposite its balanced duplicate at $500 \, \text{m s}^{-2}$ ($\equiv 51 \, g$, generated at 593 rev. min^{-1} at a distance of 13 cm from the centrifuge axis) for 5 min. Remove the tubes without disturbing the contents, and observe that the *Microcystis* colonies are floating at the surface or at the top of the inverted tube.

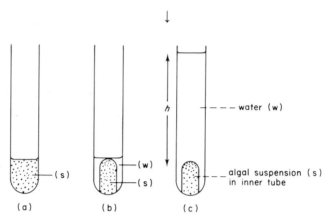

Figure 1   Diagram showing how the three centrifuge tubes should be set up. For (a) and (b) first drop in the inverted tube. Fill the outer tube with the suspension of cyanobacteria and, placing a finger over the open end, rock it over and back until the air has been displaced from the inner tube. Remove most of the suspension from the outer tube with a syringe fitted with a long needle and replace with water to the level shown.

Calculate the pressure generated at the top of the inverted tube, taking account of the depth h of liquid in the tube (*see* the method given below).

↓

Without disturbing the contents, replace the tubes in the centrifuge and spin at 1000, 1200, 1400, 1600, 1000 and 3000 rev. min^{-1} for 1 min. After each step lift out the tubes and see if the colonies have sunk or are floating.

↓

Calculate the pressure at depth h in tube (c) and thereby determine the pressure required to collapse sufficient of its gas vacuoles to destroy its buoyancy (cf. experiment 2).

Students should be asked why the colonies do not sink in tubes (a) and (b) at the highest speeds (*see* Further Information).

4. Loss of buoyancy in high light intensity *(10 min and 10 min after 4 h incubation period. Next morning: 5 min, or 30 min for filament counts)*

Place two 100 ml samples of an *A. flos-aquae* culture which has been grown at low light intensity (< 800 lux, 18–22 °C with aeration for 10–14 days) in separate 250 ml Erlenmeyer flasks. Insert cotton plugs pierced with Pasteur pipettes attached to an air-line to aerate the cultures.

↓

Place one flask at low light intensity (~ 400 lux) and the other at a high intensity (~ 4000 lux; say 2 cm from a 40 W fluorescent light tube) and leave them bubbling until the end of the practical class (preferably for at least 4 h).

↓

Turn off the bubbling and leave to stand overnight in the same position

↓

Next morning compare, by eye, the buoyant state of the *Anabaena* in the two flasks.

A more quantitative comparison of the buoyancy state can be made as follows.

Mix the contents of each flask thoroughly by swirling gently in one direction and then the next.

↓

Transfer a drop of the culture with a Pasteur pipette to the gap between a haemocytometer (depth 0·2 mm) platform and its coverslip.

↓

Leave for 10 min while the buoyant filaments float up in contact with the coverslip and the non-buoyant ones sink on to the haemocytometer platform. The position of the filaments can easily be determined by the different planes of focus under the × 10 objective; the sunken filaments will be in focus with the grid rulings on the platform.

↓

Count at least 100 filaments over the grid squares, scoring the number sinking and floating and then calculate the percentage of each.

_ _

5. *Isolating gas vesicles from* Anabaena flos-aquae *(30 min preparation, 1½ h centrifugation (over lunch hour) followed by 30 min)*

Get a microscope slide and coverslip ready to use, and a phase contrast microscope set up and in focus with the × 40 objective.

↓

Place 1 ml of a highly concentrated *Anabaena* suspension in a 10 ml round-bottomed centrifuge tube. Add 1 ml of 1·4 M sucrose solution and mix.

↓

Immediately withdraw a drop of the mixture to the microscope slide and observe the cells. They shrink in the hypertonic sucrose solution and then, after 1–2 min of mixing, are torn apart from their neighbours. The gas vesicles, seen as tiny points of light, highly agitated by Brownian movement, escape from the erupting cells.

↓

Leave the suspension to stand, with occasional gentle shaking, for 10 min and then layer over it

0·5 ml of 0·01M phosphate buffer, pH 7·5; this should be added slowly, drop by drop close to the meniscus so that it forms a clear layer floating on top of the sucrose lysate.

↓

Put the tube, balanced against a similar one, in a centrifuge with a swing out rotor and spin at 1500 rev. min^{-1} for $1\frac{1}{2}$ h (in the lunch break). (If larger quantities of gas vesicles are to be isolated a slower initial speed must be used: *see* experiment 3, in this section, on pressure developed by centrifugation.)

↓

Remove the tube very carefully. The isolated gas vesicles will have floated to the meniscus (and may be removed with a syringe needle held just touching the meniscus in the tube). The preparation will be discoloured green by unlysed gas-vacuolate cells and by packets of gas vesicles trapped in photosynthetic membrane vesicles. These can be removed by:

1. filtering twice through 0·45 μm pore size membrane filters, with only very gentle suction;
2. addition of a drop of sodium dodecyl sulphate, giving a final concentration not exceeding 0·015 %, followed by centrifugation of the gas vesicles to the surface again.

The purified gas vesicle preparation is milky white. It clears to the transparency of water on applying pressure sufficient to collapse the structures. Over 98 % of the turbidity of the suspension is lost on collapsing the structures. This is the turbidity generated by the gas-filled spaces, which have a much lower refractive index than the surrounding water.

The approximate size distribution of the isolated gas vesicles can be investigated by applying drops of the turbid suspension to membrane filters of smaller pore size. The filtrate is turbid if the vesicles pass through and clear if they do not.

B. Experiments needing Specialist Equipment

6. *Accurate measurements of gas vesicle critical pressure and cell turgor pressure (20 min)*

The collapse of gas vesicles with pressure can be measured with a pressure-nephelometer. This comprises a stout glass tube of the sort shown in Figure 2a (capable of withstanding 25 bars pressure) connected to a compressed air or gas supply. The turbidity of a gas-vacuolate suspension in the tube is measured in a nephelometer (turbidometer) or colorimeter. As pressure is applied, causing gas vesicles to collapse, the turbidity falls.

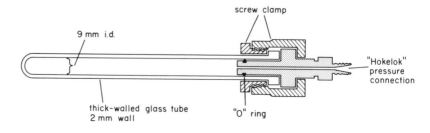

Figure 2a The pressure nephelometer tube and its connector (half actual size).

The apparatus used is shown in Figure 2b. It can be constructed from any tubing, needle valves and couplings capable of withstanding high pressure. Teflon tubing, 1 mm bore, joined with Hokelok coupling (1/8″) makes a neat arrangement. An EEL nephelometer is the ideal optical arrangement but other makes of colorimeter will do instead.

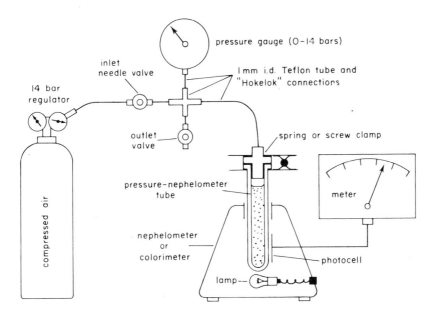

Figure 2b Diagram of the principal components of the pressure nephelometer. For a full description *see* Walsby (1973).

Fill the pressure-nephelometer tube (*see* note 3) with the suspension of gas-vacuolate cyanobacteria. If necessary, shake it thoroughly to disperse the organism, and then tilt it to and fro to remove air bubbles. Do not knock or jolt the tube as the shock may collapse the gas vesicles.

↓

Fasten on the pressure-tight coupling. Lower the tube gently into the nephelometer or colorimeter. With the instrument zeroed against a tube of water, amplify the reading to, or near, the full scale deflection.

↓

When the reading (B) is steady, open the inlet valve until the pressure reaches 0·5 bar, then close the valve. Read the turbidity on the meter ($A_{0.5}$).

↓

Take the pressure up in steps of 0·5 to 9 bars noting the steady reading (A_1 $A_{1.0}$, etc.) after each application.

↓

Finally, apply 13 bars and note the final reading (C).

----

7. Turgor pressure (additional 10 min)

Repeat the measurement made in experiment 6, but with a suspension of the cyanobacteria freshly mixed with an equal volume of 1·0 M sucrose.

↓

The readings must be completed within 4 min or the cells will lyse, generating other turbidity changes (see Walsby, 1980). The hypertonic sucrose solution abolishes the cell turgor pressure so that more pressure is required this time to collapse the gas vesicles.

↓

Plot the collapse pressure curve as before.

↓

Measure the difference between the top of the two curves on the pressure axis. This difference (in bars) is equivalent to the turgor pressure, which contributes to gas vesicle collapse in the first series of measurements with cells suspended in water (or dilute culture medium).

7a. Turgor pressure rise (10 min + 10 min after 3 h)

Experiments 4 and 7 can be combined to demonstrate the rise in turgor pressure on transferring A. flos-aquae to an increased light intensity. Take one set of measurements at the beginning of experiment 4, and another after the culture has been held at a high intensity for 3 h. The turgor pressure eventually rises to a level at which some of the weaker gas vesicles in the cells collapse. As a result, the cells lose their buoyancy at the high light intensity.

Notes and Points to Watch

1. The culture of *Anabaena flos-aquae* CCAP strain 1403/13f should be grown at 18–22 °C, a relatively low light intensity (< 1000 lux) so that it becomes highly gas vacuolate. For concentrated suspensions (experiment 5) leave a 3-week-old culture (21) to stand on the bench for 3 days and then draw off the floating cream of filaments with a wide-bore syringe needle, its tip held just in contact with the meniscus.
2. *Microcystis aeruginosa* forms waterblooms on many shallow, eutrophic lakes or reservoirs in the UK from July to late September, and large quantities may be collected using a bottle held under the water with its lip at the surface. The material may be preserved for later use, with its gas vacuoles intact, by addition of neutralized 40 % formaldehyde solution to a final concentration of 4 %. Before the practical, pour generous samples into large beakers, diluting as

necessary with more 4 % formalin, and collect the buoyant material which floats up. Do not use material which is too concentrated for the hammer, cork and bottle experiment or the colonies will trap each other in a gel and will not sink or float.

3. Test all pressure nephelometer tubes to 50 % above the maximum pressure in the experiment. The exposed top of the tube should be taped with black tape. (I have never had a tube burst. I use tubes made from Beckman column chromatography tubes, 9 mm internal diam., and 13 mm outside diam. made to withstand 35 bars and tested without failure to 70 bars).

The round-bottomed tube has a 6 mm deep, 16 mm diam. lip, to which the high pressure connection can be clamped.

4. The rubber bung should fit half-way into the neck of the McCartney bottle. If it is too large it will bounce out on being struck; if too small it will be driven inside the bottle. Use number 13 bungs with bottles having a neck of 13 mm internal diam.

5. Test each McCartney bottle before the practical class to eliminate unsatisfactory ones. Never use flat-sided bottles: they burst.

MATERIALS

1. Culture of *Anabaena flos-aquae* CCPA strain 1403/13f which is rich in gas vacuoles and suspends very evenly.
2. Samples of *Microcystis* waterbloom collected during the summer and preserved in neutral formaldehyde.
3. Fluorescent light or illuminated incubator.
4. Pressure-nephelometer with a supply of compressed air or nitrogen at 14 bars.
5. Culture medium for *A. flos-aquae* containing the salts shown in the following schedule.

Macroelements:
K_2HPO_4, 0·039 g l^{-1}, 0·218 mM
$MgSO_4·7H_2O$, 0·075 g l^{-1}, 0·304 mM
$CaCl_2·2H_2O$, 0·036 g l^{-1}, 0·245 mM
Na_2CO_3, 0·020 g l^{-1}, 0·189 mM
We prepare stock solutions at 1000 × the concentrations given above and then add 1 ml per litre of final medium.

Iron
Prepare a solution containing:
Na_2EDTA, 6·34 g l^{-1}
$FeSO_4·7H_2O$, 4·98 g l^{-1}.

Bubble the solution overnight. Keep in the dark. Use 1 ml per litre of final medium.

Trace elements

Element	Salt	$mg\,l^{-1}$ of salt	$mg\,l^{-1}$ of element	
Mn	$MnSO_4·4H_2O$	2·030	0·5	A
Mo	MoO_3	0·184	0·1	
B	H_3BO_3	2·853	0·5	
Zn	$ZnSO_4·7H_2O$	0·2199	0·05	
Cu	$CuSO_4·5H_2O$	0·0786	0·02	B
Co	$Co(NO_3)_2·6H_2O$	0·0494	0·01	

Prepare separate "superstock" solutions of group A at 10000 ×, and of group B at 100000 × the concentrations given. Take 50 ml each of group A superstocks and 5 ml each of group B and make up to 500 ml. This is the "trace element stock solution". Take 1 ml of this per litre of final medium.

Alternative: use the Allen and Arnon trace element solution of which this is based.

SPECIFIC REQUIREMENTS

Experiment 1
2 McCartney bottles with well-fitting rubber stoppers

hammer
cloth
microscope, slides and coverslips

cont.

cont.

Experiment 2
250 ml beakers
Pasteur pipettes
10 ml graduated measuring cylinder
1 litre beaker
stop-watch
apparatus from experiment 1

Experiment 3
Pasteur pipettes
six 10 ml round-bottomed centrifuge tubes
two tiny test tubes, 3 mm i.d. × 10 mm long
centrifuge with swing-out head

Experiment 4
air supply to aerate cultures
2 × 250 ml Erlenmeyer flasks with Pasteur
pipette serators
fluorescent lamp or light incubator
haemocytometer (preferably 0·2 mm depth)

Experiment 5
centrifuge and centrifuge tubes as experiment 3
1·4 M sucrose, 20 ml
0·01 M phosphate buffer, pH 7·5
1 ml syringe (disposable) with long blunt ended
    cannula (10 cm)
0·45 μm pore size membrane filters, 25 mm
    diam.
membrane filtration apparatus

Experiments 6 and 7
pressure nephelometer, with sample tube, blank
    tube, supply of compressed air or nitrogen at
    14 bars
1·0 M sucrose solution
test tube rack, for nephelometer tubes

FURTHER INFORMATION

1. The Hammer, Cork and Bottle Experiment

On striking the cork, the turbidity of the suspension decreases as the refractile gas vacuoles are destroyed. The suspension in the treated bottle immediately appears much darker than in the untreated one if viewed from the side from which it is illuminated (owing to a decrease in back-scattering). It is more transparent to light shining through it, but the difference may not be very obvious in concentrated suspensions.

The buoyant gas-vacuolate filaments of *Anabaena* float up only very slowly—a few millimetres per hour—owing to their small size (Stokes' law). Likewise, after gas vacuole collapse, they sink only slowly. The *Microcystis* colonies may have a similar density and degree of gas vacuolation, but being larger they float, and then sink, much faster (Stokes' law).

2. Proportion of Gas Vacuoles Required to Provide Buoyancy

The proportion of gas vacuoles required for neutral buoyancy is given by the expression

$$\frac{S}{S+F},$$

where F is the initial floating velocity (mm s^{-1}) of the colony and S its sinking velocity (mm s^{-1}) after the gas vacuoles are collapsed. (The average student can work this out for him/herself.)

The proportion of gas vacuoles required usually falls within the range of 0·4–0·7 (40–70 % of the initial gas vacuolation).

3. Collapse of Gas Vacuoles by Centrifugation

If buoyant colonies are selected for this experiment they probably will float up in all three tubes after the first centrifugation, as it is unlikely that even the pressure generated in tube (c) will be sufficient to collapse enough gas vacuoles to destroy buoyancy.

After the first centrifugation the colonies will therefore be at the surface (h = zero) in tubes (a) and (b), where they will not be subjected to pressure in subsequent centrifugations at higher speed. In tube (c), the colonies are prevented from rising to the surface at the first spin and when the required pressure at the depth h is generated in subsequent spins, probably between 1200 and 1600 rev. min^{-1}, the colonies will sink.

From this simple experiment students get experience in calculating the relationship between speed, acceleration and pressure in the centrifuge (invaluable for subsequent safe operation and of value in the next exercise). This exercise provides a rough alternate method to pressure-nephelometry (experiment 6) for investigating critical collapse pressures.

Centrifugation can be used to separate floating gas-vacuolate cyanobacteria from other micro-organisms and to purify isolated gas vesicles: but conditions which generate pressures that cause collapse of the gas vesicles must be avoided.

Calculation of pressure generated by centrifugation

When a column of liquid is accelerated, e.g. by gravity or by centrifugation, a pressure p is generated as given by the expression

$$p = h\rho g,$$

where h is the depth below the liquid surface (m), ρ is the density of the liquid (kg m^{-3}) and g is the acceleration (m s^{-2}). The S.I. unit of pressure is the Pascal (Pa).

$$1 \, Pa = 1 \, kg \, m^{-1} s^{-2}.$$

$$10^5 \, Pa = 1 \, bar \, (\simeq 1 \text{ atmosphere}).$$

Example 1   Find the hydrostatic pressure at a depth of 10 m in a freshwater lake.

$h = 10$ m; $\rho = 1000$ kg m^{-3} ($= 1$ g cm^{-3}).
$g = 9.8$ m s^{-2} (acceleration due to gravity at the earth's surface).
Therefore $p = 10$ m $\times 1000$ kg m$^{-3} \times 9.8$ m s^{-2}
$\qquad = 98\,000$ Pa
$\qquad \simeq 1$ atm.
Note that this is 1 atm. in excess of the atmospheric pressure pressing on the lake surface.

Example 2   Find the pressure at the base of a centrifuge tube containing 0.7 M sucrose solution of density 1090 kg m^{-3} forming a layer 0.02 m (2 cm) deep in a tube being centrifuged at 3900 m s^{-2} ($= 400 \, g$) generated by centrifuging at 1500 rev. min^{-1} at a distance of 16 cm from the centrifuge axis.
$p = 0.02$ m $\times 1090$ kg m$^{-3} \times 3900$ m s^{-2}
$\quad = 85\,020$ Pa
$\quad \simeq 0.85$ atm.
Calculations of this sort are used when employing centrifugation to purify gas vesicles or gas-

vacuolate organisms. If the pressure generated at the bottom of the cup does not exceed the minimum critical collapse pressure then all the buoyant gas vesicles or gas-vacuolate organisms will float to the surface.

4. Loss of Buoyancy in High Light Intensity

In the flask left at low light intensity the majority of the filaments will have floated to the surface overnight, and will tend to have accumulated around the edge of the meniscus (because it is the highest point); some filaments will remain suspended in the culture. Most of the filaments in the flask left at high light intensity overnight will have sunk to the bottom and very few will be left in suspension.

The quantitative assessment by haemocytometer will show that $> 80\%$ of the low-intensity treated filaments are buoyant, while fewer than 10% of the high-intensity treated filaments are buoyant. (These experiments also show that very few filaments are neutrally buoyant.)

5. Isolating Gas Vesicles from *Anabaena*

This exercise is self-explanatory. The final miracle of "turning milk into water" on pressurizing the suspension of isolated vesicles has a strong visual impact.

The gas vesicles of *A. flos-aquae* are 70 nm wide and mainly 200–700 nm long. They are increasingly retained by filters of pore size < 450 nm and quantitatively retained by one of 50 nm.

Table 1   Pressure/turbidity changes in *Anabaena*.

Pressure (bars)	In medium			In 0·5 M sucrose		
	Turbidity reading (A)	Turbidity change ($B - A$)	Percentage g.v. collapse $100(B - A)/(B - C)$	Turbidity reading (A)	Turbidity change ($B - A$)	Percentage g.v. collapse $100(B - A)/(B - C)$
0·0	100(= B)	0	0	100	0	0
0·5	100	0	0	100	0	0
1·0	98	2	2	100	0	0
1·5	87	13	16	100	0	0
2·0	68	32	38	100	0	0
2·5	50	50	60	100	0	0
3·0	35	65	78	100	0	0
3·5	25	75	90	100	0	0
4·0	21	79	95	100	0	0
4·5	19	81	98	100	0	0
5·0	17	83	100	94	6	7
5·5	17	83	100	77	23	26
6·0	17	83	100	57	43	49
6·5	17	83	100	37	63	72
7·0	17	83	100	22	78	89
7·5	17	83	100	16	84	95
8·0	17	83	100	13	87	99
8·5	17	83	100	12	88	100
12·0	17(= C)	83	100	12	88	100

6/7. Measurement of Critical Pressure and Turgor Pressure

The percentage of the gas vesicles collapsed by each pressure is given by the expression

$$\frac{B - A}{B - C} \times 100.$$

Table 1 gives typical results for changes in turbidity on applying pressure to a suspension of *Anabaena* filaments, first suspended in culture medium and then in 0·5 M sucrose.

These two sets of results are plotted in Figure 3 (solid lines). The difference between the two curves on the pressure axis gives a measure of the turgor pressure (dashed line) which, as is seen, falls slightly as the gas vesicles are collapsed and the cells consequently shrink.

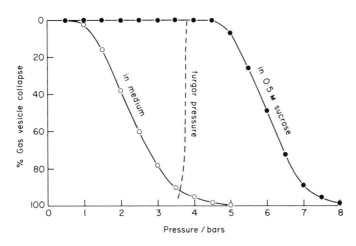

Figure 3   Collapse of gas vesicles with pressure in the planktonic cyanobacterium *A. flos-aquae.*

REFERENCES

C. C. Bowen and J. E. Jensen (1965). Blue–green algae: fine structure of the gas vacuoles. *Science*, **147**, 1460–2.

D. D. Jones and M. Jost (1970). Isolation and chemical characterization of gas vacuole membranes from *Microcystis aeruginosa* Kuetz. emend. *Elenkin. Archives für Mikrobiologie*, **70**, 43–64.

H. Klebahn (1895). Gasvakuolen, ein Bestandteil der Zellen der Wasserblulebildenden Phycochromaceen. *Flora (Jena)*, **80**, 241–82.

A. E. Walsby (1969). The permeability of the blue–green algal gas–vacuole to gas. *Proceedings of the Royal Society of London, B*, **173**, 235–55.

A. E. Walsby (1971). The pressure relationships of gas vacuoles. *Proceedings of the Royal Society of London, B*, **178**, 301–26.

A. E. Walsby (1972). The structure and function of gas vacuoles. *Bacteriological Reviews*, **36**, 1–32.

A. E. Walsby (1973). A portable apparatus for measuring relative gas vacuolation, the strength of gas vacuoles, and turgor pressure in planktonic blue–green algae and bacteria. *Limnology and Oceanography*, **18**, 653–8.

A. E. Walsby (1978). The gas vesicles of aquatic prokaryotes. *In* "Relations between Structure and Function in the Prokaryotic Cell", pp. 327–58. Symposia of the Society for General Microbiology No. 28.

A. E. Walsby (1980). The water relations of gas-vacuolate prokaryotes. *Proceedings of the Royal Society of London, B* **208**, 73–102.

71

Use of Mathematical Models in Microbial Ecology

J. I. PROSSER

*Department of Microbiology, Marischal College, University of Aberdeen,
Aberdeen, AB9 1AS, Scotland*

and

M. J. BAZIN

*Department of Microbiology, Queen Elizabeth College,
Campden Hill, London, W8 7AH, England*

> *Level*: Advanced undergraduates
> *Subject areas*: Microbial ecology
> *Special features*: Introduction to use of computers

INTRODUCTION

Many undergraduate practicals are concerned with either experimental testing of a previously constructed hypothesis or with observation of phenomena which may then lead to formulation of hypotheses. The exercise described here is concerned with the process of constructing a hypothesis within the framework of a mathematical model. As such, it involves no practical experimental work with micro-organisms, but is linked to the experimental process by generation of predictions which, in a complete investigation, would be tested experimentally.

Mathematical models are increasingly being used in the field of microbial ecology (*see* Bazin *et al.*, 1976). They have been applied to natural ecosystems, where the complexity of the system usually prohibits consideration of the growth kinetics of individual species. Such models are often used to seek out correlations between different factors and to describe experimental data in a simplified form. Models have also been used to study interactions between two species in well-defined, controlled laboratory systems, e.g. chemostats and percolating soil columns. This usually involves a description of growth of each species in terms of the Monod equation

$$\mu = \mu_\mathrm{m} \frac{S}{K_s + S},$$

where μ = specific growth rate,
  μ_m = maximum specific growth rate,
  S = substrate concentration,
  K_s = saturation constant for growth,

or some similar function, with added terms for product formation, inhibition, etc. Here the model usually represents a hypothesis about the fundamental mechanisms controlling growth and interaction between the two species.

The aim of this exercise is to introduce students to the technique of mathematical modelling by allowing them to construct a model representing an interaction, of their own choice, between two micro-organisms in a chemostat. It also leads to a greater understanding of chemostat growth and provides an introduction to the use of computers. The exercise described here requires access to the CDC simulation package MIMIC (*see* note 3).

EXPERIMENTAL

Introduction *(30–45 min)*

The following model represents "neutralism" in a chemostat, the situation where two micro-organisms coexist without any observable effects on each other. It is equivalent to a lack of interaction.

$$\frac{dA}{dt} = \mu_A A \frac{S_A}{K_A + S_A} - DA \tag{1}$$

$$\frac{dB}{dt} = \mu_B B \frac{S_B}{K_B + S_B} - DB \tag{2}$$

$$\frac{dS_A}{dt} = D(S_{A_I} - S_A) - \mu_A A \frac{S_A}{K_A + S_A} \tag{3}$$

$$\frac{dS_B}{dt} = D(S_{B_I} - S_B) - \mu_B B \frac{S_B}{K_B + S_B}, \tag{4}$$

where A = biomass of organism A,
  B = biomass of organism B,
  D = dilution rate,
  S = concentration of substrate,
  μ = maximum specific growth rate,
  K = saturation constant for growth (substrate concentration at which specific growth rate = half maximum rate),
  S_I = substrate concentration in inflowing medium,
  t = time.

Subscripts A and B refer to values for organisms A and B and their respective substrate S_A and S_B (*see* note 1).

This model is either (a) presented and explained or, preferably, (b) extracted from the students by describing a chemostat, if they are not already familiar with one, and asking them what is determining rates of change of A, B, S_A and S_B.

Individual models (*30 min*)

Arrange for each student to construct a model of an interaction of his/her own choice, between two micro-organisms (not neutralism) in a chemostat, by modifying the above equations (*see* note 2). It is preferable, but not necessary, that the student chooses an interaction that he/she knows to occur in nature.

Computing (*30 min*)

The differential equations are solved by means of a FORTRAN-based simulation package, MIMIC. Explain the following program to the students. It produces predictions from the neutralism model in the form of a table of values for A, B, S_A and S_B at different times and also produces graphical output (*see* note 3).

Statement	Comment
CON (UA, UB, KA, KB)	Define and read in the values for
CON (SAO, SBO, AO, BO)	constants.
CON (SAI, SBI, TMAX, DT)	
PAR (D)	Define and read in the values for parameters
A = INT (UA*A*SA/(KA + SA) − D*A, AO)	
B = INT (UB*B*SB/(KB + SB) − D*B, BO)	Integrate the differential
SA = INT (D*(SAI − SA) − UA*A*SA/(KA + SA), SAO)	equations
SB = INT (D*(SBI − SB) − UB*B*SB/(KB + SB), SBO)	
FIN (T, TMAX)	Finish simulation when $T \geqslant$ TMAX
HDR (T, A, B, SA, SB)	Print headings for tables.
OUT (T, A, B, SA, SB)	Print values of T, A, B, SA and SB at interval DT.
PLO (T, A, B)	Plot A and B vs T.
PLO (T, SA, SB)	Plot SA and SB vs T.
END	

·5	·2	5·	10·	
0·	0·	1·	1·	Data values for constants
100·	100·	100·	2·	
·05				Data values for parameters
·02				

Symbols for program variables, and corresponding symbols in equations (1) to (4), are given in Table 1 (*see* Further Information), along with values assigned to them by the above program.

Individual programs (2 h, but see *note 4)*
Get the students to modify this program to incorporate their own model.

Data (15 min)
Provide data values for growth constants, initial conditions, etc. (*see* Further Information).

Calculation (variable time, see *note 4)*
Run the modified program.

Results (variable time)
Examine the output and correct errors if necessary. Data values, and even the model itself, may then be changed to give more predictions, the aims being: (1) to provide theoretical predictions which could be tested by experiment, (2) to discover which parameters and constants are important in controlling the interaction and (3) to discover, by model modification, if alternative hypotheses predict significantly different behaviour.

Notes and Points to Watch

1. To increase simplicity, yield coefficients are not included in these equations. This should be pointed out and explained by stating that biomass is described in terms of substrate equivalents. If a single species is consuming more than one substrate, yield coefficients must be included.
2. Commonly chosen interactions and required modifications are given below.

Interaction	*Modification*
Competition	Single substrate S replaces S_A and S_B. Equation (4) omitted. Equation (3) contains the additional term $(\mu_B BS)/(K_B+S)$.
Predation	Replace S_B by A (the prey). Omit equation (4). Modify equation to include removal of A by predation by addition of term $-\mu_B BA/(K_B+A)$.
Product/toxin formation	Requires an additional equation $dP/dt = K_p A - DP$, where P = concentration of product K_p = rate of production by A, or $dP/dt = \mu_A A \dfrac{S_A}{S_A + K_A} \cdot K_p - DP,$ where production is growth rate dependent.
Inhibition	Modification of growth rate function to give, e.g. $\mu = \mu_A \dfrac{S_A}{K_A + S_A} \cdot \dfrac{K_I}{K_I + P},$ where P is inhibitor concentration, K_I a constant.

Growth stimulation   Modification of growth rate function to give, e.g.

$$\mu = \mu_A \frac{S_A}{K_A + S_A} \frac{P}{K_p + P}$$

$$\text{or } \mu = \mu_{A_1} \frac{S_A}{K_A + S_A} + \mu_{A_2} \frac{P}{K_P + P}.$$

These modifications are merely suggestions and many other possible functions may be used depending on how students view the particular interaction which they are considering.

3. The program above is written in a simulation language, MIMIC, which is available on a number of university computers. If it is unavailable, there will almost certainly be a suitable, alternative simulation language. An example of the use of the simulation language CSMP (available on IBM machines) using a model of the type described here is given by Curds and Bazin (1977). Programs may be written in FORTRAN or ALGOL, but such programs will be more complex and more difficult to modify.

4. The methods for running, storing and modifying programs will depend on the university, the type of computer and the simulation language used. Programs may be stored on file (and modified using visual display units) or on punch cards. This will affect the time taken to complete stage 4. Turn-round times will also vary. Some systems are interactive, providing results immediately, while others may provide turn-round overnight.

5. Throughout, interactions between micro-organisms, as opposed to bacteria, are stressed. This broadens the range of natural interactions that can be chosen, e.g. antibiotic production by *Penicillium*. It also allows consideration of prey–predator interactions, which often given the most interesting predictions.

SPECIFIC REQUIREMENTS

computing facilities

FURTHER INFORMATION

It is wise to run programs of different interactions before this practical so that sensible values for constants may be used and tested. It is also desirable, where possible, to provide references to experimental work on interactions similar to those actually chosen. This enables comparison of students' results with published experimental data and may aid in choosing values for growth constants, etc. Examples of such work are given by Curds (1971), Harder and Veldkamp (1971) and Lee *et al.* (1976).

Worked Example

To illustrate the running of this practical an example has been chosen where organism A produces a compound P, e.g. an antibiotic, which inhibits growth of organism B. It is assumed that A and B are growing in a chemostat with two limiting substrates, S_A and S_B respectively, and that P inhibits growth of B non-competitively.

Table 1 Program variables and their values.

Model symbol	Program symbol	Value assigned
μ_A	UA	0·5
μ_B	UB	0·2
K_A	KA	5·0
K_B	KB	10·0
—	SAO	0·
—	SBO	0·
—	AO	1·
—	BO	1·
S_{A_I}	SAI	100·
S_{B_I}	SBI	100·
—	TMAX	100·
—	DT	2·
D	D	·05
D	D	·2

Equations 1 and 3 are unchanged from the neutralism model. Equations 2 and 4 must be modified to take into account inhibition of growth of B by P:

$$\frac{dB}{dt} = \mu_B B \frac{S_B}{K_B + S_B} \frac{K_I}{K_I + P} - DB \tag{2}$$

$$\frac{dS_B}{dt} = D(S_{B_I} - S_B) - \mu_B B \cdot \frac{S_B}{K_B + S_B} \frac{K_I}{K_I + P} \tag{4}$$

where P = the concentration of compound P in the chemostat and K_I = the concentration of P at which specific growth rate = half maximum rate.

A fifth equation is also required describing the rate of production of P;

$$\frac{dP}{dt} = K_P A - DP, \tag{5}$$

where K_P = rate of production of P per cell per unit time.

Data values are now required for K_P, K_I and P_O (the initial concentration of P). Chosen values are ·015, 10 and 0 and the modified program is as follows:

```
      CON (UA, UB, KA, KB)
      CON (SAO, SBO, AO, BO)
      CON (SAI, SBI, TMAX, DT)
      CON (KP, KI, PO)
      PAR (D)
A     INT (UA*A*SA/(KA + SA) − D*A,AO)
SA    INT (D*(SAI − SA) − UA*A*SA/(KA + SA),SAO)
B     INT (B*UB*(SB/(KB + SB))*(KI/(KI + P)) − D*B,BO)
SB    INT (D*(SBI − SB) − B*UB*(SB/(KB + SB))*(KI/(KI + P)),SBO)
```

```
P    INT  (KP*A – D*P, PO)
     FIN  (T,TMAX)
     PLO  (T,A,B)
     PLO  (T,SA,SB,P)
     HDR  (T,A,B,SA,SB,P)
     OUT  (T,A,B,SA,SB,P)
     END
·5        ·3        5·        10·
0·        0·        1·        1·
100·      100·      ·500      5·
·015      10·       0·
·05
·2
```

The modified program is then run and results are illustrated graphically in Figure 1. Biomass of

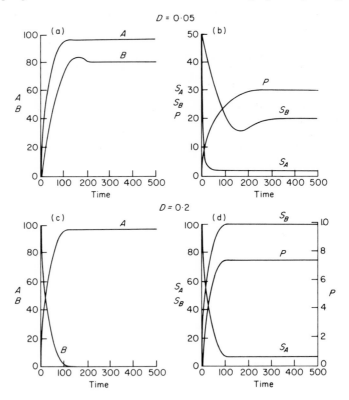

Figure 1 Simulation of a modified model (as used in the worked example) describing inhibition at two dilution rates.
(a) changes in A and B at $D = 0.05\ \mathrm{h}^{-1}$.
(b) changes in S_A, S_B and P at $D = 0.05\ \mathrm{h}^{-1}$.
(c) changes in A and B at $D = 0.2\ \mathrm{h}^{-1}$.
(d) changes in S_A, S_B and P at $D = 0.2\ \mathrm{h}^{-1}$.

A and substrate concentration S_A show expected behaviour, reaching steady state values of 99·44 and 0·56 respectively, at a dilution rate of 0·05 h^{-1}, and 96·67 and 3·33 at $D = 0·2$ h^{-1}. (It is a useful exercise for students to check the program by calculating these steady state values from the steady state solutions of the chemostat equations, i.e. by setting equations (1) and (3) equal to 0.)

B increases steadily until the concentration of P becomes significant. B then falls and eventually achieves a steady state. The steady state level is much lower than that of A despite similar values for μ_A and μ_B and K_A and K_B. (Growth of B in the absence of P can be observed by setting K_P and P_0 equal to 0.) Concentration of P increases parallel to increase in A. At the faster dilution rate, the inhibitory effect of P results in wash-out of B.

Again the program could be checked by comparing simulated steady state values with those calculated by setting equations (1) to (5) equal to 0.

REFERENCES

M. J. Bazin, P. T. Saunders and J. I. Prosser (1976). Models of microbial interactions in the soil. *CRC Critical Reviews in Microbiology*, **4**, 463–98.

C. R. Curds (1971). A computer simulation of predator-prey relationships in a single stage continuous culture system. *Water Research*, **5**, 793–812.

C. R. Curds and M. J. Bazin (1977). Protozoan predation in batch and continuous culture. *In* "Advances in Aquatic Microbiology", (M. R. Droop and H. W. Jannasch, eds). Academic Press, London and New York.

W. Harder and H. Veldkamp (1971). Competition of marine psychrophilic bacteria at low temperatures. *Antonie van Leeuwenhoek*, **37**, 51–63.

J. H. Lee, A. G. Fredrickson and H. M. Tsuchiya (1976). Dynamics of mixed cultures of *Lactobacillus plantarum* and *Propionibacterium shermanii*. *Biotechnology and Bioengineering*, **18**, 513–28.

A Taxonomic Investigation of Bacteria by Simplified Numerical Techniques

T. R. G. GRAY

Department of Biology, University of Essex, Colchester, CO4 3SQ, England

Level: All undergraduate years
Subject areas: Microbial taxonomy
Special features: Stresses logical procedure in taxonomy and separates objective and subjective thinking

INTRODUCTION

A large body of data about a set of diverse objects is difficult to appreciate unless it is organized in some way. The organization of organisms into groups based on such data is termed classification. With micro-organisms, the data employed to make such classifications are obtained from studies of structural, physiological, biochemical and immunological properties. The groups within these classifications are usually defined in terms of similarity, the similarity between any organisms within a group being greater than that between organisms in different groups. It is important to distinguish this process of classification from that of identification: the latter involves **the assignment of further unknown organisms to an already established classification.**

A crucial step in producing a classification is the estimation of resemblance between organisms, and this exercise aims to show one way of doing this using the procedures of numerical taxonomy (Sneath and Sokal, 1973). In this system, resemblance is estimated from a comparison of organisms based on a large number of properties. In the absence of *a priori* reasons for treating different properties differently, all properties are given equal weight in producing the classification. However, weighting is introduced when deciding how to produce a key to identify an organism.

In full-scale studies, a computer is necessary to make all the comparisons between all the organisms. However, in this limited study, it is possible to use punch cards instead. This has the advantage that each student can appreciate the information which needs to be incorporated in a computer program.

Five stages in the production of a classification and a key for identification are performed. These are

(1) testing a set of anonymous bacterial strains to reveal a wide range of characters (in this case 75 structural, physiological and biochemical properties);
(2) recording these characters on punch cards;
(3) calculating the resemblance between organisms by a simple matching coefficient;
(4) arranging the organisms into groups based on percentage similarity (phenons);
(5) selecting diagnostic characters for these phenons and constructing a key for identification.

At the end of the exercise, the names assigned to these bacteria by culture collections or other bacteriologists may be revealed to see how far the new classification is in accord with the existing system, and what hitherto unsuspected relationships may have been revealed.

EXPERIMENTAL (stages 1 and 2)

Each student is allocated three 24 h slope cultures of an anonymous bacterium, but students may work in groups if the time available is short. If students work individually, about $2\frac{1}{2}$–3 h are needed to initiate the testing, especially if this type of work has not been attempted before. The time taken may be shortened by use of micro-tests (*see* Further Information).

Day 1 *($2\frac{1}{2}$–3 h)*

Take one of the nutrient agar slope cultures provided and transfer bacteria aseptically to one or more vials for use by each student. Sufficient bacteria must be transferred so that after dispersion, the suspensions in the vials are faintly turbid. Use these suspensions as a source of inoculum for the physiological and biochemical tests.

↓

From the suspension in the vial, inoculate the different diagnostic media (*see* Table 1) by the following techniques:

Liquid cultures:	1 loopful of suspension
Tubed agar cultures:	1 stab with an inoculating needle
Agar plate cultures:	1 streak across the diameter of the plate
Antibiotic sensitivity tests:	Inoculate molten agar (45 °C) with 1 loopful of suspension, pour into plate, allow to solidify and then place four different antibiotic discs on each plate.

↓

Take the second nutrient agar slope culture provided and pour a small quantity of very freshly prepared 1 % tetramethyl *p*-phenylene diamine dihydrochloride onto the slope. The formation of a dark purple colour within 10s indicates the presence of an oxidase (test 62).

↓

Take the third nutrient agar slope culture. Observe the pigmentation. Streak the culture on a plate of nutrient agar to give separate colonies, incubate it until the next practical and confirm the observations on pigmentation. Record the results as follows:

Table 1 Scheme of tests for numerical taxonomy.

Type of culture	Medium	Character and character number
Liquid	Nutrient broth	Pellicle produced — 1
	Glucose ammonium salts broth	Uses ammonium as sole source of nitrogen — 2
	Glucose nitrate salts broth	Uses nitrate as sole source of nitrogen — 3
	Glucose cysteine salts broth	Uses cysteine as sole source of nitrogen — 4
	Glucose asparagine salts broth	Uses asparagine as sole source of nitrogen — 5
	Glucose urea salts broth	Uses urea as sole source of nitrogen — 6
	Glucose tryptophane salts broth	Uses tryptophane as sole source of nitrogen — 7
	Glucose uracil salts broth	Uses uracil as sole source of nitrogen — 8
	Glucose glutamic acid salts broth	Uses glutamic acid as sole source of nitrogen — 9
	Gluconate ammonium salts broth	Produces diffusible green pigment — 10
		Uses gluconate as sole source of carbon — 11
	Acetate ammonium salts broth	Uses acetate as sole source of carbon — 12
	Lactate ammonium salts broth	Uses lactate as sole source of carbon — 13
	Salicin ammonium salts broth	Uses salicin as sole source of carbon — 14
	Citrate ammonium salts broth	Uses citrate as sole source of carbon — 15
	Glucosamine salts broth	Uses glucosamine as sole C and N source — 16
	Peptone water (+ lead acetate strip *see* note 3)	Produces ammonia from peptone — 17
		Produces H$_2$S from peptone — 18
	Tryptone water	Produces indole from tryptone — 19
	Voges Proskauer medium	Produces acetoin from glucose — 20
	Nitrate broth	Reduces nitrate to nitrite — 21
	5% NaCl nutrient broth	Grows with 5% NaCl — 22
	7% NaCl nutrient broth	Grows with 7% NaCl — 23
	10% NaCl nutrient broth	Grows with 10% NaCl — 24
	12% NaCl nutrient broth	Grows with 12% NaCl — 25
	Nutrient broth	Grows at 37°C — 26
	Nutrient broth	Grows at 45°C — 27
	Nutrient broth (pH 4·5)	Grows at pH 4·5 — 28
	Nutrient broth (pH 9·0)	Grows at pH 9·0 — 29
Tubed agar	Hugh and Leifson medium (tests 30–46) with glucose (open)	Oxidative metabolism — 30
	glucose (closed)	Fermentative metabolism — 31
		gas from glucose — 32

Group	Medium/Substrate	Test	No.
	fructose (open)	acid from fructose	33
	rhamnose (open)	acid from rhamnose	34
	galactose (open)	acid from galactose	35
	xylose (open)	acid from xylose	36
	l-arabinose (open)	acid from l-arabinose	37
	lactose (open)	acid from lactose	38
	raffinose (open)	acid from raffinose	39
	sucrose (open)	acid from sucrose	40
	maltose (open)	acid from maltose	41
	salicin (open)	acid from salicin	42
	glycerol (open)	acid from glycerol	43
	dulcitol (open)	acid from dulcitol	44
	erythritol (open)	acid from erythritol	45
	mannitol (open)	acid from mannitol	46
	Aesculin agar	Hydrolyses aesculin	47
	Arginine agar	Produces NH_3 from arginine	48
Agar plate	Starch agar	Hydrolyses starch	49
	Tributyrin agar	Hydrolyses tributyrin	50
	Casein agar	Hydrolyses casein	51
	Gelatin agar	Hydrolyses gelatin	52
	Crystal violet agar	Sensitive to crystal violet	53
Antibiotic sensitivity	Nutrient agar + appropriate antibiotic disc, i.e.	Sensitive to:	
	chloramphenicol 5 µg disc^{-1}	chloramphenicol	54
	erythromycin 3 µg disc^{-1}	erythromycin	55
	polymyxin B 50 units disc^{-1}	polymyxin B	56
	novobiocin 5 µg disc^{-1}	novobiocin	57
	aureomycin 5 µg disc^{-1}	aureomycin	58
	penicillin 2 units disc^{-1}	penicillin	59
	streptomycin 2 µg disc^{-1}	streptomycin	60
	bacitracin 2 units disc^{-1}	bacitracin	61

Pigmented (on slope or as in test 10: *see also* note 2) or not pigmented   63
Yellow orange pigment or not yellow/orange pigmented   64
Colonies circular or otherwise.   65

↓

Remove some of the culture from the slope and make a faintly turbid suspension in demineralized water in a watchglass. Carry out the following staining procedures and record your results as shown.

Gram stain	Rod or coccus	66
	Single cells or filaments/clumps	67
	Gram +ve or Gram −ve	68
Spore stain	Spores present or absent	69
	Sporangia swollen or not swollen	70
	Spores oval or spherical	71
	Spores central or terminal	72
Hanging drop and	Motile or non-motile	73
flagella stain	Polar or peritrichous flagella	74

↓

When pigmentation has been checked on the nutrient agar streak plate, examine the shape of the cells again to see if they have undergone a marked change in shape.

Pleomorphic or not pleomorphic   75

↓

Incubate all cultures at 25 °C (except where stated otherwise).

Day 2 *(ideally 1 week later) (2–3 h)*

Observe all cultures for changes and make any necessary tests to detect these changes *see* note 1). The main changes should be as follows.

Test
1	Pellicle on surface of broth.
2–9, 11–16, 22–29	Presence of turbidity (dispersed or as a sediment: *see* note 5).
10	Presence of bright green colour (can occur in other media (10–15)).
17	Orange precipitate with Nessler's reagent.
18	Blackening or browning of part of lead acetate strip.
19	Pink colour, after addition of Ehrlich's reagent and a few drops of concentrated hydrochloric acid, soluble in amyl alcohol.
20	Add 5 ml 40 % KOH (*see* note 6) to 2 ml culture plus a trace of creatine; shake. Red colour within 1 h indicates production of acetoin.
21	Add Griess-Ilosvay's reagents I and II. A red colour indicates nitrite. If the test is negative, it is possible that all nitrite has been further reduced. This can be detected by testing for residual nitrate in the culture. Reduce this with Zn dust and then retest for nitrate (*see* note 4).

30–46   Examine the tubes for an indicator colour change. If a yellow colour is produced, even in only part of the tube, record it as positive. Ignore the change if the indicator has gone blue. It is helpful to compare the result with a control.

47   Hydrolysis of aesculin is shown by a blackening of the medium.

48   Production of ammonia is shown by the indicator turning blue.

49–52   Hydrolysis is indicated by a clear halo around the streak following development of the plate with Lugol's iodine (test 49), acidified mercuric chloride (tests 51 and 52: *see* note 6). A clear halo is visible on tributyrin agar (test 50) without further development.

53–61   Sensitivity is detected by a lack of growth on the agar (test 53) or around the sensitivity disc (tests 54–61: *see* note 7).

↓

Record all your results as + or − (including the morphological characteristics obtained in the previous practical) in a table. With some characteristics, no logical decision can be taken on which character state is positive, e.g. sensitivity or resistance to antibiotics. In such cases, score sensitivity as +.

↓

Transfer the results to punch cards. A useful card is the Pl Paramount card 190GSM Green CC19 $6\frac{1}{8} \times 4$ (Copeland Chatterson, Stroud, Gloucs) (Figure 1). If the character is positive, punch out the card at the appropriate numbered hole. Write the code letter for the organism clearly in the centre of the card.

↓

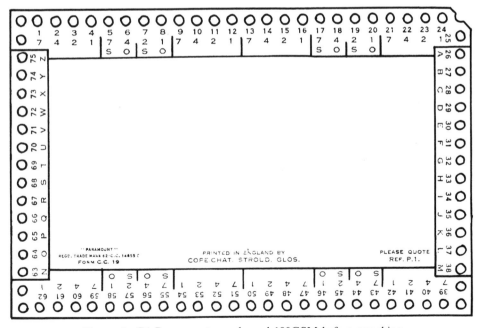

Figure 1 P1 Paramount punch card 190GSM before punching.

Then prepare as many duplicate cards as there are organisms being studied by the class. Label these with the code letter also. Exchange cards so that each group of students has a card for each of the organisms.

ANALYSIS OF DATA (stages 3, 4 and 5)

Data analysis may be done either during a practical class or during the students' own time.

Calculation of a Matching Coefficient

Determine the total number of characters being considered by subtracting from 75 the number of uniformly positive or uniformly negative characters.

↓

Stain edges of the card, corresponding to the students' own organism, with a red dye (dip it into a 0·5 cm deep trough of carbol fuchsin) and allow it to dry (*see* Brisbane and Rovira, 1961).

↓

Place the stained card on the first of the unstained cards, i.e. the cards for the organisms tested by the other groups of students. Any test which is positive and punched on the stained card, but negative and not punched on the unstained card will show as a green space in a red surround. Count these spaces and record their number.

Turn the pair of cards over so that the unstained card now lies on top of the stained card. Any differences will now show as red spaces filling gaps in the green card. Count these spaces and add them to the previous number of spaces. This shows the number of properties differing between the two strains. Subtract this figure from the total number of characters being considered to give the number of matching characters for the two strains.

↓

Calculate a matching coefficient as follows.

$$\text{Matching coefficient} = \frac{\text{No. of matching characters}}{\text{Total number of characters considered}} \times 100.$$

Calculate this coefficient for the strain investigated and all the other strains used (*see* note 8).

NB. This method of calculating the matching coefficient is simplified, for it assumes that all characters which are not positive are all directly comparable. This is not true for all characters, e.g. spores oval or spherical is a non-comparable character for organisms which do not produce spores. The effect of these errors is small in the experiment, and would be corrected if the data were analysed by computer.

↓

Arranging the Organisms into Groups

Draw up a similarity matrix, expressing the matching coefficients as percentages (to the nearest whole number). The following is a simple example.

	Strains				
	A	B	C	D	E
Strains A	100				
B	90	100			
C	55	50	100		
D	80	90	40	100	
E	40	50	95	40	100

The upper right-hand part of the diagram will be a mirror image of the lower left-hand part and is omitted.

↓

Rearrange the order of the strains so that the most closely related strains are put next to one another. This is most easily achieved by listing all the pairs of organisms in order of decreasing matching coefficients to form a table as below (Lockhart and Liston, 1970).

%M	Organism pairs
95	C–E
90	A–B B–D
80	A–D
55	A–C
50	B–C B–E
40	A–E C–D D–E

   Always cite the members of a pair in alphabetical order (e.g. A–B, not B–A). If there is more than one pair at any level, cite these in strict alphabetical order (e.g. A–B before B–D and B–C before B–E).

↓

Then arrange the organisms into groups. Select from the table the pair or pairs with the highest matching coefficient. If the pairs have organisms in common, amalgamate them, keeping the strains in alphabetical order. Repeat the exercise for pairs at the next highest matching coefficient level. If pairs at the new level share an organism in common with a pair at the preceding level, attach the pair from the current level to the end of the pair or group from the preceding level. Continue until all organisms are linked together in one group and a table is produced as follows.

%M	group
95	(C–E)
90	(A–B–D) (C–E)
80	(A–B–D) C–E
55	(A–B–D–C–E)

This procedure mimics that done by the computer and is known as single linkage clustering.

↓

Redraw the similarity matrix with the strains arranged in the new order

A	100				
B	90	100			
D	80	90	100		
C	55	50	40	100	
E	40	40	40	95	100
	A	B	D	C	E

↓

Translate the values of the matching coefficients into colours or shades (*see* note 9) and redraw a coloured or standard matrix (Sneath, 1962). A suggested scheme is 100 % (black or fully shaded), 90–99 % (violet or cross hatched), 80–89 % (blue or vertical hatching), 70–79 % (green or horizontal hatching), 60–69 % (red or dashed shading), 50–59 % (orange or dotted shading), 40–49 % (yellow or a single dot), below 40 % (no colour or shading).

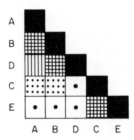

In this example, two major groups are evident, one consisting of organisms A, B and D, the other of C and E.

Construction of Dendrogram

The relationships between strains are often better shown by means of a tree-diagram or dendrogram.

To prepare the dendrogram, take a sheet of graph paper and mark a vertical axis with the % matching coefficent values.

↓

Re-examine the ordered table and, starting with the first pair (e.g. C–E), draw vertical lines from the 100 % level down to the point at which they amalgamate (e.g. 95 %). Connect the two lines at this point with a horizontal line.

↓

Then draw vertical lines for the next group (A–B–D) and join these at the 90 % level.

↓

The resulting groups link together at the 55 % level, so draw in further vertical lines, joined at this value (shown as dashed lines in the diagram.

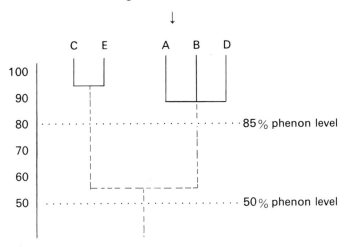

From the dendrogram, see what groups (or phenons) can be established. In the above example all five organisms would belong to one 50 % phenon, comprising two 85 % phenons.

Construction of a Key

Using the original punch cards, determine which characters are diagnostic for each of the suggested phenons, and for each of the strains studied. Use this information to construct a dichotomous key with the minimum number of steps possible. At each stage in the key, try to use more than one pair of contrasting characters, giving if possible some idea of the reliability of the character (i.e. is it possessed by 100 %, 90 %, or only 75 % of the organisms in the phenon), e.g.

1. Gram-negative (100 %), oxidative metabolism (100 %), sensitive to polymyxin (90 %)   2
   or not as above   3
2. Produces fluorescent pigment, reduces nitrate   Organism A
   Not as above   Organsim B
3. Rod shaped (100 %), with endospores (100 %), acid from fructose   Organism C
   Not as above   4
   etc., etc.

Comparison with Established Classifications

The students should then be given a list of the names of the organisms and asked to contrast the groupings they obtained with the "established" classification, and to comment on the similarities and differences between them. Useful information can be obtained from Buchanan and Gibbons (1974).

Notes and Points to Watch

1. A set of control tubes showing positive and negative test results is useful so that different members of a class can standardize their results. This is especially important where indicator changes are involved.
2. Diffusible green pigments, if they are produced, usually occur in the gluconate ammonium broth, but it is worth examining other organic acid media for the pigment.
3. When the sterile lead acetate paper is inserted in the peptone water tube, it must not come in contact with the broth.
4. Test 21 is a test for the ability of an **organism** to reduce nitrate to nitrite. The scoring of this test often confuses students who have to test for residual nitrate with zinc dust reduction.
5. All liquid cultures must be tapped gently and repeatedly to disturb growth that has sedimented.
6. Acidified mercuric chloride, used in tests 51 and 52, is highly toxic. It should not come in contact with skin (or gold objects with which it forms an amalgam). Remember to warn washing-up staff. Similar precautions should be taken with 10% KOH (test 20).
7. Some antibiotics produce wide zones of inhibition which merge with adjacent ones, especially if micrococci are being used. Results need careful interpretation.
8. If every individual or group calculates each matching coefficient, then all calculations will be done twice in the class. This serves as a useful check. All disputed results should be checked.
9. When similarity matrices are being coloured, ensure that all the colours used are either of the same intensity or that colour intensity decreases with the matching coefficient classes. If this is not done, a false picture of the groups can be obtained.

MATERIALS

1. Three replicate 24–48 h nutrient agar slopes of sufficient strains to provide each individual or group with a different strain. If only 10 strains are used, try to provide organisms to produce only 3 main groups, e.g. some pseudomonads, staphylococci and bacilli. With more strains, include a range of enterobacteria or coryneforms.
2. 2 × 200 ml 1% tetramethyl *p*-phenylene diamine dihydrochloride (made within 15 min of doing the test).
3. Differential media (1 tube per group except where marked). Useful formulae are listed by Skerman (1969) and in the SAB manual (see below):
   - 3 nutrient broths,
   - 8 basal glucose salts liquid media (with alternative nitrogen sources as listed,
   - 6 basal ammonium salts liquid media (with alternative carbon sources as listed),
   - peptone water,
   - sterile lead acetate strip (in sterile dish),
   - tryptone water,
   - Vosges Proskauer medium,
   - nitrate broth (in Universal vial, with Durham tube),
   - 4 sodium chloride nutrient broths (NaCl concentrations as listed),
   - nutrient broth, pH 4·5,
   - nutrient broth, pH 9·0,
   - 2 Hugh and Leifson stabs with glucose,
   - 14 Hugh and Leifson stabs, each with sugars as listed,
   - aesculin agar stab,
   - arginine agar stab,
   - plate of starch agar,
   - plate of tributyrin agar (Oxoid),
   - plate of 0·4% gelatin nutrient agar,
   - plate of 0·4% casein nutrient agar,
   - plate of 0·01% crystal violet nutrient agar,
   - 2 plates nutrient agar.
4. 1 tube sterile vaseline.
5. 8 different antibiotic discs (as listed).
6. Test reagents as follows:
   - Nessler's reagent,
   - Ehrlich's reagent,

concentrated HCl,
amyl alcohol,
40% potassium hydroxide,
creatine powder,
Griess Ilosvay's reagents I and II (British Drug Houses),
zinc dust,
Lugol's iodine,

Acidified mercuric chloride (10 ml 15% mercuric chloride solution plus 20 ml concentrated hydrochloric acid).
7. Pl Paramount punch card 190GSM Green CC19 $6\frac{1}{8} \times 4$ in. (Copeland Chatterson, Stroud, Gloucs, England),
   4 card punches,
   1 trough with 0·5 cm depth carbol fuchsin.

SPECIFIC REQUIREMENTS

cultures of bacteria
tetramethyl *p*-phenylene diamine dihydrochloride
differential media
vaseline
antibiotic discs
test reagents
punch cards, punches and carbol fuchsin
3 test tube racks per group to hold all the necessary tubes
1 Universal vial with 20 ml sterile distilled water per person
2 watch glasses and 3 grease-free slides per person
reagents for Gram stain (Hucker's modification), spore stain (Conklin's method, Society of American Bacteriologists, 1957), flagella stain (Rhodes, 1958)
plasticene

FURTHER INFORMATION

It is possible to minimize the preparation of the media used in this exercise by substituting API test strips for some groups of tests. The most useful test strips for this purpose are the API 20E and API 50E designed for the Enterobacteriaceae, the API ZYM system used for detecting ten different enzymes and the API STAPH system used for coagulase negative staphylococci. If these are used a modified list of tests should be issued to the students. Full instructions for the use of API test strips are given with the kits. They are obtainable from API Laboratory Products Ltd., Invincible Road, Farnborough, Hants., GU14 7QH, England.

In a large class, it is instructive to use some strains twice so that a check on errors can be made.

REFERENCES

P. G. Brisbane and A. D. Rovira (1961). A comparison of methods for classifying rhizosphere bacteria. *Journal of General Microbiology*, **26**, 379–92.
R. E. Buchanan and N. E. Gibbons (1974). "Bergey's Manual of Determinative Bacteriology," 8th edn. Williams and Wilkins, Baltimore.
W. R. Lockhart and J. Liston (1970). "Methods for Numerical Taxonomy." American Society for Microbiology, Bethesda, Md.
M. E. Rhodes (1958). The cytology of *Pseudomonas* spp. as revealed by a silver-plating staining method. *Journal of General Microbiology*, **18**, 639–48.
V. B. D. Skerman (1969). "Abstracts of Microbiological Methods." Wiley-Interscience, New York.
P. H. A. Sneath (1962). The construction of taxonomic groups. *In* "Microbial Classification," (G. C. Ainsworth, P. H. A. Sneath, eds). Symposium of the Society of Microbiology, **12**, 389–332.
P. H. A. Sneath and R. R. Sokal (1973). "Numerical Taxonomy." W. H. Freeman, San Francisco.
Society of American Bacteriologists (1957). "Manual of Microbiological Methods." McGraw-Hill, New York.

Part Seven
Micro-organism Meets Animal

73

Bactericidal Activity of Human Blood

A. C. WARDLAW

Microbiology Department, University of Glasgow,
Glasgow, G11 6NU, Scotland

> *Level*: All undergraduate years
> *Subject areas*: Immunology; mammalian defence
> mechanisms against infection
> *Special features*: Very simple to do, convincing;
> use one drop of each student's
> own blood

INTRODUCTION

One of the significant observations made during the explosive development of microbiology in the latter decades of the nineteenth century was that freshly drawn blood can kill many of the common saprophytic bacteria of dust and water. This activity is mainly localized in the serum and works best against Gram-negative bacteria. The nine components of complement are now known to be involved in the bactericidal reaction and may act either alone or in conjunction with antibodies and/or components of the properdin (alternative) pathway. Lysozyme in the serum may also contribute to the lysis of the bacteria after complement has damaged the outer membrane.

The aim of this exercise is to provide a convincing demonstration of the bactericidal activity of the student's own blood towards a special strain "Lilly" of *Escherichia coli*. *Staphylococcus aureus* which usually is not killed under these conditions is used as an insusceptible control.

EXPERIMENTAL

The exercise is very simple and economical of materials, and is designed for students working in pairs, one using the *E. coli* Lilly and the other *S. aureus*.

Day 1 *(60 min, but with intervals of free time)*

Into a 12 × 100 mm capped test tube deliver 0·9 ml of sterile 0·85 % NaCl and label it "blood", and giving the student's name.

↓

Set up a 12 × 100 mm control tube containing 1·0 ml nutrient broth. Mark each of two previously dried, nutrient agar plates into quadrants. Label one plate "blood" and the other "broth" and the quadrant spaces with 0, 10, 20, 40 min as shown in Figure 1, which also shows the results of typical tests with *E. coli* Lilly.

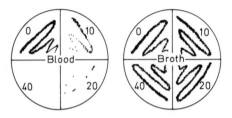

Figure 1   Layout of bactericidal test and typical results with *E. coli* Lilly.

↓

Prepare the bacterial inocula: Make a 1/100 dilution (0·1 ml + 9·9 ml) of an 18-h, 37 °C culture of *E. coli* Lilly or *S. aureus* in 0·85 % NaCl just before they are required. One student in each pair should use *E. coli* and the other *S. aureus*.

↓

Now prepare to take the sample of blood: the donor (and assistant!) should wash their hands in warm water and dry them thoroughly. Swab the tip of a finger or thumb with 70 % ethanol and allow it to dry completely without touching anything.

↓

Puncture the sterilized area near the tip of a finger or thumb with a sterile lancet (*see* note 1) and allow the blood to be drawn out by capillary action into a 0·1 ml or 0·2 ml sterile pipette (Figure 2).

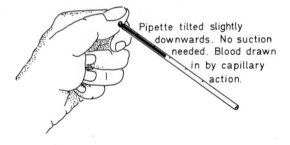

Pipette tilted slightly
downwards. No suction
needed. Blood drawn
in by capillary
action.

Figure 2   Drawing off 0·1 ml of blood after pricking the side of the thumb with a lancet.

Suction is not required. Collect 0·1 ml blood. It may be useful if the assistant massages the punctured finger of the donor to promote blood-flow while the donor holds the tip of the pipette at the puncture point. Promptly rinse the blood out into 0·9 ml saline in the "blood" tube before clotting takes place. Since blood contains approximately 50% by volume of serum, you can regard the mixture in the tube as roughly equivalent to a 1/20 dilution of serum. The red and white blood cells do not significantly contribute to, or interfere with, the bactericidal reaction and may therefore be ignored. Thus it is not necessary to remove them by centrifugation.

↓

With a sterile Pasteur pipette deliver one drop of diluted suspension of either *E. coli* or *S. aureus* to the tubes containing the diluted blood and the broth. Mix well and put the tubes in a 37 °C water-bath.

↓

At time zero (i.e. before incubation) and after 10, 20 and 40 min incubation, remove loopfuls from each tube and spread by zig-zag inoculation over the appropriate quadrant on the agar plates.

↓

Incubate the plates for 18 h at 37 °C. After incubation the plates may be refrigerated for several days without deterioration.

_____

Day 2 *(5–10 min)*

Inspect the distribution of growth.

_____

Notes and Points to Watch

1. The only significant problem is that the occasional student becomes faint at the sight of blood (even 0·1 ml of it!) so the instructor should watch out for any suspicious pallor indicative of an imminent vaso-vagal attack. Sometimes, too, students turn out to be remarkably bloodless and even 0·1 ml blood difficult to extract, like from the proverbial stone! However, this can be overcome by getting the hands well warmed before pricking with the lancet. It is probably wise to excuse anyone who shows reluctance to take part in the exercise. Proper sterile precautions should be taken in using the lancets, e.g. do not contaminate the point by handling; do not lay the lancet on the bench; do not let students use the same lancet (because of the danger of transmitting hepatitis!). Some people find it easier to get someone else to administer the lancet prick rather than doing it themselves.

MATERIALS

1. *Escherichia coli* "Lilly"—a special, very rough strain—and *Staphylococcus aureus* "Oxford" (probably almost any strain will do). The cultures may be maintained on nutrient agar slopes and inoculated into nutrient broth on the day before the experiment.
2. Sterile, disposable lancets.

SPECIFIC REQUIREMENTS

Glassware should be sterile.
12 × 100 mm capped test tubes in rack
15 × 150 mm capped test tubes in rack
0·1 ml or 0·2 ml pipette
1 ml pipettes
5 ml or 10 ml pipettes
Pasteur pipettes
0·85 % NaCl
nutrient broth

nutrient agar plates
18 h 37 °C static nutrient broth culture of *E. coli*
Lilly and *S. aureus* Oxford (one of each would
be enough for up to 20 students)
sterile lancets
sterile absorbent cotton swabs in glass Petri dish
70 % ethanol
37 °C water-bath

FURTHER INFORMATION

Typical results of the experiment with *E. coli* Lilly are shown in Figure 1. The bacteria are usually killed within 10–20 min of exposure to the diluted blood at 37 °C, while there is no loss of viability in broth. With *S. aureus*, the normal result is for the bacteria to survive incubation in both blood and broth.

Wardlaw (1962) investigated the bacteriolytic action of human serum on *E. coli* Lilly and found it to be markedly dependent on ionic strength, pH and temperature, with optima at 0·06, 8·3–8·5 and 37 °C respectively. All samples of human sera tested exhibited bacteriolytic activity, and a requirement for complement and lysozyme was demonstrated. There did not appear to be any requirement for specific antibacterial antibody. In a later study (Pruzanski *et al.*, 1972), bactericidal and bacteriolytic activity towards *E. coli* Lilly was found in human serum at birth, suggesting that IgM antibodies were not involved. In contrast to the above, there is an extensive literature showing that antibodies do play a role, in conjunction with complement, in serum bactericidal activity towards other strains. Another serum factor, properdin, is involved in the killing of certain bacterial strains (Wardlaw and Pillemer, 1956).

A bacterio**static** effect of human serum on *S. aureus* has been reported by Ehrenkranz *et al.* (1971).

REFERENCES

N. J. Ehrenkranz, D. F. Elliott and R. Zarco (1971). Serum bacteriostasis of *Staphylococcus aureus*. *Infection and Immunity*, **3**, 664–70.
W. Pruzanski, W. D. Leers and A. C. Wardlaw (1972). Bacteriolytic and bactericidal activity of maternal and cord sera. Relationship to complement, lysozyme, transferrin and immunoglobulin levels. *Canadian Journal of Microbiology*, **18**, 1551–5.
A. C. Wardlaw (1962). The complement-dependent bacteriolytic activity of normal human serum. I. The effect of pH and ionic strength and the role of lysozyme. *Journal of Experimental Medicine*, **115**, 1231–49.
A. C. Wardlaw and L. Pillemer (1956). The properdin system and immunity V. The bactericidal activity of the properdin system. *Journal of Experimental Medicine*, **103**, 553–75.

74

Measuring the Resident Bacterial Flora of the Human Hand

B. J. B. WOOD

Department of Applied Microbiology, University of Strathclyde,
Glasgow, G1 1XW, Scotland

> *Level*: Advanced undergraduates
> *Subject areas*: Industrial and medical
> microbiology: hygiene
> *Special features*: Human interest; scope for more
> advanced projects

INTRODUCTION

The bacterial flora of the animal surface may be divided into two broad groups.
  (1) The **transient** flora are organisms present on the skin surface as the result of contact with the surroundings; normally their stay is only temporary, until they are washed off.
  (2) The **resident** flora are microbes living in or on the skin surface, presumably utilizing the skin secretions and wastes as nutrient sources. Although potential pathogens such as *Staphylococcus aureus* are fairly often encountered among the resident flora, it appears that these organisms usually live at peace with their host. It has been suggested that the normal resident flora has a part to play in defending the host against microbial attack.
  Clearly there will be much interchange between the organisms of the transient and resident flora, but the resident flora of a particular individual seems to be a reasonably stable community unless subjected to fairly severe disturbance.
  Normal methods of sampling the skin flora by swabbing, applying sticky tape, etc. tend to pick up a mixture of transient and resident species. The experiment described here sets out to remove the transient organisms by a series of washes, and then enumerates a reproducible fraction of the resident flora of each student at the time of sampling. For this to be achieved reliably, strict adherence to the timings and procedures is essential.

EXPERIMENTAL

Considerable flexibility of organization is possible, including everything from each student doing his/her own determination, to a pair of students sampling from a big population. However, it is convenient for students to work in groups of four, and the protocol given is for such a group examining their own resident skin flora. A satisfactory arrangement of the work is for each member of the group to wash his/her hands, while the other three members make preparations, dilute and plate the previous subject's sample, and time the present subject's hand-washes. There is a total of four washes, the first three being to remove the transient flora and open up the pores, the fourth one being the sampling wash (*see* note 1).

Day 1 *(40 min)*

Wet the hands and lower forearms under the tap, preferably with warm water. Work up a lather with the bar of soap for 15 s.

↓

Put down the soap and lather the hands and lower forearms vigorously for 60 s.

↓

Rinse under the tap for 15 s.

↓

Repeat the soaping (15 s), lathering (60 s) and rinsing (15 s), twice more.

↓

Sterilize a stainless steel bucket or bowl and add 1 or 2 litres of sterile distilled water for the 4th washing (*see* notes 2 and 3).

↓

Just **before** the subject commences to rinse his/her hands into the sterile water, the operator pipettes 2 × 1 ml samples of the water into Petri dishes. These are the controls for assessing contamination of the sterile rinse water.

↓

For the final wash, confine the soaping and lathering to the hands only. Rinse the hands into the sterile water.

↓

After the rinsing is complete, take samples for bacterial counting as soon as possible (*see* note 4).

↓

For bacterial counting, put 1·0 ml samples into each of 2 Petri dishes.

↓

Make successive 10^{-1}, 10^{-2}, 10^{-3} dilutions of the rinse water with 1·0 ml transfer volumes into Universal bottles each containing 9·0 ml of 0·1 % peptone solution.

↓

Pipette 1·0 ml samples of each dilution into duplicate Petri dishes.

↓

Add sterile molten plate count agar (50 °C) to all dishes, sufficient to give a reasonable depth of medium: the amount required will depend on the size of Petri dish being used, but normally 10 ml is sufficient. Mix carefully and thoroughly.

↓

Incubate at 37 °C for 2–3 days.

_ _

Day 2 *(30 min)*

Examine Petri dishes and record results from dishes giving between 20 and 200 colonies. Record as a total count for the volume of wash-water used (*see* notes 5 and 6).

_ _

Notes and Points to Watch

1. It is essential that subjects avoid contact with strongly germicidal soaps or other potentially harmful materials on the day of the test. For the washing procedure a non-germicidal soap **must** be used; Lux toilet soap is suitable. Ideally, soft water should be used throughout, but normal tap-water is permissible for the first three washes.
2. The bowl or bucket should be autoclaved before beginning the day's work. If it is capped with a lid made from a sheet of thin autoclavable plastic (e.g. a bag of the type provided for the disposal of plastic Petri dishes), this can be used to protect the container from contamination when it is not in use.
3. The container used for the final hand-wash is prepared by pouring any contents down the drain and rinsing the interior **thoroughly** with water from the cold tap. As much water as possible is drained off and then approx. 10–20 ml methylated spirit is added. By carefully rotating the bucket or bowl, the alcohol is distributed over its entire inner surface. Excess alcohol is poured away, and the bucket or bowl is set down in a safe place. With careful observance of all safety procedures, the spirit remaining in the container is ignited; it burns slowly at first, but with increasing vigour as the vessel heats up. When the flames have died out, the sterile washing water is added and the vessel capped temporarily if necessary. With care the same bowl can be used many times during the day, without increase in the bacterial count of the "control" sample, and without causing a conflagration in the laboratory.
4. The free fatty acids of soap are somewhat germicidal and contact with them should be kept to a minimum. Peptone helps to protect microbes against the acids, which is why peptone water is the preferred diluent rather than Ringer's solution.
5. Further examination (e.g. identification to species) of the colonies can be made by the usual methods, if desired.

6. The total count is calculated by reference to the volume of wash-water used. A rough estimate of the area of the subject's hands can be made by treating the palm and back of the hand as rectangles, and the fingers and thumb as cylinders; an estimate of organisms removed per unit area of skin can then be made. For guidance, a medium-size pair of hands has (surprisingly perhaps) an area of about 500 cm^2.

MATERIALS AND SPECIFIC REQUIREMENTS

Day 1
sterilized stainless steel bucket or bowl
non-germicidal soap (e.g. Lux)
methylated spirits
clock (to time washing)

For four subjects:
40 Petri dishes

4×1 litre portions of sterile soft water
20×1 ml pipettes
40×9 ml portions sterile Plate Count Agar, molten, 50 °C
15×10 ml portions sterile 0·1 % peptone solution

FURTHER INFORMATION

The technique described here is a simple and basic one for use in undergraduate classes. Even so, it can produce very interesting information about the variations between individuals and between ethnic groups. For example, it has been my impression over the past 10 years that women in general have lower counts than men. Again, **in general**, Orientals and Asiatics have slightly lower counts than do Negroids, and all tend to have lower counts than do Caucasians.

A wide variation between individual subjects is to be anticipated, and there is no such thing as a "typical" set of results. A few years ago an Indian girl student, normally resident in East Africa, consistently gave a zero count although re-tested several times; it is difficult to account for this result. At the other extreme, an occasional subject, clean, healthy, with no evidence of infection on the hands, and seemingly normal in every respect, will give a sample uncountable at even the highest

Table 1 A set of results of the type which might be produced by a class. They are not typical or limiting and are included merely to show how results might be collated.

Subject	Sex	Race[a]	Dilution	Colony count	Control count	Total organisms
AN	M	C	1:10	96	0	$1·9 \times 10^6$
EJ	M	A	1:10	99	0	$2·0 \times 10^6$
NP	F	C	1:100	25	0	$5·0 \times 10^6$
JR	F	N	1:10	30	20	—[b]

2 litres of wash-water was used in each case.
[a] C, Caucasian (European); N, Negroid; A, Asian.
[b] Rejected; control count too high. An approximate total might be $(300-20) \times 2000 = 5·6 \times 10^5$.

dilution normally employed. Table 1 gives a set of class results but these must not be taken as "typical" or "limiting".

It must be emphasized that there is no simple relationship between resident flora counts and "personal hygiene"; on occasion, failure to appreciate this important fact has caused unnecessary misunderstanding and distress.

The simple basic experiment can be developed in many ways. For example, counts can be carried out on each washing, and the change in numbers and types of organisms to be investigated. The effect of different types of soaps on the skin flora can be studied. For this purpose it is necessary for subjects to use a germicide-free soap for a week, the test soap for a second week, and finally the control soap for a third week; during this time gloves must be worn for all jobs involving other detergents, etc. such as hair shampooing, dish-washing, working on a motor car, etc. A minimum of four controlled washing sessions would be needed.

Differences in total number and species composition of the skin flora between different sexes or racial types, changes in the same group of people at different times of the year, and so forth, all provide interesting areas for investigation, as does the possibility of a build-up of a resistant flora in people using one type of soap for a period of time. By suitable modifications of the test, differences in resident flora between different parts of the body might be examined.

Many branches of the food industry find a version of this test to be a valuable way to demonstrate to employees the necessity for strict hygiene codes. Even trained microbiologists are surprised at the number of micro-organisms which can be removed from the skin **after** a series of preparatory washes far more exacting than they would normally use. The magnitude of impact of the test on the microbiologically naive can be quite considerable—even alarming if they are not carefully briefed.

Finally, by repetitive sampling at regular intervals (e.g. once a week) combined with control over the soaps used by volunteer panels, industrial organizations use advanced versions of this test to assess proposed formulations for soaps, skin care and personal hygiene products.

There exists a considerable literature on the skin's microbial flora and the list appended is but a small selection thereof, incorporated for the use of students who wish to pursue this subject further.

ACKNOWLEDGEMENTS

I wish to acknowledge with gratitude and thanks the contributions of friends and former colleagues of Unilever Ltd., Colworth House Research Laboratories, who gave their time in discussion of the application and modification of this test for student laboratory use.

Thanks also to Mrs J. Winter and Mrs M. Provan for enduring an excessive number of re-typings of this manuscript.

FURTHER READING

A. Hurst, L. W. Stuttard and R. C. S. Woodroffe (1960). Disinfectants for use in bar soaps. *Journal of Hygiene*, **58**, (2), 159–76.

M. J. Latham (1979). The animal as an environment. *In* "Microbial Ecology: A Conceptual Approach", (J. M. Lynch and N. J. Poole, eds), pp. 124–5. Blackwell Scientific Publications, Oxford.

E. J. L. Lowbury, H. A. Lilly and J. P. Bull (1963). Disinfection of hands; removal of resident bacteria. *British Medical Journal*, **1963**, 1251–6.

H. I. Maibach and G. Hildick–Smith (eds) (1965). "Skin Bacteria and Their Role in Infection". McGraw-Hill, New York.

A. F. Peterson (1978). The microbiology of the hands; effects of varying scrub procedures and times. *Developments in Industrial Microbiology*, **19**, 325–34.

J. B. Williams, J. Brown, J. R. Jungermann and E. Jungermann (1976). An evaluation of the effects of antibacterial soaps on the microbial flora of the hands. *Developments in Industrial Microbiology*, **17**, 185–92.

Patricia E. Wilson (1970). A comparison of methods for assessing the value of antibacterial soaps. *Journal of Applied Bacteriology*, **33**, 574–81.

75

Bacterial Populations of Human Skin

K. T. HOLLAND

*Department of Microbiology, University of Leeds,
Leeds, LS2 9NL, England*

> *Level*: Beginning undergraduates
> *Subject areas*: Bacterial ecology, medical
> microbiology
> *Special features*: Strong personal interest

INTRODUCTION

Studies on the microbial ecology of skin have shown that most areas of the human body are populated with *Propionibacterium* species and members of the Micrococcaceae (McGinley *et al.*, 1978; Kloos and Musselwhite, 1975). In addition, various yeasts, brevibacteria and aerobic diphtheroids can be isolated in more restricted areas. Near the nose, mouth and groin many other types of micro-organisms may be found. The distribution of bacteria is determined by the local skin environment and probably by interactions of the various residents. The micro-organisms regularly isolated from the skin are known as resident flora, whilst those found intermittently and usually in low numbers are regarded as transient flora. Transients are acquired from materials with which the individual comes into contact, e.g. *Bacillus* sp. from air, dust, soil, etc. or may be acquired from other persons, e.g. *Staphylococcus aureus*.

The isolation and identification of skin bacteria in most cases presents few problems. However, accurate enumeration of the flora is impossible without destructive sampling, i.e. biopsy, which for general purposes is clearly not acceptable. The flora resides not only on the surface of the skin cells, but also in the pilosebaceous ducts of hair follicles.

Swabbing of a defined area may be used to sample the microflora but does not give 100% recovery. A scrubbing technique (Williamson and Kligman, 1965) will quantify the microflora of the surface and upper regions of the follicles but, like the swabbing technique, cannot satisfactorily deal with intrafollicular flora.

The aim of this experiment is to determine (1) whether oily, dry and moist areas of skin support different numbers of propionibacteria and members of the Micrococcaceae; (2) whether equivalent left and right positioned skin of the same individual supports different flora, and (3) whether the chosen human subjects are homogenous in the number of bacteria carried on the skin.

EXPERIMENTAL

Left and right areas of the forehead (oily), forearms (dry) and axilla (moist) are sampled either by the Williamson and Kligman method or by swabbing, as detailed below. Students should arrange to sample each other. One student can take a swab sample but two are required for a scrub sample.

Day 1 *(3h)*

Sample the specific areas by either

Scrub technique
Arrange the person under study so that the skin surface is horizontal.

↓

Place a special stainless steel cup on the area to be sampled and hold it to the skin surface with just sufficient pressure to ensure no leakage of fluid.

↓

Pipette 1 ml wash fluid into the cup.

↓

Rub the skin for 1 min with a teflon rod, lifting the rod away from the skin about every 10 s and immediately replacing it; ensure the total area of the skin enclosed by the cup is rubbed.

↓

Withdraw the wash fluid with a pipette.

↓

Pool it with a second sample taken in the same manner.

↓

Estimate the total volume of the pooled sample.

↓

Swab method
Moisten a swab in wash fluid and rub it over the defined marked area of skin for 20 s.

↓

Break the swab stick so that the cotton wool end falls into 2 ml half-strength wash fluid.

↓

Agitate the fluid and swab for 10 s with a Whirlimixer.

Treat the 2 types of washings in the same way, as follows.

↓   ↓

Gram stain the fluid sample and observe the types of bacteria and their arrangements.

Prepare decimal dilutions of the samples in half-strength wash fluid (*see* note 1).

↓

Plate the 0·1 ml of the appropriate dilutions onto RCMF and HBA media and spread evenly (*see* note 2).

↓   ↓

Incubate the HBA plates at 37 °C aerobically for 2 days.

Incubate the RCMF plates at 37 °C anaerobically for 7 days.

Day 2 *(2h, 2 days later)*

Examine the HBA plates. Count and note the different colonies (*see* note 3).

↓

Gram stain from representative colonies and correlate findings with results from Day 1.

Day 3 *(2h, 1 week later)*

Examine the RCMF plates. Count and note the different colonies (*see* note 4).

↓

Gram stain from representative colonies and correlate findings with results from Day 1.

Notes and Points to Watch

1. The number of dilutions required for a given sample depends on the site and the particular individual sampled. Table 1 gives a guide to the dilutions which are usually satisfactory.

Table 1   Dilutions required for samples from different sites.

Skin site	Dilution up to	Medium	Dilutions to be plated out
Forehead	10^{-4}	RCMF	10^{-1}–10^{-4}
		HBA	N -10^{-3}
Axilla	10^{-4}	RCMF	10^{-1}–10^{-4}
		HBA	N -10^{-3}
Forearm	10^{-1}	RCMF	N -10^{-1}
		HBA	N -10^{-1}

N = neat

2. A single autoclaved, or dry-heat sterilized, spreader is adequate for each sample, provided it is used first on the plates with the highest dilutions.
3. Aerobic flora. An indication of flora variation between sites can be obtained by noting the types of colonies and their Gram-stain appearance from the skin sites sampled. Colony types of Micrococceae on HBA are: proteolytic or non-proteolytic and coloured white, grey, golden or yellow.
4. Anaerobic flora. The three species of *Propionibacterium* found on the skin site are *P. acnes*, buff coloured, *P. avidum*, pink and mucoid, and *P. granulosum*, pink and drier with a crenated edge. No Staphylococci or Micrococci should grow on RCMF because of furoxone selection against the former and anaerobic conditions against the latter.

MATERIALS

1. Scrub samplers. The sampling cups are made from stainless steel tubing or autoclavable plastic with measurements shown in Figure 1, but other tubing of near this specification is adequate. The flange is made of such a diameter that it requires hammering into position over the tube. Alternatively the cap and flange can be made by turning on a lathe.

   The sampling cups should be individually wrapped in aluminium foil or autoclave paper for sterilization.

   The scrubbers are made from 1 cm diam. Teflon rods cut to 10 cm lengths by a fine toothed saw. The edge and face must be smoo-

thed by rubbing lightly with fine sand-paper. Sterilize them by autoclaving in capped tubes.
2. Plain white cotton wool swabs. (Medical Wire and Equipment Co. Ltd., Corsham, Wilts., England, Ref. M.W. 104.
3. Wash fluid. Add 8·5 ml 0·075 M NaH_2PO_4 to 91·5 ml 0·075 M Na_2HPO_4 and check that the pH is 7·9. Adjust it if necessary with the appropriate solution. Add Triton X-100 to give a final concentration of 0·1 % v/v.
4. Half-strength wash fluid. Dilute wash fluid with an equal volume of distilled water.
5. Reinforced clostridial medium with furoxone (RCMF). Prepare the RCM (Oxoid Ltd.,

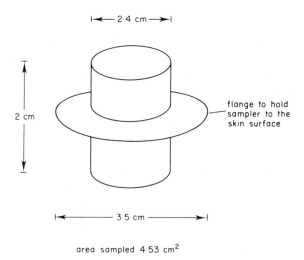

|← 2·4 cm →|

2 cm

flange to hold sampler to the skin surface

|← 3·5 cm →|

area sampled 4·53 cm²

Figure 1   Sampling cup.

CM149) as directed by the manufacturers with an addition of $14.5\,g\,l^{-1}$ lab M agar No. 2. (London Analytical and Bacteriological Media Ltd., 50, Mark Lane, London EC3 R7QJ, England). Autoclave and then bring to 50 °C. Prepare $0.3\,mg\,ml^{-1}$ furoxone (Eaton Laboratories, Regent House, The Broadway, Woking, Surrey, England) in acetone and add 20 ml of this to every 1 litre RCM at 50 °C. Leave for a few minutes to allow the acetone to evaporate before pouring.

6. Heated blood agar (HBA). Prepare blood agar base (Oxoid Ltd., CM55) as directed by the manufacturers. Autoclave it and then bring it to 50 °C. Aseptically add defibrinated horse blood (Oxoid Ltd. HD03) to give 5% (v/v) concentration. Gradually bring to 70 °C with mixing until the medium becomes chocolate-coloured and then pour the plates.

SPECIFIC REQUIREMENTS

Day 1
6 samplers or swabs per student
Gram stains
18 tubes containing 0.9 ml sterile half-strength wash fluid per student
bottle containing 20 ml full-strength wash fluid per student
20 plates RCMF medium per student
20 plates HBA medium per student
spreaders
pipettes

anaerobic jars
Whirlimixer

Day 3
Gram stains
colony counters

Day 8
Gram stains
colony counters

FURTHER INFORMATION

The data may be treated in two ways. Firstly, they should be converted to c.f.u. cm^{-2} skin to allow comparisons of the numbers of various bacteria at different sites and on the left and right side of the equivalent site on the same individual. Typical results are shown in Table 2.

Different species are frequently isolated from different sites and from the left and right sides of the same person (Kloos and Musselwhite, 1975) and up to 5 different colony types may be obtained from the forehead. *Propionibacterium acnes* should be isolated at every site, whilst *P. granulosum*

Table 2 Numbers of two types of bacteria isolated from different skin environments on the same subject.

Site	c.f.u. cm^{-2} skin	
	Micrococcaceae	*Propionibacterium* sp.
Forehead (oily)	1.5×10^4	7.5×10^6
Axilla (moist)	5.0×10^6	5.3×10^6
Forearm (dry)	4.8×10^2	1.3×10^2

may be found on some subjects but in lower numbers. *P. avidum* is found mainly in the axilla (McGinley *et al.*, 1978). In our experience, *P. avidum* may be isolated from the face and back as well as from the axilla.

Secondly the data should be transformed to $\log_{10}$, which has the effect of giving an approximately normal distribution for the populations from homologous sites of different individuals (Evans, 1975; Noble and Somerville, 1974).

Results in Figure 2 show the frequency of distribution of numbers of bacteria from the forehead of 180 subjects of mixed sex but the same age. These results clearly illustrate the large variation in bacterial populations on different subjects and also indicate a small population of people without either of the two bacterial types studied. If no organisms are isolated from the forehead of a person, either there has been a sampling failure or the person is using a topically-applied antibacterial product.

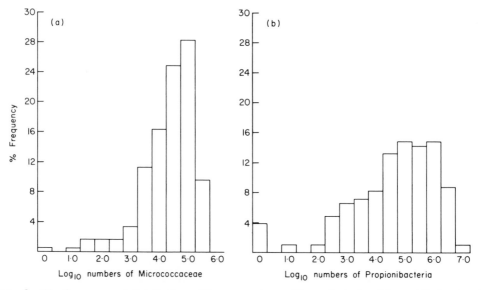

Figure 2   The frequency of distribution of $\log_{10}$ numbers of bacteria isolated from 180 foreheads. (a) Micrococcaceae; (b) *Propionibacterium* sp.

The logarithmic data of all the subjects can be used in the Student *t* test to determine any significant differences from site to site or left to right of the same site for the total population of subjects sampled. Alternatively, any significant differences may be determined using the arithmetic data by the non-parametric statistical method of Wilcoxon (1945). Although the microflora may differ from left to right with respect to species isolated, the numbers of bacteria isolated are usually not significantly different. Significant differences can be expected between forehead and forearm, and axilla and forearm. Differences in bacterial numbers between forehead and axilla are generally not significant for propionibacteria, but may be for Micrococcaceae.

This investigation could be expanded by comparing the microflora of males and females, looking at different age groups, studying more skin sites, and following up the studies with detailed identification of the bacteria isolated.

REFERENCES

C. A. Evans (1975). Persistent individual differences in the bacteria flora of the skin of the forehead: numbers of propionibacteria. *Journal of Investigative Dermatology*, **64**, 42–6.

W. E. Kloos and M. S. Musselwhite (1975). Distribution and persistence of *Staphylococcus* and *Micrococcus* species and other aerobic bacteria on human skin. *Applied Microbiology*, **30**, 381–95.

K. J. McGinley, G. F. Webster and J. J. Leyden (1978). Regional variations of cutaneous propionibacteria. *Applied and Environmental Microbiology*, **35**, 62–6.

W. C. Noble and D. A. Somerville (1974). "Microbiology of Human Skin", 53 pp. W. B. Saunders, Philadelphia.

P. Williamson and A. M. Kligman (1965). A new method for the quantitative investigation of cutaneous bacteria. *Journal of Investigative Dermatology*, **45**, 498–503.

F. Wilcoxon (1945). Individual comparison by ranking methods. *Biometrics Bulletin*, **1**, 80–3.

76

Haemagglutination by Fimbriate *Serratia marcescens*

A. I. TIFFIN

Department of Microbiology, University of Reading, London Road, Reading, RG1 5AQ, England

---

Level: All undergraduate years
Subject areas: Microbial–animal interrelationships
Special features: Student initiative in designing experiments

---

INTRODUCTION

Common fimbriae (pili) have been found on the surface of many Gram-negative bacteria, and have been grouped according to their morphology and haemagglutinating activity (Duguid, 1968; Brinton, 1965). Type I fimbriae occurring on members of the Enterobacteriaceae have been most widely studied (Duguid, 1968; Old and Duguid, 1970 and papers cited); many of their characters are common to many species and in particular they cause the agglutination of many animal and plant cells. Agglutination of animal erythrocytes may be taken as an indicator of the presence of these fimbriae. Agglutination of guinea-pig erythrocytes is inhibited by D-mannose and α-methyl-D-mannoside; haemagglutination by fimbriae is not greatly dependent upon temperature; fimbriae are heat-stable (destroyed at 80 °C for 30 min); they may be removed by violent agitation in a blender and, when isolated, will cause haemagglutination; bacterial cells grown on solid media or in shaken liquid culture are much less fimbriate than those grown in unshaken liquid culture.

Tiffin (1975) showed that a strain of *Serratia marcescens* (17A) had fimbriae with many characters common to Type I fimbriae, but which differed in being more heat-labile, not readily removed by blending and in being produced anaerobically and in shaken culture. Stable pigmented (17AR) and non-pigmented (17AW) variants of this strain were shown to differ in their haemagglutinating activity (HA). The non-pigmented variant showed little variation in HA throughout its growth in shaken liquid medium, whereas the pigmented variant showed increasing HA during the logarithmic phase of growth, with a sudden disappearance of HA towards the end of

© 1982, The Society for General Microbiology
Those wishing to photocopy this experiment must follow the instructions given on page v.

this phase. The object of the several experiments described here is to demonstrate the above properties of *S. marcescens* fimbriae.

EXPERIMENTAL

Day 1 *(30 min, see note 1)*

Inoculate nutrient broth with *S. marcescens* strains 17AR and 17AW. Incubate overnight in shaken culture at 30 °C (*see* note 2).

↓

Inoculate nutrient agar plates with *S. marcescens* 17AR and 17AW and incubate aerobically overnight at 30 °C (*see* note 2).

↓

Inoculate 1 % (w/v) glucose nutrient broth with *S. marcescens* strains 17AR and 17AW. Incubate overnight without shaking to obtain "anaerobically" grown cells (*see* note 2).

- -

Day 2 *(1 h, see note 1)*

Dilute portions of nutrient broth cultures of strains 17AR and 17AW approximately 1/10 in nutrient broth, and incubate shaken for 1 h immediately before student use to get the cultures into the logarithmic phase of growth. Retain the undiluted portions of the cultures.

- -

Day 2 *(cont.) (4 h or 3 h if the growth-curve is to be omitted)*

Apart from the growth-curve section, all of the experiments can be carried out during the course of a three hour practical. One usually needs to allow a full three hours for the growth-curve itself.

To detect HA, add three drops (60 µl) of bacterial suspension to three drops of erythrocyte suspension in one well of a WHO agglutination tray (when appropriate, add one drop of a sugar solution at this stage) and mix continuously by shaking.

View against a plain white background. Agglutination should be apparent within 20 s if there is HA (*see* note 3).

Students should be encouraged to plan their own experiments to compare the fimbriae possessed by *S. marcescens* with published accounts of the fimbriae of other Enterobacteriaceae (*see* note 4). Most of the experiments can be completed within 20 min and several can be run concurrently. Experiments may be qualitative or quantitative and can conveniently be done in the following order (*see* note 5).

Determine the haemagglutinating power (HP) of the bacterial cells for the different species of erythrocytes (*see* note 6).

↓

Determine the correlation between the HP and physiological age of the culture using both strains 17AR and 17AW (*see* note 7).

↓

Determine the extent to which HA is inhibited by the various "sugar" solutions provided.

↓

Determine whether mannose (or fructose) is competing for sites on the fimbriae or on the erythrocytes.

↓

Determine the heat lability of the *S. marcescens* fimbriae (*see* note 8).

↓

Determine whether fimbriae are produced on solid media and under "anaerobic" conditions.

↓

Determine the ease with which the fimbriae may be removed by treatment in a homogenizer.

Notes and Points to Watch

1. Although this part of the work is more conveniently done beforehand by the supervisor, there is no reason why it should not be done by students.
2. One litre each of nutrient broth cultures of 17AR and 17AW, incubated overnight should provide adequate cells for a class of 15. On Day 2, dilute 100 ml of each of these cultures to 1 litre. There will then be available 900 ml stationary phase and 1 litre logarithmic phase cultures of each strain. The composition of the nutrient broth appears to have little effect upon the HA, but if the culture fluid is to be used directly (i.e. without washing) the medium should be approximately isotonic with 0·15 M NaCl and should have a pH value between 7 and 8.
3. Methods will be found in Cruickshank *et al.* (1975, p. 188), Cruickshank (1965, p. 840), and in Duguid and Gillies (1957).
4. Scientifically, it is preferable to let students make a direct comparison between *S. marcescens* and one or more other members of the Enterobacteriaceae, but (a) it may take many sub-cultures of other genera to obtain cultures with adequate HA, (b) large volumes of culture have to be grown statically and centrifuged to obtain enough bacteria for even a small class to use, and (c) additional strains increase the students' confusion.
5. The sequence suggested has been found convenient but as long as the haemagglutinating power (HP) is determined first there is considerable flexibility in what follows. However, if the correlation between physiological age and HP is to be determined, it should be started as early as possible, since readings should be taken at intervals over about 3 h.
6. Haemagglutinating power may be arbitrarily defined as the reciprocal of the concentration (or absorbance) of bacterial cells able to cause HA in a given time. Alternatively, it may be defined as the reciprocal of the time taken for a given concentration of cells to cause HA.
7. It is convenient to use 100 ml side-arm flasks which can be incubated in a shaking water-bath

and read directly in a nephelometer. Alternatively, 100 ml conical flasks may be employed and the contents poured into a test tube for nephelometry. With a large inoculum and short incubation time there is negligible risk of contamination affecting the result.

8. It is convenient to put 1–2 ml culture into small tubes, place the tubes into water-baths at 60 and 80 °C, take three drop samples every 1 min into wells of a WHO tray (discard surplus in sampling pipettes) and after 8 min add erythrocytes and determine the time taken for HA to occur.

9. The requirement will be greatest for erythrocytes most readily agglutinated. Citrated or oxalated horse blood up to about 2 weeks old is satisfactory and 25 ml of 3 % erythrocytes should suffice per student. With other erythrocytes, 5 ml per student should suffice. Fresh human blood with anticoagulant can usually be obtained from a hospital pathology laboratory if supplies are not otherwise available.

10. Blood should be diluted in at least an equal volume of 0·15 M NaCl and centrifuged at up to 500 g. The sedimented cells should be washed once in 0·15 M NaCl and resuspended for use at 3 % (v/v) in 0·15 M NaCl. One millilitre of undiluted mammalian blood will give about 13 ml of 3 % suspension. Note that blood supplied in Alsever's solution has been diluted in an equal volume of the salt solution. One millilitre of chicken blood will give about 7 ml of 3 % suspension. If it is wished to standardize the erythrocyte suspension colorimetrically, dilute 1 ml of the suspension with 25 ml 0·04 % (w/v) ammonia solution. For a 3 % suspension, the absorbance at 541 nm should be 0·25, (1 cm light path).

11. WHO trays may be sterilized after use by brief immersion in 1 % Chloros or similar hypochlorite solution. Some phenolic disinfectants have been found to leave a film on the Perspex.

12. A number of other strains of *Serratia* has been studied. Stable colour variants can be obtained from some strains and not others. The haemagglutinating properties are not identical in the different strains (unpublished observation).

MATERIALS

1. 3 % (v/v) suspensions of erythrocytes in 0·15 M NaCl from as many species as possible, horse and/or guinea-pig are essential (*see* notes 9 and 10).
2. WHO agglutination trays (one per student)

(Manuplastics, Southdown Works, Kingston Road, London, SW 20, England) (*see* note 11).
3. *Serratia marcescens* strains 17AR and 17 AW (*see* note 12).

SPECIFIC REQUIREMENTS

Day 1
2 × 1 litre nutrient broth
orbital shaker at 30 °C
nutrient agar plates (2 per student)
1 % (w/v) glucose nutrient broth (approx. 5 ml per student)

Day 2 (*first part*)
2 × 900 ml nutrient broth
orbital shaker at 30 °C

Day 2 (*second part*)
50 ml nutrient broth (per student)

cont.

cont.

100 ml sterile 0·15 M sodium chloride (per student)	water-baths at 60 and 80 °C with racks for small tubes
1 ml 5 % aqueous solutions of sugars and sugar derivatives (including mannose and fructose) (per student)	shaking water-bath or orbital incubator at 37 °C to take 100 ml flasks
25 calibrated dropping pipettes ("50 droppers") (per student)	(bench) centrifuges to take 10–15 ml tubes, with tubes and rack
10 × 1 ml graduated pipettes (per student)	colorimeters, nephelometers or similar apparatus to measure bacterial cell density
5 × 10 ml graduated pipettes (per student)	top drive homogenizer with blades to fit 1/4 oz (or 1 oz) bottles. (M.S.E. Homogenizer or Townson & Mercer Homogenizer 04700045)
2 small tubes, 10 × 75 mm or 12 × 75 mm (per student)	
2 bottles to fit homogenizer (per student)	

FURTHER INFORMATION

Guinea-pig and horse erythrocytes are strongly agglutinated, human and rabbit less so; sheep cells are very weakly agglutinated.

Either working definition of HP may be used as appropriate to the experiment in question. The relationship between the two definitions may also be determined. Any suspension not causing HA within 150 s may be scored as negative.

Growth curves set up with strains 17AR and 17AW, initially suspended to absorbance *c.* 1 at 600 nm, show very different patterns of HA. Strain 17AW shows little change throughout the growth cycle, a slight increase in HP perhaps being observed during the logarithmic phase of growth. However, strain 17AR shows maximal activity (similar to 17AW) in the logarithmic phase of growth but this activity drops very rapidly (*c.* 30 min) to almost zero as the stationary phase is entered. After the activity is lost from the cells it cannot be detected in the supernate fluid. On transfer to fresh medium, the activity of strain 17AR is rapidly regained as growth becomes logarithmic.

Mannose, fructose and α-methyl-D-mannoside strongly inhibit HA while other "sugars" have little or no effect. More observant students will note that the inhibition is reversible, because, after prolonged shaking, some agglutination will occur. The inhibitory effects of mannose and fructose may be compared quantitatively.

Preincubation of bacterial cells or erythrocytes with mannose solution followed by washing in saline and resuspending, results in no inhibition of haemagglutination. Inhibition occurs only when bacterial cells, erythrocytes and sugar solution are mixed together.

A suspension (absorbance *c.* 1 at 600 nm) of strongly agglutinating cells of strain 17AR or 17AW is likely to have lost most of its HA in 4 min at 60 °C or in 30 s at 80 °C. The HA of some other members of the Enterobacteriaceae is destroyed after 30 min at 80 °C.

There is less HA by bacteria grown on solid media or under "anaerobic" conditions than there is by bacteria grown aerobically in a liquid medium but it is still quite marked.

Prolonged treatment in a homogenizer results in no loss of HA from the cells and no appearance of activity in the supernate fluid. With other members of the Enterobacteriaceae there is rapid loss of activity from the cells and the cell-free suspending fluid becomes haemagglutinating.

REFERENCES

C. C. Brinton (1965). The structure, function, synthesis and genetic control of bacterial pili and a molecular model for DNA and RNA transport in Gram-negative bacteria. *Transactions of the New York Academy of Sciences*, **27**, 1003–54.

R. Cruickshank (1965). "Medical Microbiology", 11th edn, 1067 pp. E. and S. Livingstone, Edinburgh.

R. Cruickshank, J. P. Duguid, B. P. Marmion and R. H. Swain (1975). "Medical Microbiology", 12th edn, Vol. 2, 565 pp. E. and S. Livingstone, Edinburgh.

J. P. Duguid (1968). The function of bacterial fimbriae. *Archivum immunologiae et therapiae experimentalis*, **16**, 173–88.

J. P. Duguid and R. R. Gillies (1957). Fimbriae and adhesive properties in dysentery bacilli. *Journal of Pathology and Bacteriology*, **74**, 397–411.

D. C. Old and J. P. Duguid (1970). Selective outgrowth of fimbriate bacteria in static liquid medium. *Journal of Bacteriology*, **103**, 447.

A. I. Tiffin (1975). Fimbriation in *Serratia*. *Proceedings of the Society for General Microbiology*, III, 11.

How Much Lysozyme is there in a Hen's Egg?

A. C. WARDLAW and J. G. McHENERY

*Department of Microbiology, University of Glasgow,
Glasgow, G11 6NU, Scotland*

Level: Beginning undergraduates
Subject area: Animal defence mechanisms against
bacterial infection
Special feature: A quantitative experiment with a
common household item

INTRODUCTION

The bacteriolytic enzyme lysozyme is widely distributed in nature, being found in such diverse organisms as animals, plants and bacteriophages (*see* Osserman *et al.*, 1974). High concentrations of lysozyme occur in the white (albumen) of the hen's egg which is the commercial source.

Activity of the enzyme may be measured either by the lysoplate technique of Osserman and Lawlor (1966) or, as done here, by the reduction in optical density of a suspension of *Micrococcus luteus* NCTC 2665.

The exercise consists in giving the students an ordinary hen's egg and getting them to measure the total amount, in milligrams, of lysozyme contained in it. Commercial, crystallized egg-white lysozyme is provided as a reference standard.

EXPERIMENTAL

The exercise is sufficiently economical and simple for students to do it individually, and a single egg will provide enough material for the whole of a large class. However, a bottle-neck is likely to occur at the spectrophotometers (*see* note 1) so the exercise is best done in alternation with others.

Day 1 *(1 h 10 min)*

With an inoculating loop, scrape off all of the growth of *M. luteus* from a lawn plate (*see* note 2) and

---

suspend it in 5 ml of buffer (*see* note 3) so as to get a thick and even suspension of bacterial cells (*see* note 4).

↓

Crack a hen's egg into a beaker or dish and, holding back the yolk intact, pour all of the egg-white into a 50 ml measuring cylinder via a funnel and note the volume.

↓

Make a 1:1000 dilution of the egg white in buffer (*see* note 5).

↓

Prepare a 4 μg ml^{-1} standard solution of commercial, crystallized (*see* note 6), hen's egg-white lysozyme in buffer.

↓

Prepare a series of dilutions of the 1:1000 egg-white and of the 4 μg ml^{-1} standard lysozyme in buffer, as shown in Table 1.

Table 1 Protocol for the assay of lysozyme in hen's egg-white. Each tube, in addition to being labelled with a number, carries the letters H or S for hen's egg-white or standard.

Tube (duplicated series)		1	2	3	4	5	6	7
Buffer (ml)		0	2[b]	3	3·5	3·75	3·88	3·94
Sample (ml)[a]		4	2[b]	1	0·5	0·25	0·12	0·06
M. luteus (ml) suspension		0·3	0·3	0·3	0·3	0·3	0·3	0·3
Time of adding M. luteus	H / S				To be recorded as done			
O.D. after 30 min	H / S				To be recorded			

[a] Either hen's egg-white or standard lysozyme.
[b] Alternatively, the stepwise decreasing amounts of lysozyme in tubes 2–7 may be prepared by making two-fold serial dilutions with 4 ml transfer volumes and discarding the last 4 ml.

↓

At accurately timed half-minute intervals (*see* note 7), add 0·3 ml *M. luteus* suspension to the tubes, one by one and alternating the egg-white and the standard; mix thoroughly and incubate at room temperature for 30 min.

↓

When the 30 min incubation period of each tube has elapsed, measure the optical density (*see* note 7) in a spectrophotometer set at 520 nm and zeroed against a buffer blank.

↓

On semi-logarithmic paper, plot graphs of O.D. against "ml of lysozyme sample per tube". Determine the concentration of the enzyme in the 1:1000 egg-white by interpolation in the central region of the curves. Calculate the number of milligrams of lysozyme in the white of the egg.

Notes and Points to Watch

1. Although best done with spectrophotometers for improved accuracy, the results can be read quite well by naked-eye inspection, and visual matching of egg-white and standard tubes of the same opacity.
2. Instead of the students making their own suspensions from freshly grown cells, freeze-dried bacteria may be used at a concentration of about 4 mg ml^{-1}. Suitable dried, *M. luteus* may be obtained as *Micrococcus lysodeikticus* from Sigma.
3. A trace of gelatin is used in the buffer to prevent possible absorption of the standard lysozyme solution from dilute solution on to the glassware.
4. The suspension should be of a concentration such that when 0.3 ml is added to 4 ml buffer, the O.D. at 520 nm should be about 0.6–0.7.
5. Since eggs may vary in lysozyme concentration, this dilution may have to be varied slightly for best results, i.e. so that the diluted egg-white has approximately the same concentration of lysozyme as the 4 μg ml^{-1} standard.
6. Even 3x crystallized lysozyme is not pure and contains small amounts of other proteins that are readily detectable by acrylamide gel electrophoresis.
7. O. D. is most conveniently measured by setting up the mixtures in optically matched test tubes. Alternatively, each tube in turn can be emptied into the spectrophotometer or colorimeter cuvette which is then rinsed and drained. If the latter is done, it may be necessary to have longer than half-minute intervals when preparing the tests.

MATERIALS

1. *Micrococcus luteus* (NCTC 2665) grown for 40 h at 30 °C on nutrient agar plates, or purchased as "*Micrococcus lysodeikticus*" from Sigma.
2. Simple spectrophotometer or colorimeter, preferably one that will take 10 × 100 mm test tubes.
3. 10 × 100 mm test tubes, preferably preselected for optical matching, or purchased as such.
4. Crystallized egg-white lysozyme (Sigma).
5. Phosphate-buffered saline containing gelatin: dissolve one tablet of Oxoid phosphate-buffered saline (Dulbecco "A") in distilled water, add 4 ml of 1 % gelatin in water (dissolved by steaming), and adjust the volume to 100 ml. This gives 0.04 % final concentration of gelatin. Alternatively the buffer may be made by dissolving 8 g NaCl, 0.2 g KCl, 1.15 g Na$_2$HPO$_4$ and 0.2 g KH$_2$PO$_4$ in distilled water, adding 40 ml of predissolved 1 % gelatin in water, and making the final volume to 1 litre.
6. One or more hen's eggs.
7. Dishes or beakers into which the egg(s) may be broken.

SPECIFIC REQUIREMENTS

Day 1
M. *luteus* culture or suspension
buffer
standard lysozyme solution
hen's egg(s)
10 × 100 mm test tubes, preferably optically matched
larger tubes, or small bottles, for preparing dilutions

test tube rack
pipettes
spectrophotometer
dishes or beakers into which eggs may be broken
50 ml measuring cylinders
funnels

FURTHER INFORMATION

Although this procedure is well suited for use as a student exercise, the lysoplate technique of Osserman and Lawlor (1966) is better for clinical-diagnostic or research purposes. Lysozymes other than egg-white may act optimally under conditions that are different to those given here. A very neat student exercise for measuring the lysozyme content of human tears (the original source of the enzyme investigated by Fleming in 1922) has been described by Fluegel (1978).

Typical curves of O.D. versus lysozyme concentration for the present exercise are given in Figure 1 which shows that 0·6 ml of a 1:1000 dilution of egg-white had the same bacteriolytic activity as 0·54 ml of 4 μg ml^{-1} standard lysozyme. Since the volume of the egg-white was 33 ml, the total

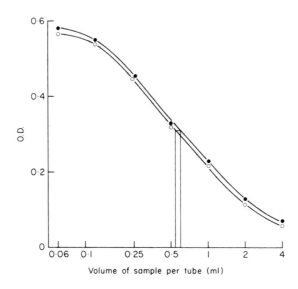

Figure 1 Lysis curves for diluted hen's egg-white (●—●, 1:1000) and standard lysozyme (○—○, 4 μg ml^{-1}) solutions.

content of enzyme in the egg (assuming that it was all localized in the white) was:

$$\frac{0.54}{0.6} \times 1000 \times 33 \times \frac{1}{1000} = 119\,\text{mg}.$$

Values considerably higher than this may be obtained.

The test as described can be altered in various ways, e.g. instead of preparing a range of dilutions of the egg-white, a single dilution such as 0·5 ml of 1:1000 could be set up in duplicate and the result interpolated on the standard curve. Also, a 4-point assay could be done (as for penicillin, *see* Experiment 58) using the central, essentially linear region of the dose–response curve.

REFERENCES

A. Fleming (1922). On a remarkable bacteriolytic element found in tissues and secretions. *Proceedings of the Royal Society, London, Series B*, **93**, 306–17.

W. Fluegel (1978). Lytic action of tears: A classroom laboratory exercise. *American Society for Microbiology. News*, **44**, 679–80.

E. F. Osserman, R. E. Canfield and S. Beychock (eds) (1974). "Lysozyme", 637 pp. Academic Press, London and New York.

E. F. Osserman and D. P. Lawlor (1966). Serum and urinary lysozyme (muramidase) in monocytic and monomyelocytic leukemia. *Journal of Experimental Medicine*, **124**, 921–52.

78

Growth of Bacteria in the Hen's Egg: Effect of Freshness, Added Fe and Mode of Incubation

P. R. HAYES

*Department of Microbiology, University of Leeds,
Leeds, LS2 9JT, England*

Level: Intermediate or final undergraduate years
Subject area: Food microbiology
Special features: Emphasizes that although the
hen's egg can readily be
infected with bacteria, it does
not behave simply like an
elaborate test tube, but
exhibits biological variation

INTRODUCTION

About ten days after a hen's egg is laid, the yolk comes to rest in its uppermost part, a movement which may have an important effect on the course of a bacterial infection of the egg. If the yolk, which is full of nutrients, comes into contact with invading organisms, the egg's antimicrobial defence system in the albumen is bypassed and bacterial growth may be permitted. Many workers have noted a lag of 10–20 days between the inoculation of fresh eggs with bacteria and the obvious presence of the bacteria or of macroscopic changes in the eggs. Board (1964) inoculated bacterial suspensions into the air sac of eggs and confirmed that there are two growth phases. In the first, bacterial growth is usually confined to the shell membranes, while the second phase only occurs after the yolk makes contact with the infected inner membrane of the shell. Thereafter a general infection of the egg contents ensues.

An important antimicrobial component in the albumen is conalbumin (ovotransferrin), first described by Schade and Caroline (1944). This protein inhibits bacterial growth by chelating iron, in particular.

The aim of this experiment is to demonstrate how the age of hens' eggs, their storage position and the addition of iron may affect their susceptibility to bacterial infection.

EXPERIMENTAL

It is desirable to examine a number of factors which may affect the resistance of eggs to bacterial infection and it is convenient for different students to study different variables, such as:

1. the age of the eggs, either fresh (3–4 days) or old (21 + days),
2. the storage position of the eggs, either upright (air sac uppermost) or inverted,
3. the type of infecting organism,
4. the size of the inoculum,
5. the presence or absence of iron in the inoculum.

It is somewhat easier if students work in pairs and, thus, such a pair could inoculate, say, 16 eggs (8 fresh and 8 old), one half (i.e. 4 fresh and 4 old) being subsequently incubated for 7–9 days and the second half for *c.* 14 days; these eggs would be incubated in an upright position. A second pair of students could repeat this procedure, but with the eggs incubated in the inverted position. Further groups could repeat these procedures with a different inoculum density, and so on.

Day 1 *(40 min, see note 1)*

Candle 16 eggs (8 fresh and 8 old) to determine the region of the air sac, the outlines of which should be marked on the shells with a pencil.

↓

Using a carborundum disc attachment, carefully drill through the shell (previously swabbed with 70% alcohol) above the air sac of the eggs to expose the outer membrane.

↓

Push the needle of a hypodermic syringe through the outer membrane and slowly introduce 0·1 ml of the bacterial suspension into each of 4 fresh and 4 old eggs (*see* notes 2 and 3).

↓

Seal the hole in the shell with either molten candle wax or with a strip of zinc oxide plaster.

↓

Add 0·4 ml 0·004 M $FeSO_4$ solution to the remainder of the bacterial suspension (*c.* 4 ml) used above, mix, and repeat inoculations into the remaining 8 eggs (*see* notes 4 and 5).

↓

Incubate the eggs at *c.* 24 °C (20–27 °C) for 7–9 days before the first examination (*see* note 6).

- -

Day 2 *(30 min)*

After incubation, 8 of the eggs (4 fresh and 4 old, with and without added $FeSO_4$) are examined.

↓

Swab the egg with 70% alcohol where the incision is to be made.

↓

Using an alcohol-flamed scalpel, crack open the egg and allow the contents (usually *c.* 40 ml) to drop into a stomacher bag or suitable macerating vessel (*see* note 7).

↓

Examine the contents and the shell membranes for signs of visible spoilage; this is particularly evident if *Serratia marcescens* is being used.

↓ ↓

Visibly spoiled eggs (*see* note 8)
Score visible spoilage by some suitable means, e.g. + + + = grossly contaminated, + + = slight/moderate contamination but with some involvement of yolk and/or albumen, + = inner membrane only affected.

Unspoiled eggs
Add 360 ml quarter-strength Ringer's (or other suitable diluent) to stomacher bag and stomach to give a 10^{-1} dilution of whole egg.

↓

Prepare 1 further decimal dilution (1 ml of the macerate → 9 ml diluent).

↓

Prepare pour plates only of this 10^{-2} dilution using nutrient agar.

↓

Incubate plates at 30 °C for 48 h.

Day 3 *(15 min)*

Count the colonies on the plates and calculate the final number per ml egg fluid.

Day 4 *(30 min)*

The eggs incubated for *c.* 14 days are examined as for Day 2.

Day 5 *(15 min)*

Perform counts as for Day 3.

Notes and Points to Watch

1. The times allowed and procedures given are for two students.
2. Difficulty is sometimes experienced with the inoculation of even 0·1 ml into the air sac, particularly with fresh eggs; it may help if the inoculum is added slowly.
3. Viable counts should be made on the bacterial suspensions used as inocula.

4. Differing $FeSO_4$ concentrations (c. 0·4–2·0 ml) may be added to the suspension.
5. It is essential to include various controls in this experiment; these should include inoculation of the suspending medium, with and without the addition of $FeSO_4$ solution, into appropriate eggs. Two sets of such controls should be prepared which should be stored in the upright and inverted positions respectively.
6. A standard bacteriological incubator should be used.
7. Care should be taken with the scalpel; a relatively light tap on the shell will suffice.
8. In the methods described above, 2 eggs are used for each experimental variation. However, since the eggs themselves are liable to variation, it is desirable in establishing trends, to duplicate all test procedures.
9. According to the experimental procedures described, this would be sufficient for 8 students; by halving the number of eggs per pair of students double the number of students would be involved.
10. Alternatively 0·1 % (w/v) peptone water can be used.

MATERIALS

1. Fresh eggs (3–4 days after laying) and old eggs (21 + days after laying).
2. Drill(s) with carborundum disc attached. Foot activated drills (Philip Harris Ltd., Lynn Lane, Shenstone, Staff., WS14 OEE, England) with discs are preferred, but any electric drill (2 mm bit) can be used.
3. Hypodermic syringes (1 or 2 ml) and needles (e.g. 31 × 0·6 mm).
4. Suitable test organisms, e.g. *Serratia marcescens*, *Pseudomonas fluorescens* and *Staphylococcus epidermidis*.
5. Egg candler (Philip Harris Ltd.).
6. Colworth "stomacher" or alternative macerating device (e.g. MSE Atomix).

SPECIFIC REQUIREMENTS

Day 1
5 ml (× 4) suspensions of test organisms in distilled water at suitable final concentrations (e.g. 100 + and 10 000 + cells ml^{-1})
egg candler
fresh (32) and old (32) eggs + additional eggs for controls (*see* note 9)
drills with carborundum disc attachments
70 % alcohol
hypodermic syringes and needles
zinc oxide plaster (e.g. 1·25 cm wide) or candle wax
scissors
sterile solution of 0·004 M $FeSO_4 . 7H_2O$

Day 2
stomacher and bags, or suitable macerating vessels

360 ml quarter-strength Ringer's solution (*see* note 10)
9 ml quarter-strength Ringer's solution
70 % alcohol
scalpels
1 ml pipettes
nutrient agar, molten for pour plates, in a bath at 45 °C
Petri dishes

Day 3
colony counters (optional)

Day 4
as for Day 2

Day 5
as for Day 3

Table 1 Typical results showing effect of age of eggs, their storage position and addition of iron on degree of spoilage by test organisms.

Organisms inoculated (cell numbers ml⁻¹) and incubation period (days) of eggs	Fresh eggs				Old eggs			
	Upright		Inverted		Upright		Inverted	
	No added Fe	Added Fe	No added Fe	Added Fe	No added Fe	Added Fe	No added Fe	Added Fe
Serratia marcescens (10000+)								
9	+ +	+ ++	500 +	12×10³ +	+ +++	+++ +++++	20×10³ +	300 +
14	+ +++	++ +++	4000 25×10³	+	++ +++	+++ ++	43×10³	++ ++
S. marcescens (100+)								
9	<100 ++	13×10³ +++	300 <100	600 7×10³	++ +++	++ ++	<100 +	9×10³ +
14		++		59×10³ +	++ ++	++ ++	16×10³ 4000	38×10³ ++
Staph. epidermidis (10000+)								
9								
14	<100 for all test results at both 9 and 14 days							

Results shown are for a pair of eggs at each incubation time.
+++ = grossly contaminated; ++ = slight/moderate contamination with some involvement of yolk and/or albumen; + = inner membrane only affected.
Total counts are expressed per ml of egg contents.

FURTHER INFORMATION

Staph. epidermidis, like many other Gram-positive bacteria, usually die out after inoculation into eggs, even under conditions most favourable to their growth; this is primarily due to the inhibitory effects of lysozyme and conalbumin (Board, 1966).

Eggs should show considerable differences in their resistance to Gram-negative bacteria depending on the age of the egg and in particular on their storage position. With fresh eggs stored upright, the bacteria inoculated into the air sacs should only grow slowly and be confined to the inner shell membrane; with heavy inocula the albumen may be invaded. The secondary, and more rapid growth phase is delayed until after the yolk makes contact with the inner membrane. However, with old eggs where yolk–albumen contact has already been made, growth should be rapid from the start. With eggs stored in an inverted position (i.e. air sac downwards) the yolk moves away from the inoculum site and hence few eggs should become contaminated (Board, 1964; Board and Fuller, 1974). Finally, the addition of iron to the inoculum should provide a growth stimulus by overcoming the chelating effects of conalbumin (*see* Experiment 79). Table 1 shows typical results obtained in the classroom.

REFERENCES

R. G. Board (1964). The growth of Gram-negative bacteria in the hen's egg. *Journal of Applied Bacteriology*, **27**, 350–64.

R. G. Board (1966). The course of microbial infection of the hen's egg. *Journal of Applied Bacteriology*, **29**, 319–41.

R. G. Board and R. Fuller (1974). Non-specific antimicrobial defences of the avian egg, embryo and neonate. *Biological Reviews*, **49**, 15–49.

A. L. Schade and L. Caroline (1944). Raw hen egg white and the role of iron in growth inhibition of *Shigella dysenteriae, Staphylococcus aureus, Escherichia coli* and *Saccharomyces cerevisiae. Science*, **100**, 14–15.

79

Infection Experiments with Hen's Egg Albumen

P. R. HAYES

*Department of Microbiology, University of Leeds,
Leeds, LS2 9JT, England*

> *Level*: All undergraduate years
> *Subject area*: Food microbiology
> *Special features*: Ease and rapidity, but ideally
> requires access to a gamma
> radiation source to sterilize
> the albumen

INTRODUCTION

The microbiology of the hen's egg has been extensively reviewed by Board (1966, 1969) and Board and Fuller (1974). It has long been recognized that the albumen contains many antimicrobial agents amongst which lysozyme is the best known. Another key component is conalbumin (ovotransferrin), a protein accounting for 13 % of the total egg-white solids, which inhibits bacterial growth by chelating iron in particular (Schade and Caroline, 1944). It has been found that changes in pH affect the stability of the conalbumin–iron complex formed. As the alkalinity increases there is a corresponding increase in the degree of association of the complex, so that less iron is made available for growth of the organism, and hence the inhibitory effect is enhanced; conversely, as the pH drops, more iron is made available and bacterial growth is stimulated. It should be noted that, in the week following laying, the pH of the albumen increases from 7·0–7·5 to 9·0–9·5 and it is at this latter pH that the chelation effect is particularly strong.

The aim of the experiment is to examine the relationship between the iron-chelating mechanism in albumen and bacterial growth, and how that mechanism is affected by changes in albumen pH and by the addition of iron.

EXPERIMENTAL

The scope of this experiment can be varied by altering the number of test organisms, by including intermediate pH levels of the albumen and by increasing the range of concentrations of iron added.

As a minimum it is recommended that two representative Gram-negative rods and a Gram-positive bacterium be used. The overall results should be combined for presentation and discussion.

It is satisfactory for each student to work with albumen (or albumen + yolk) at one pH only. The albumen, after pH adjustment, is sterilized by exposure to 500 000 rads of gamma radiation (*see* note 7).

Day 1 *(15 min)*

Take 5 × 5 ml portions of sterilized albumen (pH 9 or 7)

Add 0·5 ml bacterial suspension (*see* note 4) to first 5 ml portion of albumen.

↓

Add 2 drops (0·04 ml) FeSO₄ to a second sample, mix thoroughly, and then add 0·5 ml bacterial suspension.

↓

Add 0·1, 0·2 and 0·4 ml FeSO₄ solution respectively to third, fourth and fifth albumen samples and add bacteria as before.

↓

Take 5 × 5 ml portions of sterilized albumen (pH 5)· or albumen:yolk mixture

Add 0·5 ml bacterial suspension to first 5 ml portion of albumen or mixture.

↓

Add 1 drop (0·02 ml) FeSO₄ to a second sample, mix thoroughly, and then add 0·5 ml bacterial suspension.

↓

Add 0·04 ml, 0·1 ml and 0·2 ml FeSO₄ solution respectively to third, fourth and fifth albumen samples and add bacteria as before.

↓

Incubate all samples at *c.* 22 °C (20–25 °C) for about 24 h (*see* notes 1 and 2).

Day 2 *(45 min)*

Make viable counts on the incubated albumen samples by preparing decimal dilutions (to 10^{-6}) of each sample, e.g. 1 ml sample → 9 ml quarter-strength Ringer's solution. Prepare pour plates with 1 ml of each of the 10^{-2} to 10^{-6} dilutions with nutrient agar. Incubate the plates at 30°C for 48 h (*see* note 3).

Day 3 *(30 min)*

Count the colonies on suitable plates and calculate the number of viable bacteria per ml of substrate.

Notes and Points to Watch

1. Where a gamma radiation source is unavailable it is essential to perform sterility checks on the various samples. The inoculum level (*c.* 100 cells ml^{-1}) is likely to be greater than the level of contamination; growth of contaminants can be minimized by storing the samples at 5 °C, before and after pH adjustment, until required.
2. Care should be taken with the incubation of the inoculated albumen. Sufficient time is necessary

to elicit the different growth responses, but excessive incubation will give consistently high recoveries of the inoculated Gram-negative rods from all substrates.

3. The incubation period of the plates used for the viable counts is not critical.

4. The suspensions of the test bacteria are provided in distilled water and should be prepared as late as possible to minimize viability losses before the experiment. The viable count at the time of inoculation into the albumen samples should be determined.

5. Before breaking eggs open, they should be swabbed with 70% alcohol to minimize contamination.

6. Egg albumen can conveniently be obtained using a standard egg-yolk separator; the albumen of several eggs can then be mixed before adjustment to the required pH.

7. Adjustment of pH is difficult to perform aseptically. Thus, after adjustment of bulk albumen samples to the required pHs, they should be dispensed in screw-capped test tubes in 5 ml quantities and sterilized by gamma irradiation (c. 500 000 rads).

8. Four, or possibly 4 pairs of students, would work with one albumen sample and one bacterial suspension.

9. If, for example, 0·1 ml $FeSO_4$ solution is added to 5 ml albumen, the final concentration of available iron = $200/(5 \times 10) = 4\,\mu g\,ml^{-1}$ substrate.

MATERIALS

1. Suitable test organisms, e.g. *Serratia marcescens*, *Pseudomonas fluorescens* and *Staphylococcus epidermidis* (*see* note 4).
2. Egg-yolk separator. Metal separators are recommended (e.g. Westmark "Dotti"; Kitchen Gear, Albion Street, Leeds, England) but they should

be readily obtainable from any reputable hardware store (*see* notes 5 and 6).
3. Sterile egg albumen (pH 9, 7 and 5) and egg albumen + yolk (no pH adjustment; *see* note 7).
4. Gamma radiation source.

SPECIFIC REQUIREMENTS

Day 1
4 × 5 ml suspensions (c. 1000 cells ml⁻¹) of test organism(s) in distilled water, prepared from overnight culture(s) (*see* note 8)
5 × 5 ml quantities of egg albumen at pH 9, 7 and 5
5 × 5 ml quantities of egg albumen: yolk (1:1).
sterile solution of 0·004 M $FeSO_4$
1 ml graduated pipettes
Pasteur pipettes (0·02 ml drop⁻¹)
tube rack

Day 2
9 ml quarter-strength Ringer's solution for dilutions (30 for each pH sample)
1 ml pipettes (35)
nutrient agar, molten, for pour plates, in a bath at 45 °C
Petri dishes (25)
tube racks

Day 3
colony counter (optional)

FURTHER INFORMATION

Different organisms show different degrees of inhibition to conalbumin (Table 1). *Staph. epidermidis* is unlikely to be recovered from the albumen but this is primarily due to the presence of lysozyme rather than conalbumin, though the latter is known to have an inhibitory effect on Gram-positive forms (e.g. Theodore and Schade, 1965).

Table 1   Effect of iron on growth of *S. marcescens* (and *Staph. epidermidis*) in albumen of varying pH.

Test organism	pH	Viable counts test organism ($\times 10^6$) Amount Fe added (μg)					
		0	4	8	20	40	80
S. marcescens	9	0·5	—	1·2	2·5	11·7	18·5
	7	2·3	—	8·2	41·3	230	9·7
	5	2·1	15·3	6·8	52·0	26·0	—
	+	450	620	$> 10^3$	$> 10^3$	550	—
Staph. epidermidis		all samples typically < 100					

+ = yolk:albumen (1:1) with no pH adjustment.

In this experiment the $FeSO_4$ solution contains approximately 200 μg Fe ml^{-1} and the amount of available iron per millilitre of substrate can thus be calculated (*see* note 9). At pH 9·0–9·5, the conalbumin is saturated with about 20 μg iron ml^{-1}, a concentration which may actually inhibit growth in the absence of a chelating agent; there should thus be a progressive stimulation of growth as more iron is added. However, with decreasing pH more iron is made available due to dissociation of the conalbumin-iron complex, so correspondingly less added iron is required (Brooks, 1960; Garibaldi, 1960).

REFERENCES

R. G. Board (1966). The course of microbial infection of the hen's egg. *Journal of Applied Bacteriology*, **29**, 319–41.

R. G. Board (1969). The microbiology of the hen's egg. *Advances in Applied Microbiology*, **11**, 245–81.

R. G. Board and R. Fuller (1974). Non-specific antimicrobial defences of the avian egg, embryo and neonate. *Biological Reviews*, **49**, 15–49.

J. Brooks (1960). Mechanism of the multiplication of *Pseudomonas* in the hen's egg. *Journal of Applied Bacteriology*, **23**, 499–509.

J. A. Garibaldi (1960). Factors in egg white which control growth of bacteria. *Food Research*, **25**, 337–44.

A. L. Schade and L. Caroline (1944). Raw hen egg white and the role of iron in growth inhibition of *Shigella dysenteriae*, *Staphylococcus aureus*, *Escherichia coli* and Saccharomyces *cerevisiae*. *Science*, **100**, 14–15.

T. S. Theodore and A. L. Schade (1965). Carbohydrate metabolism of iron-rich and iron-poor *Staphylococcus aureus*. *Journal of General Microbiology*, **40**, 385–95.

80

Faecal Contamination of the Common Mussel

D. R. TROLLOPE

Department of Botany and Microbiology, University College of Swansea, Swansea, SA2 8PP, Wales

> *Level*: All undergraduate years
> *Subject areas*: Microbial ecology; public health
> *Special features*: Students collect their own material

INTRODUCTION

During the early years of this century it was realized that filter-feeding shellfish were a source of typhoid infection, and Dodgson (1928) presented evidence that molluscs had been a source of infection in several widely spread typhoid outbreaks. Shellfish thrive in many coastal waters and because of their filter-feeding mode, take up any bacteria present in the water. Many shores, and in particular estuaries, attract dense urban conurbations and the untreated sewage from the human population is dumped in the sea at varying distances from the land. This sewage, with its manifold microbial pathogens from the human population, will pollute the shellfish beds. Such bacterial and viral agents of human disease have variable survival capacities in marine sediments and water, but even small concentrations of sewage-derived organisms may become concentrated within the shellfish because each animal filters a considerable water volume every day.

The chief risk of infection from sewage-derived micro-organisms comes from oysters and mussels which are eaten raw or lightly cooked. Cockles, periwinkles, whelks and scallops are cooked by boiling or steaming with very little potential health hazard when processed commercially. However, in individual homes there will be considerable variation in the cooking and processing, whether the shellfish represent an occasional collection whilst on holiday or as a regular dietary item.

Contaminated oysters and mussels rapidly cleanse themselves (i.e. eject sewage-derived micro-organisms) if relaid in unpolluted sea-water, and this self-purification has been exploited commercially for many years. The process is monitored by bacteriological examination using standard methods to determine the viable numbers of *Escherichia coli* per millilitre of shellfish tissue. Because *E. coli* is a normal inhabitant of the animal intestine, its presence indicates recent pollution by sewage. In 1953, it was recommended that shellfish containing not more than 5 *E. coli* per ml of tissue be regarded as satisfactory for human consumption, but more recently a standard of not more than 2 *E. coli* per ml has been advocated (Thomas and Jones, 1971). This exercise yields counts of total coliform bacteria, i.e. those obtained after incubation at 37 °C on bile–salts–lactose medium, and also counts for *E. coli* itself on the same medium incubated at 44 °C. The 37 °C coliforms include various intestinal bacteria which may be present in greater numbers than *E. coli*. Thus *E. coli* and 37 °C coliforms are indicators of sewage pollution and signify a potential disease hazard; the actual disease-causing bacteria of sewage origin are not isolated in most monitoring regimes (*see* note 1).

In this experiment, the common mussel (*Mytilus edulis*) is the filter-feeding shellfish to be examined. The aim is to determine the numbers of sewage-derived bacteria in a sample of mussels and to relate this information to the recognized criteria for the consumption of mussels by man. The method closely follows that developed from the pioneer work at Conway, North Wales (Thomas and Jones, 1971).

EXPERIMENTAL

Day 1 (collection: approx. 1 h plus travelling time; laboratory: 1 h 20 min)

Arrange to arrive at the shore when the mussel bed will be exposed, i.e. around the time of low tide. Allow ample time for collection.

↓

Collect at least 5 large (5 cm long) mussels at random from each site and transport them to the laboratory in a clean plastic bag **without added sea-water**. They may be stored dry in the bag at 4 °C for several days if necessary, but should be examined within 3 h if possible (*see* notes 2, 3 and 4).

↓

Remove any barnacles from the shell with a blunt scalpel or spatula and scrub the shell under cold running water with a small nail brush. Place the mussels on tissue paper to drain.

↓

Insert a sterile **blunt** scalpel between the two component shells at the anterior end (where the off-white byssus threads emerge) (Figure 1). Force the shells apart with a slight twisting movement of the scalpel and allow the small volume of water contained within (shell water) to escape (*see* notes 5 and 6).

↓

Introduce the blades of sterile, fine-pointed scissors between the shells to cut the adductor muscles (Figure 2), and open the shells fully. With the same scissors cut the mussel flesh free from the shells, combine it in the lower shell, chop it up and drop it into a sterile measuring cylinder.

↓

Place the flesh from the 5 mussels in one cylinder and record the total volume.

↓

To the measuring cylinder add 2 volumes of sterile diluent (0·1 % peptone in 3 % NaCl); transfer the diluted flesh to a sterile polyethylene bag and homogenize in a Colworth Stomacher for 30 s (*see* notes 7 and 8).

↓

Pour the contents of the bag back into the measuring cylinder and allow the large particles to settle (2–4 min).

↓

Aseptically remove portions of the shellfish liquor (**not** the sediment) and inoculate on to MacConkey Agar No. 3 as follows:
(i) quadruplicate plates, each with 0·5 ml inoculum and incubated at 37 °C for 24 and 48 h;
(ii) quadruplicate plates, each with 0·5 ml inoculum and incubated at 44 ± 0·25 °C preferably in a brass canister in a water-bath for 24 h (*see* note 9).

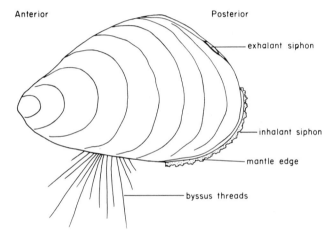

Figure 1   External features of the common mussel with the byssus threads undisturbed.

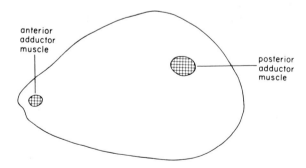

Figure 2   The interior of the common mussel after removing one shell showing the positions of the adductor muscles.

Day 2 *(30 min)*

Determine the numbers of lactose-fermenting colonies on the plates incubated at 37 and 44 °C (*see* note 10). Convert the colony counts to bacterial numbers per ml of mussel tissue by including the dilution factor of 2 resulting from the addition of 2 volumes of diluent before homogenization (*see* note 11). Interpret the results in terms of the sanitary quality of the mussels using the two recommendations for *E. coli* numbers per ml given in the Introduction.

Day 3 *(20 min)*

Recount the plates incubated at 37 °C and convert the colony count to numbers of coliforms per ml of mussel tissue. There are no recommendations for these numbers and their significance, but they are included because usually they produce a positive reading even when *E. coli* is absent.

Notes and Points to Watch

1. Shellfish may be a health-hazard for other reasons which are not of concern in this experiment. Two common examples are (i) microbial agents not derived from sewage (e.g. toxins from dinoflagellates, "red tides") and (ii) high content of heavy metals and pesticides which may not be removed by self-purification procedures.
2. Live mussels can be purchased from some fishmongers or market stalls. These could be tested to see if the bacterial content is as recommended and/or they could be deliberately polluted with *E. coli*. Students at Swansea have found locally purchased mussels to vary considerably in *E. coli* content.
3. The standard method requires, from one source, 10 mussels which are analysed as two separate batches each of 5 mussels. Therefore, two students are encouraged to work in parallel and can in effect analyse 10 mussels from one source.
4. Whilst on the shore, note the location of any drainage pipes or sewage outfalls and the presence of any solid material associated with untreated sewage.
5. During mussel preparation and opening, wear disposable plastic gloves and discard these into a sterilizable bag. In the same bag place the shell water, empty shells and any residual mussel tissue when sampling is complete. Autoclave the bag and contents and discard it.
6. Considerable force is needed to open each mussel: use the instruments with care to avoid cutting the hand.
7. Occasionally a Stomacher bag is ruptured by grit or shell fragments. To avoid this we use two bags, one inside the other, as routine.
8. The Stomacher is ideal for large numbers of separate samples (several sample bags can be stomached simultaneously if necessary). Alternatives are (i) a high-speed blender (e.g. M.S.E. Atomix) with separate jars steamed for 30–60 min beforehand or rinsed with 70 % ethanol and sterile water between samples; (ii) hand maceration in a tissue homogenizer.
9. Only *E. coli* should grow on MacConkey Agar No. 3 at 44 °C, but the temperature regulation is critical at 44 ± 0.25 °C. The accepted method requires (i) an insulated water-bath (**with lid**) employing a suitable thermostat to maintain 44 ± 0.25 °C, and (ii) the plates placed inside a heavy brass canister, with water-tight vented lid, which is submerged in the water-bath with only the vent tube protruding.

10. Typical surface colonies of *E. coli* on MacConkey Agar No. 3 (24 h, 44 °C) are 2–3 mm diam., circular, low convex, entire, glistening, butyrous and intensely violet–red.

11. There are no **statutory** standards for the sanitary quality of shellfish in the UK, but the recommendations by Sherwood and Thompson in 1953 have been accepted for many years: shellfish samples containing not more than 5 *E. coli* per ml tissue have been regarded as satisfactory for human consumption. With the increased use of improved purification techniques for shellfish, a standard of not more than 2 *E. coli* per ml tissue was suggested by Wood (1972) and has become widely accepted.

MATERIALS

1. Thermostatically controlled water-bath (with minimum working depth of 23 cm) and lid (e.g. Laboratory Thermal Equipment, Greenfield, Nr. Oldham, England).

2. Water-tight incubation case, brass, with vented lid, Burman pattern (Astell Hearson, 172 Brownhill Road, Catford, London SE6 2DL, England). Each case holds 10 standard Petri dishes. Cases can be fabricated from brass sheet, tube, etc. with a considerable saving in costs. Alternatives, which may not be suitable for complete immersion, include an anaerobic jar, a Petri dish sterilizing can, (both suitably weighted to counteract buoyancy), and a weighted plastic lunch-box sealed in plastic bags. An incubator set at 44 °C may be used but it is not an accepted method to provide and accurately maintain the close temperature control that is needed to allow only *E. coli* to grow.

3. Colworth Stomacher Model 400 (A. J. Seward, Bury St. Edmunds, Suffolk, England).

4. MacConkey Agar No. 3 (e.g. Oxoid CM115) containing 20 g peptone, 10 g lactose, 1·5 g bile salts, 5 g NaCl, 0·03 g neutral red, 0·001 g crystal violet, 15 g agar per litre of distilled water. We recommend the following: to distilled water (15 ml, sterile, in a Universal container) add MacConkey Agar No. 3 (Oxoid, CM116, 3 tablets), allow to soak for 15 min, steam at 100 °C for 15 min then cool at 45 °C until used (within 2 h of preparation).

SPECIFIC REQUIREMENTS

Day 1

scalpel, with blunt blade, (type with handle and integral blade); autoclaved individually wrapped in aluminium foil

scissors, with fine points; autoclaved individually wrapped in foil

measuring cylinder (100 ml capacity) with top covered in foil; steamed at 100 °C for 30 min

sterilized graduated pipettes, plugged, 1 ml capacity

pipette discard jars containing disinfectant

pipette bulbs for 1 ml

glass spreader; ethanol in beaker

8 plates of MacConkey Agar No. 3 with surface well dried

sterile 0·1 % peptone in 3 % NaCl, approx. 60 ml

absorbent tissue paper

disposable plastic gloves (choice of small, medium and large)

dry plastic bag for mussel collection

2 sterile polyethylene bags for stomaching

1 sterilizable bag for discarded material

materials listed in previous section

FURTHER INFORMATION

Typical variations in colony count between the quadruplicate plates and between inocula obtained from separate but related 5-mussel pools are shown in Table 1. The shores within easy access for Swansea students all receive untreated sewage so that the mussels usually contain very high

Table 1   Colony counts from quadruplicated spread plates using MacConkey Agar No. 3 and 0·5 ml homogenized mussel tissue derived from two separate pools each of 5 mussels.

Sample (date)	Mussel pool	E. coli counts					37 °C coliform counts				
		Plate				No. per ml mussel	Plate				No. per ml mussel
		1	2	3	4		1	2	3	4	
Swansea, St. Helens (24.5)	A	0	3	1	2	12	77	55	54	52	232
	B	6	7	3	1		48	45	68	65	
Swansea, St. Helens (8.6)	A	0	4	3	2	7	33	28	34	26	91
	B	2	1	1	1		17	9	14	21	
Loughor Bridge (14.6)	A	9	6	9	6	24	67	64	40	45	176
	B	4	2	5	6		44	29	39	24	
Loughor Bridge (9.11)	A	154	134	156	120	344	264	180	276	256	596
	B	21	36	36	31		64	40	52	59	

Table 2   Typical counts of E. coli and 37 °C coliforms per ml of homogenized mussel tissue as gathered from various shores near Swansea.

Location of sample	Date sampled	Bacterial no. ml^{-1} tissue	
		E. coli	37 °C coliforms
Llanelli, N. Dock	2.6.76	24	123
	2.4.76	1252	2940
Loughor Bridge	25.11.76	70	550
	25.11.76	420	432
Oxwich Point	16.2.76	11	43
	25.11.76	0	0
Swansea, Mumbles	25.11.76	20	120
	25.11.76	86	480
Swansea, St. Helens	24.5.76	12	232
	8.6.76	7	91
Worms Head	10.10.73	0	2

bacterial loads that grossly exceed the acceptable level for human consumption. However, the bacterial loads show variation from day to day, both according to the season and within a mussel bed. Table 2 lists some counts using mussels from various sites within a 15 mile radius of the University College of Swansea (Figure 3) and reflecting a wide range of sewage pollution regimes. Cleansing experiments (see below) monitored in parallel with the counts for samples St. Helens (date: 24.5) and St. Helens (date: 8.6) (Table 1) resulted in two zero counts for *E. coli* and 0 and $1\,\text{ml}^{-1}$ respectively for the 37 °C coliforms.

The shell water is not included as part of the material for bacteriological examination because it is extremely variable in volume and in bacterial content. It may contain gross particles of sediment and/or sewage that do not represent mussel tissue. If the shell water is bacteria-free, it may cause considerable dilution of bacteria derived from the mussel flesh.

The original method upon which this procedure is based stipulates a dilution factor of 2 on the assumption that, on standing, the hand-macerated tissue settles out, leaving the bacteria suspended in 2 volumes of diluent. However, from both theoretical considerations and macroscopic

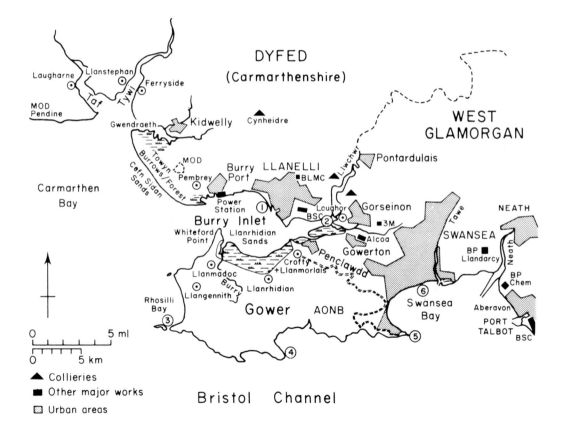

Figure 3 Swansea and surroundings, showing the sample sites of Tables 1 and 2 as locations 1 to 6. 1: Llanelli, North Dock; 2: Loughor Bridge; 3: Worms Head; 4: Oxwich Point; 5: Swansea, Mumbles; 6: Swansea, St. Helens. (After Nelson-Smith, 1977.)

observations this assumption appears incorrect. Theoretically, there are 3 volumes compared with the initial 1 volume of mussel flesh. After stomaching or machine maceration, much of the mussel tissue can be observed as a fine suspension distributed throughout the container and does not settle out. Therefore bacteria will be distributed throughout the 3 volumes and a dilution factor of 3 is more appropriate.

The mussel has marine and other bacteria associated with its tissues. Marine bacteria can be enumerated on spread plates (e.g. using ZoBell Marine Agar 2216E, Difco 0979) with 0·1 ml inocula from 10^{-2} and 10^{-3} dilutions of homogenized mussel tissue and incubation at 10 °C for 8–10 days (or 4 °C for 10–15 days). Viable counts of 10^4 to 10^6 per ml are obtained, mainly Gram-negative rods.

The sewage-derived bacteria are readily removed from the mussel alimentary canal when the shellfish filter "clean" sea-water. Such sea-water, collected at high water during flood tide, should be stored in the dark for 5–10 days to allow sedimentation; sometimes filtration using Whatman No. 1 filter paper is necessary. Alternatively prepare synthetic sea-water from a commercial powder or a standard formula. Cleansing experiments may use (i) static sea-water, 2–10 litre, (ii) running sea-water in an aquarium, or (iii) UV irradiated sea-water (e.g. a volume of 20 litre recycled past an enclosed UV sterilizer (Coast Air, Victoria works, London, NW10) into shallow plastic trays overflowing into a sump containing a cooling coil maintaining the water at 10 °C).

The bacterial counts can be portrayed as histograms. The spatial relationship between counts (e.g. within a mussel bed) and/or the time relationship between counts (e.g. from daily samples) can be displayed by print-out of the histograms using a suitable computer program (Figure 4).

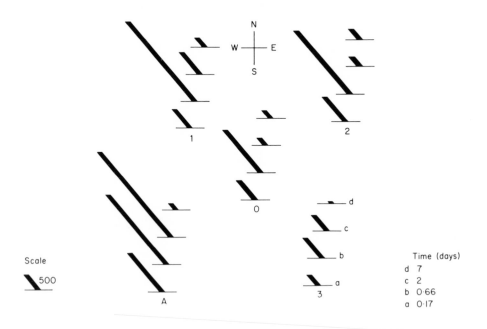

Figure 4 Histograms of the counts of 37 °C coliforms per ml mussel tissue obtained from fixed locations (A, 0, 1, 2, 3) in a mussel bed over a 7 day period.

REFERENCES

G. I. Barrow (1973). Marine micro-organisms and food poisoning. *In* "The Microbiological Safety of Food", (B. C. Hobbs and J. H. B. Christian, eds), 487 pp. Academic Press, London and New York.

R. W. Dodgson (1928). "Report on Mussel Purification". Ministry of Agriculture and Fisheries, Fishery Investigations Series 2, Volume 10, No. 1. HMSO, London.

Food and Agriculture Organization of the United Nations (1974). "Fish and Shellfish Hygiene", World Health Organization Technical Report Series No. 550, 62 pp. FAO, Rome.

A. L. H. Gameson (1975). "Discharge of Sewage from Sea Outfalls", 455 pp. Pergamon, Oxford. (See in particular Papers 8 to 11.)

A. Nelson-Smith (1977). Introduction. *In* "Problems of a Small Estuary", (A. Nelson-Smith and E. M. Bridgs, eds). University College, Swansea.

K. L. Thomas and A. M. Jones (1971). Comparison of methods of estimating the number of *Escherichia coli* in edible mussels and the relationship between the presence of salmonellae and *E. coli. Journal of Applied Bacteriology*, **34**, 717–25.

D. R. Trollope and D. L. Webber (1977). Shellfish bacteriology: coliform and marine bacteria in cockles (*Cardium edule*), mussels (*Mytilus edulis*) and *Scrobicularia plana*. *In* "Problems of a Small Estuary", (A. Nelson-Smith and E. M. Bridges, eds). University College, Swansea.

P. C. Wood (1972). The principles and methods employed for the sanitary control of molluscan shellfish. *In* "Marine Pollution and Sea Life", (M. Ruivo, ed.). Fishing News (Books) and by arrangement with the Food and Agriculture Organization of the United Nations, West Byfleet and London.

Lysozyme in the Common Mussel and Other Shellfish

T. H. BIRKBECK

Department of Microbiology, University of Glasgow,
Glasgow, G61 1QH, Scotland

> *Level*: All undergraduate years
> *Subject areas*: Microbial ecology, marine micro-
> biology
> *Special features*: Simple to perform

INTRODUCTION

The bacteriolytic enzyme lysozyme was discovered by Fleming in 1922 in human tears. Similar enzymes have subsequently been demonstrated in a wide variety of animals, plants and micro-organisms. Chicken lysozyme has been the most intensively studied because of its easy purification from the white of the hen's egg, which contains about 4 mg lysozyme per ml. Lysozyme hydrolyses peptidoglycan, the principal structural polymer of the bacterial cell wall, at the β, 1–4 glycosidic bond between *N*-acetyl muramic acid and *N*-acetyl glucosamine.

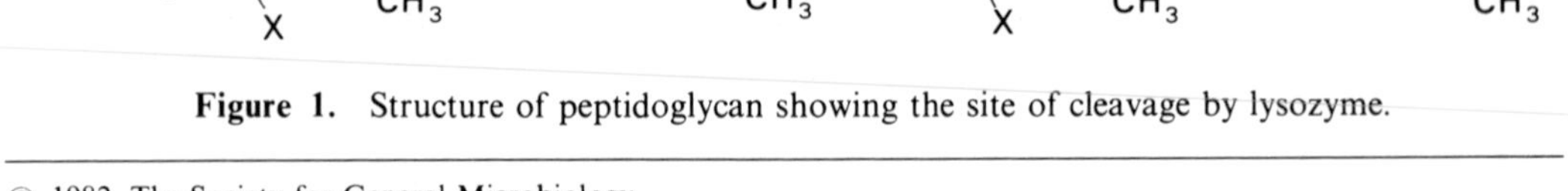

Figure 1. Structure of peptidoglycan showing the site of cleavage by lysozyme.

Few bacteria are attacked by lysozyme alone, as the outer layers of the bacterial cell frequently prevent direct access of the enzyme unless they have been altered by complement, EDTA or organic solvents such as butanol. However, some bacteria, such as *Micrococcus luteus*, are highly sensitive to the direct action of lysozyme and provide a convenient assay for it.

Lysozymes, by virtue of their bacteriolytic activity, have been assumed to play a significant role in host defence against bacterial infection. In shellfish, however, lysozyme may be involved in nutrition, since bacteria floating in the sea and in marine sediments may be a significant food source.

The aim of this experiment is to demonstrate lysozyme in a range of marine bivalves which may utilize bacteria as food.

EXPERIMENTAL

Lysozyme in shellfish such as the common mussel may be detected by applying extracts or homogenates of tissues from the animal to wells punched in an agar layer containing *M. luteus*. After overnight incubation, the resultant zones of lysis around the wells are compared to those formed by hen egg-white lysozyme standards to estimate the lysozyme content of the shellfish tissues.

Day 1 *(1½h)*

Melt the ionagar by steaming and then place it to cool in a 45 °C water-bath.

↓

Pipette 1 ml physiological saline into a 16 × 150 mm tube and, using an inoculating loop, emulsify the growth from an agar slope culture of *M. luteus* in the saline to form a dense even suspension (*see* note 1).

↓

Warm the bacterial suspension in the 45 °C water-bath for 1 min. Pour the bacterial suspension into 4 ml molten agar, mix by rolling (*see* note 2) and quickly pour into a 50 mm Petri dish.

↓

When the agar has solidified, punch up to 6 evenly spaced wells in the agar with a 4 mm diam. (no. 1) cork borer. Suck out the agar plugs with a Pasteur pipette attached to a suction pump (and trap).

↓

Prepare tissue extracts from the shellfish (*see* note 3). Open the shells with a knife and empty the mantle fluid and flesh into a Petri dish. Chop the flesh finely in an equal volume of saline and, with a Pasteur pipette, carefully fill alternate wells in the assay plate with the tissue extracts.

↓

Fill the other wells of the assay plate with hen egg-white lysozyme standard solutions of 5, 25 and 125 μg ml^{-1}.

↓

Incubate the plates overnight at room temperature.

Day 2 *(15 min)*

Inspect the plates and measure the diameter of each zone of lysis with a millimetre rule. For the standard lysozyme solutions a plot of zone diameter against log lysozyme concentration should give a straight line. It should therefore be possible to obtain an estimate of the concentration of lysozyme present in each tissue extract (*see* note 4).

Notes and Points to Watch

1. Where facilities exist for bulk preparation of bacterial cultures a suspension of freeze-dried or acetone-dried, heated (20 min at 100 °C) *M. luteus* cells is equally satisfactory.
2. The bacterial suspension and molten agar should not be mixed by shaking otherwise bubbles will be formed which are difficult to remove.
3. Suitable shellfish are, e.g. common mussel (*Mytilus edulis*), horse mussel (*Modiolus modiolus*), sand gaper (*Mya arenaria*), scallop (*Pecten maximus*), queen scallop (*Chlamys opercularis*), oyster (*Ostrea edulis*), thin tellin (*Tellina tenuis*), common cockle (*Cerastoderma edule*). Specimens should be live and fresh.
4. Different lysozymes may require different conditions for assay. Therefore estimates of concentration may be subject to large errors if the wrong conditions are used.

MATERIALS

1. Hen egg-white lysozyme (Sigma Chemical Co.).
2. *Micrococcus luteus*, NCTC strain 2665.
3. Selection of marine shellfish obtained from the seaside, a local marine station or, commercially, from the Supply Department, University Marine Biological Station, Millport, Isle of Cumbrae, Scotland, KA28 0EG.

SPECIFIC REQUIREMENTS

Day 1
nutrient agar slope culture (24 h, 30 °C) of *M. luteus* in a Universal container
4 ml 1% ionagar in 0·1 M, pH 6·5 phosphate buffer in a Universal container
No. 1 cork borer
Petri dish, 50 mm
Pasteur pipettes
1 ml pipette
16 mm × 150 mm test tubes
sterile physiological saline (5 ml per shellfish)
hen egg-white lysozyme solutions, 5, 25 and 125 μg ml^{-1} in physiological saline (0·5 ml of each)
knife for opening bivalves
scalpel
45 °C water-bath
boiling water-bath or steamer
vacuum pump and trap with connecting tubing and Pasteur pipette
tube rack

Day 2
millimetre rule

FURTHER INFORMATION

A typical standard curve is shown in Figure 2 but the zone diameters will be greatly influenced by *M. luteus* concentration, temperature and time of incubation. All the shellfish listed in note 3 have been shown to possess lysozyme-like activity, although in widely differing amounts (McHenery *et al.*, 1979). Results with other invertebrate species which may appear to lack lysozyme should be interpreted with caution as the best conditions for extraction and assay may not have been used. For example, in sea-water, or in the presence of the concentrations of Na^+ or Mg^{2+} found in sea-water (0.55 M and 0.05 M respectively), *Mytilus* lysozyme is virtually inactive. Although lysozyme can be readily detected in tissue homogenates of shellfish, one may wish to compare the lysozyme content of different tissues and structures of a single animal. The dissection of the bivalve *Anadonta* is described in detail by Rowett (1953) and only a brief guide will be given here to the dissection of the common mussel *Mytilus*. Considerable variation occurs from one species to another and for details of the comparative anatomy of bivalves, Bullough (1958) or Fretter and Graham (1976) should be consulted.

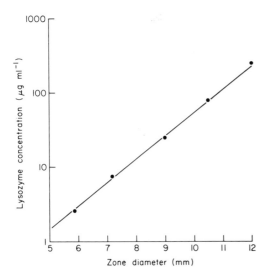

Figure 2   Linear relationship between lysozyme concentration (logarithmic scale) and diameter of zone of clearing.

Hold the bivalve in the left hand with the anterior end to the right. With a scalpel, carefully cut the anterior muscles and then the posterior muscles. Remove the upper translucent mantle tissue to reveal the tissues shown in Figure 3. Samples which may be assayed for lysozyme include mantle fluid, mantle tissue, adductor muscle, palps, gill, digestive gland and crystalline style. The latter two are usually of high lysozyme content whereas the others are generally low. The digestive gland, which is dark brown in colour, is sited close to the hinge (Figure 3). Of particular interest is the crystalline style, a gel-like rod up to 3 cm long which is associated with the digestive tract (Figure 4). In *Mytilus*, the crystalline style is buried in the gonad (cream/orange tissue) and may be released by probing with a blunt spatula or forceps. After removal, the style may dissolve quite rapidly (but not in all species); animals which are not feeding may lack a detectable style as it may dissolve at the end of a feeding cycle.

Extracts of the tissues may be prepared by homogenization or by chopping finely in saline. A style extract may be prepared by allowing dissolution in a little saline.

The lysozyme of the American oyster (*Crassostrea virginica*) has been studied by McDade and Tripp (1967), Feng (1974) Cheng and Rodrick (1975) and Cheng *et al.* (1975). This enzyme is

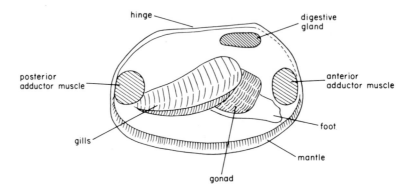

Figure 3. Anatomy of a typical bivalve after removal of the upper shell and mantle.

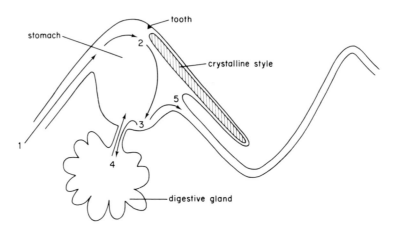

Figure 4. The digestive system of *Mytilus edulis*. Particles entering the stomach trapped in mucus (1) are wound around the style (2) and sorted on the densely ciliated wall of the stomach (3). Food particles enter the digestive gland (4) and unsuitable particles are rejected as pseudofaeces (5).

reportedly localized in the mantle or haemolymph of the oyster and a role in host defence has been proposed. However, in the common mussel (*Mytilus edulis*) and several other bivalves, McHenery *et al.* (1979) found very low levels of lysozyme in haemolymph, mantle fluid or mantle tissue. Instead the enzyme was localized principally in the digestive system and highest levels were associated with the crystalline style. This suggests that lysozyme should be considered primarily as a digestive, rather than a defensive enzyme. It probably acts in conjunction with other factors to degrade bacteria as a food source. Thus the enzyme may act on bacteria in the stomach with the debris being taken up by the digestive gland.

REFERENCES

W. S. Bullough (1958). "Practical Invertebrate Anatomy", 483 pp. Macmillan, London.

T. C. Cheng and G. E. Rodrick (1975). Lysosomal and other enzymes in the hemolymph of *Crassostrea virginica* and *Mercenaria mercenaria*. *Comparative Biochemistry and Physiology*, **52B**, 443–7.

T. C. Cheng, G. E. Rodrick, D. A. Foley and S. A. Koehler (1975). Release of lysozyme from hemolymph cells of *Mercenaria mercenaria* during phagocytosis. *Journal of Invertebrate Pathology*, **25**, 261–5.

S. Y. Feng (1974). Lysozyme-like activities in the haemolymph of *Crassostrea virginica*. *In* "Contemporary Topics in Immunobiology", Vol. 4, "Invertebrate Immunology", (E. L. Cooper, ed.), pp. 225–31. Plenum Press, New York.

V. Fretter and A. Graham (1976). "A Functional Anatomy of Invertebrates", 589 pp. Academic Press, London and New York.

J. E. McDade and M. R. Tripp (1967). Lysozyme in the hemolymph of the oyster *Crassostrea virginica*. *Journal of Invertebrate Pathology*, **9**, 531–5.

J. G. McHenery, T. H. Birkbeck and J. A. Allen (1979). The occurrence of lysozyme in marine bivalves. *Comparative Biochemistry and Physiology*, **63B**, 25–8.

H. G. Q. Rowett (1953). "Dissection Guides", Vol. V, "Invertebrates", 59 pp. John Murray, London.

82

Quantitative Electron Microscopy and Biological Activities of Influenza Virus Particles

N. J. DIMMOCK

Department of Biological Sciences, University of Warwick, Coventry, CV4 7AL, England

> *Level*: All undergraduate years
> *Subject areas*: Virology, electron microscopy
> *Special features*: Animal virology, quantitative use of the electron microscope

INTRODUCTION

Human influenza viruses were first isolated by inoculating ferrets intranasally (Smith *et al.*, 1933). Subsequently the viruses were adapted to grow in the allantoic cavity of embryonated chicken's eggs, and even today this method gives the best yield of virus—better than any cultured cell line. Influenza viruses can infect a wide range of host cells as its receptor on the plasma membrane is the ubiquitous sialic (or *N*-acetyl neuraminic) acid.

Virus which is shed into the allantoic cavity of embryonated eggs can be measured by quantitating a number of different procedures, such as the enumeration of virus particles by electron microscopy, titration of infectivity and agglutination of red (haemoglobulin-containing) blood cells. This "haemagglutination" involves the attachment of virus by the haemagglutinin proteins to neuraminic acid residues in the plasma membrane of erythrocytes. It does not depend on infectivity; completely non-infectious virus can have the same HA titre as infectious virus. The three methods first mentioned give answers which differ from 10-fold up to 10^7-fold! Each method has its advantages and disadvantages. Electron microscopy will give the absolute number of virus particles present, but this takes no account of their biological activity and furthermore requires very expensive equipment. Infectivity measurements may be necessary for some purposes, but for influenza viruses take about 3 days. The time problem is solved by the haemagglutination titration which gives an answer in under one hour and uses equipment costing just a few pence. However, as already mentioned, this test does not show if the virus is infectious.

The aim of this experiment is to measure and compare the number of particles, infectivity units (EID_{50}) and haemagglutinating units (HAU) contained in a preparation of influenza virus.

EXPERIMENTAL

Day 1 *(30 min)*

Check that each egg contains an embryo (*see* note 2) and mark in pencil the position of the air sac. Also mark a point about half way down the long axis which is free of blood vessels so that you can inoculate the egg without damaging the embryo (*see* Figure 1). Swab this area with ethanol, methanol or industrial methylated spirits to help sterilize the shell. Puncture the shell with a sterile spike (sharp forceps, etc.). Inject about 100 μl virus, taking care to insert the needle no more than 1 cm.

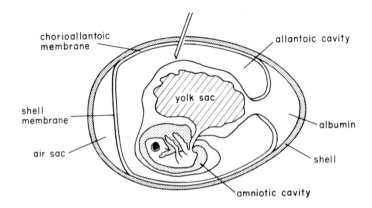

Figure 1 Method for inoculating a fertile egg.

↓

Seal the puncture with dripping wax from a lighted candle. Incubate eggs vertically at 37°C with the air sac uppermost.

Day 2 *(3½ h, but 5½ h if eggs are not chilled overnight)*

Chill the eggs at −20°C (2 h wait), **or** chill the eggs at 4°C (overnight).
  Allow free circulation of air over the eggs. Do not stack the trays. Chilling minimizes bleeding and kills the embryo.

↓

Harvest the allantoic fluid (30 min)
Break and remove the shell from over the air sac and use sterile forceps to tear the shell membrane and prevent embryonic membranes from blocking the pipette (*see* Figure 2). Push down the

embryo with forceps and avoid puncturing the fragile yolk sac. Discard the egg if this occurs. Up to 7 ml can easily be obtained but a small (2 ml) amount of **clear** allantoic fluid is sufficient. Use a bench centrifuge to remove debris.

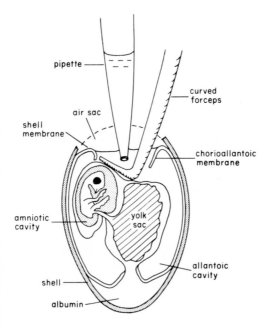

Figure 2 Method for removing allantoic fluid from an inoculated egg. NB This is not to scale; the embryo is larger at 13 days.

↓

Check by haemagglutination (HA) titration that the virus has grown.

↓   ↓

HA titration (30 min + 30 min wait)

Add 0·4 ml buffered saline to the first well of the HA tray and 0·2 ml to each of the remaining wells in the row. With a micro-pipette, add 40 μl allantoic fluid to the first well; mix thoroughly and transfer 0·2 ml to the adjacent well. Repeat the mixing and transfer to the end of the row to give a 2-fold dilution series.

Infectivity titration (see note 7) by terminal dilution in embryonated eggs (45 min)

Proceed only if the HA test is positive. Dilute allantoic fluid serially in 10-fold steps (to 10^{-10}) in buffered saline, pH 7·4, containing 1000 μ ml^{-1} crystamycin to decrease bacterial contamination (e.g. 0·1 ml virus + 0·9 ml saline). Keep dilutions cold if they are not inoculated immediately. Inoculate 100 μl per egg of the 10^{-6}, 10^{-7}, 10^{-8}, 10^{-9}, 10^{-10} dilutions and uninfected controls as described above. Use 5 eggs (or more) per dilution.

↓

Repeat on 3 rows. Add $20\,\mu l$ 5% chicken erythrocytes (*see* note 5) to each well and to wells containing no infectious allantoic fluid as negative controls. Shake the tray thoroughly and allow red cells to settle at room temperature (30–40 min). Estimate HA titre by interpolation (*see* note 6).

Electron microscopy (60 min)

Influenza virus particles are visualized by negative staining and are counted using T2 phage of known concentration as a standard.

Dilute the T2 1/10 in allantoic fluid. Add a drop of virus onto a formvar (carbon coated) EM grid. Remove excess liquid by touching with filter paper. Add to the grid a drop of 1% phosphotungstic acid (pH 7·0) and blot again.

↓

Count the number of influenza virus and T2 particles in each field until about 100 T2 particles have been scored. Calculate back to the known T2 concentration and use this factor to calculate the concentration of influenza virus particles.

Calculations

Students should calculate the following ratios (*see* note 8):
particle: EID_{50}
particle: HAU
EID_{50}: HAU
particles per red blood cell in 1 HAU.

Incubate for 3 days at 37°C.

Day 5

Chill the eggs for 2 h at -20°C or overnight at 4°C.

Day 5 or 6 *(90 min, $3\frac{1}{2}$ h if eggs not chilled overnight)*

Remove about 0·5 ml to the first well of an HA tray. Dilute serially 1/2, 1/4, 1/8 only. Add r.b.c.s to dilutions, shake and allow to settle. Calculate egg infectivity titre EID_{50} ml^{-1}) (*see* note 8).

Notes and Points to Watch

1. Fertile eggs can be obtained from local chicken breeders. These can be left at room temperature for several days providing incubation (and hence development of the embryo) has not been started. Eggs should be incubated at 37°C in humidified air (keep a wide-mouthed container full of water). To develop normally, eggs have to be moved from side-to-side. Custom-made egg incubators have a mechanism to do this. Otherwise it is necessary to do this manually each day by putting a block under the stack of egg trays (*see* Figure 3).
2. Eggs are best inoculated at 11 days after incubation at 37°C but 10–12 days is satisfactory. Before inoculation check that eggs are fertile by shining a light through each, preferably in a darkened room. Fertility can vary between 70–95%. Use a low wattage lamp (less heat) or a

torch with an adaptor tube (Fig 4) onto which an egg can sit sideways. After inoculation, egg trays may be incubated horizontally. Note that chicks hatch at 22 days!

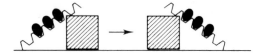

Figure 3   Method for incubating fertile eggs.

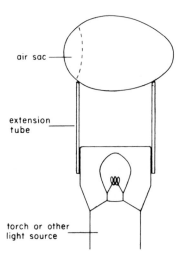

air sac

extension tube

torch or other light source

Figure 4   Method for checking eggs for their fertility.

3. Most laboratory strains of influenza virus multiply well in embryonated chickens eggs. Virus multiplies in the embryonic membranes and is released into the allantoic fluid. Use a strain of virus which gives a high yield, about $10^{3.5}$ to 10^4 HAU ml^{-1} (see below). To prepare a stock of virus, inoculate about 10^5 EID$_{50}$ per egg, which is about 10^{-3} dilution of infectious allantoic fluid. Maximum titres are reached in about 48 h at 37° C. Store virus undiluted at around $-70°$ C or below, or lyophilysed at 4° C. Influenza virus is unstable at $-20°$ C. For the class experiment use 10^{-1} dilution to obtain virus after only overnight incubation. Avoid serial low dilution passage or defective virus will accumulate.

4. Safety: to some extent, egg-adapted influenza viruses retain their infectivity for man. Most of the commonly used laboratory strains are antigenically distinct from the current wild viruses and students will have immunity only to strains current in the preceding 17 years. However, the combination of egg-attenuation, some common antigenic determinants with current strains and exposure only to low amounts of virus make the risk of infection minimal. I have seen no infections during 7 years of running 5-week influenza virus laboratory classes. However, students should use proper microbiological procedures and particularly avoid creating aerosols containing infectious material.

5. Any red blood cells (r.b.c.) may be used to titrate influenza virus. However avian r.b.c.s are nucleated and settle faster than the anucleate mammalian r.b.c.s. Settling can be speeded up by

adding 0.1% bovine serum albumin to the diluent. Use fresh r.b.c.s (up to 5 days old). Blood should be taken into an "anti-clotting" medium such as Alsever's solution and stored at 4° C without further washing. Add $100\,\mu\,\text{ml}^{-1}$ crystamycin. Shortly before use (24 h) wash r.b.c.s twice in saline (containing 0.5 mM Ca^{2+}, 0.5 mM Mg^{2+}, buffered to pH 7.4) by centrifuging with the minimum centrifugal force on a bench centrifuge. Measure concentration by centrifuging in a calibrated 1 ml (Wintrobe) tube and make up a 5% v/v suspension. Check that this contains 4×10^8 cells ml^{-1}.

6. The commonest error is to shake the tray containing red cells insufficiently which results in a low agglutination titre. Figure 5 shows what the HA titration will look like when the red cells have settled.

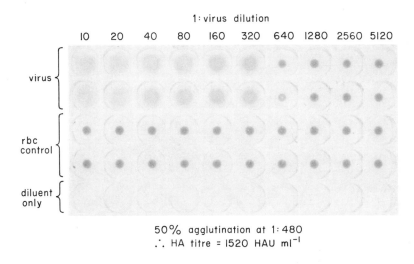

1: virus dilution

| 10 | 20 | 40 | 80 | 160 | 320 | 640 | 1280 | 2560 | 5120 |

50% agglutination at 1:480
∴ HA titre = 1520 HAU ml^{-1}

Figure 5   Example of an HA titration.

The HA titre is expressed as the reciprocal of the dilution of virus agglutinating 50% red cells

7. The infectivity titration uses about 30 eggs. If this is too extravagant, it could be done by groups of students who can then relate the infectivity titre back to the HA titre of their particular virus preparations.

8. The infectivity titre is expressed as the reciprocal of the dilution of virus which will infect 50% eggs inoculated (EID_{50}). This can be calculated according to Reed and Muench (1938) or Karber (see below).

Example (Kärber) Dilution of virus

	10^{-6}	10^{-7}	10^{-8}	10^{-9}	10^{-10}
No. −ve	0	0	3	5	5
No. inoculated	5	4	6	5	5

$$\log_{10} EID_{50}\ \text{vol.}^{-1}\ \text{inoculated} = \left[a - \frac{b}{2} \right] - \left[\frac{\text{no. of eggs } -\text{ve for virus}}{\text{no. of eggs inoculated}} \text{at } 10^{-6} + \ldots \ldots \right]$$

Where a = highest dilution of the series
  b = dilution interval

$$\log_{10} \mathrm{EID}_{50}\ 100\,\mu\mathrm{l}^{-1} = \left[10+\frac{1\cdot 0}{2}\right] - \left[\frac{0}{5}+\frac{0}{4}+\frac{3}{6}+\frac{5}{5}+\frac{5}{5}\right]$$

$$= 10\cdot 5 - [2\cdot 5]$$
$$= 8\cdot 0$$

or $10^{9\cdot 0}\ \mathrm{EID}_{50}\ \mathrm{ml}^{-1}.$

9. For electron microscopy of allantoic fluid it is essential to use unfrozen material. Fresh material has virtually no electron dense debris and the ultrastructure of the virus shows clearly. T2 is an excellent standard as it is can be clearly distinguished from influenza virus (Figure 2) and should possess similar "spreading" properties on the grid to influenza virus as they both have proteinaceous exteriors. Purified T2 is available commercially and should be diluted to

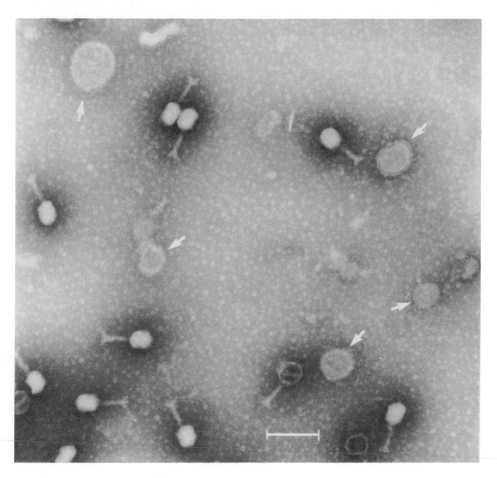

Figure 6 Electron micrograph of T2 diluted in influenza-containing allantoic fluid. The influenza virus is arrowed. The scale represents 200 nm.

about 10^{11} particles ml^{-1} in allantoic fluid.

In a fresh preparation of T2, the particle : infectivity ratio approaches unity; thus a plaque assay will give the concentration of physical particles.

MATERIALS

1. Fertile eggs: probably best obtained from a local poultry farmer. Contact the local office of the Ministry of Agriculture for breeders.
2. Influenza virus, e.g. strain A/PR/8/34. Store at -70° C or lyophilized at 4° C. Available on request from the author.
3. Erythrocytes. Preferably avian, from Tissue Culture Services, 10, Henry Road, Slough, Bucks, England.
4. Wintrobe (Haemocrit) tubes, Gallenkamp, Frederick Street, Birmingham. B1 3HT, England.

5. Alsever's solution for erythrocytes:
D-glucose (Analar),	20·5 g
NaCl,	4·2 g
trisodium citrate dihydrate,	8·0 g
citric acid,	0·55 g

   made up to 1 litre; pH 6·1–6·2.
6. Haemagglutination (WHO) trays such as those sold by Appleton Woods Limited, 313, Healy Road, Birmingham, England.
7. Purified T2 can be purchased from Miles Laboratories Ltd., Stoke Poges, Bucks, England.
8. Buffered saline.

SPECIFIC REQUIREMENTS

Day 1
incubator at 37° C
egg inspection light
buffered saline, pH 7·4
forceps
1 ml syringe + 25 gauge 16 mm needle
candle to seal eggs
industrial methylated spirits

Day 2
−20° C freezer or 4° C refrigerator
forceps
pipettes + bulb
bench centrifuge + 10 ml centrifuge tube
micro-pipettes to deliver 0–200 μl

buffered saline (50 ml)
5 % chicken erythrocytes (1 ml) (but *see* note 5)
disposable haemagglutination tray
embryonated eggs, 10–12 days old (30)
tubes for 10-fold dilutions to 10^{-10} (10) + rack
formvar E. M. grids (2)
1 % phosphotungstic acid (freshly prepared, adjusted to pH 7·0 with KOH)
EM forceps for holding grids
T2 virus

Day 5 or 6
HA tray
buffered saline (50 ml)
5 % chicken erythrocytes (3 ml)

FURTHER INFORMATION

The final results are consistent within an order of magnitude, but minor deviations from the figures below can be expected:

$$\text{particle: EID}_{50} = 10^1 - 10^2$$
$$\text{particle: HAU} = 10^7 - 10^8$$
$$\text{EID}_{50} : \text{HAU} = 10^5 - 10^6$$
$$\text{particles per red blood cell in 1 HAU} = 2{-}20$$

The disease influenza is caused by type A and type B viruses which are distinguished by having serologically different internal ribonucleoprotein antigens. Morphologically, the viruses are indistinguishable. They mature by budding from the plasma membrane of the infected cell. The virus membrane is composed of cellular lipids and viral proteins. Inside the particle are other viral proteins and the viral genomic RNA. This is in 8 segments, which are complementary to the message. The virus codes for at least eight proteins, the haemagglutinin and neuraminidase in the envelope and others situated internally: the type-specific ribonucleoprotein which is associated with small amounts of three putative transcriptase proteins and a major protein, known as the M protein. (*See* Kilbourne (1975), Stuart-Harris and Schild (1976) or British Medical Bulletin (1979) for further information.)

It is important for students to realize that different measures of biological activities of viruses do not usually measure the number of physical virus particles. For instance, in a population of virus particles many particles are not infectious. There can be many reasons, which can be divided into two categories: (1) those concerned with the host cell, and (2) those concerned with the integrity of the particle itself. In the former, either the cells may lack the appropriate receptor (unlikely for influenza virus) or not all receptors lead to a successful infection. Alternatively, in the second category some particles may be deficient in nucleic acid or some other component and hence be incapable of causing an infection under any circumstances.

ACKNOWLEDGEMENTS

I am very grateful to Dr C. S. Dow, Janis M. Wignall and Andrew S. Carver for providing photographs.

REFERENCES

British Medical Bulletin (1979). Influenza, **35**, 1–96.

G. Kärber, quoted by D. B. W. Reid (1968). *In* "Textbook of Virology", 5th Edn, (A. J. Rhodes and C. E. Van Rooyen, eds), p. 118. Williams and Wilkins, Baltimore.

E. D. Kilbourne, (ed.) (1975). "The Influenza Viruses and Influenza". Academic Press, London and New York.

W. Smith, C. H. Andrewes and P. P. Laidlaw (1933). A virus obtained from influenza patients. *Lancet*, **ii**, 66–8.

C. H. Stuart-Harris and G. C. Schild (1976). "Influenza, the Viruses and the Disease". Publishing Sciences Group, Littleton, Mass.

L. J. Reed and H. Muench (1938). A simple method of estimating 50% endpoints. *American Journal of Hygiene*, **27**, 493–7.

83

Bactericidal Activity of Coelomic Fluid from the Sea-urchin

A. C. WARDLAW and L. I. MESSER

Microbiology Department, Glasgow University,
Glasgow, G11 6NU, Scotland

> *Level*: All undergraduate years
> *Subject areas*: Marine microbiology, invertebrate immunology
> *Special features*: Suitable for doing at a marine station; needs only minimal laboratory facilities, but requires healthy sea-urchins

INTRODUCTION

This experiment arose from the observation that when large doses of marine bacteria were injected into the common sea-urchin, *Echinus esculentus*, very few of the bacteria could be recovered next day. Evidently sea-urchins possess an effective bacterial-clearance mechanism, and later work (Wardlaw and Unkles, 1978) showed that coelomic fluid, taken from the body cavity of the animal, was bactericidal *in vitro*. To demonstrate this phenomenon is the object of the present exercise.

EXPERIMENTAL

For purposes of setting up this experiment, the sea-urchin can be regarded simply as a sealed box containing a bactericidal fluid. This fluid may conveniently be sampled by syringe and needle inserted through the peristomial membrane on the underside of the animal (Figure 1), the needle being angled away from the axis to avoid the mouth parts. However, before withdrawing the coelomic fluid, it is necessary to prepare the suspension of the test bacterium, Pseudomonas strain no. 111. Students should work in pairs.

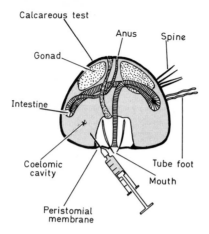

Figure 1 Diagrammatic cross-section of *Echinus*, showing the position of insertion of the syringe for withdrawing coelomic fluid.

Day 1 *(40 min)*

Agar slope culture of Pseudomonas strain no. 111 (*see* note 1).

↓

Wash the organisms off the slope (*see* note 2) into 5 ml of sterile diluent (*see* note 3).

↓

Adjust the turbidity of the suspension to about 10^9 colony-forming units (c.f.u.) ml^{-1} (*see* note 4).

↓

Dilute the suspension, in ice-cold sterile diluent, serially in 10-fold or 20-fold steps to give a concentration of bacteria of approx. 5000 c.f.u. ml^{-1} (*see* note 5) and keep in an ice bath.

↓

Add 0·2 ml of diluted bacterial suspension to:

↓

the control tube containing 1·8 ml sterile diluent in a 15

Hold (*see* note 7) a healthy (*see* note 8) specimen (*see* note 9) of *Echinus esculentus* at above eye level and with its mouth pointing downwards (*see* note 10).

↓

Insert a previously chilled 2·5 ml plastic disposable syringe, fitted with a 25 gauge $\frac{5}{8}$ in. needle (*see* note 11) through the peristomial membrane, at an angle of about 45° to the axis of the animal. Insert the full length of the needle.

↓

Without use of excessive suction (to avoid shear-rupture of the coelomocytes), withdraw about 2 ml coelomic fluid and, without delay (*see* note 12) dispense it directly from the syringe as:

1·8 ml into

↓

a tube labelled coelomic fluid and held in an ice bath.

↓

0·1 ml on to the surface of a marine agar plate (disperse with a glass spreader) to provide sterility control.

↓

× 150 mm tube in an ice bath.

↓

Incubate at 20 °C for 4 days.

Mix the inoculum into the test fluid, remove 0·1 ml and spread it over the surface of a marine agar plate. This is the zero-time sample.

Mix the inoculum into the test fluid but do not take a zero-time sample since the one taken from the control fluid suffices for this purpose.

↓   ↓

Transfer the tube to a refrigerator at about 8 °C (*see* notes 6 and 13).

_ _

Day 2 *(10 min, about 24 h from the start)*

Mix the contents of each tube by gentle swirling and plate out 0·1 ml samples (24 h-exposure samples) on to marine agar using a glass spreader to disperse the inoculum. Incubate the plates at 20 °C).

_ _

Day 3 *(10 min, about 48 h from the start)*

Repeat Day 2, labelling the plates as "48 h samples" (*see* note 14).

_ _

Day 4 *(10 min, 3 days from the start)*

Count the number of colonies of strain 111 on the zero time and "24 h" sample plates and also note any contaminants. If the colonies are too numerous to count, record the growth as AC, for almost confluent, or C for confluent.

_ _

Day 5 *(10 min bench work + tabulation and discussion time, 4 days from the start)*

Count the colonies on the 48 h sample plates and inspect the sterility control. Tabulate and compare the results from the individual urchins.

_ _

Notes and Points to Watch

1. This is a psychrophilic marine isolate that grows readily on Difco marine agar (Zobell agar 2216 E) at temperatures between 5 and 25 °C and forms highly characteristic, jet-black, agar-digesting colonies. Cultures for bactericidal tests are grown for 2 days at 20 °C.
2. Because the bacterium is an agar-digester, the surface of the slope tends to disintegrate easily. Therefore very gentle washing with a Pasteur pipette (do not scrape with a loop) is needed to avoid detaching particles of agar into the bacterial suspension. Agar particles that are inadvertently dislodged should be allowed to settle out before the suspension is used.
3. Sterile diluent is freshly membrane-filtered sea-water (99 ml) to which has been added 1 ml nutrient broth and then heated for 15 min at 100 °C as an extra precaution against contamination.

4. This may be done either spectrophotometrically or by comparison with the International Reference Preparation of Opacity. For guidance, a 10^9 c.f.u. ml^{-1} supension of strain no. 111 has an O.D. at 540 nm of about 0·72 in a 1 cm cuvette of square cross-section. It also matches the opacity of the International Reference Preparation of Opacity.

5. The instructor may find it more convenient (and reliable!) to prepare the diluted bacterial-suspension to this point rather than to let individual pairs of students wash a slope and standardize and dilute the suspension. If the whole class uses the same bacterial suspension a major source of error may be eliminated.

6. Bactericidal activity is maximum at around 8 °C. If a refrigerated, temperature-controlled incubator is not available, a satisfactory alternative is a wide-mouthed vacuum flask containing water at 8 °C, and a dial-type metal thermometer, and placed in a refrigerator which will probably be at 4–6 °C. The insulation in such flasks is so good that only infrequent manual adjustment of temperature, by adding small quantities of warm water, will be needed.

7. Normally the sea-urchins will be kept in a tank or bucket of sea-water until the time of the experiment. Although the spines of the animal will not penetrate the human skin, it is more comfortable for the operator to wear disposable plastic or rubber gloves. The animals, if healthy, will be found adhering by means of their tube-feet to the walls of the container. They may be detached without tearing the tube-feet by grasping the animal gently and rocking it to and fro for about 30 s until the tube-feet are withdrawn.

8. It is essential to use healthy animals that have not been subjected to mechanical injury, the stress of exposure to the air, or transport. Urchins are best collected by a SCUBA diver who hand-picks them from the sea bed into a net bag and the animals are either used immediately or kept in a marine aquarium with flowing sea-water for a few hours or days.

9. To allow for possible variation in bactericidal activity between individual animals, it is preferable to use at least 6 urchins for the experiment.

10. It is best if one student holds the animal while the other manipulates the syringe. The animal should not be taken out of the holding tank until after the diluted bacterial suspension of 5000 c.f.u. ml^{-1} is ready.

11. A mature specimen of E. esculentus contains well over 100 ml of coelomic fluid; therefore the removal of 2·5 ml presents no difficulty nor threat to the life of the animal. If volumes larger than 2·5 ml are needed, use a 5 or 10 ml syringe fitted with a 20 gauge needle. For still larger volumes, it is best to slit the peristomial membrane with a scalpel and let the fluid drain out into a wide-mouthed bottle. Here, however, the fluid will be more heavily contaminated with marine bacteria than if taken with a syringe.

12. Within a few seconds of removal from the animal, the cells (coelomocytes) of the coelomic fluid start to aggregate and form a large clump in the syringe. Therefore, after collection, the fluid should be dispensed as rapidly as possible. To minimize shear-rupture of coelomocytes, remove the needle before expelling the fluid from the syringe into the empty 15 × 150 mm tube.

13. Although up to this point it is essential to do the experiment at the seaside, the rest of it can be done in the students' usual classroom, since the tubes containing the test mixtures can be transported in a wide-mouthed vacuum flask, or other insulated container, containing water at 6–8 °C. Therefore it would be feasible to take a class of students to a marine station, or to the seaside, for a day trip to set up the experiment which could then be completed in the city.

14. If the class timetable does not permit the students to have access to the experiment on all the days listed, then the minimum requirement is Day 1 to set up the test and, 48 h later, access to

remove the final samples from each test mixture. The incubation time of the marine agar plates inoculated with strain 111 is not critical; about 2 days at 20°C is needed to get readily visible colonies, after which the plates can be refrigerated without deterioration.

MATERIALS

1. Pseudomonas strain no. 111 (Glasgow University Microbiology Department). This organism may be maintained conveniently on slopes of Zobell 2216 E Marine Agar (Difco) which, after inoculation, are incubated for 2 days at 20°C and then stored at 4°C.
2. Zobell 2216 E Marine Agar (Difco) is dissolved in distilled water according to the manufacturer's recommendation and dispensed mainly in Petri dishes, but also as a few slopes in Universal containers for culture maintenance.
3. Sterile diluent. With sterile precautions, add 1 ml of normal strength nutrient broth to 99 ml sea-water that has been passed through a sterile $0.22\ \mu m$ membrane filter. As an extra precaution against contamination, heat the solution for 15 min at 100°C.

4. Healthy specimens of the sea-urchin *Echinus esculentus*, preferably hand-picked by a SCUBA diver from the sea bed and either used the same day or kept for a few days in a marine aquarium.
5. Spectrophotometer or colorimeter capable of reading at 540 nm. Alternatively, the International Reference Preparation of Opacity is obtainable from the National Institute for Biological Standardization and Control, Hollyhill, Hampstead, London, NW3 6RB.
6. Plastic disposable syringes (2.5 ml or 5 ml) with 25 gauge $\frac{5}{8}$ in. needles.
7. Refrigerated incubator set at 8°C, or wide-mouthed vacuum flask, large enough to take test tubes and fitted (preferably) with a dial-type, metal-stemmed thermometer.

SPECIFIC REQUIREMENTS

These items are needed for each pair of students.

Day 1
slope culture (2 day, 20°C) of Pseudomonas strain no. 111, one for the whole class (*see* note 5) or per pair
Zobell marine agar plates, 2
sterile diluent, 3 ml, plus extra for preparing and diluting the bacterial suspension
opacity standardization tube, spectrophotometer or colorimeter (*see* note 4)
healthy specimen of *E. esculentus* (*see* note 9)
syringe and needle
plastic disposable, or rubber, gloves

15 × 150 sterile, capped test tubes
test tube rack
glass spreaders
1 ml and 5 ml sterile pipettes
Pasteur pipette and bulb
refrigerated incubator or wide-mouthed vacuum flask and thermometer

Day 2
Zobell marine agar plates, 2
1 ml pipettes

Day 3
as Day 2

FURTHER INFORMATION

Typical results of a bactericidal test, with controls, on coelomic fluid from six sea-urchins are given in Table 1. Column 2 shows that four of the six animals yielded sterile coelomic fluid at the time of sampling, while two of the fluids were slightly contaminated. The test bacteria grew in the sterile diluent but were killed in the coelomic fluids. With urchin no. 5, an initial reduction in count at 24 h was followed by regrowth at 48 h. For additional information on the bactericidal activity of *E. esculentus* coelomic fluid, *see* Wardlaw and Unkles (1978).

Table 1 Typical results of a bactericidal test, with controls, on coelomic fluid from six sea-urchins.

Sea-urchin no.	No. of contaminants per 0·1 ml coelomic fluid at time zero	No. of colonies of Pseudomonas strain no. 111				
		Sterile diluent			Coelomic fluid	
		0 h	24 h	48 h	24 h	48 h
1	0	46	98	AC[a]	10	1
2	0	48	120	AC	6	0
3	7	46	85	AC	4	0
4	0	40	102	AC	0	2
5	0	55	113	AC	7	66
6	2	60	127	AC	2	0

[a] almost confluent

REFERENCE

A. C. Wardlaw and S. E. Unkles (1978). Bactericidal activity of coelomic fluid from the sea urchin *Echinus esculentus. Journal of Invertebrate Pathology*, **32**, 25–34.

84

Staphylococcal Haemolytic Toxins

JOYCE DE AZAVEDO and J. P. ARBUTHNOTT

*Department of Microbiology, Moyne Institute, Trinity College,
University of Dublin, Dublin 2, Ireland*

> *Level*: Intermediate/Advanced undergraduates
> *Subject areas*: Exotoxin production, assay, inter-
> action with membranes and
> purification.
> *Special features*: Assay of cytolytic toxins, kinetics
> of haemolysis, use of ammonium
> sulphate in protein purification

INTRODUCTION

Staphylococcal exotoxins are thought to play an important role in the pathogenesis of staphylococcal infections, especially in relation to localized tissue damage at the site of infection. *Staphylococcus aureus* produces several toxic factors including α-, β-, γ and δ-toxins, leucocidin, epidermolytic toxin and enterotoxin. *S. aureus* α-, β-, γ and δ-toxins belong to a group classified as cytolytic toxins which act by impairing the permeability properties of various mammalian cell membranes. Most cytolytic toxins are lytic for erythrocytes of various species and are therefore often referred to as haemolysins.

In addition to being cytolytic, α-toxin is dermonecrotic, neurotoxic, lethal and causes paralysis of smooth muscle. It is a protein with a molecular weight of approximately 36 000 and a sedimentation coefficient of 3S. The toxin forms a biologically inactive aggregate ($S_w = 12S$) on contact with cell membranes. In its purified state it is electrophoretically heterogeneous: the major form has an isoelectric point of 8·5 with minor forms having pIs of 6·3, 7·2 and 9·1. It exhibits species specificity and is more lytic for rabbit erythrocytes than sheep or human erythrocytes.

Staphylococcal β-toxin has been shown to possess sphingomyelinase C activity; it catalyses the following reaction:

$$\text{sphingomyelin} + \text{water} \xrightarrow[\text{Mg}^{2+}]{\beta\text{-toxin}} N\text{-acylsphingosine} + \text{phosphorylcholine}.$$

The toxin has a mol. wt of 30 000, an isoelectric point of 9·5 and is heat-labile. The susceptibility of erythrocytes to β-toxin depends on the sphingomyelin content of the erythrocyte membrane; ovine and bovine erythrocytes having a high sphingomyelin content (40–50 % of total phospholipid) are very sensitive. One of the most interesting aspects of the action of β-toxin on sensitive erythrocytes is the phenomenon of hot–cold haemolysis. Assay of the haemolytic activity of β-toxin is usually performed by incubating toxin with erythrocytes at 37 °C in the presence of Mg^{2+} ions, followed by a period of chilling below 10 °C. After chilling, the haemolytic titre of highly purified toxin for sheep erythrocytes may increase by as much as 10^6 to 10^8 times over that determined at 37 °C with no period of chilling. In fact, little haemolysis is evident at 37 °C, even with such highly sensitive cells. As yet, there is no satisfactory explanation for this hot–cold haemolysis effect.

Production of both α- and β-toxins is enhanced by the presence of CO_2 or by incorporation of a yeast diffusate in the growth medium. However, it has been shown (Coleman and Abbas-Ali, 1977) that α-toxin production follows a biphasic pattern, i.e. the rate of exotoxin production is low during the exponential phase when the cell's biosynthetic machinery is engaged largely with cellular protein synthesis and only reaches a maximum during the stationary phase.

The objects of the three experiments described in this chapter are as follows:

Experiment 1. To measure production of α- and β-toxins by *S. aureus* strain NCTC 7121 (Wood 46) and strain BB respectively and to correlate toxin production with bacterial growth.

Experiment 2. To demonstrate a rapid and quantitative method for following erythrocyte lysis and to investigate the relationship between rate of haemolysis and α-toxin concentration.

Experiment 3. To observe the effect of different concentrations of ammonium sulphate on precipitation of α- and β-toxins and also to compare specific activities before and after concentration.

EXPERIMENTAL

1. Production of *S. aureus* α- and β-toxins in Relation to Growth

Day 1 *(9 h, but with large gaps of spare time)*

Inoculate 1·0 ml volumes of *S. aureus* strains Wood 46 and BB into 2 litre flanged conical flasks containing 500 ml Bernheimer medium. Incubate at 37 °C in an orbital shaker at 150 cycles min^{-1}.

↓

Remove 50 ml samples of culture (enough for 10 students) at 1, 3, 5, 7 and 9 h, by which time the cultures should be entering the stationary phase.

↓

Stop the growth in each 50 ml sample by adding formalin (37 % formaldehyde) to a final concentration of 1 % formalin.

↓

Centrifuge 25 ml of each formalinized sample for 30 min at 10 000 rev. min^{-1} (13 000 g) in a refrigerated centrifuge and save the supernate.

↓

Store the formalinized culture and supernate at 4 °C (*see* note 1).

Day 2 *(2–3 h)*

(a) *Measurement of bacterial growth*

Set the spectrophotometer or colorimeter at 650 nm and adjust to O.D. = 0 with Bernheimer's medium as blank.

↓

Measure the O.D. of the samples of formalinized culture removed at the different times during growth. If necessary, dilute the samples with Bernheimer's medium to bring the O.D. into the range where it can be read accurately i.e. ∼ 0·4.

↓

Plot growth curves on semi-logarithmic paper, remembering to allow for any dilutions made.

(b) *Titration of haemolytic activity*

Follow the protocol given in the following table.

Tube no.	1	2	3	4	5	6	7	8	9
PBSA or TBS Mg (ml)	0	0·5	0·5	0·5	0·5	0·5	0·5	0·5	0·5
Toxin (ml)	1·0	0·5	0·5	0·5	0·5	0·5	0·5	0·5	Discard
1% (v/v) erythrocytes	0·5	0·5	0·5	0·5	0·5	0·5	0·5	0·5	0·5
Final dilution of toxin	neat	1/2	1/4	1/8	1/16	1/32	1/64	1/128	Control

To titrate Wood 46 supernates for haemolytic activity, make nine serial doubling dilutions in duplicate in 0·5 ml volumes, with PBSA as diluent (*see* note 2).
   To titrate BB supernates, proceed likewise but use TBSMg as diluent (*see* note 3).

↓

Add 0·5 ml 1% (v/v) rabbit erythrocytes to one set of dilutions of each sample and 0·5 ml 1% (v/v) sheep erythrocytes to the other. (Use 5 ml pipettes and take care not to touch the sides of the tubes.)

↓

Incubate for 30 min at 37°C with periodic shaking, and read the 50% haemolysis end-points by comparison with a 50% haemolysis standard (*see* note 4). Express the titre of each supernate in HU_{50} ml^{-1} (*see* note 5).

↓

Transfer the supernates to 4°C for 1 h and re-read the titres.

↓

On the same graphs used to plot the growth curves, superimpose the supernate haemolytic titres against the two species of erythrocytes.

2. *S. aureus* α-toxin: the Kinetics of Haemolysis

This experiment requires a spectrophotometer with thermostatically controlled cell-holder at 37°C and a chart recorder.

(30–45 min)

Set the spectrophotometer at 650 nm with the cell holder at 37°C and a 5 × expansion scale on the chart recorder.

↓

Prepare a series of reaction mixtures in 1 cm cuvettes (*see* note 1) as set out in the following table. Mix the PBS and the rabbit erythrocytes (*see* note 2) and equilibrate at 37°C in a water-bath for 10 min before adding the toxin (*see* note 3).

Mixture	1	2	3	4	5
PBS (ml)	2·62	2·64	2·66	2·68	2·70
0·8% rabbit erythrocytes (ml)	0·28	0·28	0·28	0·28	0·28
α-toxin (ml)	0·1	0·08	0·06	0·04	0·02

Insert each mixture, without toxin, in the cell holder and start the recorder. Wait until a steady reading is obtained.

↓

Add the toxin, cover with parafilm, invert the cuvette twice to mix and replace in the cell holder. Continue the reading until haemolysis is complete.

↓

Carefully rinse and drain the cuvette against filter paper before proceeding to the next mixture.

↓

From each trace, read off the rate of haemolysis in Δ O.D. min^{-1} and plot a graph of rate of haemolysis against α-toxin concentration.

----

3. Concentration of *S. aureus* α- and β-toxins by Precipitation with Ammonium Sulphate

Students are divided into groups, each testing a different % saturation.

Each group is supplied with 50 ml of strain Wood 46 and BB supernates harvested at 9 h after culture under the conditions described in the first experiment; a 5 ml sample is also provided for Lowry protein estimation and titration of haemolytic activity against RBC of the appropriate species.

Consult the nomogram for ammonium sulphate precipitation (Figure 1) and work out the weight of ammonium sulphate to be added to give the required degree of saturation:

Group 1 . . . . 30 % saturation,
Group 2 . . . . 50 % saturation,
Group 3 . . . . 70 % saturation,
Group 4 . . . . 90 % saturation.

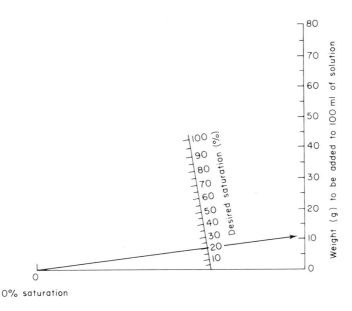

Figure 1 Nomogram for ammonium sulphate additions. A straight line drawn through zero and the desired saturation gives the amount of solid ammonium sulphate to be added to 100 ml of solution. (Modified from Dixon, 1953.)

Day 1 *(1 h)*

Place each toxin sample in a beaker in an ice bath and cool for 15–30 min.

↓

Add a magnetic stirring bar, transfer to a magnetic stirrer and add the appropriate amount of solid $(NH_4)_2SO_4$ slowly (about 3–4 g at a time) and with constant stirring to avoid denaturation of the toxin. Stir until all the $(NH_4)_2SO_4$ has dissolved (this may take up to 40 min for 70 % and 90 % saturation).

↓

Place the beaker in a cold room to allow the precipitate to settle out (*see* note 1).

Day 2 *(2–3 h)*

Transfer the contents of each beaker to two 50 ml plastic centrifuge tubes which should be balanced against each other and then centrifuged at 15 000 rev. min^{-1} for 20 min at 5°C (*see* note 2).

↓

Pour off the supernates and discard them.

↓

Dissolve the precipitate in 5 ml PBS and hold in an ice bath.

↓

Perform a Lowry protein estimation (see below) on the original culture supernate and on the redissolved $(NH_4)_2SO_4$ precipitate.

↓

Determine the haemolytic titre (*see* experiment 1) of the original culture supernate and on the redissolved $(NH_4)_2SO_4$ precipitate. Wood 46 supernates and redissolved precipitates should be titrated against rabbit erythrocytes at 37°C and BB preparations against sheep erythrocytes at 37 and 4°C.

↓

Tabulate the protein concentrations, haemolytic titres and specific activities (*see* note 3) of the original culture supernate and of the ammonium sulphate precipitated fractions.

Protein estimation by the Lowry method

The bacterial supernate and redissolved precipitate should be diluted appropriately and at least two dilutions should be incorporated in the assay (*see* note 4).

Prepare a series of mixtures in duplicate as in the following table:

Tube	1	2	3	4	5	6	7 and 8	9 and 10
BSA (200 μg ml^{-1})	—	0·05	0·15	0·30	0·45	0·6	—	—
PBS (ml)	0·6	0·55	0·45	0·30	0·15	—	—	—
Supernate (ml)	—	—	—	—	—	—	0·6	—
Redissolved precipitate (ml)	—	—	—	—	—	—	—	0·6
μg BSA per tube	—	10	30	60	90	120	—	—

↓

To each tube add 3 ml Reagent C, noting the time of addition to each tube, and mix. Let stand at room temperature.

↓

After exactly 10 min, add 0·3 ml Reagent D to each tube. Keep at room temperature for 30 min.

↓

Read the optical densities at 700 nm. Plot a standard curve of O.D. against μg BSA and read off the amount of protein in each bacterial fraction.

Notes and Points to Watch

Experiment 1

1. It may be more convenient for the instructor to do all the work outlined on Day 1 for the students beforehand.
2. PBSA is phosphate buffered saline with bovine serum albumin added to a concentration of 1 mg ml^{-1}. Seven tubes for samples up to 5 h and 12 tubes for samples between 5 and 9 h are usually adequate for the titration series, but these may have to be checked by the instructor before the practical class. When making doubling dilutions, a fresh pipette should be used after every third tube in the dilution series.
3. TBSMg is tris buffered saline containing 0·001 M MgCl$_2$. The number of tubes required is the same as detailed in note 2.
4. A 50 % haemolysis standard is prepared by mixing 1 ml 2 % erythrocytes with 1 ml distilled water until the cells are lysed, and then adding 2 ml 1 % erythrocytes in double strength saline. One millilitre of this mixture is used as a 50 % haemolysis standard.
5. The dilution causing 50 % haemolysis is accepted as containing 1 haemolytic unit (HU) and the titre is taken as the reciprocal of the dilution of toxin that causes 50 % haemolysis.

Experiment 2

1. Glass cuvettes are recommended but plastic cuvettes are adequate.
2. The concentration of rabbit erythrocytes has been chosen such that the relationship between O.D.$_{650}$ and erythrocyte concentration can be taken as linear.
3. Although it is preferable to use purified α-toxin for experiments of this sort, crude supernates from strain Wood 46 having a haemolytic titre of 300–500 HU ml^{-1} can be used. The concentration of α-toxin should be adjusted by the supervisor before the practical is held to ensure that the student obtains a representative series of haemolysis curves.

Experiment 3

1. The ammonium sulphate precipitate must be left at least overnight to settle but may be left for longer (up to one week) if necessary.
2. Instead of centrifuging toxin precipitates, they may be filtered through supercell or celite. This procedure often gives better recovery but additional equipment such as vacuum pumps and Buchner funnels are required.
3. The specific activity is defined as the haemolytic titre per milligram of protein (HU mg^{-1}).

MATERIALS

Experiment 1

1. *S. aureus* strains NCTC 7121 (Wood 46) and strain BB.
2. Bernheimer's medium is prepared as follows:
   Weigh out 2×200 g Oxoid yeast extract. Dissolve each lot in 500 ml distilled H_2O with the aid of a steam bath.
   Cut $2 \times 2\frac{1}{2}$ card lengths of 2 in. Visking dialysis tubing, soak in methylated spirit for 1 h and rinse the tubing well in sterile distilled H_2O. Fill each length with one of the 500 ml of yeast extract and dialyse each against 1600 ml distilled H_2O in a 2 litre graduated cylinder (or similar container) for 48 h.
   Discard the sac and its contents, pool the remaining fluids and make the volume up to 3200 ml.
   To this add: 64 g casaminoacids,
         8 g glucose,
         3·7 mg nicotinic acid,
         0·4 mg thiamine,
   Adjust the pH to 7·1 with NaOH.
   Distribute 500 ml amounts into 2 litre flanged Erlenmeyer flasks and autoclave at 103·4 kPa (15 lbf in^{-2}) for 15 min.
3. Phosphate buffered saline pH 7·0, 0·067 M sodium phosphate, 0·077 M sodium chloride:
   $Na_2HPO_4 \cdot 12H_2O$ (BDH), 120 g,
   NaCl (BDH), 22·3 g,
   dissolve in 3 litre distilled water. Add 11 ml conc. HCl (BDH Analar).
   Check pH and adjust volume to 5 litre with distilled water.
4. Tris buffered saline, pH 7·4, 0·01 M Tris, 0·15 M NaCl:
   Tris (BDH), 2·42 g,
   NaCl (BDH), 18 g,
   dissolve in 1900 ml distilled water, adjust pH to 7·4 with HCl (Analar).
   Make up to 2 litres.

5. 10–15 ml rabbit and sheep blood mixed with 10 ml 3·8 % sodium citrate in distilled water to prevent clotting Spin down cells (bench centrifuge, 1000–2000 rev. min^{-1} for 5 min) and wash three times in 0·9 % saline. Use packed cells to make up 1 % erythrocyte suspensions.
6. Bovine serum albumin, Fraction V. Sigma Chemical Co.
7. Magnesium chloride. BDH.
8. Formalin.
9. Gallenkamp orbital shaker.
10. Sorvall RC5 centrifuge or equivalent.

Experiment 2

1. 10 ml citrated rabbit blood (*see* experiment 1).
2. *S. aureus* α-toxin with a titre of 300–500 HU ml^{-1}.
3. Phosphate buffered saline (*see* experiment 1).

Experiment 3

1. 50 ml each of *S. aureus* strain Wood 46 and BB 9 h supernates prepared as described in experiment 1.
2. Ammonium sulphate BDH, Analar.
3. 10 ml citrated sheep and rabbit blood (*see* experiment 1).
4. Bovine serum Albumin Fraction V, Sigma Chemical Co.
5. PBS (*see* experiment 1).
6. Tris buffered saline (*see* experiment 1).
7. Lowry reagents:
   Solution A: 2 % Na_2CO_3 in 0·1 N NaOH;
   Solution B: 0·5 % $CuSO_4 \cdot 5H_2O$ in 1 % sodium potassium tartarate;
   Solution C: 50 ml A + 1 ml B;
   Solution D: Folin and Ciocalteau's Reagent (BDH) diluted 1:2 with distilled water.

SPECIFIC REQUIREMENTS

Experiment 1

Day 1

5 ml starter cultures of *S. aureus* strain Wood 46 and BB grown in Bernheimer's medium over-night in a shaking water-bath at 37 °C
sterile 1 ml and 10 ml pipettes
Gallenkamp orbital shaker or equivalent
sterile Universal bottles for storing cultures and supernates

cont.

cont.

formalin (37% formaldehyde) BDH
Sorvall RC5 refrigerated centrifuge

Day 2
Amounts given are for 1 complete experiment.
5 ml volumes of formalin-fixed cultures and supernates (prepared as described on Day 1) taken at different times in the growth cycle
30 ml of 1% (v/v) rabbit erythrocytes in PBS and sheep erythrocytes in TBS
30 ml PBS containing 1 mg ml^{-1} B.S.A.
30 ml Tris buffered saline containing 0·001 M MgCl$_2$
30 ml Bernheimer's medium
120 75 mm × 10 mm glass tubes + racks
10–20 150 mm × 15 mm glass tubes + rack
10 × 5 ml pipettes
60 × 1 ml pipettes
10 × 10 ml pipettes
water-bath at 37°C
spectrophotometer

Experiment 2
Amounts given are for 2 complete runs.
6 ml 0·8% (v/v) rabbit erythrocytes in PBS, the concentration adjusted so that the relationship between O.D.$_{650}$ and erythrocyte concentration is linear
1 ml α-toxin with a titre of 300–500 HU ml^{-1} (*see* note 2)
recording spectrophotometer with thermostatically controlled cell-holder set at 37°C, wavelength 650 nm
cuvettes
cuvette rack
35 ml PBS

water-bath set at 37°C
parafilm
5 × 5 ml pipettes
5 × 1 ml pipettes
5 × 0·1 ml pipettes

Experiment 3

Day 1
50 ml and 5 ml volumes of supernate from strain Wood 46 and strain BB
ammonium sulphate
magnetic stirrer + stirring bars
2 × 100 ml beakers
ice bath
balance

Day 2
2 × 50 ml plastic Sorvall RC5 centrifuge tubes
10 ml PBS
10 ml 1% (v/v) rabbit and sheep erythrocytes (*see* experiment 1)
10 ml PBSA and Tris buffered saline + MgCl$_2$ (*see* experiment 1)
30 75 mm × 10 mm glass tubes + rack
75 ml Lowry Solution A + 1·5 ml Lowry Solution B
2 ml BSA solution containing 200 μg protein ml^{-1}
50 150 mm × 15 mm chrome-cleaned glass tubes + rack
stop-clock
spectrophotometer
10 × 1 ml pipettes
5 × 5 ml pipettes
5 × 10 ml pipettes
water-bath at 37°C

FURTHER INFORMATION

1. Production of *S. aureus* α- and β-Toxins in Relation to Growth

Using Bernheimer's medium, McNiven (1972) showed that strain Wood 46 started to produce α-toxin at around 3 h. Toxin production reached a peak after the culture had attained the stationary phase (8–14 h), at which time toxin levels in the supernate reached a titre of 4–8 × 10^3 HU ml^{-1} (Figure 2). β-toxin production by strain BB follows a similar pattern.

Supernates from strain Wood 46 contain small amounts of δ-toxin in addition to α-toxin; this does not significantly affect the haemolysin titres of Wood 46 supernates as the α-toxin is in excess. Similarly, strain BB supernates contain some α- and δ-toxins in addition to β-toxin. For this reason haemolysis occurs in the first few tubes of the β-toxin titration series after incubation at 37 °C and accordingly the hot–cold effect is not as obvious as with purified β-toxin.

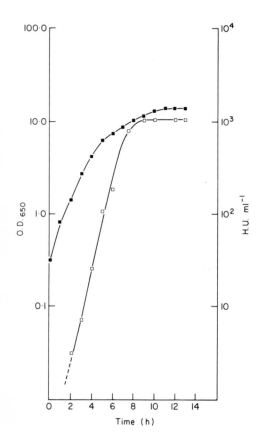

Figure 2 Growth and extracellular α-toxin production by *S. aureus* (Wood 46) in BS medium. ■—■ growth measured by $O.D._{650}$; □—□ haemolytic activity. (From A. C. McNiven, 1972.)

Typically, a 9 h supernate from strain Wood 46 would have a titre of 2–4×10^3 HU ml^{-1} against rabbit erythrocytes, and a titre of about 1–2×10^2 HU ml^{-1} against sheep erythrocytes. A 9 h supernate from strain BB would have a titre of about 4×10^3 HU ml^{-1} against sheep erythrocytes at 4 °C, and a titre of about 3×10^2 HU ml^{-1} against rabbit erythrocytes. The relative sensitivities of erythrocytes of different species of α-, β- and δ-toxins has been estimated (*see* review, McCartney and Arbuthnott, 1978) and is shown in Table 1.

The haemolytic assay as described in this experiment is useful in that it is a rapid and easy method for measuring cytolytic toxins. However, it is subject to certain limitations such as the inaccuracy of assessing 50 % haemolysis visually. This may be overcome by centrifuging the reaction mixtures and comparing the haemoglobin released (this is directly proportional to erythrocyte lysis) spectrophotometrically with a half-strength erythrocyte suspension that is fully haemolysed.

Table 1 Relative sensitivity of erythrocytes of different species to α-, β- and δ-toxins.

| | Sensitivity compared with rabbit[a] | | |
	α	β	δ
Rabbit	100	100	100
Horse	0–0·06	6	100
Human	0–0·08	400	50
Sheep	0·6–1·0	3000	20
Guinea-pig	0–0·1	6	25
Cat	9	100	10
Fowl	0–0·5	6	10
Dog	10–25	6	N.D.
Rat	10	6	N.D.
Mouse	9	6	N.D.
Pig	N.D.	50	20

[a] With individual toxins, sensitivity against rabbit is taken as 100. From McCartney and Arbuthnott (1978).

2. *S. aureus* α-toxin: the Kinetics of Haemolysis

The S-shaped curves typical of haemolysis by α-toxin are shown in Figure 3.

There is a characteristic pre-lytic lag phase, followed by a period of rapid haemolysis and, finally, a tailing off. Taking the slopes of the curves at the time of maximum rate of haemolysis as an index

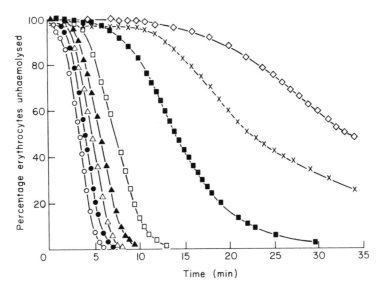

Figure 3 Relation between haemolysis and time at different concentrations of partially purified staphylococcus alpha haemolysin. The concentrations tested were: ○–○ 160 M.H.D. ml^{-1}; ●–● 80 M.H.D. ml^{-1}; △–△ 40 M.H.D. ml^{-1}; ▲–▲ 30 M.H.D. ml^{-1}; □–□ 20 M.H.D. ml^{-1}; ■–■ 10 M.H.D. ml^{-1}; ×–× 5 M.H.D. ml^{-1}; ◇–◇ 2·5 M.H.D. ml^{-1}. From Lominski and Arbuthnott (1962).

of the velocity of the reaction, it can be shown that the rate of haemolysis is directly proportional to the concentration of toxin at low α-toxin concentrations (Arbuthnott, 1970).

The primary interaction between α-toxin and erythrocytes which occurs in the pre-lytic phase results in loss of selective permeability of the erythrocyte membrane, as evidenced by leakage of potassium ions. Subsequent lysis of the damaged erythrocytes is independent of the continued presence of toxin. The addition of antitoxin very early in the pre-lytic phase prevents lysis and the release of potassium ions. However, if the addition of antitoxin is delayed, haemolysis cannot be inhibited.

The study of cytolytic toxin kinetics may provide useful information about their mode of action, since the effect of various parameters (e.g. erythrocyte concentration, pH, temperature) on the rate of haemolysis can be analysed systematically. Bernheimer (1970) showed that the kinetics of haemolysis of several bacterial cytolysins differed from that of detergents. Subsequent kinetic studies (see review, Arbuthnott, 1970) led to the concept that α-toxin may act enzymatically or else may become activated by a membrane-bound enzyme or receptor, to act enzymatically (Wiseman and Caird, 1972).

However, other concepts which may not necessarily be exclusive have been put forward. Freer et al. (1968) have demonstrated penetration of the hydrophobic region of the cell membrane by the hexameric 12S form of α-toxin. Cassidy and Harshman (1976) showed that native α-toxin and [^{127}I]-labelled α-toxin could block the binding of [^{125}I]-labelled α-toxin to the rabbit erythrocyte membrane, indicating the presence of specific membrane receptors.

One of the disadvantages of the kinetic approach in studying toxin–membrane interactions is that the event being monitored is cell lysis and therefore the effect of sub-lytic amounts of toxin on the cell membrane cannot be assessed. Thelestam et al. (1973) have shown that sub-lytic doses of toxin alter membrane permeability sufficiently to allow the release of low molecular weight compounds from cells in tissue culture.

3. Concentration of S. aureus α- and β-toxin by Precipitation with Ammonium Sulphate

Precipitation of α- and β-toxins by ammonium sulphate ("salting out") effects both concentration of the toxins (as much as fifty-fold concentration can be achieved in a single step) and removal of contaminating proteins; biological activity is also retained. The specific activity is a measure of both purity and biological activity and is therefore an important index for monitoring the purification; it should increase at each stage of the purification procedure, even though protein concentration per se may decrease.

Maximum precipitation of α- and β-toxins is achieved by 90% saturation with ammonium sulphate and the specific activity should be highest in this sample. Toxin recovered at 30 or 50% ammonium sulphate saturation has a low specific activity because proteins other than α- and β-toxins are precipitated preferentially. Precipitation by 90% saturated ammonium sulphate therefore not only effects concentration of the toxins, but also results in a degree of purification.

The specific activity for both α- and β-toxins should increase by a factor of about 4–5 during this initial step. Results of typical purification procedures for α- and β-toxins are shown in Table 2.

It can be seen that the purity of β-toxin rises from a factor of 37 to 23 000 (a six-hundred-fold increase) in a single step. This is because a β-toxin inhibitor, the nature of which has not yet been identified, is removed during purification.

It is important to ensure that most of the ammonium sulphate is removed from the concentrate

Table 2 Purification and recovery of *S. aureus* α- and β-toxins.

	α-toxin[a]			β-toxin[b]	
Purification step	Specific activity (HU mg^{-1})	Purification factor	Purification step	Specific activity (HU mg^{-1})	Purification factor
Crude α toxin	340	—	Crude β toxin	1700	—
90% $(NH_4)_2 SO_4$ precipitation	1670	5	$(NH_4)_2 SO_4$ precipitation	6800	4
Isoelectric focusing	12 800	38	Biogel P60 (after concentration)	62 000	37
Refocusing	25 000	74	C.M. cellulose eluate (after concentration)	$3{\cdot}8 \times 10^7$	23 000
			Isoelectric focusing	$6{\cdot}25 \times 10^7$	38 000

[a] A. C. McNiven (1972). [b] D. C. Low and J. Freer (1977).

before proceeding to the next stage. This can easily be achieved by dialysis against a large volume of buffer.

Precipitation of protein toxins may be achieved by other neutral salts such as sodium sulphate, magnesium and zinc salts and phosphates; the most effective region of "salting out" is at the isoelectric point of the protein. Although salt precipitation is the usual first step in purification of toxins, ultrafiltration or solvent precipitation may also be used.

REFERENCES

J. P. Arbuthnott (1970). Staphylococcal α-toxin. *Microbial Toxins*, **III**, 189–236.
A. W. Bernheimer (1970). Cytolytic toxins of bacteria. *Microbial Toxins*, **I**, 183–212.
P. S. Cassidy and S. Harshman (1976). Biochemical studies on the binding of ^{125}I-labelled alpha toxin to erythrocytes. *Biochemistry*, **15**, 2348–55.
G. Coleman and B. Abbas-Ali (1977). Comparison of the patterns of increase in α-toxin and total extracellular protein by *S. aureus* (Wood 46) grown in media supporting widely differing growth characteristics. *Infection and Immunity*, **17**, (2), 278–81.
M. Dixon (1953). A nomogram for ammonium sulphate solutions. *Biochemical Journal*, **54**, 457–8.
J. H. Freer, J. P. Arbuthnott and A. W. Bernheimer (1968). Interaction of staphylococcal α-toxin with artificial and natural membranes. *Journal of Bacteriology*, **100**, 1062–75.
I. Lominski and J. P. Arbuthnott (1962). Some characteristics of Staphylococcus alpha haemolysin. *Journal of Pathology and Bacteriology*, **83**, 515–20.
D. C. Low and J. Freer (1977). Purification of β-lysin from *S. aureus*. *FEMS Microbiology Letters*, **2**, 139–43.
A. C. McCartney and J. P. Arbuthnott (1978). Mode of action of membrane-damaging toxins produced by staphylococci. *Bacterial Toxins and Cell Membranes*, 89–127.
A. C. McNiven (1972). The nature of staphylococcal α-toxin. Ph.D. thesis. University of Glasgow, 90.
M. Thelestam, R. Mollby and T. Wadstrom (1973). Effects of staphylococcal α-, β-, δ- and γ-haemolysins on human diploid fibroblasts and hela cells: evaluation of a new quantitative assay for measuring cell damage. *Infection and Immunity*, **8**, 938–46.
G. M. Wiseman and J. D. Caird (1972). Further observations on the mode of action of the alpha toxin of *S. aureus* Wood 46. *Canadian Journal of Microbiology*, **18**, 987–92.

Assay and Serology of the α- and θ-toxins of *Clostridium perfringens*

J. G. SHOESMITH and K. T. HOLLAND

Department of Microbiology, University of Leeds,
Leeds, LS2 9JT, England

Level: Advanced undergraduates (but parts could
be used for beginning undergraduates)
Subject areas: Bacterial toxins, serology and assay
Special features: Coverage of interrelated topics,
many possiblities of use in
open-ended experiments

INTRODUCTION

Clostridium perfringens (*Cl. welchii*) causes a number of infections in man and animals, and produces a variety of toxins and hydrolytic enzymes. Some of these are the dominant lethal toxins in specific diseases, others may cause less critical tissue damage, or assist the spread of the organism, and others may be of no evident importance in the pathogenic process. In man, *Cl. perfringens* Type A may cause the dangerous wound infection gas-gangrene, in which α-toxin is the dominant and potentially lethal toxin. Much more common nowadays is *Cl. perfringens* food poisoning, caused by strains producing an enterotoxin.

Other types of *Cl. perfringens*, notably Types B, C, D and E, are mainly involved in diseases of animals. All the types excrete α-toxin, but in addition produce other lethal toxins such as β-toxin (Types B and C), ε-toxin (Types B and D) and δ-toxin (Type E), which provide a basis for identification (*see* Willis, 1977).

The α-toxin of *Cl. perfringens* is one of the few bacterial toxins with a known enzyme activity, namely that of a lecithinase C (Figure 1), also known as phospholipase C (EC 3, 1, 4, 3). It is also dermonecrotic on intradermal injection, lethal and haemolytic. The enzyme activity of α-toxin may be detected by the release of phosphoryl choline from lecithin or by the breakdown of lecithin-containing emulsions such as egg-yolk. The haemolytic effect may be measured by the lysis of red

---

cells of various animals, those of the rabbit and sheep being more sensitive than those of the horse.

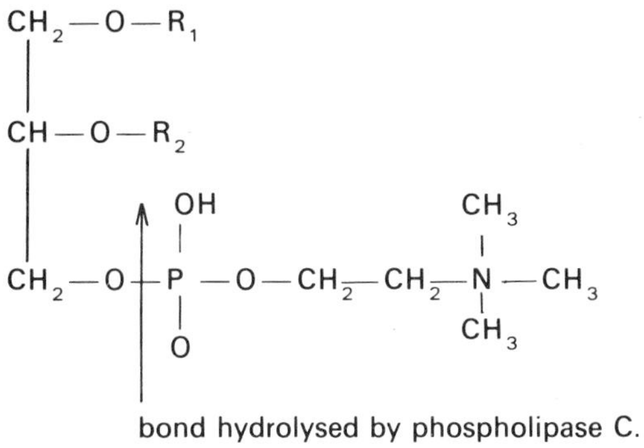

CH_2—O—R_1 R_1 and R_2 are long chain fatty acids

CH—O—R_2

CH_2—O—P—O—CH_2—CH_2—N—CH_3
with OH above P, O below P, and CH_3 above N and CH_3 below N

bond hydrolysed by phospholipase C.

Figure 1   Lecithin and the site of hydrolysis by lecithinase C (phospholipase C: EC 3, 1, 4, 3).

Other *Cl. perfringens* toxins, especially θ-toxin, are also haemolytic, though their importance in pathogenesis is less certain. With θ-toxin, unlike α-toxin, horse red-cells are as readily lysed as those of the sheep or rabbit and the typical haemolysis around colonies of the organism on horse-blood agar is usually due to θ-toxin.

The experiments that follow show methods of demonstrating and measuring haemolytic and lecithinase activities, and ways of separating the effects of α- and θ-toxins. They also demonstrate the use of "end-point" assays and the Laurell technique of crossed immunoelectrophoresis. A series of six experiments is described.

EXPERIMENTAL

1. Demonstration of Activities on Solid Media and Inhibition by Antitoxin on "Half-antitoxin Plates" (*see* note 1)

Day 1 (*20 min*)

Culture plates (dried) of (a) fresh horse-blood agar and (b) egg-yolk agar.

↓

Divide each plate into two equal parts by marking the base. Label one half "antitoxin".

↓

On to the antitoxin section deliver 1 or 2 drops of *Cl. perfringens* Type A. antitoxin and distribute evenly with a loop or glass spreader (*see* notes 2 and 3).

↓

Inoculate both halves of each plate in the zig-zag pattern as shown (Figure 2) with a culture of *Cl. perfringens* Type A.

↓

Incubate anaerobically for 18–48 h at 37 °C.

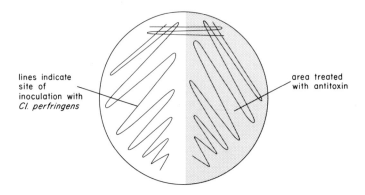

lines indicate
site of
inoculation with
Cl. perfringens

area treated
with antitoxin

Figure 2   Preparation of "half-antitoxin plate".

Day 2 *(10 min)*

Inspect the horse-blood agar plate for growth and zones of haemolysis around the colonies on the two halves of the plate (*see* note 3).

↓

Inspect the egg-yolk plate for growth and production of opalescence in the medium around the colonies on the two halves of the plate (*see* note 4).

2. Measurement of Haemolytic Activity (*see* notes 5 and 12)

(60 min)

Make nine serial 2-fold dilutions of the toxin in 2 ml volumes of tryptone diluent over the range 1/10 to 1/2560 (*see* note 6).

↓

Add 2 ml 0·20 % v/v horse red cells, mix and incubate for 20 min at 37°C in a water-bath (*see* note 9).

↓

Examine the tubes to determine the greatest dilution at which haemolysis occurs.

↓

Measure the haemolysis quantitatively in each tube either using a colorimeter with a red-filter or with a spectrophotometer set at 600–650 nm. Make sure the red cells are evenly suspended (*see* notes 7–11). Use 0·10% v/v lysed red cells as a blank.

↓

Determine the dilution at which 50% lysis occurs by plotting absorbance against the logarithm of the reciprocal dilution and interpolating if necessary.

↓

Compare this result with that obtained by visual inspection.

----

3. Measurement of Lecithinase Activity (*see* **note 3**)

(1 h 40 min)

Make dilution of toxin as above over the range 1/1 to 1/32.

↓

Add equal amounts of egg-yolk suspensions, mix and incubate for 60 min at 37 °C, taking absorbance readings at 650 nm on each dilution every 10 min. Use 1 ml egg-yolk suspension in 3 ml diluent as a blank.

↓

If a fatty layer separates, mix gently before taking a reading.

↓

Plot absorbance against time for each dilution and determine whether it would be possible to compare activities by measuring absorbance after a fixed time of incubation.

↓

Also determine the "end-point" (a) by visual inspection, as the greatest dilution at which a fatty "curd" separates after 60 min; (b) as the greatest dilution giving opalescence equivalent to an arbitrary absorbance reading (say 0·3). A graphical method, as used for the haemolysis, will allow interpolation.

----

4. Activators and Inhibitors

(2 h)

The substances tested are 0·3% hydrogen peroxide, 10 mM EDTA, 100 mM cysteine HCl, each in 20 mM Tris buffer, 0·5% NaCl adjusted to pH 7·2.

Prepare reaction mixtures in duplicate with 0·5 ml toxin, 0·5 ml tryptone diluent, and 1 ml of a solution of the substance under test.

↓

Prepare controls in duplicate for each test substance containing 1 ml tryptone diluent and 1 ml of a solution of the substance under test.

↓

Prepare two positive control tubes containing 0·5 ml toxin and 1·5 ml tryptone diluent.

↓

Mix the contents of each tube and leave for 15 min.

↓

Divide the tubes into two identical sets of controls and solutions under test.

↓ ↓

Haemolysin
Set a colorimeter to zero using 0·10 % v/v lysed cells as a blank.

↓

Add 2 ml 0·20 % v/v horse red cells to each tube and mix.

Take readings at 0, 15 and 30 min, incubating in a water-bath at 37 °C.

↓

Lecithinase
Add 2 ml egg-yolk extract to each tube and mix.

↓

Using the negative control as a blank for each substance under test, take colorimeter readings at 0, 15 and 30 min (*see* note 14).

Compare the results obtained with haemolysis and lecithinase.

----

5. Inhibition by Antitoxin (*see* note 15)

(2 h)

Make duplicate sets of four-fold serial dilutions of each of the *Cl. perfringens* antisera to Type A, B and C. Use 1 ml volumes in tryptone diluent over six tubes covering the range 1/100 to 1/102 400 (*see* note 16).

↓

Set up two positive control tubes containing 1 ml tryptone diluent alone, and two negative controls with 2 ml diluent alone.

↓

Add 1 ml toxin to all tubes except the negative controls, mix and let stand for at least 20 min.

↓

Test one dilution set of each antisera for haemolytic activity by adding 2 ml 0·20 % v/v red cells to each dilution and to the controls, mix and incubate for 30 min at 37 °C.

↓

Set the meter to zero using 0·10 % v/v lysed cells and plot meter readings against the logarithms of the reciprocal dilution.

↓

Determine the reciprocal dilution (titre) giving 50 % neutralization by interpolation.

↓

Determine antilecithinase titres similarly, but in this case either 50 % of the positive control absorbance or an arbitrary absorbance may be used as the "end-point" (see note 17).

↓

Compare the antihaemolysin and antilecithinase titres by making a table, and state whether there is evidence for more than one antigen being involved in the activities under study.

----

6. Immunoelectrophoresis

(5 h with two electrophoresis periods of 50 min and 3 h)

There are many ways in which this may be done but one useful method is the Laurell technique: crossed immunoelectrophoresis. This is carried out as follows.

Place a clean 50 × 50 mm glass slide on a coin on a horizontal bench. Dispense a mixture of 1·25 ml molten 1·25 % agarose and 1·25 ml barbitone buffer, pH 8·6, at 60 °C on to the slide, ensuring that air bubbles are not formed and all the slide surface is covered. A layer of 1 mm thickness will be formed (see note 18).

↓

When the agarose has set firmly, punch holes as indicated (Figure 3) with a 4 mm diam. stainless steel punch and remove the agarose plugs using the suction of a Pasteur pipette fitted with a rubber teat.

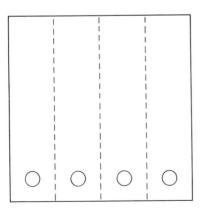

Figure 3 50 × 50 mm slide with 4 mm wells punched in the agarose, as used for the first electrophoresis.

Using a micro-syringe (e.g. Hamilton) carefully add 2 μl of filtrate to each of the four wells. Wash out immediately by filling with distilled water and emptying three times (*see* note 18).

↓

Place the slide on the bed of the electrophoresis tank (**caution**: check **power off**) (*see* note 20).

↓

Attach the wicks by gently pressing with the fingers, making sure the wicks are well soaked with buffer from the tank

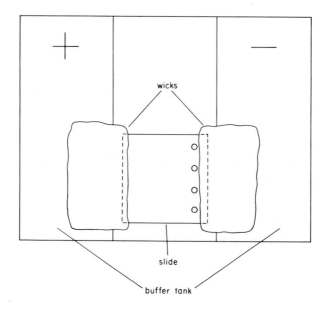

Figure 4   Slide in position for the first electrophoresis (viewed from above).

↓

Switch on the power pack and adjust the current to run at 10 mA for each slide in the tank and allow electrophoresis to take place for 50 min. Switch off, remove wicks and remove slide from the plate.

↓

Using a clean slide, cut the agarose into four strips and transfer the gels to separate clean slides as shown in Figure 5.

↓

Prepare 4 tubes of a mixture of molten 1 ml agarose and 1 ml barbitone buffer. Add *Cl. perfringens*

Type A antitoxin at different concentrations (10–40 μl) to two of these and to the other add Type B and C antitoxin (10–100 μl). The volume of antitoxin depends on the filtrate and preliminary tests should have been carried out (*see* note 21).

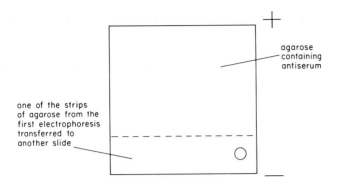

Figure 5   Slide with strip of agarose containing separated antigens and agarose-containing antiserum.

↓

Add each antitoxin–agarose mixture to a slide placed on a coin as before, allow to set and repeat the electrophoresis, but this time for 2–3 h.

↓

Switch off the current to the tank, remove the slides from the tank, rinse with distilled water, cover with a square of Whatman no. 1 filter paper and dry at 56 °C in an incubator. When dry, moisten the filter paper so that it can be removed and dry the slide completely with a hair-dryer held at a distance of about 10 cm.

↓

Stain in 0·5 % Coomassie blue for 5 min, wash in several changes of solvent until the background is clear and then dry.

↓

Examine, make a diagram of the results and interpret as far as possible. Can the precipitin patterns be directly related to haemolysin and lecithinase experiments, or would additional information be necessary?

– –

Notes and Points to Watch

1. Experiment 1 and generally: *Cl. perfringens* is capable of causing wound infections and some strains cause food poisoning. All procedures, including those which do not obviously use organisms, should therefore use good sterile technique and appropriate safety standards, especially the complete avoidance of mouth pipetting.

2. Antitoxic sera vary in potency and preliminary tests to determine the amount necessary on a "half antitoxin plate" are advised. If numerous plates are treated with antitoxin, a sterile swab soaked in antiserum may save time.

3. On fresh horse-blood agar clear zones of haemolysis are caused by θ-toxin; α-toxin is less active on horse cells and may be evident as a zone of partial haemolysis beyond the clear area. In the areas of medium treated with antitoxin no activity will be present, though growth is unaffected.

4. On the egg-yolk medium, lecithinase activity (α-toxin) is evident as a zone of precipitation within the agar around the colonies. This reaction is known as the "lecithovitellin" or "Nagler" reaction. The uninoculated medium shows a faint opalescence which may be cleared around colonies as a result of proteolytic activity. The use of "half-antitoxin plates" of egg-yolk medium is one of the simplest ways of identifying *Cl. perfringens* Type A.

5. Experiments 2, 3, 4 and 5 have been described as separate experiments but could also be carried out over a period of 5–6 h. Experiments 2 and 3 can be combined and should take 2 h. Similarly, it might be possible to carry out, say, experiment 5 during the electrophoresis time of experiment 6.

6. The quantities in the dilutions have assumed the use of 12×100 mm tubes. The use of other tubes is possible, but would involve changes in the quantities required. Check that the pipettes available will fit inside 12 mm tubes.

7. Colorimeters or spectrophotometers are used for the measurement of changes caused by the toxins. The experiments would not be possible as described if the contents of the tubes were transferred to cuvettes for readings to be taken, unless allowances were made for the time and care needed for the transfer of the tube contents and washing out of cuvettes. The tubes used therefore need to be inspected for freedom from damage or scratches, and consistency of size, so that it can be certain that they will fit into the adaptors in the colorimeters.

8. Numerous readings are taken. There should be as many colorimeters as possible, not fewer than one for two groups of students. The colorimeters should be adjacent to the water-baths, so that there is no delay in taking readings.

9. Red cells sediment faster than bacteria. Students often fail to mix suspensions properly.

10. Spontaneous lysis of red cells can be caused by mistakes in preparation of the diluent (incorrect ionic strength or pH), by detergent residues on glassware or by water dripping from racks as they are removed from the baths.

11. The method for measuring haemolysis depends on the turbidity of the red cells and detects changes in light scattering by the cells at wavelengths where absorption by haemoglobin is small. The zero is therefore set using 0·10 % lysed cells with a negative control of 0·10 % cells in the tryptone diluent. At the red-cell concentrations used (0·10%, v/v), the readings are proportional to the number of unlysed cells. Students may be confused at first because the colorimeter readings fall as haemolysis proceeds, whereas the readings rise as lecithinase activity proceeds in egg-yolk suspension.

12. If experiment 4 is omitted, experiment 3 could be repeated using tryptone diluent containing 100 mM cysteine.

13. The opalescence occurring in egg-yolk suspensions is caused by complex mechanisms and the changes may therefore be affected by the chemical composition of the medium and may vary with different samples. We have not found commercially available egg-yolk preparations to be suitable.

14. The substances under test may themselves affect the turbidity of the egg-yolk extract. In the case of cysteine, this may cause a decrease in absorbance.

15. Many toxins can be detected in more than one way (e.g. haemolysin, lecithinase, direct animal toxicity). If a single antigen of this type is present, the ratios of the amount of antibody required to inhibit each activity will be the same for different antiserum samples tested against the same toxin preparation. However, if other antigens are present with some or all of the activities, it is likely that antisera from different animals will contain different proportions of antibodies to the various antigens and the ratios will not be the same.

16. This may be done by first making a 1/100 dilution of each antiserum in tryptone diluent. Set up two rows of six tubes, one empty row, the other containing one empty tube and five with 3 ml diluent. Add 1 ml 1/100 antitoxin to the first tube of each row, 1 ml to 3 ml diluent in the row 1, mix, transfer 1 ml to the second empty tube of row 2, add 1 ml to the third tube of row 1 and discard the rest. Mix the contents of the third tube in row 1 and continue until the sixth tube, when 1 ml will be left in the tube in row 1, 1 ml transferred to row 2 and 2 ml discarded.

17. An arbitrary absorbance for the "end-point" will be essential if the toxin used is at such a high concentration that the absorbance of the positive control is unreadable on the colorimeter.

18. Students will probably require careful instruction in the preparation of the agarose slides. The coin between the slide and the bench helps to prevent overflow of the agarose. Different size wells could be used to take different volumes of toxin. If prolonged washing is required because of the timing of the experiment, the agarose may become detached from the glass. This may be avoided by smearing the slide with a thin layer of molten agarose before the main layer is poured. After preliminary drying, filter paper fibres may be removed by gentle rubbing with a moistened swab. Judgement of adequate final drying with the hair-dryer may be made by observation of changes in reflectivity at oblique viewing angles.

19. This avoids cross-contamination and jamming of the plunger caused by dried solutes.

20. Electrophoresis involves high voltages and care should be taken that the supply is disconnected when the slides are handled. This may not occur automatically in some types of equipment. Wicks prepared from 5×20 cm strips of hospital lint folded double have been found superior to filter paper.

21. Antibody concentrations in the agarose have a considerable effect on the results and it is therefore strongly advised that preliminary tests are carried out with the toxin preparation to be used. With a fixed antigen concentration the area of the peaks increases as the antibody concentration decreases and with a fixed antibody concentration the area increases as the antigen concentration increases. The intensity of the line of precipitation increases with increasing antigen or antibody concentration. For the Wellcome preparation used at 3 mg ml^{-1} and 2 μl in the well, three well-defined peaks were obtained with 20–40 μl of Type A antiserum in the gel (Figure 6). Type B antiserum required 100 μl, but use of the culture preparation required lower concentrations of Type A antiserum.

MATERIALS

1. *Clostridium perfringens* NCIB 8875 (BP6K) produces both α-toxin and θ-toxin. Maintain in a cooked meat broth, subculturing at approximately monthly intervals.

2. Horse blood for fresh blood agar plates may be obtained from a variety of sources, e.g. Oxoid Ltd., Wade Road, Basingstoke, RG24 England. Prepare the plates by cooling a molten

nutrient base (e.g. Oxoid Blood Agar Base No. 2) to 50–55 °C adding 8% (v/v) horse blood, mixing and pouring.

3. Prepare egg-yolk suspension by separating the yolk of a fresh egg, homogenizing in 20 ml 0·85% NaCl and centrifuging to remove large particles. Sterilization by filtration is not necessary if fresh eggs are used and all the other materials used are sterile. Store at 0–5 °C for not more than 5 days.

   Egg-yolk plates contain 1 ml of suspension to 20 ml nutrient agar base.

4. Antisera are available at a cost of about £15 for 10 ml from Dr P. D. Walker, Wellcome Research Laboratories, Beckenham, Kent, BR3 3BS, England. Hospital pharmacies may have stocks of therapeutic anticlostridial sera suitable for the half-antitoxin plates.

5. The toxin used may either be a dried preparation used at 3 mg ml^{-1} in 0·5% NaCl and obtained from Dr P. D. Walker (*see* note 4). Alternatively, a culture supernate from the growth of *Cl. perfringens* NCIB 8875 may be used. The organism is grown at 37 °C anaerobically in a gently stirred or shaken medium consisting of 5% tryptone (Oxoid), 1·5% soluble yeast extract (Oxoid) and 0·5% glucose. The anaerobic conditions are maintained by a slow stream of oxygen-free nitrogen (British Oxygen Company) and the culture vessel arranged so that samples can be removed. The inoculum is an 18 h culture of about one-tenth the main culture volume. Growth is established in about 2 h when samples are removed at 30 min intervals and tested for lecithinase and haemolysin by the methods described. When the lecithinase activity no longer increases (about 4–5 h), the culture is cooled and the organisms separated by centrifugation. The supernate liquid may be sterilized by filtration. The "toxin" is stored at −20 °C in 20 ml quantities.

   The dried preparation used at 3 mg ml^{-1} has a higher lecithinase activity but a lower haemolytic activity than the culture preparation. The dilution of toxin supplied to the students for experiments 2 and 3 may have to be adjusted to fit the dilution scheme, since the activities are sometimes too high. The further dilution required for experiment 4 is about 1 in 40 of the toxin used for experiment 2 and 3, and that required for experiment 5 about 1 in 10 of that required for experiments 2 and 3.

6. The isotonic tryptone diluent consists of 1% Tryptone (Oxoid) 0·5% NaCl and 0·01 M CaCl$_2$.

7. Red-cell suspensions are made by centrifuging horse blood and resuspending in 0·85% NaCl to give a 2% v/v suspension. This is diluted immediately before use to give 0·2% cells for the haemolysin substrate. The 0·10% lysed control is prepared by diluting ten-fold in water and mixing with an equal volume of tryptone diluent after lysis.

8. Egg-yolk suspension for the lecithinase substrate is produced by dilution of the original extract (*see* note 3 above) 1 in 20 in 0·85% saline.

9. Suitable colorimeters are the Cecil 404 and Pye Unicam SP15, fitted with adapters for 12 mm tubes. The standard fitting for the SP15 may not hold 12 mm tubes firmly. Although Pye Unicam supply an alternative adapter, this will not take 16 mm tubes and it is more convenient to fit a larger 13 mm roller inside the cell holder.

10. 50 × 50 mm glass slides. Manufactured for photographic use, marketed by Rowi International and obtainable from dealers in photographic supplies.

11. Agarose HSA (International Enzymes, Vale Road, Windsor, Berks, SL4 5NJ, England) is dissolved at 100 °C to give a 2% solution and 0·1% sodium azide added as a preservative.

12. Electrophoresis buffer consists of 1·38 g diethyl barbituric acid, 7·7 g sodium diethyl barbiturate and 0·05 g sodium azide in each litre of distilled water. It is diluted to half strength for use, either by agarose or by water for use in the electrophoresis tank.

13. A suitable electrophoresis system is supplied by Shandon Southern Ltd., (93, Chadwick Road, Astmoor Industrial Estate, Runcorn, WA7 1PR, England), which includes power-pack and electrophoresis tank which will accommodate four slides and must be fitted with a cooled platform.

14. The protein stain is 0·5% Coomassie blue BL dissolved in a solvent consisting of 5 vols methanol, 4 vols water and 1 vol. glacial acetic acid. The solvent is also used to wash the slides after staining. The dye is obtained from R. A. Lamb, London, NW10 6JL.

SPECIFIC REQUIREMENTS

Experiment 1
culture of *Cl. perfringens*
fresh blood plates
egg-yolk plates
antiserum
anaerobic jar
hydrogen source

Experiment 2
test tubes, 12 mm diam. (12) checked for size in
  the colorimeters
rack
pipettes (1 ml, 5 ml, 10 ml)
colorimeter
water-bath
toxin (1 ml)
red-cells (25 ml 0·2 % and 5 ml 0·10 % lysed cells)
tryptone diluent, 50 ml
graph paper and source of logarithms
discard container for excess diluted toxin

Experiment 3
as for experiment 2, except as follows:
egg-yolk suspension (25 ml, 1 in 20) replaces red
  cell suspension
10 test tubes
5 ml toxin

Experiment 4
test-tubes (20)
pipettes
racks
colorimeter
water-bath
toxin (10 ml)
red cells (25 ml 0·20 % cells and 10 ml 0·10 %
  cells)
egg-yolk suspension (25 ml)
tryptone diluent (25 ml)
EDTA, hydrogen peroxide, cysteine HC1 (5 ml
  each)
discard container

Experiment 5
test tubes (48)
pipettes
racks
colorimeter
water-bath
toxin (50 ml)
antisera to *Cl. perfringens* Types A, B and C (0·05
  ml of each)
red cells (50 ml 0·2 %) + 10 ml 0·10 % lysed cells
egg-yolk suspension (50 ml, 1 in 20)
tryptone diluent (100 ml)
graph paper and source of logarithms
discard container

Experiment 6
50 × 50 mm slides (5)
horizontal bench or levelling table
barbiturate buffer (1500 ml)
agarose (6 ml)
method for melting agarose (steamer or boiling
  water)
water-bath at 60 °C
stainless steel punch, 4 mm
Pasteur pipette and teat
pipettes
micro-syringe or pipette
electrophoresis tanks and power-pack
wicks
antitoxins, 0·02–0·2 ml, depending on toxin used
toxin 20 μl
0·5 % Coomassie blue in staining bath
solvent in baths
Whatman no. 1 filter paper
50–60 °C incubator or oven for drying
hair-dryer

FURTHER INFORMATION

The toxins responsible for the activities studied in these experiments are two among a multiplicity of soluble antigens produced by *Cl. perfringens*. The other toxins include proteases (κ- and λ-toxins), hyaluronidase (μ-toxins), deoxyribunoclease (ν-toxin), enterotoxin and lethal or dermonecrotic toxins characteristic of the different types of *Cl. perfringens*.

The two toxins examined occur in all types of *Cl. perfringens*, though food poisoning strains of Type A are sometimes deficient, especially in θ-toxin. α-toxin requires divalent cations for activity and is stable to oxidation, though readily denatured by surface tension effects (e.g. shaking of solutions). It is, however, inhibited by sulphydryl compounds, especially cysteine. On the other hand, θ-toxin is a typical oxygen-labile haemolysin, not dependent on cations. Oxygen inactivation may have occurred even in freshly prepared samples and may be reversed by sulphydryl compounds. Antibody neutralization is not completely specific, so that cross-reaction occurs with antistreptolysin O and antitetanolysin, which are other oxygen-labile haemolysins. Alouf (1977) and Arbuthnott (1978) give additional information.

Typical results obtained with a fresh culture preparation gave a lecithinase reciprocal end-point dilution of 128, and a haemolysin end-point of 640. Activation of haemolysin by cysteine could be up to 80-fold depending on the preparation, but inhibition of lecithinase may be partial. Haemolysin is little affected by EDTA, but lecithinase is completely inhibited. Hydrogen peroxide gives variable results, sometimes causing some inactivation of lecithinase as well as haemolysin. Results of this kind indicate that θ-toxin is predominant in the culture, as well as being easier to detect because of the greater sensitivity of the test for haemolysin. However, different preparations could produce other results.

There are many possibilities for variation and expansion of these experiments.

The assays could be used to examine different conditions of toxin production or methods of purification. They have also been used in conjunction with other exercises comparing the merits of various types of assay. There are many ways in which the inhibitor experiments could be expanded, such as the use of other cations or chelating agents, or the use of sulphydryl reagents.

The use of antisera is a simple example of a method which has been used ingeniously in the past (Oakley, 1954).

A typical result obtained from experiment 5 is in Table 1, from which it is evident that the Type C antiserum behaves very differently from Type A and Type B, and so more than one antigen must be involved.

Table 1

Activity	Inhibitory titres of Type sera		
	A	B	C
Haemolysin	340	260	1600
Lecithinase	5600	2100	1300

The electrophoresis experiment indicates the number of antigens precipitating with the antisera used, but does not directly relate these to the toxic activities. Three peaks are easily seen with the Wellcome toxin and Type A antisera, and one with Type B (Figure 6). The culture preparation shows fainter lines.

Experiment 5 shows that antisera A and B have similar inhibitory activities, but the precipitin patterns are quite different. It is therefore unlikely that the lines are readily associated with the activities examined.

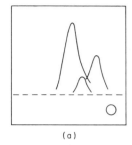
(a)

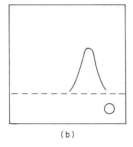
(b)

Figure 6 Results of immunoelectrophoresis experiment. (a) Wellcome Research Laboratories preparation with *Cl. perfringens* Type A antiserum; (b) with Type B antiserum.

The lack of obvious associations does, however, provide further scope for additional experimentation, such as testing sections of agarose for appropriate activity after the first electrophoresis to locate the antigen responsible. The system could also be used to examine the effects of changing antigen and antibody concentrations in the Laurell technique, or for examining changes in antigen produced by different growth conditions. Further information on the technique is given by Axelsen *et al.* (1975), and by Verbruggen (1975).

REFERENCES

J. E. Alouf (1977). Cell membranes and cytolytic toxins. *In* "Specificity and Action of Animal, Bacterial and Plant Toxins", (P. Cuatrecasas, ed.), pp. 219–70. Chapman and Hall, London.

J. P. Arbuthnott (1978) Molecular basis of toxin action. *In* "Companion to Microbiology", (A. T. Bull and P. Meadow, eds), pp. 127–53. Longman, London.

N. H. Axelsen, J. Kroll and B. Weeke (1975). "A Manual of Quantitative Immunoelectrophoresis: Methods and Applications". Oslo Universitelsforlaget.

C. L. Oakley (1954). Bacterial toxins. *Annual Review of Microbiology*, **8**, 411–28.

R. Verbruggen (1975). Quantitative immunoelectrophoretic methods. A literature survey. *Clinical Chemistry*, **21**, 5–43.

A. T. Willis (1977). "Anaerobic Bacteriology", 3rd edn, pp. 34–172. Butterworths, London.

Part Eight
Micro-organism Meets Plant

86

The Sauerkraut Fermentation

H. DALTON

*Department of Biological Sciences, University of Warwick,
Coventry, CV4 7AL, England*

> *Level*: Beginning undergraduates
> *Subject areas*: Food microbiology
> *Special features*: Minimal requirements

INTRODUCTION

Fermentation is one of the oldest means of preserving food. Sauerkraut is a preserved form of cabbage in which sugars present in the cabbage are converted to acids by lactic acid bacteria. The acids so produced serve to prevent contaminating bacteria from growing and spoiling the product. In addition, they give the sauerkraut its characteristic flavour.

During fermentation the sugars, which are drawn from the shredded cabbage tissue by salt, allow a succession of bacteria to develop until eventually lactic acid bacteria predominate. The production of organic acids causes a marked decrease in the pH value of the brine, so measuring the change in pH value and the appearance of lactic and acetic acids is a convenient way of monitoring the progress of the fermentation. Needless to say, it is important to exclude as much air as possible from the fermenting cabbage. This prevents the growth of aerobic organisms which could oxidize any organic acids produced, causing undesirable side products and impairing the flavour of the sauerkraut.

The aim of this experiment is to observe the changes in the microbial flora, pH and organic acid production during a natural fermentation of cabbage to sauerkraut.

EXPERIMENTAL

Day 1 *(2½ h)*

1. *Preparation of cabbage for fermentation*

Remove the outer leaves and any damaged or spoiled tissue from a solid head Dutch cabbage.

↓

Trim the heads and wash with water to remove surface contamination.

↓

Core the heads and shred the cabbage as finely as possible, preferably into strips a few mm in width.

↓

Mix the shredded cabbage with 3% salt (by weight) (*see* note 1).

↓

Pack into a 1 litre glass beaker or similar sized Kilner jar.

↓

Cover with muslin and a nylon or wooden disc, and press down firmly until a layer of juice is expelled from the cabbage.

↓

A weight should then be placed on the top of the disc to keep it firmly pressed down (*see* note 2).

↓

Incubate at 22° C or room temperature (*see* note 3).

2. *Sampling and examination*

Withdraw about 5 ml of the juice from the bottom of the jar using a sterile pipette (*see* note 4).

↓

Examine the juice microscopically and prepare and examine a Gram stain of the sample.

↓

Prepare appropriate dilutions of the juice and inoculate 0·02 ml drops in triplicate on to dried glucose–yeast agar plates and Rogosa agar plates.
The following dilutions and incubation times should be suitable:

Day 1	10^{-1}	10^{-2}	10^{-3}	incubate	24 h
Day 2	10^{-2}	10^{-3}	10^{-4}	incubate	24 h
Days 3–6	10^{-4}	10^{-5}	10^{-6}	incubate	24 h
Weeks 2–4	10^{-4}	10^{-5}	10^{-6}	incubate	3 days.

↓

Measure pH of the brine solution on a pH meter.

↓

Determine the total amount of acid produced by titration as follows. Transfer 5 ml of the juice to a conical flask, add 5 ml distilled water and boil for a few minutes to drive off CO_2. Cool, add 2 drops of phenolphthalein as indicator and titrate against standardized 0·1 N NaOH (*see* note 5).

↓

Determine the amount of acetic acid produced by the colorimetric lanthanum nitrate method, i.e. transfer 1 ml of the juice to a stoppered test tube (*see* note 6) and add 2 ml lanthanum nitrate (reagent A). Then add 1 ml iodine solution (reagent B), stopper the tube and place in a boiling water-bath for 5 min. After cooling on ice, immediately measure the absorbance at 625 nm against a reagent blank in which distilled water replaces the juice sample (*see* note 7).

Day 2 and subsequent days   *(1½ h)*

The plates should be examined and counted after the suggested incubation times. Record the numbers and types of colonies at each dilution, examine microscopically and do a Gram stain. A sample of juice should also be removed for pH determination, microscopic examination, total acid and acetic acid determinations, and appropriate dilutions should be prepared for plate counting (as for Day 1).

Notes and Points to Watch

1. It is important to distribute the salt evenly throughout the shredded cabbage, otherwise the soluble carbohydrates will not be removed efficiently from the tissue.
2. The shredded cabbage must be pressed down firmly—if necessary top up with more cabbage and salt—otherwise insufficient brine is produced.
3. The volume of the brine sample may have to be reduced after the first week if insufficient cabbage has been used initially. Ensure, therefore, that the cabbage is finely shredded.
4. It may be necessary to remove the brine sample with a syringe and long needle to avoid introducing too much air into the fermentation during sampling, which can occur if a pipette is used.
5. The amount of acid produced can be calculated according to the formula:

$$\% \text{ lactic acid} = \frac{\text{ml } 0.1 \text{ NaOH} \times \text{normality of NaOH} \times \text{mol. wt acid}}{\text{ml sample} \times 10}.$$

This method is fairly approximate because acids such as acetic acid may be produced, particularly if heterofermentative organisms such as *Leuconostoc* and *Streptococcus* predominate. If homofermentative organisms such as *Lactobacillus* and *Pediococcus* species predominate then the above formula will serve as an adequate measure for lactic acid formation. Acetic acid can be measured by the lanthanum nitrate method (*see* Materials and Specific Requirements) in which lanthanum reacts specifically with acetate in the presence of iodine and ammonium ions to give a jelly-like solution of basic lanthanum acetate which is blue–green in colour due to the adsorption of iodine into the complex.

6. If the tubes are poorly stoppered the colour will fade rapidly.
7. The E_{625} and acetate concentration are linear up to $250\,\mu g$ acetate ml^{-1}. The concentration of acetate is therefore read from a standard curve (up to $250\,\mu g$ acetate ml^{-1}) prepared at the same time.

MATERIALS

1. Glucose yeast agar and Rogosa agar, obtainable from Oxoid Ltd.
2. Reagents for the acetate determination, made up as follows:

   reagent A: mix equal volumes of 2·5% w/v lanthanum nitrate and 0·1 N NH_4OH. The 0·1 N NH_4OH is made up using either 7·6 ml of 0·91 ammonia or 3·4 ml of 0·88 ammonia per litre of distilled water.

   reagent B: 2·54 g I_2 and 32·2 g KI, dissolved and made up to 1 litre with distilled water and stored in the dark in a stoppered bottle. The reagent should be made up weekly because it tends to go off even in the dark.

3. The standard acetate solution contains 347·4 mg sodium acetate per litre distilled water, equivalent to 250 μg acetate ml^{-1}.

SPECIFIC REQUIREMENTS

Dutch cabbage (1 small cabbage per person)
1 litre glass beakers or equivalent size Kilner jars
burette
standard NaOH solution (0·1 N), approximately 50 ml per student, and phenolphthalein as indicator

glucose yeast agar plates (3 per student per day)
Rogosa agar plates (3 per student per day)
reagents for acetate determination
standard acetate solution

FURTHER INFORMATION

Figures 1 and 2 summarize the results obtained by a class of students.

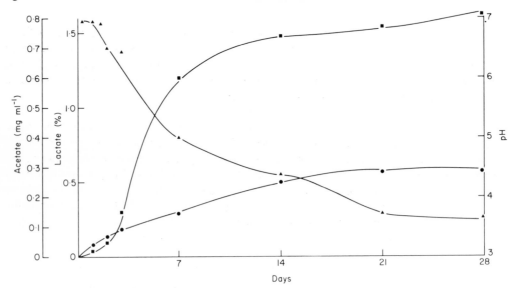

Figure 1   pH value (▲), lactate (■) and acetate (●) concentrations in the brine of the sauerkraut.

During the first 24 hours the pH of the brine was around 7 and bacteria were detected only on the glucose–yeast agar plates, and at a concentration of around $10^4 \, ml^{-1}$ at time 0 and $10^6 \, ml^{-1}$ after 24 h.

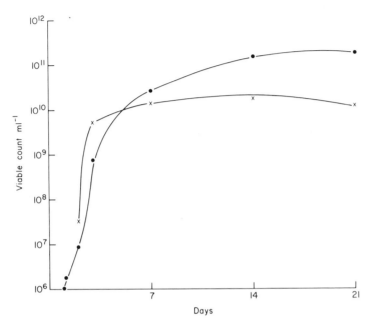

Figure 2   Viable counts of organisms from the brine on glucose yeast agar (●) and Rogosa agar (x).

During the first 24 hours (or so), the environment in the jar was aerobic and allowed the fairly rapid growth of aerobic organisms, as judged by the viable count on the GYA plates. Examination of the brine revealed both Gram-negative and Gram-positive organisms. Although the lactic acid bacteria are fermentative they were capable of growth under aerobic conditions, but only represent a very small proportion of the population in the brine at that time.

On the second day, however, there was a rapid increase in the numbers of lactic acid bacteria (as seen on the Rogosa plates), which corresponded with a drop in the pH of the brine due to the production of acid. Colonies of the Gram-positive lactic acid bacteria form white, opaque colonies on the GY agar, whereas the Gram-negative rods that are present generally form translucent colonies. The Rogosa medium, however, is fairly specific for the lactic acid bacteria (Sharpe, 1960).

After about the fourth day, the pH of the fermentation had dropped to below 6 due to the large increase in production of acid—a condition which favoured the growth of the lactic acid bacteria and inhibited growth of other heterotrophs which were present. At about this time the organisms that predominated, as judged by the Gram stain on the brine, were the Gram-positive rods and cocci.

The progressive fall in pH value down to a minimum of about 3·5 was accompanied by an increase in numbers of lactic acid bacteria which continued for another 3 weeks or so. It is important that students make some olfactory appraisal of the fermentation because a "bad" fermentation will give a very distinctive smell, as well as discolouration, low count on the Rogosa plates and lack of lactic acid production.

If the fermentation has gone well the results should be similar to those presented above, with a sweet-smelling product. The whole fermentation process could be reduced to about a week if a starter culture of selected lactic acid bacteria is inoculated at the beginning of the experiment.

REFERENCE

M. E. Sharpe (1960). *Laboratory Practice*, **9**, 223–7.

Production of Tempe, an Indonesian Fermented Food

J. N. HEDGER

*Department of Botany and Microbiology, University College of Wales,
Aberystwyth, Dyfed, SY23 2AX, Wales*

Level: All undergraduate years
Subject areas: Fun microbiology; food
microbiology
Special features: Strong visual impact; future food
resource

INTRODUCTION

Tempe (or tempeh) is a solid fermented soya bean product that is consumed widely in Indonesia. In recent years, there has been considerable interest in the West, especially the USA, in developing tempe as an alternative protein source. Hesseltine was one of the first to make a detailed description of the fermentation in his authoritative review of oriental fermented foods (Hesseltine, 1965). Only one micro-organism, *Rhizopus oligosporus* (Fungi: Zygomycotina) is required for the process and the preparation is extremely rapid, taking only two days, at most, for completion. In this it differs from other soya fermentations, such as miso and shoyu, which involve fungi, yeasts and bacteria in a multi-stage fermentation, which may take months or years to completion. Unlike these fermentations tempe is a solid "cake", which is perishable and is consumed as a meat substitute, rather than as a condiment.

Rapidity and simplicity make tempe preparation ideal for demonstrating the principle of food fermentations, and, with reasonable care, the product can be guaranteed to be free of contaminating moulds and mycotoxins.

EXPERIMENTAL

Technical requirements for the production of tempe are modest, but successful preparation depends on adherence to a number of practical points, which are outlined in the flow diagram and

---

experimental notes. Ultimate disposal of the product depends on the gastronomic predilections of student and tutor, as well as interpretation of the Health and Safety at Work Act. Tempe is delicious thin sliced and fried in soya oil until light brown. Garlic and tamarind make useful spices and the tempe can be treated as a meat substitute in the menu.

Day 1 *(1 h)*

Preparation of beans
Weigh out approximately 500 g of soya beans (previously soaked overnight in tap-water and allowed to drain) (*see* note 1).

↓

Suspend the beans in fresh tap-water and dehull by hand (*see* note 2).

↓

Discard water, drain the beans and place in a saucepan or glass beaker. Add an excess of water (*see* note 3).

↓

Bring to the boil and boil for 5 min.

↓

Immediately drain the beans **thoroughly** through a sieve (into a sink) and spread them over a clean tray (surface sterilized with alcohol).

↓

Allow to cool, stirring continuously with a spatula or spoon until a temperature of 35–40 °C is reached (*see* note 4).

↓

Add inoculum and mix the whole mass thoroughly by hand. Do not allow to cool excessively (*see* note 5).

↓

Place the inoculated beans in fermentation containers, either plastic bags or Petri dishes (*see* note 6).

↓

Preparation of inoculum
Sporulating culture of *R. oligosporus* on 3% malt agar (*see* note 11).

↓

Add 10·0 ml sterile distilled water + "Tween 80" (*see* note 12).

↓

Dislodge spores with a sterile paintbrush or strong loop.

↓

Add to a sterile centrifuge tube, centrifuge for 5 min, discard supernate.

↓

Resuspend spores in 0·5 ml sterile distilled water.

↓

Add to small quantity of beans in a clean beaker (alcohol sterilized) and mix thoroughly. Use as inoculum for main bean mass.

Incubate in the dark between 25 and 38 °C (*see* note 7).

- -

Day 2 *(15 min)*

Examine tempe and assess the state of fermentation after 24 h in incubator (*see* note 8). If in plastic bags, reverse the incubation position. Check the temperature of the incubator. If necessary, remove the tempe to a cooler incubator (*see* note 9).

- -

Day 3 *(15 min)*

The tempe should by now be ready and can be stored if necessary at 10° C for up to 1 week or can be deep frozen immediately. The product should be a firm white cake (*see* note 10).

- -

Notes and Points to Watch

1. Soaking is necessary to remove inhibitors from the soya beans and to loosen the hulls. There should be at least 5 cm depth of soak-water over the beans. If necessary, larger quantities of beans can be soaked together for class use.
2. Dehulling is a laborious process. It can be done simply by squeezing beans between thumb and forefinger. Loose hulls can be floated off with a jet of water. In fact, 100 % dehulling is unnecessary but a fairly high percentage is required for a successful fermentation, as the presence of the hulls inhibits penetration of the beans by the *Rhizopus* mycelium. A little ingenuity will soon provide more efficient ways of dehulling.
3. There should be about 5 cm depth of water over the beans.
4. This is an important stage. The beans should be vigorously stirred to promote evaporation of excess water. If possible a perforated tray should be used to promote draining of water. An excess of water will promote bacterial growth (especially *Bacillus subtilis*) and inhibit the mould, resulting in bad tempe. The surface of the beans should be almost dry without a very obvious film of water. If necessary, use a fan to promote drying. Do not inoculate until the temperature falls below 40 °C.
5. Below 35 °C, the lower the temperature of the bean mass when placed in the fermentation containers, the slower will be the fermentation, because of the insertion of a "lag phase" required for temperature build-up. In general, fermentation is most successful if the operations are carried out in a warm room.
6. The choice of fermentation container is important. In Indonesia the traditional method requires wrapping the beans in banana leaves, but fortunately plastic bags are even better.
   Place the soya beans in a new (and therefore partially sterile) plastic bag and flatten the contents out to a "cake" about 2·0 cm thick; 250 g wet weight of beans will give a cake about 24 × 12 cm. The area of the cake is not important, but the thickness should always be about 2·0 cm. Plastic bags can either be heat-sealed (with a hot needle) or the opening can be folded over tightly up to the edge of the cake, moulded straight with a ruler, and secured with metal or plastic clips. The plastic bag should then be pierced at 1–2 cm intervals all over its surface. Such holes should not be larger than the needle and a hot, but **not** red hot, needle gives excellent

results. Plastic bags can then be laid in the incubator but should never be placed on top of each other as this restricts aeration. They should be turned after the first 16–20 h of incubation.

Petri dishes provide an alternative container but produce tempe of slightly poorer quality. The new square type plastic dish is the best and used dishes, that have been thoroughly washed and sterilized with alcohol, are perfectly good. The beans should be packed down into the dishes by leaving them about "half a bean" proud of the edge and then forcing the lid down. Square dish lids can be secured with a rubber band, round dishes by a strip of sellotape. Aeration by holes is, in this case, unnecessary, though the square dishes may benefit from a small hole in the centre top and bottom.

7. The speed of fermentation is determined by the incubation temperature. Incubation temperatures above 40 °C and below 25 °C will not produce good tempe. A temperature of 37–38 °C will produce tempe within 22 h; a temperature of 28–30 °C will take up to 48 h to produce tempe.

8. Within 16–20 h there should be a covering of mycelium over the beans but the beans themselves will still be visible. Wet patches without mycelium mean that bacterial contamination has occurred and the tempe should be discarded.

9. After about 12–16 h the fermentation begins to generate heat. If a small closed incubator is used with large quantities of tempe, temperatures within the tempe cake may reach 50 °C and the growth of the *Rhizopus* is checked. If the tempe is excessively hot at the 16–24 h stage, it can be transferred to a cooler incubator or the door of the incubator can be left ajar. Ventilation appears to be important, in any case, for successful fermentation.

10. The final stage of the fermentation is reached when the soya beans are completely covered in a dense mass of white mycelium. A section through the cake reveals that the spaces between the beans are completely filled with mycelium. The cake itself should be firm and should only bend a little when held by one corner. Soft cake which tends to break up and has the interstices only partly filled by mycelium, indicates either bacterial contamination or too high a temperature in the late stages of fermentation. The tempe should have a pleasant, slightly sweet smell and nutty flavour; off odours and excessively sweet taste indicate bacterial contamination; ammonia production indicates that the fermentation has gone too far. Other problems which may occur include a black colouration around the sites of the aeration holes; this is due to sporulation by the fungus, but does not affect the quality. Heavy sporulation will occur at temperatures below 26 °C. Bright yellow patches may appear; these are due to β-carotene biosynthesis by *R. oligosporus*, and indicate that light is reaching the fermentation.

11. *R. oligosporus* should be inoculated onto 3 % malt slopes at least 7 days before the spores are required and incubated at 30 °C. Slope cultures more than 1 month old should not be used.

12. Add 1 drop of "Tween 80" to 100 ml of sterile distilled water.

MATERIALS

1. *Rhizopus oligosporus* culture. Several strains are available in this department. Alternatively, cultures are available from the Northern Regional Research Laboratory, A.R.S., Peoria, IL 61604, USA (culture no. NRRL 2710).

2. 3 % malt medium (malt extract 30·0 g, agar 15 g, distilled water 1 litre).

3. Soya beans (*Glycine max*).

4. Saucepan or glass beaker for boiling beans.

5. Bucket or large plastic beaker for soaking beans.

6. Plastic bags, e.g. 18·0 × 23·0 cm.

7. Kitchen sieve.
8. Large plastic spoon or spatula.
9. Plastic trays.

10. Artist's paintbrush (sterilize by autoclaving in grease-proof paper).
11. Gas ring to boil beans.
12. Clips or rubber bands.

SPECIFIC REQUIREMENTS

Day 1
7 day old *R. oligosporus* culture
10 ml pipettes (sterile)
250 ml sterile distilled water + "Tween 80"
autoclaved artist's paintbrush
sterile centrifuge tube
low speed centrifuge
2 clean beakers (500 ml)
balance
soaked soya beans (overnight in plastic bucket)
saucepan or glass beaker (2 litres)
gas ring

kitchen sieve
large plastic spoon or spatula
tray
plastic bags
clips
[rubber bands
Petri dishes]
incubation facilities (25–38 °C)

Days 2 and 3
refrigerator or deep freeze

FURTHER INFORMATION

In Indonesia, tempe is either sold fresh, to be cooked by slicing and frying, or as processed biscuits or crisps. It forms an important part of the diet of many poor people and indeed may supply much of their protein.

There is some controversy in the literature over the food value of tempe, as against unprocessed soya beans, but there seems to be some evidence that the PER (protein efficiency ratio) is increased and content of the B vitamins rises (growth of *Klebsiella pneumoniae* in Indonesian tempe may be responsible for this increase). In addition, the trypsin inhibitor and phytic acid present in soya bean, which reduce the nutritional value by, respectively, inhibiting gut enzymes and chelating important metal ions in the intestine, are destroyed by the fungus. *R. oligosporus* also appears to produce an antioxidant which prevents the dried product becoming rancid. These and other aspects are well reviewed by Djien and Hesseltine (1979).

This method of tempe production can be used as a basis for further experimental work. We have used the bacterium *Lactobacillus casei* var. *rhamnosus* NCIB No. 6375 to bio-assay the accumulation of riboflavine during the fermentation (Roelofsen and Talens, 1964). The effect of temperature, light level and aeration upon the fermentation also provides useful variations.

R. oligosporus can be used to ferment other substrates. Wang and Hesseltine (1966) describe wheat tempe, and it is also possible to ferment a variety of pulses other than soya to produce tempe-like products (e.g. peanuts, etc.).

A frequent question which arises is the potential problem of mycotoxin production. *R. oligosporus* is not known to produce any mycotoxin. However, fermentation at too low a temperature (25 °C) could allow growth of *Aspergillus flavus* and other mycotoxin producers to take place. In practice, even in Indonesian cottage industries, soya bean tempe is rarely reported to contain any mycotoxin.

REFERENCES

K. S. Djien and C. W. Hesseltine (1979). Tempe and related foods. *In* "Microbial Biomass", (A. H. Rose, ed.), pp. 115–40. Academic Press, London and New York.

C. W. Hesseltine (1965). A millenium of fungi, food and fermentations. *Mycologia*, **57**, 149–97.

P. A. Roelofsen and A. Talens (1964). Changes in some B vitamins during moulding of soya beans by *Rhizopus oryzae* in the production of tempeh kedele. *Journal of Food Science*, **29**, 224–6.

H. L. Wang and C. W. Hesseltine (1966). Wheat tempeh. *Cereal Chemistry*, **43**, 563–70.

A Rapid Colorimetric Method for the Estimation of Cellulose Decomposition by Micro-organisms

M. J. SWIFT

Botany Department, University of Zimbabwe,
P.O. Box MP 167, Mount Pleasant, Salisbury, Zimbabwe

> *Level*: All undergraduate years
> *Subject area*: Microbial ecology
> *Special features*: Flexible application to terrestrial
> and aquatic environments

INTRODUCTION

In studies of decomposition processes in terrestrial habitats the measurement of cellulose-degrading (cellulolytic) activity has often been taken as a useful index of decomposer activity. This is because cellulose is the most abundant component of all photosynthetic land plants and thus represents the main organic food source for heterotrophic decomposers. On land, the major agents of cellulose breakdown are fungi, with both aerobic and anaerobic cellulolytic bacteria playing a minor role. Fresh waters, such as rivers, streams and small ponds, also receive large inputs of cellulose in the form of higher plant detritus. In marine habitats, however, cellulose is much less abundant and is only a small component of primary production in the open sea. Nevertheless in estuaries, cellulose may be an important constituent of materials from three sources; (a) the cell walls of green, brown and red macrophytic algae; (b) terrestrial plant detritus washed into the estuary; (c) man-produced waste such as paper. Bacteria are probably the major agents of cellulose degradation in marine habitats. In freshwaters, however, aquatic fungi (mainly Hyphomycetes) are the dominant agents in well-aerated zones, with bacteria playing an increasing role under diminishing oxygen availability.

Remazol Brilliant Blue is a dye molecule which forms a covalent bond with a fraction of the sugars within a number of polysaccharides, including cellulose. The dye remains fast under normal environmental conditions but when subject to the hydrolytic action of microbial cellulases the dye is released proportionately to the fraction of the polymer broken down. Estimation of dye release

(Poincelot and Day, 1972) or of dye remaining (Moore *et al.*, 1979) thus gives a rapid assay of cellulolytic activity. The experiment described here has the objective of estimating the extent of cellulose decomposition over a given time in an aerobic soil held at a standard temperature in the laboratory.

EXPERIMENTAL

Many variants of the experiment are possible, e.g. investigation of the influence of environmental factors on decomposition rate (temperature, soil moisture); effects of amendments (fertilizer, pesticides); comparisons of soils and horizons, etc. Different habitats such as those of marine and freshwater ecosystems may also be investigated, with suitable alterations in experimental design and presentation of the cellulose. Additional information may be gained by carrying out microscopic examination of the films to determine the dominant micro-organisms associated with cellulolytic activity.

Day 1 *(30 min)*

Sieve soil through 2 mm mesh to remove large particles, stones, etc.

↓

Fill 250 ml beaker to about three-quarters with sieved soil. (Weigh, and adjust the moisture content to desired level if required.)

↓

Make vertical slits in the soil with a knife and insert litter bags (*see* note 1) containing RBB-dyed cellophane films. Press the soil back into place so that good contact is made between soil and films.

↓

Place in chosen conditions of incubation. Different groups of students may choose different conditions and sampling times (*see* note 2).

Day 2 *(30 min, 30 min gap, 20 min)*

The number of replicates and the period of time will vary according to the type of soil, the conditions of incubation and exact experimental objectives. However, for guidance, intervals of 2–3 weeks at 25 °C are likely to be suitable.

↓

Remove the litter bags from the soil, open them and remove the cellophane films.

↓

Wash the films free of adhering soil with a **gentle** stream of water from a wash-bottle. Use fine forceps and a fine paintbrush to remove particles not previously freed.

↓

Using fine forceps, pick up the film or film fragments from the bag and place each film in 60 ml of 0·35% KOH in a 100 ml flask capped with aluminium foil. Autoclave for 10 min at 121 °C.

↓

Make up the extract to 100 ml with distilled water in a volumetric flask. Read the absorbance (E) at 595 nm in a spectrophotometer against a water blank (*see* note 3).

↓

Perform the same procedure with control, unexposed films retained from Day 1 (*see* note 4). Calculate the percentage of weight of cellulose lost (W) from the equation:

$$W = \frac{\text{E control} - \text{E exposed}}{\text{E control}} \times 100.$$

↓

Prepare a table of results for the class and calculate the mean rate of decomposition of the films under the conditions used.

- -

Notes and Points to Watch

1. Litter bags are bags made of nylon mesh which retain the films but allow contact between soil and contents. Maximum mesh-size for microbial experiments should be about 50 μm. Larger mesh permits entry of soil micro-arthropods with resultant destruction of both film and microbes by grazing activity.
2. A preliminary experiment should be made to determine the rate of decomposition under the conditions to be used by the class. Then choose sampling times that give between 40 and 70% weight loss.
3. Read the colour immediately the extracts have cooled to room temperature, as fading sometimes occurs.
4. Ensure that control films come from the same dye-batch as those placed in the soil. The amount of dye taken up by the cellulose during dyeing (see below) varies from batch to batch.

MATERIALS

1. Cellulose film. A good material is industrially pre-cut "cellophane" (Grade 325 P, non-moisture proof) in strips of 2 × 5 cm (D. J. Parry and Co. Ltd., 7 Avon Trading Estate, Avonmore Road, London, W14 8UE). Other types of cellulose may be used, but the advantage of cellophane is that it is also suitable for microscopic work.
2. Dyeing procedure. For 100 strips of cellophane 2 × 5 cm. (NB For other materials or sizes different quantities may be needed.)

a. Before dyeing, boil the cellophane in two changes of distilled water, 15 min each time.
b. Put the cellophane in 500 ml distilled water, heat and stir until the temperature reaches 80 °C; hold at this temperature.
c. Add 1·5 g of Remazol Brilliant Blue R. (Raymond A. Lamb, 6 Sunbeam Road, London, NW10 6JL).
d. At 2 min intervals add five lots of 20 ml 30% aqueous Na_2SO_4 which has been pre-heated to 80 °C.

e. Add 15 ml solution containing 2·5 g Na$_3$PO$_4$, also pre-heated to 80 °C.

f. Continue stirring at 80° C for 20 min.

g. Pour off the supernate and wash the strips first in running tap-water and then by autoclaving in one litre of water.

h. Repeat the autoclaving until no more dye leaches out of the strips.

i. Separate the strips on a sheet of blotting paper and allow them to dry in air. Store until needed.

3. Litter bags: these can be stitched with nylon threads and made from nylon mesh 48 μm (HC grade, Pronk, Rusby and Davis, 90–96 Brewery Road, London, N7 9PD).

SPECIFIC REQUIREMENTS

Day 1
soil
250 ml beaker
sieve, 2 mm mesh
tray for receiving sieved soil
spoon for transferring soil
burette for adding water to soil (if required)
RBB-dyed film in litter bags (replicated as required)
control films (at least 5)
knife for making slit in soil

Day 2
fine scissors

fine forceps
fine paintbrush
0·35 % KOH solution
measuring cylinder
100 ml flasks plus caps (enough for exposed plus control films)
bench autoclave ready for use
100 ml volumetric flask
supply of distilled water
spectrophotometer and cuvettes
pipettes

FURTHER INFORMATION

Two to three weeks' exposure at 25° C usually produces the required extent of decomposition. A more instructive approach, but one that takes up more time than is likely to be available, is to do a time series and plot a decomposition curve. Microcosm experiments of the type described above are usually most suitable for class experiments because of the good control of variables affecting the rate of decomposition. However, exposure of films in soils or other habitats under field conditions may be of ecological interest. Films exposed in marine habitats show about 40 % weight loss in two weeks during the summer months in UK, being somewhat slower in winter. The inclusion of microscopic examination of films adds a great deal to the experiment, enabling a picture of microbial succession to be built up (Tribe, 1957). Fungi (e.g. from soil) show two distinct patterns of growth—that over the surface where hyphae are usually of "normal" appearance and that of hyphae penetrating into the film where much broader jointed hyphae with shortened cells are formed (*see* Tribe, 1957; Went and Jong, 1966 for illustrations). Bacteria (e.g. from marine habitats) erode the film from the surface forming elongated pockets of clearing which eventually coalesce into a reticulate network.

The dye technique can also be used for demonstrating cellulolytic activity in pure culture, as an assay for cellulases and for counting cellulolytic organisms (Moore *et al.*, 1979).

REFERENCES

R. L. Moore, B. B. Basset and M. J. Swift (1979). Developments in the Remazol Brilliant Blue dye-assay for studying the ecology of cellulose decomposition. *Soil Biology and Biochemistry*, **11**, 311–12.

R. P. Poincelot and P. R. Day (1972). Simple dye release assay for determining cellulolytic activity of fungi, *Applied Microbiology*, **23**, 875–9.

H. T. Tribe (1957). Ecology of micro-organism in soils as observed during their development upon buried cellulose film. *Symposia of the Society of General Microbiology*, **7**, 287–98.

J. C. Went and F. D. Jong (1966). Decomposition of cellulose in soil. *Antonie van Leeuwenhoek*, **32**, 39–56.

89

Isolation of Cellulose and Lignin Decomposing Fungi from Wood

J. N. HEDGER

Department of Botany and Microbiology, University College of Wales, Aberystwyth, Dyfed, SY23 2AX, Wales

Level: All undergraduate years
Subject areas: Ecology and environmental microbiology
Special features: Isolation of higher fungi ("mushrooms and toadstools") into laboratory culture; combination of field and laboratory work

INTRODUCTION

In a temperate forest, 200–300 g of plant material may be deposited on to each square metre of soil each year to be subsequently degraded by the micro-organisms and fauna present in the litter ecosystem. A large proportion of this material consists of lignin, hemicelluloses and cellulose (perhaps 80%) and a substantial part of this lignocellulose is degraded by the action of the Basidiomycete and Ascomycete fungi. Demonstration of the action of these fungi in colonizing lignocellulose material in the litter layers of the soil is difficult, but the principles involved can be illustrated by considering the mode of attack of these fungi on wood, which in any case represents a high proportion of the biomass in most forests. Although it is possible to isolate lignocellulose-degrading fungi from the soil litter layer, the large homogeneous mass of lignocellulose in tree trunks and branches represents a simpler system to study. There is lower species diversity, there are very obvious physical and biochemical changes as a result of decomposition, and there are fewer problems in isolating the main decomposer organisms.

During the initial stages of decomposition of wood, the substrate may be occupied by one or two species of fungi, though subsequently secondary invaders, both bacteria and fungi, increase in numbers as it is broken down. This degradation is accompanied by a change in colour, depending partly on the mode of attack on the lignocellulose complex by the wood-rotting fungi. Classically, two types of attack are distinguished: "brown rot", in which the colour of the wood darkens, eventually becoming a brown powder, and "white rot", in which action of the fungi bleaches the wood. "Brown rot" attack involves selective degradation of the cellulose and hemicellulose components of the wood, whilst in "white rot", both components of the lignocellulose of the cell walls are degraded simultaneously.

These "brown rot" and "white rot" fungi have been long considered by microbiologists to be rather difficult to isolate and manipulate in the laboratory, especially as most are Basidiomycetes. Consequently they have received little attention as teaching organisms. The object of this exercise is to obtain pure cultures of these fungi by a very simple procedure, and to combine this with field observations on their mode of attack on the wood. Cultures so obtained can be subsequently used to study cellulose and lignin degradation (*see* Experiments 90 and 91).

EXPERIMENTAL

The methods used to isolate fungi from samples of decaying wood are given below. The procedure is simple and the only potential problem is the slow rate of growth of some of the fungi. This means that the exercise must be extended over a number of weeks, but the time required for each stage is quite short. Reasonable care must be taken to exclude fast growing contaminant moulds.

Day 1 *(2–3h)*

Collection of wood samples (*see* note 1). Small samples can be collected in the field by the student, using a saw or chisel. These should be labelled with a jewellers's tag and field notes written up on the type of attack (white or brown rot), the presence of zone lines (Rayner and Todd, 1979) and the identity of any carpophores (*see* note 2). Ideally, wood specimens with carpophores attached should be collected, so that the identity of the rotting organism can be confirmed.

Samples should be stored under cover to allow some drying out to take place (*see* note 3) before the laboratory practical.

Day 2 *(20 min, 4 days later)*

Isolation of fungi from samples of decayed wood. The procedure given is for the processing of 1 sample by 1 student (*see* note 4).

Preparation of media
Add 0·5 ml of 1 % sterile streptomycin solution to each of 2 sterile Petri dishes.

↓	↓
To 1 Petri dish add approximately 15 ml melted, cooled 1·5 % malt agar. Allow to gel.	To 1 Petri dish add approximately 15 ml melted, cooled *O*-pp medium. Allow to gel.

Preparation of samples
Cut into wood or carpophore (*see* note 5) with flame/alcohol sterilized scalpel or carpenter's chisel (*see* note 6) to expose a flat, clean surface.

↓

Remove small cubes (approx. 2–3 mm^3) of material, by cutting with a sterile scalpel.

↓

Transfer 4 cubes to each of the two Petri dishes placing them equidistantly and approx. 1·5 cm from the margin.

↓

Incubate at 22 °C (*see* note 7).

Day 3 *(30 min, 7 days later)*

Make a preliminary examination of the Petri dishes. Compare growth on the two media. Make preliminary identification checks on colonies by mounting the Petri dishes directly on the microscope stage, viewing either through the lid (incubate upside down to prevent condensation) or directly. If necessary, drop a coverslip directly onto the hyphae at the colony margin for viewing under high power.

↓

Make isolations from colonies by removing portions from the margin with flame sterilized mycological spears or needles, and transferring to 1·5% malt plates or slopes (*see* note 8). Re-incubate plates and slopes at 22 °C.

Day 4 *(30 min, 7 days later)*

Repeat examinations and make any necessary further isolations from slower growing colonies as above.

↓

Discard Petri dishes.

↓

Check isolates from the previous week for contamination by mounting portions of colonies in lactophenol and examining under the microscope (*see* note 8).

Day 5 *(30 min, 7 days later)*

Check the purity of the second batch of isolates (*see* note 8).

Notes and Points to Watch

1. Wood samples can be collected at any time of the year, e.g. the carpophores of the "corky" wood-rotting Basidiomycetes are still viable in January and February, even when completely frozen. Fleshy species of fungi are best avoided since attempted cultures are more liable to bacterial contamination. Useful species are suggested in Further Information.

The exercise is most effective if students can collect their own material and make their own field notes. Samples are best transported in boxes or baskets rather than polythene bags. The wood should show evidence of fungal colonization, but should nevertheless be reasonably firm and dry; excessively decomposed or wet samples will be heavily contaminated and will not readily yield cultures of the primary decay fungus.

2. Wood-rotting fungi can be identified from the carpophores by use of one of the field guides to higher fungi (e.g. Lange and Hora, 1963; Findlay, 1967; Pegler, 1973).

3. Samples should be left exposed in a well-ventilated area to allow any surface water film to evaporate. This area should be well away from any culture facilities (danger of introduction of fungus mites). In the laboratory class the use of polythene bags and the swabbing of bench tops with a suitable disinfectant helps to reduce the risk of contaminants being introduced.

4. Students can process 2–3 samples with only a little increase in time required (30 min as against 20 min).

5. The surface of the material should be cut away and inocula removed from the interior. In the case of carpophores with a thin structure, such as *Stereum*, this may be difficult. In this case, isolations can also be made from decayed wood associated with the carpophore. In practice, we always make isolations from both carpophore and wood, and use a comparison of the morphology of the cultures produced from the two sources to emphasize the presence of the decay fungus in the wood.

6. It is important to emphasize safety procedure here. Cuts should always be made away from the wrist and **never** towards. Strong surgeon's scalpels should be employed (**not** scalpels with replaceable blades, these are not strong enough to cut into a specimen with safety, although they can be used to cut out and transfer inocula).

7. Incubation at 22–25 °C will give rapid growth of most wood-rotting fungi.

8. In practice, there should be little problem with contamination of cultures isolated from the *O*-pp medium, although occasionally bacteria (*Pseudomonas* spp.) appear on the plates. Those cultures isolated from the unamended plates are more prone to contamination by mould fungi and on this medium it is advisable to isolate from the margin of the colony of the decay fungus as soon as it has grown 2–3 mm from the inoculum. Delay may lead to overgrowth by moulds, especially *Trichoderma* and *Mortierella*. Presence of these fungi in the cultures is readily established under the microscope.

MATERIALS

1. 1·5 % malt medium (15·0 g malt, agar 20 g, tap-water to 1 litre. Adjust pH to 6·0).

2. Sterile solution of *O*-phenyl phenol (2-hydroxy-biphenyl). This should be adjusted in strength so that when added to 1·5 % malt medium it gives a final concentration in the medium of 0·003 %. This compound is not readily soluble in water. To obtain a solution, first add the weighed out quantity to a small volume of 1·0 M NaOH with **gentle** heating. When dissolved, neutralize with 0·1 M HCl drop by drop until the precipitate is **beginning** not to redissolve on shaking. Membrane filter to sterilize and add to the correct volume of sterilized melted 1·5 % malt agar for immediate use, (*O*-pp medium). In practice, 250 or 500 ml volumes are easiest to prepare in this way and can be supplied, as such, for use by the whole class.

3. Saws, carpenter's chisels or stout scalpels.

SPECIFIC REQUIREMENTS

Day 1
saw and chisel
mushroom baskets, boxes or other suitable containers for collecting samples (avoid polythene bags if possible)
jeweller's tags for labelling specimens in the field
field note books and identification texts

Day 2
1 % streptomycin solution (10 ml in McCartney bottle)
sterile 1·0 ml pipettes
sterile plastic Petri dishes
water-bath at 45–50 °C to store media
cooled, melted 1·5 % malt agar
cooled, melted 1·5 % malt agar containing 0·003 % O-phenyl phenol (O-pp medium)

alcohol
chisel
stout scalpel
mycological spear or needle (**not** loop)
22 °C incubation facility

Day 3
mycological spear or needle
lactophenol mountant
coverslips and microscope slides
1·5 % malt agar slopes or freshly poured plates
22 °C incubation facility

Day 4
as Day 3

FURTHER INFORMATION

The development of colonies on the two media, 1·5 % malt and O-phenyl phenol medium, is strikingly different and the species composition can be compared as a class exercise. Russell (1956) developed the use of O-phenyl phenol in a medium designed for the preferential isolation of Basidiomycete decay fungi from pulp-mill samples. He found that it effectively suppressed growth of *Trichoderma*, an especially aggressive competitor on agar media, although a few strains of *Trichoderma* appeared to be tolerant. *Trichoderma* and other moulds will often dominate the unamended 1·5 % malt plates, whilst the O-phenyl phenol medium may produce pure cultures of the primary decay fungus. A few of the common moulds isolated in this way are listed in Table 1, together with notes on their ecology.

The basis of tolerance to O-phenyl phenol is not clear; the compound is fungitoxic and has been used as a fungicide (e.g. Dowicide). It is tempting to assume that, since it is a phenolic and the degradation of the lignin component of the wood releases phenolic compounds, the white rot fungi, which actively degrade lignins, would be more tolerant to the compound than the brown rots. In fact, although the brown rot Basidiomycete *Serpula lacrymans* (the "dry rot" fungus) will not grow on the O-phenyl phenol medium, *Piptoporus betulinus*, another brown rot Basidiomycete, grows as vigorously as one of the most active white rot Basidiomycetes, *Coriolus versicolor*. However, an element of taxonomic distribution of sensitivity may also be involved, as with tolerance to Benomyl fungicide since *Xylaria*, an active white rot, but an Ascomycete, will not grow on O-phenyl phenol medium. This assumption underlies Russell's original use of the compound as a selective agent for Basidiomycetes.

Table 1 Moulds commonly isolated from decaying wood. These include examples of pioneer colonies such as "blue stain" fungi as well as secondary fungi associated in various ways with the primary lignocellulose degrading fungi listed in Table 2. Most are members of the Zygomycotina and Deuteromycotina. They appear almost exclusively on 1·5 % malt and rarely on the O-phenyl phenol amended medium.

Fungus	Colony description	Remarks
Deuteromycotina		
Cladosporium herborum	Slow-growing dull green colonies bearing branching chains of blastic conidia	Common pioneer colonizer of freshly exposed wood surfaces
Trichoderma species (especially *T. viride*, *T. polysporum* and *T. haematum*)	Fast-growing light green or white colonies, sometimes with a distinct smell of desiccated coconut. Conidia phialidic	*Trichoderma* shows marked antagonism to other fungi and rapidly overgrows plates. It is often a pioneer colonizer of exposed wood surfaces
Diplodia spp. (and other "Coelomycetes")	Fast-growing grey colony with scattered black pycnidia producing masses of conidia embedded in mucilage	*Diplodia* and other "Coelomycetes" are colonizers of the outer sap-wood and bark
Zygomycotina		
Mortierella isabellina	Grey colonies with little aerial mycelium, sporangiophores arising directly from substrate	*Mortierella* species are common in wood in an advanced state of decay. They probably utilize secondary products derived from the decomposition of the lignocellulose
Mortierella vinacea	Dark red colonies with little aerial mycelium	
Mortierella ramanniana	Reddish colony, with "half columella" to sporangium	

Confirmation of the identity of the cultures of wood-rotting fungi is difficult and must usually be made by correlation with the carpophore associated with the decay. In the case of Basidiomycetes, the presence of clamp connections in the cultures is a useful confirmation, but not all wood-rotting Basidiomycetes will form clamps in agar culture. Nobles (1948, 1965) gives a key to identification of Basidiomycetes in culture. One of the most easily confirmed is *Heterobasidion* (*Fomes*) *annosum*, which readily forms an *Oedocephalum* conidial state.

A selection of the most easily available wood rotting fungi are listed in Table 2, together with a brief indication of colony morphology, host, rot type and references to useful illustrations of their carpophores.

Cultures of the faster growing taxa obtained by this procedure, e.g. *Piptoporus* and *Coriolus*, can subsequently be use to study production of cellulose enzymes, and to assess degradation of lignin preparations (*see* Experiments 90 and 91).

Table 2 Examples of wood-rotting fungi, selected on the following bases: 1, common; 2, mostly woody or corky; 3, available for most of the year; 4, culture easily; 5, show clear brown or white rot attack on wood.
Useful field descriptions and illustrations of carpophores are referred to as follows: F = Findlay (1967); L and H = Lange and Hora (1963); p. = page no; pl. = plate no.

Fungus	Description and Illustration	Rot type	Usual host	Colony description
White rots ***Basidiomycotina***				
Armillaria mellea	F p. 99 pl. 27	White, rhizomorphs beneath bark	Hardwoods (standing dead and stumps)	Slow-growing brown colony; forms rhizomorphs after 2–3 weeks
Bjerkandera adusta	L and H p. 71	White, black zone lines in wood	Hardwoods, especially old stumps	Pale yellow or white colony
Coriolus hirsuta	F pl. 72 L and H p. 68	White, black zone lines in wood	Hardwoods, branches and stumps	Very easily cultured; rapidly growing white colony
Coriolus versicolor	F pl. 72 L and H p. 68	White, black zone lines in wood	Hardwoods, branches and stumps	Very easily cultured; rapidly growing white colony
Ganoderma applanatum	L and H p. 64 F pls 76, 49	White	Beech, living and dead trees	Slow-growing white colony, releasing black pigment into medium
Heterobasidion annosum	L and H p. 66	White	Conifers—dead stumps	Pure white fast-growing colony with abundant *Oedocephalum* state
Lenzites betulina	L and H p. 70	White	Birch and beech	White slow-growing colony
Panellus stypticus	L and H p. 104	White	Willow, dead trunks	Thin mycelium, white colony
Pseudotrametes gibbosa	L and H p. 768 F pl. 74 p. 144	White	Beech	White colony, medium rate of growth.

	Reference	Wood	Substrate	Colony
Stereum hirsutum	L and H p. 50	White	Oak, especially fallen branches	Yellow-brown fluffy colony, medium rate of growth
Stereum rugosum	L and H p. 50	White	Hazel, especially standing dead branches	Pale yellow fluffy colony, medium rate of growth
Ascomycotina				
Xylaria hypoxylon	L and H p. 46	White, black zone lines	Ash, sycamore (fallen branches)	White or grey colony. Little aerial mycelium
Xylaria polymorpha	L and H p. 46	White, black zone lines	Hardwoods, esp. old stumps	Grey colony with little aerial mycelium. Stromata produced after 2 weeks
Brown rots				
Basidiomycotina				
Coniophora cerebella (*puteana*)	L and H p. 52	Brown (wood tends to crack into squares)	Fallen logs (on underside) and buildings	Yellow fast-growing colony
Daedalea quercina	L and H p. 66	Brown (wood becomes pale brown and powdery)	Oak (old stumps)	Yellowish slow-growing colony
Piptoporus betulinus	L and H p. 66 F pl. 71 p. 146	Brown (wood becomes dark brown)	Birch (standing dead trees)	Fluffy white fast-growing colony
Serpula lacrymans	L and H p. 52	Brown (wood cracks into squares)	Structural timbers	White colony with yellow patches

REFERENCES

W. P. K. Findlay (1967). "Wayside and Woodland Fungi", 202 pp. Frederick Warne, London.

M. Lange and F. B. Hora (1963). "Collins Guide to Mushrooms and Toadstools", 256 pp. Collins, London.

M. K. Nobles (1948). Identification of cultures of wood rotting fungi. *Canadian Journal of Research*, **26**, 281–431.

M. K. Nobles (1965). Identification of cultures of wood inhabiting hymenomycetes. *Canadian Journal of Botany*, **43**, 1097–139.

D. N. Pegler (1973). The Polypores. *Bulletin of the British Mycological Society*, 7.

R. W. Rayner and M. K. Todd (1979). Population and community structure and dynamics of fungi in decaying wood. *Advances in Botanical Research*, **7**, 333–420.

P. Russell (1956). A selective medium for the isolation of basidiomycetes. *Nature*, **177**, 1038–9.

90

Rapid Screening of Cellulolytic Activity of Micro-organisms by the use of Acid-swollen Cellulose

J. N. HEDGER

Department of Botany and Microbiology, University College of Wales, Aberystwyth, Dyfed, SY23 2AX, Wales

> *Level*: All undergraduate years
> *Subject areas*: Ecology and environmental microbiology
> *Special features*: Visual impact

INTRODUCTION

A number of the procedures used to screen micro-organisms for cellulolytic ability are cumbersome, especially if a semiquantitative assessment of relative activity is required. The use of acid-swollen cellulose incorporated in an agar base allows such comparative studies to be made, and can be employed as a class exercise to investigate cellulolytic activity of micro-organisms isolated from soil, or indeed any other habitat where cellulose is a potential carbon source.

EXPERIMENTAL

The assay follows the procedure of Tansey (1971) and requires the growth of isolates on acid-swollen cellulose incorporated in agar deeps. Relative activity is assessed by measuring the depth of clearing of the cellulose beneath a colony growing on the surface of the deep. Depth of clearing can be measured at regular intervals.

Day 1 (*15 min*)

Inoculate the surface of the agar deep centrally with the culture to be tested and incubate at the appropriate temperature.

Days 2, 3 etc. *(5 min, at 3–4 day intervals)*

Measure the depth of clearing of the cellulose agar beneath the colony by holding the tube against light and measuring the cleared zone with a ruler. The margin of the clearing zone will be quite definite (*see* note 1).

Notes and Points to Watch

1. Active species will clear zones down to 20 mm depth. A point to watch at higher incubation temperatures is that the surface of the agar will also shrink down the tube after some days of incubation and this must be allowed for in measurement of the cleared zone.

MATERIALS

1. Plugged test tubes (2 cm diam.) each containing 6 cm depth of acid-swollen cellulose agar.
2. Acid-swollen cellulose agar (Tansey, 1971). $NH_4H_2PO_4$, 2·0 g; KH_2PO_4, 0·6 g; K_2HPO_4, 0·4 g; $MgSO_47H_2O$, 0·8 g; thiamine, 100 μg; Difco Yeast Extract, 0·5 g; adenine, 4·0 mg; adenosine, 8·0 mg; acid-swollen cellulose, 5·0 g; agar, 17·0 g; distilled water, 1 litre. Sterilize by autoclaving at 121°C for 15 min.

Method of preparation of acid-swollen cellulose
(Unless otherwise indicated, carry out in a cold room at 1°C or in an ice bath.)

Place 30·0 g of air-dried Whatman cellulose powder CF 11 in a 5 litre beaker. Add to the beaker 400 ml of 88% ortho-phosphoric acid (analar) stirring to prevent formation of lumps. After 2 h of constant stirring, add 2 litres of distilled water to the beaker with rapid stirring. Suction-filter the suspension through five layers of cheesecloth over two layers of Whatman no. 1 filter paper. Take the moist cellulose and disperse in 2 litres of distilled water, filter as before and place in 1 litre of 2% Na_2CO_3. Homogenize for 5 min in a Waring Blender and store for 12 h. Wash with 5 litres of distilled water and filter. Pellet by centrifugation at 10 000 g for 5 min and homogenize for 5 min. The resulting suspension should have a pH value of 6·5 $\pm$ 0·1. Oven dry at 80°C for long-term storage.

Quantities should be adjusted according to the amount required. Thirty grams of cellulose yields about 20 g of acid-swollen cellulose. The quantity per litre of medium is not critical and $\sim$ 5·0 g can be calculated on a **wet weight** basis from this conversion figure.

Preparation of the agar deeps
After dispensing the cellulose agar into previously sterilized plugged tubes it should be cooled as rapidly as possible in an ice bath, with constant shaking to prevent settling out of the cellulose particles.

SPECIFIC REQUIREMENTS

Day 1
cultures
loop or spear for inoculation
acid-swollen cellulose deeps
incubator

Days 2, 3, 4 etc.
millimetre ruler

FURTHER INFORMATION

It is also possible to grow isolates on agar plates of acid-swollen cellulose. In this case, clearing zones may be visible at the colony edge. Such a procedure can be used in dilution plating of soil samples to pick out cellulolytic colonies.

Ball-milled cellulose agar has been employed in a similar way, but we have found results with this medium disappointing. The cleared zones are much less well defined than on acid-swollen cellulose and considerable ball-milling may be required to produce short enough fibre lengths to obtain any result. We therefore recommend acid-swollen cellulose as the better medium for class exercises, where the easy recognition of the cleared zones makes an unequivocal demonstration of cellulase production.

Table 1 lists typical results from fungi grown on acid-swollen cellulose agar deeps for 7 and 14 days. It is interesting to note that the many isolates of the brown rot Basidiomycete *Piptoporus betulinus* which we have tested fail to clear acid-swollen cellulose, though they do form colonies on the surface. This **may** be a reflection of the different, non-enzymic mode of attack on cellulose by brown rot Basidiomycetes, which some workers believe to occur.

The rate of clearing slows down and stops after 2–3 weeks' growth of most fungi, presumably because of inactivation of the cellulases by adsorption.

Table 1   Depth of clearing of acid-swollen cellulose agar deeps by selected fungi.

Fungus	Origin	Incubation temp (°C)	Depth clearing (mm) 7 days	14 days
Coriolus versicolor	Wood (white rot)	22	8·3	15·0
Stereum hirsutum	Wood (white rot)	22	6·6	13·0
Xylaria hypoxylon	Wood (white rot)	22	1·0	3·0
Piptoporus betulinus	Wood (brown rot)	22	0	0
Trichoderma polysporum	Soil	22	8·0	12·0
Chaetomium globosum	Straw	22	7·0	14·0
Humicola grisea var. *thermoidea*	Compost	45	12·0	21·0
Mortierella isabellina	Wood	22	0	0
Mucor mucedo	Soil	22	0	0

REFERENCE

M. R. Tansey, (1971). Agar diffusion assay of cellulolytic ability of thermophilic fungi. *Archiv für Mikrobiologie*, **77**, 1–11.

Degradation of Lignin by Fungi

J. N. HEDGER

Department of Botany and Microbiology, University College of Wales,
Aberystwyth, Dyfed, SY23 2AX, Wales

> *Level*: All undergraduate years
> *Subject areas*: Ecology and environmental
> microbiology
> *Special features*: Visual impact

INTRODUCTION

Lignins are amongst the most complex known biopolymers. A schematic formula for conifer lignin is shown in Figure 1. They constitute up to 20–30 % of the plant cell wall material which enters the soil and litter ecosystem each year, and which is subsequently subject to degradative attack by micro-organisms.

Although the mechanism of depolymerization by micro-organisms of other polymers in the plant cell wall, such as cellulose and hemicellulose, is relatively well understood, the degradation of lignins remains much less clear. Many bacteria and fungi have been shown to be capable of degrading aromatic compounds derived from lignins, but the only micro-organisms that appear to be able to degrade intact lignin are the "white rot" fungi. This group of fungi obviously play an important role in litter and soil in depolymerizing lignins, and possibly making units from the polymers available for utilization by bacteria and other fungi. Recent research indicates that lignin degradation is energy expensive, but may be a requirement for exposure of the cellulose and hemicellulose components of the secondary cell wall to subsequent attack by hydrolases released by white rot fungi.

Assessment of lignin degradation by fungi is difficult, since it can range from simple demethylation of side chains to complete depolymerization of the molecule. In addition, pure preparations of lignins are necessarily highly modified during the extraction procedure, and interpretations of microbial attack in terms of intact lignins may not be valid.

However, two approaches to the demonstration lignin degradation are possible as class exercises, and the results of these can be compared usefully with what is known of the field ecology of the fungi employed in the tests (i.e. whether they are white rot fungi). Firstly, it has long been assumed that there is a correlation between the production of oxidase (phenolase) enzymes by fungi and the ability to produce a white rot in wood (Lindberg, 1948). Secondly, it is possible to grow micro-organisms on lignin–agar preparations, and to test for dephenolization of the lignins in the vicinity of the colony (Sundman and Nase, 1971). This latter approach gives direct evidence of ability to degrade lignins, at least partially. Both procedures are inexpensive of time and materials. However, pure cultures of wood-rotting fungi are a prerequisite, either from a previous class exercise (*see* Experiment 89) or from a culture collection.

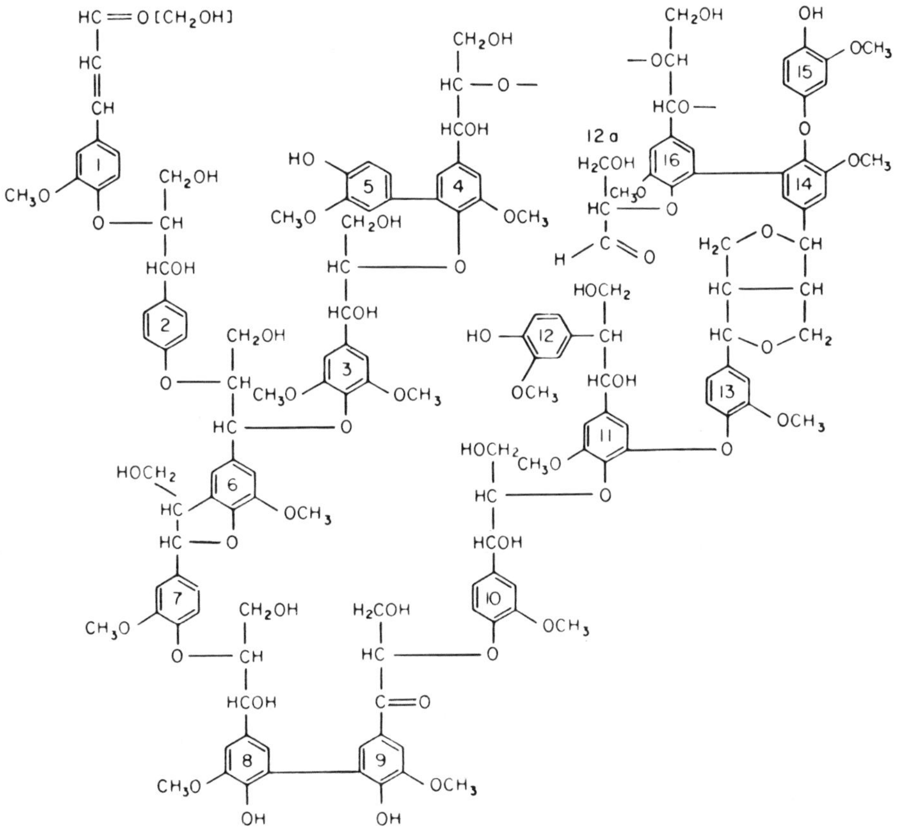

Figure 1 Formula for conifer lignin (Adler, 1966).

(From M. J. Swift, O. W. Heal and J. M. Anderson, eds (1979). "Decomposition on Terrestrial Ecosystems". Blackwell Scientific Publications, Oxford and London.)

EXPERIMENTAL

The exercise is divided into two parallel sections:
  1. demonstration of production of extracellular oxidase enzymes;
  2. demonstration of dephenolization of lignin preparations by fungi.

Ideally, students should screen several isolates, some of which should be selected to give negative results for comparison. A list of suitable test fungi is provided in Further Information.

Day 1 *(30 min)*

5·0 cm colony of fungus on 1·5% malt agar plate (*see* note 1).

↓

Cut out 6 mm agar discs from the colony margin with a cork borer (*see* note 2).

↓   ↓

Production of oxidases	*Dephenolization of lignin*
Pour one plate of approx. 15·0 ml gallic acid agar (*see* note 3) and allow to gel.	Pour two plates of approx. 15·0 ml lignin agar (*see* note 4) and allow to gel.

↓   ↓

Inoculate the centre of the plate with an agar disc from the malt–agar plate.	Inoculate the centre of 1 plate with an agar disc from the malt–agar plate. Keep the other plate as control.

↓   ↓

Incubate at 22 °C.   Incubate at 22 °C.

Day 2 *(15 min, 7 days later)*

Production of oxidases	*Dephenolization of lignin*
Examine the underside of the plates for presence of dark brown quinones in the medium around the colony (*see* note 5).	If a 1–2 cm diam. colony has formed (*see* note 6) scrape off aerial mycelium (*see* note 7), flood the test plate and control plate with ferricyanide reagent and store in the dark for 10 min (*see* note 8).

↓

Decant excess reagent. Place the plates over a sheet of white paper. The control should be dark blue–green. Dephenolization of lignin is shown by yellow–green colour under and around the colony (*see* note 9). Compare with the control.

Notes and Points to Watch

1. 1·5% malt–agar plates can be inoculated in a previous class by the student (allow about 7 days for growth at 22 °C).
2. Use a standard 6 mm cork borer. Dip in alcohol, flame to sterilize and press into agar. The discs

remain in place in the agar. Remove discs from the colony with a flame-sterilized mycological spear and place the disc mycelial side downwards on the agar plate to which it is transferred.
3. Gallic acid agar **must** be freshly prepared for the practical (*see* Materials).
4. A number of different lignins can be used (*see* Materials).
5. The activity of the phenolases can be roughly measured from the depth of colour and extent of the zone.
6. It may be necessary to allow a further growth period for some isolates.
7. A razor-blade held vertically is most effective in removing surface mycelium.
8. A bench cupboard or black sheet will suffice.
9. In some cultures it is impossible to remove all fungal hyphae, and colonies will stain dark blue with this reagent, but a lighter green zone at the edge of the colony will be seen if the culture is active (compare with control plate).

MATERIALS

1. Fungal cultures: *see* Further Information for useful isolates.
2. 0·5% gallic acid agar. This must be **freshly prepared** as follows (for 1 litre medium): make up 750 ml 3% malt–agar; (malt extract 22·5 g, agar 10 g, distilled water 750 ml) autoclave to sterilize; add 5·0 g gallic acid to 250 ml **hot** sterile distilled water (ex autoclave); add the gallic acid solution to the molten agar; finally, add a few drops of 0·1 M NaOH (sterilized by autoclaving) to adjust pH to 5–6 (test by dropping agar onto indicator paper).
3. Lignin agar. Basal medium: sucrose 30 g, $NaNO_3$ 2·0 g, KCl 0·5 g, KH_2PO_4 1·0 g,

$MgSO_4$ $7H_2O$ 0·5 g, $FeSO_4$ 0·1 g, yeast extract 0·5 g, agar 15·0 g, distilled water 1 litre. *Lignins*: add 0·5 g litre^{-1} basal medium. Mix to a paste with a little distilled water before adding to melted basal medium prior to autoclaving.

The following lignins are obtainable at low cost: *Indulin AT* from Westvaco Polychemicals, Box 5207, North Charleston, S. Carolina 29406, USA. *Peritan Na* from Norcem Ltd., Oslo, Norway. Both of these materials are industrial lignins. Peritan Na is the more modified of the two and is in fact sodium lignosulphonate.
4. 1% $K_3Fe (CN)_6$ solution.
5. 1% $FeCl_3$ solution.

SPECIFIC REQUIREMENTS

Day 1
gallic acid agar (melted, in bottles, for class use)
lignin agar (melted, in bottles, for class use)
water-bath at 50 °C, to store melted media
fungal cultures on 1·5% malt plates
plastic Petri dishes
6 mm cork borer
mycological spear
22 °C incubation

Day 7
freshly mixed ferricyanide reagent
  (mix 1% $FeCl_3$ solution with 1% $K_3Fe (CN)_6$
  solution in 1:1 ratio just prior to the practical)
caution
bench cupboard or black plastic sheet

FURTHER INFORMATION

If a number of isolates of wood-rotting fungi are examined as a class exercise, the correlation between phenolase production on gallic acid agar (the "Bavendamm reaction") and mode of attack

Table 1 Examples of class results. Only species with reasonably high growth rates are included.

+ = production of phenolase or dephenolization of lignin preparation.
− = no phenolase or dephenolization.

Rot type Isolate	Phenolase production on galic acid	Indulin AT dephenolization	Peritan Na dephenolization
A. *White Rots*			
Armillaria			
mellea	+	+	−
Coriolus			
versicolor	+	+	+
Ganoderma			
applanatum	+	+	−
Heterobasidion			
annosum	+	+	+
Stereum			
hirsutum	+	+	+
Xylaria			
polymorpha	−	+	+
B. *Brown Rots*			
Coniophora			
cerebella	+	−	−
(puteana)			
Daedalea			
quercina	−	−	−
Piptoporus			
betulinus 1	−	−	−
Piptoporus			
betulinus 2	+	+	−
C. *"Soft Rots"*			
Chaetomium			
globosum	−	−	−
Humicola			
grisea	+	0	+
Trichoderma			
polysporum	−	+	−
D. *"Non-cellulolytic"* fungi			
Mortierella			
isabellina	−	−	−
Mucor			
mucedo	−	−	−

on wood is usually found to be fairly good, although there are always some exceptions, as illustrated in Table 1. However, whether phenolase enzymes are involved in depolymerization of lignins, or whether they may be concerned with detoxification of products of lignin degradation, remains a matter of controversy. Certainly many species of fungi not thought of as white rots produce phenolases, and a few white rots show no evidence of phenolase activity (Table 1).

The lignin plate test gives direct evidence of ability to degrade lignins and is based on the fact that the green colour developed by $FeCl_3/K_3Fe(CN)_6$ reagent is an indication of phenolic groups, so that lack of staining around fungal colonies means that the lignin has been dephenolized. It must be assumed that this dephenolization is due to degradation of the lignins, although it could merely be the result of oxidation of phenolic groups on the aromatic rings of the lignin polymer by phenolases. A full discussion of lignin degradation can be found in Kirk *et al.* (1980).

If a number of species of fungi are examined during the exercise, the correlation between the two tests can be assessed. Some typical results are set out in Table 1. It is usual to find that some of the so-called non-lignolytic fungi, such as *Trichoderma* and *Piptoporus*, often show the ability to degrade lignins emphasizing the danger of over-simplifying ecophysiological classifications.

An additional exercise can be carried out by pairing up species of fungi on lignin media by inoculating them about 2–3 cm apart (Sundman and Nase, 1972). Such paired cultures often show synergistic acceleration of lignin breakdown, compared to single cultures, even if one of the partners proves to be non-lignolytic when tested in single culture. This emphasizes the types of positive interaction which must occur in microbial communities in wood.

REFERENCES

E. Adler (1966). The chemical construction of lignin. *Svensk Botanisk Tidskrift*, **80**, 279–90.

T. K. Kirk, T. Higuchi and H.-M. Chang (eds). (1980). "Lignin Biodegradation", 2 vols, pp. 241, 256. C.R.C. Press.

G. Lindberg (1948). On the occurrence of polyphenoloxidases in soil inhabiting Basidiomycetes. *Physiologia plantarum*, **1**, 196–205.

V. Sundman and L. Nase (1971). A simple plate test for the direct visualisation of biological lignin degradation. *Paper and Timber*, **53**, 67–71.

V. Sundman and L. Nase (1972). The synergistic ability of some wood degrading fungi to transform lignin and lignosulphonates on various media. *Archiv für Mikrobiologie*, **86**, 339–48.

Cellulase and Pectinase Induction and Estimation

P. J. WHITNEY

*Department of Microbiology, University of Surrey,
Guildford, Surrey, GU2 5XH, England*

> *Level*: All undergraduate years
> *Subject area*: Plant pathology
> *Special features*: Direct and simple method of
> assaying enzyme activity. Demon-
> stration of enzyme induction

INTRODUCTION

Verticillium albo-atrum, in common with several other plant pathogenic fungi, will only produce cellulase in the presence of cellulose, cellulose breakdown products or substituted celluloses. With the exception of cellobiose which is a cellulose breakdown product, low molecular weight substrates such as sugars inhibit cellulase induction. Natural cellulose is degraded first by the action of a C_1 enzyme to produce short linear polyanhydro-glucose chains and then by a C_x enzyme which breaks these chains into short lengths and ultimately to cellobiose. Carboxymethyl cellulose forms a suitable substrate for assaying C_x enzyme activity. With pectin in the medium instead of cellulose *V. albo-atrum* can be induced to produce pectinase.

The breakdown of carboxymethyl cellulose and pectin can be followed conveniently by measuring the change in flow time of a solution through a capillary tube. As the chain length of the substrate is reduced, so the viscosity is reduced and the flow time decreases. The speed with which the viscosity is reduced can therefore be used as a measure of the C_x cellulase or pectinase activity.

EXPERIMENTAL

Day 1 *(15 min*, see *note 8)*

Prepare liquid media containing cellulose powder, carboxymethyl cellulose or pectin $(1-2\%)$ and mineral salts, some with and some without the addition of sugars. Sucrose or other sugar should be used as "control" carbon source for enzyme induction.

Inoculate these media with a small piece of *Verticillium* or other fungal culture (*see* note 1). Transfer as little as possible of the agar medium on which the fungus was growing.

↓

Incubate in shaken or unshaken culture for 10–14 days at 25 °C.

Day 10–14 *(2 h)*

The procedure below is for the assay of cellulase. Pectinase activity is estimated in the same way, but with 1·5 % pectin solution as substrate.

Decant off the culture medium from the mycelial mat. Any small pieces of mycelium or conidia which remain should be removed by centrifugation. If total sterility of the medium is required it should be sterilized by filtration. This is only necessary if unusually long enzyme incubation times are to be used (many hours or even days).

↓

Dilute the culture filtrate with 5 or 6 parts of 1 % carboxymethyl cellulose in 0·1 M phosphate buffer pH 6·5 (5 ml + 25 ml gives a convenient colume for the Ostwald viscometer BS/IP/U type D) (*see* note 2). At the time of mixing a stop-clock should be started to record the incubation time.

↓

As quickly as possible after mixing, pour the incubation mixture into the viscometer to line 1 using a drawn out boiling tube as a funnel (*see* Figure 1). The mixture is drawn up to the upper mark in

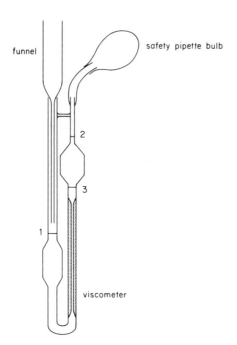

Figure 1 Ostwald viscometer and filling funnel.

the viscometer (line 2) and the flow time required for the meniscus to fall to the lower mark (line 3) should be recorded.

↓

This viscometric assay should be repeated at 10–15 min intervals during the incubation period if incubation is to be for about 1 h, less frequently if longer incubations are used.

↓

Calculate cellulase activity (C_x) as follows. The "relative decrease in flow time" after each incubation period t, (R.D.F.T.$_t$) is given by $(F.T._0 - F.T._t)/(F.T._0 - F.T._w)$, where
  F.T.$_0$ is the flow time immediately after mixing enzyme and substrate,
  F.T.$_t$ is the flow time after incubation period t,
  F. T.$_w$ is the flow time of distilled water in the same viscometer at the same temperature as the assay.

↓

Plot R.D.F.T.$_t$ against incubation time and interpolate from the resulting curve the incubation time for a 50% reduction in flow time $(t_{50}R.D.F.T.)$

$$C_x = \frac{1}{t_{50}\,R.D.F.T.} \times 100.$$

Notes and Points to Watch

1. Other suitable organisms include *Verticillium dahliae*, *Fusarium oxysporum* and *Myrothecium verrucaria* for the cellulase assay, and any soft rot organism, such as *Sclerotinia fructigena*, for the pectinase assay. Other sources of enzyme may be used, e.g. fruit infected with soft-rot fungi such as *Sclerotinia* or *Penicillium* or potato infected with the bacterium *Erwinia*, form a convenient source of pectinase. The growth of cellulose and pectin-degrading organisms in the soil can be demonstrated by inoculating a pectin or carboxymethyl cellulose solution with a few crumbs of soil and measuring the flow time before and after 24–48 h incubation.
2. An Ostwald viscometer is not essential as the following, less accurate, method can be used. The reaction mixture (culture filtrate plus substrate) is drawn into a Pasteur pipette onto which have been drawn two marks 3 or 4 cm apart. The mixture is allowed to flow freely from the pipette and the time is noted as the meniscus falls past the upper mark and again as it passes the lower mark. The difference is taken as the flow time.
3. Suction by mouth is not recommended and a rubber bulb, water pump or vacuum line should be used to draw the mixture up the viscometer or Pasteur pipette.
4. As the viscosity of solutions is very temperature-sensitive the estimations should be carried out in a temperature-controlled water-bath, and the substrate solution and culture filtrate equilibrated to this temperature before mixing if reproducible quantitative results are required.
5. When making up solutions of pectin, and more particularly carboxymethyl cellulose, there is a strong tendency for the powders to become surrounded by a viscous gel when the water is added. If the mixture is left overnight the powder will go into solution. If the solution is required in a

hurry a few minutes in a blender will break up the gelatinous lumps and allow complete solution of the powder.

6. The viscometers must be washed out thoroughly immediately after use. If a small amount of carboxymethyl cellulose or pectin is left in the viscometers it will dry to a very tiny volume and be difficult to detect. The next time the viscometer is used it will start to hydrate and progressively constrict the capillary reducing the flow rate until the particle is sufficiently hydrated to be flushed from the capillary. This is indicated by a sudden increase in flow rate. The viscometers should be dry before use to avoid dilution of the incubation mixture. If rapid drying of a viscometer is required, rinsing with acetone and drying with air from a vacuum or compressed air line is quick and effective.

7. It is possible to assay other substrates viscometrically but care should be taken in the choice of substrate. Many viscous materials are thixotropic to a greater or lesser extent; for example the flow time of a gelatine solution will depend on the degree of agitation it has received just prior to running, e.g. how vigorously it is drawn into the upper reservoir of the viscometer.

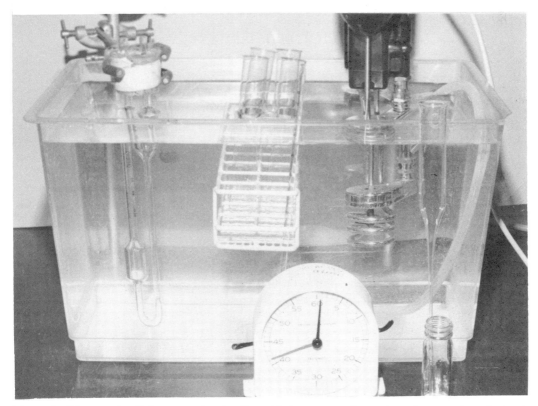

Figure 2 Viscometric cellulase assay showing Ostwald viscometer and test tubes of enzyme plus substrate in a temperature-controlled water-bath. Outside the water-bath is a viscometer filling tube and a stop clock.

MATERIALS

1. Culture of: *Verticillium albo-atrum* or (CMI Cat. No. 62, 131).
   or *Verticillium dahliae*
   (CMI Cat. No. 118, 379),
   or *Fusarium oxysporum*
   (CMI Cat. No. 141, 140),
   or *Myrothecium verrucaria*
   (CMI Cat. No. 45, 541),
   or *Sclerotinia fructigena*
   (CMI Cat. No. 162, 408).
2. Medium. Any mineral salt medium containing cellulose powder, carboxymethyl cellulose or pectin as the carbon source can be used. One suitable medium is Czapeck-Dox which contains in 1 litre $NaNO_3$, 2·0 g; KH_2PO_4, 1·0 g; $MgSO_4 7H_2O$, 0·5 g; KCl, 0·5 g; $FeSO_4 7H_2O$, 0·01 g; carbon source (Cellulose, C.M.C. Pectin or Sucrose), 10 g.
3. 1 % carboxymethyl cellulose in 0·1 M phosphate buffer pH 6·5 or 1·5 % pectin in 0·1 M phosphate buffer pH 6·5.
4. Ostwald viscometers and filling tubes made from drawn out boiling tubes (*see* Figure 1) or Pasteur pipettes.
5. Temperature controlled water-bath with transparent front, e.g. fish tank and water-bath control unit (optional) (*see* Figure 2).

SPECIFIC REQUIREMENTS

low speed centrifuge and centrifuge tubes
stop-clocks (at least 1 per group, preferably 2)
5 ml and 25 ml non-sterile pipettes/measuring cylinders (1 per group)

boiling tubes (1 per assay) and racks
acetone (to dry viscometers and pipettes or measuring cylinders) (*see* note 6)
viscometer

REFERENCES

R. N. Goodman, Z. Kiraly and M. Zaitlin (1967). The Biochemistry and Physiology of Infectious Plant Disease", Van Nostrand, Princetown, New Jersey.

E. T. Reese (1963). "Advances in enzymic hydrolysis of cellulase and related materials". Pergamon Press, Oxford and New York.

P. J. Whitney, J. M. Chapman and J. B. Heale (1969). Carboxymethylcellulase production by *Verticillium alboatrum*. *Journal of General Microbiology*, **56**, 215–25.

93

Inoculum Potential, Barriers to Infection and Host Specificity of Fruit Pathogens

P. J. WHITNEY

*Department of Microbiology, University of Surrey,
Guildford, Surrey, GU2 5XH, England*

```
            Level: All undergraduate years
       Subject area: Plant pathology
  Special features: Material easily obtained.
                    Illustrates important concepts.
                    Encourages variation in
                    experimental design by students.
                    Can be preceded by
                    experiments on isolation of fungi
```

INTRODUCTION

The ability of a potential pathogen to colonize a host depends on the vigour of the pathogen and the defence mechanisms of the host. The vigour of the pathogen will depend not only on its genetic make-up but also on the nutrient reserves on which it can draw. At its simplest, this means a large amount of infected material is more infectious than a small amount, but it also means that the material is more infectious if there are considerable nutrient reserves either within the organism or that it can draw on in the medium. In a few cases, nutrients can be transported by the pathogen from an established infection to the region of attack on a new host. These various aspects are combined into the concept of "inoculum potential". Frequently this is simply a qualitative concept, but in some instances it is refined into a quantitative measure of infection potential, e.g. the minimum infectious dose can be estimated using specified standard conditions.

The ability of the host to withstand an attack by a pathogen depends on a large number of factors that include: the presence of antimicrobial material in the tissue, the ability to synthesize antimicrobial compounds when attacked, the ability to lay down barriers in advance of the infecting organism and the presence of preformed barriers to infection.

The aim of these experiments is to investigate the ability of micro-organisms to breach barriers to infection and to colonize plant tissue. Fruit is used as a convenient form of plant tissue which is protected from microbial attack by outer barriers—"skin" or "rind". Increasing amounts of inoculum should have an increasing ability to breach these barriers (greater inoculum potential). The attack by the micro-organism can be assisted to different degrees by mechanical wounds of different severity. The role of inoculum potential of the micro-organism and the hosts' barriers to infection can therefore be tested and compared.

Some pathogens are better adapted to colonize one host rather than another or, conversely, some host defence mechanisms are not as effective against some micro-organisms as against others. This can be investigated by inoculating more than one host species with a range of potential pathogens, using different amounts of inoculum and assisting the pathogen to different extents by making wounds of different degrees of severity.

EXPERIMENTAL

Day 1 (30–45 min)

Carry out the procedures below on both apples and oranges using three different species of fungi.

Cut blocks of inoculum of different sizes ranging from very large (2–4 cm^2) to very small (2–4 mm^2) from fungal cultures on agar, or from infected fruit, and place on the surface of the fruit to be inoculated (see notes 1–3). Cover the inoculum with a piece of polythene and tape into position (see note 4).

↓

Carry out similar inoculation using only fungal spores scraped from the surface of a freely sporing culture or fruit.

↓

In the same way inoculate fruits that have been wounded to different extents, e.g. (i) wax removed with acetone, (ii) surface lightly abraded with a grater, (iii) skin cut with a scalpel, (iv) block cut from fruit and inoculum placed within fruit (see note 4).

↓

Incubate the fruits at 20 °C for approximately 14 days or 25 °C for approximately 10 days (see note 5).

Days 3–14 (15 min)

Observe and make notes on the development of infection at 2–3 day intervals.

↓

At the end of the incubation period cut the fruit so that the depth of penetration may be observed.

Notes and Points to Watch

1. Surface sterilization of fruit is not usually needed.
2. Some fruit as purchased may have been treated on the surface with antimicrobials so that washing with water or alcohol may affect results.
3. More than one treatment can be made on each fruit. The number depends on the size of the fruits and the severity of wounding used.
4. Ensure that all treatments are clearly labelled. Waterproof felt pen on the polythene covering works well.
5. To avoid excessive contamination of the incubator cultures by vigorously sporing fungi, the fruits should be enclosed in sealed polythene bags.

MATERIALS

1. Apples (*see* notes 2 and 3).
2. Oranges (*see* notes 2 and 3).
3. *Penicillium expansum* (C.M.I. Cat No. 39, 761), *Penicillium digitatum* (C.M.I. Cat No. 143, 627) and *Penicillium italicum* (C.M.I. Cat No. 143, 321). These fungi can either be obtained from a culture collection or they can be isolated from infected fruit as described below. Other suitable fungi include *Sclerotinia fructigena*, *S. fructicola*, *Trichoderma viridi* and *Botrytis cinerea*.

P. expansum is common on apples and slowly produces a brown rot of the fruit. It is easily isolated from an infected apple simply by shaking an infected fruit that is sporing over the surface of a Petri dish containing almost any sugar-based medium. The colonies that result should be greenish when sporing, but this may vary from yellow–green through grey–green or blue–green.

P. digitatum and *P. italicum* can be similarly isolated from infected citrus fruits. *P. digitatum* is olive–green or yellow–green but never blue–green and produces a rather dry rot of citrus fruits. *P. italicum* is rather more grey–green and produces a rapid soft rot of the citrus fruit, reducing it to a slimy pulp.

Sclerotinia (Monilinia) fructigena is a common soft rot of apples which, if left long enough, will produce characteristic pustules of conidia on the surface of the fruit which will finally dry out and become mummified. *S. fructicola* is similar and found on peaches and other stone fruits. Both are easily cultured by taking small pieces of infected tissue or better pustules of conidia and inoculating onto P.D.A. or malt agar.

Trichoderma viridi is slightly more difficult to isolate but grows well in culture. It is common on fallen timber and in damp soil. It grows well as a thin surface growth on most media producing patches of verdigris green spores.

Botrytis cinerea (Sclerotinia fuckeliana) is a grey mould common on many soft fruits, flower buds, and some tubers in storage. When growing on soft fruit and some media it appears light grey, becoming speckled with a darker grey as the fungus matures and finally becoming buff or light brown coloured. It is easily grown in culture but does not always spore very vigorously in culture.

SPECIFIC REQUIREMENTS

scalpels and razor blades
grater
acetone
alcohol (for sterilizing instruments)
polythene bags

sellotape or PVC tape
medical wipes or tissues
fungal inocula
fruit (*see* note 3)

Table 1 Effect on apples of *P. expansum* inoculation.

Type of wound	Small inoculum			Medium inoculum			Large inoculum		
	2 days	4 days	6 days	2 days	4 days	6 days	2 days	4 days	6 days
Large	Just beginning to soften	Slightly soft and brown in	Soft: brown in and around wound, white fungal growth	Just beginning to soften	Soft and brown in wound	Brown in and around wound, white fungal growth	Very soft, very brown, very runny	Very soft, very brown, very runny	Brown in and around wound, white fungal growth
Small	Very small, brown and soft area	Quite brown and soft	Very brown, very soft	Slightly more brown	Brown and soft	Very, very brown and very soft	Soft, brown and quite runny	Very brown and soft	Extremely brown and soft
Grated	Very soft; brown, runny	Very soft and brown; runny	Extremely wet; all of apple soft and brown	Very soft brown	Very large brown patch; quite wet	Extremely wet; all of apple soft and brown	Very soft; brown, runny	Large brown patch; quite wet	Extremely wet; all of apple soft and brown
Dewaxed	Slightly soft and brown	Brown, soft area	Brown and soft; wet	Slightly more soft and brown than with small inoculum	Quite brown and softer than at day 2	Brown and soft; a little wet	Very brown and soft; very runny	Very brown, soft and wet	Big, brown soft areas
Intact	—	—	—	—	—	Brown spotted area	—	—	Little, soft brown area

Results may be quantified by measuring the diameter or volume of infected tissue.

FURTHER INFORMATION

Typically, *P. expansum* grows vigorously on apples but very little on oranges, whereas *P. digitatum* and *P. italicum* grow vigorously on oranges but little on apples. Even on the fruits they are best able to colonize, and using the largest inoculum, it is not uncommon for no infection to occur without wounding. Wounding increases the frequency of infection and deep wounds, with the inoculum placed within the wound, regularly result in infection of susceptible fruits even when spores are used as inoculum. Typical results for *P. expansum* infection of apples are given in Table 1.

There are a large number of experimental variations possible and some of these are listed here.

1. Other fruits or tubers may be used. It is difficult to avoid contamination of some tubers even with surface sterilization. For example, carrots work well but potatoes are more prone to contamination.
2. Other organisms can be tested, e.g. *Botrytis* sp., *Sclerotinia* sp.
3. Other methods of wounding can be investigated, e.g. heat.
4. The effect of incubation at different temperatures can be investigated, e.g. from 35 down to 5 °C. Longer incubation will obviously be needed at lower temperatures. Frequently the effect is not just a change of rate, but the success or failure of the pathogen can vary at different temperatures.
5. Inocula may be taken directly from other infected fruits.
6. Competition by other fungi can be investigated by using a mixed inoculum, e.g. *Trichoderma viridi* and a fruit rotting organism.
7. Koch's postulates can be tested as a follow up to an isolation experiment.

94

Antimicrobial Effects of Seed Dressing

P. J. WHITNEY

Department of Microbiology, University of Surrey,
Guildford, Surrey, GU 5XH, England

> *Level*: All undergraduate years
> *Subject area*: Plant pathology
> *Special features*: Cheap and technically very simple

INTRODUCTION

Many seeds that are commercially available for sowing are coated with a fungicide/bacteriocide. The purpose of this dressing is to prevent infection by pathogenic micro-organisms during storage and more especially during the early stages of germination and growth, when the seed is exposed to pathogenic soil organisms.

This experiment tests the ability of a widely used seed dressing (Captan, *see* below) to protect germinating seeds from attack by two common pathogens (*Pythium* and *Penicillium*). This test of the fungicide is particularly severe as the pathogens are growing vigorously on a medium which provides all the nutrients they require, thus increasing their inoculum potential (*see* Experiment 93).

Captan

EXPERIMENTAL

Day 1 *(10 min)*

Place a pea seed dressed with Captan on one side of a Petri dish of oatmeal agar and a pea seed with no dressing on the other side. The seeds should be pressed firmly into the agar so that they do not move about when the dish is handled.

↓

Inoculate the centre of the dish with *Pythium debaryanum* (*see* note 1).

↓

Repeat the above procedure on another Petri dish and inoculate with *Penicillium expansum* (*see* note 2).

↓

Incubate at 20 or 25 °C or room temperature. Observe and record the growth of both the fungi and the seeds, inspecting them at 1 or 2 day intervals for about 7 days (5 min for each observation).

Notes and Points to watch

1. *Pythium debaryanum* does not always grow vigorously on agar, but inoculation into sterile distilled water containing sterile autoclaved hemp seeds has reliably provided vigorous growth and one, or even part of one, of these seeds forms a suitable inoculum.
2. *Penicillium expansum* grows well on a wide variety of media and no difficulty should be found in culturing it. A small cube of agar cut with a sterile scalpel from a culture forms a suitable inoculum.
3. Before use in these experiments, test the fertility of dried peas by incubating a sample of them on damp filter paper in a Petri dish for a few days.
4. Variations can be made on these experiments, such as washing the Captan-treated seeds to see if the protection is lost. Other seed dressings may be tried, such as copper sulphate, mercuric chloride or borax (sodium tetraborate) by soaking the seeds for a few minutes in a 5–10% solution before plating out. Other commercial fungicides can also be tested. Other fungi can be used as the test organism.

MATERIALS

1. *Pythium debaryanum* (C.M.I. Cat. No. 173340).
2. *Penicillium expansum* (C.M.I. Cat. No. 39761).
3. Captan-coated pea seeds (Johnson seeds, from local seed merchants).
4. Untreated pea seeds (dried peas for human consumption—carefully select undamaged seeds).

5. Oatmeal agar contains oatmeal 30 g, agar 20 g, water 1 litre. Boil oatmeal for $\frac{1}{2}$–1 h and squeeze through muslin, add agar and make up to 1 litre. Autoclave at 103·4 kPa (15 lbf in^{-2}) for 20 min.

SPECIFIC REQUIREMENTS

Petri dishes of oatmeal agar (2 per student)
scalpel
forceps
bottles of alcohol to sterilize scalpel and forceps

Captan-coated pea seeds (2 per student)
treated pea seeds (2 per student)
culture of *P. debaryanum* (1 plate per class)
culture of *P. expansum* (1 plate per class)

FURTHER INFORMATION

Growth of the *Pythium* will be enhanced on and around the untreated pea seed and, even if germination starts, the developing seedling will be rapidly killed. The *Penicillium* is not so pathogenic towards seedlings, being more of a saprophyte/opportunist pathogen, and some seedlings may survive. Both fungi will grow right up to the Captan-treated seeds and the *Penicillium* may even occasionally colonize small areas of the pea testa. *Pythium* is more sensitive to Captan and does not usually colonize the treated seed at all (Figure 1). The effect of the Captan is localized and does not diffuse through the agar because it is fairly insoluble. This is an advantage under field conditions where it would be undesirable for the dressing to be washed off the seed before germination; however, it must have some solubility for it to be toxic to the fungi.

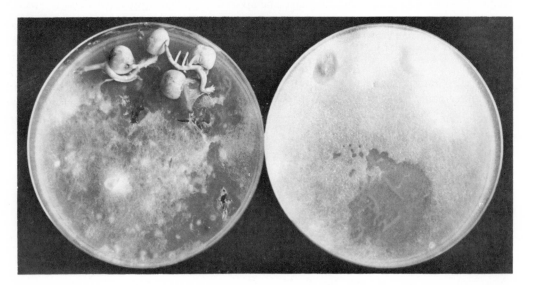

Figure 1   Protection of the pea seeds against attack by *Pythium debaryanum*. Left: Captan-treated seeds. Right: untreated seeds.

95

Antifungal Compounds in Plants

P. J. WHITNEY

*Department of Microbiology, University of Surrey,
Guildford, Surrey, GU2 5XH, England*

Level: All undergraduate years
Subject area: Plant pathology
Special features: Technically very quick and simple.
Very cheap. Many variations
possible

INTRODUCTION

Many plants contain chemical compounds that are inhibitory to the growth of micro-organisms and this may play an important part in their resistance to pathogens. Different plants produce different compounds which vary in their antimicrobial action, and micro-organisms differ in their sensitivity to these compounds. Garlic is a good example of one of the many plants containing such antimicrobial compounds (Arik and Thompson, 1959; Appleton and Tansey, 1975; Murphy and Amonkar, 1974). The presence of antimicrobial compounds can be detected by challenging the growth of micro-organisms on agar medium with homogenates or extracts of plant material, and this is the aim of the following experiment.

EXPERIMENTAL

Day 1 (15–20 min)

Using a sterile cork borer or scalpel, cut a well approximately 1 cm diam. in the agar towards the edge of a Petri dish of potato dextrose agar.

↓

Fill the well with extract or homogenate of plant material, prepared either in a blender or with a

---

pestle and mortar (*see* note 2). The homogenate may be used directly, or an extract made by squeezing the homogenate in a piece of muslin to express the liquid extract.

↓

Inoculate the agar on the other side of the Petri dish with the test micro-organism (*see* note 1).

↓

Carry out the same procedure on another Petri dish, but fill the well with sterile distilled water as a control.

↓

Incubate both the experimental Petri dish and the control at 25 °C.

Days 2–4 *(5 min)*

Note the effect of the plant material on both growth rate and form of the micro-organism.

Notes and Points to Watch

1. Suitable fungi for this experiment are *Pythium debaryanum*, the cause of "damping off" in seedlings, and a *Penicillium* sp. The latter is easy to culture but the *Pythium* may occasionally prove difficult to grow on agar. Good growth of *Pythium* can be obtained if it is inoculated into sterile distilled water containing sterile autoclaved hemp seeds. One, or even part of one, of the hemp seeds forms a suitable inoculum. Hemp seeds can be obtained from pet shops and fishing tackle shops, but care should be taken that they have not been chemically treated to stop germination.
2. Garlic cloves form suitable plant material for the experiment as they contain material which undergoes autolysis following homogenation and is converted into a form strongly inhibitory to *Pythium* but less inhibitory to *Penicillium*.

MATERIALS

1. *Pythium debaryanum* (C.M.I. Cat. No. 173340)
2. *Penicillium* sp., e.g. from mouldy bread or fruit (*see* Experiment 93).
3. Potato dextrose agar: can be made up from proprietary media or made as follows. Scrub clean, or peel, 200 g potatoes and cut into small cubes. Boil in 500–600 ml water until very soft. Break up lumps to form a slurry. Filter through a layer of cotton-wool in a large Buchner funnel using a water pump or vacuum line to speed filtration. Filter one funnel-full of potato slurry through the cotton-wool and discard compacted layer of starch and cotton-wool. Use fresh cotton-wool for each funnel-full of potato slurry as the compacted starch will stop further extract being filtered. Add 20 g agar and 20 g glucose to the extract and make up to 1 litre with water. Autoclave at 103·4 kPa (15 lbf in^{-2}) for 20 min.

SPECIFIC REQUIREMENTS

pestle and mortar or blender
muslin (to strain homogenate for extract)
pipettes and teats (sterile)
cork borer or scalpel (sterilize with alcohol and
   flame)

alcohol
potato dextrose agar plates
garlic cloves (or other plant material)
test organisms

FURTHER INFORMATION

Typical results are shown in Figure 1. Many variations are possible with this experiment and some suggestions are given below.

1. Other plant material may be used as a potential source of antifungal material, e.g. rhubarb leaves, onion bulbs and orange peel.
2. Extracts may be made at different times after homogenation to investigate the autolytic production of antifungal compounds.
3. Organic extracts may be made and dried onto filter paper discs and applied directly to the agar, rather than in a well.

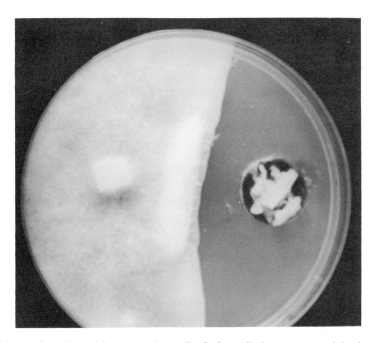

Figure 1 Inhibition of *Pythium debaryanum* by garlic. Left: garlic homogenate; right: inoculum; centre: inhibition of growth and changed morphology of mycelium. Note that the distance the fungus has grown towards the garlic is only 40 % of the distance it has grown towards the edge of the Petri dish.

4. The effect of the extracts on other micro-organisms may be investigated.
5. The effect of heat treatment of homogenate/extract can be investigated.
6. The effect of nutrient levels in the medium can be investigated using different media or different dilutions of a single medium.
7. Volatile antifungal compounds may be detected by placing the homogenate or extract on a filter paper in the lid of an inverted Petri dish and the radial growth of the fungus growing across the agar above the homogenate/extract compared with a control.

REFERENCES

J. A. Appleton and M. R. Tansey (1975). Inhibition of growth of zoopathogenic fungi by garlic extract. *Mycologia*, **67**, 882–6.

M. B. K. Murphy and S. V. Amonkar (1974). Effect of a natural insecticide from garlic (*Allied sativum*) and its synthetic form (diathyl-disulphide) on plant pathogenic fungi. *Indian Journal of Experimental Biology*, **12**, 208–9.

P. A. Arik and J. P. Thompson (1959). Control of certain diseases of plants with antibiotics from garlic (*Allium Sativum* L.). *Plant Diseases Reports*, **43**, 276–82.

96

Isolation of Tobacco Mosaic Virus from Widely Available Sources

M. A. MAYO and A. T. JONES

*Virology Section, Scottish Crop Research Institute,
Invergowrie, Dundee, DD2 5DA, Scotland*

> *Level*: All undergraduate years
> *Subject areas*: Plant virology, agricultural microbiology
> *Special features*: Students can test material of their own choice

INTRODUCTION

Plant viruses are abundant in nature and several can be isolated by simple techniques. A small proportion of the plant viruses that induce diseases in crop plants are transmitted by physical contact between plants and infected sources. Particles of most such viruses reach high concentrations in infected plants and are sufficiently stable to survive in plant sap.

One such virus, tobacco mosaic, is a significant pathogen of commercial tomatoes and much effort is expended on glasshouse hygiene to prevent infection of young plants (Broadbent, 1962; Broadbent and Fletcher, 1963). This experiment is to demonstrate some possible sources of the virus and thereby to explain horticultural practice.

One effect of many viruses is to induce necrotic spots in inoculated leaves. These local lesions represent the foci of infection and their abundance can be used to assay infectivity.

EXPERIMENTAL

Day 1 (60–90 min)

Dust carborundum lightly over fully expanded leaves of the assay host being used (*see* note 1).

↓

Grind about 1 g of material being tested in about 2 ml buffer using a small pestle and mortar (*see* note 2).

↓

Dip a muslin pad into the extract and **gently** stroke a leaf with the wet pad. Inoculate 1–2 fully expanded leaves with each suspected source of virus. Use a new plant for each virus source.

↓

Lastly repeat the whole operation using about 0·25 g of infected tobacco leaves in place of the test material.

↓

Wash hands thoroughly with soap between inoculations (*see* note 3).

↓

Rinse the inoculated leaves under running water after inoculation.

↓

Place plants in a glasshouse or culture room or under lights.

Days 3–5

Observe the appearance of lesions on the inoculated plants.

Day 5, 6, 7 or 8 *(40 min)*

Cut out lesions produced by different sources of TMV, especially those from tobacco and tomato. Use a sterile blade to cut about 5 mm outside the dead brown area.

↓

Place a single lesion in a mortar (*see* note 2) and grind in about 0·5 ml phosphate buffer. Dust *N. tabacum* cv. White Burley plants with carborundum and inoculate the extracts as for Day 1.

↓

After 2–4 days look for symptoms in the inoculated leaves.

↓

After 8–20 days look for symptoms in the uninoculated (but possibly systemically infected) leaves.

Notes and Points to Watch

1. Tobacco plants will need to be sown 7–11 weeks before use, depending on the growing conditions. The experiments are therefore best done in the spring or summer.
2. Particles of TMV are exceptionally stable and therefore present a likely source of

contamination. Glassware and pestles and mortars should be decontaminated after use by rinsing in 3% trisodium phosphate.
3. It is important that hands are washed thoroughly with soap between inoculations. This is especially important when inoculating a host which will become systemically infected.

MATERIALS

1. *Sources of virus.*
   a. A variety of cigarette, cigar and pipe tobaccos should give a range of virus isolates. Students can test their own samples. The incidence of virus may be related to the method used to cure the tobacco.
   b. Many tomato fruits will be infected with TMV. Nearly all those from Guernsey will have been inoculated deliberately with a mild strain in an attempt to protect the plants from subsequent attack by more severe TMV strains.
   c. Virus can be obtained from the authors or from the American Type Culture Collection. Infected leaves may then be stored frozen for many years. As a control for the inoculation technique, a small amount of this material should give hundreds of necrotic local lesions on the assay hosts.

2. *Plants*
   a. Assay hosts. Plants of *Nicotiana glutinosa*, *N. tabacum* cv. Xanthi-nc or cv. Samsun-NN grown to give at least two fully expanded leaves. Germination of seeds may be poor and variable. Plants are best raised in a glasshouse but can be grown on a bench under fluorescent lights. Plants should be inoculable 7–11 weeks after sowing if growing conditions are reasonable. Seed of *N. glutinosa* may be obtained from University Botanic Gardens, such as Birmingham, Cambridge or the University of London Supply Unit, Elm Lodge, Englefield Green, Surrey, England; samples of other seed for propagation can be obtained from the authors.
   b. Propagation hosts. Plants of *Nicotiana tabacum* cvs White Burley, Turkish or Samsun grown to give several mature leaves. Seed of smoking tobaccos such as these can be found in several commercial seed catalogues.

SPECIFIC REQUIREMENTS

Day 1
several pestles and mortars
several plants of the assay host(s)
labels for the plants
suspected sources of virus
pipettes
0·05 M sodium phosphate, pH 7
    carborundum (500–600 grit; British Drug Houses Ltd.) in a small glass pot covered with a piece of butter muslin; this acts as a sprinkler
pads of muslin (approx. 5 cm² pieces, folded twice)
soap

Day 7
pestles and mortars
pads of muslin
several plants of the propagation host
labels for the plants
pipettes
0·05 M sodium phosphate, pH 7
scalpel blades
a Bunsen burner for flame-sterilizing blades

FURTHER INFORMATION

Many features of plant viruses, including those shown by this experiment, are discussed in textbooks such as those by Gibbs and Harrison (1976) and Matthews (1970). Access to much basic information about tobacco mosaic virus can be obtained from the description by Zaitlin and Israel (1975) and also that of the tobamovirus group, of which TMV is the type member, by Gibbs (1977).

Deliberately inoculating tomatoes with a mild strain of TMV to protect the plants against subsequent infection with the common (and more severe) strain is discussed by Fletcher and Rowe (1975). This reference also gives access to many earlier papers describing the horticultural impact of TMV, notably those by Broadbent and co-workers.

TMV is very readily transmitted—indeed, in most laboratories working with plant viruses much effort is expended to avoid chance infection from sources such as cigarettes and infested soil. In these class experiments the main problem is likely to be avoiding this inadvertent infection.

REFERENCES

L. Broadbent (1962). The epidemiology of tomato mosaic II. Smoking tobacco as a source of virus. *Annals of Applied Biology*, **50**, 461–6.

L. Broadbent and J. T. Fletcher (1963). The epidemiology of tomato mosaic IV. Persistence of virus on clothing and glasshouse structures. *Annals of Applied Biology*, **52**, 233–41.

J. T. Fletcher and J. M. Rowe (1975). Observations and experiments on the use of an avirulent mutant strain of tobacco mosaic virus as a means of controlling tomato mosaic. *Annals of Applied Biology*, **81**, 171–9.

A. J. Gibbs (1977). Tobamovirus group. Commonwealth Mycological Institute/Association of Applied Biologists Descriptions of Plant Viruses No. 184, 6 pp.

A. J. Gibbs and B. D. Harrison (1976). "Plant Virology: The Principles". Edward Arnold, London.

R. E. F. Matthews (1970). "Plant Virology". Academic Press, London and New York.

M. Zaitlin and H. W. Israel (1975). Tobacco mosaic virus (type strain). Commonwealth Mycological Institute/Association of Applied Biologists Descriptions of Plant Viruses No. 151, 5 pp.

Bioassay of Infectivity of Tobacco Necrosis Virus in French Beans

A. T. JONES and M. A. MAYO

Virology Section, Scottish Crop Research Institute, Invergowrie, Dundee, DD2 5DA, Scotland

> *Level*: All undergraduate years
> *Subject areas*: Virology, bioassay, experimental design
> *Special features*: Simple and open to development into a wide range of projects

INTRODUCTION

An intrinsic property of viruses is their infectivity, and with some plant viruses this may be assayed by counting the number of small discrete lesions produced in inoculated leaves of a suitable host. Holmes (1929), studying tobacco mosaic virus, was the first to show the value of local lesion assays for determining the relative infectivity of inocula and subsequently many workers have developed the technique for other plant viruses.

The accuracy of the technique is influenced by a number of factors including the virus concentration of the inoculum, genetic variation of assay plants, variation in the susceptibility of leaves of the same plant, environmental factors and operator technique. In properly designed assays, however, differences in virus concentration of 10–20% have been detected (Price, 1946).

To obtain the greatest accuracy from the technique some of the variation can be minimized by following these simple procedures:

    a. assaying at suitable inoculum dilutions—the relationship of lesion number to virus concentration is sigmoid;

    b. selecting uniform assay plants from pure genetic lines grown under uniform environmental conditions;

    c. comparing inocula on opposite halves of leaves, randomizing and replicating treatments among leaves of several plants;

d. using identical techniques for inoculating all doses and leaves.

Relatively few virus host systems are suitable for local lesion assay, but that of tobacco necrosis virus (TNV) in French beans (*Phaseolus vulgaris*) is one of the easiest to handle. TNV has stable isometric particles and *P. vulgaris* sap containing TNV stays infective for several years at $-20\,°C$ (Kassanis, 1970). Usable assay plants of *P. vulgaris* cv. The Prince can be produced simply and reliably from seed within about 2 weeks of sowing. Most strains of TNV induce necrotic local lesions in leaves of this host, although TNV strain D is recommended.

The aim of the experiments is to show the relationship of lesion number to the infectivity of the inoculum and the importance of experimental design.

EXPERIMENTAL

The following experiments for bioassay can be run either concurrently with different groups or as separate experiments run at weekly intervals. However, the use of a common inoculum for the class will allow comparisons between experiments and between groups.

Preliminary

At least 2 weeks before the class is due, inoculate 40–50 *P. vulgaris* cv. The Prince plants with TNV. Inoculate as many leaves as possible. One week later harvest the inoculated leaves which show necrotic local lesions and triturate them in a sterile blender in distilled water (1 g leaf: 1 ml water). Squeeze through cheesecloth and freeze *c.* 20 ml aliquots (sufficient to supply each student or group with about $0\cdot5$–1 ml). Do not completely fill containers for freezing—sap expands on freezing!

Day 1 *(2–3 h)*

Thaw frozen sap and centrifuge for 10 min at 3000–8000 *g* before the class begins. Supply each student or group with about 0·5–1 ml of the clarified stock inoculum.

↓   ↓

1. *Effect of dilution* or *Effect of buffer*
Dilute stock inoculum in ten-fold steps up to 10^{-5} in H_2O. To compare effect of buffer as diluent, dilute in H_2O and in buffer (e.g. 0·1 ml stock + 0·9 ml H_2O = 10^{-1}, etc.).

2. *Effect of heating*
Dilute stock inoculum 1/10 in H_2O or buffer (0·1 ml stock + 0·9 ml H_2O). Pipette 1 ml samples into thin walled glass tubes and heat each for 10 min at 70, 80 or 90 °C in a water-bath. Allow to cool.

↓   ↓

Select the 4 most uniform plants in each pot and discard the remainder.

↓

Pierce the tip of one primary leaf of each plant with a pencil or pen to distinguish the 2 primary leaves.

↓

Dust plants lightly and evenly with carborundum (see note 2).

↓

Label each plant with its respective treatments (*see* note 5).

↓

Using muslin pads, inoculate half leaves with inoculum treatments (*see* notes 1, 3 and 4) starting with the most dilute inoculum and finishing with the most concentrated
  (i.e. experiment 1: $10^{-5}, 10^{-4}, 10^{-3} \rightarrow 10^{-2}$, and
  experiment 2: 90, 80, 70 °C → unheated control).

↓

Wash hands thoroughly between each inoculum treatment.

↓

Rinse inoculated leaves of plants with water.

↓

If inoculated plants are to be placed in direct sunlight, or under high light intensity lamps, cover with newspaper until the following morning to prevent the inoculated leaves from wilting.

Days 5–8 *(1 h)*

Count the lesions on each half leaf of detached leaves. It is sometimes necessary to wipe off surplus carborundum from leaves in order to see the lesions.

↓

Total the number of lesions/treatment and compare results.

↓

Simple analyses of variance can be made on the results.

Notes and Points to Watch

1. *General hygiene.* Good hygiene is essential since TNV is stable and infective for long periods. Students should be encouraged to take care to handle plants and inoculum (sap, pads and watch glasses) only with "clean" hands. Hands should be washed thoroughly in soap and water, and rinsed in running tap-water before the experiment starts and immediately prior to inoculation. Once hands are clean they should not be recontaminated before handling inoculum and plants, i.e. by handling pencils, notebooks, equipment, doors, taps, etc. Planning the work beforehand is essential—all equipment and material should be at hand and ready to use before inoculation. Keep a stream of tap-water running to avoid handling taps.

    To sterilize glassware and pestles and mortars after use, soak in 3 % trisodium phosphate and thoroughly rinse before autoclaving or heating in an oven at *c.* 130 °C.

2. Carborundum dust is messy—cover benches with newspaper before the experiment.
3. Sterile muslin (cheesecloth) pads about 50 mm² should be folded to form a suitable pad. Resoak the pad in the inoculum after inoculating each leaf. Use a fresh pad for each inoculum treatment.
4. Support the leaf from beneath with one hand, with the petiole pointing away from the operator, and with the muslin pad in the other hand make a single, slow, **gentle** stroke over the leaf half to be inoculated. Repeat to cover the entire half-leaf area—it is easy to see uninoculated areas when the leaves are dusted with carborundum.
5. Suggested randomization plans.
   Experiment 1: in a layout to compare water and buffer as diluent the following is suggested:

	Plant no.:	1		2		3		4	
Pot 1	Leaf side:	L	R	L	R	L	R	L	R
	Marked leaf:	W^2	B^2	W^3	B^3	W^4	B^4	W^5	B^5
	Unmarked leaf:	B^2	W^2	B^3	W^3	B^4	W^4	B^5	W^5

   Repeat for pot 2.
   W = water extract; B = buffer extract; 2, 3, 4, 5 = ten-fold dilutions.
   To study the effect of dilution on infectivity a better arrangement is a latin square of the 4 dilutions for each treatment on each plant (*see* experiment 2).
   Experiment 2: a latin square arrangement using a single plant for each block of treatments would be as follows:

	Plant no.:	1		2		3		4	
Pot 1	Leaf side:	L	R	L	R	L	R	L	R
	Marked leaf:	70	80	80	90	90	C	C	70
	Unmarked leaf:	90	C	C	70	70	80	80	90

   This could be duplicated for pot 2.
6. Several dilutions of inoculum are suggested to give an optimum number of lesions to count, *c.* 10–100 per half leaf are acceptable. Dilutions may need to vary depending upon the sensitivity of the plants and age of inoculum.

MATERIALS

1. Tobacco necrosis virus in clarified *P. vulgaris* sap. The D strain of TNV (available from the authors for initial propagation) is recommended as it gives large, discrete necrotic local lesions. Alternatively, TNV can often be recovered from the roots of old herbaceous ornamentals grown in their original pots of soil in heated glasshouses. To test for the presence of TNV, grind up roots that have been briefly washed in tapwater and inoculate *P. vulgaris*. Do not select material from glasshouses with a history of tomato culture as tobacco mosaic virus is likely also to be present. Alternatively, a culture of TNV can be purchased from the American Type Culture Collection (ATCC), 12301 Parklawn Drive, Rockville, Maryland 20852, USA.

2. 6–8″ pots of sterilized John Innes compost no. 1 (or similar) in which are 5–6 plants of *P. vulgaris* cv. The Prince with well expanded primary leaves; secondary leaves should just be emerging. This stage is reached 10–16 days after sowing depending on light and temperature conditions.

3. Quantities (20 ml samples) of distilled water and 0·005 M sodium phosphate buffer, pH 7.

SPECIFIC REQUIREMENTS

Day 1

Stock inoculum: *P. vulgaris* sap containing TNV either fresh or thawed from deep-freezer; centrifuged for 10 min at 3000–8000 *g* before distributing as 0·5–1 ml aliquots

Experiments 1 and 2

soap

distilled water

0·005 M sodium phosphate buffer, pH 7

carborundum, *c.* 600 grit, in shakers

wash bottles

1 ml pipettes

0·1 ml pipettes

3–5 ml test tubes for dilutions

test tube racks

muslin squares (50 mm^2)

watch glasses

pot labels

2 × 6–8″ pots of *P. vulgaris* cv. The Prince/group

newspaper

Experiment 2

water-bath

thermometer

Days 5–8

an illuminated background for lesion counting is useful but not essential

FURTHER INFORMATION

Local lesions should become evident within 4–6 days and all should be countable by one week. Usually some students are "heavy-handed" and leaves may be severely scratched and wilted, but a high proportion should have countable lesions. For the same treatments, variation in lesion number on similar half leaves should be relatively small, but in practice it often reflects the right- (or left-) handedness of the operator. From the more concentrated inocula, lesions may be too numerous to count.

The relationship of dilution to lesion number is sigmoid (Roberts, 1964) and the optimum dilution of inocula to compare treatments is within the straight line of the curve. Detailed mathematical equations have been produced to analyse data (Roberts, 1964). However, in practice, statistical analysis of data has not been widely used. In most cases workers have been concerned with major effects on infectivity where differences in lesion number are large.

Yarwood (1957) and others have demonstrated that the type of buffer, molarity and pH affect lesion number. These and other treatments are further possible experiments.

The thermal inactivation point of most strains of TNV is 85–95 °C. This and other properties of the virus are described by Kassanis (1970).

REFERENCES

F. O. Holmes (1929). Local lesions in tobacco mosaic. *Botanical Gazette*, **87**, 39–55.

B. Kassanis (1970). Tobacco necrosis virus. Commonwealth Mycological Institute/Association of Applied Biologists Descriptions of Plant Viruses No. 14, 4 pp.

W. C. Price (1946). Measurement of virus activity in plants. *Biometric Bulletin*, **2**, 81–6.

D. A. Roberts (1964). Local lesion assay of plant viruses. *In* "Plant Virology", (M. K. Corbett and H. D. Sisler, eds), pp. 194–210. University of Florida Press, Gainesville.

C. E. Yarwood (1957). Mechanical transmission of plant viruses. *Advances in Virus Research*, **4**, 243–78.

98

Demonstration of the Rhizosphere

R. CAMPBELL

Department of Botany, University of Bristol,
Bristol, BS8 1UG, England

Level: All undergraduate years
Subject areas: Microbial ecology, soil micro-biology, plant–micro-organism interactions
Special features: Simple to set up and easily replicated. Uses several different methods for counting bacteria and fungi on the same sample

INTRODUCTION

Plant roots release soluble and insoluble organic materials into the soil and slough off dead cells, consisting mostly of cellulose walls. Many micro-organisms in the soil increase in numbers near plant roots, largely because of this release of organic nutrients into a normally nutrient deficient environment, the soil (Bowen and Rovira, 1976; Brown, 1975). The volume of soil affected in this way by plant roots is called the rhizosphere and it is important for the plants themselves because the micro-organisms affect the mineral nutrient availability, may influence plant pathogens and may affect the root morphology by the production of growth factors such as indole acetic acid. The aim of this experiment is to estimate the numbers of fungi and bacteria in the rhizophere soil around wheat seedlings and to compare them with the numbers in soil without roots. Several different enumeration techniques will be used (Rovira *et al.*, 1974).

EXPERIMENTAL

Students normally work in groups of 3 or 4 and the times given and the quantities of materials are for each group. Several groups are desirable for replication of the results.

Day 1 *(15 min)*

Fill large seed trays with a mixture of unsterilized loam and sand; moisten it if necessary.

↓

Bury cleaned microscope slides on edge with the long axis horizontal, so that they are as far below the surface as possible. Take care not to get greasy fingerprints on the slides. Mark the position of each slide with a plant pot label put in alongside (Figure 1).

Figure 1   The seed tray, half-planted with wheat, with the labels marking the positions of the buried slides.

↓

Sow wheat seeds in one half of the tray (*see* note 1), taking care to put at least some of the seeds next to buried slides. Water the soil and allow the plants to grow for 4 or 5 weeks (*see* note 2) until there is a dense mass of roots, mainly in one half of the tray. Keep the bare soil weed-free. Figure 1 shows the appearance of the grown plants.

_ _

Day 2 *(3 h)*

This is devoted to the microscopical and/or cultural examination of 4 types of sample:
1. the buried slides (80 min)
2. root surfaces (60 min)
3. rhizosphere soil (20 min)
4. non-root soil (20 min)

1. *Examination of the buried slides*

Recover the buried slides so that they can be orientated in relation to plant roots, if any. Dig the soil away from one side (A) and lift out of the slide without smearing the side (B) which was in contact with the roots.

↓

Thoroughly clean side A. Pick large pieces of soil off side B then very gently wash it to remove all but very small adhering soil particles. Side B is the one to be observed. Air-dry the slide.

↓

Place the slides over a beaker of boiling water and flood with carbol erythrosin. Leave for 5–8 min. Wash, blot, then air-dry and examine microscopically. In place of carbol erythrosin the slide may be stained at room temperature in phenol acetic aniline blue for 4 or 5 min, then washed and dried.

↓

Using an eyepiece graticule with a square grid (Figure 2), count bacteria in one or more squares whose area has been calculated with a micrometer slide. Take 30 random fields for the near-root slides and another 30 for the no-root sample. Convert to numbers per square millimetre (*see* note 3).

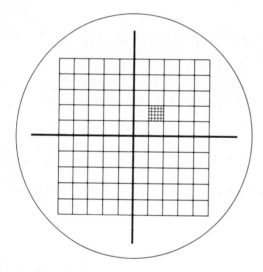

Figure 2   The eyepiece graticule used for direct observation.

↓

For hyphae, use the two central cross-lines on the graticule; measure the apparent length of the lines with the micrometer slide. Count in 30 random fields on near-root slides the number of times that hyphae intersect the lines, then use the formula $L = \pi N/2H$, where N is the total number of intersections per field, H is the total length of the cross lines in mm and L is the calculated total length of hyphae per unit area expressed in mm mm^{-2} (*see* Rovira *et al.*, 1974). Try to arrange the magnification or the length of the lines so that not more than a few zero values are recorded. Repeat for no-root slides.

2. *Examination of root surfaces*

Gently pull up 4 or 5 wheat plants with the closely attached soil. Cut off the roots with soil still attached and add to 90 ml sterile water in a flask. Shake gently to remove most of the rhizosphere soil.

↓

Recover the roots, taking care that the contents of the flask are kept free of contamination. Wash the roots gently to remove remaining soil, stain in phenol acetic aniline blue for 2–4 min, wash, mount in water.

↓

Examine several pieces of root, taking a total of about 30 fields to count the numbers of bacteria, and measure the length of hyphae per unit area of root surface using the formula given above (*see* note 4).

3. *Examination of rhizosphere soil*

Take the 90 ml of water plus the rhizosphere soil (approximately 10^{-1} dilution) and shake on a reciprocating shaker for 10 min.

↓

Take 1 ml of this suspension and transfer it to 9 ml sterile water (*see* note 5) in a test tube; shake or use a vortex mixer to give a 10^{-2} dilution. Continue the dilutions to 10^{-5} (*see* note 6). Mix thoroughly and use a new pipette or sterile tip each time: suck up, blow out again, suck up again before transferring the measured volume.

↓

Take 9 plates each of Tryptic soy agar (TSA), Martin's rose bengal agar (MRB) and *Pseudomonas* medium; label them with the medium type, inoculum source, dilution and name or group number.

↓

Plate the dilutions 10^{-3}, 10^{-4} and 10^{-5} on TSA and 10^{-2}, 10^{-3} and 10^{-4} on MRB and *Pseudomonas* medium with 3 replicates of each dilution (*see* note 6). Put 0·1 ml of the required dilution on each plate, remembering to suck the liquid up and down the pipette or tip several times on each occasion. Spread the drop with a sterile spreader.

↓

Remove the soil from the flask by filtration and use it for a dry weight measurement. Incubate the plates aerobically and inverted at 15–20 °C for one week.

4. *Examination of non-root soil*

Take and weigh about 10 g of non-root soil. Place it in 90 ml sterile water in a 250 ml flask and shake for 10 min. Prepare a dilution series down to 10^{-5} and plate as in 3 above. Do a dry weight estimation on a separate sample of the soil.

Day 3 *(2h)*

Examine the dilution plates (*see* note 7) from 3 and 4 above. All organisms on MRB are streptomycin resistant. All organisms on the *Pseudomonas* medium show multiple antibiotic resistance.

↓

Arrange the replicate plates in order of increasing dilution and check that no obvious errors have been made, i.e. that the numbers decrease with the dilution, that replicates are more or less the same, that no plates are obviously contaminated, etc. (*see* note 8). Select the dilution that gives 30–100 colonies per plate.

↓

Count the colonies, Gram-stain a sample, examine the fungi microspically and identify if possible. Classify the bacteria, e.g. it should be possible to state that "half the colonies are Gram-negative, motile rods" or whatever the case may be. Calculate the number of colony-forming units per gram dry weight of soil.

↓

From the whole experiment compare the numbers of micro-organisms in the rhizosphere with those in non-root soil by both plating and direct observation and, in the latter, consider also the numbers on the root surface.

Notes and Points to Watch

1. Sow the plants fairly densely so that there is a clear distinction between the root and non-root soil.
2. Day 1 can be set up in advance so that the students are presented with ready grown plants in the tray with the slides already in position. This saves the 4 or 5 week delay. We grow the plants over a vacation which conveniently falls in the middle of the course.
3. Direct microbial counts are difficult because of the problem in distinguishing between bacteria and organic and mineral debris. Students, and probably demonstrators as well, need some help and practice at first to pick out the bacteria. It is worth practising before starting to record the numbers. It is also worth trying slightly different staining times if necessary to obtain deeply stained organisms on an almost white background. Variability is usually high with so many unskilled observers: a single skilled worker can get quite consistent results and, for a given period of time, direct observation can give a more precise estimate of numbers than does plating (Rovira *et al.*, 1974). Comparisons of the figures within a student group should be satisfactory, but there may be considerable variation between different groups, even if their soil trays are similar.
4. In the direct examination of roots, distinctions can be made between old and young roots, or roots of different individual plants. This gives a more detailed analysis but the number of fields becomes excessive for an undergraduate class.
5. Other diluents may be used if desired (e.g. quarter-strength Ringer's or 0·1 % peptone), but we have not found that these have any advantage over water.

6. The actual numbers of colonies obtained depend on the soil used and the root density, etc. A trial run should be done so that the students can be told to plate more or less the correct dilutions. The ones suggested cover the numbers that we find in a rather poor, sandy loam.
7. Strict aseptic techniques should be used because of the undesirability of allowing any escape of antibiotic-resistant organisms.
8. The students' attention should be drawn to the problems of dilution plates in general, but particularly when used for estimating the numbers of fungi.

MATERIALS

Materials specified are for each group of students.
1. A seed tray (30 × 60 cm) filled with a 3:1 mixture of loam and sand.
2. 20 plastic plant pot labels.
3. About 15 g wheat seed.
4. Greenhouse space for growing the plants for 4 or 5 weeks.
5. Carbol erythrosin (erythrosin 1 g; 5% w/v aqueous phenol 100 ml; $CaCl_2$ 0·05 g).
6. Phenol acetic aniline blue (5% w/v aqueous phenol 15 ml; 1% w/v aqueous aniline blue 1 ml; glacial acetic acid 4 ml; filter 1 h after mixing).
7. 20 plates TSA (Tryptic soy broth, Difco 0370-01, 3 g; agar 15 g; water to 1 litre). This is 1/10 the usual strength of the nutrients.
8. 20 plates Martin's rose bengal agar (glucose 10 g; peptone 5 g; KH_2PO_4 1 g; $MgSO_47H_2O$

0·5 g; rose bengal 0·033 g; agar 20 g; water to 1 litre. 3 mg of streptomycin per 100 ml added to cooled medium just before pouring).
9. 20 plates of *Pseudomonas* medium (peptone 20 g; agar 12 g; glycerol 10 ml; K_2SO_4 1·5 g; $MgSO_47H_2O$ 1·5 g; distilled water to 1 litre. Adjust to pH 7·2 with NaOH. Autoclave. Add to cooled medium just before pouring, cyclo-heximide 75 μg ml^{-1}; chloramphenicol 12·5 μg ml^{-1}; ampicillin 50 μg ml^{-1}. (*See* Simon and Ridge, 1974).
10. Micrometer slide marked in hundredths of mm.
11. Eyepiece graticule marked in squares. A Whipple graticule (Gallenkamp Ltd., Number MNH-420-000A) is suitable (*see* Figure 2).
12. Oven at 105 °C.
13. Reciprocating shaker.

SPECIFIC REQUIREMENTS

These items are needed for each group of students.

Day 1
seed tray with soil
20 microscope slides, cleaned and degreased
20 plastic plant pot labels
wheat seed
greenhouse space

Day 2

1. Buried slides
micrometer slide

eyepiece graticule
carbol erythrosin
phenol acetic aniline blue
normal bacteriological and mycological stains, staining racks, forceps, etc.
microscopes, slides and coverslips
Bunsen burners, tripods, 250 ml beakers
calculators, preferably with a "standard deviation" key

2. Root surface
250 ml conical flask containing 90 ml tap-water sterilized by autoclaving (*see* note 5)

cont.

cont.

staining dishes, clean 9 cm Petri dishes are
  suitable

3 and 4. Soil spread plates
reciprocating shaker
10 sterile capped test tubes each with 9 ml sterile
  water
test tube racks
20 sterile 1 ml plugged serum pipettes or, better,
  automatic pipette with sterile tips to deliver
  1 ml and 0·1 ml (we find that "Selectapette"
  manufactured by Clay Adams Ltd.,
  Parsippany, N. J., USA is suitable)
vortex mixer for test tubes
Buchner funnel and Whatman no. 1 filter paper,
  vacuum source

balance and weighing dishes
oven at 105 °C
plates of TSA medium
plates of Martin's rose bengal agar
plates of *Pseudomonas* medium
marker pens
glass spreaders
250 ml conical flask with 90 ml sterile water
incubator at 15–20 °C; room temperature can be
  used if sufficient cooled incubator space is not
  available

Day 3
"Sterilin" bags for the disposal of Petri dishes
calculators

FURTHER INFORMATION

The results obtained by one group of students are given in Tables 1 and 2. The high standard errors in the fungal assessment by direct examination are caused by the number of zero values: these reflect the discontinuous growth pattern and scarcity of active hyphae in the soil.

Table 1 Numbers from direct observation; means of 30 replicate fields and standard errors.

	Bacteria (10^4 mm^{-2})	Fungi (mm mm^{-2})
Slides, no roots	0·32 ±0·05	0·58 ±0·36
Slides with roots	0·84 ±0·09	0·88 ±0·45
Root surface	2·25 ±0·05	4·42 ±0·72

Table 2 Plate counts on various media. The numbers are colony-forming units per gram dry weight, being the means of 3 plates.

	Rhizosphere	Non-root soil
Bacteria on TSA	$1·5 \times 10^8$	$5·7 \times 10^7$
% Gram-negative	45	31
Actinomycetes on TSA	$3·8 \times 10^6$	$9·2 \times 10^5$
Bacteria on *Pseudomonas* medium	$2·0 \times 10^7$	$1·6 \times 10^6$
% *Pseudomonas* compared to "total" on TSA	13·3	2·8
Filamentous fungi on MRB	$1·8 \times 10^6$	$2·6 \times 10^5$
Yeasts on MRB	$4·2 \times 10^4$	$3·9 \times 10^4$

The experiments outlined above may be modified by using different plants (comparing the effects of different species or mixtures; Christie *et al.*, 1978) or by using different culture media (Newman *et al.*, 1977) or incubation conditions for the isolation of other organisms. For example, the dilution series can be pasteurized and then plated again to get spore-forming bacteria (Newman *et al.*, 1977) or anaerobic or high temperature incubation conditions can be used. In the direct examination there are also specific stains for vesicular arbuscular mycorrhizas (Christie *et al.*, 1978).

The *Pseudomonas* medium is highly selective and generally the number of colonies is about 10–15 % of the "total" count on TSA. Occasionally some fungi grow on this medium. The TSA may have yeasts as well as bacteria and it is worth checking on a sample of colonies to determine the percentage of yeasts. MRB medium may have yeasts, filamentous fungi and some streptomycin resistant bacteria. There are more organisms in the rhizosphere soil, often an order of magnitude more, than in the non-root soil.

Direct counts give similar differences between rhizosphere and non-rhizosphere samples and the largest numbers of all occur on the root surface.

REFERENCES

G. D. Bowen and A. D. Rovira (1976). Microbial colonization of plant roots. *Annual Review of Phytopathology*, **14**, 121–44.
M. E. Brown (1975). Rhizosphere micro-organisms—opportunists, bandits or benefactors. *In* "Soil Microbiology", (N. Walker, ed.), pp. 21–38. Butterworth, London and Boston.
P. Christie, E. I. Newman and R. Campbell (1978). The influence of neighbouring grassland plants on each others' endomycorrhizas and root surface organisms. *Soil Biology and Biochemistry*, **10**, 521–7.
E. I. Newman, R. Campbell and A. D. Rovira (1977). Experimental alteration of soil microbial populations for studying effects on higher plant interactions. *New Phytologist*, **79**, 107–18.
A. D. Rovira, E. I. Newman, H. J. Bowen and R. Campbell (1974). Quantitative assessment of the rhizoplane microflora by direct microscopy. *Soil Biology and Biochemistry*, **6**, 211–16.
A. Simon and E. H. Ridge (1974). The use of ampicillin in a simplified selective medium for the isolation of fluorescent Pseudomonads. *Journal of Applied Bacteriology*, **37**, 459–60.

Nodulation of Legumes by Rhizobium

M. J. DILWORTH and A. R. GLENN

School of Environmental and Life Sciences, Murdoch University, Murdoch, Western Australia

> *Level*: All undergraduate years
> *Subject areas*: Plant-bacterial interactions, agricultural microbiology
> *Special features*: Effect of microbial inoculation on growth of leguminous plants

INTRODUCTION

The N_2-fixing root nodules of legumes are caused by invasion of the root cells by species of *Rhizobium*. The ability of the nodulated legume to fix atmospheric nitrogen is of immense agricultural importance, both to plant production by legumes and to soil nitrogen increases for exploitation by non-leguminous crops.

Legume species are usually nodulated by particular groups of *Rhizobium* although exceptions can be found. The definitive identification of a presumptive *Rhizobium* involves a plant nodulation test (Vincent, 1970).

This experiment illustrates the isolation of strains of *Rhizobium* by successful nodulation of an appropriate host plant. Since some students inevitably isolate something other than *Rhizobium*, the experiment can also serve as a simple model for the demostration of Koch's postulates without using animal or plant pathogens.

EXPERIMENTAL

Although only short periods of time are actually needed on any one day, the experiment needs to run over a total period of about 7 weeks. In the United Kingdon it may be convenient to start the isolation before Easter so that plant nodulation and growth may occur during the Easter recess.

Day 1 *(15–20 min)*

Shake excess soil from the root system of a young (6 weeks) lucerne (*Medicago sativa*) or clover (*Trifolium* spp.) plant (*see* note 1).

↓

Remove a firm, pink-coloured nodule (*see* note 2).

↓

Surface sterilize the nodule by immersion in 0·1 % (w/v) $HgCl_2$:
    1 min for small (1–5 mm long) nodules;
    2 min for large (> 5 mm) nodules.

↓

Wash four times in 4 separate batches of sterile distilled water, making transfers with alcohol-flamed forceps.

↓

After the final wash, transfer the nodule to a drop of sterile water in a sterile Petri dish and crush with alcohol-flamed forceps.

↓

Streak out the milky exudate onto a YP agar plate (*see* note 3). Incubate the plate for 4–5 days at 28 °C.

- -

Day 2 *(5 min)*

Examine YP plates. Subculture one or more isolated colonies (presumptive rhizobia) onto fresh YP plates (*see* note 4) and incubate for 3 days at 28 °C (*see* note 5). The usual appearance of *R. meliloti* on plates is white, circular, convex colonies, 2–5 mm diam. *R. trifolii* colonies have a similar appearance, but are often very gummy.

- -

Day 3 *(5 min)*

Subculture an isolated colony onto a YP agar slope. Incubate for 3 days at 28 °C.

- -

Day 4 *(15–20 min)*

Label 4 plugged, sand tubes 1 to 4. Add sterile water to each to give field capacity (*see* note 6). All other mineral nutrients except nitrogen have already been added to tubes 1 to 3; tube 4 also contains 100 mg ammonium nitrate.

↓

To the slope culture of the *Rhizobium* isolate add 2 cm^3 sterile 0·1 % sucrose, and shake gently to throughly suspend the bacterial growth. Do the same for a slope of authentic *R. meliloti* or *R. trifolii*.

↓

Sterilize a glass rod with alcohol, and make 2 holes about 1 cm deep in each tube of sand.

↓

With **cool** alcohol-sterilized forceps, transfer two aseptically germinated lucerne (or clover) seedlings to each tube (*see* note 7).

↓

Inoculate the roots of seedlings in tube 1 with 2 drops of the putative *Rhizobium* isolate; plants in tube 2 should be inoculated with 2 drops of an authentic culture of *R. meliloti* or *R. trifolii* (depending on the host plant being used). Tube 3 is a control and students should be asked what should be added to that tube. (Demostrators should work them round to using the 0·1 % sucrose solution.) Two drops of sterile 0·1 % sucrose solution should also be added to tube 4.

↓

Using a sterile glass rod cover the inoculated seedlings with sand. Place in racks in a glasshouse at about 20 °C for about 5 weeks (*see* notes 8 and 9).

Day 5 *(10 min)*

Carefully wash plants free of sand (*see* note 10).

↓

Record qualitative results on size/colour of plants and presence or absence of nodules.

Notes and Points to Watch

1. Lucerne (*Medicago sativa*) is nodulated by strains of *R. meliloti* and clovers (*Trifolium* spp.) by strains of *R. trifolii*. If rhizobia have been isolated from wild plants, it is essential to know which plant and use seedlings of that species. Both *R. meliloti* and *R. trifolii* are fast-growing on laboratory media, and will nodulate the small-seeded hosts in tubes.

   Slow-growing rhizobia typical of lupins, cowpeas, and soybeans are much less convenient, taking 10 days for colony formation.
2. Simple, round nodules are more efficiently surface-sterilized than lobed nodules, which often trap air bubbles. Some areas of the nodule surface are not then in contact with the $HgCl_2$ solution and contamination becomes more serious. Soft, green, senescent nodules must be avoided, as they will mostly yield contaminants.
3. YP medium can be replaced by a modified potato-dextrose agar.
4. Students should be left to make their own colony selection. Some will almost certainly isolate species of *Erwinia* or *Klebsiella* which are common intercellular contaminants in legume nodules. Any type of coloured colony is **not** *Rhizobium*. Isolates which do not yield nodules are

useful for making the point that testing of isolates by nodulation is vital for authentication of *Rhizobium* cultures.

5. *Rhizobium* cultures can be refrigerated for a few days between practical classes without loss of viability.

6. The water content in the sand (to give field capacity) needs to be determined. This can be done simply by finding the percentage of water remaining in a sample of sand (weighed after oven-drying at 105 °C) which has been allowed to drain overnight on a filter funnel after addition of excess water. Too much water in the sand tubes restricts plant growth.

7. The appropriate legume seed is surface-sterilized with 0·1 % $HgCl_2$ as described for nodules, and then germinated in Petri dishes at 18 °C on sterile filter pads moistened with sterile water. Radicles should not be more than 1–2 mm long (4–5 days) or they are very difficult to plant—germination can be arrested if necessary by refrigerating at the correct stage.

8. Adequate light is vital for the experiment to work. The experiment should **not** be attempted in mid-winter in the UK without supplementary lighting.

9. Tube 4 shows what normal plant growth should be with adequate nitrogen fertilizer.

10. Laboratory sink traps do not work well once filled with sand. Some trapping arrangement ahead of the sink trap is necessary.

11. Yellow sand is required. Beach sand, acid-washed sand, etc. are not suitable since they lack iron and other trace elements.

MATERIALS

1. Cultures. An authentic strain of *R. meliloti* or *R. trifolii* suitable for the test legume being used (*see* note 1). Authentic strains can be obtained from the Department of Microbiology, Rothamsted Experimental Station, Harpenden, Herts, England.

2. Sand tubes. 4 × 30 mm internal diam. × 200 mm Pyrex tubes per student, each containing 60 mm yellow sand. A tube cut off to this length can be used to dispense the sand quickly.
   To each tube add: 0·4 ml basal nutrients, and 0·12 ml dilute phosphate solution.
   To tube 4, add 0·1 g NH_4NO_3.

3. Basal nutrients, prepared as follows.
   200 g K_2SO_4 are dissolved in 2500 cm^3 distilled water.
   14 g $ZnSO_4.7H_2O$ are dissolved in 500 cm^3 distilled water.
   7 g $CuSO_4.5H_2O$ are dissolved in 500 cm^3 distilled water.
   25 mg $Na_2MoO_4.2H_2O$ are dissolved in 500 cm^3 distilled water.
   The 4 solutions are mixed and diluted to 5000 cm^3.

4. Phosphate stock solution, prepared as follows.

Dissolve 81 g $Ca(H_2PO_4)_2.2H_2O$, 243 cm^3 conc. phosphoric acid, and 100 g $NaH_2PO_4.2H_2O$ (or 216 g $NaH_2PO_4.12H_2O$) in distilled water, neutralize to pH 6·0 and dilute to 1000 cm^3.

5. Dilute phosphate solution, prepared by diluting the phosphate stock solution to one-quarter the above concentration.

6. YP medium containing per litre: sucrose, 5 g; mannitol, 5 g; K_2HPO_4, 0·3 g; $MgSO_4.7H_2O$, 0·1 g; NH_4Cl, 0·05 g; $CaCO_3$, 0·1 g; yeast extract, 2·5 g; urea, 0·03 g; NaCl, 0·1 g; $MnSO_4.4H_2O$, 0·01 g; potato broth, 200 ml. The pH is adjusted to 7·0 with NaOH, 20 g agar are added, and the medium sterilized by autoclaving for 45 min at 109 °C.
   Potato broth is made by autoclaving for 15 min at 109 °C 200 g peeled and cubed potato in 200 ml distilled water and filtering through cotton-wool and gauze.
   Commercial potato dextrose agar medium, modified to include 2·5 g yeast extract per 1000 cm^3 and with the pH adjusted to 7·0, can be used instead of YP medium.

SPECIFIC REQUIREMENTS

Day 1
nodulated lucerne or clover plants (preferably sown 6 weeks earlier in light-textured soil)
forceps
alcohol
0.1% $HgCl_2$ (2–3 cm^3 per student)
4×5 cm^3 sterile water (in 30 cm^3 wide-mouthed McCartney bottles)
sterile Petri dish
1 YP plate (or potato dextrose agar, *see* note 3)

Day 2
1 or 2 YP (or potato dextrose agar) plates

Day 3
1 or 2 YP (or potato dextrose agar) agar slopes

Day 4
slope of authentic *R. meliloti* or *R. trifolii*
10 cm^3 0.1% sucrose (sterile)

sterile water for sand tubes (*see* note 6)
$4 \times$ sterile sand tubes, one containing NH_4NO_3
$4 \times$ sterile Pasteur pipettes
$2 \times$ sterile 1-cm^3 pipettes (for addition of fertilizers)
$5 \times$ sterile 10-cm^3 pipettes (for addition of sterile water or sucrose)
glass rod
alcohol
tube racks
aseptically germinated lucerne or clover seedlings (*see* note 7)

Day 5
facilities for trapping sand after washing out

FURTHER INFORMATION

No nodules should be present on the plants in tubes 3 and 4 since neither were inoculated. The plants in tube 3 should be weak, by comparison with those in tube 4, since they have been grown in the absence of fixed nitrogen. Plants in tube 2 should have grown as well as those in tube 4 and the roots should possess pink-coloured nodules. If the plants in tube 1 exhibit healthy growth, and are nodulated, then the bacterial isolate used for the inoculation is a genuine *Rhizobium* sp. Lack of nodules in tube 1 but their presence in tube 2 indicates that the bacterial isolate used was not *Rhizobium*. However, it should be noted that *R. meliloti* will not nodulate clover plants and *R. trifolii* will not nodulate lucerne, i.e. negative results will be obtained if the wrong *Rhizobium*/legume combination is used.

If rigorous demonstration of Koch's postulates is required, the nodules formed in tube 2 can be removed and *Rhizobium* re-isolated as before.

The effectiveness of the nodules produced can be assessed using the acetylene reduction technique. The cotton wool plugs are replaced with rubber serum caps of appropriate size and 5–10% of the air removed and replaced with acetylene (welding grade is satisfactory) by syringe. Half cubic centimetre samples are taken at 10, 20 and 30 min in disposable plastic syringes, and held by jabbing them into a rubber bung.

They can then be analysed by gas chromatography on a column (1·2 m $\times$ 2·5 mm i.d.) of Porapak T run isothermally at 100 °C with N_2 as carrier, and detected by flame ionization. For further information, *see* Trinick *et al.* (1976).

REFERENCES

M. J. Trinick, M. J. Dilworth and M. Grounds (1976). Factors affecting the reduction of acetylene by root nodules of *Lupinus* spp. *New Phytology*, **77**, 359–70.

J. M. Vincent (1970). A Manual for the Practical Study of Root Nodule Bacteria". IBP Handbook No. 15, Blackwell Scientific Publications, Oxford.

A Colorimetric Method for Measuring the Ethylene Produced in the Acetylene-reduction Assay for Nitrogenase

S. B. PRIMROSE

Department of Biological Sciences, University of Warwick, Coventry, CV4 7AL, England

---

Level: Advanced undergraduates
Subject areas: Microbial physiology, agricultural microbiology
Special features: Alternative to gas chromatography

INTRODUCTION

The acetylene-reduction test for nitrogenase, the enzyme complex responsible for biological nitrogen fixation, has become widely used in laboratories all over the world because of its convenience and sensitivity. It is based on the observations of Dilworth (1966) and Schollhorn and Burris (1967) that preparations of nitrogenase specifically reduce acetylene to ethylene. The usual method of measuring ethylene in the presence of acetylene is gas chromatography. An alternative method, in which the ethylene is measured by oxidizing it to formaldehyde which is determined colorimetrically, has been described by La Rue and Kurz (1973). This latter procedure is described below. Although it is not as rapid, sensitive or convenient as the gas chromatographic assay, it has the advantage of economy and should find use in laboratories which do not have gas chromatographs.

EXPERIMENTAL

The protocol described below assumes that two samples, one without fixed nitrogen (− N) and one with fixed nitrogen (+ N) have already been incubated with acetylene. Specific experimental details

---

of how these samples are produced are not given but an example is to be found in Experiments 99 and 100 (*see* note 6). A detailed discussion of the acetylene reduction test and the uses to which it can be put, is given by Postgate (1972).

Day 1 *(2 h)*

Label a series of ten 25 ml conical flasks (*see* note 1) from 1–10. Add 1·5 ml oxidant solution to each and cap with a rubber serum cap (*see* note 2).

$\downarrow$

Prepare an ethylene "standard" by diluting a sample of ethylene gas (*see* note 3) one-hundred-fold with air. This is done by capping a litre flask or bottle (*see* note 4) with a rubber serum cap and injecting 10 ml of ethylene gas.

$\downarrow$

Inject 2, 4, 6, 8, and 10 ml of diluted ethylene into flasks 1–5 respectively, after removing an equal volume of air.

$\downarrow$

Remove 1 ml of air from flasks 6 and 7 and inject 1 ml of gas from the ($-$N) and ($+$N) samples (*see* note 5).

$\downarrow$

Remove 1 ml of air from flasks 8 and 9 and inject 10 ml of gas from the ($-$N) and ($+$N) samples. Leave flask 10 as a control.

$\downarrow$

Incubate the flasks at room temperature for 1 h, shaking them vigorously whenever possible.

$\downarrow$

At the end of 1 h, remove the serum caps (*see* note 2) and add to each flask in order: 0·25 ml 4 M sodium arsenite (**CAUTION!**), 0·25 ml 4 N sulphuric acid (**CAUTION!**) and 1 ml Nash reagent.

$\downarrow$

Leave the flasks at room temperature, preferably for 1 h, until the desired green colour develops in the standards. Measure the absorbance of each sample at 412 nm using the contents of flask 10 (control flask) as a blank.

Notes and Points to Watch

1. Flasks with ground glass necks are best for they allow the rubber serum cap to grip tightly.
2. Only new serum caps should be used and all serum caps should be discarded at the end of the experiment. The reason for this is that all natural, synthetic and silicone rubbers absorb ethylene from atmospheres containing it and release it into atmospheres free of it. If a serum cap has been used in a test which gave a strongly positive result, it will have absorbed some of the ethylene and, if next used with a negative test, ethylene will diffuse out of the closure and stimulate very convincingly the formation of ethylene by nitrogenase.

3. Ethylene can either be prepared in the laboratory, or purchased in the form of a "lecture bottle". A convenient way of handling it is to fill a football bladder with the gas from a lecture bottle. Samples of ethylene can then be withdrawn with a syringe from the neck of the bladder.

4. Ideally, the volume of the flask or bottle should be exactly 1 litre. A suitable method is to fill a 1 litre flask or bottle with exactly 1 litre of water at room temperature and then fill the flask to the brim by the addition of gravel or glass beads. After decanting the water and allowing the vessel to dry, the litre "standard" flask or bottle is ready for use.

5. If they contain grease or are damp on their internal surfaces, syringes will retain substantial quantities of ethylene. To avoid carry-over of ethylene it is best to use separate syringes for the standards, the $(-N)$ samples and the $(+N)$ samples. In addition, each syringe should be pumped out two or three times after each sample has been taken.

6. A suitable alternative is to measure the acetylene-reducing ability of legume root nodules. Remove the entire root system from a nodulated clover, pea or bean plant and wash off residual soil under running tap-water. Place the entire root system in a 25 ml screw-cap bottle (e.g. a McCartney bottle) containing 0·5 ml H_2O. Cap the bottle with a rubber serum cap and then with a syringe withdraw 2·5 ml of air. Inject 2·5 ml acetylene (from a cylinder). Allow 5 min for the gases to equilibrate and then withdraw a 1 ml gas sample for ethylene determination. Remove a second 1 ml sample after a further 30 or 60 min.

Some plants, e.g. lupins, form very large nodules and a single such nodule can be used instead of the entire root system as used for the smaller legumes. Note that only pink-coloured (nitrogen-fixing) nodules should be used; white-coloured nodules may contain ineffective rhizobia.

MATERIALS

1. Rubber serum caps can be bought in bulk from the manufacturers (Wm. Freeman and Co. Ltd., Staincross, Barnsley, Yorkshire, England) who sell them under the trade name "Suba Seal". They can be bought in smaller quantities from Gallenkamp Ltd., Frederick Street, Birmingham, England.

2. Oxidant solution should be prepared freshly by mixing 10 ml of 0·005 M $KMnO_4$ with 0·05 M $NaIO_4$, adjusting the pH to 7·5 with KOH and diluting to 100 ml with distilled water. The separate solutions of $KMnO_4$ and $NaIO_4$ should be kept in dark bottles for no more than 1 week.

3. Nash reagent contains, per litre of distilled water, 150 g ammonium acetate, 3 ml glacial acetic acid and 2 ml acetyl acetone (2, 5-pentanedione).

SPECIFIC REQUIREMENTS

These items are needed for each group of students, except where indicated.

10 × 25 conical flasks, preferably with ground glass necks (*see* note 1)
10 rubber serum caps to fit conical flasks
1 bladder of ethylene (per class)
1 × 1 litre bottle with rubber serum cap (per class) (*see* note 4)
4 × 10 ml disposable plastic syringes

2 × 1 ml disposable plastic syringes
15 ml oxidant solution
2·5 ml 4 M sodium arsenite (**caution**)
2·5 ml 4 N sulphuric acid (**caution**)
10 ml Nash reagent
2 × 1 ml pipettes (non-sterile) with pipetting aid
2 × 5 ml pipettes (non-sterile) with pipetting aid
spectrophotometer
2 cuvettes (glass or quartz) for spectrophotometer

FURTHER INFORMATION

The chemistry of the reactions involved in this colorimetric assay is described by La Rue and Kurz (1973), who should be consulted if details are required. The absorbance of the final solution is proportional to ethylene from 0·1 to 1 μmol per sample. This sensitivity is sufficient to measure nitrogenase activity over most of a legume's (e.g. pea) growth, or excised roots of a nodulated legume, e.g. clover, or a liquid culture of free-living nitrogen-fixing bacteria. However, it would not be sensitive enough to measure fixation by algae in lake water or bacteria in unamended soils nor ethylene production by plants.

REFERENCES

M. J. Dilworth (1966). Acetylene-reduction by nitrogen-fixing preparations from *Clostridium pasteurianum*. *Biochimica Biophysica Acta*, **127**, 285–94.

T. A. La Rue and W. G. W. Kurz (1973). Estimation of nitrogenase using a colorimetric determination for ethylene. *Plant Physiology*, **51**, 1074–5.

J. R. Postgate (1971). The acetylene reduction test for nitrogen fixation. *In* "Methods in Microbiology", (J. R. Norris and D. W. Ribbons, eds), Vol. 6B. Academic Press, London and New York.

R. Schollhorn and R. H. Burris (1967). Reduction of azide by the N_2-fixing enzyme system. *Proceedings of the National Academy of Sciences, USA*, **57**, 1317–23.

Section Nine
Microbial Genes

101

Transformation in Acinetobacter

M. DAY

*Department of Applied Biology, UWIST,
King Edward VII Avenue, Cardiff, Wales*

> *Level*: All undergraduate years
> *Subject area*: Bacterial genetics
> *Special features*: Extremely simple method which
> can be adapted to meet the
> requirements of advanced
> students

INTRODUCTION

Transduction and conjugation are more commonly used than transformation for practical demonstrations of gene transfer in bacteria. Transformation has been relatively little used in the teaching laboratory, despite the fact that it was the first of the three processes of gene transfer to be described (Griffith, 1928). This lack of exploitation is due chiefly to the requirement in most systems, e.g. *Bacillus, Haemophilus* and *Staphylococcus*, for both a purified DNA preparation and for recipient bacteria in a state of competence. However, a simple transformation procedure with crude cell lysates was described by Juni and Janik (1969), and was developed by Juni (1972) as a tool for the classification of *Acinetobacter*. In addition, Janik *et al.* (1976) have now proposed transformation as a diagnostic test for suspected *Neisseria gonorrhoeae* from clinical specimens. The method also has been successfully used for transformation in *Azotobacter* (Page and Sadoff, 1976a, b) which shows parallels to transformation in *Acinetobacter*. The reasons for such a marked difference in facility, in comparison to the *Bacillus*-like systems, are not understood.

The basic experiment, to demonstrate transformation simply and easily, is described in part 2. The complexity and the value of this experiment can be increased by permitting the student to isolate his own auxotrophs as described in part 1 and then by transformation to assign these into linkage groups.

EXPERIMENTAL

Part 1

There are numerous methods for the induction of mutants in bacteria and the one described is most suited to class work (*see* note 1). It involves the isolation of histidine auxotrophs but can easily be adapted to isolate other mutant types.

Day 1 *(25 min)*

Take 10 ml culture of strain P1.

↓

Centrifuge to pellet cells. Resuspend in the same volume of sterile distilled water.

↓

Aseptically remove 1 ml to a sterile test tube and add 1 ml sodium nitrite and 1 ml acetate buffer. Incubate for 15 min at room temperature.

↓

Aseptically remove 0·1 ml and inoculate 10 ml succinate/ammonium medium supplemented with 50 μg ml^{-1} histidine. Incubate overnight at 37° C (*see* note 2).

- -

Day 2 *(4 h 15 min; with two 2-h gaps)*

Culture of mutagenized cells (from Day 1)

↓

Centrifuge to pellet cells. Resuspend in 10 ml succinate/ammonium minimal medium. Incubate for 2 h (*see* note 3).

↓

Add penicillin to final concentration of 100 μg ml^{-1}. Incubate for 2 h.

↓

Centrifuge and resuspend cell pellet in 10 ml sterile distilled water. Plate out, in duplicate, 0·1 ml from this undiluted suspension, and 0·1 ml of 10^{-1} and 10^{-2} dilutions, onto succinate/ammonium plates supplemented with histidine. Incubate at 37 °C.

- -

Day 3 *(10 min)*

Inspect the plates. Select those with 50–100 colonies and replicate onto succinate/ammonium plates. Mark the master and replica plates to ensure correct orientation. Alternatively, using sterile toothpicks transfer colonies onto succinate/ammonium plates, with and without histidine (*see* note 4). Incubate at 37 °C.

- -

Day 4 *(5 min)*

Compare master and replica plates. Note the colonies absent from the minimal plates and aseptically transfer these putative auxotrophs to a nutrient agar plate and a fresh succinate/ammonium plate (*see* note 5). Incubate at 37°C.

Part 2

It is best if each student can use several auxotrophs (6–10) for linkage-group analysis. If necessary, individual students should be encouraged to exchange mutants. However, the protocol outlined below only covers the use of 3 auxotrophs supplied by the instructor.

Day 1 *(75 min with an interval of about 45 min)*

Nutrient agar plate cultures of the wild-type strain (P1) and three auxotrophs, e.g. A1, A2 and A3 (*see* note 6).

↓

Suspend a loopful of cell paste, from each culture in 0·5 ml sterile lysing solution (*see* note 7). Heat this suspension for 60 min at 60°C to produce a cell lysate.

↓

Mark out the underside of 5 succinate/ammonium plates into four equal sectors. Label three plates as shown in Figure 1. Suspend a loopful of A1 cells in 1 ml sterile distilled water. Spread 0·1 ml over the first plate. Repeat the process for A2 and A3 on the second and third plates. To individual sectors on the fourth, a control plate, add and spread suspensions of A1, A2, A3 and P1. Repeat on the fifth, a control plate, with cell lysates of the four cultures. Label all these plates as appropriate.

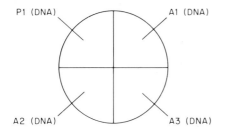

Figure 1   Minimal plate prespread with A1 cells and inoculated with cell lysates.

↓

Allow the suspensions to dry into the three experimental plates. Add a loopful of the four cell lysates to the appropriate sectors of the 1st three plates (*see* note 8). Incubate at 37°C.

Day 2 *(5 min)*

Record growth on experimental and control plates (*see* note 9).

Notes and Points to Watch

1. The choice of nitrous acid as the mutagen in this class experiment was deliberate because it is relatively safe and easy for students to use. Nitrous acid is only liberated when sodium nitrite is added to the acetate buffer (pH 4·6) and its mutagenic capacity is therefore restricted to the culture inside the container. It is inactivated on raising the pH of the medium. This and other mutagenic procedures are given by Miller (1972).

2. It is important to grow the mutagenized cells overnight since this allows segregation of the bacterial chromosomes. Plating the culture out directly will result in low auxotroph yields.

3. This interval prior to the addition of penicillin is essential to get the prototrophs actively growing. The auxotrophs will not be growing and hence will be insensitive to the action of the antibiotic. When this culture is resuspended in sterile distilled water most of those cells which have grown in the presence of penicillin will lyse.

4. It is important to ensure the correct orientation of master and replica plates when they are compared (Day 4). For both methods it is those colonies absent from the minimal plate which are required.

5. It is important to differentiate those colonies which are the result of replication errors, and are therefore presumably prototrophs, from auxotrophs. The putative auxotrophs are therefore checked for growth on succinate/ammonium minimal plates. Those that do not grow are auxotrophs and are stored on fresh medium containing histidine.

6. The experiment described here shows how to analyse three histidine auxotrophs (A1, A2 and A3) for linkage. Two of the three mutants show linkage.

7. This treatment results in a sterile crude DNA preparation. However, it is important to take care and ensure that **no** cells adhere to the wall of the test tube above the liquid. If they do it is likely that some may survive and contaminate the lysate.

8. It is critical to ensure that the student should confine the lysate or cell suspension to the appropriate sectors. This must be stressed since contamination from adjacent sectors is the most frequent cause of problems when interpreting the results.

9. Growth on any sector of the auxotroph control plate (plate 4) indicates the recipient bacterium has reverted or become contaminated. Any growth observed on the DNA control plate (plate 5) would signify that the cell lysate preparation was contaminated. It probably means that cells survived on the inside of the test tube and that these were removed as the lysate was sampled.

MATERIALS

1. *Acinetobacter calcoaceticus* BD413 (called P1 in this schedule) was isolated by E. Juni (1972). The wild type (P1) and three histidine auxotrophs (A1, A2 and A3) are streaked on nutrient agar plates.

2. Mutagenesis: 0·1 M sodium nitrite and 0·2 M acetate buffer; add 25 ml solution A (1·16 ml glacial acetic acid made up to 50 ml in water) to 24·5 ml solution B (2·73 g sodium acetate 3H$_2$O in 50 ml water). Check pH 4·6 and filter sterilize.

3. Minimal medium: dissolve in succession, KH$_2$PO$_4$ (3·89 g), K$_2$HPO$_4$ (12·50 g), MgSO$_4$ (0·19 g), sodium succinate (10 g) and ammonium sulphate (1 g) in distilled water (1000 ml) and autoclave at 103·4 kPa (15 lbf in^{-2}) for 30 min. Add Oxoid number 1 agar at 2 % w/v for plates. Add filter sterilized histidine hydrochloride solution to the sterile medium to give a final concentration of 50 μg ml^{-1}. Penicillin is prepared aseptically in sterile distilled water and added to sterile media to give a final concentration of 100 μg ml^{-1}.

4. Sterile lysing solution: dissolve 50 mg sodium lauryl sulphate in 100 ml standard saline citrate

(0·15 M NaCl and 0·051 M sodium citrate). Sterilize by filtration or by autoclaving.
5. Replica plating blocks and velvet pads can be obtained from Strand Scientific, 32 Bridge Street, Sandiacre, Nottingham, England.
6. Penicillin G can be obtained from the Sigma Chemical Co. Ltd. (Fancy Road, Poole, Dorset, England).

SPECIFIC REQUIREMENTS

Part 1

Day 1

10 ml overnight broth culture of *Acinetobacter* (P1) (the container should fit into a centrifuge)
10 ml sterile distilled water
1 ml sodium nitrite solution
1 ml acetate buffer
sterile test tube and rack
4×1 ml pipettes (sterile)
1×10 ml pipette (sterile)
10 ml succinate/ammonium medium supplemented with $50 \mu g\, ml^{-1}$ histidine; incubator (37 °C), clock, pipette-discard cylinder and centrifuge

Day 2

10 ml succinate/ammonium medium
penicillin solution (1 mg ml⁻¹) (freshly prepared, filter sterilized)
10 ml sterile distilled water
5×1 ml pipettes (sterile)
2×10 ml pipettes (sterile)
two test tubes, containing 9 ml sterile distilled water, in rack

6 succinate/ammonium plates supplemented with histidine
incubator at 37 °C

Day 3

2 sterile velvets and block
2 succinate/ammonium plates
incubator at 37 °C

Day 4

1 succinate/ammonium plate
1 nutrient agar plate
incubator at 37 °C

Part 2

Day 1

nutrient agar plate cultures of P1, A1, A2 and A3
0·5 ml sterile lysing solution in each of 4 tubes and 3 sterile test tubes containing 1 ml sterile distilled water in rack
water-bath or heating block at 60 °C
4 succinate/ammonium plates
4×1 ml pipettes
spreader and alcohol
incubator at 37 °C

FURTHER INFORMATION

The results from the transformation experiments are easy to interpret. A typical result is shown in Figure 2.

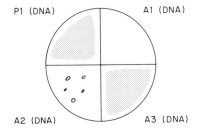

Figure 2 Growth of transformants on a plate pre-spread with A1 inoculated with the four cell lysates.

Growth of A1 auxotroph cells transformed to prototrophy, by donor DNA, will be seen for P1 (DNA) and A3 (DNA). The degree of growth in these sectors is related to the number of cells and amount of DNA used, and dilutions of the cell lysate, 10^{-1}, etc. will give single colonies (transformants). The relative lack of growth (transformant colonies) with A2 (DNA) indicates that the mutations in mutants A1 and A2 are close together (i.e. linked). The high number of transformants obtained with mutant A3 (DNA) indicates that it is unlinked. By collecting the data from various plates (Table 1) the student can assign mutants to linkage groups. The data confirm the linkage between mutations in A1 and A2.

Table 1 The linkage relationship between three histidine auxotrophs.

Recipient cells	Donor DNA				
	A1	A2	A3	P1	None
A1	0^a	5	∞	∞	0
A2	3	0	∞	∞	0
A3	∞	∞	0	∞	0
None	0	0	0	0	—

[a] An estimate of transformant numbers from none (0) to confluent growth (∞).

This simple and easy method can be used merely to demonstrate the process of transformation by the use of the prototroph and just one auxotroph. However, it is obvious that the complexity of this basic experiment can be increased to make it suitable for other levels of undergraduate work. The programme of work outlined here is one of a series of experiments carried out at UWIST designed to introduce students, through experiment, to the concept of mutant isolation, linkage groups and their subsequent analysis.

Control experiments can be included to support the thesis that DNA is the key component in transformation. These can be made by using solutions of DNA'ase, RNA'ase, lipase and protease. Parallel treatments with each of these enzymes on crude cell lysates before their use in transformation will reveal that only the DNA'ase has any effect on the process.

REFERENCES

F. Griffith (1928). Significance of pneumococcal types. *Journal of Hygiene, Cambridge*, **27**, 113–59.

A. Janik, E. Juni and G. A. Heym (1976). Genetic transformation as a tool for detection of *Neisseria gonorrhoeae*. *Journal of Clinical Microbiology*, **4**, (1), 71–8.

E. Juni and A. Janik (1969). Transformation in *Acinetobacter calco-aceticus*. *Journal of Bacteriology*, **98**, 281–8.

E. Juni (1972). Interspecies transformation of *Acinetobacter*. *Journal of Bacteriology*, **112**, 917–31.

J. H. Miller (1972). "Experiments in Molecular Genetics", 466 pp. Cold Spring Harbour Laboratory.

W. J. Page and H. L. Sadoff (1976a). Transformation of *A. vinelandi*. *Journal of Bacteriology*, **125**, (3), 1080–7.

W. J. Page and H. L. Sadoff (1976b). Transformation competence in *A. vinelandi*. *Journal of Bacteriology*, **125**, (3), 1088–95.

102

Rapid Detection of Mutagens in the Environment

S. B. PRIMROSE

Department of Biological Sciences, University of Warwick, Coventry, CV4 7AL, England

> *Level*: All undergraduate years
> *Subject areas*: Bacteria genetics, environmental sciences
> *Special features*: Topical: students screen compounds of their own choice

INTRODUCTION

It is widely recognized that many chemicals can induce cancer in animals, and the identification of potential carcinogens in the environment is an urgent requirement. However, this is no easy task, for more than 50 000 different man-made chemicals are currently in commercial and industrial use, and over 500 new chemicals are put on the market every year. The standard animal tests for carcinogenicity take two to three years and are extremely expensive, so the screening in this way of all chemicals is not feasible. Fortunately, a number of simpler, quicker and cheaper alternative systems using micro-organisms are now available. All of them rely on the correlation between mutagenicity and carcinogenicity, i.e. most chemicals which induce mutations also induce cancer. The most widely used system for screening for potential mutagens is the Ames test (Ames, 1971).

The Ames test makes use of a number of test strains of *Salmonella typhimurium*, each carrying a mutation (*his*) in histidine biosynthesis. As a result of the mutation, the test strains are unable to grow in mineral salts medium unless the medium is supplemented with histidine. Occasionally, the mutations in the test strains undergo reversion, i.e. a back-mutation restores the DNA's normal coding sequence for the needed enzyme and thereby restores the internal supply of histidine. The reversion can be scored because only the revertant bacteria form colonies on histidine-free media. Obviously, the spontaneous rate of reversion will be considerably enhanced if the tester strains are exposed to a chemical which induces mutations. This is the theoretical basis of the Ames test.

The strains chosen by Ames included one with a single base change, one with a base-pair addition and a third with a base-pair deletion. The first strain is used to detect chemicals causing base substitutions while the latter two strains are used to detect mutagens causing frame shifts. A fourth strain is used to detect chemicals causing deletions but the principle involved differs considerably from that described above. In practice, the strains used now in the Ames test carry a series of additional mutations which render the strains much more sensitive to mutagens, e.g. defects in lipopolysaccharide biosynthesis and in repair of UV-induced damage (Ames *et al.*, 1973a). In addition, the strains carry a particular plasmid which makes DNA replication more error-prone (McCann *et al.*, 1975b).

One disadvantage of the original Ames test was that some known carcinogens did not induce prototrophic revertants in the test strains. The reason for this being that some chemicals are not carcinogenic but are metabolized to carcinogens in the mammalian liver. Ames *et al.* (1973b) therefore modified the test by mixing the test bacteria with an extract of rat liver, thereby subjecting the chemical under test to mammalian metabolism. Using this system, McCann *et al.* (1975a) screened over 300 chemicals for mutagenicity and found 90% correlation between mutagenicity and carcinogenicity. The significance of these results is discussed in more detail by McCann and Ames (1976).

The aim of ths experiment is to demonstrate the Ames test, including the use of rat liver homogenate. A set of known mutagens is included in the test and the students are invited to test the mutagenicity of other compounds of their own choice.

The Ames test works well with undergraduate classes. However, the test strains are *Salmonella typhimurium* and with the advent of the Health and Safety at Work Act, many departments are reluctant to use these strains in undergraduate classes. For this reason we now use derivatives of *Escherichia coli* K12. The disadvantage of using *E. coli* is that the exact defects in the test strains are rarely known. However, this disadvantage can be offset by allowing the students to use in the test, auxotrophic mutants of *E. coli* which they have isolated themselves.

EXPERIMENTAL

Day 1 *(1 h)*

Take 10 ml overnight broth culture of *E. coli* auxotroph (*see* note 1).

↓

Wash once with normal saline and resuspend cells in 10 ml normal saline.

↓   ↓

Spread 0·1 ml portions of cell suspension on 5 mineral salts agar plates.

Add 0·1 ml cell suspension and 0·5 ml of liver homogenate to 5 tubes containing 2 ml mineral salts soft agar held at 45 °C.

↓

Mix tube contents immediately and pour over 5 mineral salts agar plates.

↓   ↓

Make no further additions to 1 plate of each type (negative controls). Add a small crystal of solid chemical or 10–20 μl of liquid chemical to the centre of the remaining plates. Two plates of each type should be seeded with known mutagens, e.g. nitrosoguanidine, 2 aminopurine. The remaining three plates of each type should be seeded with the chemicals to be tested (see notes 2 and 3.)

$$\downarrow$$

Incubate plates at 37 °C for 2 days.

Day 2 *(10 min)*

Examine all plates for the presence of revertants and, where possible, count the number on each plate.

Notes and Points to Watch

1. In this experiment it is possible to use any *E. coli* auxotroph which carries a single, revertible auxotrophic mutation. In practice, we get our students to isolate their own auxotrophs which they then use in this experiment. The method we use is described by Glover in Clowes and Hayes (1968), and every year 95 % of the class succeed in isolating auxotrophs.
2. If the entire class use the same auxotroph then there is no need for each student to set up negative and positive controls. In this way a considerable saving in media can be achieved.
3. Students should be free to choose the chemicals they wish to test. If given adequate warning many will bring in standard household items for checking. The list of compounds tested by McCann *et al.* (1975) should be available for consultation.

MATERIALS

1. Any single auxotroph of *Escherichia coli* (see note 1).
2. Mineral salts agar medium. Any of the available recipes can be used. A suitable one is M9 medium, prepared as follows. Dissolve 6 g anhydrous Na_2HPO_4, 3 g anhydrous KH_2PO_4, 0·5 g NaCl and 1 g NH_4Cl in 1 litre of water. Add agar (15 g for plates, 8 g for soft agar) and sterilize by autoclaving. Before dispensing, add 20 ml sterile 20 % glucose, 10 ml sterile 0·1 M $MgSO_4$ and 10 ml sterile 0·01 M $CaCl_2$.
3. Preparation of rat liver homogenate. All steps are performed at 0–4 °C with cold, sterile solutions and glassware. The rat liver is washed in an equal volume of 0·15 M KCl, minced with sterile scissors in three volumes of 0·15 M KCl and homogenized. The homogenate is centrifuged at 9000 g for 10 min, the supernate decanted and saved. The supernate is used to produce a "liver mix" which contains per ml, 0·3 ml homogenate, 8 mM $MgCl_2$, 3·3 mM KCl, 5 mM glucose-6-phosphate, 4 mM NADP and 100 mM sodium phosphate (pH 7·4). This mixture can be dispensed in 1 ml amounts, frozen in dry ice and stored at −80 °C. As required, sufficient mix is thawed at room temperature, kept on ice and the unused portion discarded at the end of the day.

SPECIFIC REQUIREMENTS

These items are required for each student or group of students.

Day 1

Overnight broth culture of *E. coli* auxotroph
10 plates of mineral salts agar
10–15 ml mineral salts soft agar
45 °C water-bath (can be shared by several groups of students)

25 ml normal saline
3 ml liver homogenate (stored on ice or at 4 °C)
glass spreader
5 tubes (for holding 2 ml portions of soft agar)
selection of mutagens
37 °C incubator (1 per class)

FURTHER INFORMATION

Despite the fact that no attempt is made in this experiment to optimize the sensitivity of the test strains to mutagens, good results are obtained. The expected results are as follows. On the negative control plates, any revertants which are present should be scattered randomly. On the plates seeded with known mutagens, revertants should be present. Close to the point where the mutagen was initally placed there should be a zone in which toxic concentration of the mutagen has killed all the test bacteria. As the distance from the mutagen increases, so the concentration decreases and a ring of revertants should have formed in the region where the concentration of mutagen is optimal for mutagenesis and survival. Whether or not revertants are present on the other plates will depend on the chemical selected.

A rational selection of test chemicals can give interesting results. For example our students have reported that paracetamol, surgical spirit (which contains methyl salicylate), wart remover, cigarette smoke (expirated into dimethyl sulphoxide) and the fungicide "Captan" (impregnated on filter paper discs after dissolution in $CHCl_3$) are all mutagenic. Many students, though, if given a free hand, make trivial selections. Others have good ideas, but fail to appreciate the basis of the test. For example, following recent reports in the press about the mutagenicity of certain proprietary brands of beef extract, some students have tried to evaluate similar products. Unfortunately, beef extract will provide sufficient amino acids to permit growth of the auxotroph over the entire plate!

REFERENCES

B. N. Ames (1971). *In* "Chemical Mutagens: Principles and Methods for their Detection", (A. Hollaender, ed.), Vol. 1, pp. 267–82. Plenum Press, New York.

B. N. Ames, F. D. Lee and W. E. Durston (1973a). An improved bacterial test system for the detection and classification of mutagens and carcinogens. *Proceedings of the National Academy of Sciences, USA*, **70**, 782–6.

B. N. Ames, W. E. Durston, E. Yamasaki and F. D. Lee (1973b). Carcinogens are mutagens: a simple test system combining liver homogenates for activation and bacteria for detection. *Proceedings of the National Academy of Sciences, USA*, **70**, 2281–5.

R. C. Clowes and W. Hayes (1968). "Experiments in Microbial Genetics". Blackwell Scientific Publications, Oxford.

J. McCann, E. Choi, E. Yamasaki and B. N. Ames (1975a). Detection of carcinogens as mutagens in the *Salmonella* microsome test: assay of 300 chemicals. *Proceedings of the National Academy of Sciences, USA*, **72**, 5135–9.

J. McCann, N. E. Spingarn, J. Kobori and B. N. Ames (1975b). Detection of carcinogens as mutagens: bacterial tester strains with R factor plasmids. *Proceedings of the National Academy of Sciences, USA*, **72**, 979–83.

J. McCann and B. N. Ames (1976). Detection of carcinogens as mutagens in the *Salmonella* 1 microsome test: assay of 300 chemicals: discussion. *Proceedings of the National Academy of Sciences, USA*, **73**, 950–4.

Genetics and Regulation of Amidase of *Pseudomonas aeruginosa*: Mutants with Altered Regulation and Mutants with Novel Substrate Specificity

P. H. CLARKE

*Department of Biochemistry, University College of London,
Gower Street, London, WC1E 6BT, England*

> *Level*: Advanced undergraduates
> *Subject areas*: Bacterial genetics, biochemistry
> *Special features*: Can be extended in scope

INTRODUCTION

The enzymes involved in the metabolic pathways of biodegradation are usually inducible. The synthesis of these enzymes requires the presence of the inducer and may also be affected by the presence of alternative growth substrates. The organism requires genes for the enzymes and permeability systems needed. The synthesis of these enzymes and of the permeability proteins must be switched on in response to the presence of the test compound in the environment. If the inducible enzymes convert the test compound into an intermediate of the central metabolic pathways, then that compound will be able to support growth. In the natural environment, some organisms are capable of carrying out the initial reactions of a biodegradative pathway and release intermediates which provide growth substrates for other organisms.

The experiments described here are based on the genetics and regulation of amidase of *Pseudomonas aeruginosa*. This enzyme allows the organism to utilize acetamide as sole source of carbon and nitrogen for growth. Propionamide can also be utilized but no other aliphatic amides. The amide growth phenotype is dependent on (a) the specificity of the amidase determined by the enzyme structural gene, *amiE*, (b) the specificity of the regulator protein determined by the regulator gene, *amiR*, (c) the general regulatory control of the synthesis of catabolic enzymes described as catabolite repression and (d) enzymes for the metabolism of the organic acids.

Mutants can be isolated with novel amide growth phenotypes which have mutations in *amiE* resulting in amidases with altered substrate specificities e.g. the B amidase (Brown and Clarke, 1970), mutations in *amiR* resulting in altered regulatory properties and also mutations conferring resistance to the catabolite repression normally exerted by succinate and related compounds.

The strains to be used in these experiments have the following characteristics:

PAC1	Wild type, inducible amidase	$amiR^+amiE^+$
PAC111 (C11)	Constitutive	$amiR11amiE^+$
PAC351 (B6)	Constitutive, produces B amidase	$amiR11amiE16$
PAC142 (L10)	Constitutive, resistant to catabolite repression	$amiR33amiE^+$ crp^-
PAC307 (Am7)	Acetamide-negative	$amiR^+amiE7$.

The enzyme reaction that allows growth of acetamide (or other amides) is as follows:

$$CH_3CONH_2 + H_2O \longrightarrow CH_3COOH + NH_3.$$

Acetate can be used as a carbon source and ammonia as a nitrogen source. The enzyme also catalyses a transferase reaction with hydroxylamine and this reaction is used for enzyme assays. It is sensitive, rapid and simple to carry out.

$$CH_3CONH_2 + NH_2OH \longrightarrow CH_3CONHOH + NH_3.$$

The acylhydroxamate forms a deep red-brown colour with ferric chloride.

The aims of this set of experiments are (1) to isolate mutants on amide selective media, (2) to compare the growth phenotypes of known strains, (3) to examine the synthesis of amidase by known strains and new isolates in the presence and absence of acetamide as inducer.

EXPERIMENTAL

Day 1 *(40 min)*

Experiment 1

Start mutant isolation: culture *P. aeruginosa* PAC1 in 3 Universals of nutrient broth, 5 ml in each, for 5 h.

Culture A, 5 ml

Pellet cells in bench centrifuge, resuspend in dilution buffer and dilute to 50 ml.

Culture B, 5 ml × 2

Pellet cells in bench centrifuge and resuspend in 2 × 4 ml citrate buffer 0·1 M pH 6·0. Add to each suspension 4 mg (approx.) *N*-methyl-*N*-nitro-*N*-nitrosoguanidine and shake to dissolve. **Take care. No sterilization needed and a few crystals at the end of a spatula will be sufficient. Do not heat in any way** (*see* note 2). Incubate for 45 min without shaking, centrifuge and resuspend cells in 2 × 5 ml dilution buffer.

Spread 0·1 ml of culture A on plate of succinate/formamide medium.	Spread 0·1 ml of culture B on plate of butyramide media.
↓	↓

Incubate plates for 2 days at 37° C (*see* note 4). If necessary, the plates can be incubated longer or they can be removed from the incubator when colonies have appeared and kept in a refrigerator until the next class.

Day 2 *(30 min)*

Examine mutant isolation plates. If mutant colonies have appeared pick off single well-separated colonies and streak on the same medium.

Culture A: S/F plate	*Culture B: B plate*
Pick off 8 colonies and streak 4 on each of 2 S/F plates to get single colonies.	Pick off 8 colonies and streak 4 on each of 2 B plates to get single colonies.
↓	↓

Incubate at 37 °C for 2–3 days.

Experiment 2

Begin experiment 2 by inoculating Universals, containing 5 ml nutrient broth, from slope cultures of *P. aeruginosa* strains PAC1, PAC111, PAC351, PAC142, PAC307. Incubate overnight on a mechanical shaker at 37° C.

Day 3 *(60 min)*

Continue experiment 2 by setting up patch plates to examine amide growth phenotypes of the cultures grown overnight.

↓

Centrifuge cultures. Resuspend cells in 5 ml dilution buffer.

↓

With inoculation loop spread a small patch of each suspension on each of the amide selection plates (S/F, S/L, B, A, S2). Incubate at 37 °C for 2–3 days.

Day 4 *(40 min)*

Complete experiment 2 by examining patch plates and tabulating growth results. Discuss the effects of structural gene and regulatory gene mutations on the amide growth phenotypes.

Experiment 3

Start experiment on amidase synthesis in minimal salt medium both in the presence and absence of acetamide as the inducer.

Control cultures
Inoculate 5 ml medium S3 and 5 ml medium AM with PAC1 and **one** of the following: PAC111, PAC351, PAC142, PAC307.

Mutant isolates (from experiment 1)
Inoculate 5 ml medium S3 and 5 ml medium AM with 2 of the new mutant isolates.

↓   ↓

Incubate cultures overnight on a shaker at 37 °C.

Day 5 *(2 h)*

Read the optical densities of the cultures at 670 nm to obtain the dry weight of bacteria, diluting if necessary. A value of 1·0 corresponds to 0·56 mg dry wt ml^{-1}. Centrifuge cultures to pellet cells. Resuspend cells in Tris buffer (0·1 M, pH 7·2).

↓

Assay amidase activity by the transferase reaction. For very active suspensions it will be necessary to dilute the suspensions. The optical density of the final enzyme assay mixture should be within the range of 0·2–0·9. **It is not possible to dilute the final solution and if the colour is too deep it will be necessary to repeat the assay to get accurate results.**

Amidase assay by the transferase reaction

Mixed substrates: **keep on ice** (*see* note 3).
Freshly prepare the mixed substrates by mixing 1 vol. 0·4 M acetamide, 1 vol. 2·0 M hydroxylamine hydrochloride (freshly diluted and neutralized, pH 7·2), and 2 vols 0·1 M Tris buffer, pH 7·2, 1 mM mercaptoethanol (or dithiothreitol). **Keep on ice** (*see* note 3).

Equilibrate tubes containing 0·9 ml mixed substrates in a water-bath at 37 °C. Add 0·1 ml bacterial suspension. Each assay should be carried out in duplicate and each set should contain a blank assay without enzyme added.

↓

After 10 min, add 2 ml ferric chloride reagent to stop the reaction. Shake vigorously and read the optical density at 500 nm using the mixed substrate without enzyme as blank (*see* note 3). (An optical density of 1·0 corresponds to 3·5 μmol acethydroxamate.)

↓

If no colour develops use a larger volume of the suspension and incubate for a longer period. If the colour is too deep dilute the bacterial suspension with Tris buffer.

Calculate the specific activities as μmoles acethydroxamate per min per mg dry wt of bacteria. Determine if any of the mutant isolates correspond to any of the known mutant strains supplied. Is it possible to identify the mutation that has occurred?

Notes and Points to Watch

1. No biological precautions are required other than the normal safeguards to be used with all experiments with bacteria.
2. It is suggested that the students should not handle nitrosoguanidine themselves unless they are very experienced. If a few crystals are added to the culture to be treated this will avoid the problem of making up solutions.
3. The mixed substrate solution for amidase assays will undergo spontaneous reaction if allowed to reach room temperature. The substrate blank will give a check on whether this has occurred to any serious extent.
4. It is probably advisable to put the mutant isolation plates in plastic bags to avoid drying out. The S/F plates allow much faster mutant selection than the B plates and, if possible, the latter should be incubated for a longer period.

MATERIALS

1. *Pseudomonas aeruginosa* strains PAC1, PAC111, PAC351, PAC142, PAC307 (from NCIB).
2. Nutrient broth (Oxoid No. 2), 5 ml bottles distributed in Universals.
3. Minimal salt agar and minimal salt medium (Brammar and Clarke, 1964).
   Experiment 1. Mutant isolation plates, S/F (succinate 1%, formamide 0.1%); B (butyramide 0.2%).
   Experiment 2. Amide selection patch plates, S/F (succinate 1%, formamide 0.1%), S/L (succinate 1%, lactamide 0.1%), B (butyramide 0.2%), A (acetamide 0.2%), S2 (succinate 1%, ammonium sulphate 0.1%).

Experiment 3. Minimal salt medium with additions. S3 (succinate 0.2%, ammonium sulphate 0.1%), AM (acetamide 0.2%).
4. Amides; acetamide, butyramide, formamide (BDH, Fisons); lactamide (Sigma).
5. Nitrosoguanidine (*N*-methyl-*N'*-nitro-*N*-nitrosoguanidine) (Sigma).
6. Hydroxylamine hydrochloride.
7. Ferric chloride reagent. Mix 100 ml commercial 60% ferric chloride solution (BDH Chemicals Ltd.) with 57 ml conc. HCl and dilute to 1 litre with distilled water.
8. Mercaptoethanol or DTT (Dithiothreitol) (Sigma).

SPECIFIC REQUIREMENTS

Day 1
5 h broth cultures of PAC1 in 5 ml nutrient broth in Universal bottle
sterile dilution buffer
bench centrifuge
1 plate S/F medium and one plate B medium (per student)
glass spreader and alcohol
nitrosoguanidine (**care**)

Day 2
2 plates S/F medium, 2 plates B medium (per student)
5 Universal bottles with 5 ml nutrient broth
shaker at 37°C

cont.

cont.

Day 3	***Day 5***
S/F, S/L, B, A, S media plates (one per student)	bench centrifuge
inoculation loops	Tris buffer 0·1 M, pH 7·2
bench centrifuge	0·4 M acetamide
sterile dilution buffer	2·0 M hydroxylamine hydrochloride freshly prepared from stock solution 5·0 M; **must be freshly neutralized, check pH**
Day 4	ferric chloride reagent
2 bottles S medium and 2 bottles AM medium, 5 ml in Universals (per student)	test tubes for assay
slope cultures PAC1, PAC111, PAC351, PAC142, PAC307	water-baths at 37° C
inoculation loops	spectrophotometer capable of reading in visible range

FURTHER INFORMATION

Most of the mutants isolated from S/F and B plates will be constitutive. It is possible, if a large number are obtained, that some of the S/F mutants will have altered inducer specificity and that some of the B mutants will be altered in substrate specificity.

These experiments can be used easily in various ways as the basis for student projects. For example:

1. The outline experiments can be adapted to the isolation and characterization of other classes of amidase mutants.
2. The simple enzyme characterization can be extended to study substrate specificities and enzyme kinetics with any amides that are readily available. The results from this set of experiments can be used as a basis for general discussion on the regulation of gene expression.
3. The methods can be applied to the isolation of new amidase-producers from soil or other sources.

REFERENCES

W. J. Brammar and P. H. Clarke (1964). Induction and repression of *Pseudomonas aeruginosa* amidase. *Journal of General Microbiology*, **37**, 307–19.

W. J. Brammar, P. H. Clarke and A. J. Skinner (1967). Biochemical and genetic studies with regulator mutants of the *Pseudomonas aeruginosa* 8602 amidase system. *Journal of General Microbiology*, **47**, 87–102.

J. E. Brown and P. H. Clarke (1970). Mutations in a regulator gene allowing *Pseudomonas aeruginosa* 8602 to grow on butyramide. *Journal of General Microbiology*, **64**, 329–42.

J. L. Betz, P. R. Brown, M. J. Smith and P. H. Clarke (1974). Evolution in action. *Nature*, **247**, 261–4.

104

Colicins

K. G. HARDY

*Biological Laboratory, University of Kent at Canterbury,
Canterbury, CT2 7NJ, England*

> *Level*: Advanced undergraduates
> *Subject area*: Bacterial genetics

INTRODUCTION

Colicins are antibacterial proteins produced by members of the *Enterobacteriaceae*. They are specified by genes on Col plasmids. About 40 % of *Escherichia coli* strains isolated from man or from animals have one or more types of Col plasmid.

Colicins have a narrow spectrum of action. They are effective only against certain members of the *Enterobacteriaceae* and most enterobacteria are sensitive to only a few types of colicin. Colicins and Col plasmids are classified into about 20 groups, A, B, C, etc. on the basis of their effects on a set of indicator strains. All colicins of a particular group are ineffective against a particular indicator strain. The two specific indicators used in this experiment are CL143, which is resistant to colicin I, and CL145, which is resistant to colicin E. When classifying colicins and Col plasmids, the strain which originally harboured the plasmid is also indicated. The two Col plasmids used here, ColE2-P9 and ColIb-P9, were both found in *Shigella sonnei* strain P9.

Colicins bind to proteins (receptors) in the outer membranes of sensitive bacteria. Colicins E2 and E3 bind to the same receptor but have different effects. Following adsorption to receptors, these colicins are transferred through the cell envelope into the cell. Once inside, colicin E2 acts as a DNase, whereas colicin E3 is an RNase which inactivates ribosomes by nicking 16S ribosomal RNA. The lethal effects of colicins E2 and E3 are unusual; most colicins are lethal because they make holes in the cytoplasmic membrane so that ions can pass in and out. Consequently, energy-dependent reactions in the cell are inhibited. Colicin Ib acts in this way.

Col plasmids specify immunity to the colicin they determine. For example, ColE3-CA38 specifies a protein (mol. wt 10000) which prevents colicin E3 from acting as an RNase. But the ColE3 immunity protein does not confer immunity against colicin E2, nor vice versa. Similarly, colicins Ia and Ib bind to the same outer-membrane receptor, but cells harbouring a ColIb plasmid are not immune to colicin Ia.

Some Col plasmids, such as ColIb-P9, are able to transfer copies of themselves from one bacterium to another by conjugation. The conjugative plasmid used here, ColIb-P9*drd*, is transferred at a high frequency because it is a de-repressed mutant. The genes of ColIb-P9*drd* which are necessary for conjugation are not repressed, but are apparently expressed continually so that all cells containing the plasmid are able to transfer a copy of it by conjugation. ColIb-P9 has a molecular weight of 61×10^6 and is maintained in *E. coli* at 1 or 2 copies per chromosome. ColE2-P9 and ColE3-CA38 are non-conjugative plasmids (mol.wt 5×10^6) and are maintained at about 15 copies per bacterial chromosome.

Although incapable of transferring copies of themselves by conjugation, ColE2 and ColE3 can be transferred if they are **mobilized** by conjugative plasmids. The mechanism of plasmid mobilization is unknown, but it can be a very efficient process. When ColE2 is mobilized by ColIb-P9, it is transferred at about the same rate as the conjugative plasmid (Smith *et al.*, 1963). In this experiment, ColIb-P9*drd* is used to mobilize ColE2-P9 from one strain of *E. coli* K-12 to another and colonies of colicinogenic recipients are detected by overlaying them with colicin-sensitive bacteria. Col$^+$ recipients are then tested to find out which types of colicin they are producing.

EXPERIMENTAL

The Col plasmids are transferred by incubating donor and recipient strains together in broth for 20 min. A colicin-resistant recipient is used so that it is not inhibited by the colicin released from the donor. The recipient strain is also resistant to streptomycin. Appropriate dilutions of the mixed culture are spread over the surfaces of plates containing streptomycin. The plates are covered with two layers of agar so that colonies of the recipient strain are formed underneath the surface. Col$^+$ recipient colonies are detected by overlaying the plates with agar containing a colicin-sensitive strain. After incubation, circular inhibition zones are seen around colicinogenic colonies. These are picked and tested using colicin-indicator strains (CL143 and CL145) to determine whether they are producing colicin I, colicin E2 or both types of colicin.

Day 1 *(10 min + incubation period of about 2h + 45 min)*

Dilute an overnight (unshaken) broth culture of donor strain M654 by adding 0·2 ml to 5 ml of broth in a 50 ml conical flask. Do the same with the recipient, AB1157/I/E. Incubate both flasks with shaking at 37°C until the cultures are at about 2×10^8 cells ml^{-1} (about 2 h). (The donor should not be shaken vigorously during incubation as this impairs its ability to donate plasmids.)

↓

Remove a loopful of each culture and streak separately onto a nutrient agar plate containing streptomycin at a concentration of 200 μg ml^{-1}. This is to confirm that the donor is streptomycin-sensitive and the recipient streptomycin-resistant. Also put a drop of donor and recipient separately onto two plates of nutrient agar. These two plates will be chloroformed on Day 4 and tested to confirm that the donor is Col$^+$ and the recipient Col$^-$. Add the 5 ml of recipient culture to the 5 ml of donor and incubate with gentle shaking for 20 min.

↓

Dilute the mixture 10-fold by adding 0·5 ml to 4·5 ml of diluent in a screw-capped bottle. Shake vigorously for 30 s to separate mating bacteria. Immediately make a further 10^5-fold dilution. For each step, add 0·5 ml of sample to 4·5 ml of diluent. Add 0·2 ml of the 10^{-6} dilution and 0·2 ml of the 10^{-5} dilution to each of two nutrient agar plates containing streptomycin (four plates used in all). Before each 0·2 ml dries, add 4 ml of soft nutrient agar to the plate and make sure the 0·2 ml is distributed in the agar layer over the surface. After about 5 min, add another 4 ml layer of soft agar over the surface. Incubate the plates for 18–24 h at 37 °C.

_ _

Day 2 *(15 min)*

Add 0·2 ml of an overnight (unshaken) broth culture of CL142 to 20 ml of molten soft agar held at 47 °C. Add 4 ml of inoculated soft agar onto each of the four plates prepared on Day 1. Incubate plates for 18–24 h at 37 °C.

_ _

Day 3 *(15 min)*

Using a straight wire, pick 20 random colicinogenic colonies (irrespective of the size of the surrounding inhibition zone) and point-inoculate two streptomycin plates with each colony. On each plate, make no more than 7 well-spaced inoculations. Label the plates so that duplicates can be recognized. As controls, also inoculate two plates with the recipient strain (from the nutrient agar plates inoculated on Day 1), a strain producing only colicin Ib (strain KH188) and a strain producing only colicin E2 (strain KH293). Incubate plates for 18–24 h at 37 °C.

_ _

Day 4 *(5 min + 60 min for chloroforming colonies + 15 min)*

The colonies to be tested for colicin production are first killed with chloroform. For each plate, add 5 ml of chloroform to a clock glass (in a fume cupboard) and invert the plate (lid removed) over the chloroform for 30 min. Then remove the plate and leave it with the lid partially off for at least 30 min to allow the chloroform to evaporate. Overlay one set of chloroformed plates with CL143 (colicin I-resistant) and the duplicates with CL145 (colicin E-resistant). For overlays, use the technique described on Day 2. Incubate plates for 18–24 h at 37 °C.

_ _

Day 5 *(15 min)*

Confirm that the controls gave the expected results. Determine which colicins are produced by the 20 colicinogenic recipients. Pool the class results.

_ _

Notes and Points to Watch

1. When overlaying plates which have been inoculated with 0·2 ml of culture (Day 1), the agar must, of course, be distributed rapidly over the surface and then left before it begins to set.
2. Two layers of agar should be sufficient to ensure that all colonies are submerged (Day 1). If there are one or two colonies on the surface, they should not be spread when the overlay of colicin-sensitive strain is added (Day 2).

3. No more than 7 colonies should be tested per plate when examining colicinogenic recipients since ColE2$^+$ colonies form large inhibition zones.

4. Plastic Petri dishes should not be chloroformed for more than 30 min because the edges begin to dissolve. When incubating chloroformed plates which have been overlaid with indicator bacteria, check that the chloroformed edges of the plates are not sealed to the lids as this prevents growth of the indicator.

5. The mating conditions described are optimal. But other growth containers can be used, even tubes if necessary.

6. Strains can be kept on Dorset egg medium (at 4°C or room temperature) for several years. The indicators are stable on storage, but the Col$^+$ strains must be checked for colicin production before use.

7. Several additions can be made to this experiment if time permits. The precise frequency of transfer per donor can be determined by making viable counts of donor and recipient strain. Colicin immunity can be demonstrated by overlaying chloroformed ColE2$^+$ and ColE3$^+$ colonies with a colicin E2-immune strain, a colicin E3-immune strain, a colicin E-resistant strain and a colicin-sensitive strain. The additional strains required are:

W3110 (ColE2-P9) colicin E2-immune; colicin E3-sensitive.
W3110 (ColE3-CA38) colicin E3-immune; colicin E2-sensitive.

MATERIALS

1. *Escherichia coli* K-12 strains.

Strain	Relevant features
M654	ColIb-P9*drd*$^+$, ColE2-P9$^+$, str-s
KH188	ColIb-P9*drd*$^+$, str-r (*rpsL*)
KH293	ColE2-P9$^+$, str-r
AB1157/I/E	I^R, E^R str-r
CL142	I^S, E^S str-r
CL143	I^R, E^S str-r
CL145	I^S, E^R str-r

2. Nutrient broth, nutrient agar. Soft nutrient agar (50% nutrient broth: 50% nutrient agar).

3. Streptomycin. Make up a stock solution in water at a concentration of 20 mg ml^{-1}. This is stable for several weeks at 4°C.

4. Chloroform.

5. Clock glasses (10 cm diam.).

SPECIFIC REQUIREMENTS

Day 1

about 2 ml of overnight broth culture (37°C, unshaken) of strains M654 and AB1157/I/E

2×50 ml conical flasks containing 5 ml of sterile nutrient broth

incubator for shaking 50 ml flasks at 37°C

5 nutrient agar plates containing streptomycin (200 μg ml^{-1})

2 nutrient agar plates

100 ml of sterile diluent (e.g. phosphate-buffered saline containing 0·01% (w/v) glycerol)

20 ml sterile bottle, 5 sterile test tubes in rack

5 sterile 10 ml pipettes

8 sterile 1 ml pipettes

50 ml of molten soft nutrient agar (preferably containing streptomycin (200 μg ml^{-1})) at 47°C

Day 2

about 1 ml of broth culture (overnight, unshaken) of strain CL142

cont.

cont.

20 ml of molten soft agar at 47°C	**Day 4**
sterile 1 ml and 10 ml pipettes	50 ml of chloroform
	10 clock glasses
	fume cupboard
Day 3	2 bottles of molten soft nutrient agar (20 ml) at
straight wire	47°C
eight nutrient agar plates containing strepto-	about 1 ml of broth cultures (overnight,
mycin (200 μg ml^{-1})	unshaken) of strains CL143 and CL145.
strains KH188 and KH293 streaked on plates	2 sterile 1 ml pipettes, 2 sterile 10 ml pipettes

FURTHER INFORMATION

ColE2 is efficiently mobilized by CoIIb. In this experiment about 90 % of CoII$^+$ recipients should also be ColE2$^+$. The CoIIb plasmid used is de-repressed for transfer and should therefore be transferred at a high frequency (about 10^{-1} per donor).

The mechanism of plasmid mobilization is unknown. ColE2-P9 and ColE3-CA38 are mobilized by CoIIb and by other I-like conjugative plasmids (Smith *et al.*, 1963) but not by F- or by F-like plasmids. On the other hand, ColE1-K30 is mobilized by both F- and I-like plasmids. Mobilization of ColE1 requires a ColE1-specified product (Warren *et al.*, 1978). Co-transfer of the conjugative plasmid is not required for mobilization; in the experiment described here, some Col$^+$ recipients may be found to produce colicin E but not colicin I. The proportion of ColE$^+$ CoII$^-$ recipients can be increased if mating is interrupted after 5 min. Lower dilutions should be plated.

REFERENCES

K. G. Hardy (1975). Colicinogeny and related phenomena. *Bacteriological Reviews*, **39**, 464–515.

S. M. Smith, H. Ozeki and B. A. D. Stocker (1963). Transfer of ColE1 and ColE2 during high frequency transmission of CoII in *Salmonella typhimurium. Journal of General Microbiology*, **33**, 231–42.

G. J. Warren, A. J. Twigg and D. J. Sherratt (1978). ColEl plasmid mobility and relaxation complex. *Nature*, **274**, 259–61.

105

Plasmid Compatibility

R. W. HEDGES

*Royal Postgraduate Medical School, Hammersmith Hospital,
Du Cane Road, London, W12 OHS, England*

> *Level*: Advanced undergraduates
> *Subject area*: Molecular genetics

INTRODUCTION

Plasmids are DNA molecules which replicate in bacterial cells physically separate from the chromosome. They are thus autonomous replicons. Plasmid replication usually, perhaps always, requires enzymes provided by the host but plasmids contribute towards the replication complex in different degrees.

In order to ensure its transfer from generation to generation a plasmid must ensure that its replication cycle is co-ordinated with that of the host chromosome. Many plasmids have relaxed replication control, i.e. each cell contains numerous copies of the plasmid and the number of copies per chromosome varies quite widely according to growth conditions. Other plasmids replicate under stringent control where the copy number is low, say one to five per chromosome, and relatively independent of growth conditions. Thus a plasmid has some mechanism by which it can regulate its own replication. Let us suppose that two plasmids are distinguishable by antibiotic resistance markers but are otherwise identical: most importantly, they are identical in genes involved in replication and the regulation of replication. Suppose that one cell carries both plasmids. Clearly, the replication control of each plasmid will act equally upon both to limit the multiplication of both. At each cell division the two plasmids will be randomly distributed to the daughter cells and there will be a measurable probability that a daughter will receive only one sort of plasmid and thus lack the resistance characters determined by the other. Thus, if a cell carries two very closely related plasmids, its progeny will be heterogenous and many will carry only one plasmid. We say that the two plasmids are **incompatible**. If two plasmids are incompatible extensive DNA homology is almost always found. Thus, compatibility tests are useful and relatively simple ways of determining relationships between plasmids.

Since the plasmids to be used are conjugally transmissible it is possible to produce strains carrying two plasmids by selecting for transfer of one plasmid into a recipient in which some other plasmid is already resident. If the two are unrelated they will coexist stably, i.e. be compatible. If their replication specificities are identical, the two plasmids will be unable to coexist stably. If selection is made for the resistance markers associated with the incoming plasmid then, sooner or later, the resident plasmid will be lost. It is this loss which will be investigated in this experiment.

A second phenomenon may be observed. This is **surface exclusion**. The presence of a plasmid in a bacterial strain may reduce the efficiency with which that strain acts as recipient in conjugation. The effect is highly specific. For example, in the experiments to be described, the presence of RP4 in the recipient will reduce the efficiency of transfer of R751 but have no effect on the acceptance of R388. In general, if two plasmids show initial surface exclusion they are related but the converse is not true. Two very closely related plasmids may exhibit no surface exclusion.

EXPERIMENTAL

The bacteria to be used are nutritionally distinguishable F^- derivatives of *Escherichia coli* K12. J53-2 requires proline and methionine, ferments lactose, is sensitive to nalidixic acid and carries a chromosomal mutation conferring resistance to rifampicin. J62-1 requires proline, histidine and tryptophan, does not metabolize lactose, is sensitive to rifampicin and carries a chromosomal mutation determining nalidixic acid resistance.

The plasmids to be used are R702 which confers resistance to kanamycin, streptomycin, tetracycline, sulphonamides and mercuric salts; R388 which confers resistance to trimethoprim and sulphonamides; and R751 which confers resistance to trimethoprim.

Day 1 *(2–3 h, but see note 5)*

Students should be provided with overnight shaken broth cultures of J53-2, J62-1 and J53-2(R702). In addition, half the class should be given J62-1(R751) and the other half given J62-1(R388). **These will be used as recipients, only**. The first group of experiments is designed to demonstrate the phenotypic characteristics of the bacterial strains (*see* note 1).

Dilute each culture by a factor of 10^4 and plate 0·1 ml on a lysed blood agar plate (*see* note 3).

↓

Using sterile forceps, pick up a single "Multodisk" and place it on the surface of the plate. Make sure it is in contact with the agar surface at all points. Do not place the disc absolutely centrally, but so as to leave the arms marked CT and NA rather further from the edge than the opposite arms. Then place a trimethoprim disc as close as possible to the edge of the plate, between the CT and NA discs (*see* Figure 1).

↓

To test rifampicin resistance take a nutrient agar plate and cut out a sector as shown in Figure 2. Pipette 5 ml of molten agar into a tube in a 50 °C water-bath. Add 0·05 ml of 5000 μg ml^{-1}

rifampicin solution. With a 10 ml pipette mix the agar so that the rifampicin is uniformly distributed and then pipette the agar into the trough cut in the agar plate (*see* Figure 2). There will be more agar than needed, so it should be pipetted carefully. When the agar has set, dry the plate.

↓

Using either a loop or a straight wire, streak each of the four bacterial cultures towards the ditch. Incubate the plates overnight.

↓

To check ability to metabolize lactose streak out the cultures onto MacConkey agar plates. Be careful that the plates are well dried before use. It is convenient to have *lac*+ and *lac*− strains on the same plate to facilitate comparison. So streak out all four strains on one plate but keep the zones distinct (*see* note 2). Also streak out each R+ recipient culture on a nutrient agar plate attempting to get as many well separated colonies as possible.

↓

With a straight wire inoculate each culture onto one plate of minimal agar containing proline and

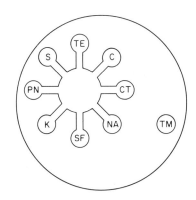

Figure 1 The plate with "Multodisk".
Each disc contains the following antibiotics:
PN = ampicillin 25 μg; S = streptomycin 25 μg; TE = tetracycline 50 μg; C = chloramphenicol 50 μg; CT = colistin 10 μg; NA = nalidixic acid 30 μg; SF = sulphonamide 500 μg; K = kanamycin 30 μg and TM = trimethoprim 30 μg.

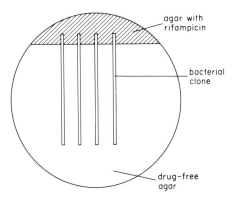

Figure 2 Construction and use of a "ditch" plate. During incubation the rifampicin will diffuse into the drug-free agar setting up a gradient.

methionine and one containing proline, histidine and tryptophan. Incubate the plates overnight and score the next day.

Having set up experiments to demonstrate the characters of the bacterial cultures, these same cultures may be used in experiments on plasmid transfer. Donor cultures should be prepared by diluting an overnight shaken broth culture (i.e. the cultures used as recipients) 10-fold into prewarmed nutrient broth. These cultures should be incubated without shaking for 2 h at 37 °C. Each student should be given two donor cultures of the same strains as the two R^+ recipient cultures he already has. Thus each will receive J53-2(R702) and either J62-1(R388) or J62-1(R751). Each student should set up four transfers. All will use the $R702^+$ culture, half will use the $R388^+$ culture and half the $R751^+$.

Place four tubes in the 37 °C bath and add 4·5 ml of pre-warmed broth. Label the tubes A, B, C and D.

↓

> To tube A pipette 4·5 ml of J53-2 (recipient culture),
> to tube B pipette 4·5 ml of J53-2 (R702) (recipient culture),
> to tube C pipette 4·5 ml of J62-1 (recipient culture),
> to tube D pipette 4·5 ml of J62-1 (R388) or J62-1 (R751) (recipient culture).

↓

To tubes A and B pipette 1 ml of **donor culture** of J62-1(R751) or J62-1(R388). Note the time.

↓

To tubes C and D pipette 1 ml of **donor culture** of J62-2(R702). Note the time.

↓

Allow the matings to proceed for 1 h. At the end of that hour place the four bottles in an ice bath.

↓

From each bottle make 6 ten-fold dilutions (0·5 ml → 4·5 ml saline).

To determine the concentration of recipients
Plate 0·1 ml aliquots from the 10^{-5} and 10^{-6} dilutions on agar appropriate to permit growth of the recipient but not of the donor. In the case of matings A and B this will be medium with proline and methionine (pro + met) (but not histidine or tryptophan). For matings C and D use medium with proline, histidine and tryptophan (pro + his + trp). Use two plates at each dilution.

↓

To determine the concentration of donors
Plate 0·1 ml of the 10^{-4} and 10^{-5} dilutions on agar appropriate to permit growth of donors but not recipients. In the case of matings A and B use plates with proline, histidine and tryptophan, and in the case of matings C and D medium with proline and methionine.

↓

To determine the concentration of transcipients (clones derived from recipient cells which have accepted incoming plasmids)
These will have the resistance markers of the donor and the nutritional characters of the recipient. Thus in matings A and B we would expect transcipients to require proline and methionine and to be resistant to trimethoprim. So, we plate on agar with proline and methionine and trimethoprim (Tmp). Plate 0·1 ml aliquots from the undiluted mating mixture, the 10^{-1}, 10^{-2} and 10^{-3} dilutions. As usual, two plates at each dilution.

↓

For matings C and D, plate the equivalent dilutions on medium with proline, histidine, tryptophan and kanamycin (Km).

↓

The resistances to nalidixic acid (nal) and rifampicin (rif) are due to chromosomal mutations and thus, unlike the other resistance genes being studied, non-transmissible (*see* note 1). In order to test for transmissibility of nalidixic acid plate 0·1 ml of the undiluted mating mixture A onto a plate containing pro + met + nal. Similarly, plate 0·1 ml of mixture C onto a plate with pro + his + trp + rif. Incubate these plates until Day 3.

Table 1. Summary of the dilution and plating schedule for first day.

Matings A and B

Plate 0·1 ml from the undiluted stock, 10^{-1}, 10^{-2} and 10^{-3} dilutions on pro + met + Tmp
Plate 0·1 ml from 10^{-4} and 10^{-5} dilutions on pro + his + trp
Plate 0·1 ml from 10^{-5} and 10^{-6} dilutions on pro + met

Matings C and D

Plate 0·1 ml from the diluted stock, 10^{-1}, 10^{-2} and 10^{-3} dilutions on pro + his + trp + Km
Plate 0·1 ml from 10^{-4} and 10^{-5} dilutions on pro + met
Plate 0·1 ml from 10^{-5} and 10^{-6} dilutions on pro + his + trp

Day 2 *(30 min)*

Characterization of the recipient strains
Inspect the "Multodisk" and ditch plates and record the resistance phenotype of the strains. Check also the amino acid requirements and ability to use lactose (*see* note 4).

Stability of resistance markers
One of the things being tested is whether entry of a particular plasmid into a recipient cell leads to the loss of the resident plasmid. This is essential to check that inheritance of the resident plasmid is stable, otherwise it is not possible to know whether any plasmid loss is due to incompatibility or to natural instability.

Take the nutrient agar plates on which the recipient cultures were streaked on Day 1. There should be plenty of well-isolated colonies.

↓

Prepare ditch plates (Figure 3) from lysed blood plates. One plate should have kanamycin in the ditch for testing J53-2(R702), the other should have lysed blood + trimethoprim. To prepare these ditch plates put two tubes each with 5 ml of molten agar in the 50 °C water-bath. To one add 0·05 ml of 2500 μg ml^{-1} kanamycin solution, mix and pipette into the kanamycin ditch. To the other tube, add 0·25 ml lysed blood and 0·4 ml of 1000 μg ml^{-1} trimethoprim solution, mix and pipette into the ditch. Dry the plates.

$\downarrow$

With a wire, stab a colony of an R$^-$ strain and make a thin streak across the plate to the ditch. Mark this streak. Then streak at least 10 separate colonies of the R$^+$ strain towards the ditch. Be careful that the streaks remain distinct. Incubate the plates overnight.

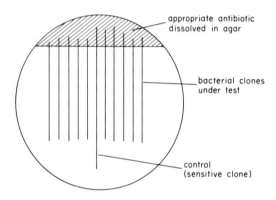

Figure 3   Ditch plate used to test resistances to kanamycin and trimethoprim.

Day 3 *(60 min)*

Count the minimal plates from Day 1. Do not touch the colonies but mark the backs of the plates. Then calculate the concentration of donor, recipient and recombinant cells in the mating mixture. Calculate the rate of transfer per donor cell in the mating and see whether the presence of an R factor in the recipient had any effect on the efficiency of transfer.

   The rest of the work will concern the fate of the resident R factor when a new one has entered. The first concern will be to purify recipient clones which have received new R factors. Since we are concerned with the fate of resident R factors we are no longer interested in transfers in which the recipient was R$^-$. We can therefore discard matings A and C.

In mating B the transfer of R751 (or R388) to J53-2(R702) was selected on pro + met + Tmp. Stab colonies from these mating plates and streak 5 on each of 4 plates of MacConkey medium plus lysed blood plus trimethoprim. Similarly streak out the products of mating D on 4 MacConkey + kanamycin plates.

$\downarrow$

Check the ditch plates of the recipient cultures and determine whether the plasmids are stably inherited in these strains.

Day 4 *(15 min)*

Inspect the plates streaked on Day 3. On the kanamycin plates there should be five streaks (total 20). From each streak select one well-isolated *lac⁻* clone, stab with a sterile wire and streak on a fresh kanamycin plate. Incubate overnight.

↓

Take the incubated trimethoprim plates from Day 3. Again there should be 20 streaked-out clones. Choose a well inoculated *lac⁺* colony of each and restreak on a fresh trimethoprim plate (*see* note 4). Incubate overnight.

Day 5 *(15 min)*

The object of this part of the experiment is to test colonies that have received R factors for retention of resident R factors. Consider the cross R751 (or R388) → J53-2(R702). Selection was made for acquisition of trimethoprim resistance (i.e. R751 or R388) and so the requirement is a check on whether R702 has survived this process. This is easy because R702 determines several resistance determinants different from those of R751 or R388, e.g. resistance to kanamycin.

Prepare ditch plates containing kanamycin and streak 10 purified recipient bacterial clones towards that ditch.

↓

Similarly test the R⁺ J62-1 strains that have acquired kanamycin resistance to see whether they have retained their resident R factors. Do this by using ditch plates with trimethoprim and lysed blood.

↓

Incubate the ditch plates overnight and examine them next day.

Day 6 *(10 min)*

Inspect plates and record results.

Notes and Points to Watch

1. Certain plasmids, notably the F factor, induce transfer of chromosomal genes. The plasmids chosen for the present demonstration do not mobilize chromosomal genes at levels detectable by the techniques chosen. Thus it can be assumed that the nutritional markers of strains J53-2 and J62-1 will be always suitable for differentiating the two strains in each conjugation and any progeny of that mating.
2. MacConkey agar is used as indicator of whether a strain can mobilize lactose. On MacConkey medium *lac⁺* colonies are stained red whilst *lac⁻* remain white (or slightly pink). It is a good idea to have *lac⁺* and *lac⁻* colonies on a MacConkey plate. This ensures reference colonies and makes it easier to decide the phenotypes of unknown colonies. It is very important to have well-

isolated colonies, formed from single cells. Otherwise, diffusion of hydrogen ions may make it very difficult to decide which colonies are *lac*+ and which are *lac*− but close to *lac*+ cells.

If plates are not properly dried before use, the free water may seal the plate after it is inverted but the bacterial growth may produce carbon dioxide sufficient to redden the complete plate. It will then be impossible to distinguish *lac*+ from *lac*− colonies. So, always dry MacConkey plates before use.

3. Trimethoprim acts as an inhibitor of the enzyme dihydrofolate reductase. The most important role of this enzyme is in the pathway producing thymine. Thus, if a *tmp*^S bacterium is treated with trimethoprim in a medium containing plenty of thymine, it will grow even though its enzyme is inhibited. To allow trimethoprim inhibition to be observed it is essential to use a medium lacking thymine. This can be done either by providing a defined (minimal) medium or by using a nutrient agar and adding lysed horse blood which contains a component capable of removing free thymine from the medium. The mechanism of this removal remains obscure.

4. After streaking out transcipients on MacConkey agar with lysed blood and trimethoprim so as to obtain isolated trimethoprim resistant colonies, the presence of the lysed blood may make it difficult to see whether a clone is *lac*+ or *lac*− by reflected light. If in doubt, hold the plate up to the light and inspect by transmitted light. A *lac*+ colony will appear dark—almost black; a *lac*− colony will be almost perfectly transparent.

5. Given a class that knew what they were doing, the time required for the work on Day 1 could be reduced to about $1\frac{1}{2}$ h by starting the plasmid transfer experiments as the first step and characterizing the strains while the transfers were proceeding.

MATERIALS

1. Minimal glucose agar contains per litre, 13·6 g KH_2PO_4, 2 g $(NH_4)_2SO_4$ and 0·5 mg $FeSO_4.7H_2O$. The pH is adjusted to 7·0 with KOH. After autoclaving, add 1 ml of 1 M $MgSO_4.7H_2O$ and 10 ml of 20% glucose. Amino acids are added to a final concentration of 25 μg ml^{-1} (or 50 μg ml^{-1} if DL form). Antibiotics are added to final concentrations of 10 μg ml^{-1} (trimethoprin) or 25 μg ml^{-1} (kanamycin, nalidixic acid or rifampicin).

2. Trimethoprim, kanamycin, nalidixic acid and rifampicin can be obtained from the Sigma Chemical Company. Rifampicin can be dissolved in 0·1 N HCl or in methanol. Nalidixic acid should be dissolved in 0·1 N NaOH.

3. Lysed blood agar—see page 709.

SPECIFIC REQUIREMENTS

These items are needed for each group of students.

Day 1
overnight cultures of bacteria
2 h dilutions of donors (described on page 698 of schedule)
Plates
4 lysed blood agar

3 nutrient agar
2 MacConkey agar
17 minimal glucose agar with proline and methionine
17 minimal glucose agar with proline, tryptophan and histidine
16 minimal glucose agar with proline, methionine and trimethoprim

---

cont.

cont.

---

16 minimal glucose agar with proline, histidine, tryptophan and kanamycin

1 minimal glucose agar with proline, methionine and nalidixic acid

1 minimal glucose agar with proline, histidine, tryptophan and rifampicin

4 "Multodisks"

5 ml molten nutrient agar

40 test tubes or small bottles

42×0.1 ml pipettes

32×10 ml pipettes

5×1 ml pipettes

1 pair forceps

saline solution for dilutions

1 scalpel

1 straight wire

1 wire loop

ice bath

37 °C water-bath

50 °C water-bath

rifampicin solution, 2500 $\mu g\,ml^{-1}$ in 0.1 N hydrochloric acid

Day 2

2 lysed blood agar plates

10 ml molten agar

water-bath at 50 °C

1 scalpel

Day 2

kanamycin solution, 2500 $\mu g\,ml^{-1}$

trimethoprim solution, 1000 $\mu g\,ml^{-1}$

lysed blood

1 straight wire

Day 3

colony counter

4 MacConkey plates again with lysed blood and trimethoprim

4 MacConkey plates agar with kanamycin

1 straight wire

1 wire loop

Day 4

as Day 3 (except for colony counter)

Day 5

4 MacConkey agar plates

4 lysed blood agar plates

water-bath at 50 °C

20 ml molten agar

1 scalpel

kanamycin solution, 2500 $\mu g\,ml^{-1}$

trimethoprim solution, 1000 $\mu g\,ml^{-1}$

lysed blood

1 straight wire

FURTHER INFORMATION

Typical results which will be obtained are described below.

Characterization of Recipient Cultures

The bacteria should have the amino acid requirements, drug resistance patterns and lactose fermenting abilities described at the start of the Experimental section.

Surface Exclusion

The assays of donor, recipient and transcipient cell concentrations in the mating mixtures should permit the calculation of plasmid transfer rates from each of the four mixtures. Each student should be able to calculate the transfer rate of R702 into a plasmid free recipient (mixture C). The transfer rate of R702 into a similar recipient carrying an R factor can be calculated from the results with mixture D. If the resident plasmid (R751 or R388) can exclude R702 this effect will lead to a reduction in the transfer rate (relative to mixture C). The presence of R751 in the recipient should reduce the transfer rate by a factor of 10^1 to 10^2. R388 should have no effect.

Similarly, the rate of transfer of the trimethoprim R factor (R388 or R751) can be calculated from the results with mixture A and if the rate observed for mixture B is lower this must be due to the presence of R702 in the recipient. The presence of R702 should reduce the rate of transfer of R751 approximately one-thousand-fold but have no effect on the transfer of R388.

Thus, R751 and R702 show mutual surface exclusion indicating relatedness but R388 and R702 show no such interaction.

Incompatibility

The entry of R751 should lead to rapid and efficient elimination of R702 from J53-2(R702). This will lead to the loss of kanamycin resistance from that strain. Reciprocally, entry of R702 will lead to the elimination of R751 and hence, loss of trimethoprim resistance. That is to say, R702 and R751 are incompatible.

It will sometimes be found that the resistance marker(s) of a resident plasmid are not lost despite entry of an incompatible plasmid. There are several common explanations.

1. The loss has been delayed till after the testing. What was tested was an unstable "double" carrying both plasmids. If allowed to grow in drug free medium such a strain will segregate progeny which have lost one of the two incompatible plasmids.

2. As described in the Introduction, two incompatible plasmids will usually contain large sequences of genetic homology. Recombination can occur between such regions producing a plasmid carrying genetic determinants from both original R factors. Thus, some transcipient cells may contain newly produced recombinant plasmids carrying resistance markers from incoming and resident plasmids.

3. Transposition. Neither the kan^R marker of R702 nor the tmp^R marker of R751 are efficiently transposable (at least, under the conditions of the experiment) but many markers are. For example the str^R and sul^R markers of R702 are highly transposable. If R751 were introduced into J53-2(R702) these markers would frequently not be lost, the str^R and sul^R determinants having been inserted into the chromosome before the R702 plasmid was eliminated.

R388 and R702 are completely compatible and the entry of one of these plasmids usually does not eliminate the other. It could be concluded that R702 and R751 are closely related and that R702 and R388 show no evidence of relatedness.

106

Experiments with Transposons

R. W. HEDGES

*Royal Postgraduate Medical School, Hammersmith Hospital,
Du Cane road, London, W12 OHS, England*

Level: Advanced undergraduates
Subject area: Microbial genetics

INTRODUCTION

Certain DNA sequences—termed transposons—can be transferred between bacterial replicons (e.g. from chromosome to plasmid) regardless of genetic homology or the presence of the enzymes involved in chromosomal recombination. Where the DNA sequence determines some recognizable phenotypic character, a transposition event can be recognized by the resultant change in location of the gene or genes involved. When the phenotypic character allows selection to be imposed (e.g. antibiotic resistance) it is easy to follow the transposition process.

This experiment starts with *Escherichia coli* K12, strain J62, which requires proline, histidine and tryptophan and is unable to metabolize lactose. Its chromosome also contains a transposon (Tn7) which confers resistance to trimethoprim and streptomycin. When a transposon has integrated into a replicon it is designated by 4 dots arranged in a square. Thus, if Tn7 has integrated into plasmid RP4 we would call this RP4::Tn7. Thus, the strain described above is termed J62::Tn7.

If a transmissible plasmid is introduced into J62::Tn7 so as to replicate autonomously it will be found that in a small proportion of cells the transposon will have inserted itself into the plasmid, i.e. transposition will have occurred. This need have no effect on the phenotype of the cell, so it would be hard to detect directly, but, if the plasmid is transferred to a drug sensitive recipient (by conjugation, transduction or transformation), it is possible to detect those plasmids which have received the transposon, to determine the proportion of these among the total plasmid population and observe any effects upon the phenotypic properties conferred by the plasmid.

The plasmid used in this experiment is RP4 which confers four distinct and easily recognizable phenotypic effects on an *E. coli* host. These are resistance to ampicillin (and other β lactam antibiotics), tetracycline and kanamycin, plus sensitivity to a variety of phages such as PR4. These phages use the RP4-determined pili, which are an essential component of the transfer machinery of the plasmid, as receptors. The recipient used is *E. coli* K12 strain J53 which requires proline and methionine for growth and is able to use lactose as carbon source.

If a transposon inserts itself into a gene, it will break the coding sequence and have a phenotypic effect. For example, RP4 carries a gene determining an enzyme which catalyses the phosphorylation, and hence inactivation, of kanamycin. If Tn7 inserts into this gene the plasmid will no longer be able to produce the protein and hence can no longer confer resistance to kanamycin. Such a plasmid will, however, still confer resistance to ampicillin, tetracycline, streptomycin and trimethoprim. If the transposon integrates into the genes involved in pilus synthesis (*tra* genes), it will abolish the ability of the plasmid to transfer itself and prevent expression of sensitivity to phage PR4.

EXPERIMENTAL

Day 1 *(10 min + 30 min separated by a gap of at least 1 h)*

Add 0·2 ml of stationary phase cultures of J62::Tn7(RP4) and J53 to the same 2 ml of pre-warmed broth in a bottle or flask. Leave for 1 h (or longer) at 37 °C.

↓

Make a series of six 10-fold dilutions and plate 0·1 ml of the 10^{-4}, 10^{-5} and 10^{-6} dilutions, in duplicate, on MacConkey agar. This will give an estimate of the final concentrations of donor (*lac⁻*) and recipient (*lac⁺*) cells.

↓

Plate 0·1 ml of 10^{-1}, 10^{-2} and 10^{-3} dilutions in duplicate on minimal agar with proline, methionine and kanamycin ($25\ \mu g\ ml^{-1}$). This will give an estimate of the concentration of recipient cells which have received RP4 and expressed kanamycin resistance.

↓

Plate 0·1 ml of the 10^{0} and 10^{-1} dilutions in duplicate on minimal agar with proline, methionine and trimethoprim ($10\ \mu g\ ml^{-1}$). This will give the concentration of recipient cells expressing the resistances encoded on transposon 7 (*see* notes 1 and 2).

Day 2 *(30 min)*

Count the colonies of the MacConkey plates (*see* note 3). Calculate the concentration of donor and recipient. Discard these plates.

Day 3 *(90 min)*

Count the colonies on the remaining plates. **Do not touch the colonies** on the trimethoprim plates. Mark the backs of the plates rather than stab the colonies. Discard the kanamycin plates after counting.

↓

If the colonies on the trimethoprim plates are well isolated three colonies may be tested for drug resistance properties by means of "Multodisks". As well as a "Multodisk", a disc of trimethoprim and a spot of phage PR4 suspension should be added to each plate, clear of the "Multodisk" and of each other.

↓

Another 10–20 colonies may be touched by a sterile wire and streaked across plates with ditches of kanamycin (25 μg ml^{-1}), ampicillin (100 μg ml^{-1}) and tetracycline (50 μg ml^{-1}) (*see* notes 4 and 8).

↓

Each colony whose antibiotic resistance pattern is being examined should also be tested for sensitivity to PR4 by inoculating a patch on a well-dried nutrient agar plate and placing a drop of phage PR4 suspension on the patch.

↓

Incubate the plates overnight. If the colonies are not well isolated, or have grown poorly, they should be streaked out on plates of MacConkey agar containing 5 % lysed blood and 100 μg ml^{-1} trimethoprim.

_____

Day 4 *(30 min)*

Inspect the "Multodisk" plates. A majority will probably display all the resistances of RP4 and be sensitive to phage PR4. Some will have lost one (or perhaps more than one?) character.

↓

Examine the ditch and phage spot plates. Estimate the proportion of ex-conjugants in which the plasmid has lost one or more of the phenotypic characters of RP4 (*see* notes 5 and 6).

There are three possible techniques for testing whether Tn7 has integrated into RP4 or whether it is independent of that plasmid, perhaps integrated into the chromosome.

(a) If Tn7 is integrated into RP4 it will be eliminated along with the rest of the plasmid if an incompatible plasmid is introduced (*see* Experiment 105 for appropriate background information). A suitable plasmid is R1033 which confers resistance to ampicillin, streptomycin, tetracycline, chloramphenicol, kanamycin, gentamicin and sulphonamides.

(b) If RP4 is transferred to a new host, selecting for transfer of one of the natural resistances of RP4, the trimethoprim resistance of Tn7 will also be transferred.

(c) After an RP4$^+$ strain is infected with phage PR4 lysis ensues and any phage resistant clones are selected. Among these will be some strains carrying mutant plasmids and some segregants which have lost the complete plasmid. If a strain carrying RP4::Tn7 is treated with phage PR4 then those survivors which have lost the PR4 markers will also have lost trimethoprim resistance.

The first technique has the advantage that it can be used to show the plasmid carriage of Tn7 in cases where the *tra* function has been lost. The first and third techniques have the disadvantage that Tn7 on the plasmid can occasionally transpose onto the bacterial chromosome and hence be retained despite loss of the plasmid.

Procedures for all three tests are described below. The instructor can choose which to use. The letters in parentheses which precede each methodology section refer to the techniques described above.

(a) Inoculate J62(R1033) and putative clone of J53(RP4::Tn7) into separate tubes of broth (10 m!). Incubate overnight at 37 °C.

(b) Inspect the phage PR4 lysis plates from Day 3 and look for colonies of resistant mutants. Use a lens if none is visible to the naked eye. Touch 10 of these, picking colonies far enough from antibiotic discs to be under no pressure to retain drug resistance, and streak out on MacConkey plates (5 per plate). Incubate the plates overnight at 37 °C.

Day 5 *(30 min + 5 min separated by at least 1 h)*

(a) Add 0·2 ml of J62(R1033) and 0·2 ml of J53(RP4::Tn7) to 2 ml pre-warmed broth. Incubate at 37 °C for at least 1 h and then streak a few drops of the mating mixture on minimal agar with proline, methionine and chloramphenicol (25 μg ml^{-1}). Incubate at 37 °C.

(b) Add 0·2 ml of J53(RP4::Tn7) and 0·2 ml of J62 to 2 ml of broth. Incubate at 37 °C for at least 1 h. Streak a few drops of the mating mixture on minimal agar with proline, histidine, tryptophan and kanamycin (25 μg ml^{-1}) (or tetracycline (10 μg ml^{-1}) if a kanamycin-sensitive RP4::Tn7 has been used). Incubate the plates at 37 °C.

(c) Test the growth of 10 PR4^R "variants" on a lysed blood agar plate with tetracycline (25 μg ml^{-1}) and trimethoprim (150 μg ml^{-1}) ditches (*see* note 7). Some of the "variants" will be *tra*$^-$ mutants which retain both kanamycin and trimethoprim resistances, whilst others will be segregants that have lost the complete plasmid, i.e. resistance to all drugs. .

Day 6 *(20 min)*

(a) Restreak 10 colonies on MacConkey agar plates with 25 μg ml^{-1} chloramphenicol. Incubate overnight.

(b) Streak out RP4$^+$ transcipients on minimal agar with proline, histidine and tryptophan plates with a ditch containing trimethoprim (50 μg ml^{-1}). Incubate overnight.

(c) Inspect the ditch plates and record results. Those clones which have regained sensitivity to tetracycline may be assumed to have lost the entire plasmid, those which retain this resistance still carry the plasmid. Consider those tetracycline sensitive strains which have lost the plasmid. Have they also lost trimethoprim resistance (i.e. transposon 7)?

Day 7 *(20 min)*

(a) Streak the colonies from the chloramphenicol plates onto lysed blood agar plates with trimethoprim ditches.

(b) Inspect the ditch plates and observe the proportion of transcipients which are trimethoprim resistant.

Day 8 *(10 min)*

(a) Inspect plates. Score the proportion which are sensitive to trimethoprim.

Notes and Points to Watch

1. Streptomycin resistance is always transposed along with trimethoprim resistance, but only trimethoprim resistance is used as a selective marker because Tn7 confers a low level of streptomycin resistance, whilst the resistance to trimethoprim is total.

2. Trimethoprim is only fully effective as an antibacterial agent if the medium contains no thymine (or related compounds). Thus, minimal agar or DST agar supplemented with lysed horse blood is used. MacConkey medium with lysed blood is not suitable as a selective medium but is adequate for purification.

3. On MacConkey plates containing lysed blood it may be difficult to see whether a clone is lac^+ or lac^- by reflected light. If in doubt, hold the plate up to the light and inspect by transmitted light. A lac^+ colony will appear dark whereas a lac^- colony will be almost perfectly transparent.

4. Both ampicillin and tetracycline solutions are unstable and freshly prepared solutions should be used.

5. The proportion of insertions of Tn7 into RP4 which lead to detectable phenotypic changes must be the result of numerous factors. Tn7 insertion is not entirely random: e.g. insertion into the β lactamase gene (producing ampicillin-sensitive mutations) are notably rare. Tra^- mutations must presumably be under represented among transcipients and any insertions into genes preventing replication must be inviable. On the other hand, certain types of mutation may possibly be over represented. Many plasmids repress synthesis of their pili (though RP4 does not). Drd mutations have lost this repression and so transfer with much higher efficiency than wild type (drd^+) plasmids. Although plausible, this prediction is untested.

6. Since the three resistance markers on RP4 are determined by quite separate genes, which are also distinct from the tra genes, insertion of a transposon at any point should affect only one of the phenotypic characters.

7. After selecting for variants of J53(RP4::Tn7), tetracycline resistance, rather than kanamycin resistance is used to distinguish between strains which have lost the plasmid and those carrying a tra^- mutant plasmid. This is because the gene conferring kanamycin resistance is located very close to the tra genes and some tra^- mutants are deletions covering the kanamycin-resistance gene. Thus some kanamycin-sensitive clones will still carry most of the plasmid including Tn7. The tetracycline-resistance determinant is located far enough from the tra genes that no tra^- deletions extending to this gene have ever been observed.

8. Ditch plates may be prepared from nutrient agar plates. The ditches are cut using a sterile scalpel and filled with molten agar containing the appropriate antibiotics. The plates should be allowed to set and then dried.

MATERIALS

1. *Escherichia coli* K12 strains J62::Tn7(RP4), J53, J62(R1033)
2. Minimal agar: Oxoid No. 3 agar L13.
3. Lysed blood agar is prepared by adding 4% lysed horse blood to Oxoid sensitest agar (CM471).
4. Lysed blood is prepared by adding 2·5 ml saponin to 100 ml horse blood and incubating at room temperature till the viscosity falls.
5. Saponin can be obtained from British Drug Houses, Ltd.
6. Phage PR4.
7. "Multodisks": Oxoid 4997E are the most suitable.
8. Trimethoprim, chloramphenicol, kanamycin and tetracycline can be obtained from the Sigma Chemical Co.

SPECIFIC REQUIREMENTS

Day 1
1 small flask or bottle
6 dilution tubes in rack
< 1 bottle of pre-warmed broth
6 plates MacConkey agar
6 plates minimal agar with proline, methionine and kanamycin ($25 \mu g\,ml^{-1}$)
8 plates minimal agar with proline, methionine and trimethoprim ($10 \mu g\,ml^{-1}$)
1 ml and 10 ml pipettes (sterile)
saline for dilutions

Day 2
colony counter

Day 3
3 lysed blood agar plates
3 "Multodisks"
about 5 MacConkey plates each with three ditches containing ampicillin ($100 \mu g\,ml^{-1}$), tetracycline ($25 \mu g\,ml^{-1}$) and kanamycin ($25 \mu g\,ml^{-1}$)
about 10 nutrient agar plates (well dried)
about 10 MacConkey agar plates (for streaking out colonies)

Day 4
2 MacConkey agar plates
3 bottles of broth

Day 5
1 lysed blood agar plate with kanamycin and trimethoprim ditches
2 small flasks or bottles
1 bottle of pre-warmed broth
1 plate of minimal agar with proline, methionine and chloramphenicol ($25 \mu g\,ml^{-1}$)
1 plate of minimal agar with proline, histidine, tryptophan and kanamycin ($25 \mu g\,ml^{-1}$)

Day 6
about 4 plates of MacConkey agar with chloramphenicol ($25 \mu g\,ml^{-1}$)
at least 2 plates of minimal agar with proline, histidine and tryptophan with trimethoprim ($10 \mu g\,ml^{-1}$) ditches

Day 7
2 lysed blood agar plates each with a trimethoprim ditch

FURTHER INFORMATION

By Day 3 it should be possible to determine the efficiency of transfer of RP4 from J62::Tn7(RP4) to J53 and the proportion of transferred plasmids which have picked up the transposon prior to transfer. The efficiency of transfer depends on many factors and will vary from laboratory to laboratory. For this reason it was suggested that a range of dilutions be plated. In any one laboratory, however, it is usually possible to attain consistent results and, if so, it will be possible to decide which dilutions will be useful and which can be omitted.

Generally about one plasmid in 10^3 will carry the transposon. Hence there should be a thousand-fold difference between the numbers of kanamycin-resistant transcipients and those resistant to trimethoprim.

On Day 4 it should be possible to observe cases where Tn 7 has integrated into a gene, giving rise to a new phenotype. The most abundant class will be *tra*⁻ mutants (resistant to phage PR4), for there are numerous genes into which insertion can occur to give this phenotype. Tetracycline-sensitive and kanamycin-sensitive variants are common but ampicillin-sensitive mutants seem to be rare.

On Days 6–8, in part (a) of the experiment, the establishment of R1033 should lead to elimination of the resident plasmid and the transposon. Thus chloramphenicol-resistant clones will be trimethoprim-sensitive. If any clones retain trimethoprim resistance the most likely explanation is that Tn 7 has integrated into the bacterial chromosome. The frequency of such integration can be determined by observations on the proportion of RP4::Tn7$^+$ strains in which trimethoprim resistance can no longer be eliminated by entry of R1033.

In part (b) of the experiment, all transcipients which have acquired kanamycin resistance by conjugation with an RP4::Tn7$^+$ donor should also acquire trimethoprim resistance, i.e. the genes responsible for the two characters should be physically associated. Note that a supplemented minimal agar plate was used for preparation of the ditch plate to avoid the need to purify transcipients. Although some donor strain cells will undoubtedly be picked up and streaked on the ditch plate, they cannot grow.

In part (c) of the experiment, the loss of sensitivity to PR4 will often be accompanied by loss of all other plasmid-determined characters. Phage-resistant variants may be cells which have lost the entire plasmid (what is wanted) or clones carrying *tra*$^-$ mutant plasmids (not wanted). Cells of the first type will have lost tetracycline resistance whereas those of the latter class will retain it. Of the tetracycline-sensitive clones the large majority should have lost trimethoprim resistance as well as the known plasmid-coded determinants.

The schedule would be more impressive if the students were given two separate strains, J53(RP4) and plasmid-free J62::Tn7 and allowed to construct the J62::Tn7(RP4). However, this adds an extra three or four days to the experiment.

The interaction between R1033 and RP4::Tn7 could be used as a main experiment demonstrating incompatibility. However, this could present difficulties. First, there is extensive overlap of resistance markers which blurs the impact of the demonstration of elimination of the resident plasmid. Second, the chloramphenicol and gentamicin resistances of R1033 (the most suitable for selection) are effective only against low levels of the drugs and may not be impressively displayed. Third (and most important) the Tn7 will transfer from RP4 onto the bacterial chromosome in a minority of cells. Where this happens, entry of R1033 will not eliminate trimethoprim resistance. This fairly infrequent occurrence should not cause confusion if the marker is known to be transposable but it is best to use markers not subject to transposition in formal demonstrations of incompatibility.

Restriction and Modification of Bacteriophage Lambda in Strains of *Escherichia coli*

S. W. GLOVER

Department of Genetics, University of Newcastle,
Newcastle upon Tyne, NE1 7RU, England

Level: Advanced undergraduates
Subject areas: Molecular genetics
Special feature: Insight into some of the machinery
of genetic engineering

INTRODUCTION

The classical experiments of Werner and Arber (Arber and Dussoix, 1962; Dussoix and Arber, 1962) established that many strains of bacteria possess restriction and modification (R-M) systems which are responsible for the differences in plating efficiencies often observed when the same preparation of bacteriophage is plated on different strains of the same species.

In a given R-M system, restriction is caused by a **restriction endonuclease** which cleaves the bacteriophage DNA molecule (and other DNA molecules) not specifically protected against its activity. Modification is caused by a **DNA methylase** which specifically protects DNA from cleavage by the endonuclease.

Many strains of bacteria possess both these activities which form the basis of the R-M systems. The activities may be coded by chromosomal genes, e.g. in *Escherichia coli* K12 or *E. coli* B; or by plasmid genes, e.g. R factors, or by bacteriophage genes, e.g. phage P1.

R-M systems are widespread among prokaryotes and play an important role in protecting bacteria against invasion by foreign DNA and, it can be argued, act as a genetic isolation mechanism. In recent years restriction endonucleases have been the focus of much attention because many of them cleave DNA molecules into precise fragments by cutting them at defined nucleotide sequences. Some of them produce DNA fragments with short single-stranded termini which can be used to generate recombinant DNA molecules *in vitro*.

The aim of this experiment is to use bacteriophage lambda grown on an *E. coli* strain C, which

---

does not possess an R-M system, to detect the presence of R-M systems in other *E. coli* strains and to demonstrate the specific protection afforded to bacteriophage lambda after growth in strains which do possess R-M systems.

EXPERIMENTAL

Students are provided with a suspension of bacteriophage lambda (approx. 10^9 particles ml^{-1}) previously grown on an *E. coli* strain lacking R-M systems, e.g. *E. coli* C. Using dilutions of this suspension, the R-M systems in *E. coli* K, B, C(P1) and K(P1) can be detected by measuring the efficiency of plating of the suspension on each strain. Phage from plaques produced on each strain is then used to show the specificity of the protection afforded by each R-M system. Other *E. coli* strains may be used in addition, e.g. *E. coli* strain 15 or strains carrying appropriate R factors.

Day 1 *(1 h)*

Test for restriction
Prepare nutrient agar plates overlaid with soft agar containing each of the 5 *E. coli* strains provided (*see* notes 1 and 2).

↓

Prepare 10^{-2}, 10^{-4} and 10^{-6} dilutions of the suspension of bacteriophage lambda ($\lambda.0$) (*see* note 3).

↓

With a 0·1 ml pipette or a standard loop carefully spot 0·01 ml of undiluted $\lambda.0$ and the 10^{-2}, 10^{-4} and 10^{-6} dilutions on to marked quadrants of the agar plates (*see* note 4). **Exercise care to avoid spots running into one another and allow time for the spots to dry before incubating the plates at 37 °C.**

Day 2 *($1\frac{1}{2}$ h)*

Through the back of each plate score the number of plaques in each spot. Calculate the approximate efficiency of plating of phage $\lambda.0$ on each of the 5 strains. Assume that the number of plaques obtained for $\lambda.0$ on *E. coli* C is equivalent to an efficiency of plating of 1·0.

↓

With a sterile straight wire select and stab 1 well-defined plaque from each of the 5 *E. coli* strains (*see* note 6). Resuspend the phage in 1·0 ml of buffer (this phage suspension will contain approximately 1×10^6 particles ml^{-1}). Lable each tube with the source of the phage, e.g. $\lambda.$K, etc. (*see* note 3).

↓

Prepare 10^{-1}, 10^{-2}, 10^{-3} dilutions of the 5 phage suspensions.

↓

Prepare 3 nutrient agar plates overlaid with each of the 5 *E. coli* strains C, C(Pl), K, K(Pl) and B.

↓

With a 0·1 ml pipette or a standard loop spot each phage suspension and the 3 dilutions on to plates containing the 5 indicator strains. Two sets of spots can be accommodated on a single plate with care. **Exercise caution to avoid spots running into one another and allow time for the spots to dry before incubating the plates at 37 °C.**

Day 3 *(30 min)*

Count the number of plaques produced by each phage suspension on the 5 *E.* coli strains.

↓

Calculate the efficiency of plating of each modified phage suspension. Assume that the number of plaques obtained on *E. coli* C is equivalent to an efficiency of plating of 1·0.

↓

From the results of the restriction tests and the modification tests, determine the restriction and modification systems present in the 5 *E. coli* strains.

Notes and Points to Watch

1. In the preparation of all soft-agar overlay plates it is advisable to use 3 ml of soft agar and 0·2 ml of an overnight culture of indicator bacteria to ensure a uniformly smooth lawn of growth.
2. Soft-agar should be held at 45 °C in a water-bath before use, poured directly over the agar plates and rocked gently to ensure that the surface is evenly covered.
3. A stock of λvir gives the best results because it gives clear plaques and can be used with *E. coli* strains which are lysogenic for lambda. Nomenclature: The host specificity of a phage is indicated by a symbol representing the phage followed by a symbol representing the last host in which the phage was grown (Arber and Dussoix, 1962), e.g. λ.K represents λ which was last grown on *E. coli* K 12 and λ.0 represents λ which was last grown on a host which neither restricts nor modifies DNA.
4. In spotting phage dilutions, a 0·1 ml pipette can be used, taking care not to damage the soft agar surface and without serious risk of contamination from one plate to another. The alternative method of spotting with a standard loop takes more time and students are apt to damage the soft agar surface. In laboratories where phage-typing is a routine procedure, the mechanical multiloop innoculator commonly used can be adapted for the spotting of phages for restriction and modification tests.
5. In the restriction test the suggested dilutions 10^0, 10^{-2}, 10^{-4} and 10^{-6} should give end-points on the *E. coli* strains used. If not, the test dilutions should be adjusted as required.
6. In the modificaion test, students should select one well-isolated plaque from each *E. coli* strain and stab it accurately through the centre. If plaques are not stabbed centrally, the modified phage suspension will be seriously contaminated with bacteria. If an area of confluent lysis is stabbed rather than a single plaque the modified phage suspension may contain significant amounts of unmodified phage particles.

7. The experiment can be made more interesting for students if the *E. coli* strains used in the restriction test are numbered rather than labelled with the strain designation which reveals the R-M system characteristic of the strain. The strains used in the modification test should be correctly labelled and from the results students can determine which of the original 5 strains restrict lambda and which do not and, from the modification test, they will be able to identify the R-M systems present in each of the 5 strains.

MATERIALS

1. *Escherichia coli* strains C, C(P1), K, K(P1) and B.
2. A stock of λvir previously grown on *E. coli* C (i.e. $\lambda vir.$O) at about 10^9 particles ml^{-1}.
3. Nutrient agar containing Oxoid No. 2 nutrient broth powder 25 g; agar 12·5 g; distilled water to 1 litre.
4. Nutrient broth containing Oxoid No. 2 nutrient broth powder 25 g; distilled water to 1 litre.

5. Bacteriophage buffer containing Na$_2$HPO$_4$ (anhydrous) 7 g; KH$_2$PO$_4$ (anhydrous) 3 g; NaCl 5 g; 0·1 M MgSO$_4$ 10 ml; 0·01 M CaCl$_2$ 10 ml; distilled water to 1 litre.
6. Soft agar containing Difco "Bacto" agar powder 6 g; water to 1 litre.

SPECIFIC REQUIREMENTS

Day 1
5 ml broth cultures of *E. coli* C, C(P1), K, K(P1) and B
1 ml $\lambda vir.$O at 10^9 phage ml^{-1}
100 ml phage buffer
5 nutrient agar plates
20 ml soft agar at 45 °C
tubes for dilutions
tubes for soft agar
tube racks
water-bath at 45 °C
pipettes for dilutions

pipettes or standard loop for phage spotting
37 °C incubator

Day 2
as for Day 1 + the following:
15 nutrient agar plates
75 ml soft agar at 45 °C
straight nichrome wire for plaque stabbing

Day 3
a colony counter will help students to see and to count plaques more accurately

---

FURTHER INFORMATION

Colson *et al.* (1965) describe in detail methods for scoring restriction and modification and illustrate the results of a rapid method which avoids dilutions. A plaque of λ grown on the strain to be tested is stabbed and the phage resuspended in 0·3 ml of buffer in a small tube. A straight wire is dipped vertically into the tube and used to stab repeatedly (about 12 times) along a radius of a thick agar plate overlaid with soft agar containing the indicator bacteria. A phage that is not restricted by the indicator bacteria will show a zone of lysis at each of the stabbing points, a phage which is restricted either lyses the indicator bacteria at the point of the first stab only, or not at all. However, some experience is necessary before this rapid method can be used reliably.

An alternative method for scoring restriction semiquantitatively is described by Piekarowicz and Glover (1972) which relies on cross-streaking critical dilutions of phage and bacteria. Vertical streaks of phage are made on nutrient agar plates and allowed to dry before cross-streaking with cultures of bacteria. The titre of the phage suspension used for streaking can be adjusted so that bacteria unable to restrict the phage are completely lysed at the point of intersection, while bacteria able to restrict the phage are not lysed or show the presence of a small number of plaques at the point of intersection.

Table 1 shows a typical set of results for the restriction test (Day 1).

Table 1   *E. coli* strains.

Dilution of $\lambda.O$	C	K	B	C(Pl)	K(Pl)
10^0	L	L	L	L	P
10^{-2}	L	P	P	P	O
10^{-4}	L	O	O	O	O
10^{-6}	P	O	O	O	O

(L = lysis; P = countable number of plaques (about 10); O = no lysis.)

Approximate efficiencies of plating can be calculated by assuming the plaque count of $\lambda.0$ on *E. coli* C to be 1·0. Ten plaques at 10^{-6} on *E. coli* C and 10 plaques at 10^{-2} on *E. coli* K means that the plating efficiency of $\lambda.0$ on *E. coli* K is 1×10^{-4}.

Table 2 shows a typical set of results for **part** of the modification test (Day 2).

Table 2   Indicator strains.

Dilution of $\lambda.K$	C	K	B	C(Pl)	K(Pl)
10^0	L	L	P	P	P
10^{-1}	L	L	O	O	O
10^{-2}	L	L	O	O	O
10^{-3}	P	P	O	O	O

(L = lysis; P = countable number of plaques (about 10); O = no lysis.)

REFERENCES

W. Arber and D. Dussoix (1962). Host specificity of DNA produced by *Escherichia coli*. I. Host controlled modification of bacteriophage λ. *Journal of Molecular Biology*, 5, 18–36.

C. Colson, S. W. Glover, N. Symonds and K. A. Stacey (1965). The location of the genes for host-controlled modification and restriction in *Escherichia coli* K-12. *Genetics*, 52, 1043–50.

D. Dussoix and W. Arber (1962). Host specificity of DNA produced by *Escherichia coli*. II. Control over acceptance of DNA from infecting phage. *Journal of Molecular Biology*, 5, 37–49.

A. Piekarowicz and S. W. Glover (1972). Host specificity of DNA in *Haemophilus influenzae*. The two restriction and modification systems in strain Ra. *Molecular and General Genetics*, 116, 11–25.

108

Linked Transduction in *Salmonella typhimurium* with Bacteriophage P22

D. E. S. STEWART-TULL

*Department of Microbiology, University of Glasgow,
Alexander Stone Building, Garscube Estate, Bearsden, Glasgow, Scotland*

and

P. SMITH-KEARY

Department of Genetics, Trinity College, Dublin 2, Eire

> *Level*: All undergraduate years
> *Subject areas*: Bacterial genetics
> *Special features*: Linked transduction is easily
> observed by difference in
> colour of recombinants
> plated on selective media

INTRODUCTION

In transduction a fragment of bacterial chromosome incorporated into bacteriophage DNA is transferred to a recipient strain. This fragment may be recombined into the recipient chromosome to form a stable transductant. The bacteriophage is grown lytically on the donor strain and allowed to lysogenize the recipient strain. The frequency of generalized transduction is about 1 per 10^5 phage-infected cells. Joint transduction of two genes is unlikely to be the result of two independent transductions since this would have a probability of $10^{-5} \times 10^{-5}$, i.e. 10^{-10}. Thus if two genes are co-transducted it is likely that they are linked together on the same chromosomal fragment. It is possible to use this transduction method for genetic mapping. For example, take genes P Q and R.

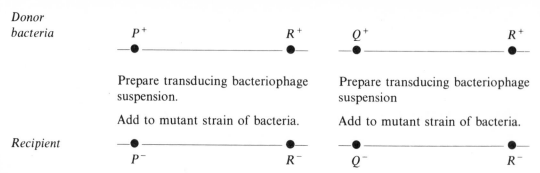

*Donor
bacteria*

Recipient

In these crosses select for bacteria carrying the markers P^+ and Q^+. Subsequently, the selected bacteria are tested to determine whether they are R^+ or R^-. The percentage of linked transductions, i.e. P^+R^+ or Q^+R^+ are calculated and these values are used as mapping distances. Note that the closer together two genes are on the chromosome the higher the

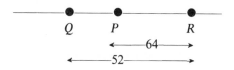

percentage of linked transduction.

This principle is conveniently shown with the group of genes controlling the biosynthesis of leucine in *Salmonella typhimurium* (Table 1).

Table 1   Linked transductions obtained from crosses with a *leu*⁺ *ara*B9⁺ donor and various *leu*⁻ *ara*B9⁻ recipients.

Progeny obtained from *leu*⁺ *ara*B9⁺ × *leu*⁻ *ara*B9⁻	*leu*⁻ *ara*B9⁻ recipients used in crosses			
	*leu*A121	*leu*B129	*leu*C126	*leu*D128
leu⁺	2890	3024	3205	2951
leu⁺ *ara*B9⁻ transductants	1607	1565	1615	1315
leu⁺ *ara*⁺ transductants	1283	1459	1590	1636
Percentage of *leu*⁺ *ara*⁺	44·4	48·3	49·6	55·4

Transductants *leu*⁺ *ara*⁺ involve cross-overs in regions I and III and *leu*⁺ *ara*B9 involve cross-overs in regions I and II. Region II will be small if the *leu* marker is close to *ara*B9 and large if the *leu* marker is distant from *ara*B9. If region II is small there will be fewer cross-overs occurring and more *leu*⁺ *ara*⁺ transductants will be generated by cross-overs in regions I and III.

Thus from Table 1, the following linkage map can be deduced:

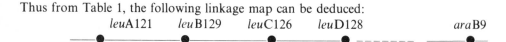

The aim of this experiment is to map the position of two genes controlling the biosynthesis of leucine in *S. typhimurium* in relation to the *ara*B9 marker. The percentage of linked transduction values are used as mapping distances.

Safety precautions: Care should be exercised in handling the pathogenic *S. typhimurium* used in this experiment. All pipettes should be discarded into a suitable disinfectant solution. Petri dishes and tubes containing this organism should be autoclaved. Wash your hands thoroughly before leaving the laboratory.

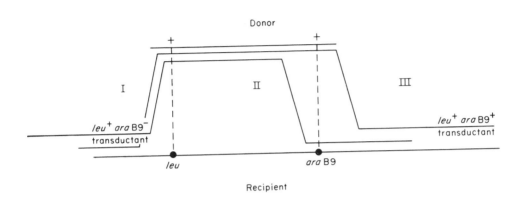

EXPERIMENTAL

Day 1 *(10 min + 60 min separated by 2 h)*

Inoculate 20·0 ml amounts of nutrient broth with 0·2 ml of overnight broth culture of *S. typhimurium leu*39⁻ *ara*B9⁻ or *leu*125⁻ *ara*B9⁻. Incubate at 37 °C for 2 h with shaking.

↓

Remove 5·0 ml of culture aseptically into a sterile Universal bottle and centrifuge at 3000 rev. min⁻¹ for 20 min to deposit cells (*see* note 3). Carefully remove the culture supernate into disinfectant with a sterile Pasteur pipette and resuspend the pellet in 5·0 ml phage buffer.

↓   ↓

Transfer 1·0 ml of this suspension to a sterile 4 × ½″ test tube.

Plate out loopfuls of the mutant suspensions on enriched minimal agar.

↓

Add 0·1 ml of phage P22 lysate, obtained by infection of *S. typhimurium met*A22 (*see* note 2), and incubate at 37 °C for 15 min. Inoculate two enriched minimal agar plates (*see* note 4) with 0·2 ml of each transduction mixture and two plates with 0·1 ml of each mixture. Spread

the inoculum over the surface of the agar with
a sterile glass spreader. Incubate the plates at
37 °C for 48 h.

Day 2 *(30 min)*

Examine the minimal agar plates for leucine-
independent colonies (*leu*⁺). Make streak inoculations from a total of 40 of these colonies
on to the surface of each of two eosin methylene blue arabinose agar plates as shown.
Incubate the plates at 37 °C for 2 days.

Examine the enriched minimal agar plates to
check that the mutant strains did not grow.

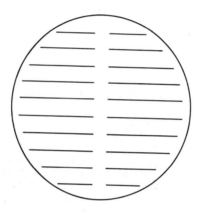

× 2 plates for each cross

Day 3 *(20 min)*

Count the number of arabinose-utilizing (*ara*⁺) and arabinose-non-utilizing (*ara*⁻) streaks on the
plates. The two types are easily recognized as *ara*⁺ colonies are dark purple with a greenish sheen
and *ara*⁻ are pale pink.

    Calculate from the results the percentage of linked transductions, i.e. transductants which are
leu⁺ *ara*⁺ and determine which of the two *leu* markers is closer to *ara*B9.

Notes and Points to Watch

1. Do not assume that the production of a single lysate will give good transducing activity—this
   must be checked.
2. The donor wild-type strain LT2 will grow on enriched minimal agar and therefore the P22
   phage lysate must be checked for bacterial sterility before use by the student.

This problem could be avoided by using a donor (*leu*⁺ *ara*⁺) which has a requirement for another amino acid, e.g. *met* A22⁻ which never reverts to *met*⁺.

3. It is advisable to remove the nutrient broth from the mutant strains (*leu*39⁻, *leu*125⁻) and to resuspend in buffer to prevent carry-over of sufficient nutrient broth to allow growth of the mutants on minimal agar. There must be no growth on the enriched minimal agar plates inoculated with the buffer suspension of mutant.
4. Best results will be obtained if the surface of the enriched minimal agar plates and eosin methylene blue arabinose plates are dried prior to use.
5. An *leu*⁻ *ara*⁻ check by laboratory staff on the mutant strains each year is desirable.
6. Other *leu*⁻ mutants may be used according to availability.

MATERIALS

1. Bacteriophage P22.
Salmonella typhimurium met A22.
*S. typhimurium leu*39 *ara*B9⁻ (68% linked)
*S. typhimurium leu*125 *ara*B9⁻ (46% linked) *see* notes 5 and 6 and Further Information

These strains and bacteriophage P22 can be obtained from Dr Peter Smith-Keary, Dublin. Please include one Irish punt or equivalent to cover the cost of handling.

2. Enriched minimal agar.

NH₄Cl,	5·0 g
NH₄NO₃,	1·0 g
Na₂SO₄,	2·0 g
K₂HPO₄,	3·0 g
KH₂PO₄	1·0 g
MgSO₄.7H₂O,	0·1 g
Agar (Oxoid L28),	15·0 g

Dissolve each salt in cold water in the order shown. Make volume to 1 litre. Autoclave at 103·4 kPa (15 lbf in⁻²) 15 min and add 20 ml 10% glucose (sterilized by filtration) before pouring.

3. Phage buffer.

Na₂HPO₄, anhydrous,	7 g
KH₂PO₄, anhydrous,	3 g
NaCl,	5 g

Dissolve these salts in water in order and add:
0·1 M Mg.SO₄,	10·0 ml
0·01 M CaCl₂,	10·0 ml

Make up the volume to 1 litre. Autoclave at 103·4 kPa for 15 min.

4. Phage broth.

Difco Bacto-peptone,	15·0 g
Oxoid tryptone,	8·0 g
NaCl,	8·0 g
D-glucose,	1·0 g
water,	1·0 litre

pH 7·2. Autoclave at 103·4 kPa for 15 min.

5. Eosin methylene blue arabinose agar.
(a) Eosin yellow 4 g, water to 100 ml.
(b) Methylene blue 0·65 g, water to 100 ml.
(c) L-arabinose 20 g dissolved in water and sterilized by filtration. Add 8·0 ml (a) and (b) to 750 ml nutrient agar base. Sterilize at 103·4 kPa for 15 min.

Before pouring plates add 40 ml sterile L-arabinose solution aseptically.

6. Suspension of bacteriophage P22 (*see* note 1). Grow the donor strain of *S. typhimurium met* A22 in 20 ml nutrient broth overnight at 37°C. Inoculate 20·0 ml phage broth with 0·5 ml starter culture and place in shaking incubator at 37°C; leave until the $E_{600\,nm}$ is approximately 0·6. Add 0·5 ml P22 suspension (10^{8-10} p.f.u. ml⁻¹) and continue the incubation for 5 h; clearing should be visible. Heat the culture at 60°C for 30 min and leave overnight. Centrifuge to remove bacterial debris. Retain the supernate containing phage particles.

The phage suspension can be used in this form but it would be worthwhile to determine the number of p.f.u. ml⁻¹ by standard titration methods.

SPECIFIC REQUIREMENTS

Day 1
overnight broth cultures of *S. typhimurium*
  *leu*39⁻ *ara*B9⁻ and *leu*125⁻ *ara*B9⁻
P22 phage lysate
2 × 20 ml nutrient broth in 100 ml Erlenmeyer
  flasks
2 sterile Universal bottles
2 × 5·0 ml phage buffer
two 4 × ½″ test tubes and rack
6 enriched minimal agar plates

2 sterile glass spreaders
5·0 ml pipettes
1·0 ml pipettes
sterile Pasteur pipettes
centrifuge
shaking water-baths

Day 2
4 eosin methylene blue arabinose agar plates

FURTHER INFORMATION

The results obtained from a series of class experiments are tabulated below:

	*leu*125⁺ *ara*B9⁺ × *leu*125⁻ *ara*B9⁻	*leu*39⁺ *ara*B9⁺ × *leu*39⁻ *ara*B9⁻
Total number of *leu*⁺ transductants tested	800	800
Number of *leu*⁺ *ara*B9 transductants	436	282
Number of *leu*⁺ *ara*⁺ transductants	364	518
Percentage of *leu*⁺ *ara*⁺ linked transductants	45·5	64·75
Percentage of *leu*⁺ *ara*B9 recombinants	54·5	35·25

It is apparent that the percentage of linked transduction is higher and the percentage recombination is lower with *leu*39 than with *leu*125. The linkage map can be represented as:

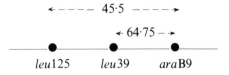

The linkage order can be established with a three-point test cross. Single mutants (*leu*39⁻ and *leu*125⁻) are used as donors with the double mutants (*leu*125⁻ *ara*B9⁻ and *leu*39⁻ *ara*B9⁻ respectively) as the recipients.

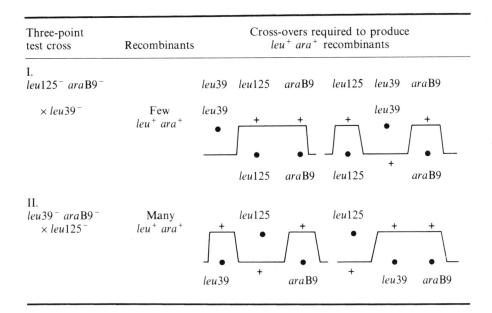

| Three-point test cross | Recombinants | Cross-overs required to produce *leu*+ *ara*+ recombinants |

If the order of genes is *leu*39 *leu*125 *ara*B9 only two cross-overs are required to yield *leu*+ *ara*B9+, whereas four cross-overs would be required if the order was *leu*125 *leu*39 *ara*B9. Quadruple cross-overs will be rare compared with double cross-overs. In cross I, few *leu*+ *ara*+ were scored consistent with the requirement for four cross-overs and in cross II many *leu*+ *ara*+ were scored which is indicative of the order *leu*125 *leu*39 *ara*B9.

REFERENCES

W. Hayes (1968). "The Genetics of Bacteria and their Viruses", pp. 127–57. Blackwell Scientific, Oxford.

E. Lennox (1955). Transduction of linked characters of the host by bacteriophage P1. *Virology*, **1**, 190–206.

P. F. Smith-Keary (1975). "Genetic Structure and Function." Macmillan, London.

B. A. D. Stocker, N. D. Zinder and J. Lederberg (1953). Transduction of flagellar characters in *Salmonella. Journal of General Microbiology*, **9**, 410–33.

N. D. Zinder and J. Lederberg (1952). Genetic exchange in *Salmonella. Journal of Bacteriology*, **64**, 679–99.

Induction of Thymidine Kinase in Bacteriophage T4-infected *Escherichia coli*

D. A. RITCHIE

*Department of Genetics, University of Liverpool,
Liverpool, L69 3BX, England*

> *Level*: Advanced undergraduates
> *Subject areas*: Microbial genetics and virology
> *Special features*: Combines biochemical and genetical techniques

INTRODUCTION

The enzyme thymidine kinase phosphorylates thymidine (TdR) to form thymidine monophosphate (dTMP) which is subsequently converted to thymidine triphosphate (dTTP) for incorporation into DNA. This synthesis of dTMP from exogenously supplied TdR is not normally essential for DNA metabolism since bacteria can synthesize dTMP endogenously from dUMP catalysed by the enzyme thymidylate synthetase. This can be deduced from the observation that mutants of the bacterium *Escherichia coli* which are unable to synthesize active thymidine kinase (*tdk* mutants) are viable under normal conditions. Thymidine kinaseless mutants, however, are unable to incorporate

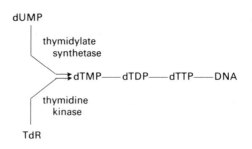

TdR into DNA (Hiraga *et al.*, 1967; Ritchie *et al.*, 1974). A functional thymidine kinase becomes essential for DNA synthesis in cells lacking an active thymidylate synthetase.

tdk mutants of *E. coli* infected by the wild type of phage T4 will incorporate TdR into DNA suggesting that the phage also specifies a thymidine kinase activity which is induced after infection. Proof of this comes from the isolation of phage T4 mutants which fail to incorporate TdR into DNA following infection of a *tdk* mutant host (Chace and Hall, 1973; Ritchie *et al.*, 1974).

The aim of this experiment is to demonstrate this interrelationship between phage and bacterial thymidine kinases. It also demonstrates the kinetics of appearance of a phage-induced function and provides a technique by which phage or bacterial DNA synthesis can be measured.

EXPERIMENTAL

The method used here for detecting thymidine kinase activity is indirect since it measures the incorporation of the precursor, TdR, into DNA which is the end-product of the pathway. The assay depends on the measurement of the incorporation of radioactivity from ^{3}H-labelled TdR into a form insoluble in trichloracetic acid (TCA). DNA is insoluble in TCA whereas TdR, dTMP, dTDP and dTTP are soluble.

Day 1 *(2 h)*

To each of three 5 ml cultures of both the wild type and the *tdk* mutant of *E. coli* add 0·1 ml of TdR and 0·1 ml of ^{3}H-TdR and aerate at 30 °C (*see* notes 1 and 4).

$$\downarrow$$

Immediately infect one wild type culture with T4$^+$ phage and a second with T4tk phage to give a multiplicity of 2–5 phage per cell. The third culture is left uninfected. This time marks the beginning of the experiment, i.e. 0 min.

$$\downarrow$$

Immediately repeat this procedure for the three *tdk* mutant cultures (*see* note 2).

$$\downarrow$$

At 0, 10, 20, 30, 40, 50 and 60 min after infection remove a 100 μl sample from each culture on to a labelled filter paper disc and immediately immerse in ice-cold TCA solution (*see* note 5) **caution**.

$$\downarrow$$

At 70 min pour off the TCA solution from the samples, wash them with three changes of tap-water and set out to dry (*see* note 3) **caution**.

$$\downarrow$$

Transfer each dried filter paper disc to a scintillation vial, add 10 ml scintillation fluid and assay the radioactivity.

$$\downarrow$$

The radioactive counts for each culture should be drawn as a graph which plots time on the abscissa against counts per minute on the ordinate.

- -

Notes and Points to Watch

1. Radioactive samples must be handled with care. This involves the use of disposable plastic gloves, protective bench covering (Benchkote) and provision for disposal. Samples should be taken with a hand-operated automatic pipette and **not** mouth-pipetted.
2. It would be possible to divide the experiment into two halves, one half for wild type bacteria (3 cultures) and the other for *tdk* bacteria (3 cultures). The class could then pool results.
3. The work could be spread over 2 days by interrupting the schedule after the filter paper discs have been washed and set to dry. Drying can be done in a hot oven for a few minutes or overnight at 37 °C or on the bench.
4. Bacterial cultures can be aerated by either bubbling or shaking.
5. The filter paper discs can be impaled on pins arranged on a board or simply laid flat on a sheet of foil prior to adding the sample. It is best if the discs are labelled with a pencil.

MATERIALS

1. *Escherichia coli* strains W3110 (wild type) and W3110 *tdk*32 (*tdk* mutant).
2. Stocks of phage T4 wild type (T4$^+$) and T4 *tk*26 (T4*tk*) at a titre of about 5×10^{10} plaque-forming units per ml.
3. TCG medium containing 0·023 g Na$_2$SO$_4$, 0·12 g MgSO$_4$, 0·5 g NaCl, 0·01 g CaCl$_2$, 0·09 g KH$_2$PO$_4$, 1 g glucose and 0·5 g casamino acids (Difco) per litre of distilled water, buffered to pH 7·2 with 0·1 M Tris.
4. L-tryptophane, 2'-deoxyadenosine and thymidine (Sigma Chemical Co.).

5. Methyl ^{3}H-thymidine can be obtained from the Radiochemical Centre, Amersham, England, and is diluted to 250 μCi ml^{-1} with distilled water.
6. Filter paper discs, 2·5 cm diam. The grade of paper is not critical.
7. Scintillation fluid—any standard non-aqueous scintillant is suitable.
8. Vials for a liquid scintillation spectrophotometer.

SPECIFIC REQUIREMENTS

3 × 5 ml cultures of W3110 and 3 × 5 ml cultures of W3110 *tdk*32 (Cultures should be grown in TCG medium with aeration to the logarithmic phase and have a titre of about 2×10^8 cells ml^{-1}. At this stage add 20 μg ml^{-1} *l*-tryptophane (required for phage T4 adsorption) and 250 μg ml^{-1} 2'-deoxyadenosine to promote thymidine incorporation.)
aqueous solutions of thymidine (125 μg ml^{-1}) and ^{3}H-thymidine (250 μCi ml^{-1})
250 ml of a 10 % solution of trichloracetic acid

(TCA) standing in an ice bath (**caution**)
pipettes, but an automatic pipette with disposable tips is very useful
filter paper discs labelled in pencil
forceps for transferring discs
scintillation fluid and tubes or vials for a scintillation spectrophotometer
water-bath with tube rack and aeration supply or shaking incubator with small culture flasks suitable for 5 ml cultures
protective gloves and bins, etc. for radioactive disposal

FURTHER INFORMATION

The results from this experiment should be striking and clear-cut. No incorporation of ^{3}H-TdR should be observed for the uninfected W3110 tdk32 culture or following infection with T4 tk phage. Infection of W3110 tdk32 by T4 wild type phage should show ^{3}H-TdR incorporation increasing from about 5 min after infection (Figure 1). This is evidence for the induction of a T4-specified thymidine kinase function. The three W3110 wild type cultures should all incorporate ^{3}H-TdR. However, following T4 wild type infection the level of incorporation should be enhanced compared with the uninfected control. This results from the appearance of the T4-specified enzyme. This enhanced incorpration will not, therefore, be observed following infection by T4 tk phage.

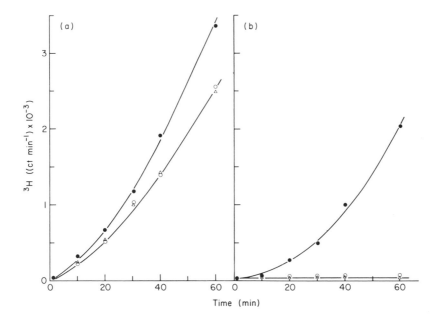

Figure 1 The effect of T4 infection on the incorporation of ^{3}H-TdR by W3110 wild type (a) and W3110 tdk32 mutant (b) bacteria. ^{3}H-TdR incorporation was determined for uninfected bacteria ($\triangle$) and for bacteria infected with T4 wild type ($\bullet$) and T4 tk26 mutant ($\bigcirc$) phage.

This experiment could be developed or modified in two ways. The first would be to make an extract from each culture and separate the various thymidine compounds—TdR, dTMP, dTDP and dTTP—by chromatography. The activity of the thymidine kinases could then be determined directly by examination of the nucleotides for the presence of radioactivity. The second development would be to extract the thymidine kinase activity from each culture and assay the amount of activity by the ability to convert TdR to the nucleotide.

REFERENCES

K. V. Chace and D. W. Hall (1973). Isolation of mutants of bacteriophage T4 unable to induce thymidine kinase activity. *Journal of Virology*, **12**, 343–8.

S. Hiraga, K. Igarishi and T. Yura (1967). A deoxythymidine kinase-deficient mutant of *Escherichia coli*. I. Isolation and some properties. *Biochimica et Biophysica Acta*, **145**, 41–51.

D. A. Ritchie, A. T. Jamieson and F. E. White (1974). The induction of deoxythymidine kinase by bacteriophage T4. *Journal of General Virology*, **24**, 115–27.

110

Extracellular Morphogenesis of Phage T4

U. WINKLER, W. RÜGER and W. WACKERNAGEL

Lehrstuhl Biologie der Mikroorganismen, Ruhr-Universität,
Postfach 102148, D-4630 Bochum, FRG

Level: Advanced undergraduates
Subject areas: Phage genetics
Special features: Demonstration of self-assembly of
infectious phage particles from
suspensions of phage heads
and tails

INTRODUCTION

Genetic experiments have shown that about one half of the more than 100 known genes of *Escherichia coli* phage T4 participate in its morphogenesis (Wood and Revel, 1976). However, as demonstrated by biochemical methods and electron microscopy, only between 15 and 20 different structural proteins appear in mature phage particles. This discrepancy can be explained by the assumption that various steps in the assembly of the structural components require proteins with catalytic functions. Some of these proteins act as "scaffolding devices"; others may activate individual subunits, e.g. by specific cleavage, enabling their subsequent spontaneous assembly (Showe and Kellenberger, 1975; Wood, 1979).

With nonsense mutants of phage T4, some steps of phage morphogenesis can be studied. When non-suppressing host bacteria (*su*) are infected with phages having amber (*am*) mutations in one of the genes involved in morphogenesis, only parts of phages, such as heads, tails, tail fibres or other structures accumulate. If two cell extracts prepared from *su* bacteria after infection by different T4 *am* mutants are mixed, then the accumulated phage subunits may complement mutually, so that infectious phage particles are formed extracellularly (Edgar and Wood, 1966). This proves that at least some of the unassembled phage components visible in the electron microscope are intermediate products of normal phage formation rather than defective subunits unable to assemble. Based on the extracellular complementation of phage mutants, the pathway of T4 assembly has been established (Figure 1). At first the heads, tails and tail fibres are synthesized

---

independently. Then the head and tail subunits react with one another and, only after this, are the tail fibres able to attach. This sequential self-assembly of the phage particles indicates that some gene products must react with certain intermediate structures to condition them to be a substrate for the next step in the reaction scheme.

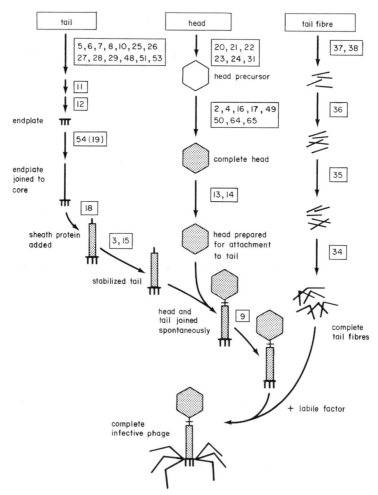

Figure 1 Pathway of the formation of infectious phage T4 by sequential assembly of phage subunits. The numbers refer to the various genes involved in each step of the assembly process. (Reproduced with permission from J. D. Watson, "Molecular Biology of the Gene", 3rd edn, W. A. Benjamin, 1976.)

The morphogenesis of phage T4 is one of the best understood models of sequential self-assembly of supramolecular structures starting out from a large number of gene products (Kushner, 1969; Wood, 1979). The aim of this experiment is to demonstrate T4 morphogenesis in a cell-free system. Cell extracts will be prepared from *su* bacteria infected separately with two different T4 mutants which have *am* mutations in genes, no. 23 and 27 (*see* Figure 1) required for morphogenesis. After incubation of a mixture of both extracts the increase of the number of phage particles able to form plaques on su^+ bacteria will indicate spontaneous extracellular self-assembly of phages.

EXPERIMENTAL

In the first part of the experiment cells are infected with phages and extracts prepared. These may be frozen at $-70\,°C$ and the experiment interrupted for several hours or even days. In the second part the phage titres in the extracts and in the mixture of extracts are determined by plating.

Day 1 (about 7 h)

1. *Preparation of cell extracts*

Take 30 ml culture of *E. coli* BA in Hershey broth.

↓

Inoculate 190 ml of Hershey broth with 10 ml of culture and aerate at $37\,°C$ to 4×10^8 cells ml^{-1}.

↓   ↓

Add 2 ml of tryptophane solution and 0·3 ml of T4 *am* N120 (*see* notes 1 and 4). Aerate vigorously at $30\,°C$ for 30 min. | Add 2 ml of tryptophane solution and 0·3 ml of T4 *am* B17. Aerate vigorously at $30\,°C$ for 30 min.

↓   ↓

Chill the culture quickly by pouring it into a 1 litre Erlenmeyer flask standing in ice water.

↓   ↓

Transfer culture to a centrifuge bottle and sediment the cells at 5000 g for 8 min at $+4\,°C$.

↓   ↓

Discard supernate and remove any residual liquid with a Pasteur pipette. Dry the inside of the bottle with sterile absorbent paper (*see* note 2).

↓   ↓

Take up sediment in 1 ml phosphate buffer containing DNaseI; the DNase reduces the viscosity of the suspension. Cells are now concentrated about 200-fold. Transfer cell suspension to a sterile plastic tube holding 5 ml.

↓   ↓

Quickly freeze the cells in a methanol dry ice bath at $-70\,°C$ (*see* note 3) and thaw at $30\,°C$. Repeat this procedure. Then 99 % of the cells should be broken. Check under the microscope after diluting a sample 1 : 100 in phosphate buffer.

2. *Determination of phage titres in the extracts*

Label plates according to Table 1 and prepare the dilution series with phosphate buffer. Thaw both of the cell extracts at 30 °C and mix 0·5 ml of each. Immediately take 0·1 ml samples from the N120 and B17 extracts and from the mixture, dilute according to Table 1 and plate on *su* and *su*$^+$ indicator using the soft agar layer technique. Samples of 0·1 ml of the dilutions are plated with

Table 1   Plating schedule for phage extracts.

Sampling time and extracts	Dilution for plating	Plate no. and indicator bacteria		Plaque titre: typical results (per ml)	Symbol
		BA (*su*)	CR63 (*su*$^+$)		
Time: 0 *min*					
am N120	10^{-2}	1	—	$5·6 \times 10^4$	A_0
	10^{-2}	2	—		
	10^{-5}	—	3		
	10^{-5}	—	4	$1·2 \times 10^8$	B_0
	10^{-6}	—	5		
am B17	10^{-2}	6	—	$9·6 \times 10^4$	C_0
	10^{-2}	7	—		
	10^{-5}	—	8		
	10^{-5}	—	9	$6·4 \times 10^7$	D_0
	10^{-6}	—	10		
am N120 +	10^{-2}	11	—		
am B17	10^{-2}	12	—	$2·0 \times 10^5$	E_0
	10^{-3}	13	—		
	10^{-6}	—	14		
	10^{-6}	—	15	$8·0 \times 10^7$	F_0
	10^{-7}	—	16		
Time: 200 *min*					
am N120	10^{-2}	17	—	$5·2 \times 10^4$	A_{200}
	10^{-2}	18	—		
	10^{-5}	—	19		
	10^{-5}	—	20	$1·0 \times 10^8$	B_{200}
	10^{-6}	—	21		
am B17	10^{-2}	22	—	$1·0 \times 10^5$	C_{200}
	10^{-2}	23	—		
	10^{-5}	—	24		
	10^{-5}	—	25	6.2×10^7	D_{200}
	10^{-6}	—	26		
am N120 +	10^{-2}	27	—		
am B17	10^{-2}	28	—	$2·2 \times 10^5$	E_{200}
	10^{-3}	29	—		
	10^{-6}	—	30		
	10^{-6}	—	31	$3·1 \times 10^9$	F_{200}
	10^{-7}	—	32		

0·2 ml of a culture of indicator cells. While this is being done, incubate the 3 cell extracts at 30 °C for 200 min (*see* note 5). Then again dilute samples of 0·1 ml according to the Table 1 and plate. Incubate the plates at 30 °C overnight.

Day 2 *(about 2 h)*

Count the plaques on plate nos. 1 to 32 and calculate the phage titres. The titres A, C and E represent revertants (wild-type) in the three extracts and the titres B, D and F the total of infectious phages in the three extracts.

Calculate the ratios of the different titres according to Table 2.

Table 2   Evaluation of data.

Extracts/ indicator bacteria	Ratio	Frequency of wild-type revertants at 0 and 200 min: typical results
am N120	A_0/B_0	$4\cdot7 \times 10^{-4}$
	A_{200}/B_{200}	$5\cdot2 \times 10^{-4}$
am B17	C_0/D_0	$1\cdot5 \times 10^{-3}$
	C_{200}/D_{200}	$1\cdot6 \times 10^{-3}$
am N120 + *am* B17	E_0/F_0	$2\cdot5 \times 10^{-3}$
	E_{200}/F_{200}	$7\cdot0 \times 10^{-5}$
am N120 + *am* B17	Ratio	Frequency of infectious phages, formed by spontaneous extracellular assembly: typical results
Indicator CR63	$2 \times F_0/(B_0 + D_0)$	0·87
	$2 \times F_{200}/(B_{200} + D_{200})$	38·7
Indicator BA	$2 \times E_0/(A_0 + C_0)$	2·6
	$2 \times E_{200}/(A_{200} + C_{200})$	2·9

Notes and Points to Watch

1. Phage T4 requires tryptophane for proper adsorption to the host cell. A concentration of 10 μg ml^{-1} in the medium is sufficient. Apparently, tryptophane acts by bringing the phage tail fibres into the "active" conformation necessary for adsorption. Broth media generally contain tryptophane, but the amount is usually low and most of the tryptophane may be consumed by the cells before addition of phage T4. It is therefore advisable to add tryptophane to cultures in both complete and minimal medium immediately before infection with T4.
2. For the preparation of extracts, the cells are infected with *am* mutants of T4. It is important to remove as many of the non-adsorbed phages from the sedimented cells as possible (Pasteur

pipette, absorbent paper). These phages, if not removed, will be present in the extracts (background) and may obscure extracellular morphogenesis.

3. Breakage of *E. coli* cells by repeated freezing and thawing is a relatively mild method. Freezing and thawing breaks phage-infected cells effectively since lysing enzymes coded by the phage DNA weaken the cell envelope from inside. Non-infected cells require the additional application of egg-white lysozyme and/or a detergent (e.g. Brij 58, Triton X-100, sodium dodecyl sulphate) for lysis.

4. Phage stocks of *am* mutants are propagated in *su*⁺ bacteria. During several passages the frequency of revertants may increase since *am*⁺ phages often have selective advantages (suppression of *am* mutations rarely restores 100% of gene function). In order to reduce the background of revertants to the minimum value it may be necessary to start a new phage stock from a single plaque.

5. The given incubation time of 200 min for the extracts may be reduced to 100 min with only minor loss of extracellularly formed infectious phages.

MATERIALS

1. *Escherichia coli* strains BA *su* and CR63 *su*⁺.
2. Phage mutants. T4 *am* N120 (gene No 27; no phage tails) and T4 *am* B17 (gene No 23; no phage heads).
3. Hershey broth containing 8 g nutrient broth, 5 g peptone, 5 g NaCl and 1 g glucose per litre of distilled water.
4. Enriched Hershey (EH) agar containing 15 g agar, 13 g tryptone, 8 g NaCl, 2 g sodium citrate . 2 H_2O, 1·3 g glucose per 1 litre distilled water. EH soft agar is EH agar but contains 6·5 g instead of 10 g agar and 3 g instead of 1·3 g glucose.
5. Methanol bath cooled to −70 °C. Alternatively a mixture of dry ice and methanol may be used.
6. DNase I (pancreatic DNase; commercial grade).

SPECIFIC REQUIREMENTS

Day 1

30 ml of an overnight culture in Hershey broth of *E. coli* BA (*c.* 2×10^9 cells ml^{-1})

10 ml of an overnight culture in Hershey broth of *E. coli* CR63

0·5 ml phage suspension of T4 *am* N120 with a titre of 1×10^{12} ml^{-1} on indicator *E. coli* CR63

0·5 ml phage suspension of T4 *am* B17 with a titre of 1×10^{12} ml^{-1} on indicator *E. coli* CR63

2·5 ml of ice cold phosphate buffer (see below) containing 10 μg DNaseI ml^{-1}

2 Erlenmeyer flasks (500 ml), each with 190 ml Hershey broth and aeration device

5 ml aqueous tryptophane solution (1 mg ml^{-1})

2 Erlenmeyer flasks (1000 ml)

1 bucket with crushed ice

2 centrifuge bottles (250 ml)

2 plastic tubes holding about 5 ml (to withstand freezing at −70 °C and thawing at 30 °C)

2 Pasteur pipettes

sterile absorbent paper

refrigerated centrifuge with rotor for 250 ml centrifuge bottles

2 water-baths, at 37 °C and at 30 °C

1 water-bath at 47 °C with a rack for soft-agar tubes

32 small tubes containing molten EH soft-agar (3 ml each)

32 plates with EH agar

300 ml sterile neutral phosphate buffer (60 mM) containing 0·4% NaCl and 2 mM $MgSO_4$

pipettes

tubes for dilutions

tube rack

FURTHER INFORMATION

Typical results of the experiment are shown in the Tables 1 and 2. Some comments on the experimental data may be helpful for the evaluation.

Cell extracts for extracellular morphogenesis

The phage-infected cells were concentrated to about

$$200\,\text{ml} \times (4 \times 10^8\,\text{ml}^{-1})\,\text{ml}^{-1} = 8 \times 10^{10}\,\text{ml}^{-1}.$$

Assuming that (1) all cells were infected, (2) 99 % of the cells were disintegrated by freezing and thawing and (3) each cell produced on average 20 phage heads or phage tails alternatively, then each of the two extracts should have contained phage equivalent substructures at a concentration of

$$0.99 \times (8 \times 10^{10}\,\text{ml}^{-1}) \times 20 \approx 1.6 \times 10^{12}\,\text{ml}^{-1}.$$

However, in addition to this, the extracts also contained infectious phages as shown by plating on indicator CR63 su^+. The ratio of infectious phages to phage equivalents was, for example, $1.2 \times 10^8/1.6 \times 10^{12} \approx 10^{-4}$ indicating a large excess of phage subunits over complete particles.

Since a phage titre of 3 to 4 orders of magnitude lower was observed on BA su the majority of the infectious phages consisted of am mutants. They originated possibly as:

(1) Phages which did not adsorb to BA su cells during the infection and which were trapped by the sedimented cells during centrifugation.

(2) Progeny of am phages which adsorbed to BA su cells and multiplied as a result of so-called transmission, which means that the am mutation did not completely abolish the respective gene function of the phage (leakiness).

Evident extracellular morphogenesis

When the mixture of am N120 and am B17 extracts was incubated, the phage titre on indicator CR63 su^+ increased to a value which was about 40 times higher than the control (mixture with no incubation). As a corresponding increase of the plaque titre on BA su was not observed, the phages formed by extracellular morphogenesis must be am mutants as expected.

REFERENCES

R. S. Edgar and W. B. Wood (1966). Morphogenesis of bacteriophage T4 in extracts of mutant-infected cells. *Proceedings of the National Academy of Science, USA* **55**, 498–505.

D. J. Kushner (1969). Self-assembly of biological structures. *Bacteriological Reviews*, **33**, 302–45.

M. K. Showe and E. Kellenberger (1975). Control mechanisms in virus assembly. *In* "Control Processes in Virus Multiplication", (D. C. Burke and W. C. Russell, eds), pp. 407–38. Cambridge University Press, Cambridge.

W. B. Wood (1979). Bacteriophage T4 assembly and the morphogenesis of subcellular structure. *The Harvey Lectures Series*, **73**, 203–23.

W. B. Wood and H. R. Revel (1976). The genome of bacteriophage T4. *Bacteriological Reviews*, **40**, 847–68.

111

Radiotracer Studies with Phage-infected Bacteria

M. IOSSON

Department of Microbiology, University of Reading,
London Road, Reading, RG1 5AQ, England

> *Level*: Advanced undergraduates
> *Subject area*: Microbial biochemistry; virology
> *Special features*: An introduction to labelling studies for nucleic acid metabolism and hybridization techniques

INTRODUCTION

Synthesis of nucleic acids in phage-infected bacteria may be examined by labelling with a radioactive nucleotide precursor, e.g. ^{3}H-uridine for RNA and ^{3}H-thymidine for DNA. When a phage such as λ infects a host bacterium, the host DNA is not broken down, and synthesis of host DNA may continue for a short period after infection.

Phage nucleic acid metabolism commences with the synthesis of "early" mRNA, followed several minutes later by phage DNA and "late" mRNA (which codes for structural proteins). Thus infected bacteria may be synthesizing both host and phage DNA. Phage λ DNA has the same average base composition (approx. $50\% \, G + C$) as *Escherichia coli* DNA, and cannot readily be distinguished from it by chemical means. The techniques of nucleic acid hybridization can be used to identify the species of nucleic acid: immobilized single-stranded DNA molecules of known specificity are incubated (under conditions promoting annealing of single strands) with a radioactively labelled mixture of single-stranded nucleic acids. Any of the radioactively labelled molecules which "finds" a complementary sequence of bases on an immobilized molecule will form a stretch of double-stranded, base-paired nucleic acid and itself become immobilized. The resulting hybrid molecules will be radioactively labelled in proportion to the frequency of complementary species of nucleic acid in the incubation mixture. Removal of the non-annealed molecules allows the retained hybrid molecules to be counted, and thus gives a measure of the amount of nucleic acid of similar specificity in the incubation mixture. Slightly different procedures are used for DNA–RNA hybridization (Gillespie and Spiegelman, 1965) and DNA–DNA hybridization (Denhardt, 1966), but the principle is the same: DNA of known type is immobilized on a nitro-cellulose membrane filter, and incubated with the labelled extract from infected cells. After incubation, the filter is washed thoroughly to remove non-annealed nucleic acids, dried and counted.

In addition to determining the specificity of the nucleic acids synthesized in a given time, by a slight modification of technique, it is possible to study the **kinetics** of synthesis, and hence obtain information on some of the events occurring after infection by phage. Kinetics of incorporation are studied by **pulse-labelling**: a sample of culture is incubated with a radioactive precursor of DNA or RNA synthesis for a short time (1–2 min) then incorporation is stopped by the addition of ice-cold trichloracetic acid (TCA). The **incorporated** labelled material is determined by collecting the TCA precipitate on a filter, washing to remove non-incorporated material, drying and counting. By taking samples of culture for pulse-labelling at different times through the latent period, it is possible to determine the instantaneous **rate** of DNA or RNA synthesis.

The purpose of the experiments included here is to attempt to get an overall view of the nucleic acid metabolism of phage-infected bacteria: the kinetics of RNA and DNA synthesis, and the nature (whether host or phage) of the nucleic acids synthesized at different times after infection.

EXPERIMENTAL

The techniques and experiments given are designed for five inter-connected experimental groups. The overall scheme is given below.

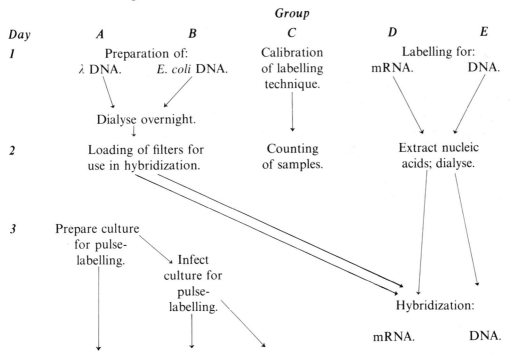

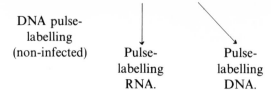

DNA pulse-
labelling
(non-infected)

Pulse-
labelling
RNA.

Pulse-
labelling
DNA.

Day 4 Washing, drying and counting of samples of all groups.

Days 5, 6 Analysis of results.

Group A

1. Preparation of unlabelled λ DNA

Day 1 *(1 h: 1h + 30 min separated by 1h)*

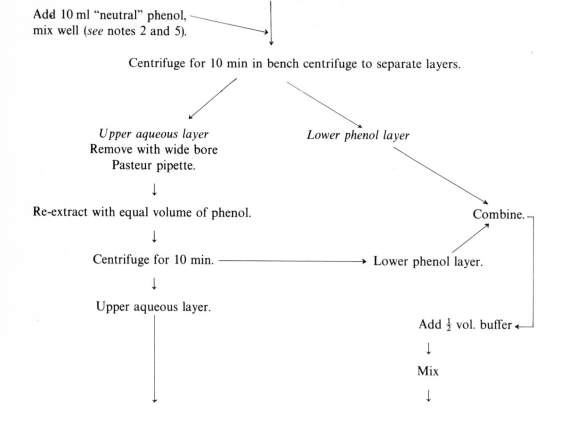

Take 10 ml λ phage *(see* note 4).

Add 10 ml "neutral" phenol,
mix well *(see* notes 2 and 5).

Centrifuge for 10 min in bench centrifuge to separate layers.

Upper aqueous layer
Remove with wide bore
Pasteur pipette.

Lower phenol layer

Re-extract with equal volume of phenol.

Combine.

Centrifuge for 10 min. ⟶ Lower phenol layer.

Upper aqueous layer.

Add ½ vol. buffer

Mix

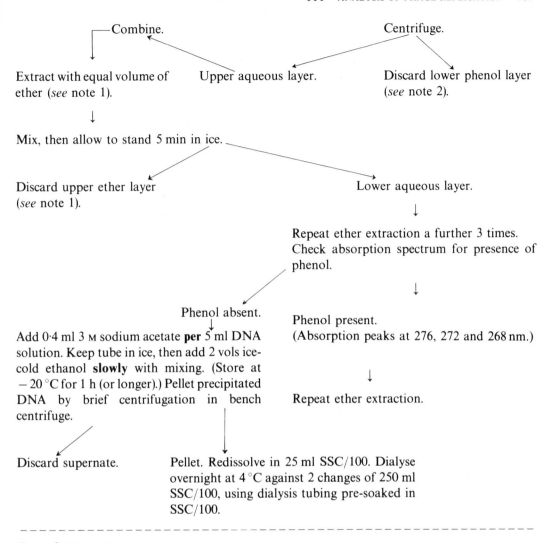

┌─ Combine.   Centrifuge.

Extract with equal volume of   Upper aqueous layer.   Discard lower phenol layer
ether (*see* note 1).   (*see* note 2).

Mix, then allow to stand 5 min in ice.

Discard upper ether layer   Lower aqueous layer.
(*see* note 1).

                                          Repeat ether extraction a further 3 times.
                                          Check absorption spectrum for presence of
                                          phenol.

                        Phenol absent.   Phenol present.
Add 0·4 ml 3 M sodium acetate **per** 5 ml DNA   (Absorption peaks at 276, 272 and 268 nm.)
solution. Keep tube in ice, then add 2 vols ice-
cold ethanol **slowly** with mixing. (Store at
− 20 °C for 1 h (or longer).) Pellet precipitated   Repeat ether extraction.
DNA by brief centrifugation in bench
centrifuge.

Discard supernate.   Pellet. Redissolve in 25 ml SSC/100. Dialyse
                       overnight at 4 °C against 2 changes of 250 ml
                       SSC/100, using dialysis tubing pre-soaked in
                       SSC/100.

- -

Day 2 *(3 h + 4 h)*

Record absorption spectrum of DNA, and measure A_{260} to estimate DNA concentration.
(Assume A_{260} of 1·0 = 50 μg DNA ml^{-1}.) Dilute in SSC/100 if necessary to less than 100 μg ml^{-1}.
Heat in boiling water-bath for 20 min to denature DNA.

Pour into sufficient 6 × SSC (in an ice bath) to give a 5-fold dilution, or 10 μg DNA ml^{-1},
whichever is the greater dilution.

Re-check absorption spectrum; compare with double-stranded DNA and confirm denaturation
(*see* note 15a). Keep DNA solution on ice. (Retain 4 ml for groups D and E, Day 3.)

2. Loading of filters

Soak 24 nitrocellulose filters in $6 \times$ SSC for 10 min (*see* note 6).

$$\downarrow$$

Set up filter(s) in filtration unit.

$$\downarrow$$

Load each filter with 5 ml of denatured DNA solution; adjust vacuum for flow rate of approximately 1 ml min^{-1}.

$$\downarrow \qquad\qquad\qquad \downarrow$$

(First sample only.) Collect filtrate. Check absorption spectrum for absence of A_{260} material. If material present, **repeat** "loading" of filter at lower flow rate.

Pass 50 ml cold $6 \times$ SSC through filter, then remove filter, invert and repeat with a further 50 ml $6 \times$ SSC.

$$\downarrow$$

Air dry (1 min).

$$\downarrow$$

Remove filter and place in a vacuum desiccator over P_2O_5.

$$\downarrow$$

Leave on bench (4 h).

Place vacuum desiccator in 80 °C oven (overnight, or for a minimum of 2 h) (*see* note 7).

- -

Day 3 *(approx. 5 h; including 3 h incubation)*

Return desiccator to room temperature.

3. Pulse labelling of non-infected E. coli

Groups **must** be co-ordinated, to enable all three cultures to be pulse-labelled at the same time.

Take 50 ml PN medium.

$$\downarrow$$

Add 2·5 ml overnight culture *E. coli* K12. Incubate in shaking water-bath at 37 °C until $A_{610} = 0·5$ (1 cm cell).

$$\downarrow$$

Harvest by centrifugation (15 min, 10 000 rev. min^{-1}).

$$\downarrow$$

Wash pellet in 50 ml sterile glass distilled water. Resuspend in TCM buffer to $A_{610} = 0.8$.

Non-infected culture   *Culture for group B*
Store on ice.

↓

Warm 10 ml suspension to 37 °C (2 min).

↓

Add 5·0 ml TCM at 37 °C (control for groups B and C).

↓

Incubate 20 min, 37 °C.

↓

Pipette 5·0 ml into 250 ml flask containing 45 ml 1·1 × concentrated K medium at 37 °C. Start timer and shake culture gently.

↓

At 2 min intervals (0–20 min) then 5 min intervals (25–50 min) remove 0·5 ml for pulse labelling.

↓

Add each sample to tube containing 0·2 μCi (0·01 ml) ^{3}H-thymidine. Incubate for 1 min (*see* note 8) at 37 °C.

↓

Add 0·3 ml ice-cold 20 % TCA containing 5 mg thymidine ml^{-1}. Store tubes in ice for 30 min (or overnight).

Day 4 *(approx. 3 h)*

Collect filtrates on Whatman glass fibre filters (*see* note 9), taking care to wash all precipitate out of each tube.

↓

Wash each filter with 5 × 3 ml vols 5 % TCA; follow by 5 × 5 ml vols 70 % ethanol.

↓

Dry by air passage (2 min).

↓

Dry filters for 5 min in open glass Petri dishes under an infra-red lamp.

↓

Place filters in scintillation vials; add scintillator solution and count (*see* Appendix).

Group B

Day 1 *(5 h, including 2 × 1 h incubations)*

Preparation of E. coli *DNA*

Pellet of frozen *E. coli* K12 (*see* note 10).

↓

Add 10 ml lysozyme (300 μg ml^{-1}) containing 1 % SDS and 50 μg Pronase ml^{-1}. Incubate at 37 °C for 30 min.

↓

Add equal volume of phenol saturated with Tris–SDS buffer, pH 9. Extract DNA as for group A (*see* note 5).

↓

Precipitate DNA with ethanol (as for group A). (**Omit** ether extraction.)

↓

Keep at −20 °C for 1 h.

↓

Redissolve pellet in 20 ml SSC/10. Add 2·0 ml × 10 × SSC.

↓

Add RNAse to final concentration of 50 μg ml^{-1}. Incubate for 30 min at 37 °C.

↓

Re-extract with phenol–Tris–SDS as before.

↓

Ether extraction and ethanol precipitation as for group A.

↓

Store pellet overnight at, −20 °C.

Day 2 $(3\frac{1}{2} h + 4 h)$

Redissolve pellet in 20 ml SSC/100.

↓

Add 2·2 ml 3 M sodium acetate containing 0·001 M EDTA, pH 7.

↓

Add 0·54 volumes (12 ml) ice-cold *iso*-propanol. Centrifuge for 10 min in bench centrifuge.

Pellet containing DNA.   Discard supernate which contains RNA.

↓

Dissolve in SSC/100 to 100 μg DNA ml^{-1}.

↓

Denature, shock – cool and load filters as described for group A, Day 2. (Retain 4 ml for groups D and E, Day 3.)

- -

Day 3

Pulse-labelling of infected cells (co-ordinate with groups A and C)

Take 10 ml of washed, log phase *E. coli* (*from group A*).

↓

Add 5 ml λ phage (*see* note 4) in TCM buffer.

↓

Incubate 20 min, 37 °C to allow adsorption.

Remove 5 ml. Add to flask containing 45 ml   Balance of culture to group C.
1·1 × concentrated K medium at 37 °C. Mix
quickly, allow to shake gently at 37 °C.

↓

Take samples for pulse-labelling as described
for group A.

- -

Day 4

Collection of samples, drying and counting, *as described for group A*.

- -

Group C

Day 1 *(4 h including 3 h incubation)*

Standardization of labelling period (see note 11)

Prepare a culture of non-infected *E. coli*, as described for group A, Day 3.

$$\downarrow$$

Pipette 5·0 ml culture into 45 ml 1·1 × concentrated K medium at 37 °C. Shake gently for 5 min.

$$\downarrow \qquad\qquad\qquad \downarrow$$

0·5 ml samples

Add to 12 tubes containing 0·2 µCi (50 µl) ³H-thymidine at 37 °C. At times of 0, 30, 60, 90, 120 and 180 s, and 0·3 ml ice-cold TCA (20 % TCA containing 5 mg thymidine ml⁻¹) to 2 tubes to stop isotope incorporation.

0·5 ml samples

Add to 12 tubes containing 0·2 µCi (50 µl) ³H-uridine at 37 °C. At times of 0, 30, 60, 90, 120 and 180 s, add 0·3 ml ice-cold TCA (20 % TCA, containing 5 mg uridine ml⁻¹) to 2 tubes to stop isotope incorporation.

Place tubes in ice (30 min).

$$\downarrow$$

Collect precipitate; wash and dry filters as described for group A, Day 4.

$$\downarrow$$

Count and calculate results.

Day 2

Results calculated to determine optimum labelling time for DNA and RNA labelling. Other groups informed of results.

Day 3 *(90 min)*

Pulse-labelling of infected bacteria: RNA synthesis
(co-ordinate timing of events with groups A and B)

Take infected culture from group B.

$$\downarrow$$

Pipette 5 ml into 45 ml 1·1 × K medium at 37 °C.

$$\downarrow$$

At times indicated for *group A* (Day 3), remove 0·5 ml samples for pulse-labelling.

$$\downarrow$$

Add each sample to a fresh tube containing 0·2 µCi ³H-uridine. Incubate for 1 min (or optimum

time as shown by results from Day 2) at 37 °C, then stop incorporation by addition of 0·3 ml 20 %
TCA containing 5 mg uridine ml^{-1}.

$$\downarrow$$

Place in ice for 30 min or overnight.

_____

Day 4

Collect precipitate, wash, dry and count as for group A.

_____

Group D

Day 1 *(5 h, including 3 h incubation)*

Preparation of culture (co-ordinate with group E after preparation of culture)

Inoculate 100 ml PN medium with 4 ml overnight K 12 culture. Incubate in shaking water-bath at
37 °C to A_{610} of about 0·5.

$$\downarrow$$

Harvest by centrifugation (15 min, 10 000 rev. min^{-1}).

$$\downarrow$$

Wash pellet in 100 ml sterile glass distilled water.

$$\downarrow$$

Resuspend pellets in sterile TCM to A_{610} of 0·8 (65 ml required).

$$\downarrow$$

Add 2 ml λ phage at 10^{12} p.f.u. ml^{-1} *(see note 4)*.

$$\downarrow$$

Incubate for 20 min at 37 °C to allow adsorption.

$$\downarrow \qquad\qquad\qquad\qquad\qquad \downarrow$$

6 × 5 ml samples.   Balance of culture to *group E*.

$$\downarrow$$

Add each 5 ml volume to 45 ml 1·1 × concentrated K medium in 250 ml flask at 37 °C.
Start timer.

$$\downarrow$$

At time zero, add 20 μCi ^{3}H-uridine to flask A: incubate with gentle shaking.

$$\downarrow$$

At $t = 3$ min, terminate the "pulse" by plunging the flask into a freezing mixture (ice-salt) bath and
adding 0·5 ml 1 M sodium azide *(see note 16)*.

$$\downarrow$$

Keep flask in ice for 30 min (minimum). Add 20 μCi ^{3}H-Uridine to flasks B, C, D, E and F at times 5, 10, 15, 20 and 30 min respectively. Terminate each "pulse" after 3 min incubation (as above for flask A).

$\downarrow$

Collect cells by centrifugation (15 min, 10 000 rev. min^{-1}).

$\downarrow$ $\qquad\qquad\qquad\qquad\qquad\qquad\quad$ $\downarrow$

Discard supernate (*see* notes 3 and 16). $\qquad$ Wash each pellet in 50 ml TCM/0·01 M NaN$_3$.

$\downarrow$

Discard supernate. $\qquad$ Resuspend pellet 1 ml TCM/0·01 M NaN$_3$.
Transfer to Piccolo (10 ml) centrifuge tubes.

$\downarrow$

Freeze overnight.

Day 2 *(2½ h, including 1 h incubation)*

Thaw each pellet of cells at 37 °C (5 min).

$\downarrow$

Add 1 ml lysozyme at 300 μg ml^{-1} in Tris–SDS buffer.

$\downarrow$

Incubate for 10 min at 37 °C, or until lysis occurs.

$\downarrow$

Add DNAse and Pronase to give final concentration of each of 50 μg ml^{-1}. Incubate for 1 h at 37 °C.

$\downarrow$

Extract with an equal volume of phenol saturated with 0·05 M sodium acetate, ph 5·2 (*see* note 5), using procedure as for group A, but omit the ethanol precipitation.

Samples for total radio-activity (50 μl) (*see* note 12). $\qquad$ Dialyse overnight against three changes (250 ml) SSC buffer at 4 °C. Monitor extent of dialysis (*see* note 12).

Samples for dialysed radioactivity (1 ml) (*see* note 12).

Day 3 *(15 h/overnight incubation)*

Measure and note the volume of each dialysed extract.

↓

Take samples of each extract for hybridization and controls.

Samples for total radio-activity (50 μl) (*see* note 12).

Controls

1. *TCA-precipitable radioactivity*
Pipette 50 μl of each extract into duplicate tubes containing 1 ml 10% TCA, in ice. Add 0·1 ml bovine serum albumen (1 mg ml⁻¹) to act as "carrier" for the precipitate. Keep in ice for 30 min.

↓

Collect each precipitate on a Whatman GF/C or GF/F filter paper presoaked in 5% TCA. Wash, dry and count the filters as in part A, Day 4.

2. *Non-specific binding of labelled material to membrane filters*
Use an "unloaded" membrane filter in the hybridization system.

3. *RNAse contamination in the DNA preparations used in loading filters*
Incubate 50 μl of labelled extract (in duplicate) with 1 ml vols each of λ DNA and *E. coli* (denatured) DNA preparations from group A and B respectively. Incubate for 1 h at 37 °C. Precipitate DNA with TCA, collect, wash and dry as in (1) above.

Hybridization
For each "time point", 6 filters and vials are used: 2 "λ-DNA"; 2 "*E. coli* DNA" and 2 blank (*see* note 13). Place each filter flat on the bottom of a glass scintillation vial. Add either 0·1 ml or 0·3 ml labelled, dialysed extract (total: 1·2 ml per time-point).

↓

Add 6 × SSC to give a final volume of 2·0 ml. Cap each vial tightly.

↓

Incubate for 15 h (or overnight) at 66 °C.

- -

Day 4 *(4 h approx. including 30 min incubation)*

Take vials from 66 °C incubator.

↓

Cool in ice bath. Remove filters (*see* note 13).

↓

Wash by passage of 5×10 ml vols $2 \times$ SSC.

↓

Combine into sets (λ, *E. coli* and blank), place the 3 filters in a vial and add 5 ml $2 \times$ SSC $+ 20 \mu$g RNAse (heat-treated).

↓

Incubate at $37\,^\circ$C for 30 min then chill in ice.

```
        ↙                          ↘
     Filter.                    Solution.
        ↓                          ↓
```

Wash 5 times with 10 ml portions of $2 \times$ SSC.   Discard (count a sample if required).

↓

Dry (air passage: 2 min; infra-red: 5 min).

↓

Count (*see* Appendix).

Group E

(Co-ordinate operations with group D to ensure identical timing)

Day 1 *(2 h)*

Culture provided by group D; 6×50 ml cultures for labelling. Use $20\,\mu$Ci ^{3}H-thymidine per culture (methods as for group D).

Day 2 *($2\frac{1}{2}$ h)*

Extract DNA. Purify as for group D using "*alkaline phenol*" (*Tris/SDS/NaCl*), and *RNAse* in place of *DNAse*.

Day 3 *(6 h incubation; 1 h)*

Hybridization and controls:

```
        ↙                          ↘
```

Controls (as in Part D, Day 3).   Hybridization (6 vials per time point as in D).

↓

"Loaded" and "unloaded" filters supplied by   Measure A_{260} of each DNA preparation.
groups A and B (36 in total).   Adjust to $5\,\mu$g ml^{-1} with SSC.

↓   ↓

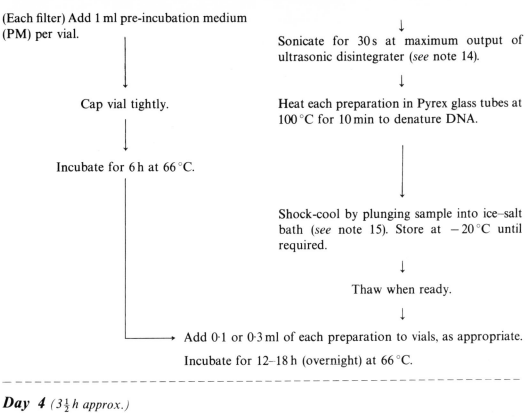

(Each filter) Add 1 ml pre-incubation medium (PM) per vial.

Cap vial tightly.

Incubate for 6 h at 66 °C.

Sonicate for 30 s at maximum output of ultrasonic disintegrater (*see* note 14).

Heat each preparation in Pyrex glass tubes at 100 °C for 10 min to denature DNA.

Shock-cool by plunging sample into ice–salt bath (*see* note 15). Store at − 20 °C until required.

Thaw when ready.

Add 0·1 or 0·3 ml of each preparation to vials, as appropriate.

Incubate for 12–18 h (overnight) at 66 °C.

- -

Day 4 $(3\frac{1}{2}h$ *approx.)*

Chill vials.

Wash, dry and count filters as in part D and Appendix.

- -

Notes and Points to Watch

1. Ether is highly flammable, and the vapour is heavier than air. **Do not use** ether in the presence of naked flames, i.e. Bunsen burners, matches, etc. **Do not** pour waste ether down the sink, but put into waste bottles provided.
2. **Phenol** is toxic by skin absorption, and corrosive. Wear disposable gloves and safety glasses when handling phenol; wipe bottles before handling them and discard waste phenol into the appropriate waste bottles. All phenol used should be redistilled before use, and stored in either dark glass bottles or in bottles covered with aluminium foil. When distilling phenol, operations should be performed in a fume cupboard, and an air condenser used. The distillate should be collected under the appropriate buffer: the lower phase is the phenol phase. Distillation removes impurities whose formation is induced by light. Freshly redistilled phenol may be either frozen or kept at 0–4 °C for 2–3 weeks before use.

3. Normal precautions for the use of radioactive materials should be employed: no eating, smoking or drinking in any room where such materials are used or stored. Do not use with open cuts on the hands, etc. Disposable gloves should be worn by all users. Water soluble waste may be discarded into a **designated** sink and flushed away with plenty of running water (designated sinks should, if at all possible, have a short pipe run to a main sewer, and be of such a construction that cleaning is easy and efficient). The amounts of radioactivity used are well within the amounts permitted for disposal by most institutes/laboratories, but this should be checked before proceeding. Solid waste is likely to be of relatively low activity (e.g. filter papers after counting) and may be discarded into normal waste-paper bins **after** sealing into plastic bags. Contaminated glassware and apparatus should be decontaminated as soon as possible after use (Decon 90 or Decon 75 for simple radioactive decontamination; RBS50 if there is also microbiological contamination).

4. (a) The λ phage used for DNA extraction (group A, Day 1) may be any strain capable of being produced to high titre, e.g. λ cI60. Phage produced should be purified by differential centrifugation, then treated with RNAse and DNAse to remove exogenous DNA. Further purification by banding on velocity (sucrose) gradients is adequate for most purposes, but isopycnic (CsC1) centrifugation will give a "cleaner" product. Minimum titre required 5×10^{12} p.f.u. ml^{-1}.

   (b) For infection (group B, Day 3); 5 ml λcI phage at 10^{10} p.f.u. ml^{-1} is required; group D, Day 1: 2 ml at 10^{12} p.f.u. ml^{-1}. The experiments could be adapted to study events leading to the establishment of lysogeny, when λ^{+} at the same titre(s) would be required; infection conditions might need modification.

5. (a) Three different phenol preparations are used: neutral phenol, saturated with TCM buffer, pH 7·2, for DNA extraction from phage; "alkaline phenol", saturated with Tris–SDS, pH 9, for DNA extraction from bacteria (this prevents RNA contamination of DNA by reducing solubility of RNA in the aqueous phase) and "acid phenol", saturated with 0·05 M sodium acetate, pH 5·2, for mRNA extraction.

   (b) Extraction with phenol may be conveniently done either by inversion or by mixing on a vortex mixer. For either method tubes should be fitted with tightly fitting silicone rubber stoppers, or cotton-wool plugs for vortex mixing only where tubes are less than half-full. Plugs should not be wetted by the mixture.

   For centrifuge tubes, mixing should be for $3 \times 1\frac{1}{2}$ min periods, with 2 min in ice between mixing periods; for inversion, mixing should be for 5 min at the rate of 1 inversion cycle per 5 s. Samples should be kept as cold as possible during mixing. Centrifugation should be performed in the cold, using either a refrigerated centrifuge, or, more conveniently, a bench centrifuge in a cold (4 °C) room.

   Extraction from bacterial lysates may prove difficult, in view of the high viscosity of the aqueous phase caused by released nucleic acids. Care should be taken to transfer as much aqueous phase as possible into the fresh container for re-extraction after the first phenol extraction.

6. Nitrocellulose filters, 25 mm diam., Millipore type "HA" or Sartorius type SM11306/25 are used for hybridization assays. They should be soaked in 6 × SSC for 10 min before use.

7. Two-stage drying in the vacuum desiccator (room temperature, 4 h; 80 °C, 2 h) reduces renaturation of DNA on the surface of the filter.

8. Duration of labelling for groups A, B and C should be such that the rate of uptake is constant. Time should be deduced from results of group C, part 1: calibration.

9. Whatman glass fibre filters (GF/C or GF/F) are used for collection of TCA precipitates. Nitrocellulose membrane filters can be used, at considerably greater cost, but there is no particular advantage in doing so. Filters should be soaked in 5% TCA before use.

10. The *E. coli* K12 used for *E. coli* DNA extraction may be any strain of K12 not lysogenic for λ^+ phage. Bacteria should be grown to mid-log phase in nutrient broth; harvested and washed in SSC, then concentrated to 5×10^{10} cells ml^{-1} in SSC. The cells are frozen quickly and stored frozen until required. Approximately 2 litres of culture will provide sufficient DNA for these procedures.

11. Labelling periods for pulse-labelling must be standardized to ensure that the rate of uptake of label is constant, i.e. the amount of uptake is proportional to the duration of uptake. This allows the assumption that the **amount** of uptake during a period gives an indication of the **rate** of uptake during that period.

12. Extent of dialysis for radioactively labelled preparation may be monitored by measuring the volume of material to be dialysed, and counting the activity present in a small (50 μl) volume.

    The volume of dialysis fluid is measured at each change, and a sample (1 ml) kept for radioactive counting. The activity in aqueous samples may be counted with the aid of scintillator-containing dispersal agents, e.g. "Triton X-100" or "Beckman Biosolv BBS-3". Dialysis may be assumed to be complete when the total activity in the dialysis fluid is less than 30 000 d. min^{-1} for 300 ml volume (i.e. 100 d. min^{-1} ml^{-1} above background).

13. Dried nitrocellulose membrane filters are extremely fragile and should be handled with extreme care. They are liable to carry heavy charges of static electricity. Where differently loaded filters are to be present in the same vial, they may be marked appropriately (use 2B pencil) on the edge of the filter.

14. DNA extracted from *E. coli* is likely to be in the form of a mixture of relatively large fragments (mol.wt $1-50 \times 10^6$); the λ-DNA should have a molecular weight of $30 \cdot 6 \times 10^6$. Such high mol.wt DNA will not hybridize readily with stabilized DNA due to the problem of complementary sequences "finding" each other. Sonication will break the DNA into shorter pieces which hybridize much more readily when denatured.

15. (a) When double stranded DNA is denatured, a **hyperchromic shift** occurs, i.e. the absorbance at 260 nm increases by up to 50% because of the conformational changes (double stranded rod to two random coils) taking place in the molecules on denaturation. Comparison of the absorption spectra before and after heat treatment for the introduction of a hyperchromic shift should confirm whether denaturation has taken place.

    (b) Shock-cooling of denatured DNA prevents renaturation of separated complementary strands in solution. Glass vessels used for heating DNA should obviously be able to withstand thermal shock, i.e. borosilicate glass.

16. Sodium azide is poisonous and reacts with copper in plumbing to form explosive compounds. **Do not** discard azides down the sink. Dilute to 10 mм (or less) then react with a 10-fold (volume) excess of dilute (5%) ceric ammonium nitrate in an ice bath until the reaction is complete. Flush down the sink with plenty of water.

MATERIALS

A. (1) Phage λcI60.
   (2) *E. coli* K12 F$^+$.

B. *Media/Buffers*
   (1) SSC at various relative concentrations (i.e. $1 \times$; $6 \times$; $10 \times$, etc.) "$1 \times$" contains 0·15 M NaCl; 0·015 M trisodium citrate.
   (2) TCM buffer is 0·01 M Tris/0·01 M CaCl$_2$/0·01 M MgCl$_2$, pH 7·2.
   (3) Tris–SDS buffer is 0·1 M Tris/1 % sodium dodecyl sulphate/0·1 M NaCl, pH 9·0.
   (4) K medium: 100 ml of a solution containing 3 g KH$_2$PO$_4$; 15 g Na$_2$HPO$_4$; 12 H$_2$O; 0·96 g NH$_4$Cl; 0·5 g NaCl. Sterilized at 121 °C for 20 min. Add:
       100 ml 15 % w/v casamino acids (sterile);
       20 ml 25 % w/v glucose (sterile);
       1 ml FeSO$_4$. 7H$_2$O at 50 mg 100 ml^{-1} 0·1 N H$_2$SO$_4$ (sterile);
       1 ml 1 % w/v CaCl$_2$ (sterile);
       1 ml thiamine hydrochloride (5 mg ml^{-1}, sterile);
       25 ml 0·1 M MgSO$_4$ (sterile, added last).
       Add sterile water to 900 ml for 1·1 $\times$.
   (5) PN medium contains 1 % peptone, 1 % NaCl in water, pH 7·0.
   (6) Pre-incubation medium (PM) contains:
       0·02 % w/v Ficoll–400 (Pharmacia);
       0·02 % w/v polyvinyl pyrrolidone–360 (Sigma);
       0·02 % w/v bovine plasma albumen (Armour Pharmaceuticals, Eastbourne, England);
       in 3 $\times$ SSC containing 0·01 M Tris–HCl, pH 7.

C. *Radiochemicals and materials for scintillation counting*
   (1) ^{3}H-uridine (sp. act. 25–30 Ci mmol^{-1}) and methyl–^{3}H-thymidine (sp. act. 40–60 Ci mmol^{-1}); Amersham International Ltd., Amersham, England.
   (2) *p*-terphenyl (primary fluor); Koch-light Laboratories Ltd.
   (3) 1, 4-Di {2 – (5 phenyloxazolyl)}-benzene (POPOP) (secondary fluor): Koch-Light Laboratories Ltd.
   (4) Beckman "Biosolv BBS-3" solubilizing agent: Beckman—R.I.I.C. Ltd.
   (5) Scintillation grade xylene: BDH Chemicals Ltd., Poole, Dorset, England.
   (6) Scintillator contains 4 g *p*-terphenyl; 125 mg POPOP per litre of xylene.

D. *Enzymes*
   (1) Ribonuclease (pancreatic); 1 mg ml^{-1} in TCM; heated at 90 °C for 10 min to destroy DNAse (Sigma (London) Ltd.).
   (2) Deoxyribonuclease, Type 1: 1 mg ml^{-1} in TCM (Sigma (London) Ltd.).
   (3) Pronase: 1 mg ml^{-1} in TCM (Calbiochem B grade; nuclease free).
   (4) Lysozyme: 300 μg ml^{-1} in Tris–SDS buffer with or without the further addition of Pronase (to 50 μg ml^{-1}).

E. *Materials: filters*
   (1) Nitrocellulose membrane filters, 0·45 μm pore size, 25 mm diam.: Millipore Ltd.
   (2) Glass-micro fibre filters: Type GF/C or GF/F, 25 mm diam. (Whatman Ltd., Springsfield Mill, Maidstone, Kent, England).

F. *Apparatus*
   (1) Bench centrifuge capable of holding 25 ml glass tubes.
   (2) Bench centrifuge capable of holding 10 ml glass tubes (e.g. Piccolo Centrifuge, H. Christ GmbH, Germany).
   (3) Filter-holding apparatus, to take 25 mm diam. filters. Millipore type XX1002530 will hold one filter (several units required) or type XX27 02550 will hold 12 filters, allowing simultaneous processing of 12 samples (Millipore Ltd.).
   (4) Liquid scintillation counter, capable of counting in at least 2 channels, either sequentially or simultaneously.
   (5) Pipetting unit capable of delivering 20–200 μl. A suitable instrument is the Gilson Pipetteman (Gilson, France S.A.).
   (6) Vortex mixers.
   (7) Shaking water-bath with 250 and 500 ml flask holders.
   (8) Ultrasonic cell disrupter, e.g. M.S.E. 150 Watt Model PG 100 (MSE Ltd., Manor Royal, Crawley, Sussex, England).
   (9) Spectrophotometer capable of measuring in both the ultra-violet and visible ranges. Scanning model preferable.
   (10) Quartz cuvettes (1 ml).
   (11) High-speed centrifuge (e.g. MSE "18"; MSE Ltd., Manor Royal, Crawley, Sussex, England).

SPECIFIC REQUIREMENTS

Day 1

Group A

bench centrifuge + 25 ml glass tubes

λ phage suspension, 10^{12} p.f.u. ml^{-1}, purified to be free of *E. coli* RNA and DNA

35 ml "neutral" phenol

pipettes (10 ml; Pasteur) and appropriate pipette bulbs

ether, peroxide-free, 100 ml ⎫
absolute ethanol, 100 ml ⎭ stored in ice

3 M sodium acetate, 5 ml

"SSC/100", 750 ml

dialysis tubing, narrow bore

1 litre glass beaker

magnetic stirrer

safety glasses

disposable gloves

bottles for waste ether and phenol

timer

Group B

No dialysis tubing or magnetic stirrer required, otherwise as for group A, but with additions and substitutions:

substitute λ phage by frozen pellets *E. coli* K12 (*see* note 10)

lysozyme, in Tris–SDS buffer + Pronase

37 °C water-bath

substitute "neutral" phenol by 70 ml "alkaline" phenol (in Tris–SDS buffer)

30 ml SSC/10; 5 ml 10 × SSC

RNAse

Group C

50 ml PN medium

250 ml sterile flask

2·5 ml overnight culture *E. coli* K12 in PN medium

shaking water-bath at 37 °C

(use of) high speed centrifuge; 100 ml (plastic) tubes

100 ml sterile glass-distilled water

100 ml sterile TCM buffer

(use of) spectrophotometer and cuvettes

sterile pipettes; bulbs as appropriate

45 ml "1·1 ×" K medium

timer

24 test tubes (3 × $\frac{1}{2}$ in.) + test tube racks

pipetting unit set to 50 μl

2·4 μCi (0·6 ml) each of ^{3}H-uridine and ^{3}H-thymidine

5 ml (each) of 20% trichloracetic acid (TCA) containing either 5 mg uridine ml^{-1} or 5 mg thymidine ml^{-1}

ice bucket and ice

filter unit(s) to hold 25 mm diam. filters

450 ml 5% w/v TCA (in ice)

750 ml 70% w/v ethanol (in ice)

250 ml scintillation fluid

24 scintillation vials and caps

container with Decon 90 solution for contaminated glassware

(use of) scintillation counter

Group D

100 ml PN medium

500 ml sterile flask

4 ml overnight *E. coli* K12 culture in PN medium

sterile pipettes; bulbs as appropriate

2 ml λ phage at 10^{12} p.f.u. ml^{-1}

100 ml sterile glass distilled water

100 ml sterile TCM buffer

(use of) spectrophotometer and cuvettes

timer

7 × 250 ml sterile flasks

(use of) shaking water-bath with 500 and 250 ml flask holders

300 ml "1·1 ×" K medium

120 μCi ^{3}H-uridine (1 mCi ml^{-1}, i.e. 20 μl per flask)

pipetting unit set to 20 μl

ice–salt (freezing) bath

3 ml 1 M sodium azide

(use of) high-speed centrifuge; 100 ml sterile (plastic) tubes

350 ml TCM containing 0·01 M NaN$_3$

6 × 10 ml (Piccolo) centrifuge tubes

use of −20 °C freezer

cont.

cont.

Group E
6 × 250 ml sterile flasks
10 ml + 5 ml (sterile) pipettes and bulb
300 ml "1·1 ×" K medium
120 μCi-^{3}H-thymidine (at 1 mCi ml^{-1}, i.e. 20 μl per flask)
pipetting unit set to 20 μl
(use of) shaking water-bath, holder for 250 ml flasks
3 ml 1 M sodium azide
freezing bath
(use of) high speed centrifuge; 100 ml sterile tubes
350 ml TCM containing 0·01 M NaN$_3$
use of $-20\,°$C freezer

Day 2
Group A
(use of) spectrophotometer and 1 ml quartz cuvettes
(500 ml) glass beaker, wire gauze and tripod stand
ice bath, 250 ml beaker
24 nitrocellulose membrane filters, 25 mm diam.
holder(s) for membrane filters
2·5 litres 6 × SSC
vacuum desiccator; P$_2$O$_5$ in base
oven/incubator set to 80°C
50 ml measuring cylinder
plastic forceps for membrane filters

Group B
pipettes (non-sterile: 10 ml, 5 ml, 1 ml, Pasteur as required)
3 ml 3 M sodium acetate, containing 1 mM EDTA, pH 7
ice bath
20 ml propan-2-ol (isopropanol) (in ice)
(use of) spectrophotometer and 1 ml quartz cuvettes
50 ml SSC/100
other requirements as for group A

Group D
water-bath, 37°C
non-sterile pipettes (1 ml; 10 ml; 5 ml; Pasteur; 0·1 ml; 0·2 ml)

7 ml lysozyme solution (in Tris–SDS buffer **without** pronase)
timer
1 ml each of DNAse and Pronase solutions
30 ml "acid phenol"—saturated with 0·05 M sodium acetate, pH 5·2 (kept in ice)
Piccolo (10 ml) centrifuge tubes and centrifuge
vortex mixer
ice bath
100 ml peroxide-free ether
dialysis tubing (narrow bore)
1 litre 1 × SSC buffer

Group E
As for group D with certain exceptions: substitute "alkaline" phenol for "acid" phenol, i.e. saturated with Tris–SDS buffer
substitute RNAse solution for DNAse

Day 3
Group A
50 ml PN medium
2 × 250 ml sterile flasks
shaking water-bath with 250 ml holder, set at 37°C
sterile pipettes (10 ml, 2 ml, 1 ml) plus bulbs as appropriate
2·5 ml overnight culture *E. coli* K 12 in PN medium
(use of) spectrophotometer and cuvettes
(use of) high speed centrifuge and 100 ml sterile tubes
50 ml sterile glass-distilled water
50 ml sterile TCM buffer
ice bath
timer
45 ml "1·1 ×" K medium
16 tubes, 3 × $\frac{1}{2}$ in., and rack
pipetting unit set to 100 μl and tips
^{3}H-thymidine solution, 1·6 ml at 20 μCi ml^{-1}
5·0 ml 20 % w/v TCA containing 5 mg thymidine ml^{-1}

Group B
sterile 10 ml, 5 ml, 1 ml pipettes plus bulbs as appropriate
5 ml λcI phage at 10^{10} p.f.u. ml^{-1} in TCM buffer
timer
2 × 250 ml sterile flasks

cont.

cont.

45 ml "1·1 ×" K medium
(use of) shaking water-bath at 37°C
16 tubes, 3 × ½ in., and rack
ice bath
pipetting unit set to 100 μl and tips
^{3}H-thymidine solution, 1·6 ml at 20 μCi ml^{-1}
5·0 ml 20% w/v TCA containing 5 mg thymi-
    dine ml^{-1}

Group C

As per group B, but substitute
(a) ^{3}H-uridine at 20 μCi ml^{-1} for ^{3}H-
    thymidine solution
(b) 20% w/v TCA + 5 mg uridine ml^{-1} for
    TCA-thymidine

Group D

non-sterile pipettes (1 ml; 0·1 ml; 0·2 ml) plus
    bulbs as appropriate
pipetting unit set to 50 μl and tips
100 scintillation vials (glass, with foil-lined caps)
1 litre scintillation fluid
solubilizer (Beckman "Biosolv–BBS-3")
24 tubes 3 × ½ in., and rack
ice bath
30 ml 10% w/v TCA
5 ml bovine serum albumen (1 mg ml^{-1} in TCM
    buffer)
24 glass-fibre filters (25 mm diam.) in 5% TCA
filter holding unit(s)
400 ml 5% w/v TCA ⎤
600 ml 70% v/v ethanol ⎦ stored in ice
infra-red lamp
glass Petri dishes
12 nitrocellulose membrane filters
(*Group A* to supply 12 "λ-DNA" filters;·
Group B to supply 12 "*E. coli*–DNA" filters)
30 ml 6 × SSC

Group E

As per group D, with the *addition of*:
12 ml pre-incubation medium (PM)
(use of) ultrasonic disrupter
boiling water-bath tripod and gauze; 250 ml
    beaker)

6 Pyrex test-tubes 6 × ⅝ in., and rack
freezing (ice–salt) bath

Day 4

Groups A, B and C

16 Whatman GF/C or GF/F 25 mm diam.
    filters
filtration unit(s)
300 ml 5% w/v TCA ⎤
450 ml 70% v/v ethanol ⎦ in ice until required
infra-red lamp (20 cm above bench)
glass Petri dishes
16 scintillation vials (glass)
160 ml scintillator solution in dispensor
plastic-tipped forceps for filter-handling

Group D

ice bath
plastic-tipped forceps
3 litres 2 × SSC
48 scintillation vials (glass)
RNAse, 1 mg ml^{-1} in TCM
pipetting unit set to 20 μl and tips
37°C water-bath
10 ml pipettes, non-sterile
dispensor set to 10 ml, containing 2 × SSC
infra-red lamp, set to 20 cm above bench
glass Petri dishes
360 ml scintillation fluid in dispensor (set to
    deliver 10 ml)

Group E

As for group D except:
no RNAse; pipetting unit or water-bath and
    only 36 vials
1·5 litres 2 × SSC

Days 5 (6, 7)

No specific requirements (counting in progress).
    Some of the requirements may be best sup-
    plied in the form of a general "lab. stock" of
    glassware which is recycled as necessary.

FURTHER INFORMATION

Group A

With a starting volume of 10 ml λ phage (5 × 10^{12} p.f.u. ml^{-1}) the yield of λ-DNA is typically
800–1000 μg. To prepare 24 filters, a total of 600 μg is required. The phage yield from 5 litres of
culture should suffice to provide enough phage for the experiments.

Pulse-labelling of non-infected cultures for DNA synthesis typically shows a logarithmic rate of increase in rate of thymidine uptake with exponentially growing cultures. Dependent on the delay between preparing the washed suspension of bacteria, and the final growing culture, a variable delay in the start of incorporation may be seen. The curve shown (Figure 1) demonstrates the synchronization of the culture and completion of a round of replication (plateau 17–20 min).

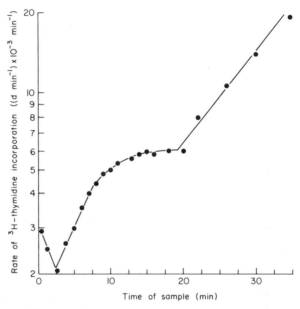

Figure 1   Incorporation of ^{3}H-thymidine into non-infected *E. coli*. This is a typical result obtained when a non-infected, exponentially growing culture of *E. coli* was prepared and pulse-labelled as described in the text. Cell density approx. 7×10^7 cells ml^{-1}.

Group B

Phage-infected bacteria typically show a higher initial rate of DNA synthesis than non-infected bacteria, reaching a peak by about the 7–10th min after infection (Figure 2). A second peak may be observed during the stage of the latent period when phage maturation is occurring (i.e. 20–40 min).

Similar results were obtained by Young and Sinsheimer (1967) when studying the kinetics of intracellular DNA synthesis in λ phage-infected bacteria, the pattern of incorporation being noticeably different for λ^+ phage under conditions giving 90–95 % lysogeny.

Group C

Calibration experiments indicate that a labelling period of 1 min or more for DNA or RNA typically gives a **linear** graph when the **amount** of uptake is plotted against the **time** allowed for uptake. After a short time to achieve a steady state, the **rate of uptake is constant** (Figure 3). It may be assumed that the lag in achieving the steady state represents the time required to saturate the pools of precursors with the labelled molecules.

Pulse-labelling of infected bacteria to show RNA synthesis typically shows peaks in the rate of synthesis soon after infection (2–4 min) and late after infection (25–35 min), corresponding to the synthesis of "early" and "late" messengers (Figure 4).

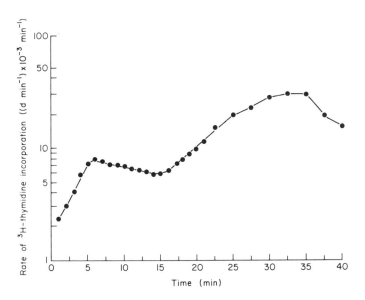

Figure 2 Incorporation of ³H-thymidine into *E. coli* infected by phage λcI 60. This is a typical result obtained when *E. coli*, prepared and infected as described (multiplicity of infection of approx. 10; cell density approx. 6×10^7 ml^{-1}) was pulse-labelled with ³H-thymidine as described in the text.

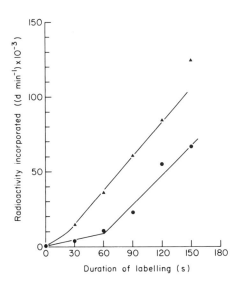

Figure 3 Uptake of radioactive material by exponentially growing non-infected *E. coli*. This is a typical class result obtained for incorporation of ³H-uridine (●) and ³H-thymidine (▲) into TCA–insoluble material. Conditions for labelling as described in the text. Cell density approx. 5×10^7 cells ml^{-1}.

Group D

Hybridization studies show that early in infection the proportion of mRNA species hybridizeable with *E. coli* DNA is about 0·5 % of the total TCA-precipitable material. The level of λ specific messengers increases rapidly from zero at infection to about 2·0 % at 10 min and subsequently the level of *E. coli* specific messengers decreases to zero over this time. Few if any "counts" are adsorbed to blank filters.

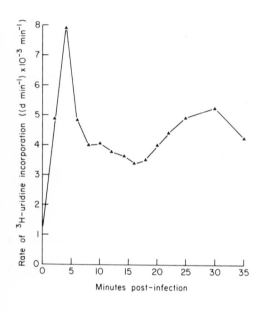

Figure 4 Rate of incorporation of ³H-uridine into TCA-insoluble material after λcI phage infection. This is a typical class result obtained when *E. coli* cells were infected and pulse-labelled as described in the text. Cell density approx. 5 × 10⁷ ml⁻¹; multiplicity of infection approx. 10 p.f.u. cell⁻¹.

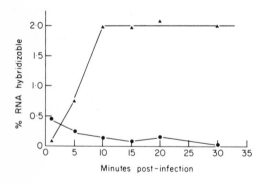

Figure 5 Specificity of RNA synthesized in λcI infected *E. coli*. This is a typical class result obtained when pulse-labelled, infected *E. coli* were broken to release RNA, and the labelled mRNA incubated with filters carrying denatured *E. coli* DNA (●) or denatured λ-DNA (▲) as described in the text.

Results normalized as percentages of total radioactive material incorporated into TCA-insoluble material at each time point.

Group E

Results obtained by the procedures used typically show a higher incorporation of ³H-thymidine into DNA than ³H-uridine into RNA. The pattern of synthesis of DNA shows a similar effect of a rapidly increasing synthesis of λ-specific DNA, from 1·1 % at 0–3 min; 16 % at 5–8 min, reaching

27% at 30 min after infection (Figure 6). The level of *E. coli* DNA synthesis stays constant at about 2·5% (at 0–3 min) to 2·3% at 30 min, confirming that host DNA synthesis is unaffected by λ-DNA synthesis.

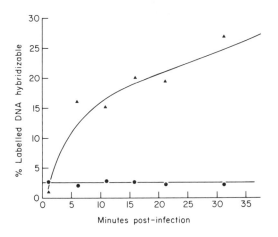

Figure 6 Specificity of DNA synthesized after infection of *E. coli* by λcI phage. This is a typical class result obtained after extraction of DNA from infected pulse-labelled *E. coli*. Extracted nucleic acids were purified and incubated with filters loaded with either λ-DNA (▲) or *E. coli* DNA (●) as described in the text. Retained radioactivity was counted and calculated as a percentage of the total radioactivity present in DNA at each time-point.

The level of activity adsorbing to "unloaded" filters under such conditions varied from 0 to 0·3%.

BIBLIOGRAPHY

Methods in Enzymology XIIB pp. 641–68 (Academic Press, London and New York) for discussion (by Gillespie) on DNA–RNA hybridization techniques.
Methods in Enzymology Volume XIIA pp. 543–50, articles by Miura and Smith on DNA extraction from bacteria.

REFERENCES

D. T. Denhardt (1966). A membrane-filter technique for detection of complementary DNA. *Biochemical and Biophysical Research Communications*, **23**, 641–6.
D. Gillespie and S. Spiegelman (1965). A quantitative assay for DNA–RNA hybrids with DNA immobilised on a membrane. *Journal of Molecular Biology*, **12**, 829–42.
M. H. Green (1966). Inactivation of the prophage Lambda repressor without induction. *Journal of Molecular Biology*, **16**, 134–48.
E. T. Young and R. L. Sinsheimer (1967). Vegetative λ-DNA: II Physical characterisation and replication. *Journal of Molecular Biology*, **30**, 165–200.

APPENDIX

Radioactive Counting

Results will be obtained in the form of counts of radioactivity. The **count rate** is related to the true disintegration rate by the relation:

$$\frac{\text{ct min}^{-1}}{\text{d. min}^{-1}} = \text{efficiency of counting,}$$

where ct min^{-1} = counts per minute and d. min^{-1} = disintegrations per minute. To convert ct min^{-1} to d. min^{-1}, a calibration series of samples must be used. These all contain the same amount of radioactivity, but different amounts of "quenching" agents. Quenching agents interfere with energy transmission in the vials, by preventing transfer of energy from activated solvent molecules to solute molecules. As solute molecules fluoresce, and emit their activation energy as light quanta (photons), quenchers reduce the number of photons produced per decay event, and hence the pulse of voltage in the photomultiplier tubes. If the activity in each sample is known the efficiency can be readily calculated.

For homogeneous (liquid) samples, such a standard set may be used to construct a curve relating efficiency to a parameter known as the "external standard ratio". This parameter is obtained by irradiating each sample with γ-photons from a Cs^{137} source. This greatly elevates the count rate, and the increased rate is measured in each of two (high and low) energy channels. The ratio of such counts is computed: quenching will "divert" counts from the higher to the lower energy channel, hence the ratio can give a measure of the efficiency. It is assumed that quenchers affect the induced energy spectrum in the same way as the natural energy spectrum.

For a calibration set, a curve of external standard ratio (ESR) vs efficiency can be constructed. When an unknown sample is being counted, the external standard ratio is measured and the efficiency can be read from the graph.

There is a second method of calibrating the liquid scintillation counter: the so-called "sample channels ratio method". This method is more appropriate for the evaluation of non-homogeneous samples, such as the use of a filter disc at the bottom of the counting vial. It may also be used for the evaluation of homogeneous samples, but is more tedious than the external standard method, since two counts of the activity in each sample are required. Like the ESR method, it requires the use of a set of standards of known activity and varying amounts of quenching. The energy of the radioactive decay events in the vial is spread over a spectrum from 0 to Emax, where Emax is the maximum energy of a decay event for the isotope in question. For Tritium, Emax is 18·6 keV, but very few β-particles are emitted with this energy. Most are emitted at an energy of about 5 keV (Emean). The energy of any β-particle determines the number of photons it will produce in the scintillation vial, the photon yield in most organic solvent/scintillator combinations in common use being about 5 photons keV^{-1}. Most modern scintillation counters are coincidence counters, i.e. they have two photomultiplier tubes looking at the sample from opposite sides, and a pulse in **both** is required before the counter will register one count. For an event with energy of 1 keV, there are 5 photons produced: there will be a 1/6 chance that the event will not be detected because all of the photons travel to **one** of the photomultiplier tubes, i.e. the count rate will be 5/6 of the disintegration rate, or the efficiency is 83·5 %. However, this assumes that the photomultiplier tubes are 100 % efficient in converting incident photons to pulses of electrons: they have a photocathode efficiency of the order of 25 %, i.e. they only detect/convert 1 photon in any 4. For two photomultiplier tubes operating in coincidence, a minimum of 8 photons (4 to each PM tube) are required for an event to be detected. The effect of these constraints is that events of low energy (below 2 keV) are not detected at all, and events less energetic than about 12 keV will not be detected with 100 % efficiency. Since for tritium the Emean value is about 5 keV, the maximum efficiency in the absence of quenching is of the order of 60 %. Any event in the vial that interferes with energy transmission from the solvent to the solute (i.e. quenching) will reduce the photon yield per decay event, and hence reduce the efficiency of counting. If the energy spectrum is examined, then quenching has the effect of shifting the whole spectrum downwards to the low energy end. If any of the events are shifted to an energy below 2 keV, then they will not be detected at all. If the counter is set to count the total number of events in the normal energy spectrum (i.e. unquenched), the effect of quenching will be to reduce the count rate, as the less energetic events are shifted to below detection level of the instrument. In the sample channel ratio method, the sample is also counted a second time, in an energy channel at the high end of the normal spectrum for the isotope. In an unquenched sample, the count rate in the first channel (Ch. 1) might be (say) 10000 ct min^{-1}, and in the second channel (Ch. 2) (say) 5000 ct min^{-1}. (The best results obtained with a channel ratio of about 0·5 for Ch. 2/Ch. 1 for an unquenched sample.) If a quenching agent is added, the counts in channel 1 might now be 8000 min^{-1} and 3500 min^{-1} in channel 2 (Ch. ratio of 0·438): addition of more quenching agent might alter these counts to 7500 and 3150 (Ch. ratio = 0·42), 7200 and 2975 (Ch. ratio = 0·413), etc. in a progressive manner. If each sample has a known amount of activity present, then by knowledge of the count rate over the whole spectrum (Ch. 1) then the efficiency of counting could be determined. This method of efficiency determination reflects the composition of the sample in the vial, as only the activity present is being counted. If, for instance, quenching takes place at the surface of a filter, the external standard method tends to give an incorrect result since that method looks at the solvent phase of the sample. The disadvantage of the method is that it requires two counts of each sample, which in many scintillation counters means sequential counts. (More sophisticated instruments can count in up to 4 channels simultaneously.) To obtain good results, the lowest number of counts in the low count rate channel should be 10000 or greater (i.e. the 2 × s.d. value is 200, or the "error" is 2%). Obviously, for low activity samples, this could require very long counting times. **In use**, a set of "quenched standards" containing equal amounts of activity (known, by the use of a standard radioactive solution) is counted before the addition of varying amounts of quenching agent, to ensure that the samples do contain equal amounts of activity, then varying amounts of the quenching agent are added. The samples are counted again, in two channels, to determine the channel ratio values for the set. The absolute amount of activity/sample (in d. min^{-1}) is known from the known activity of the radioactive standard. The counts per minute in channel 1 for each sample will then give a value for the efficiency, $E\%$, of each sample (since $E\%$ = ct min^{-1}/d. min^{-1}). A graph is plotted of $E\%$ vs channel ratio; any unknown sample with the same scintillator system may then be counted under the same conditions to give a value of channel ratio, and by reference to the graph, the efficiency may be calculated, and hence the absolute activity in disintegrations per minute.

Computation of "d. min^{-1}" values may be facilitated by the use of simple computer programs. A set of standards is used, composed of equal (10 ml) volumes of scintillator plus standard radioactive solution (e.g. ^{3}H$-n-$Hexadecane. Amersham International Ltd. Catalogue No. TRR6) and varying amounts of quenching agent, (e.g. chloroform) to obtain a set of data relating the sample channels ratio to efficiency of counting. The efficiency is determined from knowledge of the total amount of radioactive standard used per vial (e.g. 100000 d. min^{-1}) and the (measured) count rate in the **total energy spectrum**. If a set of, say, 30 such standards is prepared, with differing amounts of quenching agent, the data is suitable for computation of an equation by a computer program for multiple regression (equation of the form $E = a_0 + a_1 x + a_2 x^2 + a_3 x^3 + \ldots$, with coefficients a_0, a_1, a_2, etc. being determined by the computer, where E = efficiency of counting and x = count rate in the total energy channel. The use of a third order equation can give a fit to within $\pm 0·1\%$ of the points determined experimentally. The coefficients may then be used in another computer program to determine the efficiency of counting, and hence the d. min^{-1} values of unknown samples. Both programs used at Reading are in BASIC language (Reading Version); further information may be obtained from the author on request.

Culture Collections

ATCC   American Type Culture Collection,
12301 Parklawn Drive,
ROCKVILLE,
Maryland 20852,
USA

CBS   Centraal Bureau voor Schimmelncultures,
Oosterstraat 1,
BAARN,
The Netherlands

CCAP   Culture Centre of Algae and Protozoa,
36, Storey's Way,
CAMBRIDGE,
CB 3 ODT,
England

NCIB   National Collection of Industrial Bacteria,
Torry Research Station,
P. O. Box 31,
ABERDEEN,
AB9 8DG
Scotland

NCMB   National Collection of Marine Bacteria,
Torry Research Station,
P. O. Box 31,
ABERDEEN,
AB9 8DG,
Scotland

NCTC   National Collection of Type Cultures,
Central Public Health Laboratory,

Colindale Avenue,
LONDON NW9,
England

NCYC   British National Collection of Yeast Cultures,
ARC Food Research Institute,
Colney Lane,
NORWICH,
NR4 7UA
England

Subject Index